CW01572544

Petroleum Geology of Libya

Petroleum Geology of Libya

Don Hallett

Page preparation sponsored by

2002

ELSEVIER

Amsterdam • London • New York • Oxford • Paris • Shannon • Tokyo

ELSEVIER SCIENCE B.V.
Sara Burgerhartstraat 25
P.O. Box 211, 1000 AE Amsterdam, The Netherlands

First edition 2002

Library of Congress Cataloging in Publication Data
A catalog record from the Library of Congress has been applied for.

ISBN: 0 444 50525 3

⊗ The paper used in this publication meets the requirements of ANSI/NISO Z39.48-1992 (Permanence of Paper).
Printed in The Netherlands.

'Africa semper aliquid novi affert'

– Pliny

CONTENTS

Chapter 4 STRATIGRAPHY: MESOZOIC 144

Chapter 5 STRATIGRAPHY: CAINOZOIC 201

Chapter 6 STRUCTURE 265

Chapter 7 PETROLEUM GEOCHEMISTRY 322

Chapter 8 PETROLEUM SYSTEMS 355

Chapter 9 POSTSCRIPT: WHERE ARE THE REMAINING UNDISCOVERED RESERVES? *417*

LIST of FIGURES

8

10

PREFACE

Libya has the largest petroleum reserves of any country in Africa and since production began in 1961 over 20 billion barrels of oil have been produced. It has been calculated that Libya will reach the mid-point of reserves depletion in 2001 and this provides a timely point at which to review the state of petroleum exploration in Libya. It is perhaps surprising that no comprehensive account of the petroleum geology of Libya has yet been published. This is partly attributable to the fact that it is only within recent years that oil companies have released the information necessary to compile a book of this kind, and partly because of the onerous task of collecting the data. On average around fifty articles are published each year on the petroleum geology of Libya. A recent bibliographic search revealed over 5000 publications dealing directly or indirectly with Libyan geology, and a large proportion of these publications relate to hydrocarbons.

This book is intended as a synthesis of the petroleum geology of Libya. It is based exclusively on published data, supplemented by the author's experience gained during ten years work with Sirte Oil Company in Libya. There is no lack of available data, but much of the older material is now out of date, and some of the publications are difficult to find, especially those published in Libya. The idea of a synthesis of Libyan petroleum geology was suggested by the book by Arthur Whiteman on the *Petroleum Geology of Nigeria* (Graham and Trotman, 1982) which had similar aims, and which successfully synthesised a large body of information within the confines of a single volume. Whilst this book was in the final stages of preparation a book by Ed Tawadros was published on the *Geology of Egypt and Libya* (Balkema, 2001) which presents an excellent synthesis of the stratigraphy and structural evolution of the two countries, and which in many respects complements the data presented in the present volume.

The aim of this book is to present a systematic review of the petroleum geology of Libya in which the plate tectonics, structural evolution, stratigraphy, geochemistry, and petroleum systems are examined, along with data on fields, production, and reserves. It is hoped that it will provide a ready source of reference for individuals and companies who wish to obtain an overview of the petroleum geology of Libya, and that it will save them the laborious task of sifting though hundreds of publications to find the data they require.

I have drawn heavily on the series of symposium volumes published under the general title of *Geology of Libya* between 1971 and 1991, and on the symposium volumes for the Sirt Basin and Murzuq Basin published in 1996 and 2000. These volumes contain invaluable data. The present book could not have been written without them and I wish to acknowledge my debt to the contributing authors. The latest volume on the western basins of Libya is scheduled for publication in 2002/3. In addition, the Industrial Research Centre in Tripoli produced a geological map of Libya in sixty-one sheets, each with an explanatory booklet, plus maps at other scales, culminating in a 1:1,000,000 map published in four sheets in 1985. They also issued a series of bulletins which included a stratigraphic lexicon of Libya and a bibliography of Libyan geology. These publications represent a further valuable source of information. The Earth Sciences Society of Libya (formerly the Petroleum Exploration Society of Libya) published a series of excellent field guides mostly in the period 1962 to 1976, and stratigraphic lexicons for the Sirt Basin and offshore areas. In 1998 the Geological Society of London issued a Special Publication on the Petroleum Geology of North Africa based on a symposium held in London in 1995, which contains some useful contributions, not only on Libya but on adjacent parts of Algeria, Tunisia and Egypt. Data on production, reserves, and the economics of the Libyan oil industry were obtained from the statistical reviews published by BP, the Oil and Gas Journal, World Oil, OPEC, the US Energy Information Administration and the American Association of Petroleum Geologists. There are also a number of books which deal with the economics of the oil industry in Libya, the most recent of which was an excellent account by Judith Gurney entitled *Libya, the political economy of oil* (Oxford University Press, 1996). Data sources have been cited throughout and are listed in notes at the end of the book, and a full list of references is provided.

It is impossible to ignore the impact of political events on the oil industry, as illustrated vividly in Daniel Yergin's book *The Prize* (Simon and Schuster, 1991). Nowhere is this more true than in Libya, particularly during the 1970's, when Libya played a pivotal role in OPEC. The development of oil exploration and production in Libya cannot be understood without a knowledge of the political events of the time, and for this reason the chapter on the history of oil exploration in Libya includes an outline of the events which shaped the oil industry in Libya during those turbulent times.

Finally, I wish to acknowledge with gratitude the help I have received from numerous people and organisations during the preparation of this book. In particular I am deeply indebted to the organisation which sponsored the cost of drafting, and to Lynx Information Systems Ltd and JFA Global Ltd who contributed generously towards the the cost of page preparation. I am also most grateful to IHS Energy Group for kindly providing the satellite image for the cover, in addition to supplying a set of well location maps. I owe a special debt of gratitude to my former colleagues at Sirte Oil Company in Marsa al Brayqah and at NOC for discussions on many aspects of Libyan geology extending over ten years. The search for data was greatly facilitated by the library staff at the Geological Society of London, particularly Wendy Cawthorne, who went to endless trouble to obtain obscure and rare publications. Many publications were supplied from the extensive collections of the British Library and the excellent Libyan Collection at the School of Oriental and African Studies of London University.

I am indebted to Mike Ridd, Reza Sedaghat, Kent Wallace and Bindra Thusu for proof reading the typescript and for suggesting many improvements to the text and figures. It should be emphasised however that the opinions expressed in this book are mine alone. I thank my wife Irina for undertaking the tedious job of checking the index. I am also grateful for the dedication of the drafting team of Cathy Hickey, Pat Jarvis and Tony Stephens. I wish to thank the following for information or comments on particular topics, Sebastian Lüning, Raphael Unrug, Khaled Echikh, Mike Anketell, Haydon Bailey, Andrew Racey, Ralph Burwood, Muhammad Ibrahim, Chris Cornford, Ranald Kelly, Syed Rasul, Gertrud Baric, Jonathan McQuilken and Bob Newport. I have received encouragement from several Libyan colleagues including Prof. El Hawat from Garyounis University in Binghazi, and from Dr Mustafa Salem of Earth Sciences Society of Libya who kindly provided transliterations of many Arabic geographical names. My thanks also to Elsevier Science, particularly Femke Wallien, for their patience and professionalism in producing this volume. It is inevitable in a book of this kind that some errors will have passed undetected. The author would be grateful to receive notification of errors or inaccuracies found in this book.

Don Hallett
Richmond, UK
August 2001

NOTES and DEFINITIONS

Units of Measurement

Units of measurement employed in the oil industry in Libya are in a state of transition from fps to SI units. SI units have generally been adopted for units of length and area, but for oil and gas volumes barrels and cubic feet are still widely used. This mixed system is used in the current work since it conforms to current work practices in Libya.

Units of time are conventionally expressed as: Ga billion years, Ma million years, m.y. time interval in millions of years.

Billion is used in the sense of 10^9, and trillion as 10^{12}. The oil industry is inconsistent in abbreviations for volume and flow rates. M is conventionally used for thousand and MM for million. For gas volumes TCF is used for trillion cubic feet, BCF for billion cubic feet, MMCF for million cubic feet, and MCF for thousand cubic feet. For oil volumes however the corresponding abbreviations are GB for billion barrels, MMB for million barrels, and MB for thousand barrels. CFGPD represents cubic feet of gas per day, and BOPD represents barrels of oil per day.

Geographical Names

As in other Arabic countries the method of transliterating geographic names has changed with time. Publications before independence used transliterations based on Italian orthography. Between 1951 and 1978 transliterations moved towards a standard Arabic system, as reflected on the maps published in Libya during this period. In 1978 a National Atlas of Libya was published which introduced a new system of Libyan transliterations. As an example of these changes the Italian rendering of Bengasi was subsequently replaced by Benghazi and is now rendered as Binghāzī in the National Atlas. The 1978 system has been formally adopted in Libya, and has been used in most Libyan publications since then including the proceedings of the all the geological symposia held in in Libya since 1978. The 1978 transliteration system is used in the current work but without the accents, which are of little significance to non-Arabic speakers. Transliterations of names not shown in the National Atlas have kindly been provided by Dr Mustafa Salem of Al Fatah University in Tripoli. In some cases the changes are startling. El Gheriat ech Cherguia is now Al Qaryat ash Sharqiyah, Sceleidima is now Ash Shulaydimah and Bu Ngem is now Abu Njaym. In other cases, names have changed completely. Serdeles is now Al Awaynat, Barce is Al Marj, and amongst oil and gas fields Zelten has been renamed Nasser and Ora (Awra) has been renamed Tibisti. In an effort to assist the reader, a glossary of old and new transliterations is provided as an appendix. In his book on the *Geology of Egypt and Libya* Tawadros gave a fascinating insight into place-name changes in post-Revolutionary Libya. Jardas al Abid (place of the slaves) is now Qasr al Ahrar (fort of the free).

Tripolitania, Fezzan and Cyrenaica are no longer administrative regions, and have been replaced by the regions of Tarabulus (Tripoli), Sabha, Binghazi and Al Khalij. However the old regional names are still well known, and are used in this work without any administrative significance. Libya is formally named the Great Socialist People's Libyan Arab Jamahiriya, but in the interests of brevity is referred to as Libya throughout this book.

Stratigraphic Names

If the geographic name changes in Libya are complex, the stratigraphic nomenclature is byzantine. There are three principal 'rule books' for stratigraphic nomenclature: the *International Stratigraphic Guide*, edited by H.D. Hedberg (1977) published by John Wiley, New York, the *Code of Stratigraphic Nomenclature* published by the North American Committee on Stratigraphic Nomenclature (last full version 1983), and the *International Stratigraphic Guide* (Salvador, 1994) published by the International Union of Geological Sciences and the Geological Society of America. All three guides have much in common. A fundamental principle is that a stratigraphic name should not be changed, unless there are sound and justifiable reasons. But what if a name is given a new transliteration? What if the geographic name of a type locality is changed – as in the case of Jardas al Abid as mentioned above? Would it be logical to have the type section of the Jardas al Abid Formation at Qasr al Ahrar? The codes give no guidance on these problems, and opinion amongst geologists is divided. Barr and Weegar and Tawadros argued for strict retention of the original name, but many geologists have preferred to modify the stratigraphic nomenclature to match the geographic name. This approach was adopted for the Geological Map of Libya and has become more and more common in recent publications. The latter approach is adopted in the present work – not without some misgivings, but in the interests of practicality. It is surely not logical to have the type section of the Gialo Formation on the Jalu field, or the type section of the Jefren Marl at Yifran or the Scecsiuc Formation at Shakshuk. A further complication arises with the subsurface stratigraphy. Many formations have been defined with reference to subsurface type sections in wells. Most of these conform to the rules of stratigraphic nomenclature and are valid names, but some are very confusing such as the relationship between the Lutetian Al Jir Formation defined at outcrop, and the Ypresian Al Jir (Gir) Formation defined in the subsurface. These problems are discussed in the text but there is an urgent need to resolve these anomalies. The organisers of the Sirt Basin Symposium established a Stratigraphic Committee in 1993 to address these issues, but the committee has yet to issue its recommendations.

Well Terminology

For the purposes of this book a wildcat well is an exploration well designed to test either a new structure or new reservoir. An appraisal (or delineation) well is designed to define the size and extent of a hydrocarbon pool. A development well is planned with the specific intention of producing hydrocarbons from a proven pool. The date on which drilling commenced is known as the 'spud' date. In this book all the well statistics relate to spud date. For example a well spudded in December 1985 which was not completed until March 1986 is regarded as a 1985 well. Wells which yield no sign of hydrocarbons are 'dry holes'.

Well Nomenclature

The system of well numbering in Libya is unique, and requires some explanation. Three types of concession have been awarded – numeric licences in the period up to 1968 (no prefix), joint venture concessions from 1968 to 1973 (prefix LP), and production-sharing concessions from 1974 to the present time (prefix NC, for new concession). Prospects on each concession are identified alphabetically, and wells are identified numerically. The first well on the first prospect on concession 11, for example, is A1-11. Subsequent wells on the same structure are A2-11, A3-11 etc. The first well on the next prospect is B1-11. After prospect Z the next prospect is AA, and the next after ZZ is AAA, and so on. Thus C10-6 is the tenth well on the third structure drilled on concession 6, and F4-NC 100 is the fourth well on the sixth structure drilled on new concession 100. This distinction is necessary because there is still an original concession 100. On concession 59 the alphabet has already been scrolled through five times, so a recent well has the sequential number FFFFFF1-59, usually shortened to 6F1-59 (the 131st prospect to be drilled on concession 59). Significant discoveries are usually named, but names have sometimes changed. The Zaltan field was discovered by the C1-6 well, but the name was subsequently changed to Nasser in honour of President Nasser of the U.A.R. Most minor discoveries have not been named and are still identified by their prospect number.

Reserves

It is usual to distinguish between oil (or gas) in place (hydrocarbons in the ground under reservoir conditions), original recoverable reserves (the amount of hydrocarbons recoverable, at surface conditions, prior to production start-up), and remaining recoverable reserves (the amount of hydrocarbons remaining, at surface conditions, at the date of the estimate). Estimates are also made of undiscovered reserves (reserves yet-to-be- discovered). Produced oil, remaining reserves plus undiscovered reserves are known as ultimate potential, the intention of which is to estimate the total amount of hydrocarbons recoverable from an area by the end of production.

Field Discoveries

For the purposes of this book 'discoveries' are wells which have tested some amount of hydrocarbons. Commercial discoveries are those which have either been developed and put on stream, or are planned to be brought on stream. For the sake of convenience, the date of discovery of a field is the same as the spud date of the discovery well. Giant fields, as normally defined, are those with recoverable oil reserves greater than 500 million barrels of oil. Super giants, of which there are none in Libya, have reserves greater than 10 billion barrels.

Chapter 1

HISTORY of LIBYAN OIL EXPLORATION and PRODUCTION

1.1 Before Independence

It is one of the ironies of history that the North African campaign of the Second World War was won and lost largely due to the ability of the opposing sides to supply their forces with petroleum which had to be transported by sea at great risk and at enormous cost to the ports of Libya and Egypt. Less than twenty years later oil was being produced from the Libyan desert in quantities which could have supplied both armies many times over.

Italian geologists were active in Cyrenaica, the Tripoli area and the coastal belt from 1901, and by 1912 the first geological map of Libya was published.[1] In the 1920's a second generation of Italian geologists systematically investigated the more remote regions of Kufrah, Fezzan and the Tibisti Mountains under the leadership of Professor Ardito Desio who went on to become the doyen of Libyan geologists, publishing over 120 papers on the geology of Libya during the period 1927 to 1975.[2] The mapping programme showed the presence of four principal sedimentary basins, three of which are filled with dominantly Palaeozoic sediments, and one, the Sirt Basin, filled with mainly Mesozoic and Cainozoic sediments (Figure 1.1).

Traces of methane were reported in a water well drilled near Tripoli in 1914[3] and subsequently gas was noticed in several wells drilled on the coastal plain and on the Jabal Nafusah. In 1926 Crema found traces of oil in a water well near Tripoli.[4] This led to a two-year reconnaissance campaign by Agip in Tripolitania and the

drilling of one dry hole. Further reconnaissance work in 1940 was cut short by the outbreak of war and the Italian invasion of Egypt. As a result of these investigations Desio identified the Sirt Basin as being a favourable area for hydrocarbon generation and accumulation.[5]

The war took a heavy toll on Libya. Cyrenaica, which had been the site of some of the fiercest fighting, was largely destroyed and most of the Italians had fled. The economy was in ruins and the population totally impoverished. The Allies established a presence in Libya as early as 1943 and after the war the strategic value of Libya quickly became apparent. Fezzan was regarded by the French as an appendage to Algeria and French Equatorial Africa, the United States established an air base close to Tripoli, and the British regarded Libya as a valuable addition to their ring of Mediterranean bases. The Libyans themselves had aspirations for independence, but with no clear programme on the form that independence should take. Wrangling between the various powers remained unresolved, and in 1948 the matter of Libya's future was referred to the United Nations. The resolution for independence for Libya was approved by the General Assembly in November 1949, and independence became a reality on 24th December 1951 when Libya was established as a federal constitutional monarchy.[6]

1.2 The Fledgling Libyan Oil Industry

Within two years of independence a Minerals Law was enacted which permitted the allocation

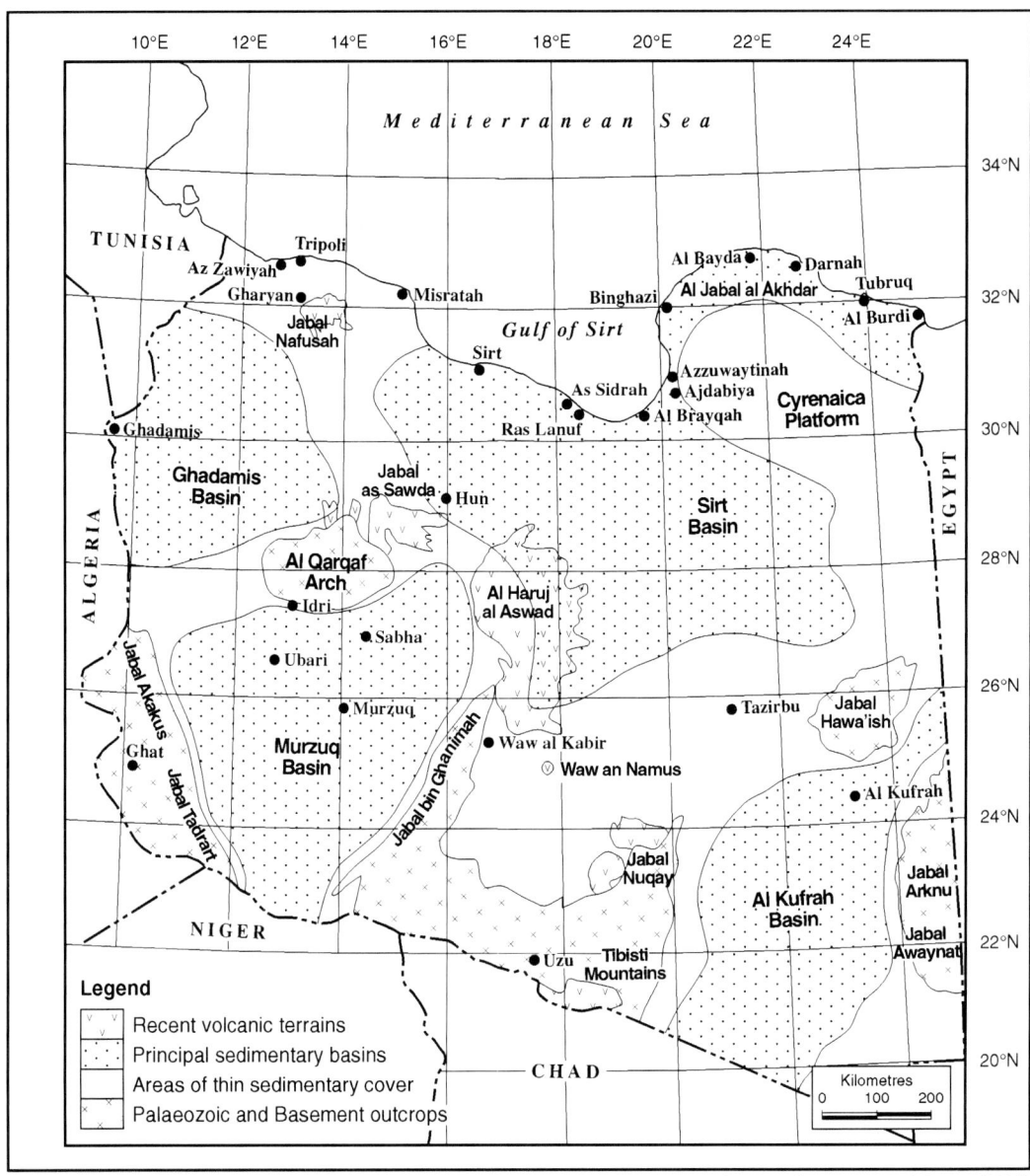

Source: *Geological Map of Libya (1985), Futyan and Jawzi, (1996)*

Figure 1.1   Libya: Principal Physiographic Features

The Ghadamis, Murzuq and Al Kufrah Basins are principally Palaeozoic basins whereas the Sirt Basin is a Mesozoic and Cainozoic basin. The volcanic rocks are Neogene to Recent in age.

of preliminary prospecting permits to foreign oil companies. Eleven major international companies obtained permits including Esso, Shell, BP and CFP. Drilling operations were not permitted, and the areas awarded for prospecting were not exclusive.[7] In June 1955 the first Libyan Petroleum Law was published. The law declared that all sub-surface minerals were state property, and established the rules under which concessions could be awarded. The intention of the drafters of the law was to encourage an open-door policy in regard to exploration and to develop a competitive attitude between companies. The country was divided into four petroleum zones with each zone carrying different rental charges and work commitments. Almost certainly there was an underlying intention to prevent the major oil companies from gaining a position from which they could exercise effective control over the Libyan oil industry. The terms of the law in respect of royalties, tax and rents were regarded as favourable by the industry and such as would encourage foreign participation. The law however contained two items which were to provoke major problems in the future. Royalities were defined as being payable on the value of crude oil 'on the field', and were allowable as a tax credit. There was no reference anywhere in the law to posted price. Secondly, the law provided specific protection against retrospective legislation being applied to a concession agreement without the consent of both parties.[8]

The areas which appeared most favourable to the oil companies in 1955 as a result of the reconnaissance programmes were northern Cyrenaica where major structures had been observed, and adjacent to the Algerian border where exploration by French companies in the Illizi Basin was very active. The first two concessions, awarded in November 1955, were in these areas: concession 1 was awarded to Esso adjacent to the Illizi Basin, and concession 2 was awarded to Nelson Bunker Hunt in northern Cyrenaica.[9] In December a French consortium operating in Algeria discovered the large Edjeleh field on the very boundary of the Esso concession. Forty-five more concessions were awarded in December 1955 and January 1956, and a further 42 concessions were awarded in the period up to

June 1960 (Figure 1.2).[10] As a result of these awards 70% of the land area of Libya was placed under licence, including all but the most unprospective or inhospitable areas. In west Libya the entire Ghadamis and Murzuq Basins were licensed, and further east all of Cyrenaica and the Sirt Basin were covered by concessions. Even at this early date some of the concessions were offshore, mostly in the Gulf of Sirt, but also off the coast of Cyrenaica. The principal concession holders at this time were Esso, Mobil, Oasis (a consortium formed by Amerada, Continental and Marathon), Amoseas (a consortium of Texaco and Socal), Gulf, BP, Shell, and CFP. Among smaller American companies were Libyan American (a subsidiary of Texas Gulf), Nelson Bunker Hunt and W.R Grace. European companies were represented by Deutsche Erdoel, Wintershall, Elwerath, CORI, Ausonia Mineraria and SNPA.[11] It could be fairly claimed that the Libyan Government had attained its objective of a diverse portfolio of licence holders with no one company holding a dominant position.

1.3 Exploration Activity, 1956 to 1958

The first wildcats to be drilled in Libya under the new Petroleum Law were located mostly on structures identified by surface mapping. Libyan American spudded well A1-18 in April 1956, only four months after the licence was granted.[12] The well tested a large anticlinal structure in northern Cyrenaica which proved to be dry. In 1957 four companies were involved in drilling campaigns. Libyan American drilled two more dry holes on concession 18, and BP drilled a dry hole in the area near Tripoli investigated by Agip before the war. Esso began operations in concession 1 close to the recent French discoveries in Algeria. Their second well made a gas discovery which, on appraisal, turned out to be a significant find containing both oil and gas in multiple reservoirs. Mobil drilled the first well in the Sirt Basin, A1-57, which was drilled primarily to evaluate the Cainozoic section on the Az Zahrah Platform, but proved to be dry.[13]

Drilling began in earnest in 1958. Twenty-eight wildcat wells were spudded and the first major discoveries were made. In west Libya Esso continued to evaluate concession 1 with six further wells, several of which had gas shows, and Shell began evaluating the area north of concession 1. Gulf drilled three dry holes in concession 67 on the edge of the Murzuq sand sea, and tested oil in well A1-68 in an area where major discoveries were to be made in later years. CFP made a small oil discovery in Devonian sandstones at Tahara in concession 49, and Shell made the first discovery in the northern Ghadamis Basin at Tlakshin in the Akakus reservoir of late Silurian age. Further east BP drilled two wells in the area between west Libya and the Sirt Basin without success.[14]

Drilling continued in the Sirt Basin and in April 1958 the first undoubted commercial discovery was made when the A1-32 well of Oasis discovered the Bahi field which contained oil in Palaeocene carbonates in a gentle low-relief structure on the western margin of the Az Zahrah Platform. This was followed in September by a much larger discovery in the B1-32 well which Oasis named the Az Zahrah (Dahra) field. On the adjacent block Mobil drilled the A1-11 well which encountered good shows in the same reservoir interval, and which on appraisal, proved another major discovery, which Mobil named the Al Hufrah (Hofra) field. Amoseas was active on the western flank of the Sirt Basin, but without success. Esso began operations in the central Sirt Basin and reported a small discovery in Eocene carbonates in concession 5.[15] The 1958 drilling programme had resulted in seven discoveries of which three were commercial (Bahi, Az Zahrah and Al Hufrah).

1.4   Bonanza, 1959-1961

By the end of 1958 three factors were becoming clear: firstly, in west Libya several small oil and gas accumulations had been found in Palaeozoic clastic reservoirs, secondly it was apparent that the major commercial oil accumulations were being found in the Sirt Basin, and thirdly, the large structures mapped in Cyrenaica had proved disappointing. Desio's prediction that the Sirt Basin appeared favourable for hydrocarbons was beginning to be borne out.

The discovery which changed the face of the Libyan oil industry came in 1959 when Esso drilled their C1-6 well. The well was located on an anomaly on the Zaltan Platform which proved to be a Palaeocene carbonate complex with excellent reservoirs and a 105m oil column. Oil was tested at a rate of 17,500 barrels per day.[16] Subsequent appraisal showed that the discovery was a giant field with reserves of 2.2 billion barrels. In addition the field was shown to contain a number of subordinate oil-bearing reservoirs. The field was named Zaltan, but was renamed Nasser in honour of the President of the United Arab Republic when he died in 1970. As a result of this discovery drilling activity increased dramatically and whilst some drilling continued in west Libya the focus of attention now swung firmly to the Sirt Basin.

The period from 1959 to 1961 were the formative years of the Libyan oil industry. Other successes were announced in rapid succession. Oasis, exploring in the area south of Zaltan, made two major discoveries with their first two wells in concession 59. The A1-59 well discovered the giant Al Wahah field in which the reservoir was a late Cretaceous nearshore carbonate, whilst the B1-59 well found the Dayfah (Defa) field, a huge Palaeocene carbonate build-up, differing in many respects from Zaltan, but with equally good reservoir characteristics. Oasis also continued exploration

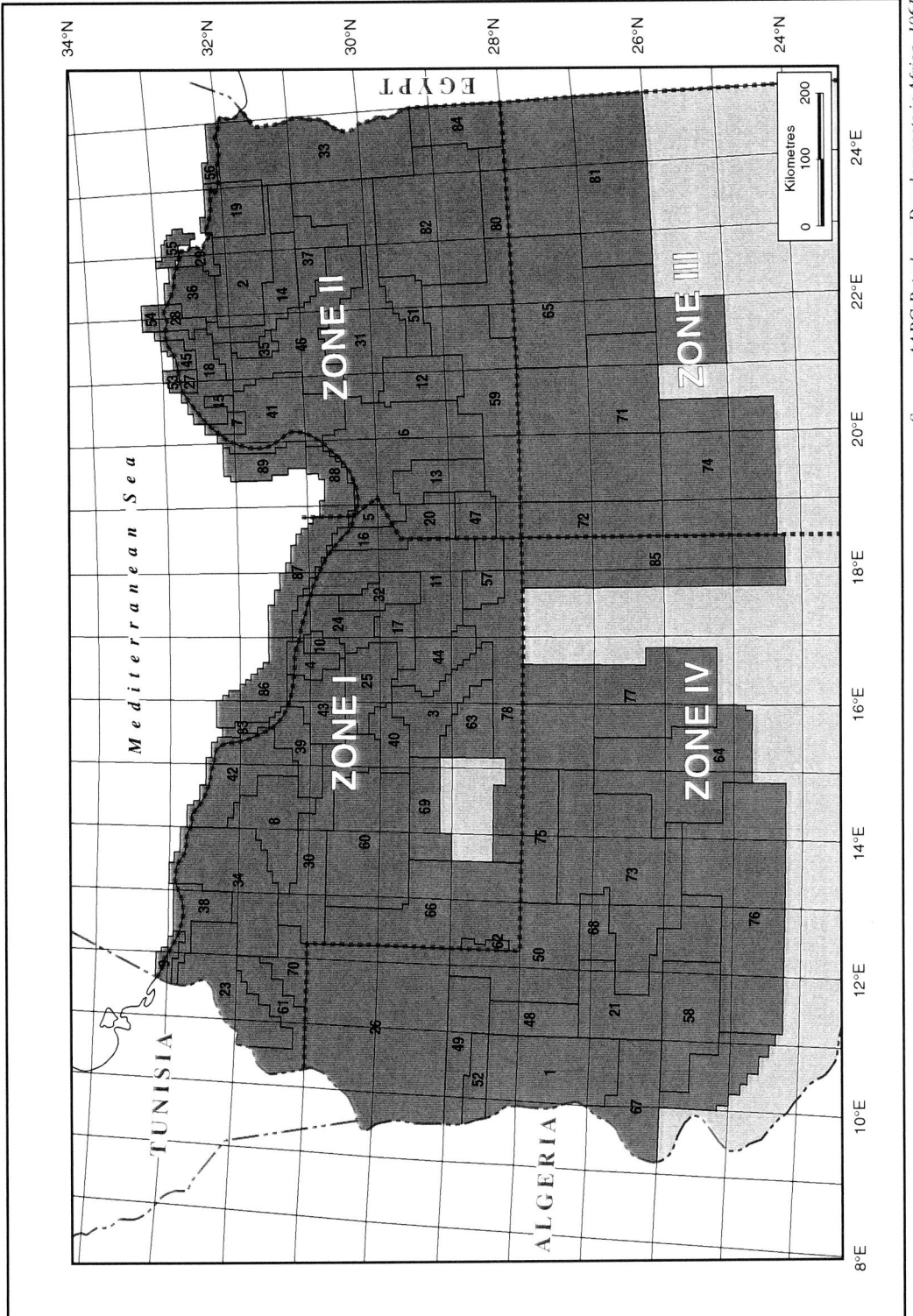

Figure 1.2   Concession Map, 1960

Source: AAPG Petroleum Developments in Africa, 1961

This map shows the maximum extent of concession awards prior to the first relinquishments. The area covered by concessions was 12,430km², or around 70% of the entire area of the country.

on the Az Zahrah Platform in concession 32 where their F1-32 well discovered the Az Zahrah East pool, which subsequently proved to be an extension of Mobil's Al Hufrah field. Mobil began evaluation of concession 12 in the eastern Sirt basin and discovered the giant Amal field with their second well, with oil reservoired in basement quartzites and Cretaceous basal sands. Amoseas had their first success in concession 47 when their second well found the Al Bayda (Beda) field with oil in several Palaeocene carbonate reservoirs. Esso, who had taken over the operatorship of concession 17 from Libyan American, discovered a large oil pool at Mabruk, but the reservoir was thin and heavily faulted. In west Libya CFP in concession 23, Gulf in concession 66 and Oasis in concession 26 all reported discoveries, but on a much smaller scale than those in the Sirt Basin.[17]

Of the sixteen discoveries announced in 1959 (from a total of 39 exploration wildcat wells), six were giant commercial discoveries (Al Bayda, Zaltan, Amal, Az Zahrah East, Al Wahah and Dayfah). The year 1959 can be said to mark the coming of age of the Libyan oil industry. Libya was rapidly becoming a significant force on the international oil scene. The Sirt Basin was developing as a major new oil province, and one with a wide variety of oil reservoirs and structural styles.

Exploration drilling continued on an upward trend throughout 1960 with 68 exploration wells being drilled. In the Sirt Basin Amoseas continued exploring concession 47 and discovered the medium-sized Al Kotlah field, BP commenced operations in concessions 65 and 80 in the southeastern part of the basin, and Esso continued work in the central Sirt Basin. Their H1-6 well found a gas field which was named Mn Berber (subsequently renamed Assumud), and in concession 20 Esso discovered another giant oilfield at Ar Raqubah (Raguba), with oil present in late Cretaceous

calcarenites. Shell began exploration in concession 41 and made a small discovery at Antlat near Binghazi in Eocene carbonates.[18] The tally for 1960 was two commercial oil discoveries (Ar Raqubah and Al Kotlah), and one gas discovery (Assumud) which was developed at a later date.

In 1961 BP's third well on concession 65 discovered the giant As Sarir field, and in so doing opened up the whole of the south-east Sirt Basin.[19] Oasis made an equally spectacular discovery at Jalu in the eastern Sirt Basin with multiple oil-bearing reservoirs of Eocene and Oligocene age. Oasis had further successes in concession 59 at Samah, Masrab, Harash, Zaqqut, and Khalifah, and Esso had a small discovery at Al Meghil in concession 6. Elwerath, one of the small German independents, reported the first indication of oil in the Zallah Trough which was subsequently to become a major producing area.[20]

In west Libya CFP and Gulf announced further small discoveries in the Ghadamis Basin on concessions 23 and 66.

Companies which had made commercial discoveries rapidly set about developing them. Development wells were drilled and pipelines and pumping stations were installed. Esso laid a 36-inch pipeline from Zaltan to Al Brayqah where they established a port and export facility, and the first cargo of crude oil from Libya was shipped in September 1961. Oasis established a similar facility at As Sidrah, and Mobil at Ra's Lanuf. By the end of 1961 ten giant fields had been discovered, over 200 development wells had been drilled and production was underway.[21]

1.5 Events leading to the Petroleum Law of 1965

By 1960 the first relinquishments took place under the terms of the 1955 law which required 25 per cent surrender within five years. In 1961 six of the surrendered blocks were reassigned to Phillips and Pan American (later Amoco). No further concessions were awarded for the next five years. The reaction of the government to the discovery of major oil accumulations was to question the liberal terms which had been offered to companies before any oil had been discovered. In November 1957 companies were persuaded to voluntarily accept new conditions not covered by the Petroleum Law of 1955, but there was a limit to what could be achieved within the framework of the Law. The new conditions primarily concerned the waiver of depletion allowances and increase of royalty payments, which under the 1955 law, were deductible against tax. These changes represented minimal concessions on the part of the companies since they did not involve any increase in payments to the government.[22]

By 1960 the government felt that the 1955 law required revision in order to secure the national interest and bring the law more into line with practices in the Middle East. Amendments were published in 1961 after lengthy discussions with the operators. They gave the government greater freedom in the award of concessions, and they attempted to modify the way in which profits were taxed by establishing posted price as the basis for calculating tax charges. The government was anxious to obtain the approval of the operating companies since the original Petroleum Law specified that changes could only be implemented by mutual consent. The companies were able to extract major concessions from the government in return for their agreement which effectively nullified the government's objective.[23] In the next four years, during which production rapidly increased, companies calculated their tax liability on the basis of posted price. The posted price was established by the companies on the 'free market price' of actual cargoes shipped from Libyan ports. Under the new agreements they were then permitted to deduct expenses and

rebates which had the effect of substantially reducing their tax liability. At a time of falling oil prices it was felt that several operators in Libya were taking unfair advantage of the posted price rebate system.[24]

OPEC was established in 1960 to protect the interests of petroleum exporting countries and Libya became a member in 1962. OPEC was at that time concerned to establish a standard level of royalties and to ensure that royalties could not be offset against taxes, as was the case in Libya. In 1964 the international majors agreed to OPEC's proposals in return for a temporary reduction in the level of posted price. In Libya however the operators were divided. The majors were in favour of the OPEC proposals, subject to certain concessions, but the independents were not. In 1965, following a meeting of OPEC ministers in Tripoli, the Ministry of Petroleum Affairs set up a committee to look into the possibility of introducing the OPEC formula on royalties to all operators in Libya. The committee recommended implementation of the OPEC proposals, and the new law was promulgated by the Libyan Parliament in December 1965. The new Petroleum Law incorporated both the principle that royalties could not be used as credits against tax, and that expenses and rebates on the posted price were limited to the amounts set out in the OPEC formula. In return the government agreed to write-off all disputed tax discounts and rebates for the period up to the end of 1964.[25] Companies were given a deadline within which to accept the new law. Several of the independents resisted, but under threat of sanctions from OPEC they eventually gave their assent. The Petroleum Law of 1965 took immediate affect and applied to all concession holders. The new law removed the worst of the tax-avoidance abuses which had been possible under the old law and brought Libya more closely into line with her Middle Eastern neighbours.

1.6 Exploration and Production Activity, 1962-1965

There is no evidence to suggest that the posted price dispute had any significant influence on drilling activity. Libya still represented a highly profitable environment in which the oil was of excellent quality with low sulphur content, and transportation was not dependent on long tanker voyages or transit through the Suez Canal. Exploration activity developed rapidly following the major discoveries in 1959 to 1961. Wildcat drilling peaked in 1963 with 133 exploration wells totalling over one million feet of exploration footage. This marked the culmination of drilling activity on the original concessions. Thereafter the annual total declined steadily as more effort was diverted into development drilling and as acreage relinquishments began to reduce the area available for exploration. The fact that no new concessions were awarded between 1961 and 1966 during the posted-price dispute was also a contributory factor.[26]

The early pace of oil exploration in Libya was phenomenal. Ten billion barrels of reserves were discovered in the first three years, and a further ten billion barrels within the next two years. By 1962 most of the major producing reservoirs in the Sirt Basin had been discovered, the Eocene and Oligocene at Jalu, the Palaeocene at Zaltan, Dayfah, and Al Bayda, the Cretaceous at Al Wahah and Ar Raqubah, the Nubian at As Sarir, the basal Cretaceous and basement at Amal. The next phase, from 1962 to 1965, was one of consolidation in which the skills and expertise gained during the initial exploration phase were applied to finding more oil accumulations of the same type. Companies were also anxious to fully evaluate their acreage before the statutory relinquishments required under the terms of the original licences.

During the period from 1962 to the end of 1965 447 exploration wells were drilled, mostly in the Sirt Basin although significant efforts still continued in west Libya (Figure 1.3). In the western Sirt Basin Elwerath made a second discovery in the Zallah Trough (F1-78), and on the Az Zahrah-Al Hufrah Platform Mobil had four small discoveries in concession 11 (Abu Maras, Al Facha, Adh Dhi'b and Al Furud). The tectonic elements of the Sirt Basin are shown on Figure 6.9. Pan American and Phillips, two of the companies which were awarded relinquished acreage in 1961, reported small finds at Umm al Furud, Al Kuf and E-92 on acreage which had been relinquished by Mobil. Further south on the Al Bayda Platform Amoseas discovered the Al Kotlah, and Haram pools, and Mobil found the multi-reservoir Al Awra (Ora) field (later renamed Tibisti). Esso's efforts concentrated on concession 6 where further important oil discoveries were made at Al Jabal, Lahib and Ar Ralah, and gas at Al Hutaybah and As Sahel. In the eastern Sirt Basin Amoseas had a major success with their G1-51 well which found the giant An Nafurah field, with reserves in multiple reservoirs, and Mobil made further discoveries in the Amal area (Ar Rakb, Amal E, and Amal N). Oasis continued work, particularly in concession 59, and reported discoveries at Bilhizan, Kalanshiyu, and in well YY1-59. BP continued exploring the Nubian province in concessions 65 and 80, and discovered another major pool at As Sarir L, and the Italian company CORI had their first success in concession 82 when they discovered a deep Nubian reservoir in their R1-82 well. This well was drilled in the Hameimat Trough and is significant in indicating that oil is still preserved even at depths greater than 4200m. These discoveries brought the total number of commercial fields by the end of 1966 to thirty-six.[27]

In west Libya Gulf, CFP and Oasis reported a string of small discoveries in the Ghadamis Basin in Palaeozoic clastic reservoirs, and the government tried to persuade Gulf to develop its

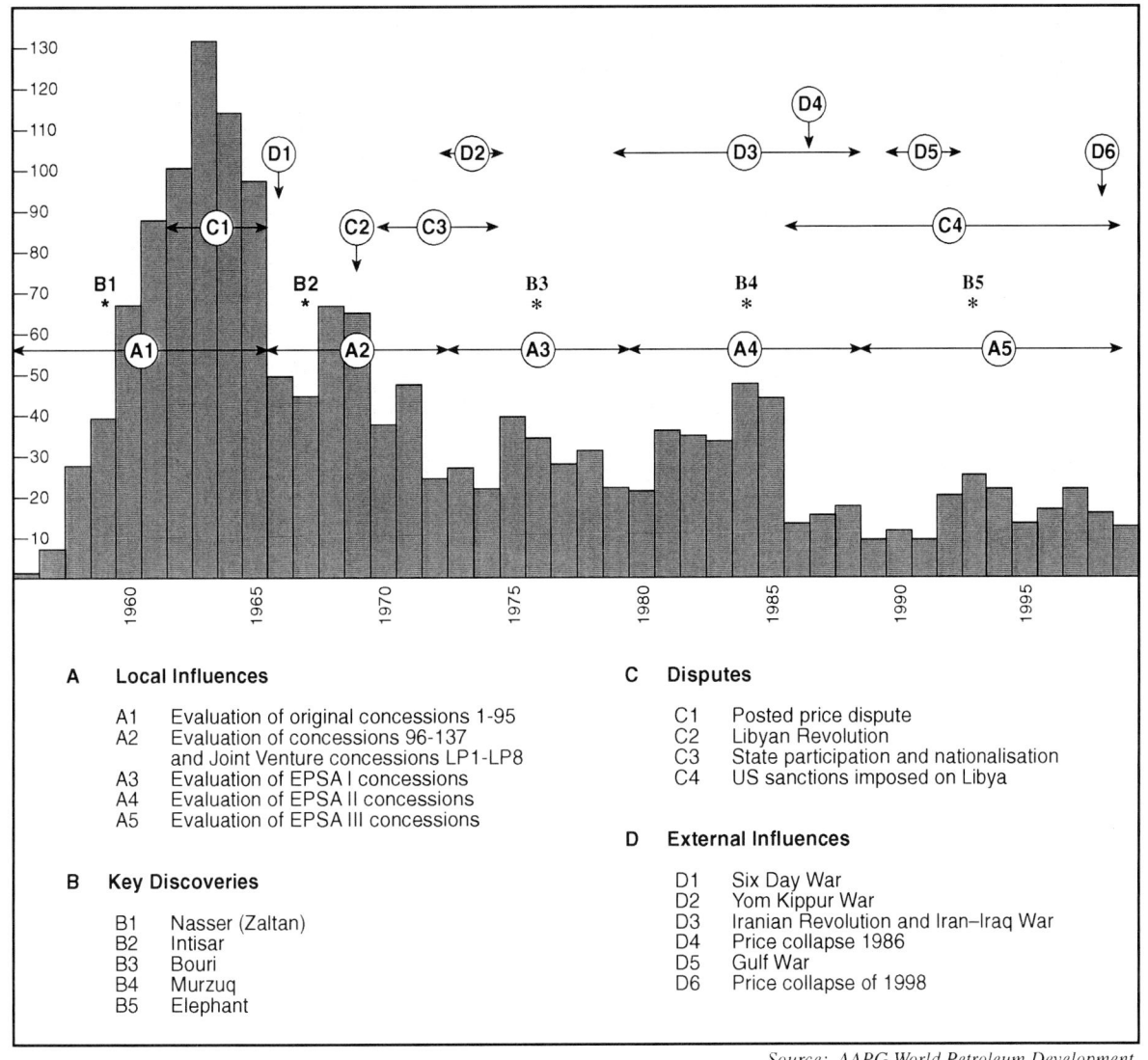

Source: AAPG World Petroleum Development

Figure 1.3   Number of Wildcat Exploration Wells Spudded per Year

The level of exploration drilling has been affected by a succession of adverse factors since 1966. The local peaks in 1968, 1975, 1984 and 1993 indicate evaluation of newly-awarded acreage.

discoveries on concession 66. Gulf however regarded them as sub-commercial and was unwilling to fund the development without financial assistance.[28] Offshore drilling began in the Gulf of Sirt with Libyan Atlantic spudding their first well in 1963. Thirteen offshore wells were drilled during the following five years. No

major discoveries were made but fair gas shows were reported in two wells offshore from Al Brayqah.[29]

Development drilling reached a peak in 1964, and fields were progressively brought on stream: Zaltan in 1961, Az Zahrah in 1962, Al Hufrah, Ar Raqubah, Samah and Al Wahah in

1963, Jalu, Dayfah, Al Bayda and Al Kotlah in 1964, and Al Awra, Jabal, Al Meghil, Al Furud, Umm al Furud and Ar Ralah in 1965. The pipeline systems were extended to link these fields to the respective marine terminals, and by the end of 1965 production had reached 1.22 million barrels per day, 95% of which was exported.[30] Companies which chose not to build their own pipeline and export facilities were obliged to make arrangements with other operators who had spare capacity, as in the case of Amoseas who made an arrangement to use Mobil's facilities (Figure 1.4).[31] In west Libya where most of the discoveries were small, no company was willing to finance the construction of a pipeline, despite the fact that about 200 million barrels of reserves had been discovered, and it was not until 1980 that a pipeline was eventually constructed to tap these reserves (Figure 1.5).

Most of the discoveries made to this date were oil accumulations with varying quantities of associated gas. There was no ready market for gas, particularly from the more distant fields, and for the most part the associated gas was flared. However in the northern part of Esso's concession 6 several non-associated gas fields had been discovered which, with the gas from Zaltan and Ar Raqubah, had the potential to produce large volumes of gas. In 1965 the decision was taken to build a gas liquefaction plant on the coast at Al Brayqah, and a gas pipeline was constructed to gather gas from these fields. The Al Brayqah facility was further expanded by the addition of a small oil refinery which came on stream in 1966, the purpose of which was to supply cheap refined products for Libya's small domestic market.[32]

1.7   New Concession Awards and Joint Ventures, 1966 to 1969

In 1965 the government invited applications for new concessions mostly on acreage relin-quished by the original concession holders. Awards were delayed until the new Petroleum Law came into effect, but in February 1966 forty-one new concessions were awarded. Much to their disappointment neither Esso nor Oasis were awarded new concessions, although Mobil and Shell both received blocks. The major winners were Occidental, at that time a small Californian independent, Phillips, Aquitaine and Agip. Most of the new concessions were awarded on acreage which had been relin-quished under the terms of the 1955 Petroleum Law. Occidental was awarded two prime blocks in the eastern Sirt Basin, and Wintershall and Agip also picked up good quality acreage in the same area. In the central and western Sirt Basin the main beneficiaries were Aquitaine, Shell and Union Rheinische. The last of the licences was an offshore block awarded to Aquitaine in 1968 adjacent to the Tunisian border. A number of awards were made to small speculative companies who, it subsequently transpired, could not meet the eligibility criteria of the Petroleum Law, and whose concessions were either revoked or reassigned.[33]

The situation in Libya was soon over-shadowed by events on the global stage. In June 1967 the Six Day War between Israel and its Arab neighbours led to the closing of the Suez Canal and an embargo by Arab producers on exports to western countries. When shipments resumed from Libya the price of crude oil had risen significantly and the Libyan government requested producing companies to adjust their posted price accordingly. The companies did not respond to this request and posted prices remained unchanged.[34]

In 1968, following the lead of several Middle Eastern countries, Libya began negotiations with ERAP-Aquitaine to set up a new type of joint venture exploration agreement. To achieve this objective the Libyan government established a national oil company, the Libyan General Petroleum Corporation (Lipetco). It was

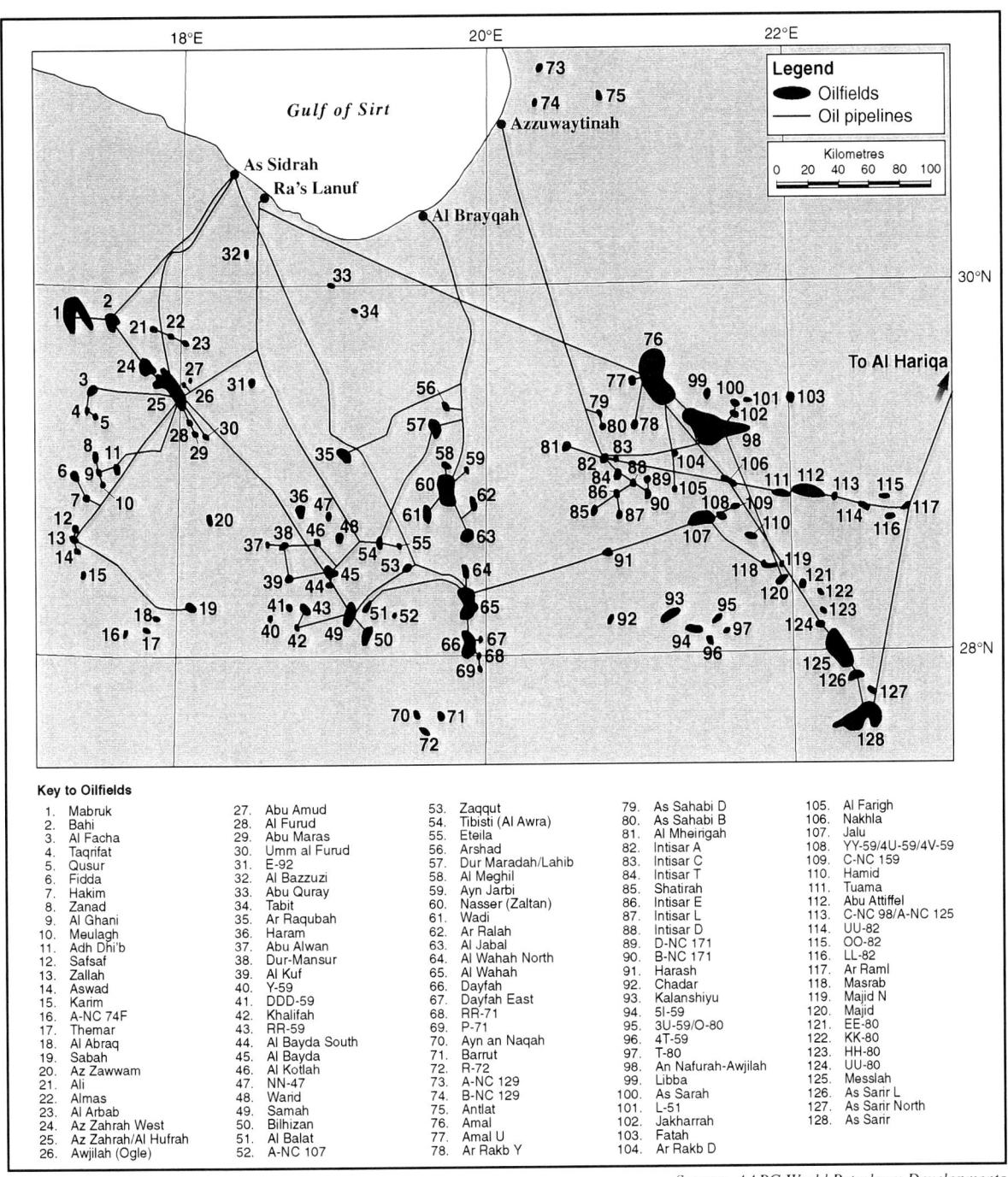

Key to Oilfields

1. Mabruk	27. Abu Amud	53. Zaqqut	79. As Sahabi D	105. Al Farigh
2. Bahi	28. Al Furud	54. Tibisti (Al Awra)	80. As Sahabi B	106. Nakhla
3. Al Facha	29. Abu Maras	55. Eteila	81. Al Mheirigah	107. Jalu
4. Taqrifat	30. Umm al Furud	56. Arshad	82. Intisar A	108. YY-59/4U-59/4V-59
5. Qusur	31. E-92	57. Dur Maradah/Lahib	83. Intisar C	109. C-NC 159
6. Fidda	32. Al Bazzuzi	58. Al Meghil	84. Intisar T	110. Hamid
7. Hakim	33. Abu Quray	59. Ayn Jarbi	85. Shatirah	111. Tuama
8. Zanad	34. Tabit	60. Nasser (Zaltan)	86. Intisar E	112. Abu Attiffel
9. Al Ghani	35. Ar Raqubah	61. Wadi	87. Intisar L	113. C-NC 98/A-NC 125
10. Meulagh	36. Haram	62. Ar Ralah	88. Intisar D	114. UU-82
11. Adh Dhi'b	37. Abu Alwan	63. Al Jabal	89. D-NC 171	115. OO-82
12. Safsaf	38. Dur-Mansur	64. Al Wahah North	90. B-NC 171	116. LL-82
13. Zallah	39. Al Kuf	65. Al Wahah	91. Harash	117. Ar Raml
14. Aswad	40. DDD-59	66. Dayfah	92. Chadar	118. Masrab
15. Karim	41. Khalifah	67. Dayfah East	93. Kalanshiyu	119. Majid N
16. A-NC 74F	42. RR-59	68. RR-71	94. 5I-59	120. Majid
17. Themar	43. Al Bayda South	69. P-71	95. 3U-59/O-80	121. EE-80
18. Al Abraq	44. Al Bayda	70. Ayn an Naqah	96. 4T-59	122. KK-80
19. Sabah	45. Al Kotlah	71. Barrut	97. T-80	123. HH-80
20. Az Zawwam	46. NN-47	72. R-72	98. An Nafurah-Awjilah	124. UU-80
21. Ali	47. Warid	73. A-NC 129	99. Libba	125. Messlah
22. Almas	48. Samah	74. B-NC 129	100. As Sarah	126. As Sarir L
23. Al Arbab	49. Bilhizan	75. Antlat	101. L-51	127. As Sarir North
24. Az Zahrah West	50. Al Balat	76. Amal	102. Jakharrah	128. As Sarir
25. Az Zahrah/Al Hufrah	51. A-NC 107	77. Amal U	103. Fatah	
26. Awjilah (Ogle)	52.	78. Ar Rakb Y	104. Ar Rakb D	

Source: AAPG World Petroleum Developments

Figure 1.4   Libya, Sirt Basin, Principal Oilfields and Pipelines

About 100 oil pools in the Sirt Basin have been developed and put on production, and there are probably a further 15 to 20 which may be developed in future.

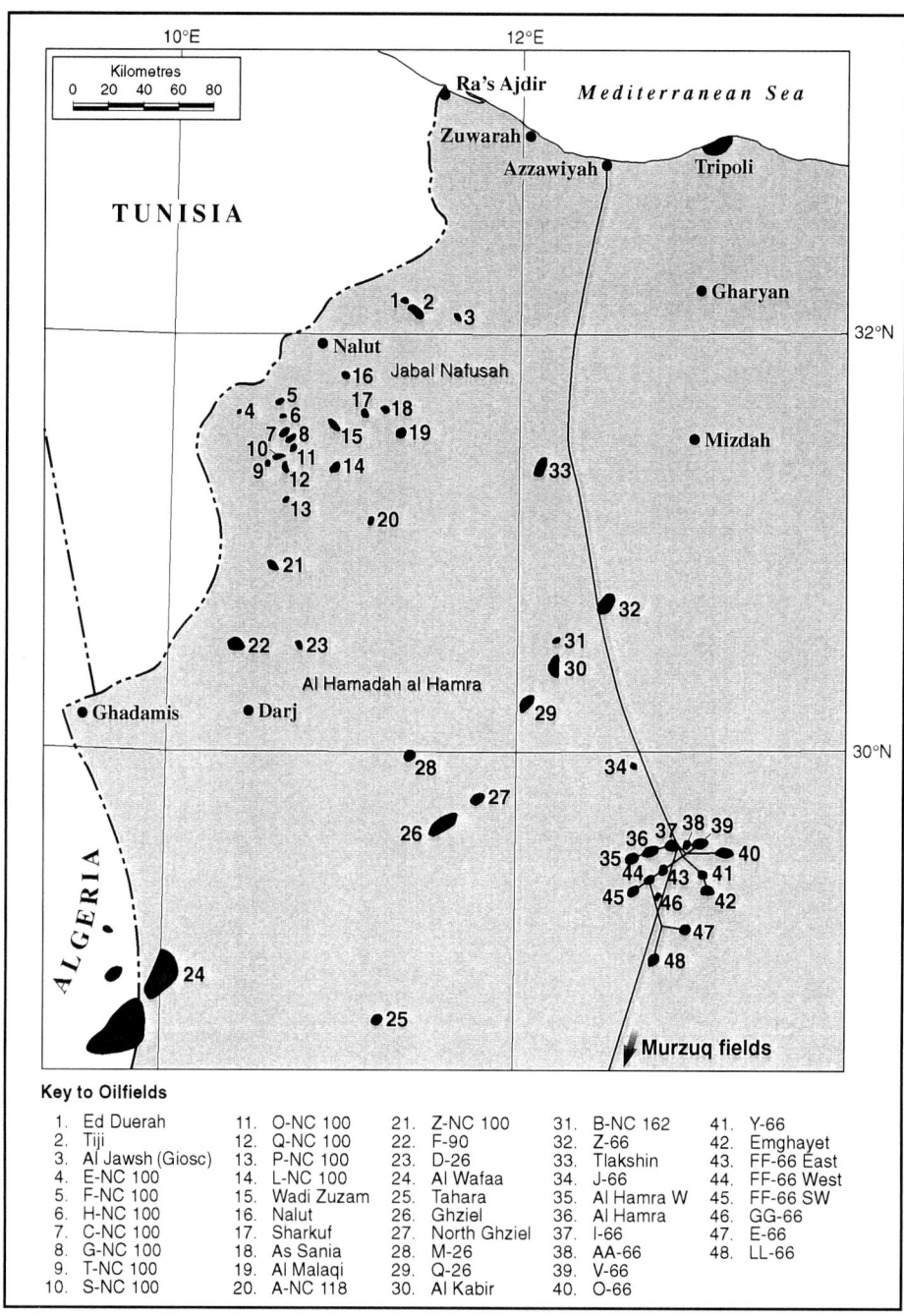

Source: *AAPG World Petroleum Developments, Futyan (1996)*

Figure 1.5   Libya, Ghadamis Basin, Principal Oilfields and Pipelines

At present only the old concession 66 fields have been developed and put on production. There are numerous undeveloped fields, but these are generally small. The Al Wafaa gas-condensate field, near the Algerian border, was being developed by Agip in 2001.

given a broad remit to participate on behalf of the Libyan government in exploration, drilling operations, production, refining and transportation of oil and gas, both domestically and internationally. The joint venture agreement with Aquitaine signed in April 1968 (LP 1) marked a new beginning and after that time no further licences were issued under the 1965 Petroleum Law. Further joint venture agreements were quickly made with Shell (LP 2 and LP 3), Agip (LP 4) and Ashland (LP 5). The terms under which these agreements were signed varied, but in general they involved substantial work commitments on the part of the foreign partner. Exploration costs would be borne entirely by the foreign operator and Lipetco would only become involved after a commercial discovery. Its level of involvement would depend on the production rate. The foreign partner would finance the appraisal and development costs and Lipetco's share of these costs would be recovered from oil sales.[35]

1.8 Exploration and Production, 1966-1969

Rapid progress was made in the three years following adoption of the 1965 Petroleum Law. Exploration activity increased as companies began to evaluate the new concessions granted in 1966. Occidental in particular was very aggressive. The company, under the dynamic leadership of Armand Hammer, was awarded two blocks in March 1966 in areas which had been relinquished by Oasis and Mobil. After three dry holes in concession 102 Occidental struck oil in well D1-102 which flowed at a rate of 14,860 barrels per day, second only to Zaltan in flow-rate. The discovery was named Awjilah (Augila). This success was shortly over-shadowed by a much more spectacular series of discoveries in concession 103. The first well, drilled on the site of an old Mobil reconnaissance camp, discovered the Intisar A reef (originally named Idris) which flowed oil at a rate of 43,000 barrels per day. The next four wildcats were all successful in discovering similar pinnacle reefs. Intisar D, the fourth and largest discovery flowed at a rate of 74,867 barrels per day, by far the highest flow-rate ever measured in Libya.[36] These discoveries, made by a small California independent, on acreage relinquished by one of the majors, caused a sensation, and incidentally transformed the fortunes of Occidental within a space of a few short months. Independent evaluation by De Golyer MacNaughton in November 1967 showed that the four reefs contained recoverable reserves of over three billion barrels of oil. Occidental constructed a 40-inch pipeline to Azzuwaytinah where they built a port and export facility. First oil arrived at the terminal on February 16th 1968, less than one year after the date of the first discovery.[37]

Other operators also had major successes. Agip discovered the Abu Attiffel field deep in the Hameimat Trough in the same Nubian play as CORI's earlier discovery at Ar Raml. Abu Attiffel produces oil from a depth of more than 4200m. The small German independent Wintershall had two discoveries in concession 97 at Hamid and Tuama, and a much larger discovery in a Triassic reservoir in the Maragh Graben in concession 96 which was named Jakharrah. Aquitaine made four substantial discoveries in concessions 104 and 105 at Mansur, Majid, D-104 and East Masrab. Pan American discovered the As Sahabi fields, and Amoseas the Dur field. Mobil were successful at Al Farigh, Chadar and Dur Maradah, and BP had further successes in the Nubian province. Offshore, Libyan Atlantic reported gas in their B1-88 well. In total twenty-five discoveries were made during this period. Further pipelines were completed. Mobil connected the Amal field to Ra's Lanuf in January 1966, Amoseas put the An Nafurah field on-stream in June 1966, and BP completed the 514km

34-inch line from As Sarir to Al Hariqa near Tubruq in December 1966. The pipeline network was extended to tie in several other fields during the next three years, including Lahib (1967), Awjilah, Intisar A, and Intisar D (1968), Majid, Mansur and Dur Maradah (1969), and by the end of 1969 forty-one fields were on production.[38]

Production increased steadily from 1.57 million barrels per day in 1966 to 3.38 million barrels a day in 1970, putting Libya amongst the top ten oil producing countries (Figure 1.6).[39]

Nevertheless a significant milestone was reached in 1967. Discovered reserves passed 32 billion barrels in 1967, but production was gaining momentum and by the end of 1967 cumulative production had reached 2.1 billion barrels. From this date production generally exceeded the discovery of new reserves, and the level of remaining reserves has steadily decreased since 1967.

1.9   The Revolution and its Aftermath, 1969-1974

The monarchy in Libya was overthrown by a military coup on 1st September 1969, led by Colonel Mu'ammar al Qadafi. A Revolutionary Command Council was established to take over the administration of the country, and new ministerial appointments were announced. The new government made it clear that existing agreements with foreign oil companies would be honoured, and that there was no intention of nationalising the Libyan oil industry. One of the first priorities of the new government was to negotiate the withdrawal of British and American forces from Libya. Agreement was reached surprisingly quickly, and British forces were withdrawn by 28th March, and the American evacuation of the Wheelus Field was completed by 30th June 1970.[40] Henceforth the British and American oil companies would no

longer shelter under the protection of their respective armed forces. The new government was anxious to secure a more equitable posted price for oil produced in Libya, and when negotiations failed to make significant progress the government responded by imposing draconian production cuts on the grounds that wells were being overproduced and were causing damage to the reservoirs. The cuts were first applied to Occidental which by 1969 was producing over 800,000 barrels of oil per day, mainly from the Intisar fields. In June 1970 their production was cut to 500,000 barrels per day and in August to 440,000 barrels per day. Similar cuts were imposed on Amoseas, Oasis, Mobil and Esso. Occidental sought help from the other producers, particularly Esso, to make up the shortfall, but no company was willing to assist.[41] Further pressures were applied by the government. In July the government nationalised the marketing of oil products in Libya, a port tax was imposed on all oil shipped from Libyan ports, and Esso was prevented from starting up its LNG plant at Al Brayqah.[42] This confrontation had an immediate effect on exploration drilling which fell by 50% within a few months. Because Libyan oil accounted for ninety percent of Occidental's total worldwide production it was forced to concede not only an increase in posted price of 30 cents cents per barrel, but an increase in tax to 58% and an adjustment to the base gravity on which the posted price was calculated. It was plain for all to see that the new government was going to be much more ruthless in its actions than the previous one. Within a month all of the producing companies had been forced to agree to similar increases, with the exception of Gulf and Phillips, who chose to relinquish their concessions rather than agree to the new measures. The production cuts which had been imposed on Occidental in the summer were lifted, but the cuts in the other companies production remained in force. Other OPEC

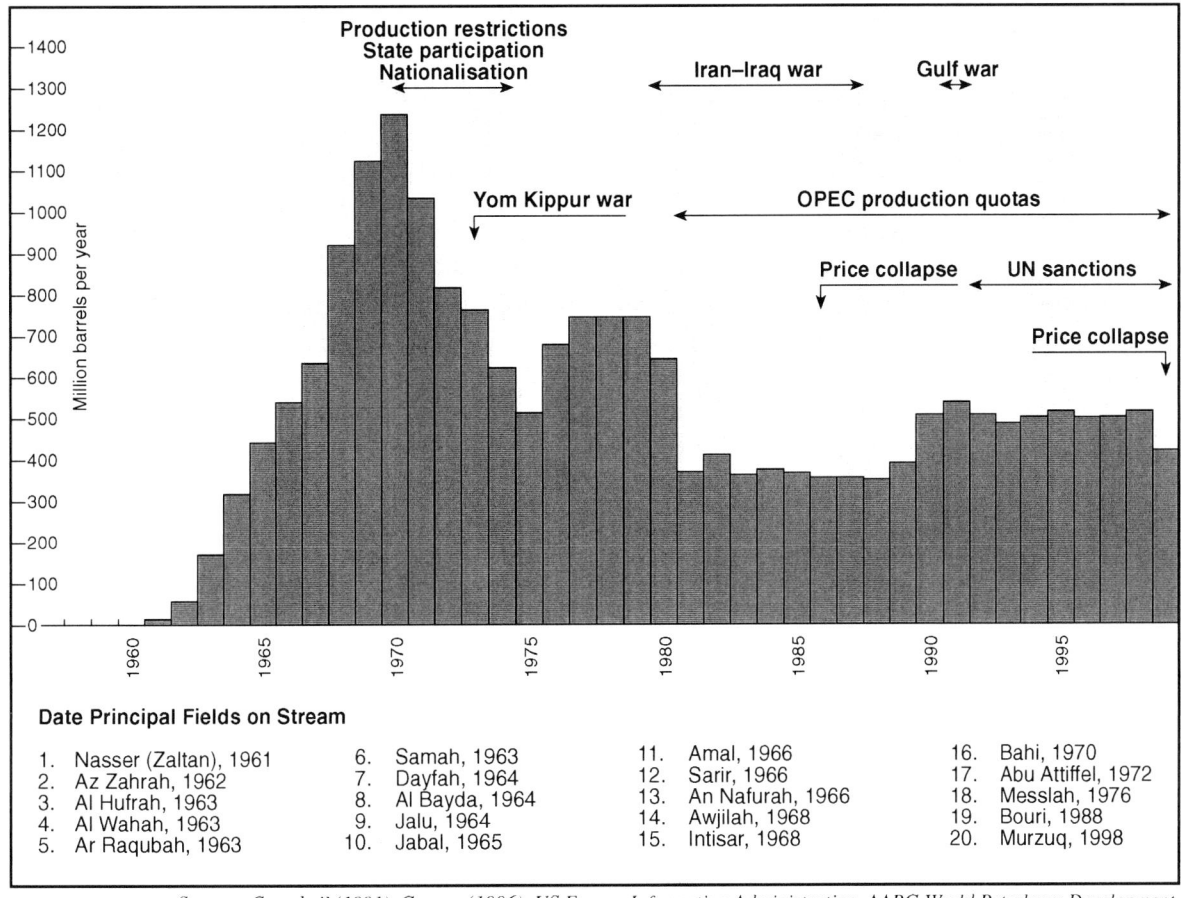

Date Principal Fields on Stream

1. Nasser (Zaltan), 1961	6. Samah, 1963	11. Amal, 1966	16. Bahi, 1970
2. Az Zahrah, 1962	7. Dayfah, 1964	12. Sarir, 1966	17. Abu Attiffel, 1972
3. Al Hufrah, 1963	8. Al Bayda, 1964	13. An Nafurah, 1966	18. Messlah, 1976
4. Al Wahah, 1963	9. Jalu, 1964	14. Awjilah, 1968	19. Bouri, 1988
5. Ar Raqubah, 1963	10. Jabal, 1965	15. Intisar, 1968	20. Murzuq, 1998

Source: Campbell (1991), Gurney (1996), US Energy Information Administration, AAPG World Petroleum Development

Figure 1.6   Libyan Oil Production (million barrels per year)

This figure shows peak oil production in 1970 with a marked decline due to government production restrictions during the period 1970 to 1974, and OPEC quota limitations after 1980.

members were not slow to follow Libya's example and in December 1970 at a meeting in Caracas OPEC ministers resolved to adopt the Libyan measures as standard.[43]

Belatedly the operators in Libya decided that action must be taken to stop the government applying pressure against the weakest members. As a result they established a secret agreement which came to be called the Libyan Producers Agreement in which the producing companies undertook to support each other in the event of government action against a member company. They also agreed that in the event of forced production cuts the other members would make up the difference at cost. In the discussions with the government which followed the companies for the first time formed a united front, and a further general agreement was signed in April 1971 which applied standard conditions to all the producing companies.[44]

The next crisis was not long in coming. In December 1971 the Libyan government unexpectedly nationalised the interests of BP in retaliation for Britain's failure to prevent three islands in the Arabian Gulf from being occupied by Iranian troops following the removal of

British military presence from the area. BP's production at this time – mostly from As Sarir – was about 430,000 barrels per day. The decision included provision for compensation, the amount to be decided by the Libyan Courts. BP immediately appealed against the nationalisation decision under the terms of the Petroleum Law, but the government did not respond. BP attempted to embargo the sale of As Sarir crude at European destinations, but the Libyan government circumvented this problem by exporting As Sarir cargoes to the Soviet Union. After a judgement in favour of BP at the International Court of Justice, the Libyan government finally paid BP £17.4 million in compensation.[45]

The question of state participation had been discussed by OPEC since 1968 and Libya strongly supported these objectives. The award of joint venture licences had begun before the revolution, and between May 1969 and 1973 twenty-four concessions, covering most of the available open acreage, were awarded to NOC's wholly-owned subsidiary Agoco, which had succeeded Lipetco as the government's operating company. In addition the new government was keen to extend state participation into existing licenses. Their first opportunity came in 1972. The Abu Attiffel field had been developed by Agip, but production start-up was prevented by the government, ostensibly because of concerns about lack of provision for the associated gas. After many months of negotiation Agip offered the government a 50% interest in two concessions as a means of breaking the deadlock. The government accepted and the state interest was transferred to Agoco.[46]

State participation in all concessions became a stated goal during 1972 with the government demanding a 51% interest in all existing licences. Firstly it was imposed on the independents Occidental and Oasis, and then on the majors Amoseas, Esso and Mobil. All of the

companies protested, but their protests were overtaken by external events.

Between April 1971 and September 1973 the posted price of Libyan crude increased from $3.45 per barrel to $4.60 per barrel, largely due to external forces, particularly the suspension of dollar convertibility to gold in August 1971, and the subsequent effective devaluation of the dollar. OPEC members however were becoming increasingly impatient with the gap which still existed between posted price and actual market price which none of their measures had succeeded in closing. The Yom Kippur War of October 1973 provided the excuse they needed. Shortly after hostilities began the OPEC Gulf States announced that in future posted price would be established by the host nations based on actual market prices, and in December, following rapid price rises caused by the war, they announced that the posted price would be calculated so as to yield a $7 per barrel income for the host government. By January 1974 the posted price for Libyan crude had risen to $15.76 per barrel (Figure 1.7).[47]

The Arab-Israeli conflict of October 1973 had profound consequences on the Libyan oil industry. The Libyan government cut production by 5% and embargoed oil shipments to the United States and the Netherlands. In June 1973 the government nationalised the assets of N.B. Hunt, in February 1974 the holdings of Amoseas, and in April those of Shell. The Amoseas operations were transferred to an NOC operating company called Umm al Jawaby. Esso and Mobil acquiesed to 51% state participation in April.[48] By the summer of 1974 the government had largely achieved its participation objectives. It had taken over BP, Hunt, Amoseas and Shell's holdings completely, it had a 50% interest or more in the concessions of Esso, Mobil, Oasis and Occidental, it had joint venture agreements with Aquitaine and Agip, and it had taken possession of all of the available open acreage. As a result of these

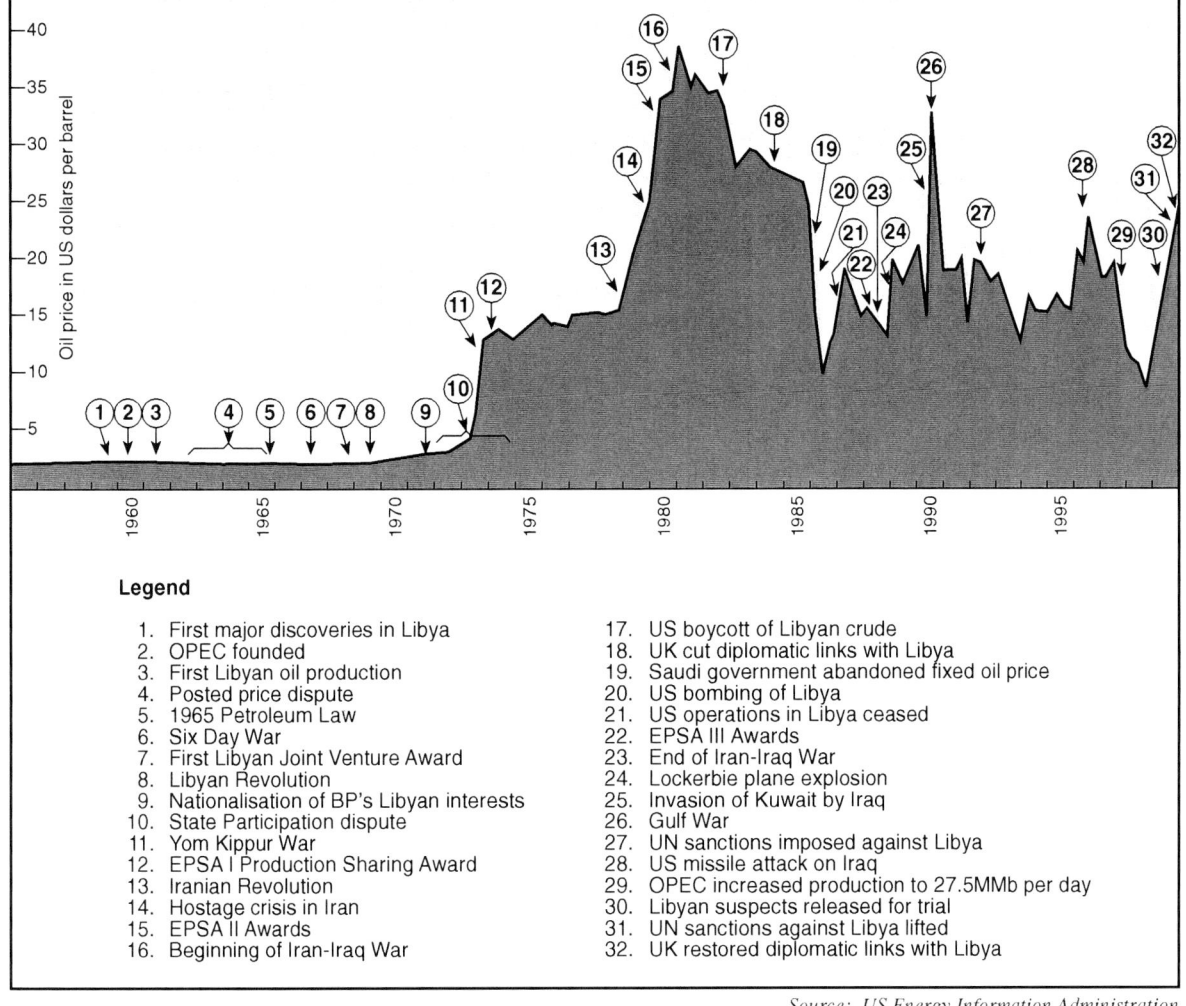

Figure 1.7   Oil Price, 1958 to 1999 (money of the day)

This figure shows the principal events which have influenced the oil price between 1958 and 1999.

1.10 The Decline in Exploration Activity, 1969-1974

The disputes over posted price, and the actions taken by the new government led to a drastic reduction in exploration activity. Seismic activity was greatly reduced and rig counts measures the government controlled approximately 70% of production.

dropped to alarming levels. The reduction in drilling inevitably led to a reduction in the rate of new discoveries, and to a decline in remaining reserves. From levels of over 100 wildcats per year in the period 1962 to 1964 the level fell to less than 30 wildcats per year in the period 1972 to 1974 (Figure 1.3). Exploration footage drilled in 1974 was 155,000 feet compared with 1.08 million feet in 1963.[49] Apart from the giant Messlah field which was discovered by BP just before nationalisation the discoveries made during this period were small. The first well on a joint venture block, A1-LP2E, was drilled by Aquitaine in 1969, and in the following year the company made a modest Palaeocene reef discovery at Al Mheirigah, northwest of Occidental's Intisar reefs.[50] Amoseas made two further discoveries on concession 47 at Abu Alwan and Warid which passed into the control of Agoco in 1974, and in 1973 Oasis made two Nubian discoveries in concession 59. For the first time the amount of reserves added fell below the amount of oil produced. Production during the five-year period totalled just over 5.5 billion barrels. Reserves added were little more than 2 billion barrels. From a reserves peak of 29.8 billion barrels in 1967 the level of reserves fell to 25.9 billion barrels in 1974 (see Figure 1.12).

Aquitaine began evaluation of offshore concession 137 adjacent to the Tunisian border in 1970 and made a number of promising discoveries which proved the presence of an excellent hydrocarbon-bearing reservoir in the western Libyan offshore area. Exploration in this area however was hampered by boundary disputes with both Tunisian and Malta, both of which were eventually submitted to the International Court of Justice for arbitration, but were not finally resolved until 1985.[51]

Further fields were brought on stream including Bahi and Intisar C in 1970 and Abu Attiffel and Abu Alwan in 1972. Oil production peaked in 1970 at a rate of 3.32 million barrels of oil per day with Oasis as the major producer and Occidental in second place.[52] Thereafter production fell in response to the production cuts and political pressures outlined above. By 1974 total production was down to 1.5 million barrels per day. Nevertheless government revenues from oil quadrupled during the same period to $6 billion in 1974 due to the dramatic increases in oil price and the increased government share of revenues.[53] Nor was the government unduly concerned about the lower levels of production. To them it represented a convenient means of conservation.

1.11 EPSA I, 1974

The government was alarmed by the decline in exploration and development drilling and in an effort to stimulate new exploration the government abandoned the joint venture licensing system begun in 1968, in favour of a new system of concessions based on exploration-production sharing agreements, which came to be abbreviated as EPSA I. At the same time a new concession numbering system was introduced, with the first group of licences being reserved for NOC and the state company Agoco (NC 1 to NC 28). The first awards under the EPSA I terms were made to Occidental in February 1974 (NC 29 to NC 34), less than one year after the nationalisation upheaval of 1973. The terms of the new agreement involved an equity split of 85:15 in favour of the government, in return for guaranteed levels of exploration expenditure, and repayment of NOC's share of development costs in the event of a commercial discovery. This neatly reversed the roles established under the previous joint venture agreements.[54]

This award was quickly followed by others. Esso and Agip received offshore concessions close to where Aquitaine had been awarded acreage in 1968. The equity split for offshore blocks was established by the government at 81:19. Both of these blocks were to yield discoveries. Mobil, CFP, and Braspetro also obtained new concessions. Braspetro obtained a large block deep in the Murzuq Basin in an area which was almost totally unexplored. Since many of the old concessions were still active three types of concessions now co-existed side by side – the original concessions first issued in 1955, the joint-venture concessions issued between 1968 and 1973, and the production sharing concessions first issued in 1974.

1.12 Consolidation, 1975 to 1979

The successes of 1974 marked the end of the Libyan government's campaign to take control of the upstream end of the domestic oil industry. By the end of that year they had essentially achieved their objectives. They were content to leave the high-cost exploration effort to foreign operators whilst their own financial exposure was limited to such field developments as might take place with ultimate reimbursement of their share of development costs. Amoco, which had taken over Pan American, was unsettled by the events of 1974, and surrendered its holdings the following year.

Occidental was awarded a further group of concessions in the Zallah Trough in 1976 in an area which had originally been assigned to Elwerath in 1959 and relinquished by them in 1972. Elwerath had tested oil in three wells but had taken no steps to develop these finds. Occidental was to have considerable success in this area. Mobil was awarded ten new blocks in 1977 and Esso was granted a further offshore block in the same year adjacent to its existing offshore block.[55] Incidentally, the name change from Esso to Exxon in 1972 was not reflected in Libya where the Esso name remained in use.

Foreign operators however were wary. Esso drastically scaled down its exploration drilling after 1974, preferring to assign its resources to development drilling and production. Oil shipments to the United States were resumed at the beginning of 1975, but the economic recession following the Arab-Israeli war led to a reduced demand for oil and the price of Libyan crude fell from $16 per barrel at the time of the war to less than $12 per barrel in mid-1975. Thereafter prices slowly recovered to about $14 per barrel and remained stable during the

next four years.[56] Production in Libya gradually increased to around 2 million barrels per day.

Having achieved its goal in upstream operations the government then turned its attention to downstream activities. The Al Brayqah refinery was too small to satisfy domestic demand for refined products so in 1969 the government commissioned a new refinery to be built at Az Zawiyah with an ultimate capacity of 120,000 barrels per day. The refinery came on stream in 1974. Crude was initially shipped from Al Hariqa near Tubruq, until the oil pipeline to the Ghadamis Basin fields was completed in 1982. A further refinery was subsequently constructed at Ra's Lanuf with a capacity of 220,000 barrels per day. The refinery started production in 1984.[57]

Occidental's bid for new concessions in 1966 included an offer to construct an ammonia plant at Azzuwaytinah to be operated as a joint venture with the Libyan government. After several changes of plan a petrochemical complex was eventually constructed at Al Brayqah where it had access to abundant supplies of natural gas to produce methanol, urea and ammonia. The plant came on stream in 1977, and Occidental sold its interest to the government in 1979.[58] An ethylene plant was subsequently built at Ra's Lanuf, and a PVC and caustic soda plant at Bu Kammash near the Tunisian border.[59] The government also wished to have an involvement in transportation, and by 1980 Libya had a fleet of 15 tankers which gave it the capability of transporting a fair proportion of its oil products to market under its own flag.[60]

The first well on a production sharing licence was drilled by Occidental in the eastern Sirt Basin in 1974, but exploration drilling remained sluggish. Only 155 wildcats were drilled in the five years from the beginning of January 1975 to the end of December 1979, an average of only 31 per year.[61] However there was a marked increase in offshore drilling. Aquitaine drilled ten wells on concession 137

36

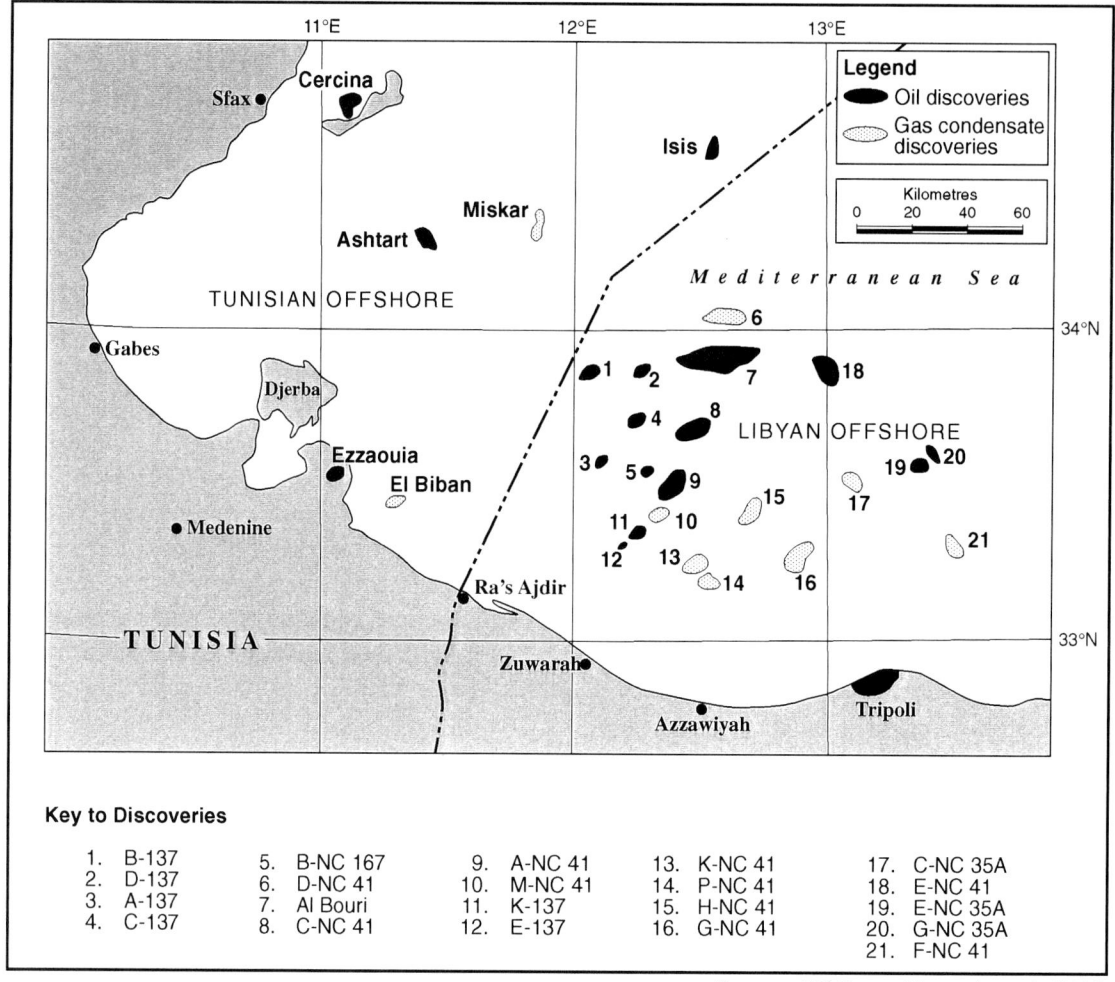

Figure 1.8   Libya, Sabratah Basin Offshore Discoveries

These fields are rich in gas and are either gas-condensate fields or oil fields with a gas cap.
Only Al Bouri has been developed in Libyan waters, and is equipped with tanker-loading facilities.

in the period up to the end of 1979 with several wells testing oil and gas from Eocene carbonates. Esso drilled two wells in block NC 35 without success, and Agip began evaluation of block NC 41. The first well drilled on block NC 41 in 1976, tested modest amounts of oil from the Eocene reservoir, but the B1-NC 41 wildcat well tested 4857 barrels of oil per day, resulting in the first commercial discovery in the Libyan offshore.[62] The discovery was

named Al Bouri and was subsequently appraised by eight delineation wells. These wells proved Al Bouri to be a giant oil pool, the largest offshore field in the Mediterranean, with quoted oil in place of 5 billion barrels and 670 million barrels recoverable (Figure 1.8).[63] Encouraged by this success Agip drilled seven more wildcats up to the end of 1979, many of which tested oil or gas, but none of which was regarded as commercial. Al Bouri was the twenty-first giant field to be discovered in Libya (Figure 1.9).

Onshore, Occidental had several notable successes. On concession NC 29 in the northern Sirt Basin the company made three small discoveries named Ali, Almas and Arbab. In the Zallah Trough, in an area previously explored by Elwerath, four discoveries were made in Eocene reservoirs: Zallah, Sabah, Hakim and Fidda. Mobil had an excellent discovery at Ghani in the Zallah Trough, plus a string of smaller discoveries on the Al Bayda and Az

Zahrah Al Hufrah Platforms. The northward extension of the Messlah field into concession 80 was proved by Agoco, and several other Nubian discoveries were made in concessions 65 and 80. During this period NOC operated a number of blocks independently from Agoco, mostly on relinquished acreage which was regarded as still having good potential. In the Hameimat Trough, they discovered oil at Qadeem, on trend with the Jalu field, and in the Ghadamis Basin the Kabir field was discovered, close to the former Gulf Q-26 field. The

Name	Discovery Date	Date on Stream	Original Oil in Place (MMb)	Original Reserves (MMb)
Bahi	1958	1970	1875	600
Az Zahrah B	1958	1962	715	229
Al Hufrah	1958	1963	681	218
Nasser (Zaltan)	1959	1961	6470	2200
Amal B	1959	1966	14655	4250
Dayfah North	1959	1964	5217	1800
Az Zahrah F	1959	1962	790	253
Al Wahah	1960	1963	3870	1200
Ar Raqubah	1960	1963	3571	1000
As Sarir C	1961	1966	15964	5428
Jalu	1961	1964	11428	4000
Samah	1961	1963	1612	500
As Sarir L	1964	1966	3152	1072
An Nafurah	1965	1966	4472	1297
Awjilah	1966	1968	700	203
Intisar A	1967	1968	1875	750
Intisar D	1967	1968	3750	1500
Abu Attiffel	1967	1972	3449	1103
Messlah South	1971	1976	3051	1038
Messlah North	1976	1977	1356	461
Al Bouri	1976	1988	5000	670
Elephant	1994	U/D	2003	561
Total			**95656**	**30333**

Source: Campbell (1991), Thomas (1995), PESGB Newsletter (May 1999), Author's estimates

Figure 1.9   Giant Field Discoveries

Discoveries are shown in chronological order. Az Zahrah F/Al Hufrah, An Nafurah/Awjilah, and Messlah N/ Messlah S subsequently proved to be single oil pools. Reserves have been prorated between the component parts. In terms of original total reserves these 22 discoveries account for 73% of all the oil discovered in Libya. U/D = Under development.

operating divisions of NOC, Umm al Jawaby and Agoco, were merged in December 1979 under the Agoco banner. Oasis reported a number of small successes, mostly in concession 59. In the Murzuq Basin Braspetro tested oil from two zones in their A1-NC 58 well which was the first report of oil from the central Murzuq Basin.[64]

Production steadily increased from 1.4 million barrels per day in 1975 to 2.04 million barrels per day in 1979. New fields brought on stream during this period include the giant Messlah field, Masrab, Khalifah, Al Mahiriqah and several of Occidental's recent finds: Zallah, Aswad, Almas and Ali. The Al Hutaybah gas field also came on stream in 1977.[65]

1.13 Turmoil, 1979-1986

By late 1978 political events were again threatening to affect the relative stability of the last three years. The international oil price had stabilised at around $14 per barrel during the period 1975 to 1978. During 1978 protests in Tehran against the regime of the Shah of Iran gradually gained momentum and in November Iranian oil production began to decline sharply. In January 1979 the Shah left Iran and on 1st February the exiled Ayatollah Khomeini returned to Tehran to take over the government. In April OPEC set the oil price for 1979 at $14.56 per barrel, but events spiralled rapidly out of control. In November the Iranians seized western hostages and President Carter imposed a ban on imports of Iranian crude to the United States. By the end of the year the international oil price had reached $24 per barrel.[66]

In September 1980 Iraqi troops invaded Iran and occupied the Shatt al Arab waterway. OPEC's pricing structure collapsed in turmoil, and by December 1980 the oil price peaked at $36 per barrel. During 1981 Saudi Arabia flooded the market with cheap oil and the price

stabilised. By October it was hovering uncertainly at $32 per barrel, but the threatening political situation led to a general recession and prompted a gradual move by Western governments towards alternative fuel sources. An oil surplus developed and the oil price progressively fell.[67]

Relations between Libya and the United States continued to deteriorate, provoked by Libya's support for anti-American causes. In 1978 the USA imposed a ban on the sale of aircraft and electronic equipment to Libya and in 1981 they asked all US citizens to leave.[68] As a result of these pressures Esso decided to close down its operation in Libya. Their assets were transferred to NOC in January 1982 in exchange for modest compensation. NOC established a new state-owned company, Sirte Oil, to take over management of Esso's operations.[69] By the time of Esso's departure 2.7 billion barrels of oil had been produced from their fields on concessions 6 and 20.[70] Three months later the United States imposed an embargo on the import of Libyan crude and placed further bans on exports to Libya. Mobil was the next to go, handing over operatorship of the Amal field and other areas to Veba, who had been partners of Mobil since 1978.

Instability continued through 1983 and 1984 as the Iran-Iraq war intensified and OPEC's market share fell as western countries reduced consumption and switched to more accessible North Sea oil. In order to maintain income some OPEC countries, including Libya, resorted to discounting, selling their oil at prices below the official OPEC price. The principal loser was Saudi Arabia who at this time acted as swing producer, maintained the OPEC price by adjusting its own output as required. Saudi Arabia's oil earnings fell from $119 billion in 1981 to $26 billion in 1985, and King Fahd made it clear that he was not willing for this situation to continue.[71] In an effort to restore a more equitable balance of production Sheikh

Yamani, Saudi Arabia's oil minister, introduced netback dealing. This ingenious mechanism worked by guaranteeing refiners a fixed profit on the sale of their products. After agreeing an acceptable profit margin with the Saudis the refiner would market his products, deduct the agreed level of profit, and remit the balance to Saudi Arabia. At a stroke this did away with the fixed OPEC oil price, and created a situation favourable to Saudi Arabia insofar as refiners were insulated from highly volatile crude prices. The guaranteed profit also encouraged refiners to maximise production and several customers signed contracts immediately. The Saudi's were deliberately trading price for volume, but it was a gamble, which ultimately failed. The Saudi government reasoned that when the oil price fell below $20 a barrel high cost areas such as the North Sea would become uncompetitive. They were wrong; the actual production cost of North Sea oil was only about $6 per barrel. The situation of 1979-81 was reversed. Now it was a question of producing countries seeking markets rather than purchasers seeking suppliers. The price of oil tumbled from over $30 per barrel in November 1985 to less than $10 a barrel in July 1986.[72] The effect on exploration was acute, as companies cut their budgets, and shut down their exploration programmes. Libya's response to this, the third oil crisis within thirteen years, was to advocate reduced quotas for OPEC producers. They had cause for concern; their oil revenues had fallen by 42% in one year. By August OPEC members had been forced to agree to a new quota system designed to ease the oil price back up to around $18 per barrel.[73]

1.14 EPSA II and New Discoveries, 1979-1986

During the late 1970's the government took the view that companies were holding acreage without making sufficient effort to evaluate it,

and they demanded that inactive acreage should be relinquished. Several companies gave up acreage in 1979 and this acreage formed the basis for EPSA II. The terms of EPSA II were significantly less favourable to foreign operators than EPSA I.[74] Nevertheless many companies showed an interest in the new blocks on offer. Awards under the EPSA II formula were made during the period 1980 to 1982.

Several newcomers received blocks including the Bulgarian and Romanian State Oil Companies who were granted concessions in the Ghadamis and Murzuq Basins, and Deminex who received concessions in the southern Sirt Basin. Two US companies, Sun Oil and Coastal, received a total of ten blocks in the Sirt Basin. Shell, Agip, Occidental and Aquitaine also received additional blocks. At the same time the state company Agoco added further blocks to its portfolio.[75]

Despite these efforts exploration remained at a disappointingly low level. Only 21 wildcat wells were spudded in 1980, rising to 48 in 1984 before declining again to 13 in 1986. The Bulgarian State company, Boco, began evaluation of its acreage in 1982 and almost immediately announced a string of discoveries in concession NC 100 near the border with Tunisia. The individual discoveries were small, and have not been developed. However on block NC 101 on the northern edge of the Murzuq Basin more significant finds were made in an area where Gulf had reported oil shows in 1958. These discoveries, along with the shows in Braspetro's A1-NC 58 well, signalled the presence of significant oil accumulations in the previously neglected Murzuq Basin. This was spectacularly borne out when Rompetrol struck oil on concession NC 115. Their first two wells on the block both found large oil pools with combined reserves of around 450 million barrels, and a third discovery in 1986 made this into the largest group of onshore discoveries in fifteen years (Figure 1.10).[76]

Elsewhere, almost as their last act before departure, Esso tested oil on their offshore block NC 35. Esso's successor Sirte Oil, drilled a further seven wells on the block and had two non-commercial discoveries, whilst Agip continued to evaluate the area around Al Bouri. Onshore, Mobil had a further discovery on their Al Bayda Platform block, and Occidental discovered the Safsaf, Abraq and Themar oil pools in the Zallah Trough.[77]

Production during this period was limited by OPEC quotas and averaged just over 1 million barrels per day. The western pipeline from the Ghadamis Basin fields to Az Zawiyah was completed in 1982 and about twenty small fields were put on production. Occidental put their remaining Zallah Trough fields on stream, and a number of small fields in the central Sirt Basin were also tied-in to existing pipelines.

1.15 Sanctions and EPSA III, 1986 to 1999

The oil-price collapse of 1986 had the effect of forcing both OPEC and non-OPEC producers to cooperate in limiting production. OPEC producers established much tighter quotas on member states, and non-OPEC producers introduced voluntary production limitations. The oil price oscillated between $12 and $18 per barrel as the Iran-Iraq War ground to an end in 1988, and OPEC countries were gradually able to increase their production ceiling to 19.5 million barrels per day. However another crisis erupted in August 1990 with the invasion of Kuwait by Iraqi forces, and the subsequent panic forced the oil price up to $33 per barrel during the autumn of 1990. The ensuing Gulf War resulted in the torching of Kuwait's oil wells by the retreating Iraqi forces, and a UN ban on the export of Iraqi oil. Ironically these unforeseen events largely achieved OPEC's

goal, and by October 1992 OPEC production had risen to over 25 million barrels per day, the highest level since 1979.[78]

In 1984 Britain severed diplomatic links with Libya following the shooting of a policewoman in London, and in 1986 the United States government broke off relations. Extensive trade sanctions were imposed by the US government on Libya, and in April US planes bombed Tripoli and Binghazi in retaliation for terrorist attacks in Europe. In June 1986 a Presidential decree ordered US companies to close down their operations in Libya and made it unlawful for US citizens to work in Libya. The Libyan government established the state oil companies Waha and Zueitina to take over operatorship of the Oasis and Occidental interests, initially on a caretaker basis for three years, but failure to lift the US sanctions has left Waha and Zueitina as continuing operators of these blocks.

The combined effect of sanctions, low oil price and the cost of the $25 billion dollar Great Man-Made River project[79] led to the government introducing new and more attractive, production sharing contract terms in 1988, under the name of EPSA III. For the first time cost recovery was allowed, bringing Libya into line with most other countries which operate production sharing contracts. The intent was to provide rapid payback of costs, and an acceptable rate of return thereafter. These terms encouraged a number of new companies to apply for blocks, notably Petrofina from Belgium, Lasmo from UK, International Petroleum Company from Canada, INA from Croatia and OMV from Austria. In addition Shell was tempted back into Libya and Braspetro was awarded a further block. Nevertheless exploration activity remained at a very low level. An average of only 13 exploration wildcat wells per year was drilled in the six years from 1986 to the end of 1991, and discoveries in this

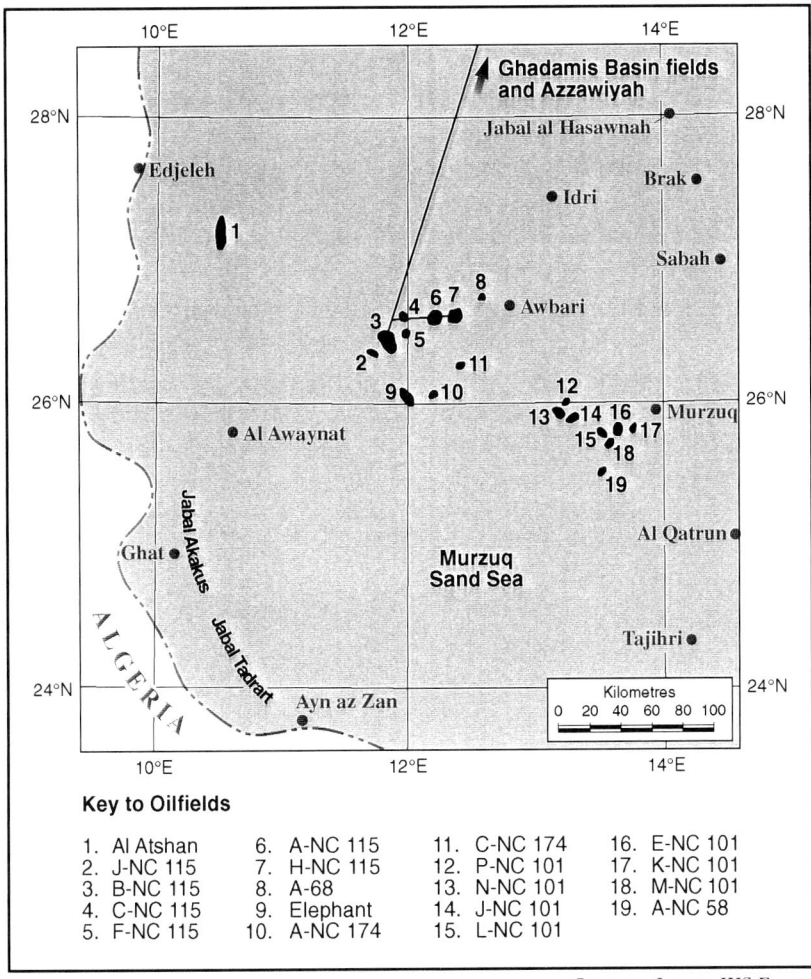

Figure 1.10   Libya, Murzuq Basin, Oilfields and Pipeline

A 30-inch pipeline has been completed to the Murzuq B field and satellite fields are being tied-in. The Elephant field was scheduled to begin production in 2000. No decision has yet been reached on the viability of the NC 101 fields.

period were proportionally few. Sirte Oil, which took over Esso's interests in 1982, developed the giant Attahadi gas field and the large Wadi oilfield both in the Sirt Basin, and Wintershall discovered a second large Triassic oil pool on concession 96, which was named As Sarah. In west Libya both Rompetrol and Boco discontinued their exploration effort, but Sirte Oil found a significant gas-condensate accumulation at Al Wafaa close to the Algerian border.[80]

In December 1988 Pan Am flight 103 was blown up over Lockerbie in Scotland and in September the following year a French plane crashed in Niger. In 1991 indictments were submitted in French, US and British courts against Libyan personnel for both of these airline incidents, and in 1992 the United

Nations imposed sanctions against Libya until such time as the suspects were handed over for trial. The sanctions prohibited all airline traffic with Libya, froze Libyan assets overseas, and banned weapons sales. Further sanctions were applied in subsequent years banning the sale of equipment for oil and gas terminals and refineries. These sanctions were to cost Libya more than $24 billion dollars, including $5 billion dollars in lost oil revenues.[81]

The years 1992 to 1999 were difficult for the Libyan government. Exploration activity picked up to a small extent as companies evaluated the acreage awarded under the EPSA III regulations, with wildcat drilling averaging about 25 wells per year. The sanctions had a severe effect, particularly on the activities of the

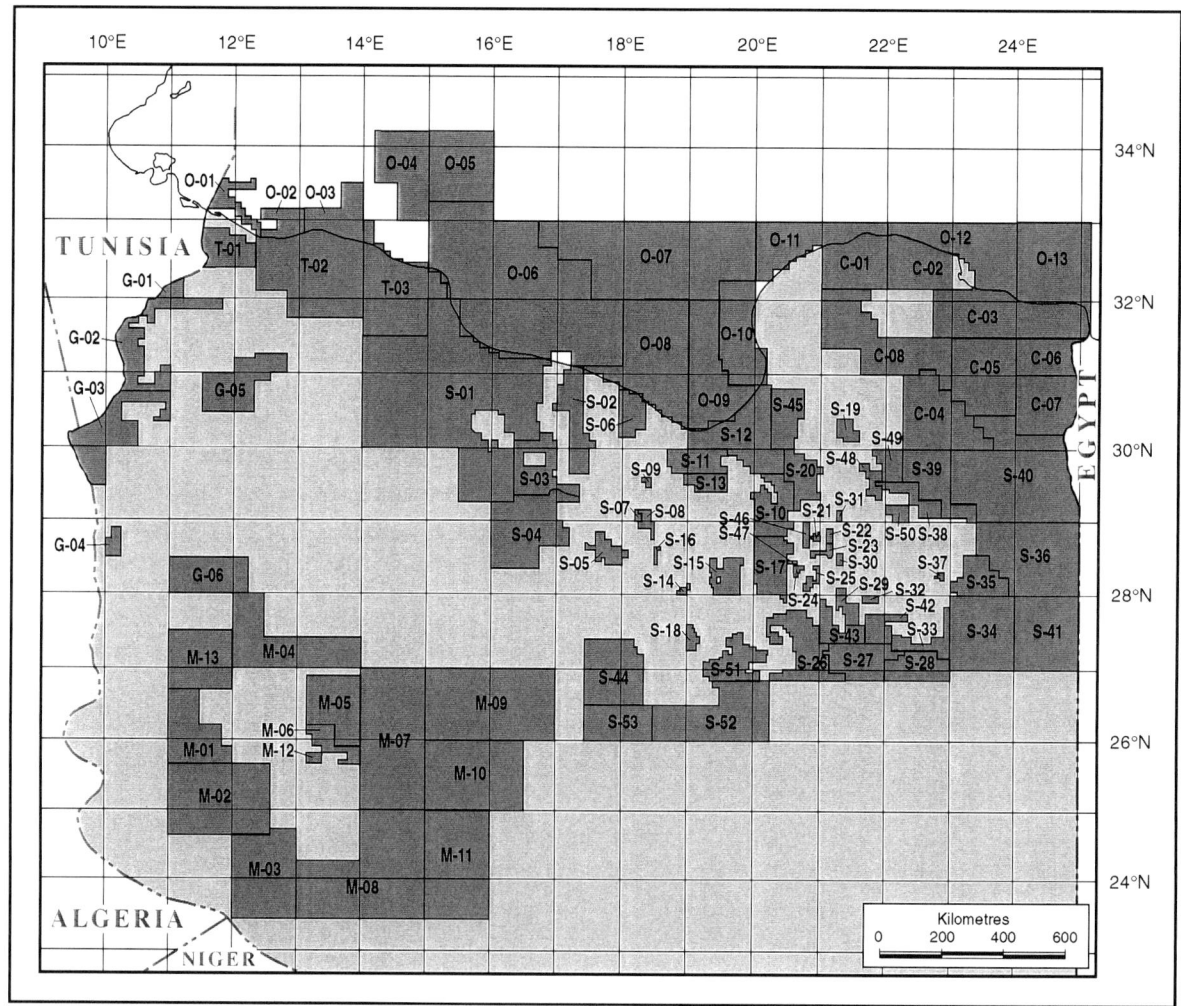

Source: NOC

Figure 1.11   Open Acreage (designated late 1999)

Areas designated as open at the end of 1999 include recently relinquished acreage and new areas not previously offered for licensing.

state-owned companies, and several field developments, including Mabruk and Al Wafaa, were transferred from state companies to foreign operators. Similarly the Rompetrol discoveries in the Murzuq Basin were transferred to the Spanish company Repsol for development.[82] Two interesting discoveries were made during this period. Lasmo discovered the large Elephant field in their EPSA III block NC 174, which lies adjacent to Rompetrol's discoveries, and in the Sirt Basin the Ayn an Naqah discovery, in the poorly-explored Abu Tumayam Trough south of the Zallah Trough, tested oil at a rate of 6517 barrels per day.[83] Production increased in line with OPEC quotas and averaged 1.3 million barrels per day during the period 1989 to 1996. The giant offshore Al Bouri field came on production in 1988, and Wintershall's As Sarah field in 1991. The Murzuq fields discovered by Rompetrol and developed by Repsol began production in December 1997 via an extension to the western pipeline to Az Zawiyah.[84]

At the end of 1996 Iraq resumed oil production under UN supervision. OPEC production increased in the wake of the Gulf War, first to 25 million barrels per day, and then in November 1998 to 27.5 million barrels per day. This last increase coupled with economic recession in the Far East led to a price collapse which in June 1998 fell below the $10 per barrel threshhold. The pattern of 1986 was repeated, and OPEC was forced to reduce production levels twice during 1998 and again in March 1999.[85]

The depressed state of the industry in Libya during the 1990's and the price collapse of 1998, in which Libya's oil revenues fell to $5.6 billion dollars, perhaps had some influence on the government's decision in April 1999 to hand over the Lockerbie suspects for trial. In response the United Nations suspended the sanctions which had been applied in 1992, and Britain re-established diplomatic relations with Libya in July 1999. The US Sanctions Act, which applies to both Libya and Iran, remained in force. Libyan production reached 1.45 million barrrels per day in 1998, but the OPEC quota reductions reduced this level to 1.22 million barrels per day.[86]

Towards the end of 1999 the Libyan government designated extensive new areas for licensing, both onshore and offshore in an effort to stimulate exploration (Figure 1.11). These tracts include acreage not previously designated, and blocks recently relinquished by both state and foreign companies. A number of companies, mostly from Europe and Canada, entered into negotiation with NOC on several of these areas.[87]

1.16 Reserves

Determination of petroleum reserves is a highly contentious issue. The official reserves claimed by the Libyan government in 1999 were 29.8 billion barrels.[88] This is a significant increase over the official government estimates of 22.8 billion barrels which remained unchanged from 1986 to 1994, and is almost certainly due to adjusting the overall recovery rate from 30% to 38% on the basis that enhanced recovery techniques will increase ultimate recovery. The following key figures are agreed by most authorities. Total oil-in-place discovered to the end of 1999 is about 134 billion barrels.[89] Assuming an overall recovery factor of 30% this gives original recoverable reserves of 40 billion barrels, or at 38% recovery 50 billion barrels. Total production to the end of 1999 was 21 billion barrels leaving remaining reserves of 19 billion or 29 billion barrels depending on which recovery factor is used (Figure 1.12).[90] However these estimates contain assumptions which merit further examination.

The reserves are made up of three basic categories, giant fields, smaller commercial

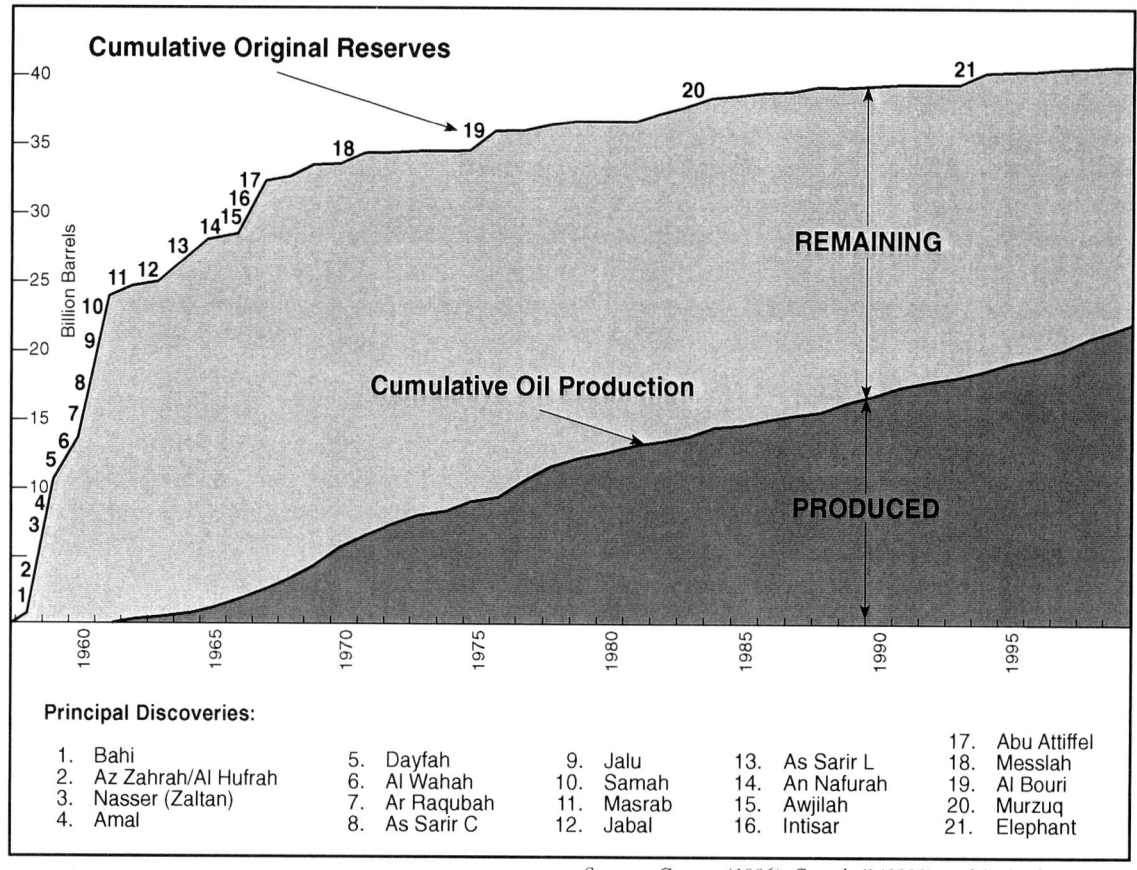

Figure 1.12   Cumulative Reserves and Cumulative Oil Production

The figure shows the early discovery of large volumes of reserves, and the gradual depletion of those reserves since 1967. Estimated remaining reserves at the end of 1999 are 19.2 billion barrels.

fields which are on production or scheduled for production, and marginal fields which are not yet developed or on production – either because they are too small, too remote or have some reservoir problem. The giant fields make up the bulk of the reserves. In 1967 giant fields accounted for 86% of the remaining reserves, but by 1999 that figure was down to 60% because most of the oil produced during this period was from giant fields. In fact by 1999 the giant fields collectively were 66% depleted.[91] Consequently in 1999 a greater proportion of the remaining reserves is contained in smaller

fields, and reservoir quality in some of these fields is poor (Figure 1.13). It is true that enhanced recovery techniques can dramatically increase recovery from difficult fields, and Total's exploitation of the Mabruk field is a good example, but it is not reasonable to assume that EOR techniques can be applied to all fields, or even to a majority of fields. Indeed NOC has currently earmarked only 35 sites for EOR tests.[92] There are many discoveries which will never be developed simply because they are too small to be economic. As a result of these considerations it is questionable whether an increase in global recovery rate to 38% is justified, and it is probably more prudent to leave it at 30%.

There is then the question of yet-to-be-discovered reserves. This figure is the difference between reserves already discovered and the ultimate potential – the final amount of oil produced at the end of all production. Mathematical models based on production history have been developed to predict ultimate potential, and several of these were discussed by Campbell.[93] Using the parabolic fractal method of Laherrere (1996) Campbell assigned an ultimate potential to Libya of 45 billion barrels.

Given cumulative production of 21.3 billion barrels and remaining reserves of 19.2 billion barrels, the yet to find figure at the end of 1999 would therefore be 4.5 billion barrels. Whether these figures are justified remains for the future, but it is perhaps significant to note that only one giant field has been discovered in Libya since 1976, and although it took only 10 years to discover the first 30 billion barrels of reserves, it has taken 27 years to find the next 10 billion barrels.

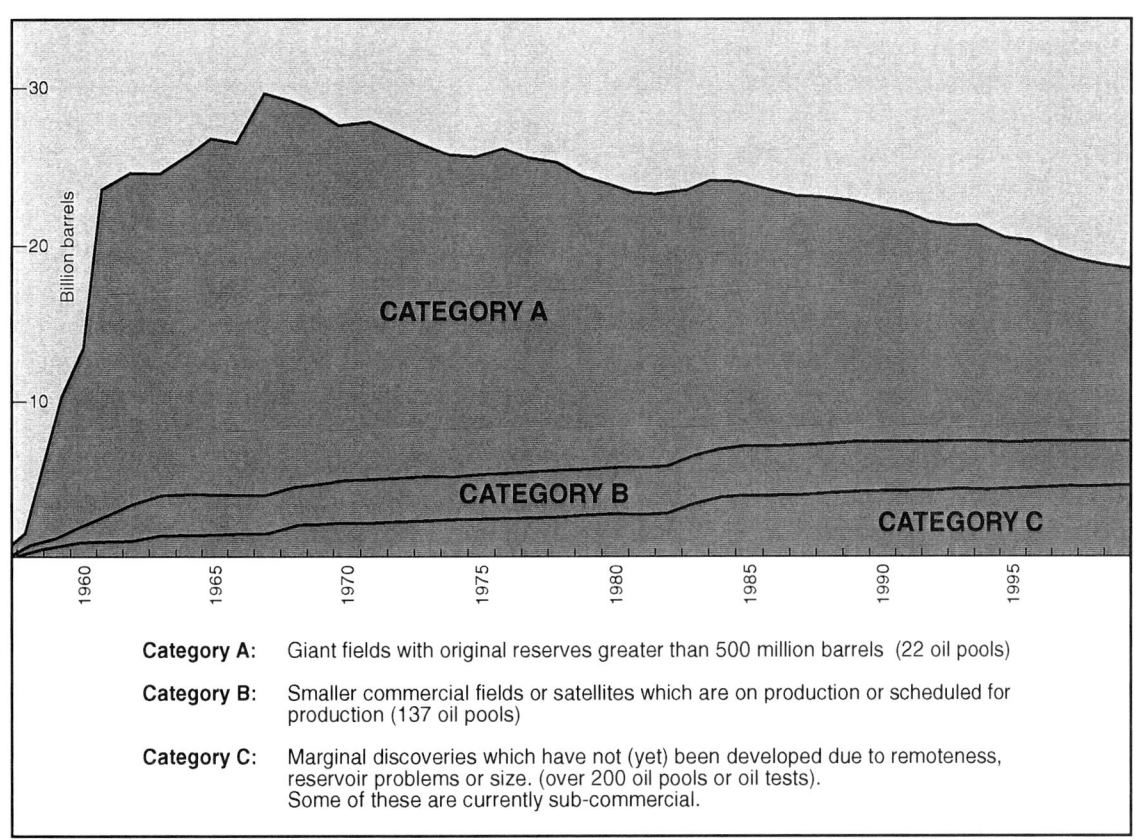

Category A: Giant fields with original reserves greater than 500 million barrels (22 oil pools)

Category B: Smaller commercial fields or satellites which are on production or scheduled for production (137 oil pools)

Category C: Marginal discoveries which have not (yet) been developed due to remoteness, reservoir problems or size. (over 200 oil pools or oil tests). Some of these are currently sub-commercial.

Source: Campbell (1991), Gurney (1996), US Energy Information Administration, Author's estimates

Figure 1.13   Remaining Reserves (billion barrels)

Remaining reserves reached a peak of 29.8 billion barrels in 1967 and by the end of 1999 had declined to 19.2 billion barrels. Category A reserves, by the end of 1999, were about 66% depleted.

1.17 Natural Gas

Until recently Libya's natural gas reserves have been regarded as relatively unimportant, both as regards volume and value. The number of non-associated gas discoveries is small compared to the number of oil discoveries, and the gas fields are of modest size compared to Algeria. Associated gas has traditionally been used for reinjection or has been flared. Only Esso, which discovered several non-associated gas fields on concession 6 and its successor Sirte Oil, have had any significant involvement in gas production and export. A gas pipeline was laid from Nasser and Ar Raqubah to Al Brayqah where a LNG plant was constructed, and production began in 1970.[94] The pipeline was extended to recover non-associated gas from Al Hutaybah in 1977. Gas production rose in line with oil production and peaked in 1979 at a rate of 2.2 billion cubic feet per day. By 1994 production had fallen to 1.2 billion cubic feet per day. About one-third of total production is for industrial use or is reinjected. Flaring was reduced from 21% in 1975 to 14% in 1994. The remaining 55% was exported as LNG.[95]

However during this same period reserves of natural gas have increased dramatically, and production should increase considerably in future years. Drilling by Agip in offshore concession NC 41 in the late 1970's and early 1980's proved the presence of ten gas pools in the Sabratah Basin, with reserves which are undisclosed but claimed to be huge. Since 1982 Sirte Oil has developed the As Sahel and Assumud gas fields, and the development of the large Attahadi field in concession 6 was scheduled for completion in 2000. The same company discovered the Al Wafaa gas field in west Libya, which was under development by Agip in 2001. These discoveries increased gas reserves from 20 TCF to 46 TCF by 1995.[96] The gas-gathering pipeline system has been extended to tap associated gas from Al Wahah,

Dayfah, Intisar, Amal. Nafurah/Awjilah and Abu Attiffel, and further extensions are planned. As a result of these developments gas production is likely to increase rapidly, and there are plans to supply natural gas to Binghazi and to extend the western branch beyond Tripoli to the petrochemical plant at Bu Kammash. Agreement was reached in October 1999 between Agip and NOC for the development of the gas and condensate fields in onshore concession NC 169 and offshore concession NC 41. The Al Wafaa field will be connected by a 550km pipeline to a gas processing plant at Melitah on the coast and an offshore platform on concession NC 41 will be connected to the same facility. It is planned to export 280 bcf of gas per year to Italy via Sicily.[97]

1.18 Summary

The above review serves to illustrate a number of key points. First of all it shows that politics and the oil industry are inextricably linked. Action in one sphere has an inevitable consequence in another. From the time when it became clear that Libya was a major petroleum province the government determined to gain control over Libyan oil reserves. In the disputes on posted price, taxation and government-participation, by acting in a determined, some might say ruthless fashion, Libya effectively blazed the trail which other OPEC members were happy to follow. It took until 1974 for the Libyan government to achieve its objectives, and subsequently Libya's influence in OPEC began to wane, in favour of Saudi Arabia which took over leadership of OPEC in the 1980's due to its position as the major producer. But the victory of the government over the foreign operators was not without cost. The companies were alienated by forced nationalisation and the arbitrary measures taken against them, and relations became confrontational and wary. Several companies withdrew completely, and most of the others reduced their exploration efforts to a minimum and concentrated on production instead. This led to an inadequate level of exploration drilling in terms of reserves replacement which has not recovered to the present day. The government tried many different methods to encourage exploration and it is interesting to note that after each major issue of new concessions there has been an increase in exploration activity. One option, which has not been tried, is that of allowing state companies to enter into farm-out arrangements with foreign operators. The government appears to prefer to appropriate inactive acreage from state companies and offer this acreage to foreign companies under production sharing terms. The December 1999 offer of new exploration acreage was based partly on appropriation of inactive acreage from state companies. It is also worth noting that the government does not hold licensing rounds, as in many other countries. Companies are free to enter into negotiations with the government at any time on the basis of current legislation.

Secondly, it is clear that oil price is a highly volatile factor which is extraordinarily difficult to control. In the 1970's it appeared that OPEC had established control over the pricing of the bulk of the world's oil supply, but events were to prove that OPEC could not retain control in times of crisis. Nor could they maintain a balance between market share, oil price and levels of production. Repeatedly one or other of these variables slipped from their grasp, and they had to impose measures which in many cases financially hurt their own members, particularly countries such as Libya which depend on oil sales for over 90% of their overseas revenue. It seems inevitable that in the long run OPEC's market share will again become dominant if only because their members control most of the world's remaining reserves. Saudi Arabia's 1986 experiment was a disaster, but fourteen years later OPEC has regained the market share which it sought at that time.

The growth of Libya's oil industry took place very quickly. In the first ten years 80% of the country's current oil reserves were discovered, 80% of the petroleum systems were found, and 50% of the exploration wells were drilled. Since 1968 the industry has operated at a much reduced level. The pace of exploration has been much slower and it has suffered from underinvestment. For considerable periods it has also been deprived of the latest technology. The practical consequence of these factors, whether intended or not, has been to leave Libya underexplored, in comparison with areas such as the North Sea, which developed at more or less the same time. In effect this has led to a conservation of oil reserves in Libya, and now, on the threshhold of the twenty-first century there are numerous exploration opportunities waiting to be evaluated. The rest of this book attempts to provide the geological framework against which the remaining potential of the country can be assessed, leading to a final chapter whose purpose is to suggest where some of the undiscovered reserves might be found.

Chapter 2

PLATE TECTONIC HISTORY of LIBYA

2.1 Introduction

It is only within the last twenty years that a coherent picture of the plate tectonic history of Libya has emerged. This is partly because two disparate aspects are involved. In broad terms, the evolution of Gondwana and Pangaea controlled the development of Libyan tectonics during the Palaeozoic and early Mesozoic, whereas the evolution of Tethys and the Mediterranean were the primary controls on the tectonics of Libya during the later Mesozoic and Cainozoic. Most plate tectonic studies have concentrated on one or the other, but north Africa, located on the leading edge of Gondwana, requires input from both. Progress has beeen further hampered by the complexity of Mediterranean tectonics, and the fact that much evidence was obliterated during the closure of the Tethys Ocean, and by subduction during the Cainozoic. As a result of this convergence there are very few areas in the Mediterranean where sea-floor magnetic data have been preserved. Nevertheless, thanks to intensive research in magnetics, deep seismic profiling, and deep-sea drilling, by the 1990's the tectonic history of the Mediterranean had been established with a fair degree of unanimity.

The main evidence for the evolution of the northern margin of Gondwana has come from studies on the ancient cratons and the intra-cratonic mobile belts exposed in Algeria, southern Libya, Chad, Niger and Sudan. Given the diverse origins of north Africa, it is not surprising that even after the break-up of Pangaea, Africa has not operated as a single plate. The inter-plate tensions within the African continent have had a significant influence on the tectonic development of Libya. Evidence from polar-wandering paths and from hot-spot tracks has also proved useful in reconstructing the tectonic development of north Africa.

By applying these methods it has become possible to trace the history of plate movements from the shadowy history of the Rodinia supercontinent during the Proterozoic, through the reorganisation of plates to form Pannotia in the late Proterozoic, and Gondwana during the early Palaeozoic. During the Palaeozoic Gondwana moved northwards from a southern high-latitude position and collided with Laurasia during the mid-Palaeozoic to form the Pangaea supercontinent. The later history of north Africa was dominated by a rifting phase associated with the establishment of the Tethys Ocean, and a post-rift phase in which the Tethys Ocean closed, and compressive tectonism became the dominant force both in north Africa and southern Europe. This chapter will review the sequence of plate movements which resulted in the formation of present-day north Africa.[1]

Important summaries of plate tectonics in relation to Libya have been made by Anketell (1996), Guiraud (1998), and Morgan et al. (1998). The relation of magmatism to plate tectonics was reviewed by Wilson and Guiraud (1998), and the connection between the break-up of Pangaea and the collapse of the Sirt Arch was examined by El Makhrouf (1996). Fouad (1987) related hot-spot tracks to plate tectonic activity, Schäfer, et al. (1980) applied stress-pattern analysis to plate movement, and

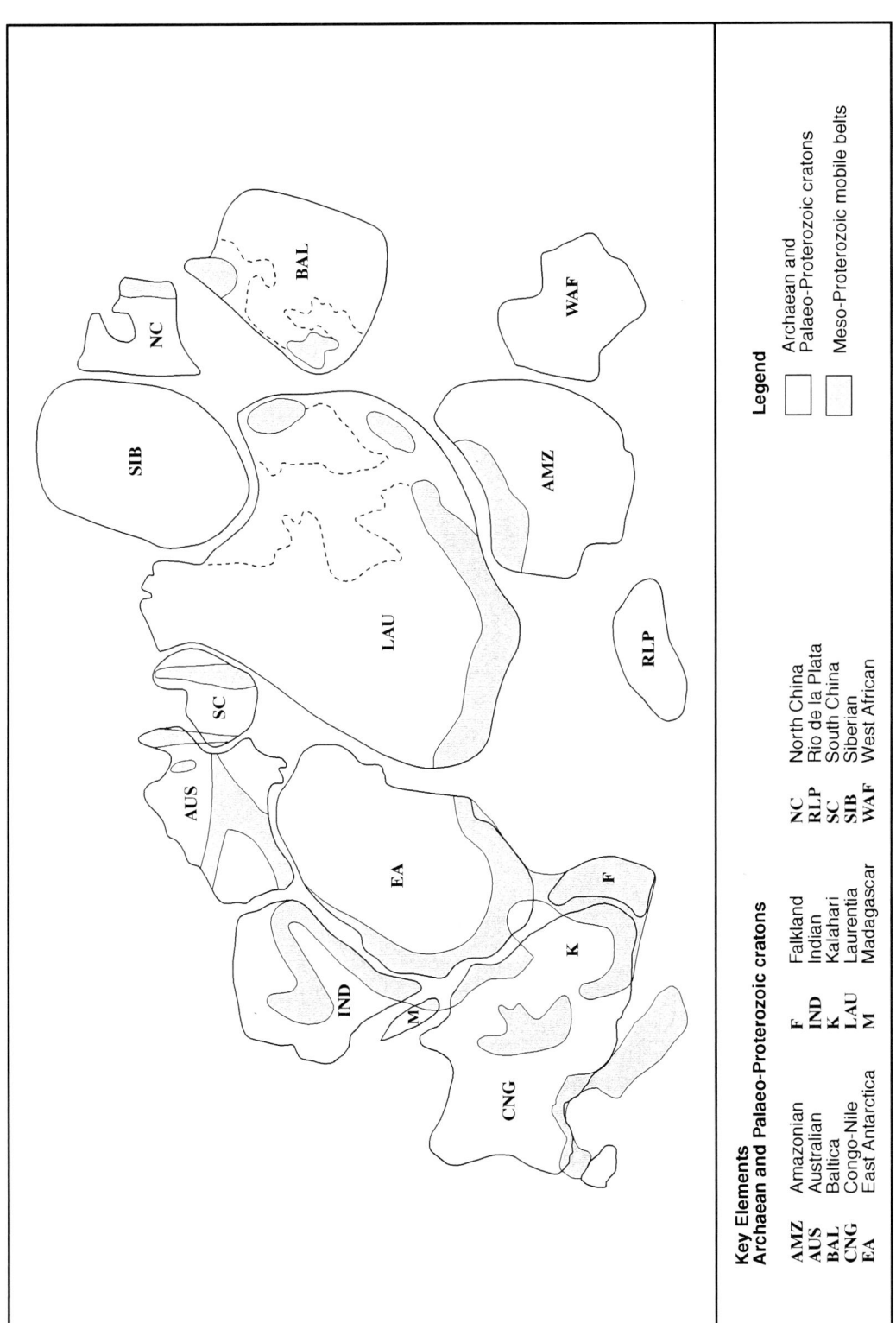

Key Elements
Archaean and Palaeo-Proterozoic cratons

AMZ	Amazonian	**F**	Falkland	**NC**	North China
AUS	Australian	**IND**	Indian	**RLP**	Rio de la Plata
BAL	Baltica	**K**	Kalahari	**SC**	South China
CNG	Congo-Nile	**LAU**	Laurentia	**SIB**	Siberian
EA	East Antarctica	**M**	Madagascar	**WAF**	West African

Legend

☐ Archaean and Palaeo-Proterozoic cratons

▨ Meso-Proterozoic mobile belts

Source: Redrawn from Unrug, et al. (1996)

Figure 2.1   Plate Tectonic Reconstruction of Rodinia at 1000 Ma

The West African craton was totally remote from the Congo/Kalahari craton at this time, and the elements which came to form North Africa had not yet evolved.

Jongsma (1987) discussed the plate tectonic origin of the Medina wrench offshore Libya. As a result of these studies the plate tectonic evolution of Libya can be established with some degree of confidence.

2.2 Rodinia

The Rodinia Supercontinent is the ancestral supercontinent which was assembled during the Meso-Proterozoic from about 1600 to 1000 Ma. Available evidence suggests that it was composed of about twenty identifiable Archaean and Palaeo-Proterozoic cratons. The core of this assemblage was formed by Laurentia which was surrounded by the cratons of South China, Australia and East Antarctica, and by the cratons of Siberia, North China, Baltica and Amazonia. The West African craton and the Congo-Kalahari cratons at that time were totally separated. The Congo-Kalahari craton formed part of 'western' Rodinia, adjacent to East Antarctica, whilst the West African craton formed part of 'eastern' Rodinia, adjacent to the Amazonia craton (Figure 2.1). The terranes which later came to form the Touareg, Nubian and Arabian shields had not yet evolved.[2]

In north Africa remnants of the West African craton are found in the Eglab Mountains of Algeria, and in Mali and Upper Volta. These areas have yielded radiometric age dates attributable to the Archaean and early Proterozoic. Cratonic fragments of similar age form the Nile craton, which extends from southern Libya into eastern Chad, and forms the northern extension of the Congo-Kalahari craton. These rocks give Rb-Sr ages of 2900 to 2600 Ma, with some younger overprinting. In Libya rocks belonging to this suite form the Jabal Awaynat, on the border with Sudan, and have also been found in the eastern Tibisti Mountains. It is likely that these rocks form basement over the entire Al Kufrah Basin (Figure 2.5).[3]

A much younger cratonic fragment is preserved in the Hoggar Mountains of Algeria which dates from the late Proterozoic, but which appears to predate the Pan-African orogeny. This has been named the East Saharan craton, and a small patch of these rocks forms the border region with Libya, to the southwest of Ghat (Figure 2.5). The basement rocks of the western Murzuq Basin are believed to belong to the same suite. In the Tibisti Massif the Bin Ghanimah granite batholith dates from the late Neoproterozoic, and also appears to have escaped remobilisation during the Pan-African orogeny.[4]

2.3 The Break-up of Rodinia

Rodinia began to break apart during the early Neo-Proterozoic, with Laurentia and the 'eastern' terranes moving away from 'western' Rodinia (Figure 2.2). Available evidence suggests that the East Sahara craton and its attendant terranes was located on the periphery of 'western' Rodinia, adjacent to the Congo and Indian cratons, but its precise location in relation to these areas is unquantified. Magmatic arc conditions were present in the terranes which later formed the Arabian-Nubian shield whilst the terranes of the Touareg shield were represented by oceanic microplates in the Pharusian Ocean. The Congo-Nile craton maintained its position on the 'south-western' margin of 'western' Rodinia bordering the Adamastor Ocean, and the West African craton remained attached to the Amazonia craton in 'eastern' Rodinia.[5]

The Pharusian microplates, which exhibit a wide variety of oceanic island arc suites of sedimentary and volcanic rocks, are now preserved as accreted terranes in the Touareg shield of Algeria and Mali. The tectonic events which produced these rocks began at about 880 Ma and persisted until about 550 Ma. Evidence from deep oil wells shows that rocks

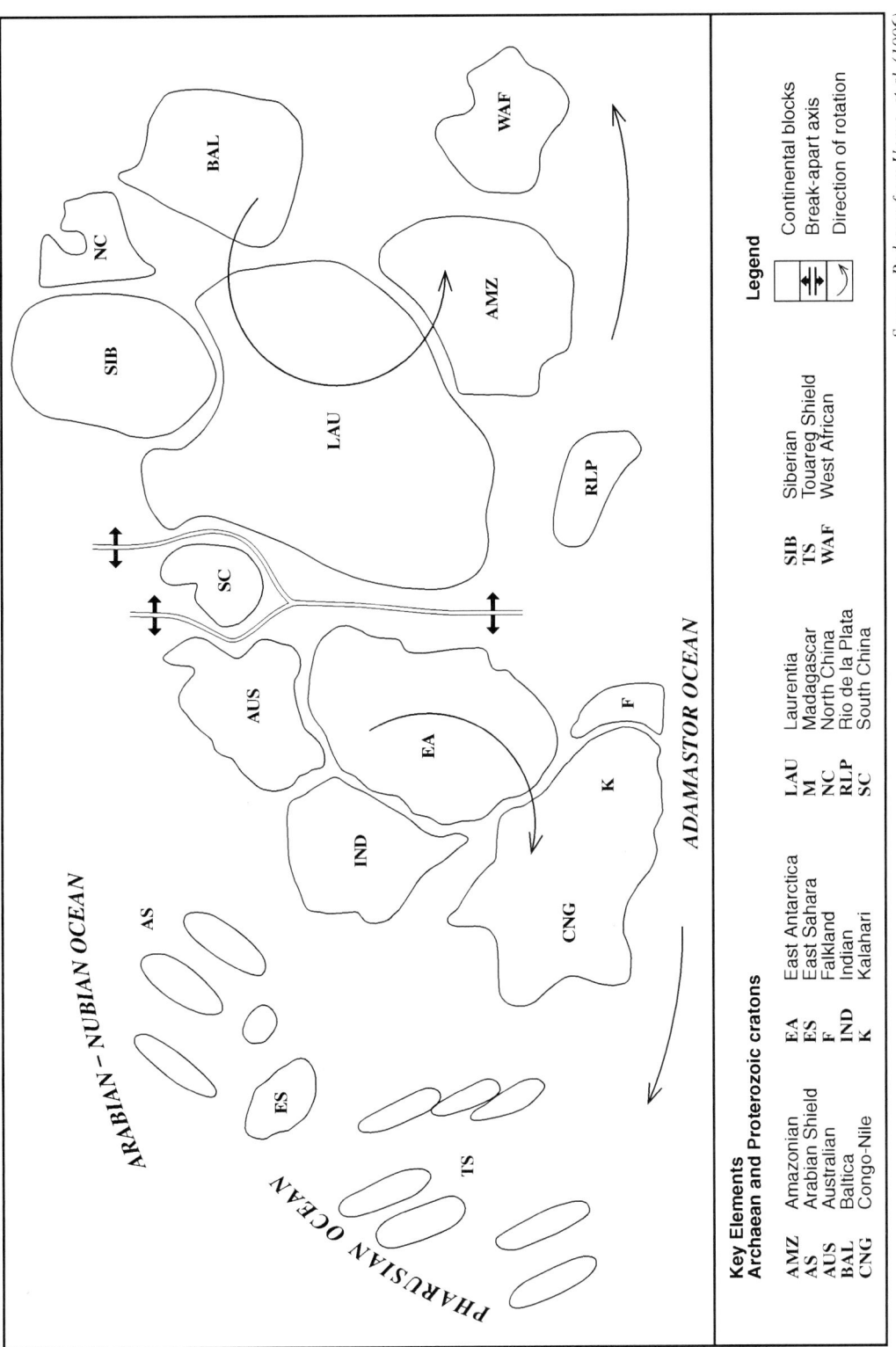

Source: Redrawn from Unrug, et al. (1996)

Key Elements
Archaean and Proterozoic cratons

AMZ	Amazonian	**EA**	East Antarctica	**LAU** Laurentia
AS	Arabian Shield	**ES**	East Sahara	**M** Madagascar
AUS	Australian	**F**	Falkland	**NC** North China
BAL	Baltica	**IND**	Indian	**RLP** Rio de la Plata
CNG	Congo-Nile	**K**	Kalahari	**SC** South China

SIB	Siberian	
TS	Touareg Shield	
WAF	West African	

Legend

Continental blocks
Break-apart axis
Direction of rotation

Figure 2.2   Plate Tectonic Reconstruction: the Break-up of Rodinia 1000–700 Ma

The West African craton was still remote from the Congo/Kalahari craton. The terranes which came to form the Touareg Shield of North Africa developed as oceanic microplates in the Pharusian Ocean, and the terranes which came to form the Arabian Shield developed as magmatic arcs in the Arabian–Nubian Ocean.

belonging to this assemblage form basement across much of northern Libya, including the Ghadamis Basin, northern Sirt Basin and Cyrenaica Platform (Figure 2.5).[6]

The period from 700 to 500 Ma was marked by the final break-up of Rodinia and the reassembly of the component plates into a new supercontinent which was given the name Pannotia by Powell. The magnitude of this reorganisation cannot be over-emphasised. In effect Rodinia was turned inside out. During this phase the entire assemblage of 'eastern' Rodinia was rotated in an anticlockwise direction and translated relatively 'eastwards', whereas 'western' Rodinia was rotated clockwise and displaced relatively towards the 'west' (Figure 2.2).[7]

2.4 The Pan-African Orogeny and the Assembly of Gondwana

The rotated and realigned cratonic blocks were reunited during the late Neo-Proterozoic, but in a totally new configuration. The re-assembly was accompanied by intense deformation, called the Pan-African orogeny, in which the cratonic nuclei of present-day Africa were fused together. The South American and Amazonia cratons were fused onto the western margin of the Kalahari craton, and the components of north Africa were emplaced along shear zones on the northern edge of the Congo-Nile craton accompanied by extensive remobilisation (Figure 2.3).[8]

This reconstruction demonstrates that Africa is made up of six components of diverse origin: the West African, Congo-Nile, Kalahari and East Saharan cratons of Archaean and Proterozoic age; the Arabian/Nubian shield, of magmatic arc origin; and the Touarag shield, made up of marine terranes derived from the Pharusian Ocean. The shear zone along which the components of north Africa were emplaced is marked by a broad belt of collisional deformation. The Pan-African orogeny was accompanied by intense metamorphism during which some of the old cratonic margins were remobilised. These Neo-Proterozoic mobile belts include the Tibisti, Trans-Saharan, Anti-Atlas and Mauritanide belts, and the Arabian-Nubian shield (Figure 2.4).[9]

The remobilised belts have been metamorphosed to green-schist facies. In north Africa this belt forms the eastern part of the Touareg shield in the Hoggar Mountains of Algeria, and the Aïr Massif in Niger. Similar remobilised cratonic rocks are found in Sudan. In Libya the rhyolites of the Upper Tibistian in the western Tibisti Mountains belong to the same suite. The small inliers of Precambrian metamorphic rocks on the Al Qarqaf Arch and west of Al Haruj al Aswad, which have radiometric ages of about 520 Ma, are also assigned to the Pan-African suite of rocks. Evidence from deep oil wells suggests that this facies extends across much of the Murzuq, and southern Sirt Basins, and the Tibisti Arch (Figure 2.5). The Pan-African metamorphism has been dated at about 550 Ma, but some of these rocks retain evidence of relict ages of 2200 Ma, giving an indication of their earlier cratonic origin.[10]

The Pannotia supercontinent had a relatively brief existence. During the earliest Palaeozoic it broke apart along the line of the Neoproterozoic suture and Laurentia, Baltica, Siberia and North China split from the new supercontinent. The smaller 'northern' group constituted Laurasia which, for the next 300 million years, had a separate history. The 'southern' continental assemblage, comprising Africa, Antarctica, South America, Australia and India, was given the name Gondwana, after a region in India where the distictive palaeofloral assemblage which characterises Gondwana was first described.[11]

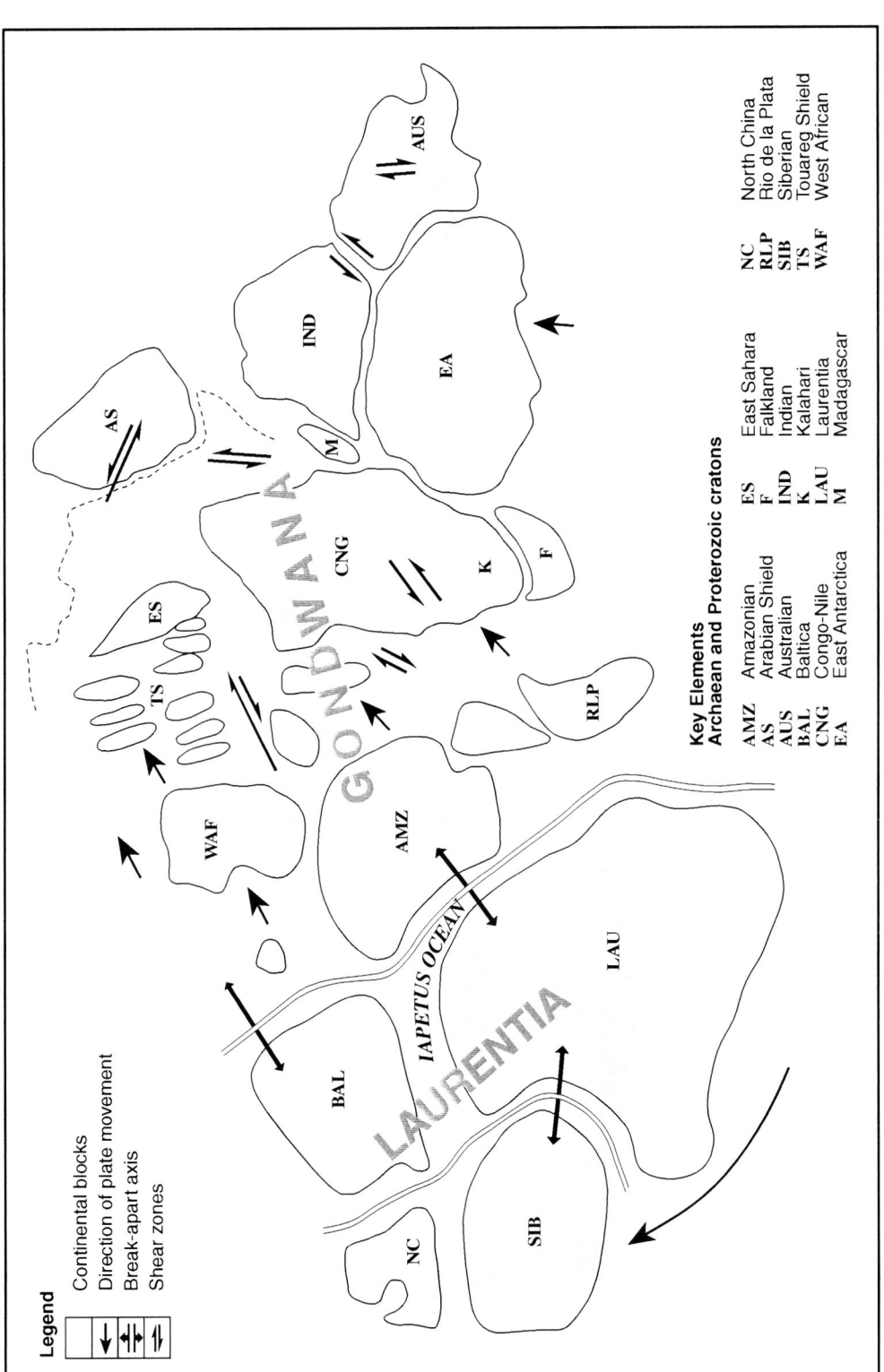

Legend

↓	Continental blocks
↔	Direction of plate movement
	Break-apart axis
⇊	Shear zones

Key Elements
Archaean and Proterozoic cratons

AMZ	Amazonian	**ES**	East Sahara	**NC**	North China
AS	Arabian Shield	**F**	Falkland	**RLP**	Rio de la Plata
AUS	Australian	**IND**	Indian	**SIB**	Siberian
BAL	Baltica	**K**	Kalahari	**TS**	Touareg Shield
CNG	Congo-Nile	**LAU**	Laurentia	**WAF**	West African
EA	East Antarctica	**M**	Madagascar		

Source: Redrawn from Unrug, et al. (1996)

**Figure 2.3   Plate Tectonic Reconstruction: Rodinia turned inside-out;
the Assembly of Pannotia and the Pan-African Orogeny, 700-500Ma**

Rodinia broke apart between 700 and 500Ma and the two halves drifted apart and rotated as shown in Figure 2.2. Their violent reassembly into the short-lived supercontinent of Pannotia (shown here) resulted in the Pan-African orogeny. Separation into the continents of Laurentia and Gondwana occurred in the earliest Palaeozoic.

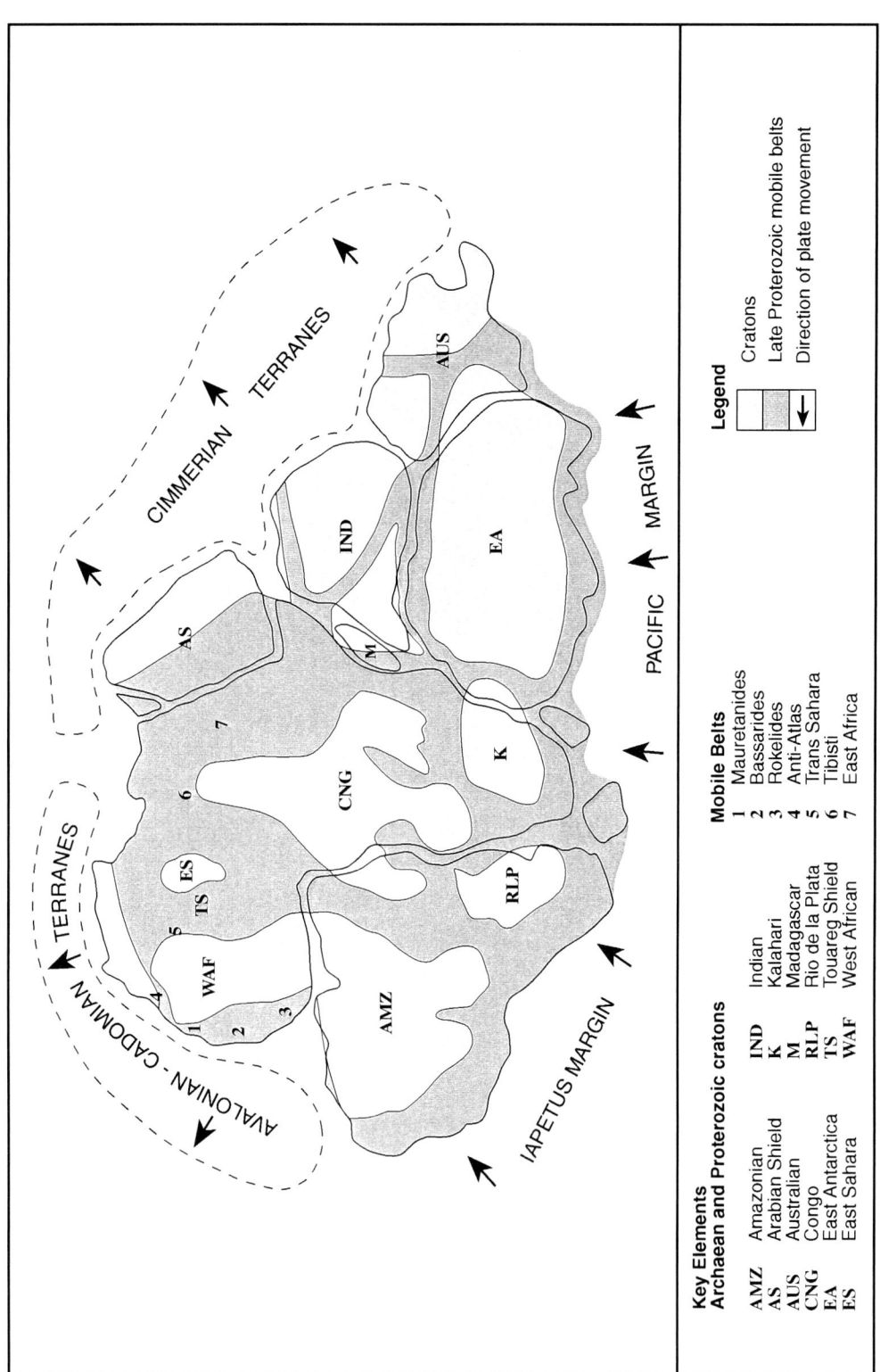

Figure 2.4   Plate Tectonic Reconstruction: Gondwana assembled, 450 Ma

Following the separation of Laurentia the Pan-African orogeny fused together the cratons and mobile belts of 'eastern' Pannotia to form Gondwana. The southern margin of Gondwana became an active margin with subduction of oceanic crust beneath the continental edge. The northern margin became a tensional passive margin. The Avalonian, Cadomian and Cimmerian terranes broke away during the early Palaeozoic and became united with Laurentia, Baltica and Asia respectively.

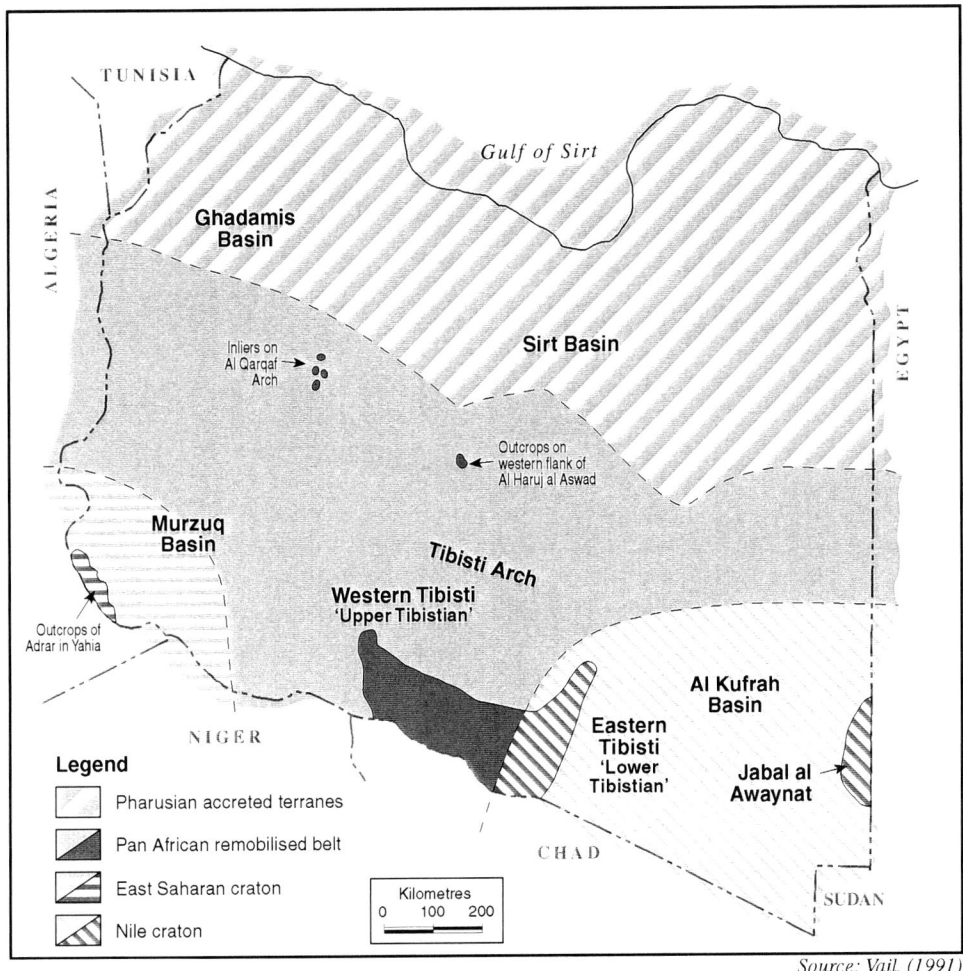

Figure 2.5   Libya: Basement Terranes

Basement in Libya is formed of four distinct Precambrian terranes. The outcrops of Jabal Awaynat and the eastern Tibisti Mountains belong to the Archaean and early Proterozoic Nile craton. The outcrops at Adrar in Yahia, which form part of the Hoggar Massif of Algeria, belong to the Neoproterozoic East Saharan craton. The western Tibisti Mountains, the isolated exposures on the Al Qarqaf Arch and west of Al Haruj al Aswad belong to the Pan-African remobilised belt, and the Precambrian rocks found in deep oil wells in northern Libya mostly belong to the Pharusian suite of oceanic accreted terranes.

2.5   Gondwana during the Palaeozoic

The assembly of Gondwana was completed by 540 Ma, and was followed by post-orogenic magmatism which persisted until 440 Ma during which granites and syenites were intruded into the mobile belts which separated the principal cratonic blocks of western Gondwana. Crustal thickening took place, accompanied by uplift, rifting and erosion. The southern margin of Gondwana developed as an active margin, characterised by mountain building, convergence and subduction tectonics, whereas the northern

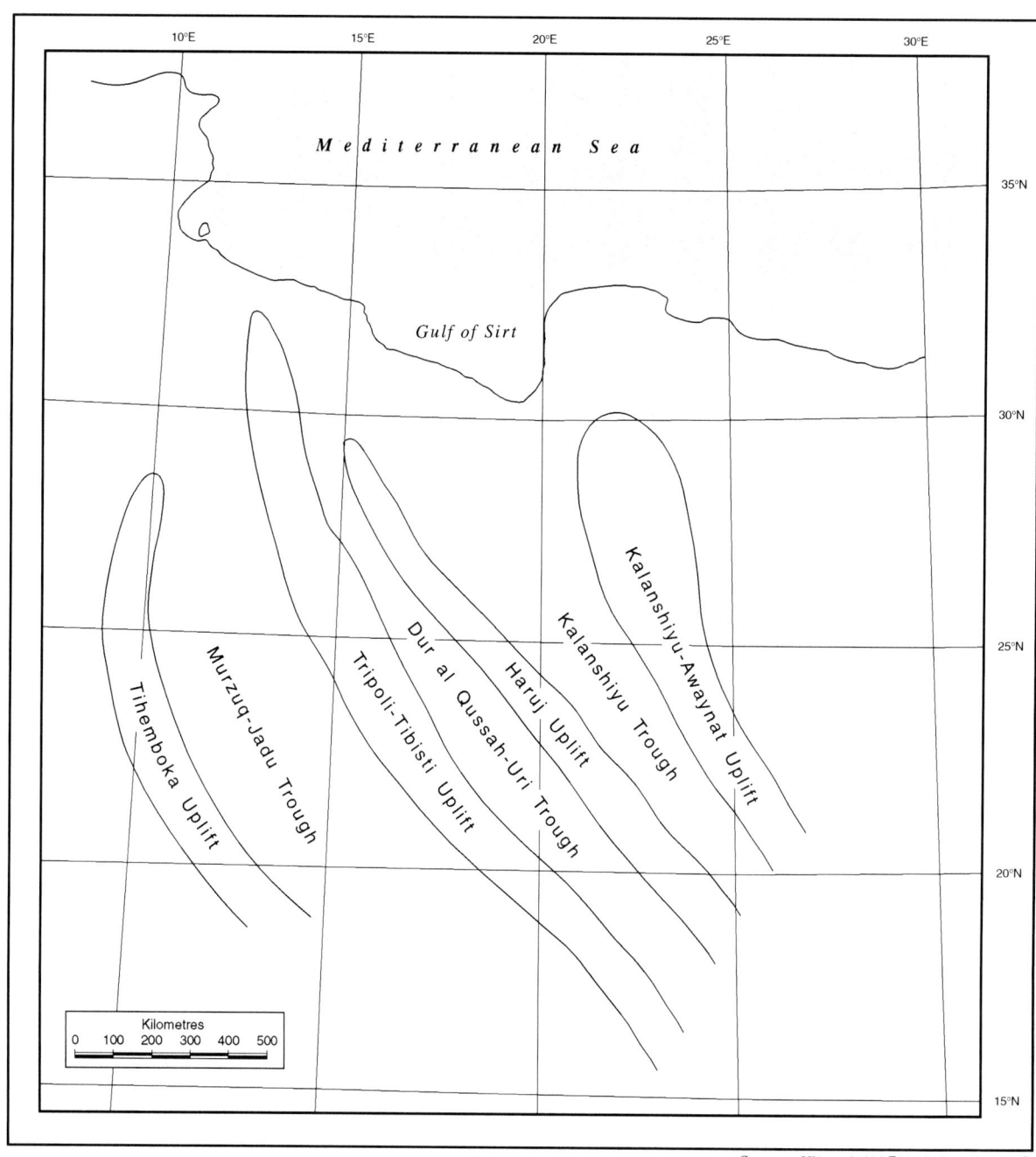

Source: Klitzsch (1971), Anketell (1996)

Figure 2.6   Post Pan-African Structural Trends in Libya

Pan-African tectonism welded together the cratonic blocks of Africa as they are known today with remobilised belts separating the cratonic nuclei. Following the de-coupling of Laurentia from Pannotia Libya was located on the passive margin of West Gondwana. The final phase of the Pan-African orogeny in the Cambrian produced a series of north-south to northwest-southeast uplifts and troughs which controlled sedimentation in the early Palaeozoic.

margin was passive, and was bordered by a broad and shallow shelf.[12]

The northern margin was unstable and fragments began to calve and drift away to the 'north'. The Avalonian and Cadomian Provinces, which include Neo-Proterozoic and early Palaeozoic magmatic arc deposits, broke away during the Ordovician and drifted northwards and collided with Laurentia and Baltica respectively, resulting in the Caledonian orogeny, which did not directly affect Gondwana. The Cimmerian Province, comprising fragments of northern Gondwana, similarly drifted away and was accreted to the Siberian plate (Figure 2.4). The seaway which developed between Laurasia and Gondwana, and which extended between Laurentia and Baltica was named the Iapetus Ocean by Harland and Gayer (Figure 2.3). This ocean reached its maximum width in the late Cambrian-early Ordovician.[13]

North Africa formed part of the passive margin of western Gondwana during the early Palaeozoic. The first sediments to be deposited after the Pan-African orogeny were continental sandstones, which are usually assigned to the Infracambrian. These rocks were formed by local erosion of Precambrian topography prior to the Cambrian marine transgression. Such rocks are found around the Tibisti Massif and in Cyrenaica, and are described in more detail in chapter 3.

The early Palaeozoic rocks in north Africa can be divided into three megacycles. The lower sequence consists of transgressive, fluvial and estuarine sands which pass upwards into marine shales (Ash Shabiyat Formation in Libya), which represent a maximum flooding event. A second cycle, dominently regressive in character, was terminated by another maximum flooding event (the Melaz Shuqran Formation in Libya), and the third cycle is terminated by the basal Silurian marine transgression. The Ordovician sequence is affected by local unconformities which reflect the final expression of Pan-African tectonism. In Libya these late tectonic events produced a series of broad, northwest-southeast to north-south troughs and swells which include the Tihemboka High, the Murzuq-Jadu Trough and the Tripoli-Tibisti Uplift which controlled deposition during the early Palaeozoic (Figure 2.6). It was during this period that the Avalonian and Cadomian terranes became detached from Gondwana and began their northward migration.[14]

During the early Palaeozoic western Gondwana was located close to the south pole and there is abundant evidence of glacial conditions in these regions during the late Ordovician. Both glacial and periglacial phenomena have been described from many areas in southern Algeria and Libya, and the periglacial rocks form major oil reservoirs in the Murzuq Basin.[15]

The glacial episode was very short lived, and was followed by a major flooding event during which early Silurian seas spread across much of north Africa. Thick black shales were deposited in these seas which form the major source rocks for the Palaeozoic oil accumulations of north Africa.[16]

Thereafter western Gondwana drifted northwards towards more temperate latitudes and the Iapetus Ocean separating Gondwana from Laurentia gradually closed. Throughout the Palaeozoic much of the interior of Gondwana was the site of widespread erosion and continental deposition, but shallow-shelf seas continued to occupy the northern margin. The black shales of the early Silurian are followed in Libya by pro-delta and deltaic high-stand deposits of the Akakus Formation, and the cycle was brought to an end by an episode of rifting and crustal separation along the margin of Western Gondwana, which produced a regional unconformity at the top of the Silurian.[17]

58

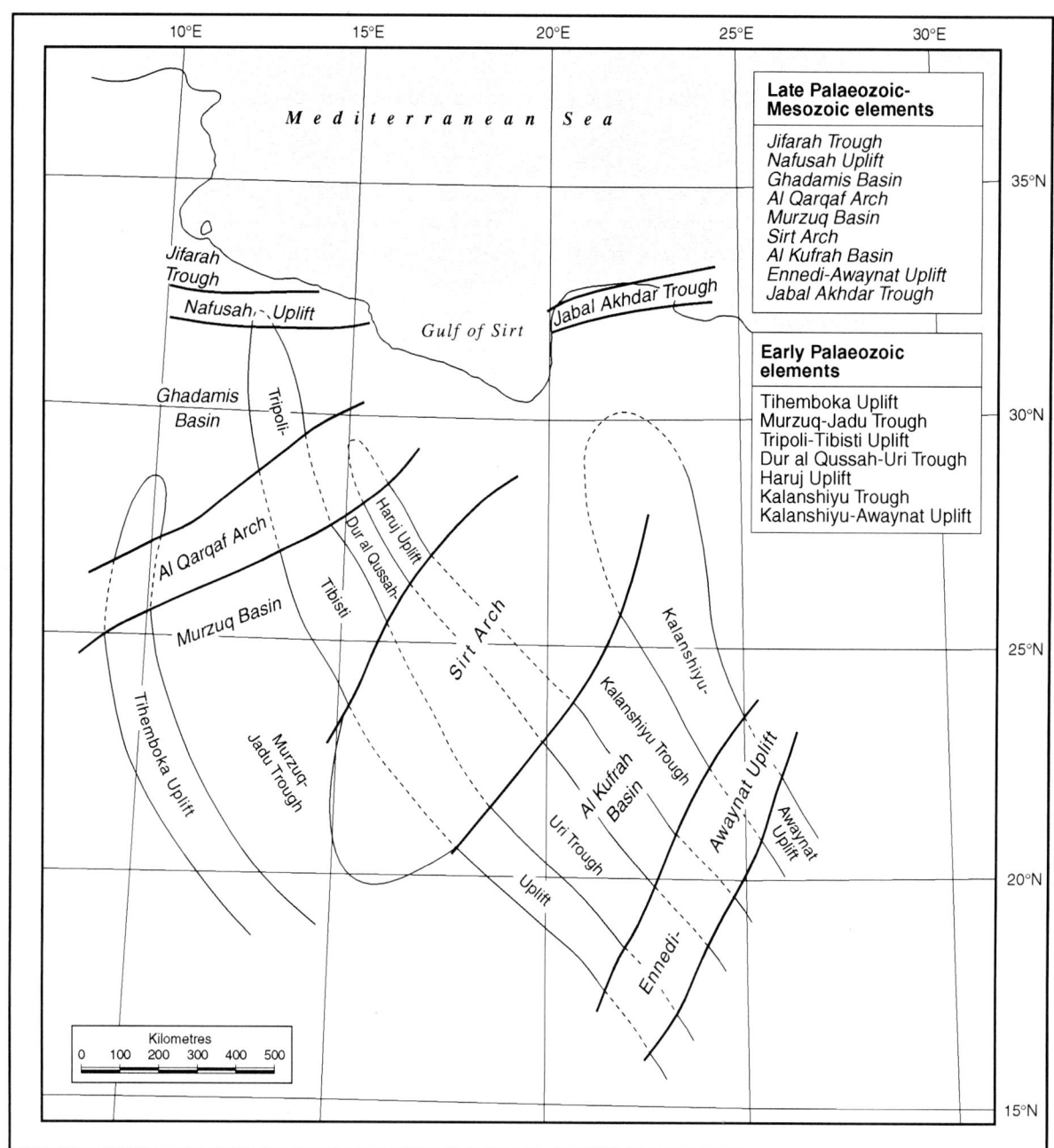

Source: Klitzsch (1971), Anketell (1996)

Figure 2.7   Hercynian Structural Trends

Gondwana collided with Laurasia in the mid-Devonian and in north Africa deformation increased in intensity throughout the Carboniferous, reaching a peak in the late Carboniferous. This is a reflection of the Hercynian orogeny. Much of Libya (and Algeria) was uplifted, eroded and deformed, and a new set of east-west to northeast-southwest structural elements was superimposed over the earlier Pan-African structural grain. Continental conditions were established over much of Libya which persisted until the mid-Cretaceous.

The early Devonian in Libya is dominated by four transgressive sequences which form a widespread deltaic complex, which were terminated by uplift and erosion in the mid-Devonian. During the mid- to late Devonian, the northwestern margin of Gondwana collided with Laurasia and the effects of the collision became progressively evident throughout north Africa. The initial effects were confined to Morocco and Algeria, but minor effects are visible in the late Devonian sequences in Libya, particularly the base Frasnian unconformity, which reflects an episode of extensive erosion on the flanks of Al Qarqaf Arch. A flooding event followed the Frasnian unconformity during which organic-rich shales were deposited in eastern Algeria and western Libya. This unit is a source rock for most of the middle and late Devonian and Carboniferous reservoirs in the area. The late Devonian in Libya is dominated by deltaic sands which include sedimentary oolitic ironstones in the Wadi Ash Shati area.[18]

Fluvial, deltaic and littoral deposits characterise the Carboniferous in north Africa. The increasing intensity of the collision between Gondwana and Laurasia led to the development of a fold belt in the Atlas Mountains by mid-Carboniferous times, with associated foreland basins on the platform to the south. In Libya a final flooding event in the mid-Carboniferous produced the carbonates of the Dimbabah Formation, after which much of the north African margin was uplifted and deformed as the Hercynian orogeny reached its culmination in the late Carboniferous. Continental conditions were established, extensive erosion took place, and the north African platform was deformed into a series of swells and sags extending from Morocco to western Egypt. In Libya the Nafusah Uplift, Al Qarqaf Arch, Sirt Arch, Ennedi-Al Awaynat Uplift and their associated troughs were formed. The superimposition of the Hercynian trend over the Lower Palaeozoic trend led to the separation of the Ghadamis from the Murzuq Basin, and the former Kalanshiyu Trough was largely destroyed by the formation of the Sirt Arch and its subsequent deep erosion (Figure 2.7). The late Palaeozoic tectonism caused a major marine regression in North Africa. Most of Libyan territory became emergent and extensive erosion and continental deposition took place in the interior.[19]

2.6 Pangaea

The collision between western Gondwana and Laurasia resulted in the formation of a supercontinent which included virtually all of the world's continental plates. This supercontinent, which was named Pangaea by Wegener in 1912, persisted until the Jurassic. The surrounding 'world ocean' was called Panthalassa (Figure 2.8). The collision involved a significant strike-slip component, and a major dextral shear-zone developed along the line of contact. The configuration of several of the Triassic basins along the north African margin is controlled by the pull-apart geometry propagated by the shear-zone.[20]

Pangaea, unlike Gondwana, was a relatively unstable assemblage. A large oceanic gulf (Palaeo-Tethys) extended deep into Pangaea from the east, along the line of the dextral shear zone, and from this gulf the Tethys Ocean ultimately developed. The gulf was occupied by oceanic crust which acted as a transit plate on which fragments of Gondwana were transported across the Palaeo-Tethys gulf and accreted onto the Siberian and southeast Asian plates. This process was in operation during the early Permian and perhaps represents a continuation of the calving process which had operated in the same area during the Palaeozoic. Even during Permian time rifting extended westwards from the gulf along the line of the Gondwana-Laurasian suture, and oceanic crust was probably present as far west as the future eastern Mediterranean (Figure 2.8).[21]

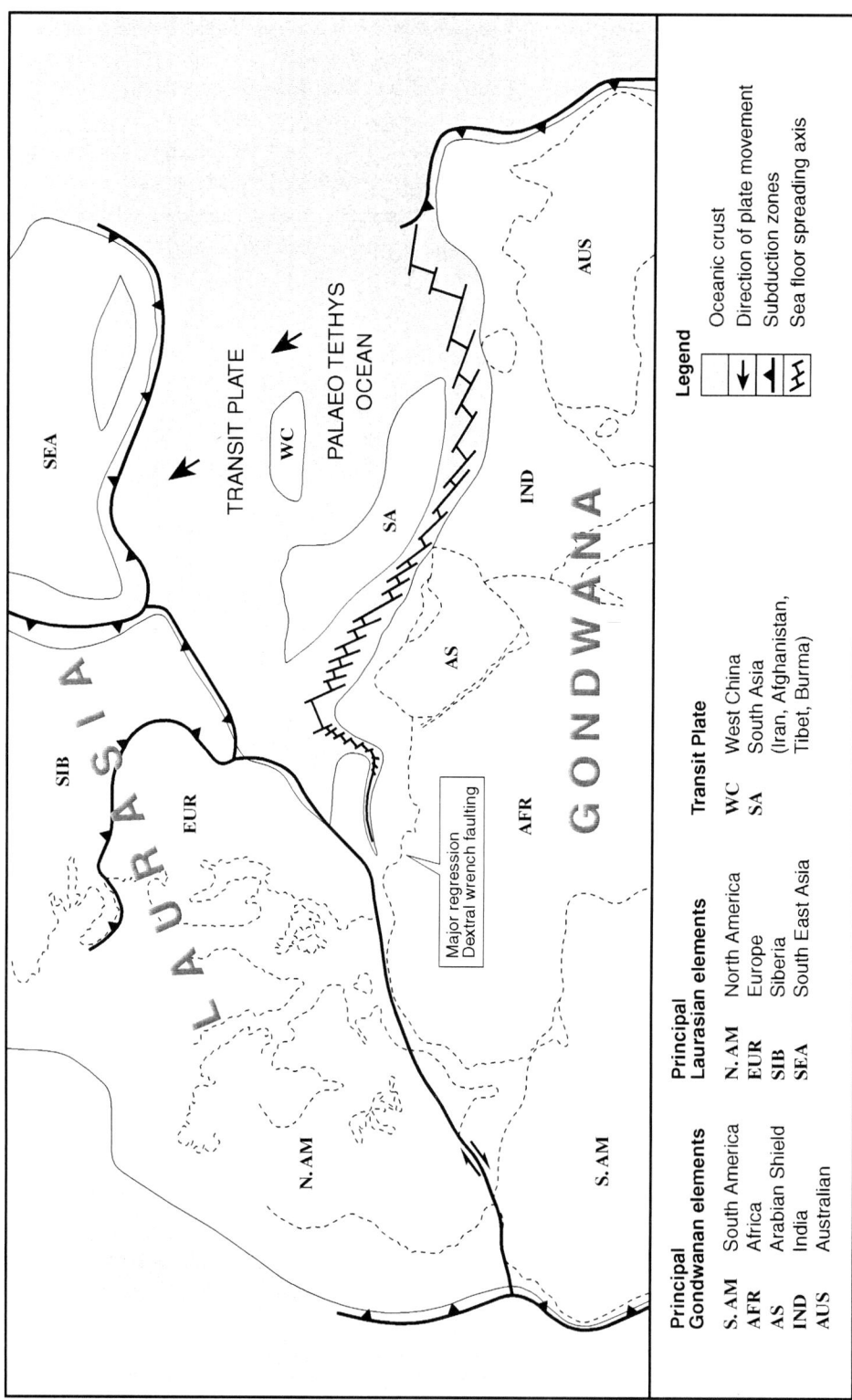

SEA

TRANSIT PLATE

WC

PALAEO TETHYS
OCEAN

AUS

Legend

	Oceanic crust
↓	Direction of plate movement
◄	Subduction zones
⋈	Sea floor spreading axis

SIB

EUR

SA

IND

AS

Major regression
Dextral wrench faulting

AFR

GONDWANA

L A U R A S I A

N. AM

S. AM

Principal Gondwanan elements		Principal Laurasian elements		Transit Plate	
S. AM	South America	N. AM	North America	WC	West China
AFR	Africa	EUR	Europe	SA	South Asia
AS	Arabian Shield	SIB	Siberia		(Iran, Afghanistan,
IND	India	SEA	South East Asia		Tibet, Burma)
AUS	Australian				

Figure 2.8   Plate Tectonic Reconstruction: Pangaea at 265 Ma. (early Permian)

From its southern high latitude position in the Ordovician Gondwana moved northwards during Silurian, Devonian and Carboniferous times, and collided with Laurasia in the late Palaeozoic to form the Pangaea Supercontinent. By the early Permian a proto-Tethys Ocean was developing. Microplates became detached from Gondwana and were transported northwards on the Transit Plate. A spreading ridge became active along the margin of eastern Gondwana and extended into the eastern Mediterranean region. A shear zone began to develop between western Gondwana and Laurasia.

Magmatic evidence in Libya gives some indication of the instability of this period. Mid- to late Permian basalts have been encountered offshore, and late Permian granite, with a radiometric age of 256 Ma, has been penetrated on the Waddan horst. In the Amal area microsyenite sills have been dated at 245 Ma. No extensive metamorphism or orogenic events are associated with these and subsequent intrusions quoted in this chapter, and they should probably be regarded as anorogenic events, or 'resetting the clock' on earlier magmatic events. This interval corresponds to the time of the great dextral wrench between Laurasia and Gondwana, the aborted rifting in the central Mediterranean region and the fragmentation of many small plates in the eastern Mediterranean.[22]

Permian marine rocks are present offshore Libya and in Tunisia, but thin rapidly from the Jifarah Basin onto the Jabal Nafusah Uplift which probably represented the southern margin of the Tethyan Basin at this time. In Algeria Permian marine carbonates are known north of the Talemzane Arch. On the emergent platform to the south continental rocks accumulated which were named the 'Continental Post Tassilien' by Kilian.[23]

The suture between western Gondwana and Laurasia was subjected to severe dextral wrenching during the Permian with dis- placement of appoximately 400km, which resulted in the formation of a patchwork of fragmented plates at the eastern end of the wrench by the mid-Triassic. This area formed a complex triple junction between the diverging plate margins of east and west Gondwana and the converging boundary along the southern margin of the Siberian and southeast Asian plates. The wrench-fault zone remained active until the commencement of rifting in the mid- Triassic (Figure 2.8).[24]

Break-up of Pangaea began in the Triassic. In Appalachia rift grabens are filled with late Triassic clastics, and in north Africa there is evidence of Triassic rifting in Tunisia and Libya. The north African continental margin at this time was characterised by extension and crustal thinning. In Algeria a blanket of Triassic fluvial sands and shales was deposited on the eroded Palaeozoic surface and these sands are overlain by extensive evaporites, which in turn are overlain by shelf carbonates. This sequence illustrates a progressive deepening of the sea due to post-rift thermal subsidence. Break-apart also began in the late Triassic between the former components of East and West Gondwana, with rifting along the southern margin of the Arabian plate, and along the eastern margin of Africa.[25]

The break-up of Pangaea involved the establishment of a spreading axis throughout the Mediterranean region. Associated igneous activity in Libya has been reported from the Waddan Platform (granodiorite 230 Ma) and from the Amal oilfield (granodiorite 207 Ma). These events correspond to the initial rifting phase in the Mediterranean region. Tensional faults of Triassic age are present both offshore and onshore in Libya, and several unconformities are present within the Triassic succession. Evidence from eastern Libya shows the presence of Triassic sediments which may represent the earliest deposits in incipient syn- rift grabens. Triassic rocks are present in the offshore Libyan basins and thin onto the Jabal Nafusah Uplift which again marks the effective Tethyan shoreline. Over the rest of Libya Triassic rocks are almost entirely continental in character.[26]

2.7 Tethys

The concept of Tethys as an equatorial ocean located along the line of the Pangaean suture was introduced by Suess in 1893. It is now generally regarded as the Mesozoic ocean which developed along this axis as Pangaea broke apart, and which persisted until the eastern Tethys Ocean was closed by the northward movement of Africa. Western Tethys is preserved as the peripheral margin of the present-day Atlantic Ocean between Morocco and Senegal and between Nova Scotia and Florida.

The earliest oceanic crust, marking the opening of the eastern Tethys Ocean was formed during the Bathonian-Callovian in the area of the future eastern Mediterranean. In Libya several Jurassic magmatic events occurred during this period. On the Waddan Platform granodiorites continued to be intruded until 180 Ma, and on the Amal oilfield basalts, rhyolites and syenites have been dated from 164 Ma to 159 Ma. A rhyolite from the Zaltan Platform gives a date of 160 Ma. Elsewhere oceanic crust of Toarcian age is present to the south of the Arabian plate, and Callovian oceanic crust is present in the western Tethys along the margins of northwest Africa and eastern North America.

By Oxfordian times Pangaea no longer existed. Total separation had been established between Laurasia and western Gondwana along the line of the Pangaean suture, and the Tethys Ocean formed a seaway from Asia to the Pacific (Figure 2.6). At the same time breakapart between east and west Gondwana was completed and separation was established between Africa and India/Antarctica. The relative motion between Africa and Europe is critical to the development of Tethys, and to the tectonic history of north Africa. It was initially believed that three main phases could be distinguished. An early phase from the Jurassic to early Cretaceous in which only the central Atlantic (Senegal to Morocco) was opening resulted in an apparent eastward movement of Africa relative to Europe. A second phase, from late Cretaceous to early Cainozoic, during which the north Atlantic between Europe and north America opened faster than the central

Atlantic, produced an apparent western motion of Africa relative to Europe. The third phase was dominated by the northern motion of Africa and the closure of Tethys. Subsequent palaeomagnetic work has shown that this model is not correct, and it has now been demonstrated that since the mid-Jurassic the motion of Africa relative to Europe has been, with one or two perturbations, in the form of a simple arc swinging from ESE in the mid-Jurassic, to east in the early Cretaceous, northeast from the mid-Cretaceous, and north from the late Eocene. It should be emphasised that these are relative motions. Attempts to show absolute motion in relation to fixed-mantle hotspots give significantly different results. Studies have been conducted on hotspot tracks both in Libya and from the Atlantic which show a southward motion of Africa during the Jurassic, a northward motion during most of the Cretaceous, and a northeasterly motion since the Campanian.[27]

2.8 The Development of Tethys

There are abundant publications on the evolution of the Mediterranean sector of Tethys. Two syntheses by Dercourt et al. (1986) and Ricou (1994) contain excellent maps illustrating the plate tectonic history, and Dercourt's maps were later elaborated into a palaeoenvironmental atlas of Tethys (1993). Dewey et al. (1989) reviewed the kinetic aspects of plate movement, and Livermore and Smith (1985), presented palaeomagnetic evidence which they used to derive relative plate motions. Fouad (1991) related plate movement to hot spot tracks. Guiraud (1998), and Mengoli and Spinicci (1984) gave overviews of the northern Tethyan margin, Woodside (1991) studied the eastern Mediterranean, and Fischer (1970), and Boccaletti and Guazzone (1970) provided data on the central Mediterranean. Geiss (1991) presented a significantly different view,

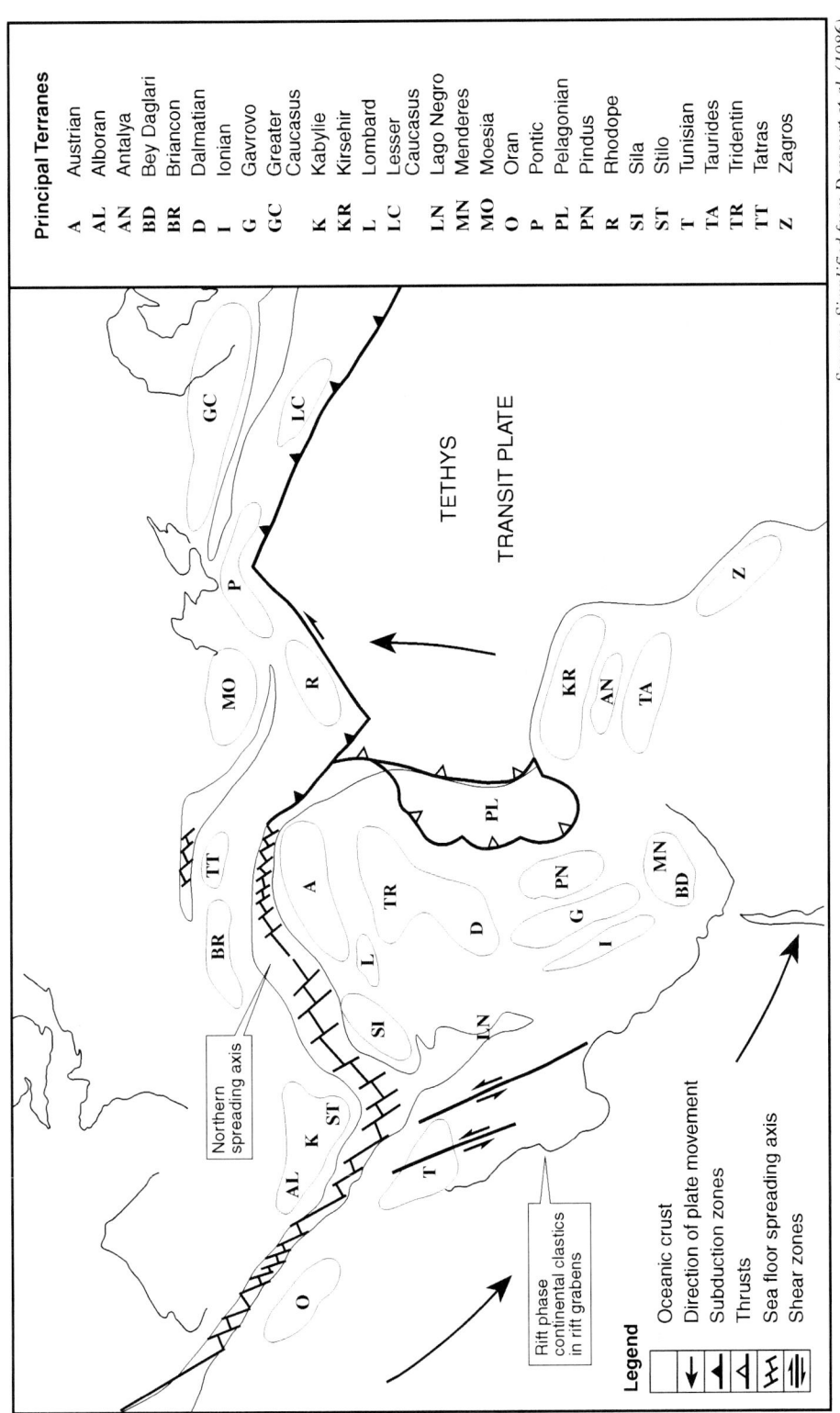

Principal Terranes

A	Austrian
AL.	Alboran
AN	Antalya
BD	Bey Daglari
BR	Briancon
D	Dalmatian
I	Ionian
G	Gavrovo
GC	Greater Caucasus
K	Kabylie
KR	Kirsehir
L	Lombard
LC	Lesser Caucasus
LN	Lago Negro
MN	Menderes
MO	Moesia
O	Oran
P	Pontic
PL	Pelagonian
PN	Pindus
R	Rhodope
SI	Sila
ST	Stilo
T	Tunisian
TA	Taurides
TR	Tridentin
TT	Tatras
Z	Zagros

TETHYS

TRANSIT PLATE

Northern spreading axis

Rift phase continental clastics in rift grabens

Legend

↓	Oceanic crust
◄	Direction of plate movement
◁	Subduction zones
⋏	Thrusts
⋏	Sea floor spreading axis
⇉	Shear zones

Source: Simplified from Dercourt et al. (1986)

Figure 2.9   Plate Tectonic Reconstruction: Tethys at 150 Ma. (late Jurassic)

This reconstruction shows the disposition of terranes in the Mediterranean and adjacent areas during the late Jurassic. A seaway became established throughout the Mediterranean region with a spreading axis extending from Arabia, via a northern route, to the newly opening Atlantic. Africa had an eastward motion relative to Eurasia, producing sinistral shear zones in the western Mediterranean. Oceanic crust continued to move northward and was subducted beneath the Pontide and Caucasus terranes. Thrusting and ophiolite obduction occurred in the Pelagonian terrane (Macedonia).

favouring mobile zones between more stable regions, but the following review largely follows the work of Dercourt and Ricou.

In the early Jurassic, before the breakthrough of Tethys in the Mediterranean region, the eastern Tethys gulf was flanked on its northern side by a subduction zone north of which were located the Cimmerian terranes of central Afghanistan, northern Iran, the Caucasus, northern Turkey, and Bulgaria/Romania. The southwestern flank of the gulf represented the unstable margin of Gondwana from which fragments were detached at intervals and transported northwards on the transit plate. This margin included terranes which were ultimately to form western Iran, southern Turkey, and the Balkans.[28]

Sea floor spreading in the Mediterranean began in the Bathonian-Callovian with a spreading axis developing between the Briancon terrane (French Alps) to the north and the Austrian and Lombard terranes to the south. In the western Mediterranean a spreading axis developed between the Iberian terranes and the Morocco-Oran-Tunisian terranes (Figure 2.9). By the late Jurasssic these two spreading axes had joined in the central Mediterranean and Pangaea was split into two. The initial spreading axis followed a sinuous path through the western Mediterranean (the African-European Rift Zone of Morgan, et al.), turning northeast, and then southeast, through the present Alpine region, to join the eastern Tethys gulf to the south of the Bulgarian terranes. Tentative evidence from the Lago Negro terrane in southern Italy suggests that a further area of oceanic crust developed along the northern margin of the African plate at this time, and represents the failed arm of a short-lived triple junction (Figure 2.9). To the east subduction continued beneath the Pontic and Caucasus terranes, but further west strike slip faulting affected the Rhodope terrane (Bulgaria), and major thrusting occurred in the Pelagonian

terrane (Macedonia) with obduction of ophiolites.[29]

In Tunisia the opening of the Tethys seaway induced a sinistral transtensional regime which broke up the Triassic and Jurassic platform sequences and produced flowage within the Triassic salt deposits. In Libya evidence of continental Jurassic is scarce, but the incipient rifts of the Triassic probably continued to develop. Jurassic rocks in the Jabal Nafusah have a more marine aspect than in the Triassic, but quickly pass into continental equivalents to the south. In Algeria evaporite deposition ended in the early Jurassic and was succeeded by a marine transgression which spread carbonates and shales over the platform margin.[30]

The establishment of seafloor spreading in the central Atlantic in the early Cretaceous produced major changes in the Mediterranean segment of Tethys. The change from an ESE to an easterly drift of Africa relative to Europe induced a major shift in the position of the main Mediterranean spreading axis (Figure 2.10). The former spreading axis in the Alpine region was obliterated by the northward movement of the Apulian plate comprising the Lombard, Tridentin and Dalmatian terranes (the Adriatic prong of Livermore and Smith), and a new spreading axis opened further south, adjacent to the north African margin. This new axis termed Mesogea by Dercourt et al. connected to the eastern Tethys gulf via a major transcurrent displacement zone. This event caused major thrusting and deformation in the Austrian and Sila terranes, and the obduction of ophiolites between the Stilo and Sila terranes (Calabria). In the eastern Tethys the Kirsehir and Pelagonian terranes (central Turkey/Macedonia) were detached from the Arabian shield and displaced northwards.[31]

During the early Cretaceous the eastward displacement of Africa continued at a rate calculated by Dercourt, et al., of about 2.5cm/year. There is evidence of subsidence and

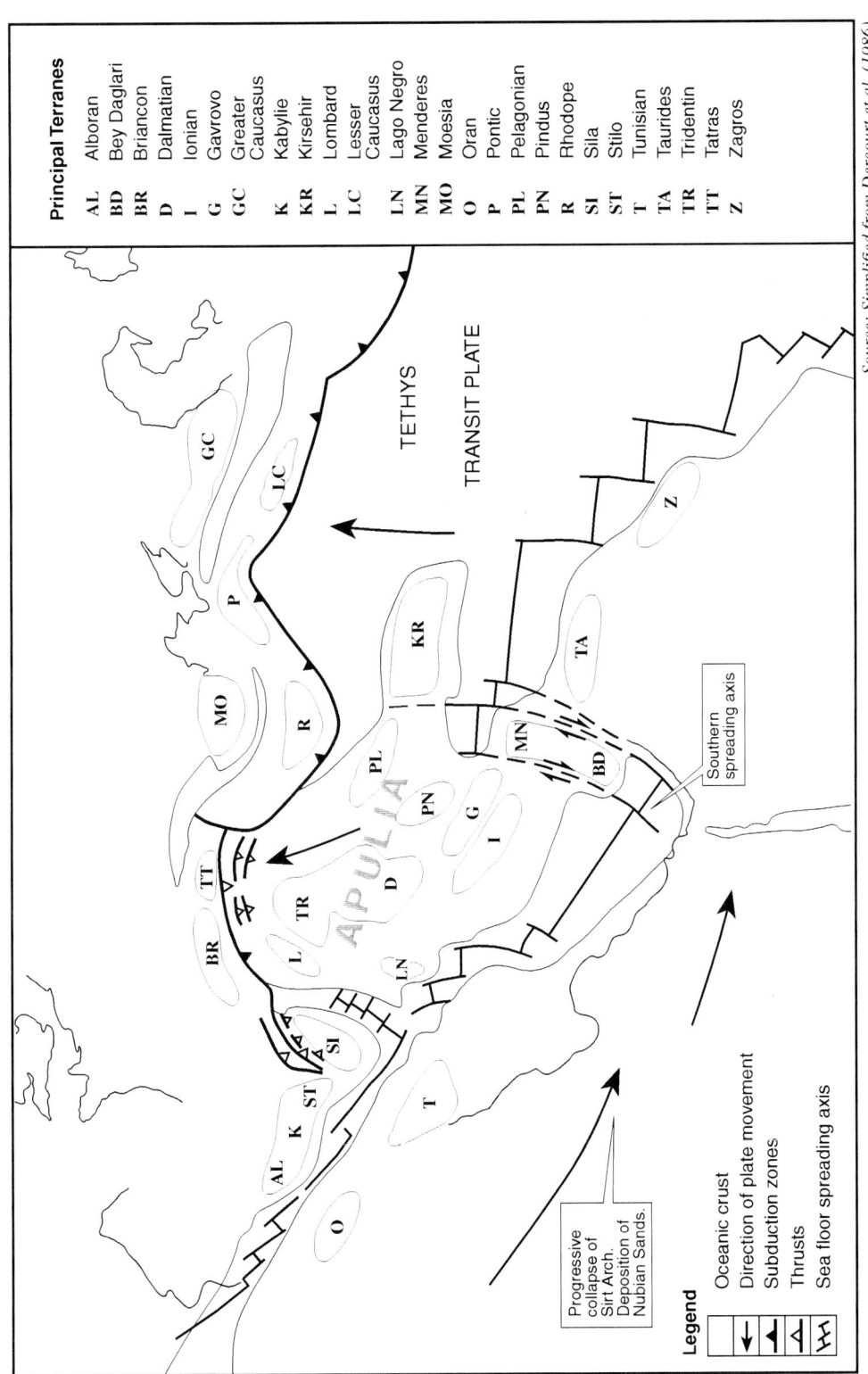

Principal Terranes

AL	Alboran
BD	Bey Daglari
BR	Briancon
D	Dalmatian
I	Ionian
G	Gavrovo
GC	Greater Caucasus
K	Kabylie
KR	Kirsehir
L	Lombard
LC	Lesser Caucasus
LN	Lago Negro
MN	Menderes
MO	Moesia
O	Oran
P	Pontic
PL	Pelagonian
PN	Pindus
R	Rhodope
SI	Sila
ST	Stilo
T	Tunisian
TA	Taurides
TR	Tridentin
TT	Tatras
Z	Zagros

Legend

	Oceanic crust
↓	Direction of plate movement
◄	Subduction zones
△	Thrusts
⋏⋏	Sea floor spreading axis

Progressive collapse of Sirt Arch. Deposition of Nubian Sands.

Southern spreading axis

TETHYS

TRANSIT PLATE

APULIA

Source: Simplified from Dercourt et al. (1986)

Figure 2.10   Plate Tectonic Reconstruction: Tethys at 120 Ma. (Aptian)

By the mid Cretaceous the composite terrane of the Balkans, Greece and Turkey had moved northwards and closed the original northern spreading centre, causing widespread deformation in the Alpine collision zone. A new, southern, spreading centre opened up in the wake of the composite terrane in the eastern Mediterranean. Africa continued its relative eastward motion and further 'Turkish' terranes were moved northwards on the Transit Plate.

66

pull-apart in the Hameimat and As Sarir troughs in the Neocomian and Barremian, and extensive areas of continental sands were deposited which form one of the most important hydrocarbon reservoirs in the country. Marine early Cretaceous rocks are limited to the northern margin of Libya. The stresses produced by the detachment of the Turkish, Balkan and Lombard terranes (Apulia Plate), active sea-floor spreading adjacent to the African margin, and the eastwards movement of Africa led to the stretching and collapse of the Sirt Arch in the mid-Cretaceous, and seaward tilting of the northern margin of the African plate (Figure 2.10). The southern prong of the Apulia Plate may have originally been located in the

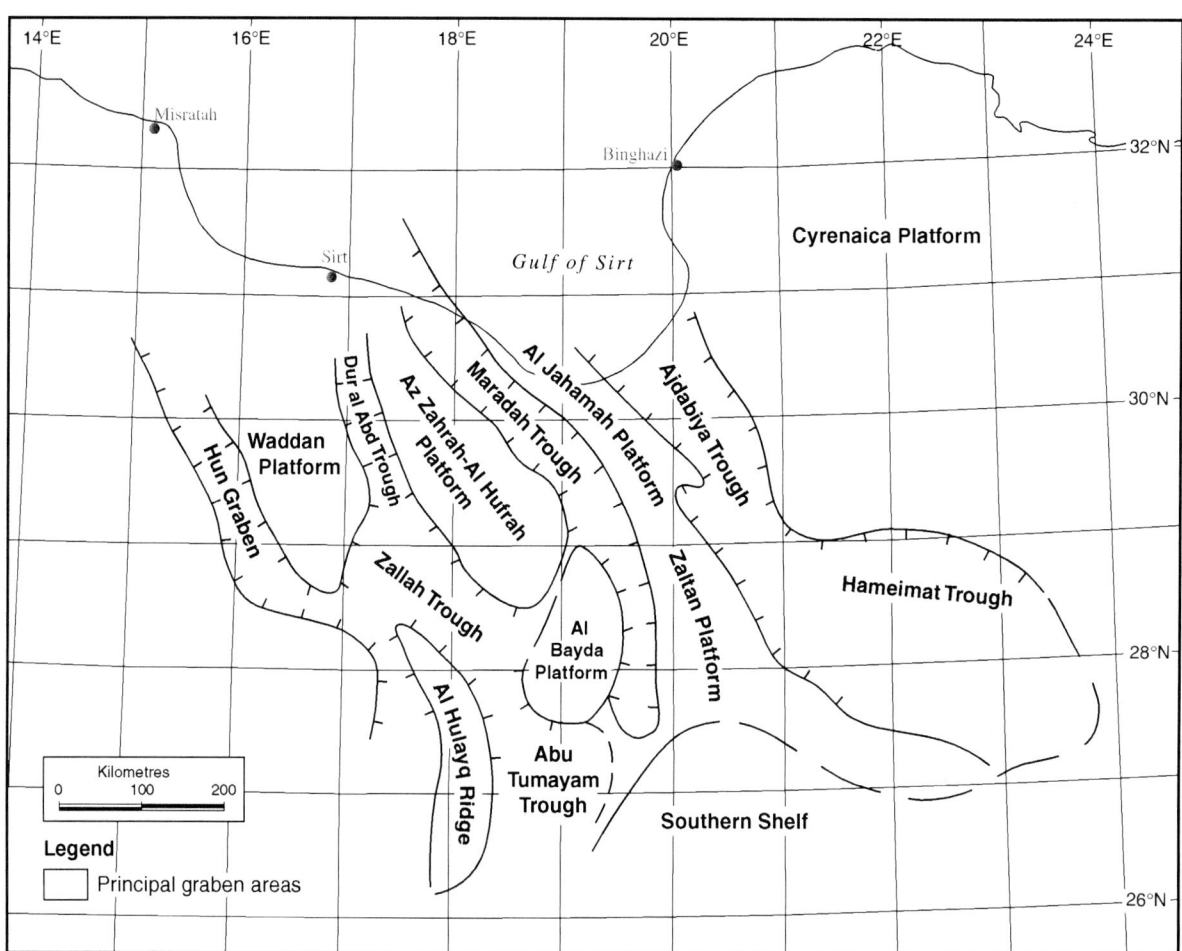

Source: Modified from Anketell, 1996

Figure 2.11   Collapse of the Sirt Arch

Rifting on the Sirt Arch began in the Triassic, and Triassic and Jurassic sediments have been found in several rift troughs. The syn-rift phase ended in the mid-Cretaceous with the collapse of the Sirt Arch. The collapse was caused by the detachment of the Turkish, Balkan and Lombard terranes from the north African margin, the establishment of an active sea floor spreading centre adjacent to the plate margin, and the relative eastwood movement of the African plate. A series of major horsts and grabens was formed which were flooded by a marine transgression in the Cenomanian.

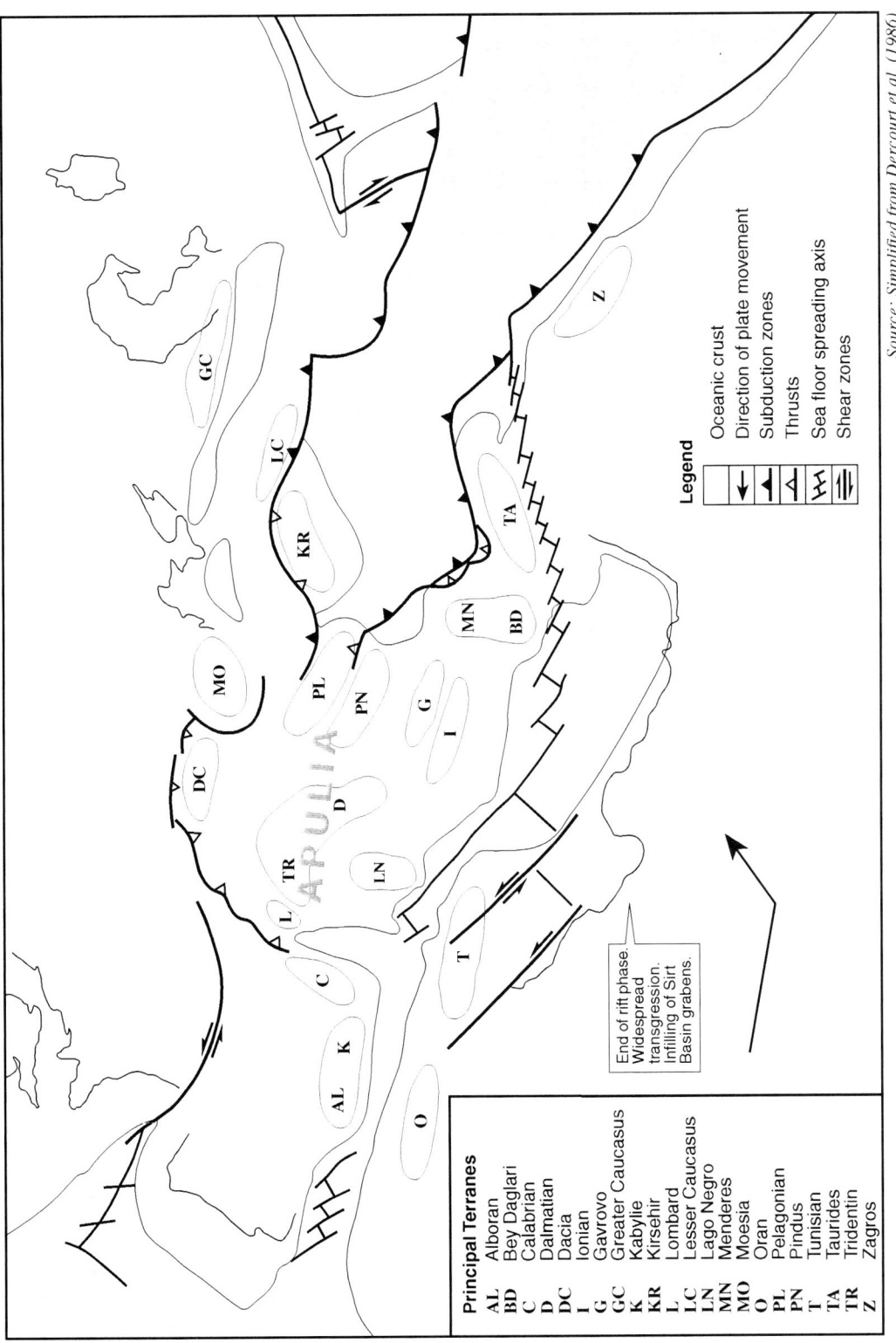

Source: Simplified from Dercourt et al. (1986)

Figure 2.12   Plate Tectonic Reconstruction: Tethys at 80 Ma. (Campanian)

This reconstruction shows Tethys at the point where seafloor spreading was ending in the Mediterranean region. The direction of movement of Africa changed from east to northeast during the Campanian. This produced sinistral shearing in North Africa and a termination of seafloor spreading in the western Mediterranean. Subduction of oceanic crust dominated the Turkish-Iranian sector.

Principal Terranes

AL	Alboran
BD	Bey Daglari
C	Calabrian
D	Dalmatian
DC	Dacia
I	Ionian
G	Gavrovo
GC	Greater Caucasus
K	Kabylie
KR	Kirsehir
L	Lombard
LC	Lesser Caucasus
LN	Lago Negro
MN	Menderes
MO	Moesia
O	Oran
PL	Pelagonian
PN	Pindus
T	Tunisian
TA	Taurides
TR	Tridentin
Z	Zagros

Legend

Oceanic crust

Direction of plate movement

Subduction zones

Thrusts

Sea floor spreading axis

Shear zones

End of rift phase. Widespread transgression. Infilling of Sirt Basin grabens.

Gulf of Sirt. The entire northern area of the Arch foundered to form the Sirt Basin, and a series of northwest-southeast horsts and grabens was established, the direction determined by shearing stresses along the northern margin of the African plate (Figure 2.11). Thermal sag led to marine conditions which flooded the basin, and by Cenomanian times much of northern Libya was covered by a shallow epicontinental sea. The syn-rift phase of tectonic development was brought to a close by the marine incursion, and the subsequent tectonic development can be assigned to the post-rift phase. Van Houten proposed a different explanation for the collapse of the Sirt Arch. He suggested that the collapse was caused by crustal thinning due to the region residing for a long period of time over a fixed mantle hotspot. Burke proposed that the Sirt Basin represents the failed southern arm of a triple junction between the Eurasian, African and Tethyan plates. It has been suggested that the Sirt Basin itself could represent a triple junction formed by the northwestern trend of the Cretaceous horsts and grabens, the southwestern trend of the Abu Tumayam Trough and the east-west trend of the Hameimat Trough, resulting from crustal thinning over a fixed mantle hotspot during the late Jurassic-early Cretaceous.[32]

These events are also reflected in magmatic activity in the Sirt Basin. Granite intrusions dated between 152 and 122 Ma (Tithonian to Aptian) and volcanic rocks from the interval 148 to 127 Ma correspond to this period. Magmatic activity increased in the late Cretaceous in response to extensive rifting and sinistral strike-slip faulting on the offshore Pelagian Shelf. Volcanic rocks of Aptian to Palaeocene age have been encountered in offshore wells, and in the Sirt Basin trachytes have been dated at 103 Ma (Albian).[33]

Separation was achieved between South America and Africa by the Cenomanian, leaving Gondwana fragmented into four continental plates: South America, Africa, India, and Australia/Antarctica. By contrast, Laurasia at this time formed a single continental plate from Mexico to China, although rifting in the North Atlantic had already begun. India began its northern migration towards Asia on the Transit Plate during the Cenomanian.[34]

During the Santonian-Coniacian period the motion of Africa relative to Europe changed to a northeasterly direction. In the western Mediterranean region seafloor spreading came to an end, marking the beginning of the end of the Tethys seaway, and Tethys began to close (Figure 2.13). The width of the eastern Tethys between Arabia and the northern Iranian terranes was dramatically reduced. Similarly in the Mediterranean, compression of the Apulian plate against the European plate led to thrusting and metamorphism in the western Alps, signalling the effective beginning of the Alpine orogeny. Further east the Kirsehir terrane was thrust against the Pontic terranes of northern Turkey, and the detachment of the Taurides terrane completed the separation of the Turkish terranes from Arabia. During this period seafloor spreading occurred in the Black Sea, as a result of crustal stretching in this area (Figure 2.12). In the central Mediterranean extensive strike-slip faulting developed, particularly on the Pelagian shelf and in the Sabratah Basin.[35]

In Libya the Cenomanian marine transgression was controlled by the rugged horst and graben topography left by the collapse of the Sirt Arch. Initially shallow marine rocks were deposited in the grabens, but as the transgression progressed deeper water conditions invaded the troughs whilst shallow water sediments gradually onlapped the horsts. It was in these troughs that the euxinic Sirt Shale was deposited, forming the source rock for most of the hydrocarbons in the Sirt Basin. Gentle uplift in western Libya led to a progressive withdrawal of marine conditions from this area. The Turonian-Coniacian

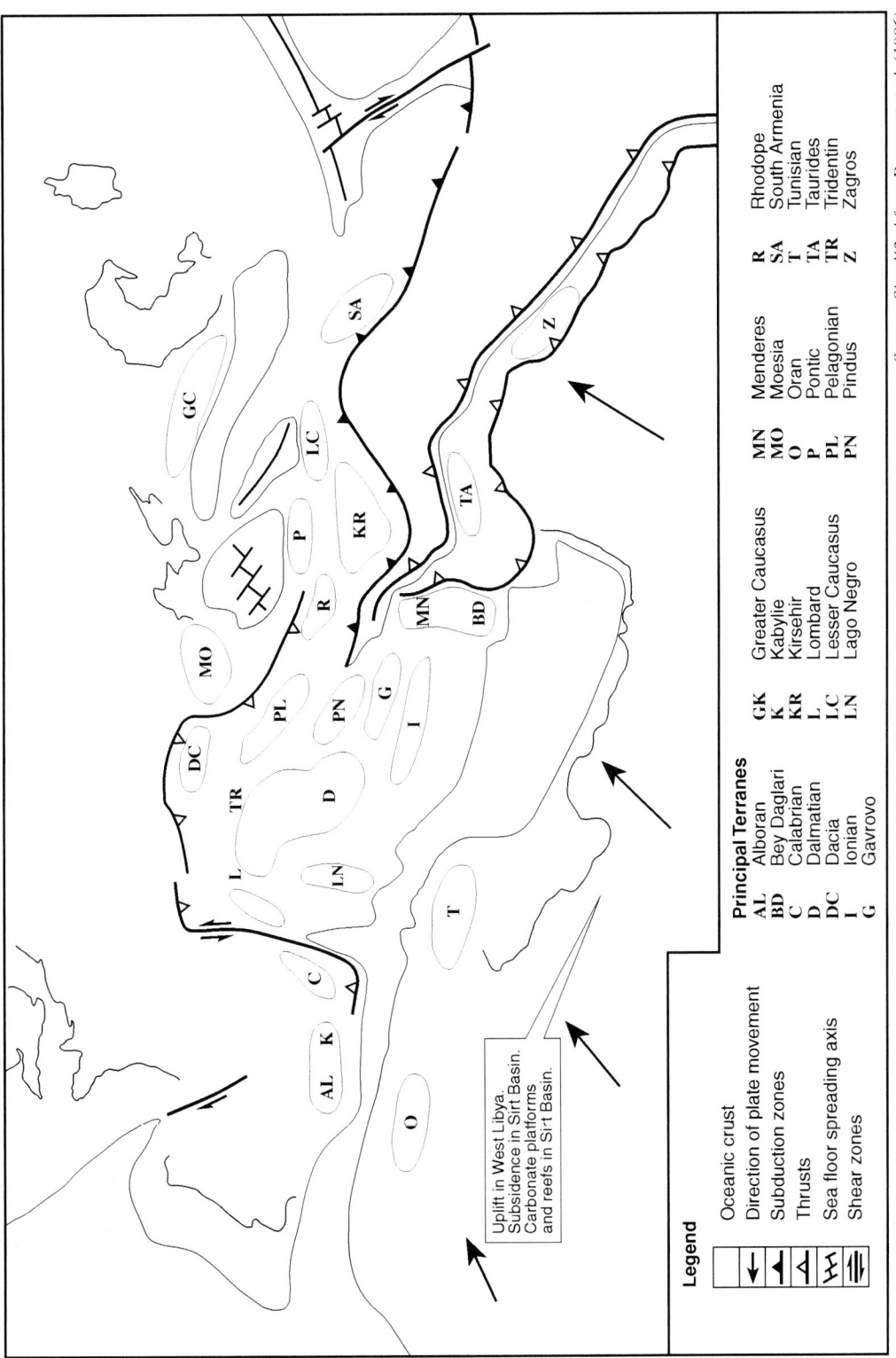

Legend

↓	Oceanic crust
	Direction of plate movement
◄	Subduction zones
◁	Thrusts
⋝	Sea floor spreading axis
⇈	Shear zones

Uplift in West Libya.
Subsidence in Sirt Basin.
Carbonate platforms
and reefs in Sirt Basin.

Principal Terranes

AL	Alboran	**GK**	Greater Caucasus	**MN**	Menderes	**R**	Rhodope
BD	Bey Daglari	**K**	Kabylie	**MO**	Moesia	**SA**	South Armenia
C	Calabrian	**KR**	Kirsehir	**O**	Oran	**T**	Tunisian
D	Dalmatian	**L**	Lombard	**P**	Pontic	**TA**	Taurides
DC	Dacia	**LC**	Lesser Caucasus	**PL**	Pelagonian	**TR**	Tridentin
I	Ionian	**LN**	Lago Negro	**PN**	Pindus	**Z**	Zagros
G	Gavrovo						

Figure 2.13   Plate Tectonic Reconstruction: Tethys at 65 Ma. (early Palaeocene)

Seafloor spreading ceased in the Mediterranean sector of Tethys due to the continued northeastward movement of Africa. Compression continued in the eastern sector with subduction of oceanic crust beneath the terranes of Turkey, Iran and South Armenia, and obduction of ophiolites in the Taurus-Zagros terranes. (Troodos and Oman ophiolites).

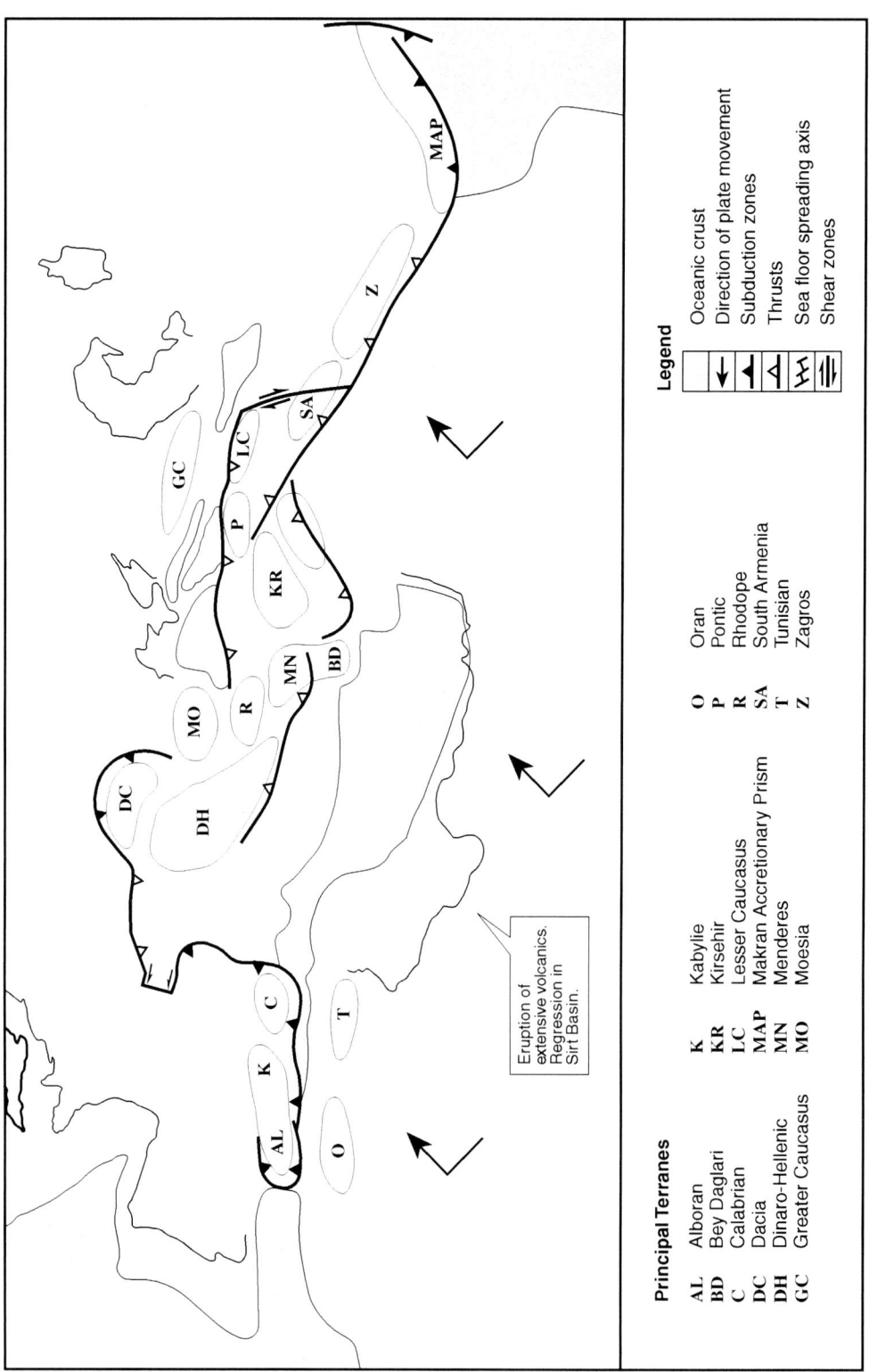

Principal Terranes

AL.	Alboran	K	Kabylie	O	Oran
BD	Bey Daglari	KR	Kirsehir	P	Pontic
C	Calabrian	LC	Lesser Caucasus	R	Rhodope
DC	Dacia	MAP	Makran Accretionary Prism	SA	South Armenia
DH	Dinaro-Hellenic	MN	Menderes	T	Tunisian
GC	Greater Caucasus	MO	Moesia	Z	Zagros

Legend

- Oceanic crust
- Direction of plate movement
- Subduction zones
- Thrusts
- Sea floor spreading axis
- Shear zones

Source: Simplified from Dercourt et al. (1986)

Eruption of extensive volcanics. Regression in Sirt Basin.

Figure 2.14 Plate Tectonic Reconstruction: Tethys at 20 Ma. (early Miocene)

The motion of Africa relative to Europe changed towards the northwest from 36 to 20 Ma, before returning to a northeastward trend. The southern terranes of Turkey and Iran were welded onto the Eurasian plate along a thrust front which extends from the Black Sea to southern Iran, finally closing the eastern sector of the Tethys Ocean. The continued northward motion of Africa closed the western Mediterranean, with subduction of oceanic crust beneath the terranes of Alboran, Kabylie and Calabria.

represents a tectonically quiet period, but in the Santonian a phase of compression, associated with the continued northeastward movement of the African plate, produced basin inversion in the Jabal Akhdar, in the Atlas mountains and in Egypt. Tension again became the dominent regime in the late Cretaceous along the southern margin of Tethys. By the end of the Cretaceous the grabens of the Sirt Basin had been largely infilled, and most of the horsts had been covered by shallow marine sediments, some of which form important hydrocarbon reservoirs.[36]

2.9   Tethys to Mediterranean

By the end of the Cretaceous, seafloor spreading had entirely ceased. Compression contined to be the dominant force, particularly in the eastern Tethys, causing extensive thrusting on the northern margin of Arabia with ophiolite obduction from the Troodos terrane (Cyprus) in the west to the Hawasina terrane (Oman) in the east.[37]

Throughout the Palaeocene and Eocene the northeastward movement of Africa continued at rates between 2.2 and 2.7 cm/year. In order to explain the more rapid movement of east Africa relative to west Africa since the mid-Cretaceous, it has been suggested that Africa did not behave as a single plate but as two, with a sinistral strike-slip zone between them. This is particularly evident in the early Tertiary where evidence of sinistral strike slip faulting has been found in the western Sirt Basin, and studies on stress fields in Libya support this view. The junction between the two plates extends from the Gulf of Guinea, via the Benue Trough, Chad Basin, and Tibisti Arch to the Sirt Basin. Other workers have suggested that the east African plate is composed of two sub-plates. As a result of the northeastward movement oceanic crust in the western Mediterranean was obliterated by the fusion of the Alboran terrane with the Oran terrane, which established a connection between Africa and Europe. This marks the effective end of the Tethys Ocean. In the central Mediterranean continued compression in the Alpine region led to the fusion of several terranes to produce the composite Dinaro-Hellenic terrane, and in the eastern Mediterranean the southern Turkish terranes were finally emplaced in their present position with considerable deformation along the line of contact from central Turkey to Crimea. This movement put an end to the crustal stretching in the Black Sea and the short-lived seafloor spreading in that area came to an end. The eastern Tethys gulf between Arabia and Iran finally closed along a subduction front which extended from the eastern Taurides to southern Iran.[38]

In Libya the early Tertiary was marked by uplift and withdrawal of marine conditions from western Libya and gentle subsidence over the earlier Cretaceous grabens in the Sirt Basin. Major subsidence continued in the Ajdabiya Trough. During this period broad carbonate platforms became established in the Sirt Basin with reefal conditions around the seaward margins. The Palaeocene and Eocene carbonates represent major hydrocarbon reservoirs in Libya. In Algeria the inversion of the Atlas Basin began in the Eocene as a result of the compressional events in the western Mediterranean. Volcanic activity began along the line of the Tripoli-Tibisti axis during the Eocene and ring-intrusions at Jabal Awaynat on the border with Sudan have been dated as Eocene.[39]

The Oligocene saw a brief change in the direction of movement of Africa. From Priabonian to Burdigalian times the dominant movement was towards the northwest. Subsidence in the Hun Graben occurred during this period. The change of drift direction led to oceanic subduction in the western Mediterranean beneath the Alboran, Kabylie and Calabrian terranes and deformation and thrusting in the Alps and eastwards as far as the

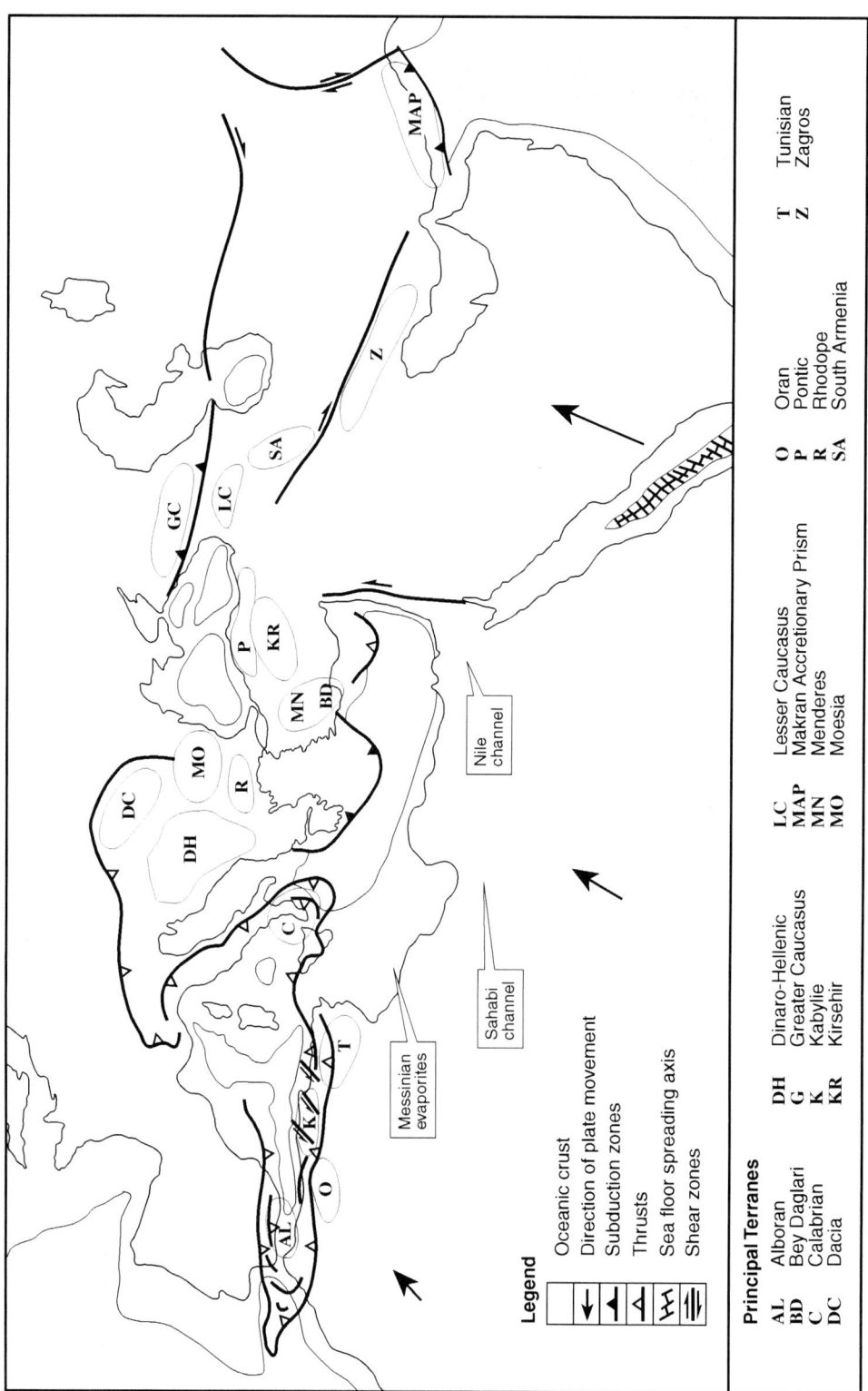

Figure 2.15   Plate Tectonic Reconstruction: Mediterranean Present Day

The northeastward movement of Africa is continuing with a greater moment of rotation in the east which has led to the opening of the Red Sea as a new spreading axis, and strike-slip faulting in the Dead Sea rift and in the Zagros Mountains. Subduction is continuing beneath the Makran terrane and the accretionary prism is continuing to build. In the west, intense thrusting in the western Mediterranean welded the Kabylie terrane to North Africa, emplaced Calabria in its present position, and caused deformation in the Rif, Atlas and Betic fold belts. Subduction of oceanic crust is active beneath the terranes of southern Greece.

Dacian terrane, marking the climax of the Alpine orogeny (Figure 2.14). A further axis of thrusting extended from the Dinaro-Hellenic terrane into the Turkish terranes. Further east, subduction along several fronts led to the emplacement of the terranes of Iran and Afghanistan into their present positions.[40]

Since the end of the Oligocene several major changes have taken place in the western Mediterranean. With a return to northeast drift of the African plate at about 20 Ma, major compression and thrusting occurred in the Alboran and Kabylie terranes. The Kabylie terrane was thrust southwards and compressed against the Oran and Tunisian terranes, forming the Atlas thrust belt, whilst the Calabrian terrane was detached and displaced eastwards with oceanic subduction along its eastern margin. The thrust front continued northwards through the Appennine belt, which in turn connected with the Alpine thrust belt (Figure 2.15). In the east, thrusting continued in Turkey and Iran, and subduction of oceanic crust continued beneath the Makran terrane of Pakistan and a major accretionary prism was formed along this margin. Rifting in the Red Sea began in the Aquitanian and strike-slip faulting in the Dead Sea rift began in the Tortonian. In Libya these events are reflected in ENE tilting, subsidence and gentle folding in the Sirt Basin, and dextral wrenching on the Medina and Sirt Wrenches offshore.[41]

Since the Tortonian, Africa has continued its northeastward drift at a rate of about 1 cm per year. The rate of drift in the Arabian plate is about 3 cm per year, as a result of which compression is increasing along the boundary between Arabia and Iran, whereas the Red Sea rift is continuing to open and has developed into an active spreading centre. In the western Mediterranean the Calabrian terrane attained its present position accompanied by extensive thrusting and wrench faulting in the Libyan and Tunisian offshore areas, and thrusting has continued in the Betic and Rif areas of Spain and Morocco (Figure 2.15). Thrusting continues around the Alpine arc, and in the eastern Mediterranean subduction of oceanic crust is continuing beneath southern Greece and Crete. Thrusting is active in southern Turkey and in the Caucasus, and subduction of oceanic crust contines beneath the Makran area of Pakistan.[42]

In north Africa the Neogene is marked by the emplacement of nappes in northern Morocco, and wrench faulting along the South Atlas fault system. In Libya marine sediments were deposited in the Ajdabiya Trough and in Cyrenaica, and large volumes of basaltic and basanitic lavas were erupted from volcanic centres along the axis of the ancient Tripoli-Tibisti Uplift. Indeed in the Tibisti Mountains and Al Haruj al Aswad volcanic activity has only recently ceased. The Tripoli-Tibisti axis marks the line of a Proterozoic suture and stress-field studies suggest that this line also represents the junction between the west African and east African plates. The Messinian salinity crisis, during which sea level in the Mediterranean fell by more than 500m and produced evaporites in the deep basins of the Mediterranean, is reflected in Libya by the presence of thin evaporites in the Sahabi Formation and in wells in the Sabratah Basin. Further evidence of the Messinian drop in base-level is found in the deep, buried valleys at Sahabi near Ajdabiya, and in the Nile Valley, excavated to more than 500m below present sea level.[43]

The present-day stress fields in Libya have been studied by Shäfer, et al. They identified three distinct stress domains, all marked by compression at the present time. The western Libyan domain, extending from Tunisia as far east as the Waddan Platform, exhibits a northwest-southeast stress regime, the central Libya area, including all of the Sirt Basin, has a northeast-southwest orientation and the coastal area of Cyrenaica shows a northwest-southeast pattern. Shäfer, et al. interpreted these findings

ERA	AGE (Ma)	PLATE TECTONIC EVENTS	LIBYA AND ADJACENT AREAS
CAINOZOIC	6.7 - 5.2	Messinian salinity crisis. Sea level in Mediterranean lowered by 500m. Evaporites formed in deep Mediterranean basins.	Messinian evaporites present in Sahabi Fm. onshore and Marsa Zouaghah Fm. offshore. Sahabi Channel cut to 500m below present sea level. Volcanics, Tibisti Mountains.
	15 - 6	Thrusting in Rif and Betic areas. Calabrian and Kabylie terranes emplaced in present location.	Eruption of Al Haruj al Aswad volcanics. Regression in Ajdabiya Trough and Cyrenaica.
	30 - 15	Main Alpine Orogeny. Thrusting in Alps and western Mediterranean. Subduction of ocean floor beneath Alboran, Kabylie and Calabrian terranes.	Continental, littoral and shallow marine facies in Sirt Basin. Volcanic eruptions in Jabal Sawda and Gharyan.
	50 - 30	Beginning of Alpine Orogeny. Closure between Iberia and north Africa marks end of Tethys.	Ring intrusions in Jabal Awaynat. Shallow water carbonates in Sirt Basin. Inversion in Atlas Mountains.
	65 - 50	Progressive closure of Tethys Ocean due to northeast movement of African plate.	Uplift in west Libya. Subsidence in Sirt Basin. Carbonate platforms and reefs in Sirt Basin.
MESOZOIC	80 - 65	Relative movement of African plate changed from eastwards to northeastwards. End of seafloor spreading in Mediterranean	Infill of grabens with marine shales and micrites. Shallow marine carbonates and clastics on horsts. Deposition of organic rich shales in Sirt Basin.
	125 - 80	Seafloor spreading axis switched to southern axis (south of Turkish, Balkan and Lombard terranes).	Major flooding episode in Sirt Basin. End of rift phase. Subsidence in Sirt Basin. Formation of horsts and grabens.
	165 - 125	Seafloor spreading in Tethys Ocean along northern axis (north of Sila, Lombard and Austrian terranes).	Progressive collapse of Sirt Arch. Deposition of Nubian sands.
	165 - 155	Establishment of Tethys seaway from Asia to the Pacific	Tensional regime in Libya. Continued rifting and deposition of continental clastics.
	240 - 155	Break up of Pangaea along Caribbean - North Atlantic - Mediterranean axis.	Beginning of rift phase. Triassic continental clastics found in rift grabens in Libya.
PALAEOZOIC	290 - 260	Calving of terranes from West Gondwana and translation northwards on transit plate.	Major regression. Establishment of continental conditions over much of Libya. Dextral wrenching in north Africa.
	330 - 290	Hercynian Orogeny. Culmination of collision between Gondwana and Laurasia to produce Pangaea.	Uplift and deformation. Formation of E-W/NE-SW tectonic elements. Carboniferous deltaic deposition.
	380 - 360	Initial collision of Gondwana with Laurasia.	Deposition of stacked, deltaic and littoral, clastic sequences. Major transgression, organic rich shales over much of north Africa.
	435 - 380	Northward drift of Gondwana.	
	440	West Gondwana located close to South Pole	Brief glacial episode over large part of north Africa.
	500 - 440	Detachment of Avalonian and Cadomian terranes from passive margin of West Gondwana.	North Africa formed passive margin of West Gondwana. Deposition of three transgressive clastic cycles.
	550 - 450	Separation of Laurentia from Pannotia to produce Gondwana.	Late Pan African activity produced broad north to south troughs and swells in Libya and eastern Algeria.
PRE CAMBRIAN	600 - 550	Pan African Orogeny brought together the dis-assembled plates of Rodinia into the new supercontinent of Pannotia.	Evidence of Pan African remobilised belt in Tibisti Mountains, on Al Qarqaf Arch and in wells in southern Libya. Period of major tectonism and metamorphism.
	700 - 600	Rodinia dis-assembled, rotated and turned inside out.	
	880 - 550	Pharusian microplates formed as island arcs in Pharusian Ocean.	Evidence of Pharusian accreted terranes in wells in northern Libya and on Touareg Shield of Algeria and Mali.
	1000 - 800	East Sahara craton formed on periphery of Rodinia.	Evidence of east Saharan craton in Hoggar Mountains of Algeria and at Adrar in Yahia near Ghat.
	1600 - 1000	Assembly of Rodinia from cratonic nuclei.	Evidence of Nile craton at Jabal Awaynat and eastern Tibisti massif, West African craton in Eglab massif, Algeria.
	2900 - 2600	Formation of cratonic blocks including West African, Nile and Congo-Kalahari.	

Sources: Plate tectonics: Vail (1991), Dercourt, et al. (1986), Ricou, (1994), Unrug, et al. (1996), Boote, et al. (1998). Events in Libya: Numerous sources.

Figure 2.16   Plate Tectonic Summary

This chart shows the main plate tectonic events which have shaped the development of north Africa and the reflection of these events in the structural and sedimentological history of Libya and adjacent areas.

as a reflection of northwestward drift of the African Plate during the last 10 million years. The western and eastern areas are actively indenting the European Plate but the central belt occupies a more passive position between the two and has acquired a different stress pattern as a result. They deduced that the central area was in a tensional regime through most of the late Mesozoic and Cainozoic but entered a compressional mode during the late Neogene. Westaway developed a different model. Based on plate tectonic considerations he concluded that the Sirt Basin and Pelagian Shelf form a zone of active oblique extension at the present time caused by the continuing rotation of the Calabrian terrain. He calculated the rate of extension at about 1mm per year and dated the start of the present regime at 0.7 Ma.[44]

A summary of the main plate tectonic events relating to north Africa, and the local expression of these events in Libya is shown on Figure 2.16.

Chapter 3

STRATIGRAPHY

PART ONE: PRECAMBRIAN and PALAEOZOIC

INTRODUCTION

3.1 The Development of Libyan Stratigraphy

There are more publications on stratigraphy than any other aspect of Libyan geology, so it may not be out of place to give a brief outline of the development of Libyan stratigraphy before embarking on the detailed review of the Palaeozoic. Early north African explorers, such as Beyrich in 1852, collected samples and fossils from Fezzan which were subsequently identified and described in the scientific literature, and von Bary published several papers describing the geology and scenery along the Tripoli to Ghat route in the 1880s. Probably the earliest geological map covering part of Libya (western Fezzan) was published by Rolland in 1880. The map correctly identified Devonian and Cretaceous strata in the Jabal Akakus and Hamadah al Hamra areas, and Krumbeck identified and described Carboniferous fossils from the southern Ghadamis Basin in 1906. In Cyrenaica the earliest major stratigraphic investigation was conducted by Gregory in 1911.[1] He examined the stratigraphy and structure of Cyrenaica and subdivided the Cainozoic sequence into seven formations, several of which are still in use today. Italian geologists however did most of the early work.[2] Vinassa de Regny conducted widespread investigations in the period 1901 to 1913, and produced a geological map of Libya at a scale of 1:6,000,000 in 1912. The outcrops of the Jabal Nafusah were studied by Parona and

Zaccagna. Stefanini published several papers on Cyrenaica between 1913 and 1935, and Franchi studied the coastal area of Tripoli in 1912/1913. Crema examined hydrological problems between 1913 and 1936, and made what was probably the first reference to oil, encountered in a water well at Sidi al Masri near Tripoli in 1926. Libyan palaeontology was investigated by Parona between 1906 and 1933, and by Checchia Rispoli from 1913 to 1928. The famous Miocene/Pliocene fossil locality at Sahabi was found by Italian soldiers in the 1920s and was scientifically studied by Petrocchi and others from 1934.[3]

The early work was mostly concerned with the coastal areas of Cyrenaica and Tripolitania, but starting in the 1920's a series of missions was organised to examine the geology of the interior, particularly Kufrah, Fezzan and the Tibisti Mountains.[4] These missions were mostly under the leadership of Prof. Ardito Desio who published his first paper in 1927. He went on to publish over 120 papers on Libya during the next forty-five years on a wide variety of topics including stratigraphy, palaeontology, hydrogeology, volcanic rocks and economic geology. In the period between the two world wars further contributions to Libyan stratigraphy were made by de Agostini (eastern Libya), Amato (Cyrenaica, Fezzan and Tripolitania), Chiesa (Sirt Basin and Fezzan), Dalloni (Tibisti and western Libya), and Marchetti (Cyrenaica), and by the palaeontologists Lipparini (foraminifera and molluscs), Serra (echinoids and pelecypods), Silvestri (large foraminifera), and Zuffardi Comerci (corals).[5]

The end of the Second World War led to the effective replacement of Italian personnel by geologists from the victorious powers. In Fezzan French geologists such as Lelubre, de Lapparent, Lefranc, Bellair, Boureau and Freulon published the first detailed descriptions of Palaeozoic outcrops in the period 1943 to 1956.[6]

During the period of United States technical assistance to Libya Christie published an account of the stratigraphy of the Garian area in 1955, and Goudarzi produced a series of reports on the economic geology of Libya for the Libyan Ministry of Industry and the U.S. Geological Survey, and in 1964 in collaboration with Conant produced a geological map of Libya at a scale of 1:2,000,000 which, after several revisions, is still in use today. Another American, J.R. Jones, produced a series of reports on hydrogeology. British geologists during this period were concerned primarily with water supply (Addison and Shotton, Curran and Dixey, Hill, Murray), geomorphology (Bagnold, Sandford, Grove), vertebrate palaeontology (McBurney, Savage, Bate, Harris), and igneous geology (Peel and Tomkieff).[7]

With the arrival of oil company geologists in 1955 a new phase of investigation began, with the acquisition of large amounts of subsurface data. In 1956 Professor Desio was invited by the Stratigraphic Commission of the Intenational Geological Congress to prepare a Stratigraphic Lexicon for Libya. A draft was prepared, but in an effort to include some of the more recent data, the newly formed Petroleum Exploration Society of Libya asked Burollet to revise and update the draft. The Lexicon was published under the auspices of the International Geological Congress in 1960, but with the benefit of hindsight it was published prematurely.[8] It was based partly on pre-war data, some of the entries were derived from unpublished company reports and thereby contravened the Code of Stratigraphic Nomenclature, and it relied exclusively on surface data.[9] As a result it quickly fell into disfavour and each operating oil company set up its own system of stratigraphic nomenclature.

French geologists continued to be active in western and southern Libya. Massa and Collomb published on the stratigraphy of the Al Qarqaf Arch, followed by Jacque on the Jabal bin Ghanimah and by Freulon on the Murzuq Basin and de Lestang on the Kufrah Basin.[10]

The Petroleum Exploration Society of Libya emerged as the focus for oil company activity, and played a key role in publishing new data during the 1960's. The Society organised the First Saharan Symposium in 1963 which included notable stratigraphic papers by Klitzsch (northeast Murzuq and Dur al Qussah), Burollet (Kufrah), Jordi and Lonfat (northwest Libya), Mennig (Jifarah Basin), and Vittimberga and Cardello (Kufrah Basin). The Society also organised annual field excursions. Three meetings in particular produced important new stratigraphic data: South-Central Libya and northern Chad was published in 1966, Northern Cyrenaica in 1968, and Southwest Fezzan in 1969. The Society issued publications on the Miocene rocks of Maradah, Gemini Space photographs of Libya, Libyan microfacies, and the stratigraphy of the Jabal Nafusah.[11]

The first oil company data appeared in 1967 with the publication of reviews on the Zaltan and As Sarir fields, followed by others on Intisar, Amal and Awjilah.[12] The Petroleum Exploration Society planned an updated Stratigraphic Lexicon as early as 1963, specifically designed to include as much subsurface data as possible. The new publication however did not appear until 1972, and was finally published, not as an updated lexicon, but as a volume on the stratigraphic nomenclature of the Sirt Basin. Compiled by Barr and Weegar the book incorporated detailed new subsurface data, including log traces for the newly defined formations.[13] The book defined over sixty new lithostratigraphic units, many of which were dated on the basis of foraminifera and macro-fossils. The principal criticism was that it was based largely on data provided by Oasis and Mobil, with little input from Esso, BP, Amoseas or Occidental. This produced some curious anomalies such as the type section of the Zaltan Limestone being selected in the atypical Oasis well AA1-59 rather than from the Zaltan field. Another questionable practice was to assign formation names based on geographic names located many hundreds of kilometers from the type section. The Taqrifat Limestone was perhaps the most stiking example. However the book satisfied a major need; it became the principal guide to subsurface stratigraphic nomenclature in the Sirt Basin, and is still widely used today.

The period of the large symposia began in 1969 when the first Geology of Libya symposium was held in Tripoli. The proceedings were published in 1971 and included ground-breaking papers on the structural history of Libya, the pre-Cretaceous stratigraphy of the Jabal Nafusah and the Wadi ash Shati sedimentary ironstones.[14]

In 1974 the Industrial Research Centre (IRC) in Tripoli embarked on an ambitious geological mapping programme designed to map the whole of northern and western Libya at a scale of 1:250,000. The geologists involved were mostly from eastern Europe, and the mapping was limited to surface data. In all sixty sheets were eventually produced in the period 1974 to 1985, with a final sheet on the Al Haruj al Aswad volcanics being added in 1993.[15] These publications contained much useful data, but they were patchy in quality and there were some serious inconsistencies from one sheet to another. About one third were published prior to the appearance of the National Atlas of Libya in 1978, and geographical and stratigraphic names were inconsistent between the earlier maps and the later ones. The stratigraphic terminology was reasonably consistent, and age dating was supported by palaeontological control. The stratigraphic terminology however was totally different from the subsurface nomenclature of Barr and Weegar. Between 1978 and 1980 IRC published four key bulletins: a Bibliography of the Geology of Libya, a synthesis of the Upper Cretaceous and Tertiary formations of northern Libya, a synthesis of the pre-Mesozoic Stratigraphy of Libya, and finally, in an effort to bring all the new data together, a Stratigraphic Lexicon of Libya, compiled by Banerjee. For the first time both surface and subsurface stratigraphic data were combined in one volume.[16]

The second symposium on the Geology of Libya was held in Tripoli in 1978 and over 70 papers from the symposium were published in 1980. These included major new data on the southern basins of Libya, biostratigraphy in Cyrenaica, the Messinian salinity event and post-Eocene stratigraphy in the eastern Sirt embayment.[17] The National Atlas of Libya was published in 1978, and for the first time standardised the spelling of geographic names.[18]

Offshore, sufficient data had accumulated for the Earth Sciences Society of Libya (successor to the Petroleum Exploration Society) to issue a publication on stratigraphic nomenclature for the northwestern offshore of Libya in 1985. This followed a detailed review of Tunisian stratigraphy, including the offshore, published in 1978.[19] At about the same time the Energy Resources Department of the Organization of Arab Petroleum Exporting Countries, based in Kuwait, issued a preliminary report on stratigraphic correlation amongst member countries, including Libya, Algeria and Egypt.[20]

By 1987 oil companies had begun to release large amounts of data and this was reflected in the third symposium on the Geology of Libya which presented key well information from the Kufrah and Murzuq Basins, and from the Cyrenaican offshore. In addition the proceedings included an important revision of the Upper Cretaceous/Palaeocene stratigraphy of the Hamadah al Hamra, and an excellent modern synthesis of the Precambrian outcrops in north Africa.[21]

The Earth Sciences Society of Libya subsequently launched a series of thematic conferences on the sedimentary basins of Libya. The first symposium, on the Sirt Basin, was held in 1993 with the proceedings issued in 1996. The publication included many high quality papers on Libyan stratigraphy including a detailed biostratigraphic analysis of the eastern Ghadamis Basin, a review of the Palaeocene in the Sirt Basin, and a major revision of the Amal Formation in eastern Libya. Several major syntheses were also presented. A second symposium on the western Libyan basins was held in 2000, and the proceedings are scheduled for publication in 2002. In 1998 a conference on geological exploration in the Murzuq Basin was held in Sabha, and the proceedings were published in 2000. This volume included new data on the Silurian hot-shales, the Mamuniyat glacial episode and the stratigraphy of the Murzuq Basin. In the UK, two conferences on north Africa in 1995 and 2000 produced further useful stratigraphic data.[22]

It is striking to compare the number of stratigraphic publications available to Barr and Weegar in 1972 with the number available now. Barr and Weegar listed thirty-three relevant publications; there are now over three hundred.

Most of the stratigraphic terms defined in the publications listed above are lithostratigraphic terms. Many of the lithostratigraphic definitions include palaeontological data which permit the formations to be dated to a greater or lesser degree. In addition the Palaeozoic in Libya is rich in trace fossils which are useful both as age and facies indicators. The diachronous nature of the Silurian Akakus Formation can be proved exclusively by the use of ichnostratigraphy. Publications on biostratigraphy are less common. The multi-well studies conducted in Cyrenaica,

and the eastern Ghadamis Basin are two of the best examples.[23] Publications applying the principles of sequence stratigraphy in Libya have begun to appear only during the last five or six years. Radiometric age dating of igneous samples has permitted an outline chronstratigraphy to be established for Libya.[24] The review which follows is principally concerned with lithostratigraphy, since it is distinctive and mappable units which are of primary concern to petroleum geologists, although sedimentological and sequence stratigraphy data is given where available.

PRECAMBRIAN

3.2 Archaean and Proterozoic

Major advances have been made in the study of the Precambrian geology of north Africa in recent years. The earlier subdivision into three series, the Suggarian, Pharusian and Nigritian, is not supported by recent radiometric analysis, and it now appears that the Precambrian rocks have been subjected to repeated cycles of tectogenesis, overprinting and rejuventation which have in many cases erased their earlier history. In view of the impact of plate tectonic concepts it is probably more useful to to examine these rocks in terms of their tectonic setting, which include both cratonic nuclei and mobile belts. An excellent summary of these various elements was given by Vail and a review of geochronology and isotopic age determinations was provided by Cahen, et al.[25]

From west to east across north Africa six tectonic domains can be distinguished (Figure 3.1).

1. The Phasurian mobile belt in the Hoggar Mountains of Algeria, which forms the western half of the Touareg Shield, is dominated by oceanic volcanic and sedimentary rocks, including ophiolites, which originated as island arc sediments which were subsequently intruded by subduction-related gabbros and diorites.

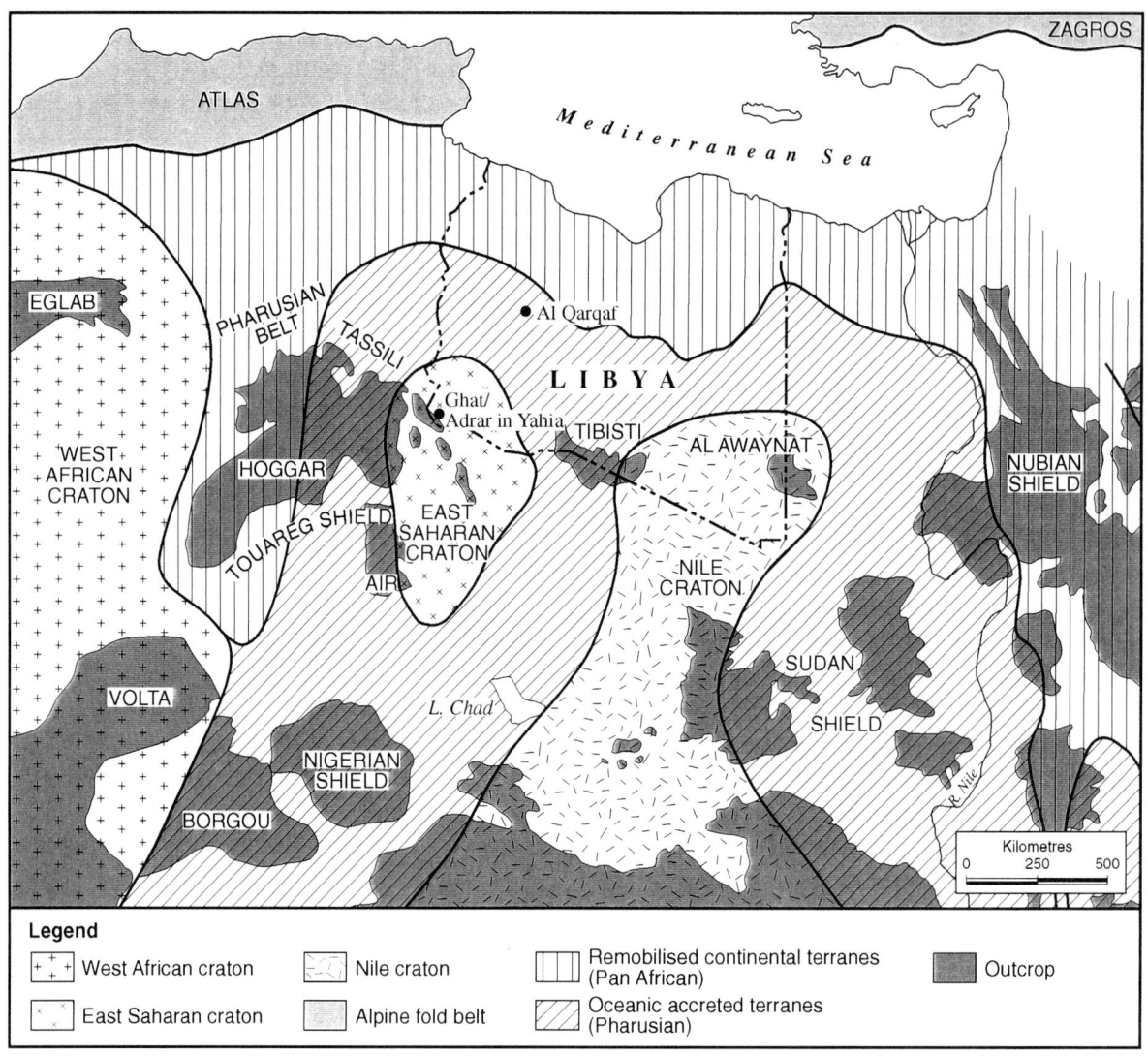

Figure 3.1   Precambrian Tectonic Domains in North Africa

The Precambrian rocks in Libya belong to four main groups: the cratonic nuclei of the East Saharan Craton (Ghat, Adrar in Yahia), and the Nile Craton (Al Awaynat and eastern Tibisti), the Pan African remobilised continental terranes (western Tibisti, Al Qarqaf, subsurface Illizi Basin), and the Pharusian oceanic accreted terranes (Hoggar Mountains of Algeria, subsurface northern Libya, subsurface Western Desert of Egypt).

Radiometric ages of 880 to 600 Ma indicate a Riphean to Vendian age.

2. The eastern Touareg Shield in southeastern Algeria and northern Niger, comprises remobilised gneisses overprinted by Pan-African metamorphism. Relict ages of 2200 Ma have been obtained which suggest that it originated as part of an Archaean continental terrane, but was subsequently remobilised during the Pan-African orogeny. The remobilised gneisses contain substantial reserves of uranium both in veins and stratiform

deposits in Algeria. Pan-African metamorphic rocks have been cored in several wells in the northern Murzuq Basin where they comprise a range of lithologies from quartzites to serpentinites.[26]

3. The Touareg shield terminates eastwards against a major north-south shear, the Tiririne fold belt. To the east lies the East Saharan craton mostly located in Niger, but also extending into southwestern Libya. The shield is made up of clastic sediments, migmatites and gneisses which have been metamorphosed to amphibolite grade. This area appears to be largely unaffected by the late Pan-African tectonic events. The small outcrops of Precambrian rocks near Ghat, Adrar in Yahia and South Anay in southwest Libya, belong to this assemblage.[27]

4. The Tibisti Massif in southern Libya and northern Chad has been studied by Wacrenier, Klitzsch, Ghuma and Rogers, and Assaf, and most recently by El Makhrouf and Fullagar.[28] Sedimentary rocks and rhyolites metamorphosed to greenschist grade are dominant in the western part of the massif (the Upper Tibistian of Wacrenier), and the rhyolites gave isotopic ages ranging from 526 to 790 Ma. The sedimentary rocks appear to indicate a progression from shelfal conditions in the west to basinal in the east. The overall sequence suggests a remobilised continental terrane similar to that of the eastern Touareg shield. The Bin Ghanimah granodiorite batholith in the northwest of the Tibisti Massif has been dated at 550 Ma by Ghuma and Rogers, and it may be related to an underlying subduction zone. The small inliers of anorogenic granites and metasediments on the Al Qarqaf Arch, which have yielded K/Ar dates of 551 to 476 Ma, belong to this suite, as do the granites and granodiorites exposed on the western flank of the Al Haruj al Aswad volcanic centre.[29]

5. The Tibistian sequence in the eastern part of the massif (the Lower Tibistian of Wacrenier), is different in character to the western Tibistian sequence. It shows a much higher grade of metamorphism, to amphibolite grade, and is intruded by granodiorites. It has closer affinity to the Al Awaynat inlier than the western Tibisti massif, and was interpreted by Vail as part of the Nile craton.

6. The Al Awaynat inlier on the frontier of Libya, Sudan and Egypt has been described by Menchikoff, Sandford, Mahrholz, Klerkx, Schurmann and Cahen.[30] It forms part of the Nile craton and contains the oldest rocks in Libya. Granulites and gneisses have given isotopic age dates from 2900 Ma to 2500 Ma with a later intrusive event at 1784 Ma. Klerkx divided this group into a lower Karkur Murr series and an upper Ayn Daw series distinguished by their metamorphic grade. The area does not appear to have been affected by Pan-African deformation. Similar rocks occur in Chad and the Central African Republic. The Al Awaynat inlier is also well known for its Tertiary ring intrusions.

The Arabian shield further to the east need not concern us here, except to note that in the Red Sea and Sinai areas oceanic island-arc rocks are present, with ophiolites, which show a remarkable similarity to those of the Pharusian belt of the Touareg shield.

Basement rocks in northern Libya are buried beneath a thick sedimentary cover, but have been reached in several deep oil wells. Recent work in Cyrenaica has proved the presence of Precambrian acritarchs in sediments previously regarded as Cambro-Ordovician in age. Wells A1-33, B1-31 and C1-82, located on the Al Jaghbub High, have all yielded sphaeromorph acritarchs of late Riphean age from reddish felspathic sandstones, interbedded with micaceous shales and siltstones. The thickness of the late Riphean section in well A1-81 is believed to be 1060m. El-Arnauti and Shelmani suggested that late Riphean rocks may also be present in wells D1-31, J1-81A, A1-81 and C1-125, but the suggestion is made on the basis of correlation alone, and is without palynological confirmation.[31]

Many other penetrations of Precambrian igneous rocks have been noted by Schurmann and by Shackleton and Grant. The principal lithologies encountered are andesites, rhyolites, green schists and intrusive igneous rocks. The northernmost wells show an affinity with the accreted terrains of the Pharusian and Red Sea belts, whereas the more southerly penetrations are principally of gneisses which suggest an affinity to the remobilised rocks of the Tibesti Massif. Isotopic ages have been determined for over a dozen wells which have penetrated igneous basement in the Sirt Basin, and almost all give ages within the range 670 to 460 Ma. Several wells incidentally have encountered Jurassic and early Cretaceous granites which have been dated isotopically.[32]

In summary, central North Africa comprises two ancient cratonic blocks, the East Saharan and Nile cratons, which are flanked by Pan-African remobilised continental terranes, (the eastern Touareg, western Tibesti and Sudan shields), which in turn are bordered by oceanic accreted terranes (western Touareg shield, northern Algeria/Libya/Egypt, and the Red Sea shield) (Figure 3.1). This interpretation is consistent with the plate tectonic model described in chapter two, and with the radiometric and petrographic data presented above.

The Proterozoic would not normally be of much interest to petroleum geologists were it not for the fact that at least one oilfield has significant production from these rocks. The D2-102 well on the Awjilah field in the eastern Sirt Basin tested 7,627 barrels of oil per day from devitrified rhyolite and weathered and fractured granophyric granite. The D8-102 well flowed 18,000 BOPD from an open hole completion in basement. The reservoir is deeply weathered and intensely fractured. Seismic data shows the field to represent a broad eroded granite high with relief of over 650m. Potassium-argon dating gives an age between 568 and 402 Ma, which suggests a Pan-African age for these rocks.[33]

PALAEOZOIC

3.3 Cambro-Ordovician

There is a very extensive literature on the Palaeozoic of North Africa due in large measure to the abundant petroleum reserves contained within these rocks in Algeria and Libya. The classic work of Beuf, et al. on the Lower Palaeozoic sandstones of the Sahara is well known, and Burollet's Gargaf Group has long been used as a general term for the Cambro-Ordovician succession in Libya. A synthesis of the Palaeozoic rocks in southern Libya was given by Bellini and Massa.[34] In recent years there has been a flood of publications reflecting the continued importance of these rocks. A review by Boote, et al. contained a comprehensive bibliography. A map showing the outcrops of Precambrian and Palaeozoic rocks in Libya is shown on Figure 3.2, and the pre-Mesozoic subcrop map is shown as Figure 3.3.[35]

The Pan-African orogeny, as previously described, had the effect of welding together a number of diverse cratonic and mobile belts into a new configuration which formed part of west Gondwana. It is worth making the point that the thick blanket of quartz-rich sandstones, spread over much of north Africa from Morocco to Oman, represents material eroded from the Pan-African mountains to the south, which was deposited on the stretched and unstable continental margin of Pannotia and subsequently Gondwana, between 550 and 450 Ma. The distal, fine-grained equivalents of these rocks have not been found, and may have been deposited on crustal fragments since detached from Gondwana. The varying thickness of this enormous sand wedge suggests the presence of rifts in the underlying basement. In any event this sand pile may be unique in extent and

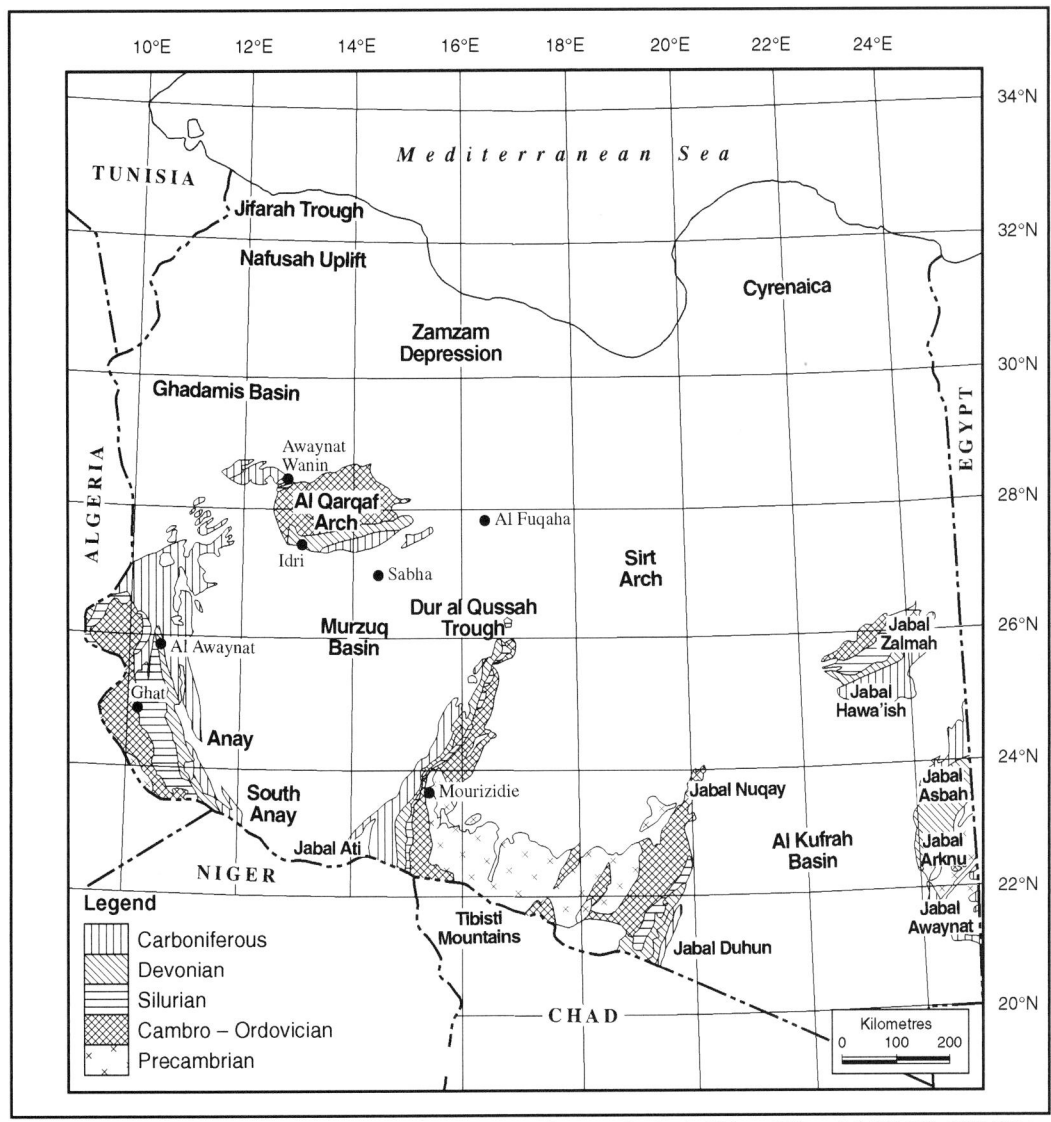

Source: Geological Map of Libya, 1:1,000,000, 1985 Edition

Figure 3.2   Precambrian and Palaeozoic Outcrops in Libya

Precambrian and Palaeozoic rocks are present throughout Libya but are mostly concealed beneath younger rocks. The principal outcrops occur around the periphery of the Murzuq Basin, on the northern flank of the Tibisti Mountains, on the western and eastern flanks of the Al Kufrah Basin, and on the Al Qarqaf Arch.

thickness. North Africa was located on the northern margin of Gondwana, and formed a passive margin during most of the Palaeozoic, from which the Avalonian and Cadomian terranes broke away during the Ordovician and migrated northwards towards Laurentia and Baltica. During the early Palaeozoic western Gondwana was located much further south than at present, and glacial conditions prevailed over much of the region during the Ordovician.[36]

In several areas in southern Libya unfossiliferous continental sandstones are present

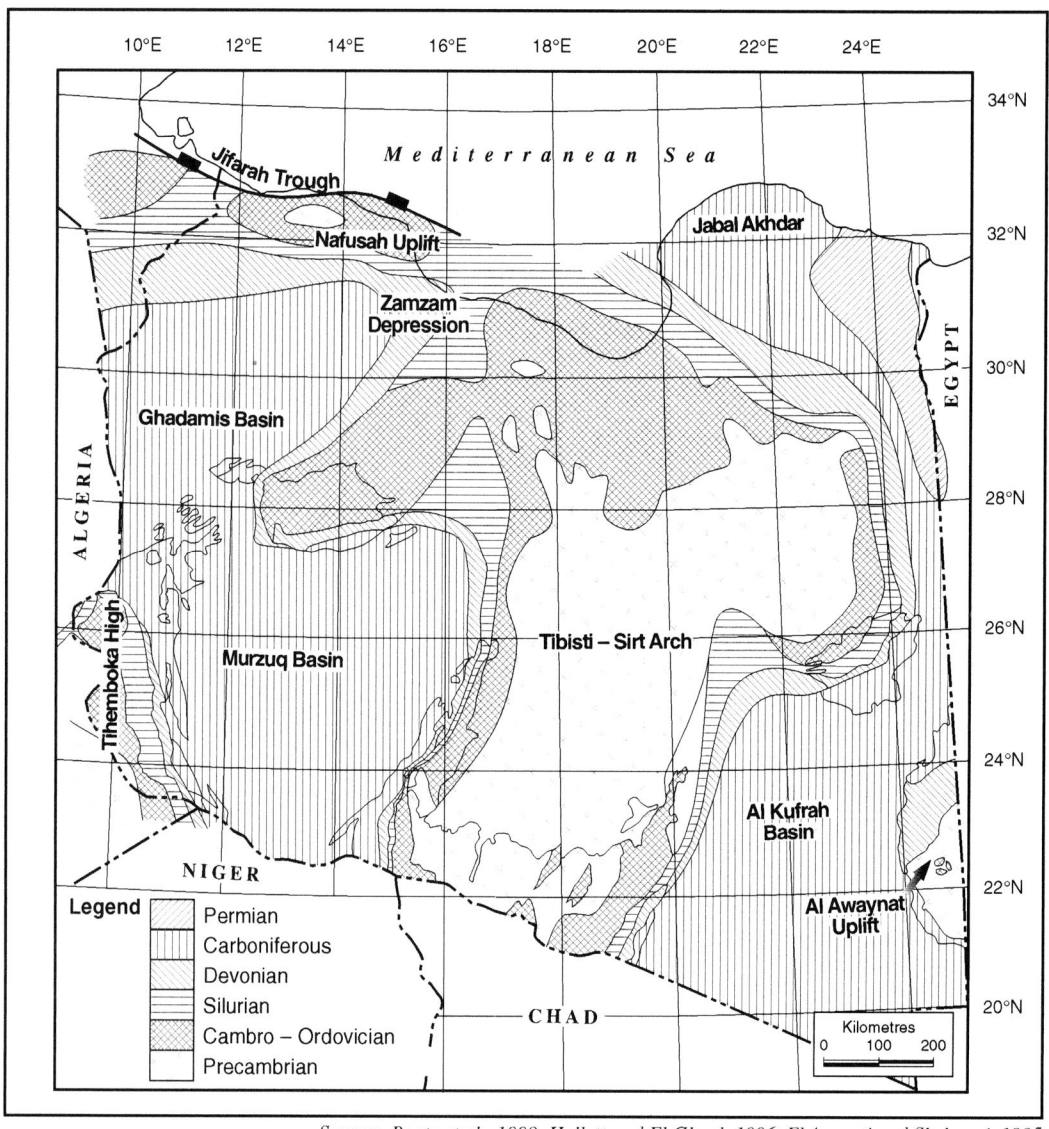

Source: Boote et al., 1998, Hallett and El Ghoul, 1996, El Arnauti and Shelmani, 1985

Figure 3.3   Pre-Mesozoic Subcrop Map

This map shows the age of the formations subcropping the Mesozoic unconformity. It represents a view of the deformed post-Hercynian surface, prior to the deposition of Mesozoic sediments. The deeply eroded Tibisti-Sirt Arch dominates the map, and other important uplifts are evident at Al Awaynat, Nafusah, Al Qarqaf, and Tihemboka. The data has mostly been obtained from oil wells. Small inliers on the Sirt Arch have not been shown. The extent of the Permian subcrops in Jabal Akhdar and in the Jifarah Trough is uncertain.

between metamorphosed basement and the conglomerate which is taken to mark the base of the Cambrian. These include the Olochi Sandstone, the Mourizidie Formation, and the Bi'r Bayai Formation, which comprise mostly red and violet sandstones, arkoses and quartzites, showing ripple marks and desiccation cracks. The Mourizidie Formation was penetrated in well D1-NC 115 in the northern Murzuq Basin where it is 45m thick. These deposits represent local erosion of Precambrian topography prior to the Cambrian marine incursion, and have been assigned an Infra-Cambrian age. Sinha et al. however regarded the Mourizidie Formation as glacial in origin. It may be that these rocks are equivalent to the late Riphean rocks of Cyrenaica mentioned above. Radiometric dating on igneous rocks associated with some of the so-called Infra-Cambrian sediments in Algeria has subsequently yielded a Cambrian age, and the Tassilian discordance therefore can not be regarded as synonymous with the Proterozoic-Palaeozoic boundary.[37]

The Palaeozoic succession in Libya can be divided into two megasequences, a lower sequence extending from Cambrian to Silurian, which represents passive margin sedimentation at high latitudes, and an upper sequence from Devonian to Permian, representing deposition at lower latitudes during the northward drift of Gondwana and the collision between western Gondwana and Laurasia to form Pangaea. The Palaeozoic as a whole was assigned to the Gondwana megacycle by Boote, et al., and subdivided into a lower and upper sequence. Cross-sections of the three principal basins are shown as Figures 3.4, 3.5 and 3.6.[38]

The transgression of the Cambrian seas was controlled by the configuration of the Precambrian residual topography, which had a northwest-southeast trend. Deposition took place on an undulating, low-angle shelf on the passive margin of west Gondwana. Klitzsch showed the transgression to have developed from the northwest into the Ghadamis-Murzuq-Jadu Trough, the Dur al Qussah-Uri Trough and the Kalanshiyu Trough, with the intervening uplifts remaining uncovered until the Ordovician.[39] The Cambrian deposits are everywhere unconformable on the underlying Precambrian or Infra-Cambrian strata. A chart of lithostratigraphic nomenclature for the Lower Palaeozoic is shown on Figure 3.7.

3.3.1 Hasawnah Formation

In the 1960 Stratigraphic Lexicon, Burollet proposed the name Qarqaf (Gargaf) Group for the thick sequence of sandstones lying between Precambrian basement and the base of the Silurian shales on the Al Qarqaf Arch. The sequence is characterised by massive, cross-bedded quartz sandstones containing the trace fossil *Harlania*. The group was subsequently subdivided into four formations by Massa and Collomb.[40]

The earliest clastic sequences overlying Precambrian basement in western Libya were named the Hasawnah Formation by Massa and Collomb based on extensive outcrops covering 16,000km[2] on the Jabal Hasawnah on the Al Qarqaf Arch (Figure 3.8). No type section was designated. Nor did Jurak define a type section during the remapping of the Jabal al Hasawnah sheet for the Geological map of Libya, probably because no complete section is present.[41] The Hasawnah Formation unconformably overlies Precambrian granites on the Al Qarqaf Arch and can be subdivided into three parts. The lower unit begins with a basal conglomerate of quartz pebbles about 2m thick, but reaching a maximum of 13m. The conglomerate is overlain by massive banks of cross-bedded quartz sandstones each up to 10m thick, with a clay or siltstone matrix. The middle unit is fine-grained and more silty, and the upper unit is again composed of massive cross-bedded quartz sandstones with abundant *Tigillites*. The sedimentology of the Hasawnah Formation on the Al Qarqaf Arch was studied by Cepek,[42] who concluded that fluvial and deltaic conditions were dominant in the lower part, with intertidal and subtidal conditions in the middle section and offshore bars in the upper part. Palaeohydraulic studies have been conducted on

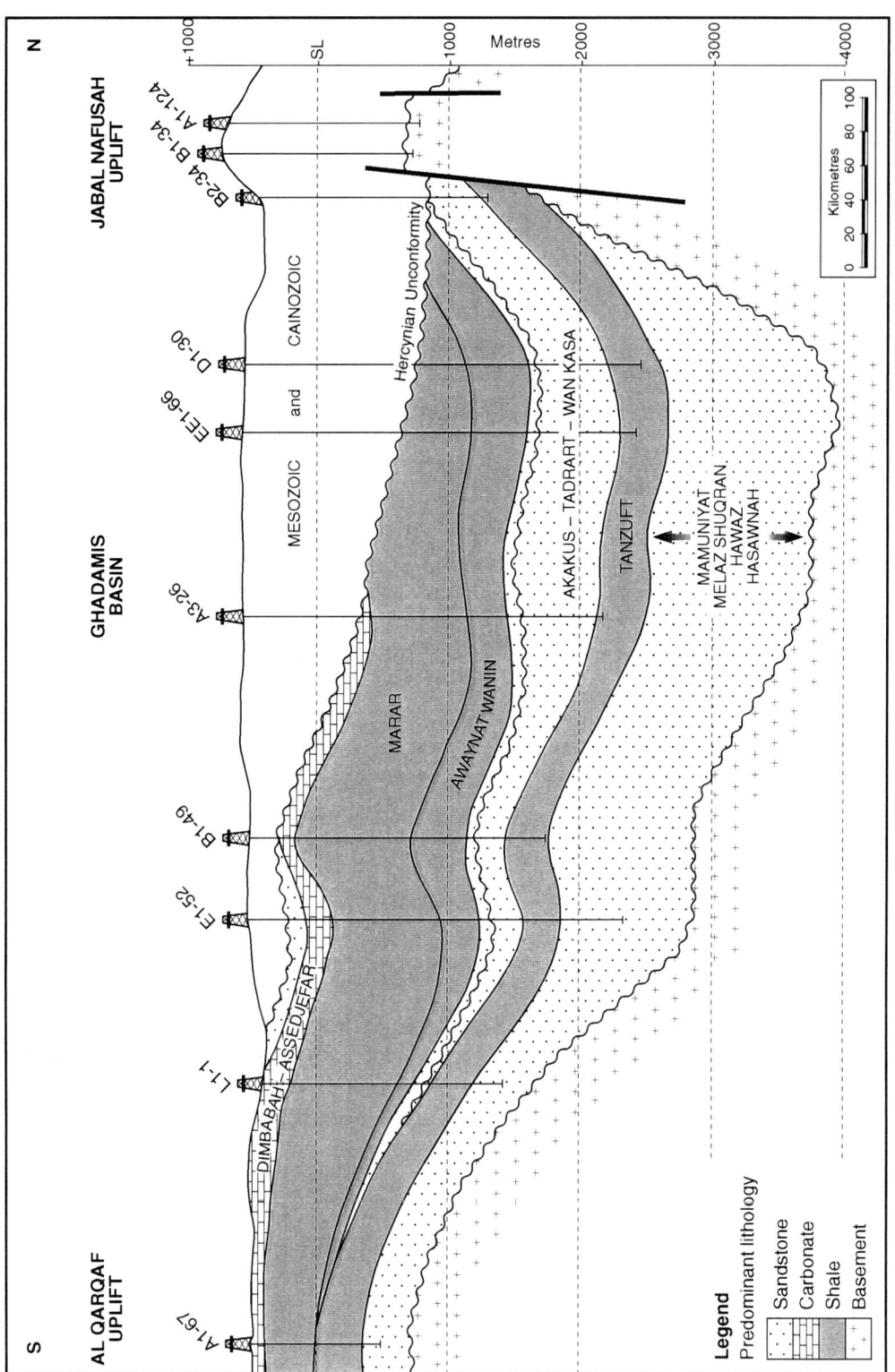

Figure 3.4   **Ghadamis Basin, North-South Cross Section, Al Qarqaf Arch to Jabal Nafusah Uplift**

The section shows the progressive truncation of the Palaeozoic strata by the Hercynian unconformity from south to north, and the abrupt northern termination of the basin against the Jabal Nafusah Uplift. Southwards the Devonian pinches out on the subsurface extension of the Al Qarqaf Arch indicating significant uplift on the arch during the early Devonian.

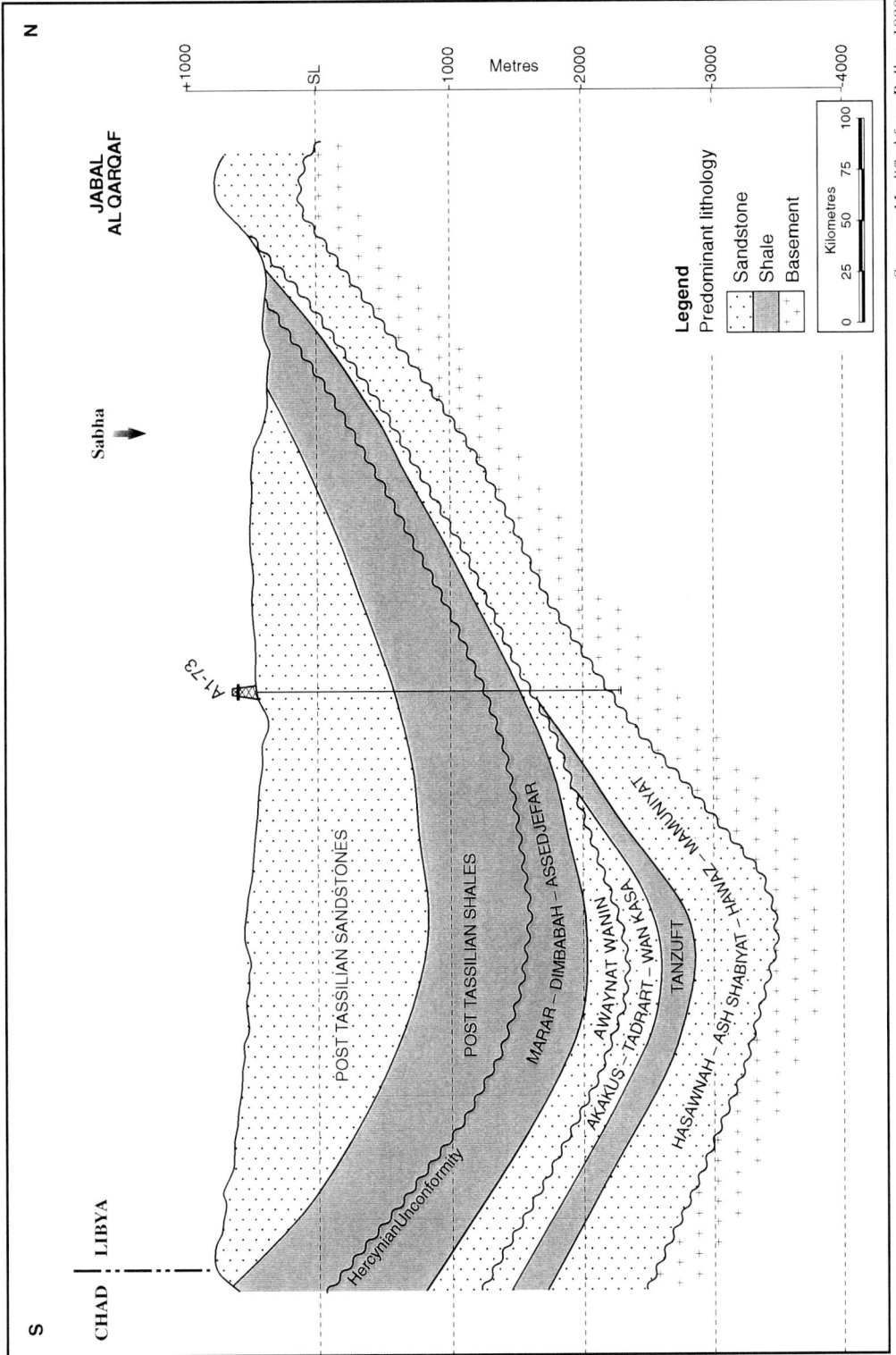

Figure 3.5   Murzuq Basin, North-South Cross Section, Al Qarqaf Arch to Southern Border of Libya

This section shows the effect of the mid-Devonian tectonism and erosion on the Lower Palaeozoic rocks on the southern flank of the Al Qarqaf Arch. The entire Silurian and early Devonian section is missing to the north of well A1-73.

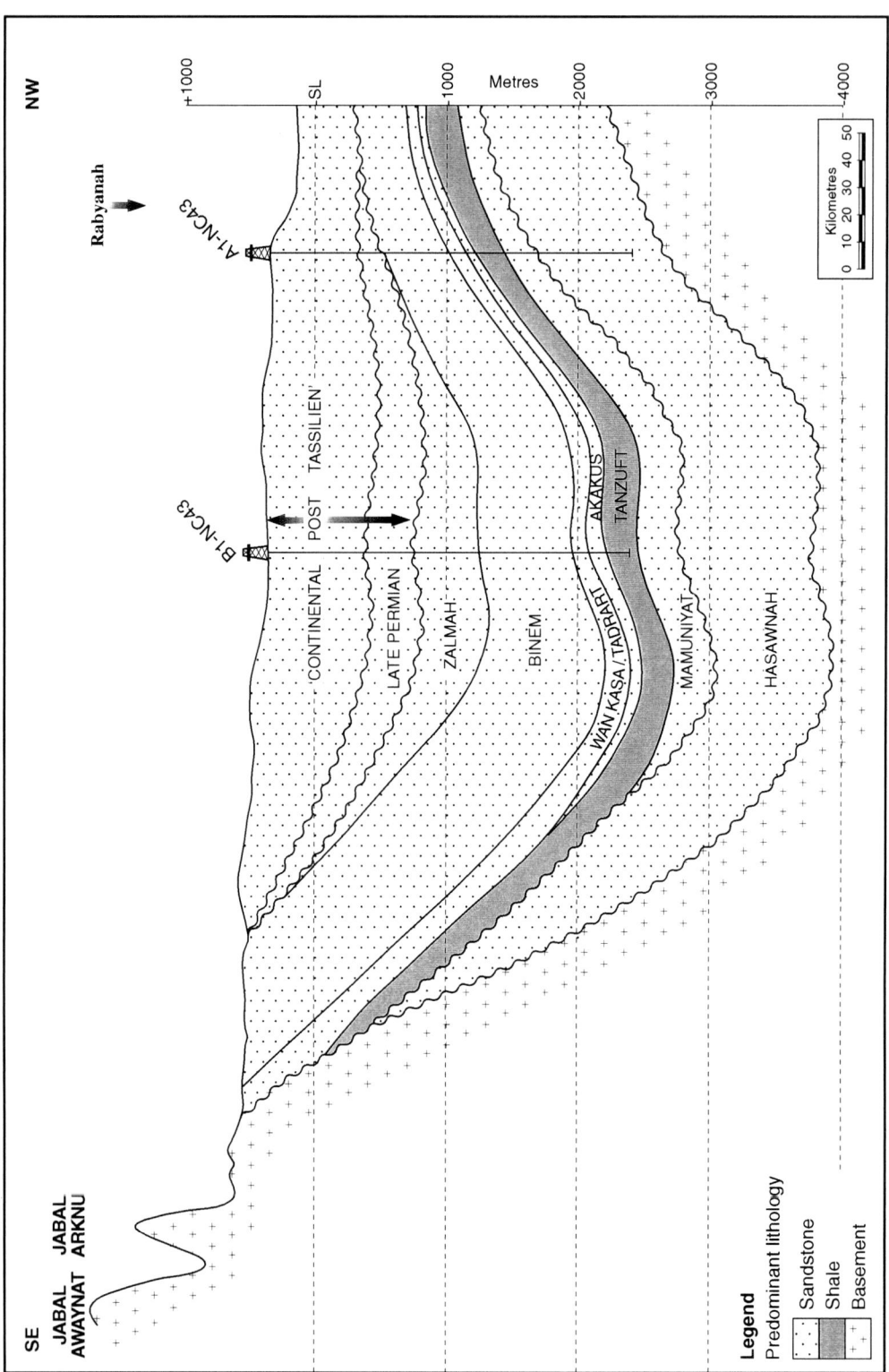

Source: Pallas, 1980, Well data from Grignani et al., 1991. Outcrops from Geological Map of Libya 1:1,000,000, 1985

Figure 3.6   Al Kufrah Basin, Northwest-Southeast Cross-Section, Jabal al Awaynat to Rabyanah

The Cambro-Ordovician and Silurian sequences are overlapped by Devonian and Carboniferous rocks on the Jabal Awaynat flank, but outcrop on the western flank of the basin. In the Carboniferous no rocks younger than Visean have been preserved. Late Permian palynomorphs have been recovered from both wells and Lower Triassic forms from well A1-NC43 within the Continental Post Tassilien sequence.

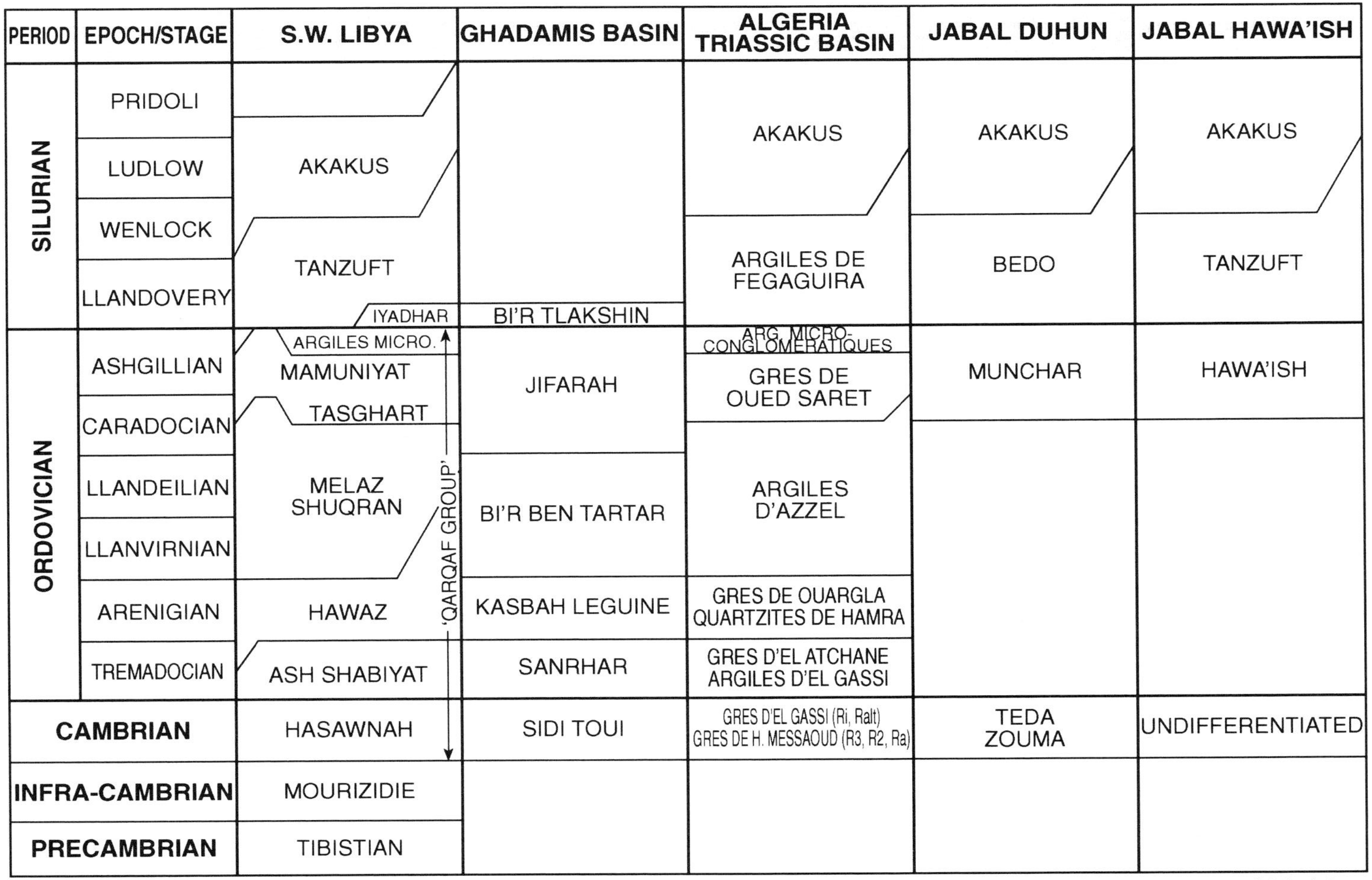

PERIOD	EPOCH/STAGE	S.W. LIBYA	GHADAMIS BASIN	ALGERIA TRIASSIC BASIN	JABAL DUHUN	JABAL HAWA'ISH
SILURIAN	PRIDOLI				AKAKUS	AKAKUS
	LUDLOW	AKAKUS		AKAKUS		
	WENLOCK					
	LLANDOVERY	TANZUFT	BI'R TLAKSHIN	ARGILES DE FEGAGUIRA	BEDO	TANZUFT
ORDOVICIAN	ASHGILLIAN	IYADHAR / ARGILES MICRO. / MAMUNIYAT	JIFARAH	ARG. MICRO-CONGLOMERATIQUES / GRES DE OUED SARET	MUNCHAR	HAWA'ISH
	CARADOCIAN	TASGHART				
	LLANDEILIAN	MELAZ SHUQRAN	BI'R BEN TARTAR	ARGILES D'AZZEL		
	LLANVIRNIAN					
	ARENIGIAN	HAWAZ	KASBAH LEGUINE	GRES DE OUARGLA QUARTZITES DE HAMRA		
	TREMADOCIAN	ASH SHABIYAT	SANRHAR	GRES D'EL ATCHANE ARGILES D'EL GASSI		
CAMBRIAN		HASAWNAH	SIDI TOUI	GRES DEL GASSI (Ri, Ralt) GRES DE H. MESSAOUD (R3, R2, Ra)	TEDA ZOUMA	UNDIFFERENTIATED
INFRA-CAMBRIAN		MOURIZIDIE				
PRECAMBRIAN		TIBISTIAN				

('QARGAF GROUP' spans the Ordovician formations of S.W. Libya)

Source: Numerous published charts

Figure 3.7   Precambrian and Lower Palaeozoic Lithostratigraphic Nomenclature

This chart shows the current lithostratigraphic nomenclature for the Lower Palaeozoic of west Libya and equivalents in the subsurface of the Ghadamis Basin, and in the Triassic Basin of Algeria. The oblique lines represent differing views on the age of the formations, as discussed in the text. No attempt has been made to show unconformities or gaps in the succession on this chart.

the Hasawnah Formation to determine flow parameters in meandering and braided stream environments, but the results have yet to be published. The eroded thickness on the Al Qarqaf Arch is from 250 to 300m, and the complete thickness is estimated to have been 400m. Nowhere on the Al Qarqaf Arch does the formation contain age-diagnostic fossils. This lack of definitive age-diagnostic fossils in the type section takes on added importance when attempting correlations from the type section into the subsurface. This is discussed further below.[43]

The formation has been mapped along the western rim of the Murzuq Basin in the areas of Tikiumit, Wadi Tanzuft, Ghat, Adrar in Yahia, Anay and South Anay. The exposures are incomplete in most of these areas and the contact with the underlying Precambrian is seen only in the extreme south. In this area the three-fold subdivision of the Al Qarqaf Arch is still evident although only patchy exposures are preserved. The upper unit is dominated by several cycles of massive sandstones showing abundant ripple cross-lamination. Several conglomeratic horizons are present. A section of 150m is exposed in the Tikiumit area but only the upper 50-60m is visible further south. The Hasawnah Formation is characterised by extensive cross-bedding and distinctive weathering into pillars and 'sphinxes'. The middle unit, exposed at Ghat, is more silty and fine grained than in the Al Qarqaf area, and shows ripple marks and boudinage structures. Clark-Lowes and Ward drew analogies between the Hasawnah Formation and the Mesozoic Nubian Formation, on the basis of similar transgressive facies associations, particularly braided-river and estuarine deposits laid down on a tectonically stable continental margin.[44]

In the subsurface of the Murzuq Basin the Hasawnah Formation is present, and a complete thickness of 175m has been penetrated in the C1-NC 58 well. The well shows a series of fining-upwards cycles which have been interpreted as fluvial plain/braided-stream deposits laid down during periods of high sediment discharge. Channel scour features characterise the base of each cycle with oblique and festoon cross-bedding features higher up. Soft sediment deformation structures attributed to dewatering are common.[45]

The Hasawnah Formation is well developed in the Dur al Qussah Trough. On the flank it has a thickness of 600m and is extremely conglomeratic and contains abundant kaolin derived from weathered feldspars, but it thickens rapidly into the trough where it reaches a thickness of 1700m. In the centre of the trough it is much less conglomeratic than on the flanks. It thins towards the Mourizidie Horst, but reappears further south where the thickness reaches 300 to 400m, and can be followed into Chad where it is known as the Bardai Sandstone in the southern Tibisti Mountains. In the subsurface it is assumed to be present on the Tripoli-Tibisti Arch which extends north-northwestwards from Mourizidie, although in the absence of well data this is speculative.[46]

In the Kufrah Basin the Hasawnah Formation is exposed around the periphery: in the northwest around Jabal Qardabah, in the southwest at Jabal Duhun, with a few isolated inliers at Jabal Asbah and Jabal Arknu in the southeast (Figure 3.8). Vittimberga and Cardello proposed new formation names in each of these areas, but these names are local synonyms of the Hasawnah Formation. The Hasawnah Formation in the Kufrah Basin varies from 350m in thickness in the northwest to 1150m in the southwest. A conglomerate overlies the Precambrian basement and is succeeded by quartzitic, coarse-grained and conglomeratic sandstones containing *Tigillites*. The Hasawnah Formation was encountered in the A1-NC 43 and the B1-NC 43 wells, comprising medium to coarse-grained quartz sandstones with interbedded gravels, but it is not possible to

91

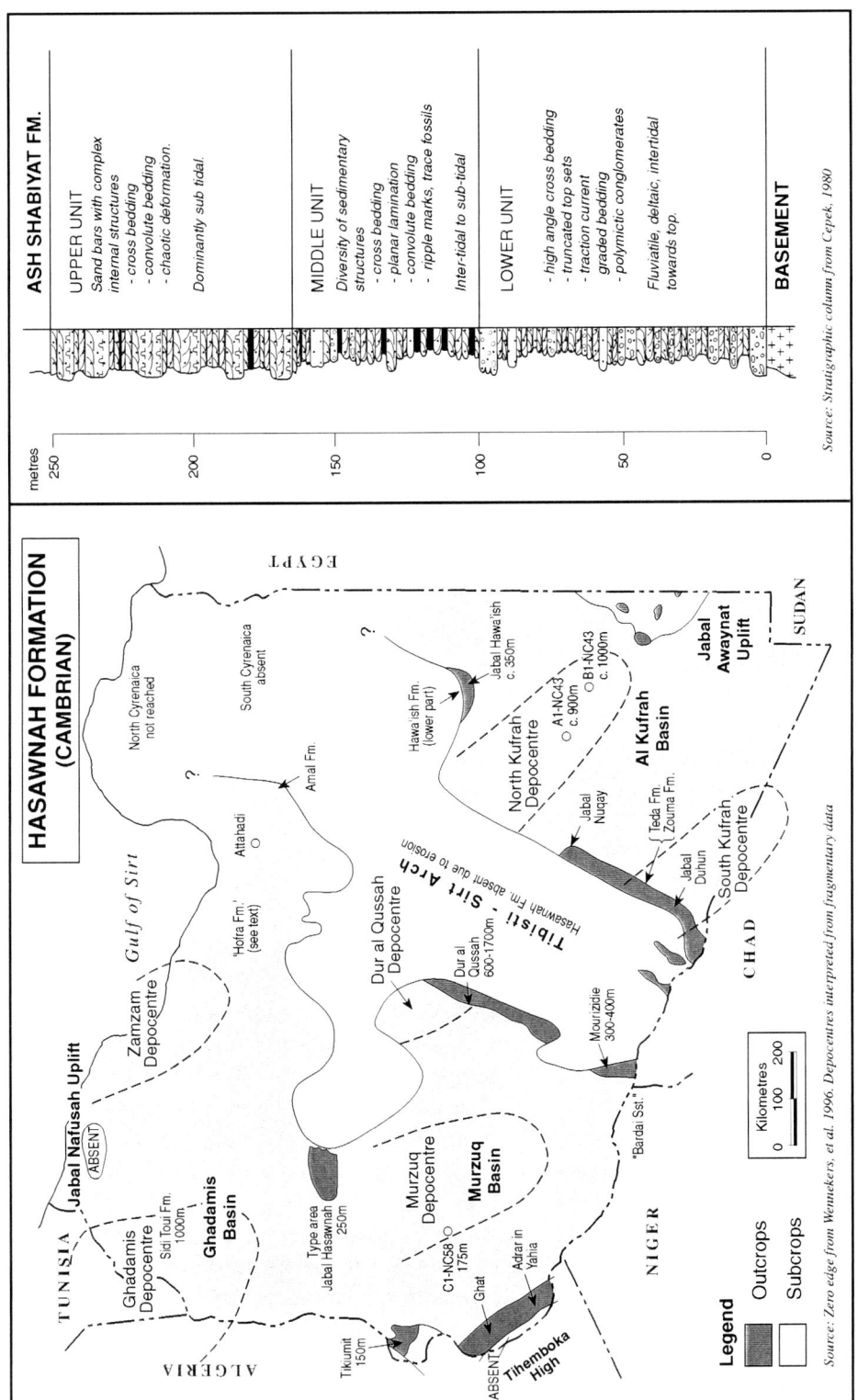

Figure 3.8   Hasawnah Formation, Distribution and Stratigraphic Column

The Hasawnah Formation is present at outcrop and in the subsurface over much of Libya, with the exception of the Tibisti-Sirt Arch, Jabal Awaynat Uplift, Tihemboka High and Jabal Nafusah Uplift. No type section has been defined but the Jabal Hasawnah on the Al Qarqaf Arch is the area in which the formation was first described. The stratigraphic column is based on a composite section compiled by Cepek in the Jabal Hasawnah area. The Amal and Hofra Formations are now highly compromised by recent palynological findings (see text).

separate it from the overlying Mamuniyat Formation. Sedimentological analysis of the Hasawnah Formation in the eastern Kufrah Basin suggests deposition in a lower shoreface environment.[47]

In the subsurface of the Ghadamis Basin Deunff and Massa established a new formation which they named the Sidi Toui Formation. This is a coarse quartzitic sandstone up to 1000m thick which is commonly believed to be equivalent to the Hasawnah Formation, although equivalence is assumed rather than proved. The Sidi Toui Formation contains Cambrian spores and acritarchs. No evidence of Cambrian palynomorphs has been found in wells to the northeast of the Al Qarqaf Arch, and the Hasawnah Formation is absent on the Tibisti-Sirt Uplift. In Algeria the equivalent of the Hasawnah Formation are the Gres de Hassi Leila.[48]

The situation in the subsurface of the Sirt Basin is more complex. In the absence of any diagnostic criteria it became common practice for oil company geologists to apply Burollet's term Qarqaf Group to any quartzitic sequence underlying the Mesozoic formations in the northern Sirt Basin. However Barr and Weegar introduced two new formation names, the Amal and Hofra Formations, with designated type sections on the respective fields, which were intended to define these sequences more precisely.[49] Both resemble the Qarqaf Group quartzitic sandstones petrographically, and both overlay what was assumed to be basement. Neither contain diagnostic fossils, but the Hofra Formation was thought to underlie Silurian graptolitic shales. However, subsequent studies have raised major questions about the age of the Sirt Basin quartzites. The first doubts were raised by Bonnefous who published an analysis of the D2-104A well in the same year that Barr and Weegar's book was issued.[50] He showed that the quartzites in the well were of two different ages, separated by a major uncon-

formity: an upper early Cretaceous unit overlying a lower Cambro-Ordovician unit. Subsequently Wennekers, et al. reported the results of a study carried out by The Robertson Group in 1991 on thirteen wells from the Sirt Basin including A1-11, the type well for the Hofra Formation, and a well from the Amal field. The results were startling. Revised age dates, based primarily on palynology, were obtained for quartzite sections in nine wells. The revised ages ranged from early Carboniferous to early Cretaceous, and include both the Hofra type section and the Amal well. In another study El Hawat et al. confirmed the age dating by Bonnefous on the Wadi field quartzites based on palynology. Analysis of the quartzites of the Attahadi field however, also based on palynology, indicated a late Cambrian age, and a ?Precambrian to Cambrian age has been established on the basis of palynology for the pre-Upper Cretaceous (PUC B) quartz arenites at Jakharrah in the Maragh Graben. Petrographic analysis of the PUC B unit, which reaches an incomplete thickness of 300m, suggests that it is a non-marine sequence deposited in a tectonically active graben during a climatically cool period. Clearly there are major problems with the quartzite section in the Sirt Basin many of which remain unresolved. The assumption that the quartzites can be assigned to a generalised Cambro-Ordovician Qarqaf Group is no longer tenable.[51]

In the subsurface of Cyrenaica palynological analysis showed no evidence of the presence of Cambrian rocks. In fact no rocks have been found at all from the interval between the late Riphean sediments and rocks of Caradocian age.[52]

In summary, the Hasawnah Formation is an extremely widespread and lithologically uniform deposit overlying eroded and folded basement and extending over large areas of Libya. It also extends across the western desert of Egypt, and over much of Algeria where it is called the Gres de Hassi Leila in the Illizi Basin, and the R1, R2 and R3 units in the Triassic Basin which form the reservoir in the Hassi Messaoud, El Gassi, El Agreb and Rhourde el Baguel fields.[53]

The petrographic and sedimentological evidence suggests that the Hasawnah Formation represents a continental or fluvial deposit over much of its extent, with deltaic and shallow marine shelfal conditions in some areas. No fossils have been found in the Hasawnah Formation, apart from trace fossils. Its age attribution as Cambrian is based on its stratigraphic position between Pan-African basement and overlying rocks which are generally believed to be of Tremadocian age, and the supposed equivalence with subsurface sequences which contain late Cambrian palynomorphs.

3.3.2 Ash Shabiyat Formation

The Ash Shabiyat Formation was established by Havlicek and Massa[54] for a silty sequence on the Al Qarqaf Arch occurring between the Hasawnah and Hawaz Formations (Figure 3.9). It represents a maximum flooding interval and is essentially a silty marine sandstone containing abundant *Tigillites*, *Cruziana* and *Harlania* trace fossils. Seilacher has demonstrated that trace fossils, particularly *Cruziana* and *Arthophycus*, can be used to date otehwise barren Lower Palaeozoic sequences over much of western Gondwana, including Libya. It originally formed part of the overlying Hawaz Formation. In the area of the Al Qarqaf Arch the contact with the underlying Hasawnah Formation is conformable.[55] It can be divided into a lower and an upper unit. The lower unit comprises thin-bedded, flaggy sandstones, often ferruginous, and containing *Tigillites* and *Harlania*. The thickness is about 25m. Cross-bedding is rare, which is taken to indicate a deeper depositional environment. The upper unit, which reaches 40m in thickness, is more thickly bedded and more extensively cross-bedded, and bears some similarity to the Hasawnah Formation. The formation has also been mapped on the southwestern flank of the Al Qarqaf Arch in the Idri area where it has similar characteristics.[56]

The Ash Shabiyat Formation is also present on the western flank of the Murzuq Basin. In this area the two-fold subdivision is less evident, but the friable, ferruginous sandstones rich in *Tigillites* are regarded as characteristic. The thickness in the Tikiumit/Ghat area is about 70m. It has not been differentiated from the overlying Hawaz Formation in the C1-NC 58 well in the subsurface of the Murzuq Basin.[57]

Nor has the formation been differentiated as a distinct unit on the eastern flank of the Murzuq Basin, but may be present, as a deeper water pulse, between the Hasawnah and Hawaz Formations. In the Kufrah Basin however an unconformity cuts out not only the Ash Shabiyat, but also the younger Hawaz and Melaz Shuqran formations.

In the subsurface of the northern Ghadamis Basin, Deunff and Massa established the Sanrhar Formation for a 100m argillaceous sandstone overlying the Sidi Toui Formation.[58] The age has been determined as Tremadocian based on brachiopods and microplankton which are similar to those of the Argiles d'El Gassi in Algeria. It is generally assumed that the Sanrhar Formation of the Ghadamis Basin is the approximate subsurface equivalent of the Ash Shabiyat Formation.[59]

In the area northeast of the Al Qarqaf Arch, Belhaj showed the presence of quartzitic sandstone of inner neritic character passing northeastwards into siltstones and mudstones of an outer neritic facies which reaches a thickness of 1357m in the A1-10 well. These rocks have been dated as Tremadocian on the basis of palynology, and are assumed to be equivalent to the Ash Shabiyat Formation.[60]

94

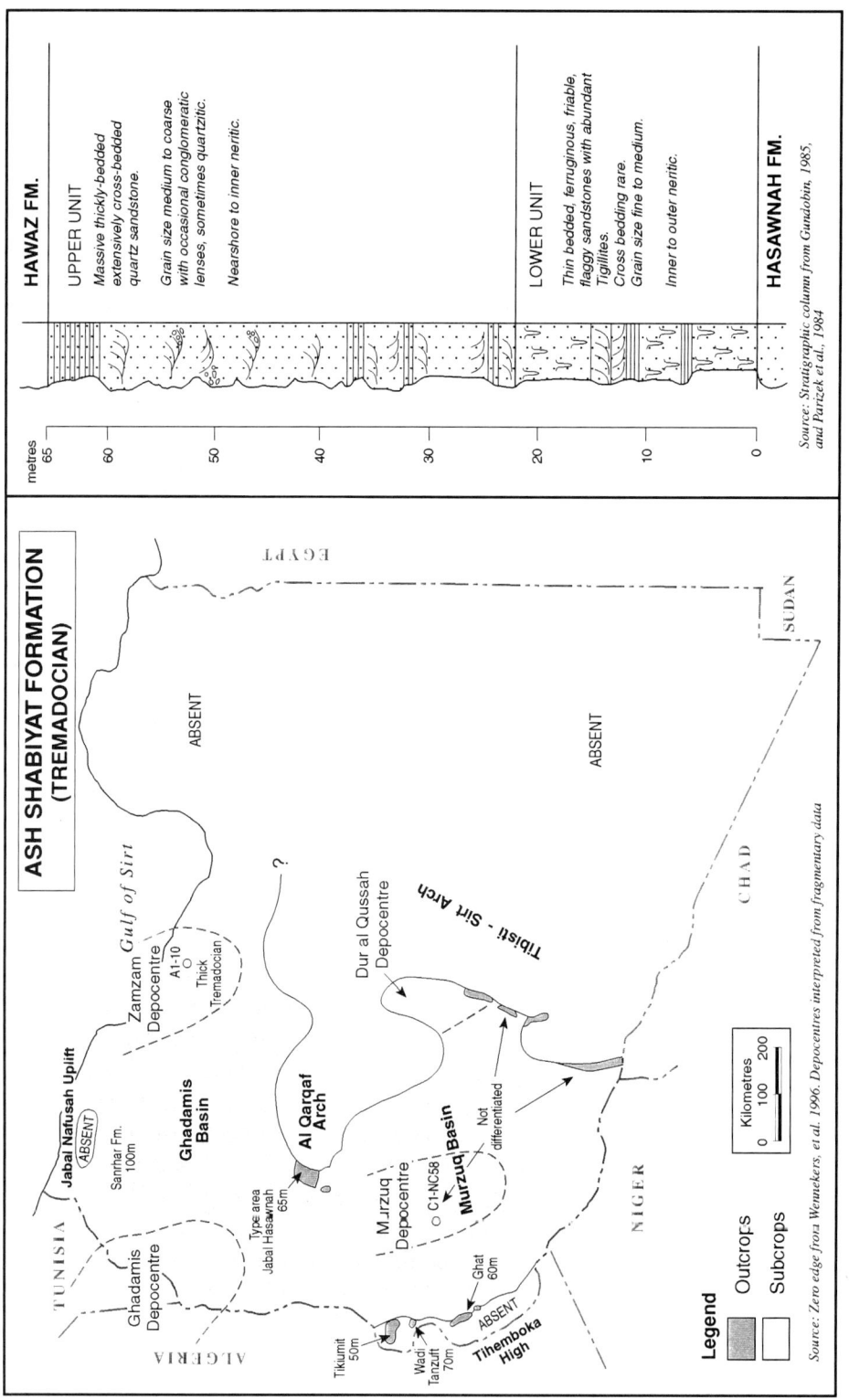

Figure 3.9 Ash Shabiyat Formation, Distribution and Stratigraphic Column

The Ash Shabiyat Formation was first described from exposures on the Al Qarqaf Arch. No type section was defined, since no complete exposure is available, but the section shown represents a composite section from the western part of the Arch. The formation has been mapped along the western flank of the Murzuq Basin, but it has not been differentiated from the overlying Hawaz Formation in other areas. In the Ghadamis Basin the Sanrhar Formation is probably equivalent to the Ash Shabiyat Formation and the thick sequence dated as Tremadocian in well A1-10 is probably also a partial equivalent. There is no evidence of equivalent sequences in eastern Libya.

During this period Cyrenaica was evidently still emergent and there is no evidence of any marine sequences equivalent to the Ash Shabiyat Formation.[61]

In the Algerian Triassic province the equivalents of the Ash Shabiyat Formation are represented by the Zone des Alternances, the Argiles d' El Gassi and the Gres d' El Atchane, whilst in the Illizi basin the equivalent is the In Kraf Formation.[62]

The Ash Shabiyat Formation represents a marine transgressive unit which in southern Libya shows a generally deeper water aspect than the Hasawnah Formation. The age is assumed to be Tremadocian, on the basis of stratigraphic position and supposed correlation with Tremadocian dated rocks in the subsurface.

3.3.3 Hawaz Formation

The second major unit of Burollet's Qarqaf Group is the Hawaz Formation named by Massa and Collomb from a type section on the western part of the Al Qarqaf Arch (Figure 3.10). Collomb later divided the formation into three subdivisions, an upper 'Tigillites superieurs', a middle 'Gres intermediaires' and a lower 'Tigillites inferieurs'.[63] The subsequent separation of the Ash Shaybiyat Formation from the Hawaz left only the upper unit attributable to the Hawaz Formation. The Hawaz Formation on the Al Qarqaf Arch, as redefined, comprises 120m of alternating siltstones and thinly bedded, cross-bedded sandstones with abundant *Tigillites*. Vos interpreted this sequence as a coarsening-upwards delta front/delta top highstand deposit.[64] The contact with the underlying Ash Shabiyat Formation is conformable. The formation has been mapped on both the northern and southern flanks of the Al Qarqaf Arch in the Qararat al Marar and Idri areas. It has not been mapped on the western flank of the Murzuq Basin, but it is known to be present in the C1-NC 58 well in the subsurface

of the Murzuq Basin where it is represented by 85m of very fine-grained sandstones and siltstones showing sigmoidal cross-bedding which are interpreted as offshore tidal sands. It is also present in the NC 115 wells where it reaches 150m in thickness. Palynological analysis of this sequence gives a probable mid-Ordovician age.[65]

The Hawaz Formation was reported by Klitzsch as reaching 50m in thickness in the Dur al Qussah Trough, but this may also include Ash Shabiyat Formation from which it is undifferentiated in this area. Its presence on the Tripoli-Tibisti Uplift in the subsurface is unquantified, due to lack of data. It is totally absent from the Kufrah Basin where it is cut out by an unconformity.[66]

No diagnostic fossils have been recovered from the Hawaz Formation, except for casts of brachiopods and pelecypods, and the trace fossils *Tigillites* and *Harlania*. In the subsurface of the Ghadamis Basin Deunff and Massa named a unit of fine-grained sandstone containing ferruginous oolites as the Kasbah Leguine Formation.[67] It contains microplankton which indicate an Arenigian age for this unit. It has long been assumed that this formation is the subsurface equivalent of the Hawaz Formation, but this equivalence has not been proved, and the correlation is circumstantial.

Further east sandy conglomerates and mudstones, which reach a thickness of 730m in well A1-8, are taken to be approximate subsurface equivalents of the Hawaz Formation and they contain Arenigian palynomorphs.[68]

In Algeria the Quartzites de Hamra and the Grès d'Ouargla, which have a wide distribution in both the Triassic and Illizi Basins, occupy an equivalent stratigraphic position.[69]

In summary, the Hawaz Formation is missing in the Kufrah, and southwestern Murzuq Basins, probably due to uplift or marine regression in these areas, possibly related to tectonic adjustments following the break away of the

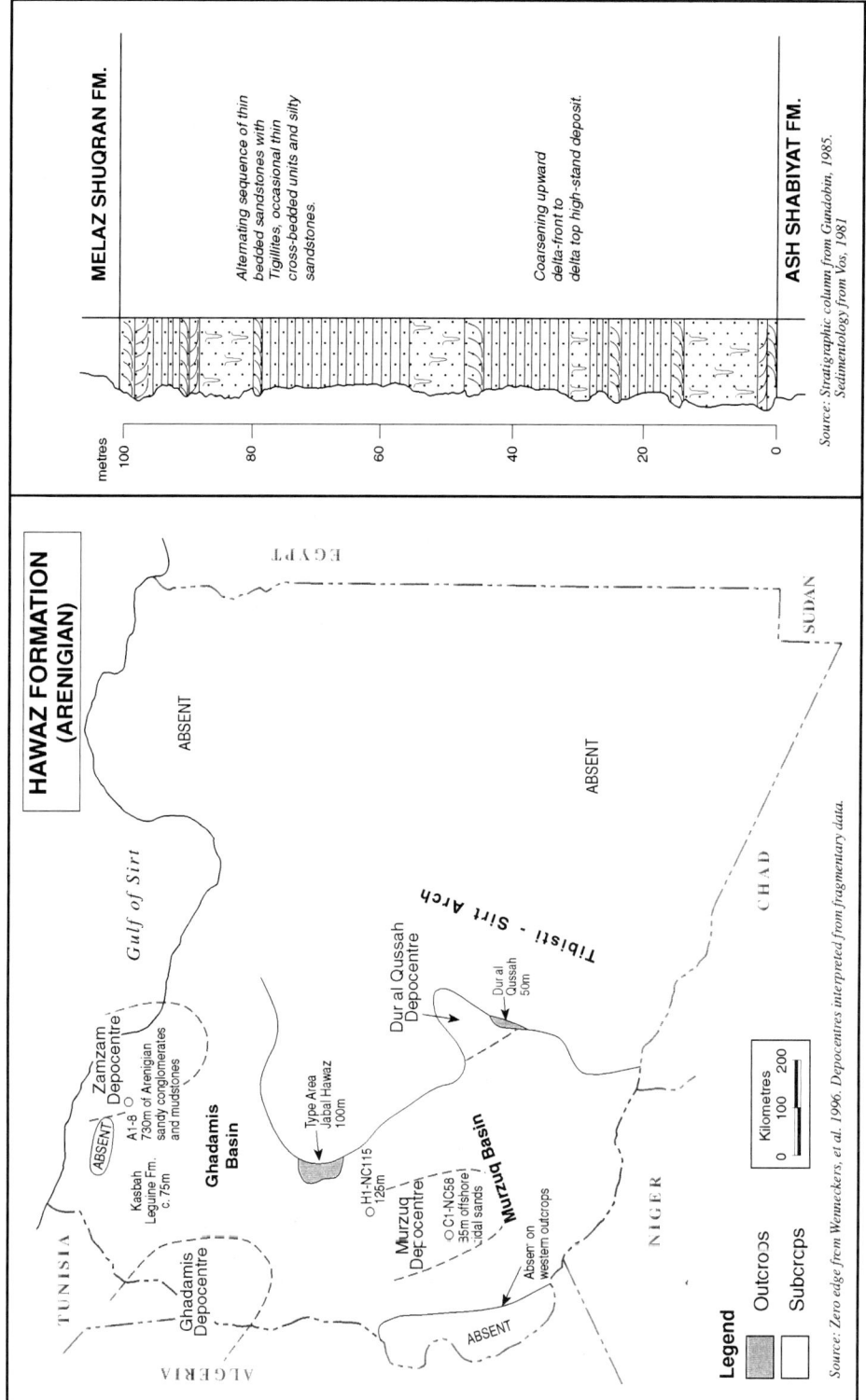

Figure 3.10 Hawaz Formation, Distribution and Stratigraphic Column

The Hawaz Formation was first described from the western Al Qarqaf Arch which is regarded as the type area. Originally it included the Ash Shabiyat Formation but this was subsequently established as a separate formation. The formation is not found along the western margin of the Murzuq Basin where it is overstepped by younger formations, but it is present in the C1-NC 58 well and in the Dur al Qussah Trough. There is no evidence of equivalent sequences in eastern Libya.

Avalonian and Cadomian terranes which occurred at about this time.[70] The formation is present in the Dur al Qussah Trough and around the Al Qarqaf Arch and in the subsurface of the Ghadamis Basin and at least into the northwestern part of the Sirt Basin. It is absent over the Tibisti-Sirt Arch and in Cyrenaica.

3.3.4 Melaz Shuqran Formation

Unconformably overlying the Hawaz Formation on the Al Qarqaf Arch is a much more argillaceous unit which Massa and Collomb called the Melaz Shuqran Formation (Figure 3.11).[71] It comprises about 60m of claystones and siltstones which are varicoloured green, brown, red and grey, chloritic and thin bedded. It represents a significant flooding event. Collomb reported the presence of abundant tracks in the formation, and the presence of trilobites, bryozoa, brachiopods and graptolites which suggest a Llanvirnian-Llandeilian age for the formation. Havlicek and Massa preferred a Caradocian age based on the brachiopods.[72] Collomb also reported the presence of boulders and cobbles of granite, gneiss and quartzite which are polished and striated, and which he took to indicate periglacial conditions and iceberg rafting. Blanpied, et al. described an association of large boulders and striated pebbles in a silty matrix, graded beds and large scale cross-bedding in the Melaz Shuqran Formation which they attributed to a proglacial delta. At outcrop on the western flank of the basin McDougall and Martin found an association of glacial and non-glacial facies representing distal-shelf and ice margin deposits. The brachiopods also indicate a cold water environment. El Hawat and Bezan regarded the Melaz Shuqran Formation as the lowest unit of their late Ordovician glacial sequence. They took the Melaz Shuqran to represent a transgressive unit reflecting the first major deglaciation event. The formation is characterised by mass-movement, liquifaction and turbidity flow features.[73]

The Melaz Shuqran Formation is well developed in the Idri area on the Al Qarqaf Arch, and has also been mapped along the western flank of the Murzuq Basin. In the Tikiumit area the Melaz Shuqran Formation exhibits good examples of ice rafted blocks of granite, schists and conglomerate up to 0.5m in diameter randomly distributed within a fine grained sandstone matrix. The sandstones show extensive subaqueous slump features. The thickness of the formation in this area is about 45m. In the subsurface of the Murzuq Basin the Melaz Shuqran Formation contains radioactive shales with high thorium content which are interbedded with minor siltstones and sandstones. The thickness of this unit is 25m in well C1-NC 58 and in other wells ranges from 15m to 75m. Its absence in well H1-NC 115 and reduced thickness in well D1-NC 115 may be indicative of the effect of Caradocian tectonism in this area.[74]

The Melaz Shuqran Formation is reduced to a thickness of 25m in the western Dur al Qussah area where it consists of reddish silty and sandy shale with sandstone stringers. It is assumed to be present on the Tripoli-Tibisti Uplift although this is unconfirmed. It is absent in the Kufrah Basin where the Mamuniyat Formation rests directly on the Hasawnah Formation.[75]

In the subsurface of the Ghadamis Basin Deunff and Massa proposed a new formation called the Bi'r Ben Tartar Formation which comprises about 100m of fine-grained sandstones, with minor red and brown claystone intervals, and containing bands of chloritic oolites.[76] The formation has a conformable relation to the underlying Kasbar Leguine Formation. It contains a rich association of trilobites, graptolites and acritarchs which collectively suggest a Llanvirnian to Llandeilian age. The Bi'r Ben Tartar Formation is usually assumed to be the subsurface equivalent of the

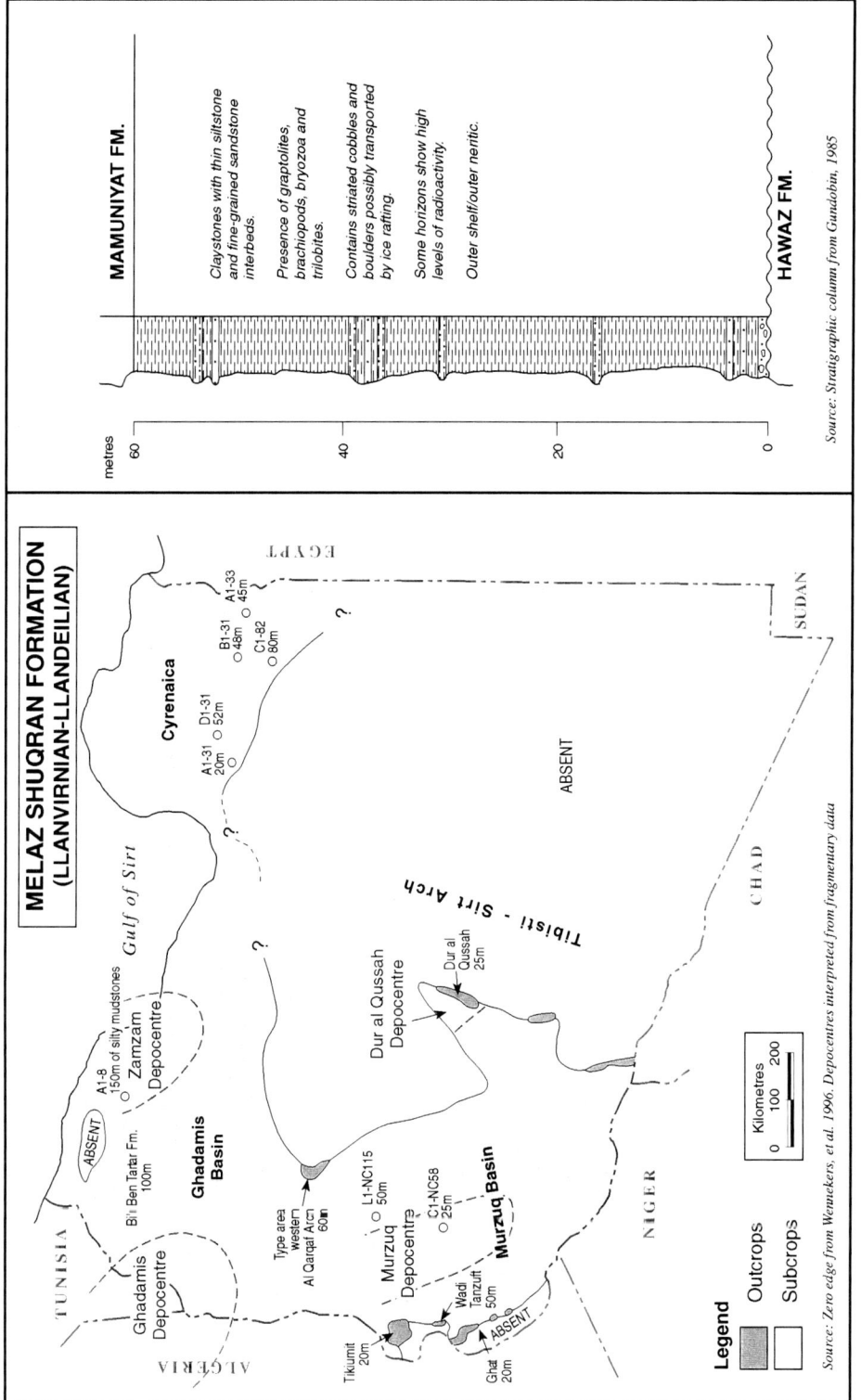

Figure 3.11   Melaz Shuqran Formation, Distribution and Stratigraphic Column

The Melaz Shuqran Formation was first described from the western Al Qarqaf Arch. It is found throughout the Murzuq Basin, and Dur al Qussah Trough, and the Bi'r Ben Tartar Formation is regarded as the subsurface equivalent in the Ghadamis Basin.

Melaz Shuqran Formation. Wells A1-42, A1-8 and B2-34, in the Zamzam Depression contain about 150m of neritic sandstones and silty mudstones which probably represent the extension of the Melaz Shuqran Formation into this area. The sediments contain Llanvirnian to Caradocian palynomorphs. The Caradocian in Libya is remarkable for containing the earliest indication of the appearance of land plants anywhere in the world. In 1982 Gray described sheets of cells associated with cutinized spore tetrads of Caradocian age from wells in the Ghadamis Basin, which resemble cuticle from vascular plants.[77]

In the Triassic Basin of Algeria the lithostratigraphic equivalents of the Melaz Shuqran Formation are the Gres d'Ourgla and the Argiles d'Azzel.[78]

In Cyrenaica marine sandstones and shales overlying late Riphean 'basement' have been found to contain Caradocian-Ashgillian acritarchs and chitinozoa in wells E1-81, C1-82 and C1-125. The maximum thickness encountered is 207m in well C1-125. The environment of deposition is shallow-water, nearshore marine, and there are similarities with the glacio-marine deposits of Morocco of similar age. These rocks are probably equivalent to part of the Melaz Shuqran Formation of western Libya.[79]

The glacial hypothesis for the dropstones in the Melaz Shuqran diamictite was reviewed by Radulovic. Continental ice sheets are believed to have developed in the Hoggar-Tibisti area with glaciers extending northwards along existing valleys. The flooding episode which gave rise to the Melaz Shuqran Formation permitted a marine transgression to extend to the edge of the ice cap, from which icebergs carrying rocks derived from the Precambrian terranes became detached, floated northwards and deposited their dropstones as they melted. This view has been challenged by Abugares and Ramaekers who proposed the alternative view that these deposits are lowstand fluvial deposits reworked during the following transgression. There is however no doubt that western Gondwana was located at high latitudes during the Ordovician, and there is an extensive literature on the Ordovician glaciation of the area. This matter is discussed further in the section on the Mamuniyat Formation.[80]

3.3.5 Tasghart Formation

In mapping the Wadi Tanzuft area on the western flank of the Murzuq Basin Radulovic found a distinctive unit between the Melaz Shuqran Formation and the Mamuniyat Formation which he called the Tasghart Formation. The unit is about 60m thick and is characterised by cyclic sequences with erosive bases associated with flute casts, load casts and slumps, individually about 2m thick. The cycles begin with a conglomeratic unit followed by fining-upward sandstone sequences, and capped by fissile, fine-grained to silty sandstones with parallel laminations. Radulovic interpreted these rocks as shallow-water turbidites within a small local depocentre. The formation is extemely conglomeratic at Ghat but shows more distal features northwards in the Tikiumit area. It has not been found outside this small area. The formation is barren of fossils but is assumed to be Caradocian in age.[81] Abugares and Ramaekers disputed the turbidite origin, and regarded the Tasghart Formation as simply the lower part of the Mamuniyat Formation. It is probably more appropriate to regard the Tasghart as a member of the Mamuniyat Formation rather than a separate formation.[82]

3.3.6 Mamuniyat Formation

The Mamuniyat Formation was first described by Massa and Collomb from outcrops on the Al Qarqaf Arch (Figure 3.12). A type section was selected by Parizek et al. in the Idri

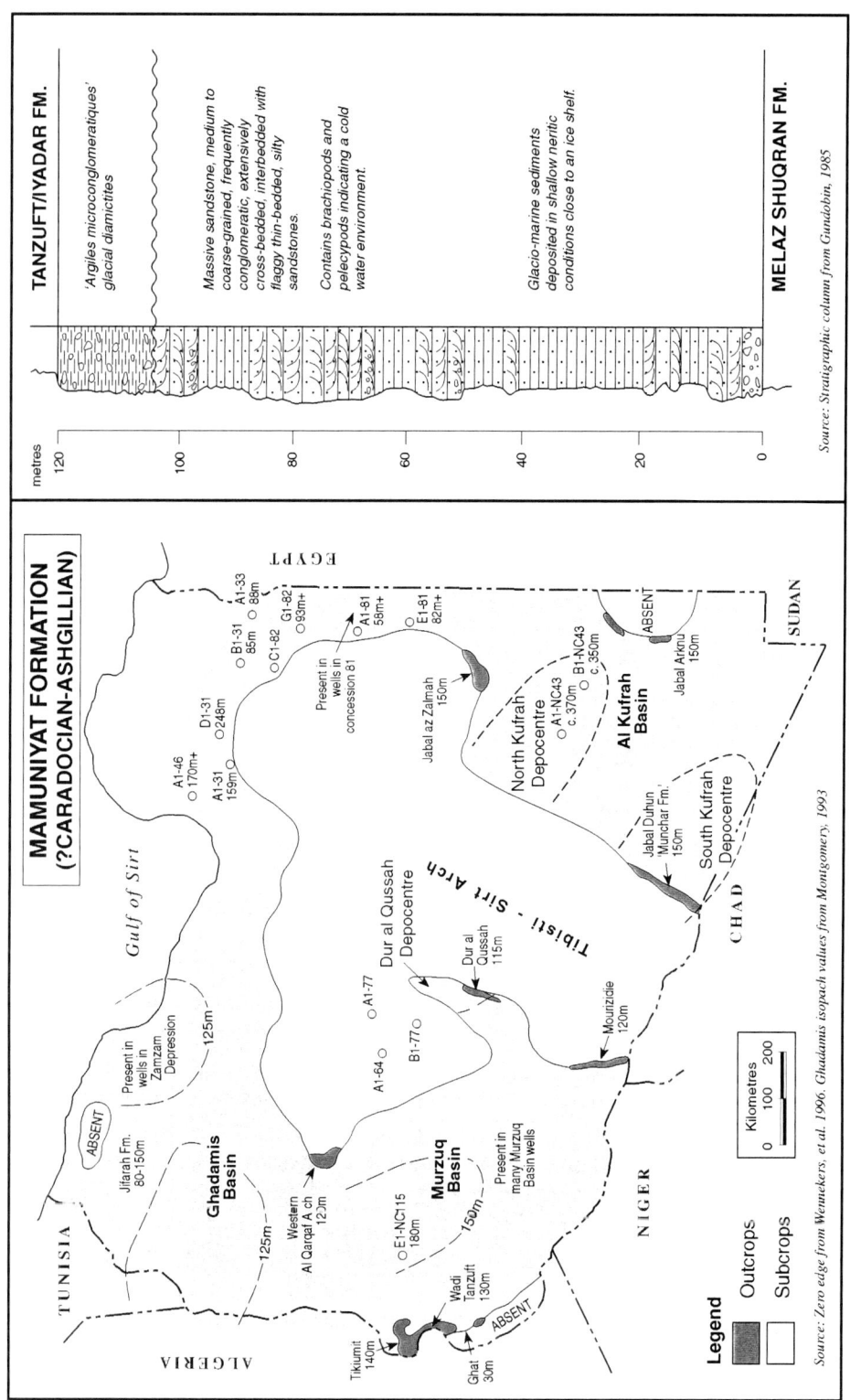

Figure 3.12 Mamuniyat Formation, Distribution and Stratigraphic Column

The Mamuniyat Formation was first defined on the western Al Qarqaf Arch. It outcrops around the margin of the Murzuq Basin and has been penetrated in many wells in the northern part of the basin where the formation is oil bearing. Its equivalents are present in the Ghadamis Basin and in the Zamzam Depression, and equivalents are also present in the Al Kufrah Basin. It is present on the southern part of the Cyrenaica Platform but its presence in northern Cyrenaica has yet to be proved. The Mamuniyat Formation is probably, in part, a glacio-marine deposit, containing evidence of glacially striated drop-stones in the Ghat area. In western Libya and Algeria it is often capped by a distinctive diamictite unit, called 'Argiles Microconglomeratiques' by French geologists, which was deposited on an irregular, possibly glacially-eroded surface.

area.[83] It has a wide distribution in the Murzuq, Al Kufrah and Ghadamis Basins. The Mamuniyat Formation is a major oil reservoir in the northern Murzuq Basin, and contains large reserves in concessions NC 115, NC 174 and NC 101. It comprises 100 to 140m of massive, cross-bedded sandstones, representing a basal lowstand system. Rubino, et al. reported a complex meander belt in the lower part of the Mamuniyat Formation which can be traced over a distance of 12km. The lower contact is unconformable and often ferruginous. The sandstone is generally medium- to coarse-grained, and frequently conglomeratic. It is friable, with a high percentage of kaolinitic cement. It contains occasional fine-grained sandstone and siltstone stringers. The Mamuniyat Formation contains rare *Tigillites* and fragmentary brachiopods and pelecypods. Palynological evidence suggests a Caradocian age for the Mamuniyat Formation whilst the brachiopod evidence indicates an Ashgillian age.[84] Additionally, the brachiopod fauna suggests a periglacial environment, and other evidence shows that at this period western Gondwana was located very close to the South Pole. Boote et al. believed that the Mamuniyat Formation was deposited towards the end of the glacial cycle as the icecap retreated. Blanpied, et al. came to a similar conclusion, and felt that the Mamuniyat Formation on the Qarqaf Arch represented a deltaic complex separated from the ice-front by a broad subaerial periglacial terrain. However on the westernmost margin of the Qargaf Arch, Deynoux et al. found two deeply incised valleys up to 4km wide and 150m deep, which cut the lower Mamuniyat, Melaz Shuqran and top of the Hawaz Formations, which they attributed to glacial incision, implying a dramatic fall in sea level during early Mamuniyat deposition. Smart and Whittington and Walker reported similar deeply incised erosional channels filled with fluvioglacial sediment from seismic data in the Awbari area which in some cases may represent

sediment-filled ice-tunnel channels. The western Qarqaf Arch has also yielded evidence of soft-sediment deformation in the form of a large gravity slide of unconsolidated Mamuniyat sands over a *decollement* surface in the shales of the Melaz Shuqran Formation. El Hawat and Bezan emphasised the major erosional event which separates the lower from the upper Mamunyiat sequences. They regarded the lower Mamuniyat as a high stand systems tract associated with the Melaz Shuqran cycle, and sharing similar debris flow and soft sediment deformation features. The upper sequence was deposited on a highly eroded surface caused by a rapid fall in sea level which marks the main glacial episode. Deep channels were cut into the underlying sequence which were filled during the rapid deglaciation phase with braided stream periglacial deposits. These sediments originated from the south, and pass northwards into microconglomeratic shales. The top of the upper Mamuniyat sequence is marked by a regional unconformity. Rapid melting of the ice sheets led to a major transgression, which marks the base of the Silurian, during which transgrssive nearshore sands were deposited, followed by the shales of the Tanzuft Formation. The channel-fill sequence forms an important oil reservoir in the northern Murzuq Basin and the Tanzuft shales are the principal source rock.[85]

The Mamuniyat Formation has been mapped in the Qararat al Marar area on the northern flank of the Al Qarqaf Arch and in the Idri and Sabha areas to the south. It has also been mapped in the Tikiumit-Ghat area on the western flank of the Murzuq Basin where it conformably overlies the Tasghart Formation or unconformably overlies the Melaz Shuqran Formation. In this area the formation contains microconglomerates with granite and gneiss pebbles, and the upper surface shows evidence of subaerial erosion. Acritarchs recovered from the Mamuniyat Formation in the Wadi Tanzuft area give a late Ordovician age. McDougall and Martin recognised two unconformities in the Mamuniyat Formation at outcrop separating three sequences: a lower proximal shelf facies deposited during glacial advance, a middle shoreface/delta-plain nearshore sequence, and an upper braided-stream fluvial sequence. Thicknesses ranging from 20m to 170m have been reported from wells in the Murzuq Basin and it is present in the NC 115 wells where its thickness varies dramatically. In wells A1 and B1-NC 58 the upper part is intensely silicified. The sediments exhibit water-escape structures which may be suggestive of permafrost conditions. Echikh and Sola recognised four facies, a fine-grained quartzitic facies in the centre of the basin (concessions NC 101 and NC 58), a silty facies in wells F1, E1, G1 and H1-NC 58, and a shallow marine sand facies in the NC 115 area. The fourth lithofacies, a coarse, poorly-cemented sandstone, known as the periglacial horizon has only been found on the B-NC 115 field. Aziz presented a sedimentological model for the periglacial horizon in the B-NC 115 field which showed a deltaic complex passing northwards into delta-front deposits. This unit, which ranges from 15 to 45m in thickness, has an unconformable relationship with both the underlying Mamuniyat Formation and the overlying Tanzuft Formation. The acritarch content suggests a late Ordovician age whilst the chitinozoans are more Silurian in character. The relationship of this unit to the Iyadar and Bi'r Tlakshin Formations is unclear. A sedimentological and petrographic study by Fello and Turner suggested that the Mamuniyat Formation in the NC 115 area was derived from the tectonically active granitic basement ridge of the Tihemboka Uplift to the southwest.[86]

In the Dur al Qussah area the Mamuniyat Formation is represented by 115m of cross-bedded sandstones fine- to coarse-grained, and often conglomeratic, containing *Tigillites*, and with a highly ferruginous basal sandstone. The unit rests with an angular unconformity on eroded Melaz Shuqran Formation. The presence of the Mamuniyat Formation on the Tripoli-Tibisti Uplift beneath the Devonian unconformity must remain speculative in the absence of well data. It is however absent in wells A1-77, B1-77 and A1-64 in the Dur al Qussah Trough. South of the Mourizidie Horst 120m of cross-bedded Mamuniyat sandstones rest directly on Hasawnah Formation.[87]

In the Kufrah Basin the Mamuniyat Formation lies unconformably on Hasawnah Formation (Figure 3.12). The relief on the unconformity surface may be due to glacial action, and conglomerates overlie the unconformity in the Jabal Arknu area. The sediments are mainly fine- to medium-grained sandstones with festoon trough cross-beds indicating an upper shoreface environment in this area. These rocks contain a rich assortment of trace fossils including *Tigillites*, *Arthrophycus*, *Rusophycus* and *Cruziana*. The Jabal Duhun outcrops were originally assigned to the Munchar Formation, but this name is synonymous with the Mamuniyat. The thickness is relatively constant with 150m at Jabal Arknu in the southeast and a similar figure at Jabal Duhun in the southwest. The Mamuniyat Formation was penetrated in both the A1-NC 43 and B1-NC 43 wells where no unconformity is detectable. It comprises medium- to coarse-grained quartz sandstones indistinguishable from the underlying Hasawnah Formation. A sample from near the top of the sequence has yielded a Caradocian-Ashgillian age based on chitinozoa.[88]

The Mamuniyat Formation probably correlates with the Jifarah (Djeffara) Formation in the subsurface of the Ghadamis Basin. The term was introduced by CFP geologists in southern Tunisia and was later extended into the Libyan subsurface by Deunff and Massa. It comprises 80 to 150m of microconglomerates and stratified clays of glacio-marine origin containing brachiopods, ostracods and acritarchs, which indicate a Caradocian age. Overlying this unit is a carbonate sequence of bryozoan biostromes which contains conodonts of Ashgillian age. The carbonate sequence is well developed in well C1-NC 143 where it reaches a thickness of 140m. Conodonts have been obtained from wells A2-70, B2-34 and C1-34 which confirm an Ashgillian age for the carbonate unit. Carbonate mud-mounds have been reported from the Jifarah Formation in the northern Ghadamis Basin along an east-west trending palaeoridge extending for 300km. The mounds are rich in bryozoans and individual mounds reach 100m in thickness. The mounds appear to pass laterally into periglacial micro-conglomerates. Massa and Bourrouilh compared these mounds with Quaternary bryozoan mounds south of Australia which developed in glacial upwelling currents originating from Antarctica. Echikh published a cross section showing the complexity of the Mamuniyat reservoir in the Ghadamis Basin, and its relationship with the overlying Bi'r Tlakshin Formation which is discussed in section 3.4.1.[89] The formation can be traced eastwards into the Zamzam Depression where both carbonates and clastics are found. The formation is present in wells B2-34, A1-43, E1-60, D1-8, and as far east as D1-32, where it has been confirmed as Caradocian-Ashgillian in age. In well A1-NC 40 sedimentological analysis indicated beach/barrier island to tidal-flat conditions.[90]

In southern Cyrenaica a widespread sequence of fine-grained sandstones, which reach 240m in thickness, is present in wells, A1-31, B1-31, D1-31, A1-33, A1-46, A1-81, E1-81, J1-81, C1-82, and G1-82 (Figure 3.12). An apparent microconglomerate has been described from well E1-81. Ashgillian chitinozoa have been recovered from well J1-81A, and it is likely that this sequence correlates with the Mamuniyat Formation of west Libya.[91]

In the Triassic Basin of Algeria the Caradocian-Ashgillian section is interrupted by an unconformity. The early Caradocian is represented by the Gres de l'Oued Saret, which is abruptly truncated and overlain by the Argiles microconglomeratiques and the Dalle de M'Kratta.[92] The Gres de l'Oued Saret is regarded as the equivalent of the Mamuniyat Formation of Libya, and the overlying units to the Bi'r Tlakshin Formation. In the Illizi Basin this grouping forms the Unit IV reservoir which is productive at Tin Fouye-Tabankort, and has been interpreted as glacial in origin.[93]

The Mamuniyat Formation is the principal oil reservoir in the Murzuq Basin, and is productive in the NC 101, NC 115 and NC 174 concessions on the northern edge of the basin. The oil is reservoired in the upper units of the formation including the periglacial horizon, which is a prolific reservoir unit on the B-NC 115 field. In all of these areas the Mamuniyat Formation is in direct contact with the overlying Silurian source rock. Oil has not been found in areas where the Bi'r Tlakshin Formation intervenes between these two formations.

Late Ordovician Glaciation

It is necessary to make a brief digression to consider the late Ordovician glaciation, since there is some disagreement on the interpretation of the field data, and because the rocks associated with this event are important reservoir rocks in the northern Murzuq Basin. The late Ordovician glaciation in north Africa is not a unique event. Glacially derived rocks have been described from across Gondwana at various times from the Proterozoic to the Permian. Late Proterozoic glacial rocks are known from the Taoudeni Basin of west Africa, the Anti-Atlas mountains of Morocco, the Touareg shield in the central Sahara, the western Hoggar in Algeria, from Sierra Leone, the Volta Basin, Zaire, Uganda and Eritrea. Late Ordovician glacial rocks have been reported from the Tindouf Basin in Morocco, the

Taoudeni Basin of west Africa, and the central Sahara. Evidence of Silurian glacial activity has been reported from Namibia and the Cape Fold Belt and from the Argentinian-Peruvian basins. Late Devonian glacial activity is well known in northern Brazil and the Amazon Basin, and early Carboniferous glacial activity has been reported from Bolivia, eastern Brazil and from the Congo Basin in west Africa. The late Carboniferous glaciation in the Karoo Basin of South Africa, southern India and Antarctica is also well documented. It might reasonably be concluded that glaciation was almost a regular feature of Gondwanan geology throughout the Palaeozoic. Caputo and Crowell developed a model to explain these occurrences by producing an apparent polar wandering path of Gondwana across the southern pole of rotation during the Palaeozoic. In Libya only the late Ordovician glaciation had any real impact and the remainder of this discussion will concentrate on this event.

The diamicrites of the Melaz Shuqran Formation have already been described. The evidence from the Mamouniyat Formation is even more impressive. These rocks were first described from the Tamajert Formation of Algeria in the 1950s and 60s, and have subsequently been described in detail by numerous workers, including Freulon, Beuf, Deynoux, Bennacef and Biju-Duval. These authors have described the presence of U-shaped valleys, glacially striated pavements extending over many kilometres with chatter-marks and crescentic gouges which indicate ice movement from south to north; varve-like clays with dropstones, and diamictites which probably represent tillites. Deformed sandstone units are interpreted as outwash fans overlying stagnant ice which collapsed and deformed as the ice melted. Towards the north there is a general passage into glacio-marine deposits. Graptolite evidence suggests an age between Caradocian and Ashgillian.

In the northern Murzuq Basin the surface of the Mamuniyat Formation is deeply incised with glacially eroded palaeovalleys which are filled with fluvio-glacial or glacio-marine sandstones and gravels, which appear in the wells as micro-conglomerates and diamictites. Abugares and Ramaekers however repudiated a glacial origin for the features exposed in the area around Ghat, attributing these features to fluid flow, soft sediment slumping and dewatering phenomena. The dropstones were interpreted as low-stand fluvial deposits reworked during the following transgression. It may be that Ghat is located too far north to show evidence of direct glacial activity and that the phenomena in the northern Murzuq Basin are more attributable to periglacial or glacio-marine conditions.

There is an extensive literature on the late Ordovician glaciation and two papers provide an interesting insight on the problem. Semtner and Klitzsch provided evidence of two glacial advances with continental ice sheets extending into southern Libya during the late Ordovician and a more resticted advance during the early Silurian which did not extend north of central Chad and Sudan. Brenchley, et al. studied bathymetric evidence and carbon and oxygen isotopic anomalies in late Ordovician carbonates, and concluded that the late Ordovician glaciation was of very short duration spanning no more than a few hundred thousand years during the Hirnantian stage. A map showing the extent of the Ordovician ice-sheet has been published by Sutcliffe, et al., which showed it to compare in size with the present-day Antarctic ice sheet. Sea-level fall during the glacial episode has been calculated at between 50 and 100m, and was accompanied by a major mass-extinction event. Sutcliffe, et al. recognised six glacial facies associations in the late Ordovician. The diamicitite was interpreted as the result of marine deposition along the grounding line of the ice sheet, the cross-bedded sandstones were formed from meltwater plumes during glacial retreat, and the hummocky cross-bedded sandstones represent reworking of older glacial sediments by storms. Some of the sediments show evidence of over-pressure and fluidisation which are characteristic of a sub-glacial environment. Stacked striated surfaces are believed to have been generated by shear in soft sediments caused by the movement of the overlying glacier.[94]

The periglacial deposits and diamictites which form part of the oil reservoir in the NC 115 and NC 174 areas and in southern Tunisia and Algeria have been described in several publications. They are very variable in thickness. and overlie an extremely irregular surface which may be due to glacial erosion. In Algeria a study of the late Ordovician sequence in wells on the Tiguentourine field revealed the dominent process in this area to be density flow. Massive debris flows are frequently overlain by turbidites and pelagic muds containing dropstones and graptolites. This sequence is taken to indicate deposition during glacial retreat, and appears to be analogous with the situation found in the northern Murzuq Basin.[95]

3.4 Silurian

Regional uplift and erosion took place at the end of the Ordovician. This was associated with the retreat of the ice sheets at the end of the Ordovician, and was followed by a widespread marine transgression, perhaps partly caused by melting of the icesheets, which spread over much of North Africa and Arabia. The transgression produced thin, locally developed, basal silty sandstones in some areas, after which thick, black, radioactive, graptolitic shales were deposited. These shales are rich in organic debris and form one of the principal petroleum source rocks in Libya and Algeria. In Cyrenaica it has been possible to demonstrate an uninterrupted succession of palynozones from the late Ordovician into the early Silurian.[96]

3.4.1 Iyadar and Bi'r Tlakshin Formations

The Iyadar Formation was split from the earlier-defined Tanzuft Formation by Massa and Jaeger in 1971, for a sequence of hematitic siltstones, claystones and minor sandstones in the Ghat area, which corresponds to three graptolite biozones at the base of the Silurian. The sequence reaches a thickness of 200m but pinches out northwards. Southwards it extends into Niger. The contact with the Mamuniyat Formation is unconformable, but it grades into the Tanzuft Formation without break. The graptolites indicate a Rhuddanian age. The Iyadar Formation was not recognised as a mappable formation by the geologists of the Geological Map of Libya, who regarded it as a basal unit of the Tanzuft Formation. In effect it is equivalent to the hot-shale zone at the base of the Tanzuft Formation. Deunff and Massa claimed to be able to recognise the Iyadar Formation in the subsurface of the Ghadamis Basin, where it contains early Silurian chitinozoa and acritarchs. The Iyadar however, as defined by Massa and Jaeger, is a biostratigraphic unit which cannot easily be mapped without biostratigraphic control. The designation is therefore questionable as a valid formation name.

The Bi'r Tlakshin (Bir Tlacsin) Formation is an informal subdivision which is probably roughly equivalent to the Iyadhar Formation. The name has never been formally described, and was not used by IRC, but has been discussed in some detail by Echikh and Sola. The unit was first recognised on the Tlakshin field (A1-70) in the Ghadamis Basin, and has subsequently been found in several other areas. The Bi'r Tlakshin Formation, unlike the Iyadhar Formation is easily recognisable on electric logs, and has a similar log response to the Argiles microconglomeratiques of Algeria and southern Tunisia. It unconformably overlies the

Mamuniyat Formation, and it appears to infill topographic lows on the eroded and irregular Mamuniyat surface, and thins over topographic highs. It also has an unconformable relationship with the overlying Tanzuft Formation which truncates the Bi'r Tlakshin Formation. The unit comprises black mudstones containing abundant coarse sand grains, granules and pebbles, which may have originated as a debris flow deposit, although this view is not universally accepted. Where the formation is complete the shales are capped by thin argillaceous sandstones. The Bi'r Tlakshin Formation varies rapidly in thickness and its distribution is very patchy (Figure 3.13). In the Ghadamis Basin it reaches 100m in thickness in well C1-NC 100 and in the Murzuq Basin it has a thickness of 130m in well D1-NC 115. The age of the Bi'r Tlakshin Formation is unclear. It has usually been regarded as a basal unit of the Silurian, and if it is accepted as an equivalent of the Iyadhar it can be assigned to the Rhuddanian stage. It may therefore be a time equivalent of the Rhuddanian hot shales described in the next section. Whether the hot shales pass laterally into the Bi'r Tlakshin Formation is debatable, but both appear to represent different aspects of the first stage of the Silurian transgression onto the eroded Ordovician surface. Others, however, have suggested that it represents an upper Mamuniyat clastic unit, implying an Hirnantian age. No age-diagnostic fossils have yet been recorded from the formation. The Bi'r Tlakshin Formation is important from a petroleum point of view since it forms a barrier between the Mamuniyat reservoir and the Tanzuft source rock.[97]

3.4.2 Tanzuft Formation

The Tanzuft Formation is an important stratigraphic unit from the standpoint of petroleum exploration. It has been proven to be one of the major source rocks of north Africa and its distribution and extent are therefore of prime importance. The Tanzuft Formation was first defined by Desio[98] from a location 65km north of Ghat on the western margin of the Murzuq Basin. The base of the sequence is not exposed at this location and Desio did not define the top of the formation. In 1965 Klitzsch established a parastratotype section 35km south of Ghat where a complete sequence is present (Figure 3.14).[99] At this location the Tanzuft Formation comprises 370m of dark-grey, marine graptolitic shales with thin siltstone and sandstone stringers in the lower part (the Iyadar Formation of Massa and Jaeger). The contact with the underlying Mamuniyat Formation is unconformable, and is marked by a thin bed of ferruginous sandstone, but the transition into the overlying Akakus Formation is gradual.

On the western margin of the Murzuk Basin it has been mapped from Tikiumit to South Anay on the border with Algeria and Niger, varying in thickness from 475m near Ghat to 350m near Al Awaynat (formerly Serdeles) to only 80m at Tikiumit. In a sedimentological study based on both outcrop and core studies de Castro interpreted the lower Tanzuft Formation as representing a predominantly turbiditic environment, whilst the upper part is characterised by thick siltstone members showing hummocky cross-stratification separated by hard grounds which he believed represent tempestite deposits. The lower part of the Tanzuft Formation contains anomalously high levels of radioactivity (hot-shales), reflecting the presence of uranium, which probably originated as wind-transported fine volcanic ash. The Tanzuft Formation outcrops on the eastern flank of the Murzuq Basin where it has been dated as early Llandoverian. In the Mourizidie area it directly overlies basement rocks.[100]

According to Bellini and Massa the graptolite zones within the Tanzuft Formation and the overlying Akakus Formation demonstrate that these formations are diachronous. They produced evidence to show

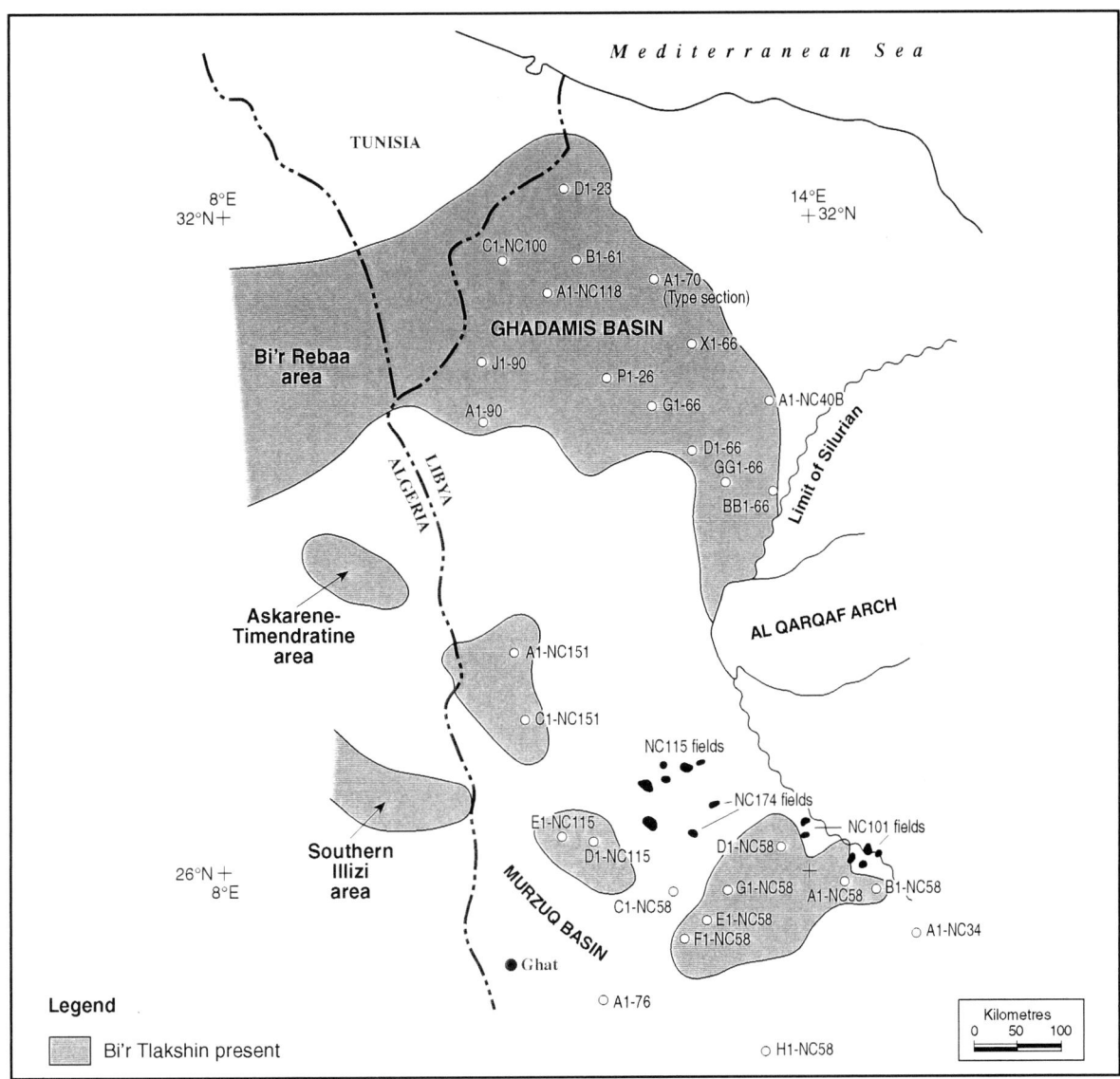

Figure 3.13   Bi'r Tlakshin Formation, Distribution

The informal Bi'r Tlakshin Formation of probable Rhuddanian age, has a patchy development in western Libya, southern Tunisia and eastern Algeria. It unconformably overlies the Mamuniyat Formation, and unconformably underlies the Tanzuft Formation. The Bi'r Tlakshin Formation forms a barrier between the Mamuniyat reservoir and the Tanzuft source rock. All of the Mamuniyat discoveries in the Murzuq Basin have been in areas where the Bi'r Tlakshin is absent.

that in the Murzuq Basin the Tanzuft Formation was deposited during the early and mid-Llandoverian, in the southern Ghadamis Basin it is mid-Llandoverian to Wenlockian in age, and in the northern Ghadamis Basin it is largely Wenlockian and Ludlovian in age.[101] This interpretation has subsequently been questioned. Abugares and Ramaekers suggested, on the basis of well-log correlations, that multiple unconformities are present within the Tanzuft Formation, and that recent palynological data does not support the simple south to north diachroneity suggested by Bellini and Massa. Lüning et al. were able to show that after the glacial retreat the pre-Silurian surface was highly irregular. According to their interpretation the Silurian transgression first flooded the topographic lows, and black shales were deposited within these closed basins (see Figure 7.3). As sea-level rose the transgression spread over the remaining topographic highs, depositing normal outer-shelf shales. There is evidence that the Rhuddanian is characterised by a low-diversity graptolite fauna which perhaps indicates an oxygen-poor environment. Evidence from wells in concession NC 174 and elsewhere shows that the black shales are exclusively Rhuddanian in age. The diachronous nature of the formation can be attributed to a general marine regression which was probably less rapid than was previously believed.[102]

In the Ghat area the association of black organic shales, metallic sulphates and the present-day arid climate has led to the formation of alum within the Tanzuft Formation. Alum is used as a mordant in dyeing and was highly prized in the pre-industrial age, being exported from Ghat throughout northern Africa and Arabia. The name Tanzuft is the Tamacheq word for the yellow alum found near Al Awaynat.[103]

In the wells of the Murzuq Basin the Tanzuft Formation ranges in thickness from 45m to 320m and its age has been determined as Llandoverian. It is present in wells H1-NC 58 and F1-NC 58 where it shows evidence of turbidite flows. It is missing in the NC 101 wells, on the Brak-Bin Ghanimah Uplift, at Sabha and in the A1-73, A1-64, A1-77 and B1-77 wells, due to mid-Devonian erosion. The Tanzuft Formation becomes more sandy towards the Brak-Bin Ghanimah Uplift, and towards the south of the basin where the sand content reaches 20%. However well C1-NC 101, close to the eastern limit of the Tanzuft Formation, shows a graptolitic black-shale facies which supports the concept of an erosional eastern limit rather than a depositional pinch-out. Wennekers et al., and Abugares and Ramaekers indicated that the Wenlockian is missing in the subsurface of the Murzuq Basin, presumably based on unpublished micro-palaeontological data. The Rhuddanian black shales are present in only about half the wells in the Murzuk Basin. Lüning, et al. published an isopach map of the hot shales in the area of concessions NC 115 and NC 174, on the northern edge of the basin, which showed the hot shales to be patchily developed in palaeotopographic lows on the pre-Silurian surface. A distinct linear trend from well C1-NC 115 to E1-NC 174 is interpreted as a palaeo-valley. On a local scale hot shales are absent in well A1-NC 115, but seismic evidence shows the structure to be surrounded by hot shales (see Figure 7.5). Similar seismic techniques may be able to clarify the distribution of Tanzuft hot shales in other areas.[104]

The Tanzuft Formation is present in the Dur al Qussah Trough where 200 to 300m of silty sandstones and shales are present. Graptolites of Llandoverian age are present at the base, but the higher sequences become progressively more brackish and graptolites are replaced by ostracods in the upper part of the section. The Akakus Formation cannot be separated from the Tanzuft Formation in this area.[105] The Tanzuft Formation wedges out south of Dur al Qussah which suggests that it is absent from the Brak-Bin-Ghanimah Uplift due to mid-Devonian uplift and erosion. South of the uplift, in the Mourizidie area, marine shales with graptolites reappear with a thickness of about 15m.[106]

The Tanzuft Formation outcrops around the periphery of the Kufrah Basin showing an unconformable contact with underlying

Cambro-Ordovician sandstones (Figure 3.14). The formation comprises varicoloured micaceous shales with minor siltstone intervals which become more silty towards the top. The thickness ranges from about 80m in the southeast to over 100m in the southwest and reaches 165m in the northeastern sector. On the eastern flank of the basin planar bedding with ripple cross-lamination are the dominant sedimentological features, suggesting a nearshore shelf environment. The Tanzuft Formation is present in both the A1-NC 43 and B1-NC 43 wells drilled by Agip in 1978 and 1980. In A1-NC 43 the Tanzuft Formation comprises grey-green shales with minor sandstones, with a thickness of 54m. In the B1-NC 43 well the Tanzuft Formation has not been separated from the overlying Acacus Formation. Shales occupy the lower 17m but above that the section is predominantly sandstone. In the Kufrah Basin it appears that the early Silurian is represented by predominantly deltaic sedimentation. In the Agip wells the contacts with the underlying Mamuniyat Formation and with the overlying basal Devonian are unconformable. The lower part of the Tanzuft Formation in both wells, and in the A1-71 and KW2 wells, has been dated as late Ordovician based on chitinozoa, and a sample near the top in the KW2 well gave an early Llandoverian age based on graptolites. This implies that in this area deposition of the Tanzuft shales began in the late Ordovician. At Jabal Zalmah the uppermost shales of the Tanzuft Formation have been dated as Rhuddanian-Aeronian in age.[107]

Returning to the Al Qarqaf Arch, the Tanzuft Formation has been mapped in the Qararat al Marar area, but only about 30m is preserved due to mid-Devonian erosion. The age of the Tanzuft Formation in this area is mid-Llandoverian, based on graptolites.[108]

In the Ghadamis Basin the Tanzuft Formation has been penetrated in many oil wells where it is characterised by dark-grey to black graptolitic shales with either one or two highly radioactive zones near the base of the formation. Significant thicknesses of hot shales have been reported from wells EE1-NC 7 and GG1-NC 7. The thickness of the Tanzuft Formation in the Ghadamis Basin reaches about 450m of which commonly 20 to 25m show high levels of radioactivity. The Tanzuft Formation in this area was first described from well data by Massa and Jaeger under the informal name 'Formation des argiles principale'. A comparison with the Silurian rocks in Tunisia was given by Jaeger, et al. As mentioned above, the Iyadhar 'Formation' has been identified in the Ghadamis Basin where it has been dated as Rhuddanian, and rocks of this age have been identified in well F1-66. The overlying Tanzuft Formation extends from the upper zone of the early Llandovery to mid-Ludlovian based on rich graptolite assemblages.[109]

Further east the Tanzuft Formation is missing on the northeastern extension of the Al Qarqaf Arch, but is present on both flanks of the ridge. In the Al Fuqaha Trough to the east 600m of Tanzuft mudstones are present in a water well drilled at the Al Fuqaha oasis, where the age is reported as being late Llandoverian and early Wenlockian. The formation is also present in the C1-44 well further north. To the west, in the Zamzam Depression, Tanzuft Formation shales of the same age, reported by Belhaj as containing some highly radioactive zones, are present in wells A1-43 and C1-25. These hot shales are evidently younger than the Rhuddanian hot shales found elsewhere. The detailed palynological studies carried out in this area have permitted the pinchout limits of the successive Silurian stages to be mapped in detail. In the Agip concession NC 40 the highly radioactive shales were interpreted as having been deposited in a marsh environment, whilst the less radioactive zones are interpreted as tidal flats. In the Sirt Basin most of the Silurian was removed during the Hercynian orogeny.[110]

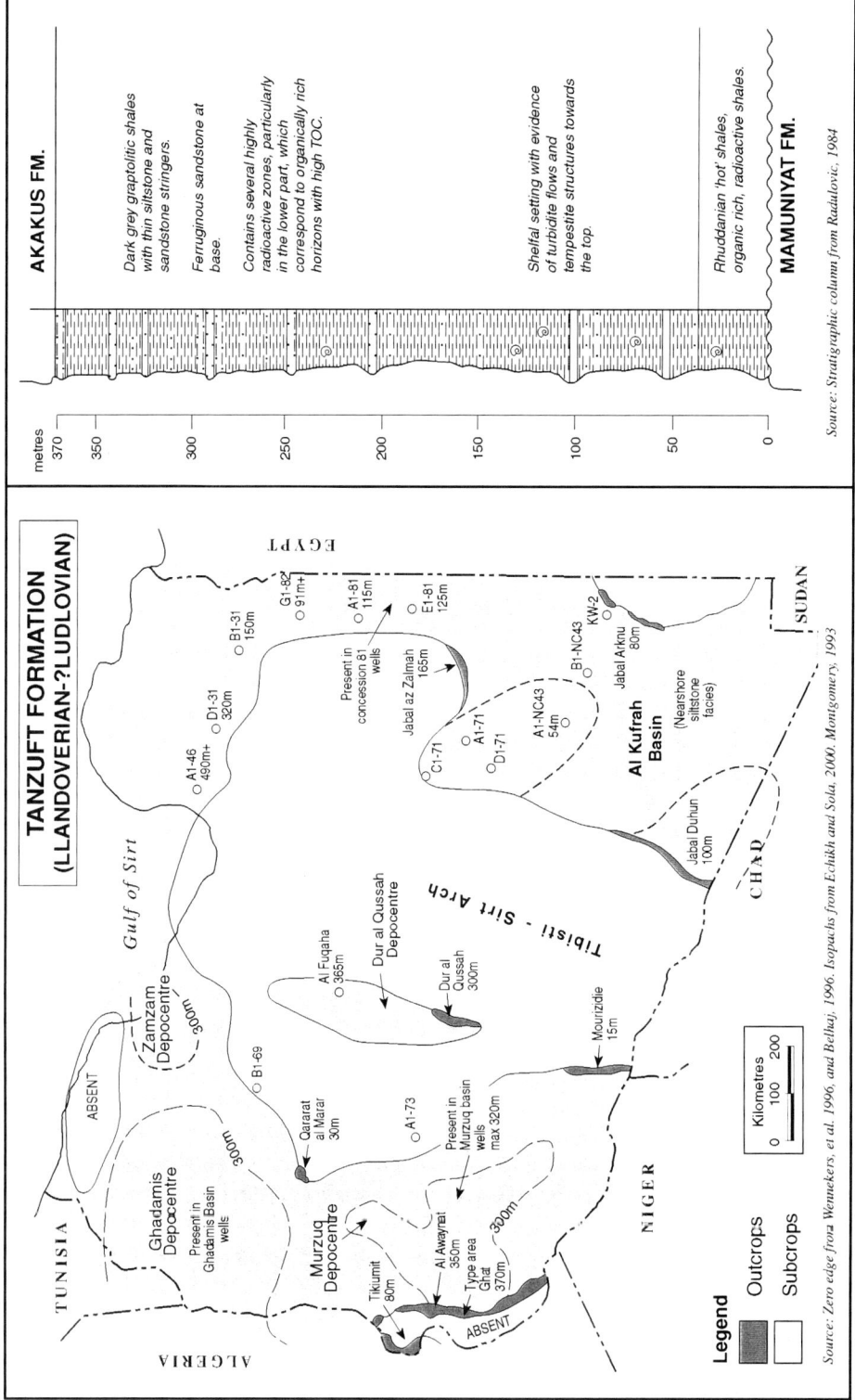

Figure 3.14   Tanzuft Formation, Distribution and Stratigraphic Column

The Tanzuft Formation represents a major flooding event over much of North Africa. The type area is defined south of Ghat where the thickness reaches almost 400m. The formation comprises dark grey graptolitic shales with thin siltstone and sandstone partings, deposited in a shelfal environment. The formation shows evidence of turbidite flows and storm disturbances. It contains one, or rarely two, highly radio-active horizons which correspond to organically rich intervals with high TOC values which represent the principal Palaeozoic oil source rocks in North Africa.

In Cyrenaica the Tanzuft Formation is present on the the Cyrenaican Platform where it is represented by black graptolitic shales passing up into siltstones and sandstones (Figure 3.14). It has been penetrated in wells A1-46, D1-31, B1-31, G1-82, A1-81 and E1-81 where age determinations based on acritarchs and chitinozoa give a Llandoverian age. The formation was probably deposited in a relatively shallow, marginal-marine environment. The thickness decreases from 490m in well A1-46 to 120m in E1-81.[111]

The Tanzuft Formation is remarkable in providing evidence of some of the earliest known land plants. The cuticle-like sheets of cells found by Gray in the Melaz Shuqran Formation have already been mentioned. In the lower part of the Tanzuft Formation conclusive proof of the presence of land plants has been found in the form of trilete spores belonging to the genus *Ambitisporites* in the A1-1 well, southwest of the Al Qarqaf Arch, which could be as old as Llandoverian, and studies on two wells in the northern Ghadamis Basin, C1-34 and B2-34, have yielded miospores from over fifty species of land plants from rocks dated as uppermost Wenlockian to Ludlovian. Actual plant remains of lycophytes and psilophytes are preserved in the overlying Akakus Formation.[112]

In Algeria the equivalent unit to the Tanzuft Formation is the 'Gothlandien argilleux', which forms the principal oil source rock in the Illizi and Triassic Basins.

The key features of the Tanzuft Formation are that it represents a major flooding event following the glacial episode of the late Ordovician. It has a wide distribution throughout North Africa, and was deposited on a very irregular surface. The oldest Tanzuft sediments yet recorded are in the Al Kufrah basin where they have been dated as Ashgillian. In other areas, depending on the pre-Silurian topography, deposition began much later. The Tanzuft Formation forms the principal

Palaeozoic petroleum source rock in north Africa. The organic rich shales, which contain TOC values up to 16.7% are largely restricted to the Rhuddanian and occur in isolated topographic lows. The black shale facies has not yet been found in the Al Kufrah Basin. The source rock aspects of the Tanzuft 'hot shales' are discussed further in chapter 7. The Tanzuft Formation is also remarkable in containing some of the earliest evidence of the emergence of land plants.[113]

3.4.3 Akakus Formation

The type section of the Akakus Formation has a similar history to that of the Tanzuft Formation. The original description by Desio from a location in the Jabal Akakus, north of Ghat was incomplete, and a replacement type section was established by Klitzsch from an area to the south of Ghat where a complete section is present (Figure 3.15). Subsequently Klitzsch, having studied the eastern margin of the Murzuq Basin, indicated that the type section may represent only the lower part of the formation.[114] At the type locality the Akakus Formation comprises 240m of fine to medium-grained silty sandstone, thinly bedded, frequently cross-bedded, with ripple and flute marks. The top is marked by a conspicuous horizon of ferruginous sandstone with convolute bedding and stromatolitic structures which suggest a period of emergence. It contains the trace fossil *Arthrophycus alleghaniensis*, up to 40cm in length, plus *Corophiodes*, *Cruziana*, and *Tigillites dufrenoyi*. The Akakus Formation provides an excellent example of ichnostratigraphy, and is characterised by the presence of *Cruziana acacensis* which Seilacher identified as trilobite burrows. Ichnostratigraphy is sufficiently precise to be capable of demonstrating the diachronous nature if the Akakus Formation from south to north in western Libya. The environment of deposition is interpreted as

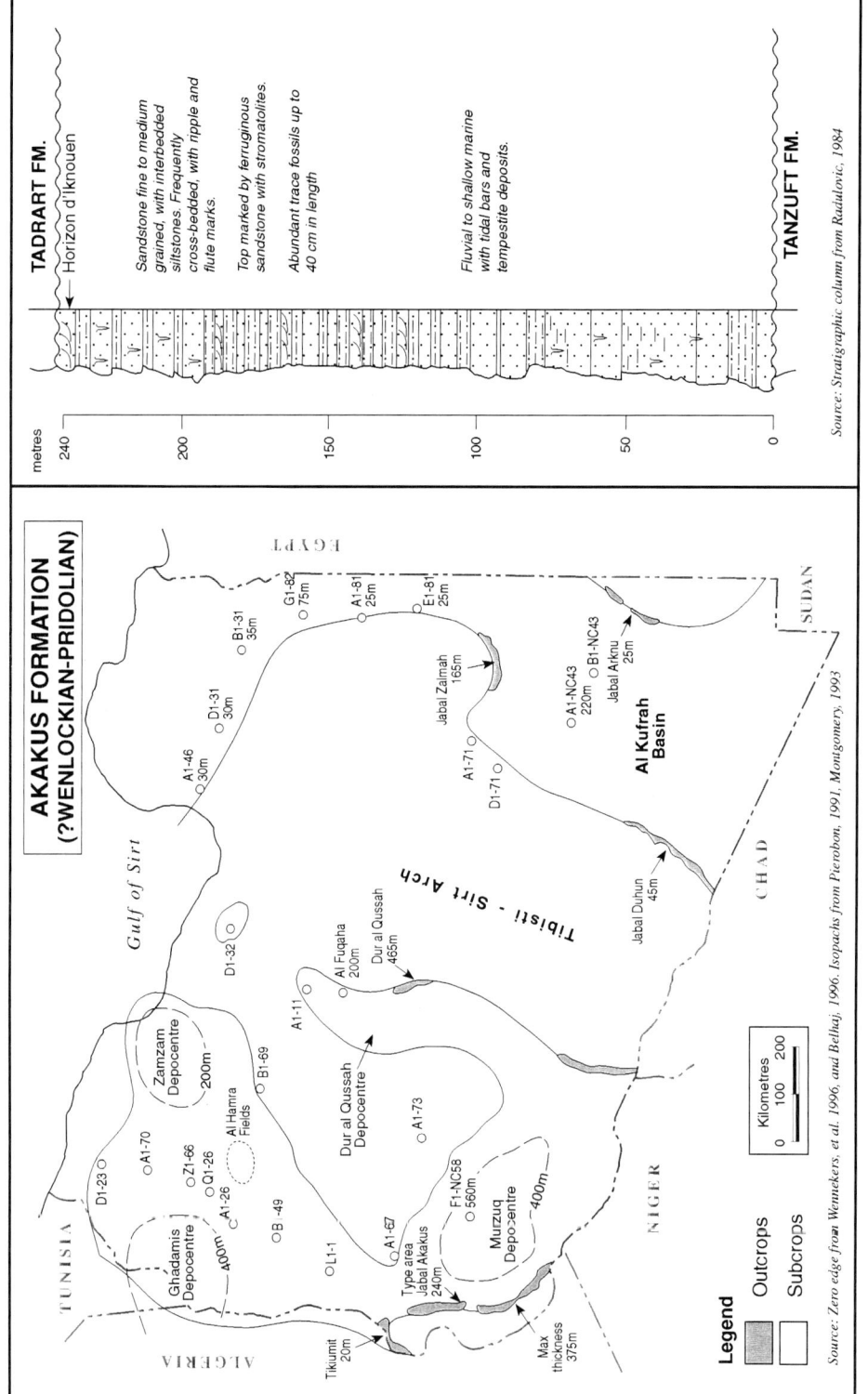

Figure 3.15   Akakus Formation, Distribution and Stratigraphic Column

The Akakus Formation represents a shallowing of the marine conditions which prevailed during Tanzuft deposition. Sedimentary structures indicate a shallow marine to fluvial setting. The top is marked by a distinctive ferruginous sandstone called the 'horizon d'Iknouen' which marks a period of emergence. The Akakus Formation is a significant petroleum reservoir in the Ghadamis Basin.

ranging from fluvial to shallow marine with tidal bars and tempestite deposits. The lower contact is apparently conformable with the underlying Tanzuft Formation, but the upper contact with the Tadrart Formation is distinctly unconformable. Bellini and Massa showed that the Akakus Formation is strongly diachronous from south to north. In the Ghat area they suggested a mid-Llandoverian to Wenlockian age although no graptolites have been found in the Ghat area. Subsequent work however suggests that in this area the Wenlockian is missing, and Akakus sandstones of late Llandoverian age are unconformably overlain by beds of Ludlovian and possibly Pridolian age. This observation can be extended to suggest that the Akakus Formation prograded northwestwards from the Murzuq Basin to the Ghadamis Basin during Ludlovian and Pridolian times.[115]

The Akakus Formation has been mapped along the western margin of the Murzuq Basin from Tikiumit in the north to South Anay on the Niger border. In this area Pfluger traced a transition from well-oxygenated proximal Akakus sandstone down the depositional palaeoslope into basinal anoxic Tanzuft shales. This clearly implies that the two formations are at least partially time-equivalent. He also recognised the presence of matgrounds formed by microbial action which are found only in areas where grazing and burrowing organisms are absent. The formation wedges out rapidly along the outcrop north of Tikiumit, due to early Devonian uplift. Southwards it thickens to 180m on the Wadi Tanzuft map sheet, and reaches a maximum of 375m south of Ghat, before thinning again to 90m on the border. The lithology remains similar throughout this area with the ferruginous stromatolitic top being a conspicuous feature. This horizon was called the 'horizon d'Iknouen' by Freulon, and represents a period of subaerial exposure.[116]

The Akakus Formation is present in the subsurface of the Murzuq Basin where it reaches a thickness of 560m in the F1-NC 58 well with depositional environments ranging from fan deltas with local tubidite flows, to offshore bars and fluvial plain deposits. Palynological evidence from A1-NC 58 and E1-NC 58 gives a Wenlockian to Ludlovian age for the Akakus Formation. The formation is also present in wells A1-76 and H1-NC 58, but is absent further east and on the Brak-Bin Ghanimah Uplift. Echikh and Sola published a map showing the extent of the Akakus Formation in the Murzuq Basin.[117]

The formation outcrops on the eastern margin of the basin where its thickness varies greatly along the outcrop edge. In the Dur al Qussah Trough it reaches a thickness of 465m, but thins progressively to the south onto the Tripoli-Tibisti Uplift. South of the Mourizidie Horst it ranges in thickness from 35m to 190m. The lithology in this area is principally thin-bedded sandstones, silty and shaly in the lower part, but more massive, compact, cross-bedded and ferruginous towards the top. In the Dur al Qussah Trough the upper part of the sequence contains psilophyte and lycophyte plant remains in what are probably deltaic or fluvial deposits. No precise age is determinable, but in general terms they certainly fall within the mid to late Silurian range. Like the miospores recorded in the Tanzuft Formation these remains represent some of the earliest vascular plants yet recorded. They were first studied by Daber in 1971, and have been the subject of much palaeobotanical interest. This significance of this flora was reviewed by Lejal-Nicol. The Al Fuqaha water well penetrated about 200m of sandstones and mudstones which have been assigned to the Akakus Formation, and the formation is present in the C1-44 well further north. Age determinations in near-by wells suggest a late Wenlockian to early Ludlovian age.[118] The Al Fuqaha occurrence may represent a northward extension of the Dur al Qussah Trough, but this remains to be proved (Figure 3.15). The Akakus Formation is absent in the subsurface on the Brak-Bin Ghanimah Uplift between Sabah and the A1-73 well, and is also absent on the subsurface WSW-ENE ridge which extends through the A1-67 well, from which it has been removed by mid-Devonian erosion.[119]

In the Kufrah Basin the Akakus Formation is recognised on the southeastern margin where is does not exceed 25m in thickness. The lithology comprises grey siltstones, with thin stringers of fine-grained sandstones and micaceous shale with abundant trace fossils. The contact with the overlying Tadrart Formation is very sharp. The outcrops on the northeastern and southwestern margins are similar, with maximum thicknesses of 165m and 45m. The Akakus Formation in this area is dominated by abundant ripple cross-laminations which indicate a subtidal setting. In the A1-NC 43 well the Akakus Formation comprises 220m of fine to coarse-grained quartz sandstones, with interbeds of micaceous, silty shales. In B1-NC 43, where it has not been separated from the underlying Tanzuft Formation, it comprises sandstones with rare shaly intervals. No age diagnostic fossils have been recognised in the Akakus Formation in the Kufrah Basin.[120]

In the subsurface of the Ghadamis Basin the Akakus Formation is represented by a series of sandstones and siltstones which were called the Formation des alternances argilo-greseuses by Massa and Jaeger. The sequence can be subdivided into two predominantly sandstone units separated by a shale. The upper Akakus has a ferruginous top and contains psilophyte plant remains and is interpreted as a continental deposit. Fossil soils have been reported from some wells. Towards the southwest the Akakus Formation is progressively truncated by pre-Devonian uplift and erosion. A full sequence is preserved in the O1-26 well, but only the middle shale and lower Akakus sandstone are preserved in the A1-90 well. The lithofacies of the Akakus Formation in the Ghadamis Basin determined from subsurface data confirms the gradual reduction in sand content and this has been interpreted as a probable indication of a ramp setting. Petrographic analyses of Lower Akakus sandstones showed a transition from fluvial channel sandstones in the south, through coastal deltaic sandstones and siltstones, to offshore marine siltstones in the north. These zones exhibit differing diagenetic histories depending on the nature of the cement. The best secondary porosity is found in the coastal/deltaic sandstones which had an original calcareous cement. In southern Tunisia the uppermost Akakus has been dated as Pridolian based on chitinozoa. In the Bani Walid water well the Akakus Formation has been dated as Ludlovian, Pridolian and possibly earliest Lochkovian in age, and in wells in the Hamadah al Hamra the Lower Akakus has been dated by palynology as Ludlovian. The Akakus Formation is oil and gas productive from numerous small fields in the Libyan sector of the Ghadamis Basin.[121]

In the Zamzam Depression the Akakus Formation has been encountered in many wells. The three-fold division becomes less clear and the lithology ranges from muddy siltstones with minor sandstones at the base, to fine-grained sandstones with minor mudstones in the upper part. The entire section is marine, ranging from deltaic to tidal flat, and to subaerial in the iron-rich layers. The thickness is variable, but reaches around 400m in the A1-8, C1-60 area. Abundant acritarchs and common miospores permit the Akakus Formation in this area to be dated as late Wenlockian to Pridolian.[122]

Evidence from the subsurface in Cyrenaica shows a general transition from the Tanzuft Formation into a more silty and sandy facies which reaches a thickness of about 40m. This section is barren, but the sequence has litho-logical similarities to the Akakus Formation of west Libya, and is believed to represent a deltaic

PERIOD	EPOCH/STAGE	EUROPEAN EQUIVALENTS	WEST LIBYA	IDRI-SABHA AREA	ALGERIA ILLIZI BASIN	JABAL DUHUN	JABAL HAWA'ISH
PERMIAN	LATE	BUNTER	BI'R AL JAJA			'CONTINENTAL POST-TASSILIEN'	
	MID	ZECHSTEIN	BI'R AL JAJA				
	EARLY	ROTLIEGENDES	AL WATYAH (CONTINENTAL EQUIVALENTS)				
CARBONIFEROUS	GZELIAN	STEPHANIAN	TIGUENTOURINE				
	KASIMOVIAN						
	MOSCOVIAN	WESTPHALIAN	DIMBABAH		'A' RESERVOIR		
	BASHKIRIAN	NAMURIAN	ASSEDJEFAR		'B' RESERVOIRS		
	SERPUKHOVIAN						
	VISEAN		COLLENIA BEDS / MARAR		'D' RESERVOIRS	OUNGA	ZALMAH
	TOURNAISIAN						
DEVONIAN	FAMENNIAN	STRUNIAN	TAHARA AO IV	ASHKIDAH / TARUT / DABDAB / QUTTAH	'F2' RESERVOIR / CUES HORIZON		
	FRASNIAN		AO III				
	GIVETIAN		AO II	IDRI	'F3' RESERVOIR / 'F4' RESERVOIR	BINEM	BLITA
	EIFELIAN	COUVINIAN	AO I / EMGHAYET	BI'R AL QASR			
	EMSIAN		WAN KASA		'F5' RESERVOIR		
	PRAGIAN	SIEGENIAN	TADRART		'F6' RESERVOIR	TADRART	TADRART
	LOCHKOVIAN	GEDINNIAN					

(West Libya: AWAYNAT WANIN)

Source: Numerous published charts

Figure 3.16   Upper Palaeozoic Lithostratigraphic Nomenclature

This chart shows the current lithostratigraphic nomenclature for the Upper Palaeozoic of west Libya and equivalents in the Idri-Sabha area, and in the Illizi Basin of Algeria. The oblique lines represent differing views on the age of the formations, as discussed in the text. No attempt has been made to show unconformities or gaps in the succession on this chart.

or nearshore regressive deposit. It is present in wells A1-46, D1-31, B1-31, G1-82, A1-81 and E1-81. The age is assumed to be Wenlockian-Ludlovian (Figure 3.15).[123]

In Algeria the equivalent of the Akakus Formation is the sequence usually called the Gothlandien argilo-greseux which contains the lower part of the F6 reservoir which is oil productive from several fields in the Illizi Basin, including Mereksen, Stah and Ohanet. Alem et al. published a detailed review of the F6 reservoir in the Illizi Basin.[124]

The Akakus Formation represents a prodelta and delta-top highstand system which developed from southeast to northwest from the late Wenlockian to the Pridolian. The delta top and neritic shelfal environments provide fair to moderate reservoirs in the Ghadamis Basin of Libya and the Illizi Basin of Algeria.

3.5 Devonian

The stratigraphic nomenclature for the Upper Palaeozoic is shown on Figure 3.16. It has long been believed that the Silurian-Devonian boundary in Libya is almost everywhere unconformable, as reflected by the 'horizon d'Iknouen', and that the Lochkovian stage is absent in all areas. This hiatus was believed to represent a major episode of uplift, folding and erosion reflecting major tectonic events on the northern margin of west Gondwana associated with the early stages of closing of the Iapetus Ocean. These events had a major impact on the configuration of the pre-Devonian topography, large areas were left emergent, and the sedimentary basins shrank in size. Most of these features have a north-northwest orientation. However Weyant and Massa presented evidence to show that in the central Ghadamis Basin at least the Lochkovian is present, and deposition across the Silurian-Devonian boundary was continuous. A synthesis of the Devonian in the Murzuq and Jadu Basins has been published by

Mergl and Massa, and a review of the sequence stratigraphy of the Devonian on the Al Qarqaf Arch and in the Ghadamis Basin by Rubino and Blanpied.[125]

The passive continental margin of the early Palaeozoic was replaced during the Devonian by a sheared margin, reflecting the first episode of the Hercynian orogeny. The clastic sequences of the Devonian were deposited on a low-angle continental margin, on which deposition was controlled by eustatic and local tectonic events. The effects of local tectonism can be seen to affect Devonian sedimentation in several areas, particularly on the Brak-Bin Ghanimah Uplift and on the Tihemboka Arch. The repeated transgressions and regressions migrated rapidly over large areas of the Gondwanan marginal shelf. The succession can be divided into three megacycles, each containing several cycles of continental, deltaic and shallow marine sequences.

3.5.1 Tadrart Formation

The Tadrart Formation was first described by Burollet in 1960. A type section was designated by Klitzsch from the Tadrart Mountains, 35km south of Ghat (Figure 3.17). At the type locality the Tadrart Formation forms a conspicuous scarp which weathers into large isolated pillars and blocks. The 25m high pillar in Wadi Attullah is one of the most spectacular. It comprises massive continental to marginally-marine sandstones, ferruginous, cross-bedded, with plant remains in the lower part and trace fossils in the upper part. The environment of deposition ranges from subtidal to intertidal and continental. The thickness is 140m, and the contact with the underlying Akakus Formation is distinctly unconformable.[126]

The Tadrart Formation has been mapped along the western flank of the Murzuq Basin from Tikiumit to the border with Niger. In this area the Tadrart Formation is made up of

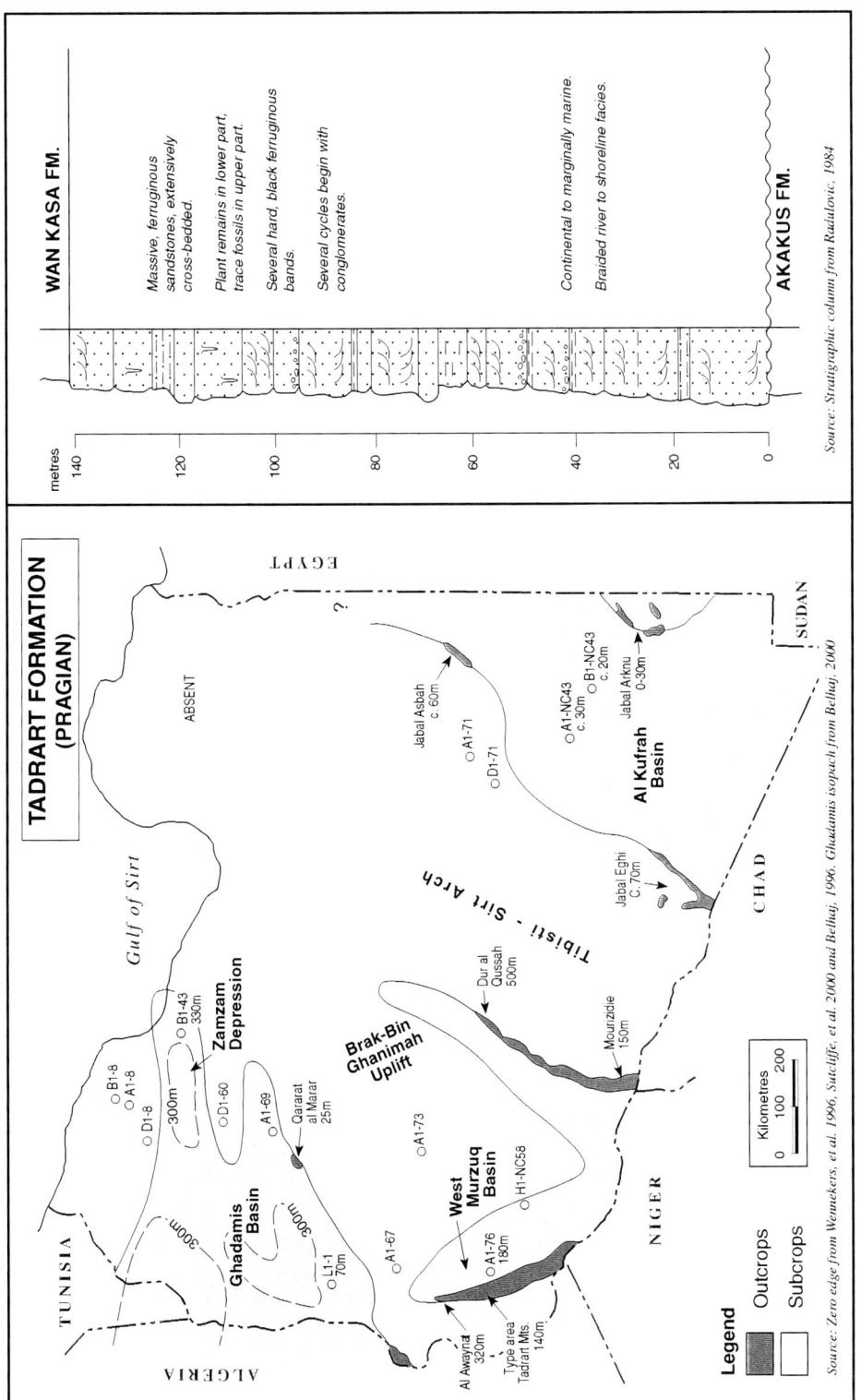

Figure 3.17   Tadrart Formation, Distribution and Stratigraphic Column

The Tadrart Formation unconformably overlies the Akakus Formation except in the deepest part of the Ghadamis Basin where the sequence is continuous. The type area is defined in the Jabal Tadrart southeast of Ghat. It represents a continental to marginally marine unit which is present in the Ghadamis Basin, but is missing over much of the Murzuq Basin due to erosion. The Tadrart Formation contains a significant oil fairway in the Al Kufrah Basin. Clastic sequences equivalent to the Tadrart Formation are present in the Al Kufrah Basin.

lenticular sandstone bodies representing filled channels within a braided-river sequence. This facies association led Clark-Lowes and Ward to suggest strong similarities between the Tadrart and Nubian Formations. Grain size ranges from fine-grained to conglomeratic. The thickness reaches 320m at Al Awaynat. Sparse fossil evidence and characteristic trace fossils suggests an early Devonian age. On the basis of palynology Belhaj indicated a range from Gedinnian to late Emsian. Clark-Lowes examined the Tadrart Formation in this area and concluded that the lower part represents a braided-river environment whilst the upper part represents shallow marine conditions with shoreline facies cut by tidal channels.[127] The Tadrart Formation exhibits festoon cross-bedding representing fluvial sandstones, and herring-bone cross-bedding usually associated with tidal currents. The tidal current influence increases upwards. These sandstones represent excellent petroleum reservoir rocks. No age diagnostic fossils have been found in the Murzuq Basin wells, but the age is assumed to be early Devonian. Sutcliffe, et al. recognised four facies associations within the Tadrart Formation, representing braided-fluvial channels, braided-deltaic channels and tidally dominated shoreface environments.[128]

The distribution of the Tadrart Formation within the Murzuq Basin has been greatly affected by Devonian tectonism. The work of Pierobon and Meister has demonstrated that the Tadrart Formation, although thick in the Dur al Qussah Trough and in the Al Awaynat Trough, is absent over much of the central part of the basin. This was confirmed by Sutcliffe, et al. who presented an isopach map showing the formation to be absent over a wide area. There is clear evidence of mid-Devonian inversion, and the Tadrart Formation was removed during the subsequent erosion. Echikh and Sola also published an isopach map of the Lower Devonian and a subcrop map on the base of the mid-Devonian. These maps confirm the presence of Lower Devonian rocks in wells A1-76 (483m) and H1-NC 58, but they are absent further east. The eastern limit can be clearly defined by well evidence. The subcrop map shows the formations underlying the mid-Devonian unconformity (Figure 3.19). On the eastern flank of the Murzuq Basin the Tadrart Formation has a thickness of 500m in the Dur al Qussah Trough but pinches out onto the Haruj Uplift. Its thickness on the southern part of the Brak-Bin Ghanimah Uplift is about 150m, but further north it is absent due to mid-Devonian erosion. In the Dur al Qussah Trough it is composed of massive ferruginous sandstones with plant fragments resembling *Lepidodendron*. Poorly preserved fossil fish have been found within interbedded siltstones. South of the Tripoli-Tibisti Uplift the sequence becomes thinner and condensed, but friable sandstones probably equivalent to the Tadrart Formation are present in the Uri Trough near the border with Chad.[129]

Tadrart Formation has been mapped on the north side of the Al Qarqaf Arch at Qararat al Marar where 25m of friable, cross-bedded, coarse-grained ferruginous sandstone containing wood fragments is present. On the south side of the arch the Tadrart Formation has not been recognised in the Idri/Sabha area.[130]

In the Al Kufrah Basin the Tadrart Formation cannot be differentiated from the overlying Wan Kasa Formation, although Pragian microfloras have been identified at Jabal Zalmah. At outcrop the combined unit comprises continental sandstones, very thick-bedded, with cross-bedding, foresets, and slump features, which suggest a braided-river environment. The combined unit reaches about 60m in thickness. A similar sequence is present in the A1 and B1-NC 43 wells. The sequence contains a rich microflora which indicates a Pragian age, confirming the absence of the Lochkovian in this area.[131]

In the Ghadamis Basin the Tadrart Formation has been penetrated in many wells where it is represented by widespread sheet sands cut by numerous sand-filled channels. An isopach map was published by Belhaj showing three separate depocentres, in the Zamzam Depression, southern Ghadamis and northern Ghadamis areas, and a structure map of Tadrart Formation in the southern Ghadamis Basin was presented by Hammuda. The thickness averages 130m. Porosity is good, particularly on the southern flank of the basin, and the formation is a major oil reservoir in the area from Al Hamra into the Illizi Basin of Algeria. The presence of numerous plant remains and miospores has allowed the Tadrart Formation in this area to be dated as Pragian to early Emsian, but palynological analysis of well D1-8 in the centre of the Ghadamis Basin demonstrated that Tadrart deposition began in the Lochkovian.[132]

In the Zamzam Depression thick sandstones attributable to the Tadrart Formation have been encountered in several wells. The environment of deposition ranges from continental to inner neritic, tidal-flat, and the thickness reaches 330m in well B1-43. On the basis of numerous acritarchs and miospores the age of this unit has been determined as Pragian. The formation is absent on the northern flank of the Al Qargaf Arch, the Jabal Nafusah Uplift, and in the Al Fuqaha Trough.[133]

The Tadrart Formation has not been reported from the Sirt Basin, nor from Cyrenaica, where the oldest Devonian sediments appear to be Emsian in age. [134]

In Algeria the upper F6 reservoir of the Devonien inferieur greseux' is equivalent to the Tadrart Formation. This unit is oil productive from many fields including Tin Fouyé, Ohanet, Guelta, Askarene and Hassi Mazoula, plus several recently discovered fields such as Bir Rebaa and Rhourde Messaoud. A review of this area has been given by Alem et al.[135]

The Tadrart Formation essentially represents a transgressive, fluvial to shallow-marine formation of wide areal extent deposited on the uplifted, and eroded pre-Devonian surface. In the centre of the Ghadamis Basin and in Cyrenaica deposition may have been continuous with the underlying Akakus Formation, but in all other areas a major hiatus separates the two formations.

3.5.2 Wan Kasa Formation

The second megacycle within the Devonian succession comprises a marine transgressive sequence comformably overlying the much more massive sandstones of the Tadrart Formation. This sequence was first described by Borghi and Chiesa in 1940 and was named the Wan Kasa Formation. It was defined from the Wadi Wan Kasa in the Jabal Tadrart. A formal type section was established by Klitzsch in 1965 in the same area, and amplified in 1969 (Figure 3.18). The Wan Kasa Formation is composed of a series of alternating grey to reddish ferruginous siltstones and shales with thin layers of gypsum, containing brachiopods, tentaculitids and trace fossils. The siltstones and thin sandstones coarsen upwards and show wavy and flaser-bedding, grading upwards into more massive cross-bedded sandstones at the top. The formation accumulated in a low-energy setting ranging from brackish lagoon to shallow sublittoral and offshore bar environments. The formation is 55m thick at the type locality. It is unconformably overlain by sandstones of the Awaynat Wanin Formation. Borghi and Chiesa and Klitzsch suggested a mid-Devonian age for the Wan Kasa Formation, and this is supported by a rich fauna of brachiopods found during the recent mapping in this area which indicates an Eifelian age. Evidence from other areas however suggests an Emsian age.[136]

The formation has been mapped along the western margin of the Murzuq Basin from

120

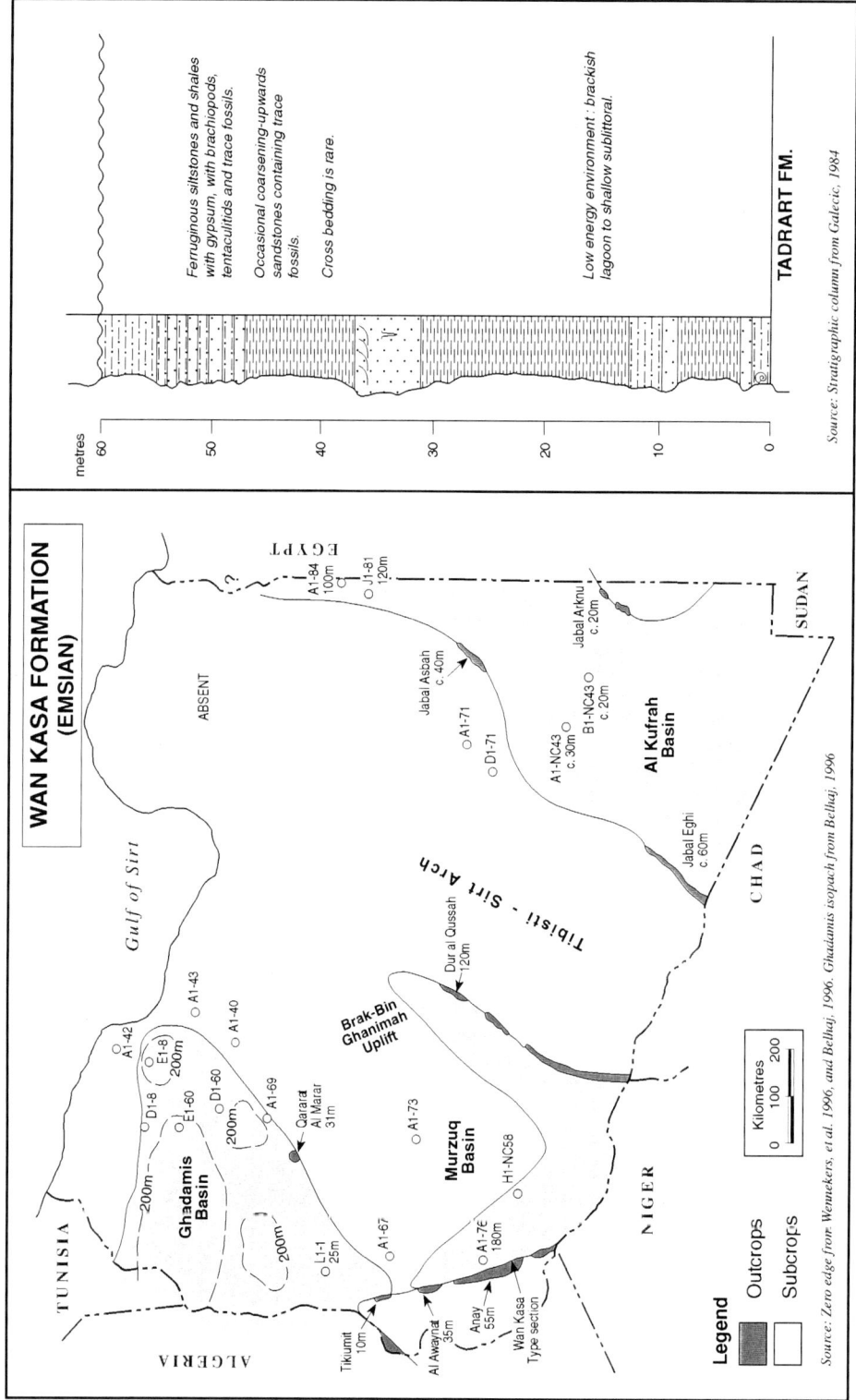

Figure 3.18 Wan Kasa Formation, Distribution and Stratigraphic Column

The Wan Kasa Formation represents a cycle of low-energy deposition in a lagoonal to shallow sub-littoral environment. It was first described from the Wadi Wan Kasa in the Jabal Tadrart, southeast of Ghat. It is present throughout the Ghadamis Basin but is absent over much of the Murzuq Basin due to mid-Devonian erosion. An equivalent sequence of ferruginous siltstones is present in the Al Kufrah Basin.

Tikiumit to South Anay. In this area the Wan Kasa Formation shows similar characteristics to the type area. Sedimentological studies by Clark-Lowes indicated an environment dominated by strong tides, with evidence of deposition from sediment-laden sub-tidal currents. Sutcliffe, et al. recognised facies associations indicative of oolitic shelf and muddy shoreface to tidal-flat environments, confirming the transgressive nature of these rocks. The thickness of the formation in this area ranges from 50m at Anay to 25m at Al Awaynat and less than 10m at Tikiumit. The top is progressively truncated by mid-Devonian erosion.[137]

In the Murzuq Basin the Wan Kasa Formation has been encountered in the A1-76 and H1-NC 58 wells, but Echikh and Sola showed that it is absent to the northeast of these wells. Well logs show a rapid transition from the estuarine Tadrart Formation into the shallow marine tidal bars of the Wan Kasa Formation. The formation is not present in the centre of the basin or on the Brak-Bin Ghanimah Uplift in the area affected by the Devonian tectonic inversion and mid-Devonian erosion. It is also absent in well A1-67 in the northwestern part of the basin. The Wan Kasa Formation can be traced for 120km on the eastern flank of the Murzuq Basin. In the Dur al Qussah Trough it is represented by 120m of silty grey shale with siltstone and thin sandstone stringers containing crinoids, corals, pelecypods, brachiopods and plates from placoderm fishes. Northwards the formation pinches out onto the Haruj Uplift, and towards the south it thins and becomes condensed towards Mourizidie.[138]

In the Al Kufrah Basin it is difficult to separate the Wan Kasa from the Tadrart Formation, although the upper part of the combined sequence is a clayey siltstone which suggests an affinity with the Wan Kasa Formation. The thickness of the combined unit ranges from 143m in the Jabal Nuqay area to 102 m at Jabal Zalmah and 60m north of Jabal Arknu. In the A1 and B1-NC 43 wells the combined Tadrart-Wan Kasa section comprises about 150m of medium to coarse-grained sandstones with minor siltstones. The section has been dated as Emsian from abundant palynological data.[139]

The Wan Kasa Formation has not been mapped on the flanks of the Al Qarqaf Arch but it is present in the subsurface of the Ghadamis Basin where Massa and Moreau-Benoit recorded a sequence of 170m of fine- to medium-grained sandstones, siltstones and dolomites of Emsian age in several wells. Later studies by Belhaj showed that the Wan Kasa Formation reaches 365m in thickness in the E1-8 well, where it comprises poorly-sorted sandstones and mudstones deposited in a neritic environment, with an upper more persistent sandstone member. An upper iron-rich layer probably indicates a period of emergence. Palynological evidence permits the sequence to be dated as Emsian to early Eifelian. The Wan Kasa Formation is well developed in wells E1-8, E1-60, C1-30, B1-39, D1-60 and several other wells in the same area. The friable sandstones within this sequence represent good oil reservoirs in the Ghadamis Basin.[140]

The Wan Kasa Formation is not present in the Sirt Basin, but in Cyrenaica there is evidence that the Devonian transgression reached the eastern part of the area during Emsian times. Dark grey shales grading up into fine-grained sandstones containing Emsian palynomorphs have been found in wells A1-33, A1-84 and J1-81. These shallow-marine sediments, which have a thickness of 90m to 115m, are regarded as approximate equivalents of the Wan Kasa Formation. They rest unconformably on sandstones of late Silurian age.[141]

In Algeria the Wan Kasa Formation correlates to part of the 'Devonien inferieur greseux' which contains the F5 reservoir. This unit is oil productive from several fields including El Adeb Larache, Taredert, Tan Emellel and Assekaifaf.[142]

122

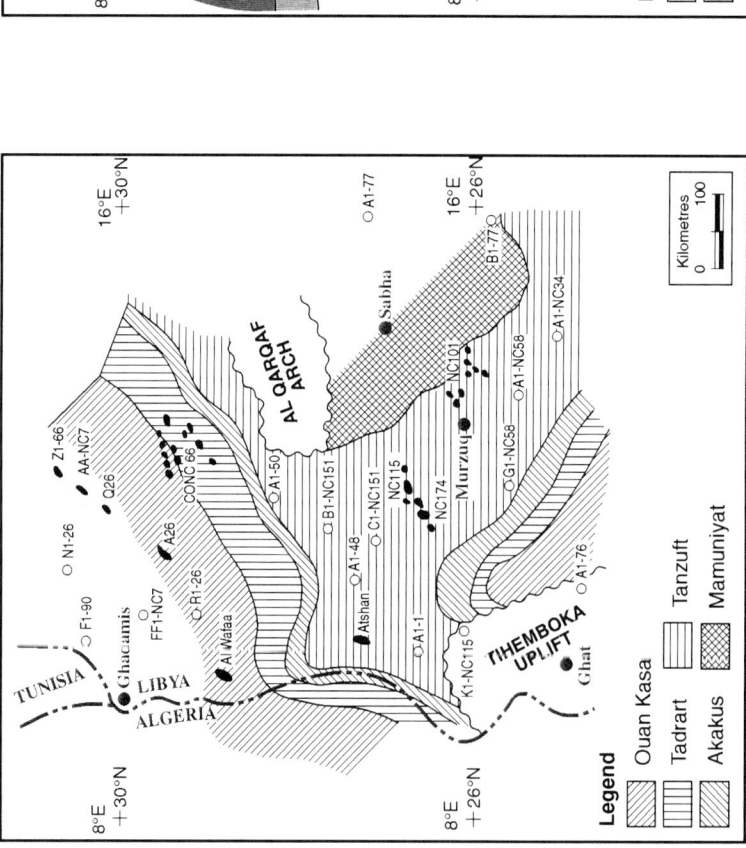

Source: Echikh and Sola, 2000

Figure 3.20   West Libya, Mid-Devonian Supercrop Map

Figure 3.19   West Libya, Mid-Devonian Subcrop Map

These maps illustrate the mid-Devonian unconformity in west Libya. The subcrop map shows that Lower Devonian rocks are preserved only in the Ghadamis Basin and in the area east of well A1-76. In the basin centre Middle Devonian rocks overlie Tanzuft Shales, and on the Brak-Bin Ghanimah Uplift Middle Devonian rocks overlie Mamuniyat Formation. The supercrop map shows the transgression of Middle and Upper Devonian formations over the unconformity surface. The last area to be covered was the Tirirene High between well K1-NC 115 and the Atshan field.

3.5.3 Awaynat Wanin Formation

Rocks of Middle Devonian age were recognised by Borghi in the Awaynat Wanin area in 1939, but it was Lelubre who first proposed the name Awaynat Wanin Formation, and Massa and Collomb who gave the first systematic description. The formation comprises the entire sequence between the top of the Wan Kasa Formation and the base of the Marar Formation (Figure 3.21). Regional studies have shown that west Libya was affected by extensive tectonism at the end of the Emsian which led to uplift, erosion and peneplaination. Erosion on the Brak-Ben Ghanimah Uplift was particularly marked and resulted in the removal of the entire section down to the Mamuniyat Formation (Figure 3.19). For the most part the Awaynat Wanin Formation was deposited on this peneplained surface. Transgression onto the eroded surface was gradual and some areas, such as the Tiririne Uplift in the Murzuq Basin, were not covered until Famennian times (Figure 3.20).[143]

The type area of the Awaynat Wanin Formation is on the northwest flank of the Al Qarqaf Arch. According to the most recent study the type section shows six repeated 'cyclothems' of claystones, sometimes ferruginous or gypsiferous, siltstones and sandstones, each cycle averaging 15-30m in thickness, with a total thickness of 162m. The sequence, according to Vos, represents a deltaic environment ranging from delta front to fluvial-distributary channels, all reworked by tides and storm waves. The sandstones are frequently cross-bedded and contain a diverse fauna of brachiopods, sometimes concentrated into coquinas. The base is apparently conformable with the underlying Wan Kasa Formation and the top has a conformable contact with the overlying Marar Formation. The formation has been dated on the basis of its brachiopod fauna, and spans the interval from Eifelian to Famennian.[144]

Subsequent attempts to revise the stratigraphy have led to considerable confusion. In 1962 Collomb divided the entire sequence into eight cycles (Figure 3.22). He restricted the name Awaynat Wanin Formation to the lower four cycles and named the upper four cycles the Chatti Formation. In 1974 Massa, Termier and Termier established a new formation – the Tahara Formation – for the transitional series immediately underlying the Marar Formation. Two years later Massa and Moreau-Benoit proposed upgrading the Awaynat Wanin to group status and dividing it into four formations: Awaynat Wanin I to IV, corresponding to the stages Eifelian, Givetian, Frasnian and Fammenian which they were able to identify on the basis of faunal assemblages. This subdivision has been further refined by Weyant and Massa. In mapping for the Geological map of Libya Gundobin accepted the original definition of Massa and Collomb in the type area of the formation, but on the south side of the Al Qarqaf Arch, on the Idri, Sabha and Al Fuqaha sheets, most of the subsequent revisions have been accepted (Figure 3.22).[145]

A partial section of the Awaynat Wanin Formation has been mapped in the Tikiumit area where 20m of ferruginous sandstones with silty and calcareous stringers has been dated as Frasnian. Fürst and Klitzsch claimed that the formation can be followed southwards, but for the most part it is overstepped by the succeeding Lower Carboniferous. Bellini and Massa illustrated a 147m section of shales, siltstones and sandstones, containing brachiopods and ferruginous oolites at Gour Iduka on the Anay sheet, which they attributed to the Awaynat Wanin Formation. However it was not recognised at outcrop or mapped in this area by the surveyors of the Geological Map of Libya. The formation is absent in well A1-76 (Figure 3.21).[146]

The south flank of the Al Qargaf Arch was mapped first by Goudarzi and then in detail by a

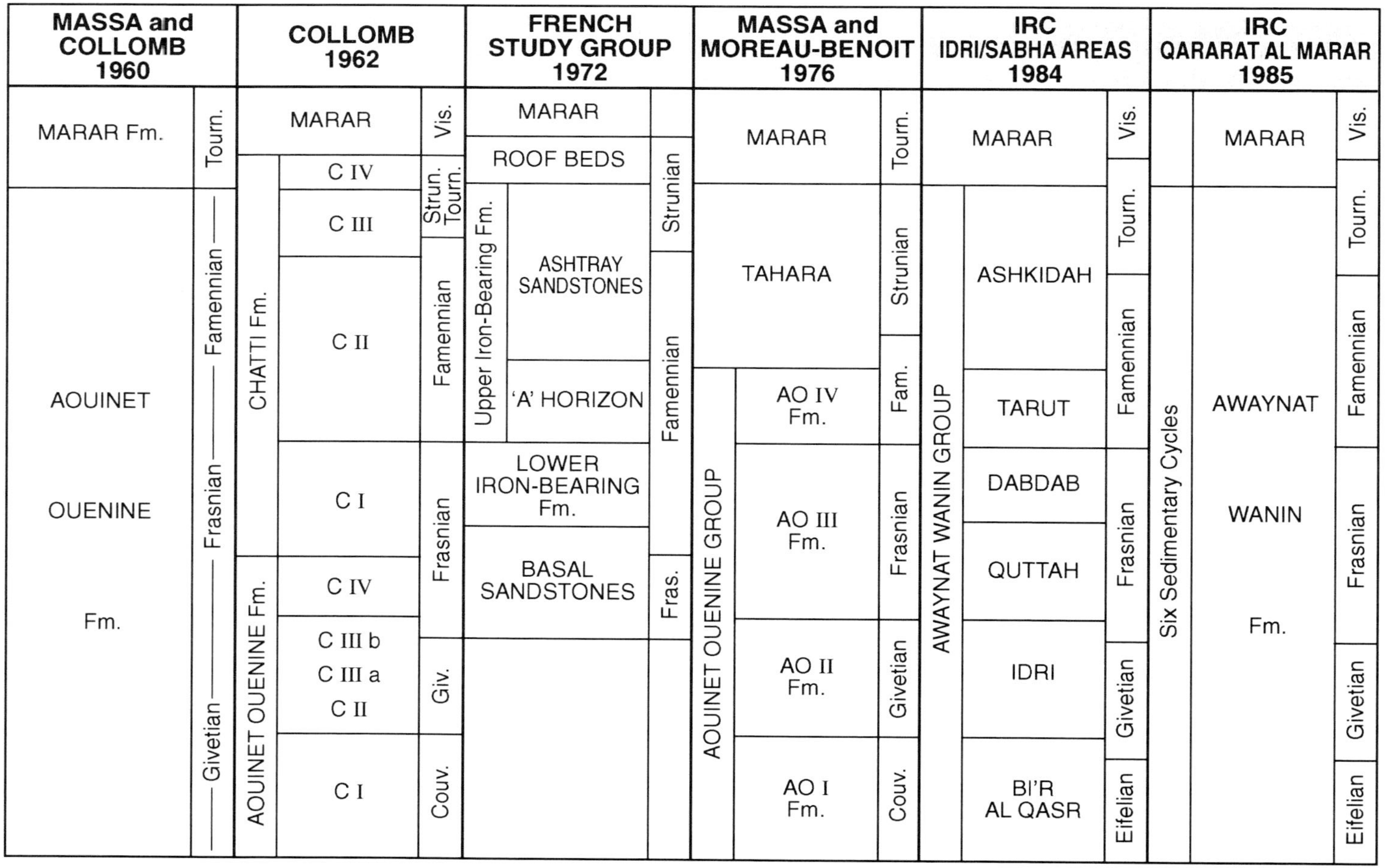

MASSA and COLLOMB 1960	COLLOMB 1962	FRENCH STUDY GROUP 1972	MASSA and MOREAU-BENOIT 1976	IRC IDRI/SABHA AREAS 1984	IRC QARARAT AL MARAR 1985
Tourn. / Famennian / Frasnian / Givetian	Vis. / Tourn. / Strun. / Famennian / Frasnian / Giv. / Couv.	Strunian / Famennian / Fras.	Tourn. / Strunian / Fam. / Frasnian / Givetian / Couv.	Vis. / Tourn. / Famennian / Frasnian / Givetian / Eifelian	Vis. / Tourn. / Famennian / Frasnian / Givetian / Eifelian
MARAR Fm.	MARAR; C IV, C III, C II, C I (CHATTI Fm.); C IV, C IIIb, C IIIa, C II, C I (AOUINET OUENINE Fm.)	MARAR; ROOF BEDS; ASHTRAY SANDSTONES; 'A' HORIZON; Upper Iron-Bearing Fm.; LOWER IRON-BEARING Fm.; BASAL SANDSTONES	MARAR; TAHARA; AO IV Fm.; AO III Fm.; AO II Fm.; AO I Fm. (AOUINET OUENINE GROUP)	MARAR; ASHKIDAH; TARUT; DABDAB; QUTTAH; IDRI; BI'R AL QASR (AWAYNAT WANIN GROUP)	MARAR; AWAYNAT WANIN Fm. (Six Sedimentary Cycles)
AOUINET OUENINE Fm.					

Figure 3.22   Successive Nomenclature of the Awaynat Wanin Formation

The Awaynat Wanin sequence was established as a formation by Massa and Collomb in 1960. Subsequently Massa and Moreau-Benoit raised it to group status and subdivided the group into four formations. IRC geologists redefined the group and proposed six new formations in the Idri/Sabha area. However in their subsequent resurvey of the type area, IRC reverted to the view of the Awaynat Wanin as a formation divisible into 'six member cycles'.

Sources: As shown in the caption

French Study Group between 1973 and 1976 in a survey of the Wadi ash Shati iron deposits. They adopted an informal stratigraphic scheme which deliberately emphasised the iron-bearing horizons. The area was subsequently remapped for the Geological Map of Libya in 1984. On the basis of field observation the surveyors of the Geological Map sheets opted to accept Collomb's view that the entire sequence should be elevated to group status - the Awaynat Wanin Group - which was then divided into six formations which were given new names and new type descriptions (Figure 3.22). The formations correspond closely to the divisions proposed by Massa and Moreau-Benoit, and the uppermost of the six formations corresponds to the Tahara Formation. In more recent studies of these outcrops, Sutcliffe et al. recognised numerous incomplete cycles of shoreface, distributary-channel and intertidal depositional environments, and Blanpied and Rubino defined nine tidal and wave-dominated transgressive and high-stand sequences.[147]

Local Stratigraphy of the Awaynat Wanin 'Group', South Flank of the Al Qarqaf Arch

The Bi'r al Qasr Formation was defined in the Idri area as the basal formation of the Awaynat Wanin 'Group'. It corresponds to cycle I of Collomb, and to the Awaynat Wanin AO I Formation of Massa and Moreau-Benoit. The formation rests on eroded Tanzuft Formation. The cycle is unusual in having a substantial sandstone unit at the base, fine-grained, ferruginous and containing shell fragments (transgressive system tract), before the establishment of a silty claystone sequence with occasional sandstone stringers. The cycle becomes progessively more sandy, ending in a massive, iron-stained quartzitic sandstone, containing brachiopods and trace fossils (high stand system tract). Poorly preserved plant remains with *Lepidodendron* affinities appear to

have been washed into the area from a considerable distance. The thickness of the formation is 42m at the type section. The brachiopod fauna gives an Eifelian age, possibly extending into the early Givetian.[148]

The second cycle was named the Idri Formation by Parizek et al. It corresponds to cycles II and III of Collomb and to the Awaynat Wanin AO II of Massa and Moreau-Benoit. It contains two sequences, beginning with a short claystone sequence which passes rapidly into sandy siltstone and then into thick massive sandstone units. The sandstones are fine- to medium-grained, and show ripple-marks and burrows in the lower part, and cross-bedding towards the top. The uppermost sandstone unit contains abundant brachiopods. Some of the shales within the Idri Formation show euxinic characteristics. The total thickness of the formation is 70m of which eighty percent is sandstone. The lower 60m has been dated as Givetian, but the upper 10m contains an early Frasnian fauna.[149]

Cycle number three of the Awaynat Wanin 'Group' was defined in the Sabha area, and was named the Quttah Formation by Seidl and Rohlich. It corresponds roughly to cycle IV of Collomb and to the lower part of the Awaynat Wanin AO III of Massa and Moreau-Benoit. It is equivalent to the Basal Sandstones of the French Study Group. Its base is formed by a maximum flooding surface and the cycle begins with a claystone sequence which rests on the uneven surface of the Idri Formation. The claystone contains brachiopods and the trace fossil *Bifungites fezzanensis*. The top of the claystone forms a sequence boundary. The claystone is followed abruptly by ferruginous massive sandstones with a conglomeratic base. The uppermost unit is a quartzitic sandstone and contains abundant pelecypods and brachiopods which give a Frasnian age for the Quttah Formation. The thickness of the formation at the type locality is 28m.[150]

126

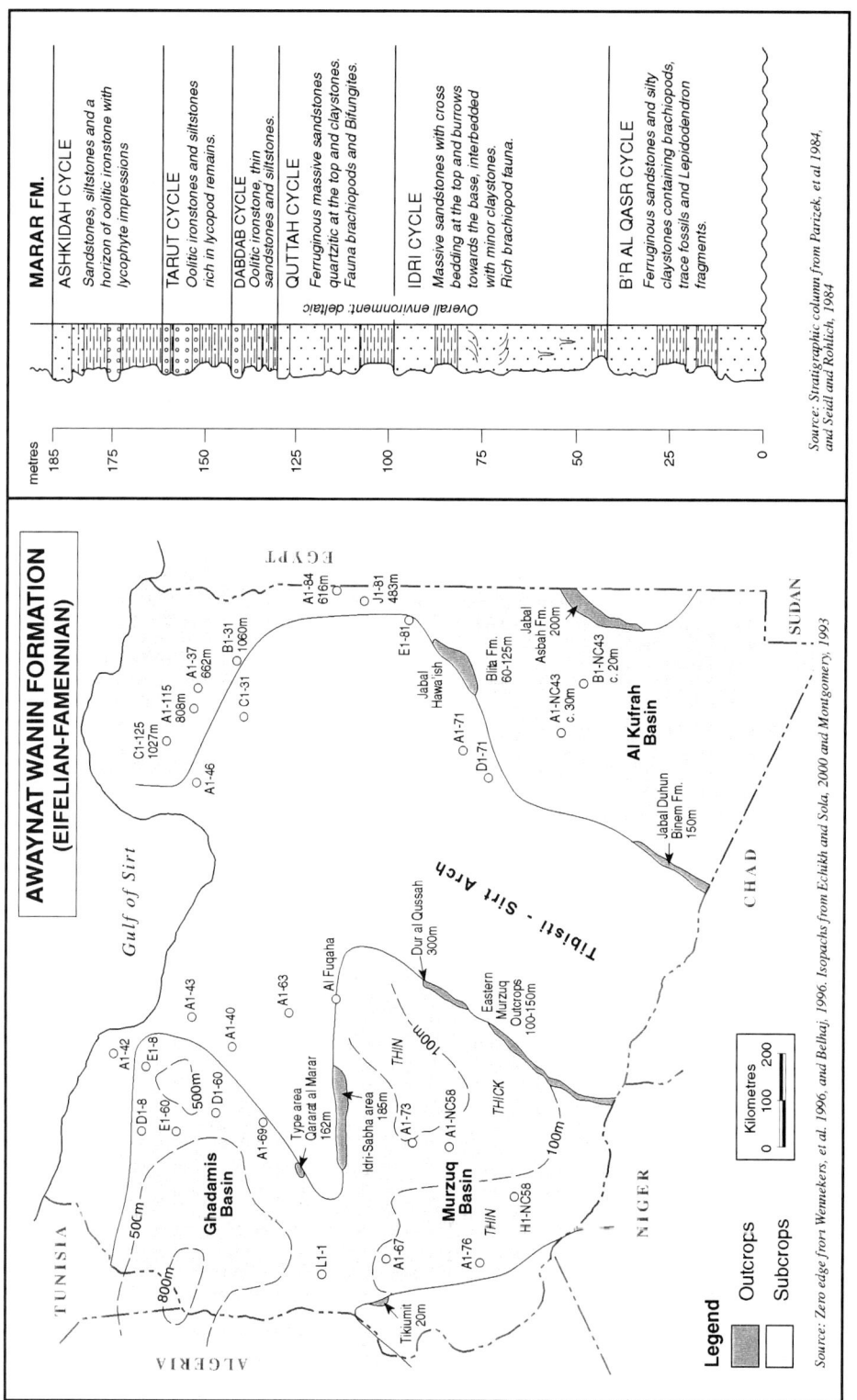

MARAR FM.

ASHKIDAH CYCLE
Sandstones, siltstones and a horizon of oolitic ironstone with lycophyte impressions

TARUT CYCLE
Oolitic ironstones and siltstones rich in lycopod remains.

DABDAB CYCLE
Oolitic ironstone, thin sandstones and siltstones.

QUTTAH CYCLE
Ferruginous massive sandstones quartzitic at the top and claystones. Fauna brachiopods and Bifungites.

IDRI CYCLE
Massive sandstones with cross bedding at the top and burrows towards the base, interbedded with minor claystones. Rich brachiopod fauna.

B'R AL QASR CYCLE
Ferruginous sandstones and silty claystones containing brachiopods, trace fossils and Lepidodendron fragments.

Overall environment: deltaic

Source: Stratigraphic column from Parizek, et al 1984, and Seidl and Rohlich, 1984

AWAYNAT WANIN FORMATION (EIFELIAN-FAMENNIAN)

Source: Zero edge from Wennekers, et al. 1996, and Belhaj, 1996. Isopachs from Echikh and Sola, 2000 and Montgomery, 1993

Figure 3.21 **Awaynat Wanin Formation, Distribution and Stratigraphic Column**

The Awaynat Wanin Formation was first defined in the Qararat al Marar area where the type section was established. The formation has been studied in detail in the Idri-Sabha area where it contains important deposits of sedimentary ironstones. It is present in the Ghadamis Basin, and as a transgressive unit in the Murzuq Basin, although there are no outcrops on the western flank of the Basin. In the Al Kufrah Basin the Binem and Blita Formations are equivalents of the Awaynat Wanin Formation. The formation contains hydrocarbons in the F4 and F3 reservoirs in the Ghadamis and Triassic Basins, and a local oil source rock, the so-called Cues Limestone.

In the Sabha area the Quttah Formation is overlain by the fourth cycle of the Awaynat Wanin 'Group', which was named the Dabdab Formation by Seidl and Rohlich. The Dabdab Formation is equivalent to the Lower Iron-bearing Formation of the French Study Group, to cycle I of Collomb's Chatti Formation, and to the upper part of Massa and Moreau-Benoit's Awaynat Wanin AO III Formation (Figure 3.22). The Dabdab Formation comprises a sequence of alternating claystones and thin ferruginous sandstones, capped by a 2m bed of ferruginous oolite (high stand system tract) which represents one of the main iron ore units (unit L) of the Wadi ash Shati iron ore deposit. The total thickness of the formation is 12.5m and the age has been determined as late Frasnian on the basis of brachiopods.[151]

The fifth cycle was named the Tarut Formation by Seidl and Rohlich. It corresponds to the lower part of cycle II of the Chatti Formation of Collomb, to the Awaynat Wanin AO IV Formation of Massa and Moreau-Benoit, and to the lower part of the Upper Iron-bearing Formation of the French Study Group. The lower part of the formation comprises laminated claystones containing pelecypods, followed by siltstones, which pass upwards into a 4m bed of ferruginous oolites representing the 'A' ore bearing unit of the Wadi ash Shati iron ore deposit (high stand system tract). The total thickness of the Tarut Formation at the type locality is 17.5m, and the age has been determined as Famennian. The Tarut Formation contains a rich palaeoflora of tree-sized lycophytes, and at a locality south of Tarut, which has attracted the name 'Paul's Garden' these fossils are preserved with trunks and roots still in growth position. A Famennian age has been assigned to the Tarut Formation.[152]

The final cycle of the Awaynat Wanin 'Group' on the south flank of the Al Qarqaf Arch was named the Ashkidah Formation by Seidl and Rohlich. It is equivalent to the upper part of cycle II plus cycle III of the Chatti Formation of Collomb, to the Tahara Formation of Massa, Termier and Termier, and Massa and Moreau-Benoit, and to the upper part of the Upper Iron-bearing Formation of the French Study Group. At the type locality in the Sabha area the Ashkidah Formation is 27m thick, and its junction with the Tarut Formation is unconformable. The cycle begins with silty claystones with thin ferruginous siltstones, followed by a 2m bed of ferruginous oolite (the 'B' ore bearing unit of the Wadi ash Shati iron ore deposit). The oolites contain impressions and compressions of lycophyte plants. A second sub-cycle overlies the oolite, passing from siltstone to a massive non-ferruginous sandstone containing brachiopods. The formation contains an assemblage of miospores and acritarchs which indicate an uppermost Devonian age and probably spans the Famennian-Tournaisian boundary. It is overlain by claystones of the Marar Formation.[153]

The Awaynat Wanin Formation in Other Areas

The localised subdivision of the Awaynat Wanin 'Group' described above is traceable along the southern flank of the Al Qarqaf Arch from Idri to Al Fuqaha, but it has not yet been extended into any neighbouring areas. The remainder of this review reverts to the original concept of the Awaynat Wanin as a formation extending from the top of the Wan Kasa Formation to the base of the Marar Formation.

The centre of the Murzuq Basin was affected by widespread inversion and peneplanation during the mid-Devonian, and the Awaynat Wanin Formation was deposited on the peneplained surface. The absence of the Awaynat Wanin Formation on the western margin of the basin has already been mentioned, and Sutcliffe, et al. published an isopach map which showed that in the subsurface the formation is absent over a large area of the western part of the basin, which corresponds roughly with the Tiririne Uplift. Echikh and Sola gave a more detailed picture showing the supercrop of Middle and Upper Devonian formations over the mid-Devonian unconformity (Figure 3.22). The Eifelian section (Awaynat Wanin I) is absent over much of the basin, but the Givetian Awaynat Wanin II extends as far south as well H1-NC 58 and westwards into Algeria to the north of the Atshan field. The Frasnian formations onlap the Brak-Bin Ghanimah Uplift, and the last remaining island of the Tiririne Uplift was not covered until Famennian times. In the Braspetro wells of concession NC 58 sedimentological analysis by de Castro on the Awaynat Wanin Formation demonstrated the presence of features indicative of a deltaic environment including suspension deposits, traction sediments, tidal channels and bars, tempestite deposits and hardgrounds formed during calm-water periods.[154]

The Awaynat Wanin Formation outcrops on the eastern flank of the Murzuq Basin from the Dur al Qussah Trough to Mourizidie and southwards into Chad. The thickness is fairly constant at about 100 to 150m but it thickens to 300m in the Dur al Qussah Trough. In this area the Awaynat Wanin Formation comprises massive cross-bedded sandstones with trace fossils and plant remains. The lower part is continental, but the upper part has some marine intervals.[155]

In the Kufrah Basin the Awaynat Wanin Formation is represented by the Binem Formation established by Vittimberga and Cardello in 1963 for a siltstone and sandstone sequence with kaolinitic and ferruginous cement occurring between the Tadrart Formation (which here cannot be separated from the Wan Kasa Formation) and the basal Carboniferous in the Jabal Duhun area. The name Blita Formation was given to the same sequence exposed at Jabal Hawa'ish. The thickness in the type area is 150m. It contains rare plant remains and the trace fossil *Zoophycus*. In the Jabal Hawa'ish area the thickness varies from 60m to 125m. The environment of deposition in this area is tidal-flat crossed by tidal channels, becoming subtidal towards the top. The formation also outcrops extensively at Jabal Asbah to the north of Jabal Arknu.[156]

The Binem Formation was penetrated in the A1-NC 43 well with a thickness of 555m and in the B1-NC 43 well with 626m. The lithology comprises a series of quartz siltstones, thin sandstones and shales with occasional coals. The formation contains a rich microflora which has allowed subdivision into eight microfloral zones, which demonstrate a continuous sequence from Eifelian to the latest Famennian.[157]

In the subsurface of the Ghadamis Basin Massa and Moreau-Benoit found well developed Awaynat Wanin Formation in many wells with rich microfloras, and they too were able to subdivide the formation into eight microfloral zones which match those in the Al Kufrah Basin almost exactly. Over much of the basin a 30m shale is present at the base of the formation which has been termed the Emghayet Shale by various authors. This unit has minor source-rock potential, but is important as a regional seal. One remarkable feature is the presence of a highly radioactive zone within the Frasnian which corresponds to the Cues Limestone horizon, which is a significant oil source rock in the Ghadamis Basin. It is present in many wells across the basin. The radioactivity is caused by uranium levels approaching 40ppm in the shales. In Algeria low-diversity miospore assemblages and *Tasmanaceae* blooms in this interval have been interpreted as indicating poorly-oxygenated conditions. The Cues Limestone horizon is the source rock for the Alrar field in Algeria and the Al Wafaa field in Libya. In both cases oil and gas are contained within a stratigraphic trap of the F3 reservoir which has been dated as Givetian. The total thickness of the Awaynat Wanin Formation in this area reaches 635m and comprises five sedimentary cycles similar to those in the Idri-Sabha area. An isopach maps of the Awaynat Wanin Formation in the Ghadamis Basin has been published by El Rweimi which shows a maximum thickness of 683m in the B1-NC 100 well.[158]

A similar picture emerges for the eastern Ghadamis Basin/Zamzam Depression where Belhaj showed a complete sequence of Awaynat Wanin cycles which can be assigned to eight palynozones extending in age from Eifelian to late Famennian. The area of deposition becomes progressively more restricted with time. The lower cycles are present from B1-42 in the north to A1-69 in the south, whereas the upper cycles are only present in the centre of the basin from C1-30 to D1-60. The maximum thickness in this area reaches 700m in well B1-60.[159] In the Ghadamis Basin the Awaynat Wanin Formation does not contain such good oil reservoirs as the Lower Devonian. The reservoirs are usually thinner and porosities and permeabilities are lower. Nevertheless Awaynat Wanin reservoirs produce oil and gas in the Al Hamra area and at Al Wafaa, as already mentioned.[160]

In Cyrenaica the equivalent of the Awaynat Wanin sequence is represented by a complex sequence of mainly shallow marine micaceous shales, sandstones and calcilutites which can be divided into several sequences. The equivalent of the Awaynat Wanin I is seen in southeastern Cyrenaica in wells J1-81, A1-84, B1-31 and as far north as C1-125, where up to 730m of nearshore sandstones and shales of Eifelian age have been penetrated. Awaynat Wanin II equivalents are represented by predominently shaly rocks which contain Givetian palynomorphs in wells B1-31 and A1-37. Awaynat Wanin III is represented by shales yielding Frasnian assemblages in wells B1-31, A1-37 and A1-115, and Awaynat Wanin IV equivalents

have been found in wells A1-37, A1-115 and C1-125 containing Famennian palynomorphs. The sequence appears to be complete, and the total thickness of Awaynat Wanin equivalents in Cyrenaica is in the order of 1200m.[161]

In the Illizi Basin of Algeria the equivalent of the Awaynat Wanin Formation is the 'Devonien Serie argileuse' which includes the F4 and F3 reservoir units. The F4 is Eifelian in age and the F3 is Givetian. The F4 is a major oil reservoir in the Illizi Basin at Zarzaitine, Edjeleh, Tiguentourine, Assekaifaf and El Adeb Larache. The F3 reservoir contains gas and condensate at both West and East Alrar on the Libyan border.[162]

3.5.4 Tahara Formation

The Tahara Formation was established by Burollet and Manderscheid in 1967, and was refined by Massa, Termier and Termier in 1974. The formation was defined as a 60-70m shale-sandstone cycle from a type section in the B1-49 well in the southern Ghadamis Basin. It outcrops in a small area to the northwest of Awaynat Wanin where two coarsening upwards cycles are exposed separated by a hard ground. The lower cycle shows typical tempestite features. The hard ground represents normal fair weather conditions and the upper unit represents a tidally dominated facies. Sutcliffe, et al. recognised similar high-energy, shallow shelf facies passing into deeper-water, outer shelf siltstones and mudstones. Hassi recognised a sequence of sixteen shallowing-upwards sequences within the Tahara Formation representing a shelf-shoreface succession in a storm-dominated environment. It is characterised by a palynological assemblage which marks the transition from the Devonian to the Carboniferous (Late Famennian-Strunian age), although Mergl and Massa attributed the formation to the Lower Tournaisian. Analysis of core samples from well A1-NC 40B yielded a rich palynomorph assemblage which gave a latest Famennian age. An isopach map of the Tahara Formation in the Ghadamis Basin was presented by El Rweimi which shows a maximum thickness of 83m in well A1-90, and a separate depocentre in the Zamzam Depression in which the thickness reaches 73m in well E1-60. The formation has been recognised as a separate entity in the Ghadamis Basin, but it is equivalent to the uppermost cycle of the Awaynat Wanin Formation as originally defined, and in areas other than the Ghadamis Basin it is included within the Awaynat Wanin Formation. It is an exact equivalent of the Ashkidah Formation of the Sabha area. In reservoir terms it represents the F2 reservoir of the Algerian classification. The formation has also been recognised in the Murzuq Basin where it occurs as a thin unit overlying the Awaynat Wanin Formation over much of the basin, except for the western margin. The Strunian nowadays is regarded as equivalent to the uppermost Famennian.[163]

3.6   Carboniferous

Deposition of Carboniferous rocks in Libya was controlled by tectonic events associated with the collision of western Gondwana with Laurasia to form Pangaea. These orogenic events resulted in compression, uplift, and the gradual emergence of a series of southwest-northeast intercratonic sags and arches which increasingly affected sedimentation during the Carboniferous. The early Carboniferous is dominated by deltaic and inner neritic facies. Late Namurian carbonate rocks reflect the last marine incursion of the Palaeozoic, and the late Carboniferous is characterised by continental rocks. By the end of the Carboniferous most of Libya was emergent and was being actively eroded and peneplained.

3.6.1 Marar Formation

The Marar Formation was first defined by Lelubre in 1948 by reference to a type section at Qararat al Marar northwest of the Al Qarqaf Arch (Figure 3.23). At the type location the Marar Formation comprises about fifty stacked cycles of silty claystones, micaceous siltstones and feldspathic quartz-sandstones. The claystones are greenish-grey, silty and gypsiferous, the sandstones and siltstones are ferruginous, flaggy and often ripple-marked. The sandstones are sometimes conglomeratic at the base, frequently cross-bedded, and contain brachiopods and pelecypods. The fauna indicates an Upper Tournaisian-early Visean age. The extension of the Marar Formation into the late Serpukhovian as suggested by Belhaj in the southwestern Ghadamis Basin is difficult to justify given the palaeontological evidence provided by El Harbi who demonstrated a late Visean age for the upper part of the upper member of the Marar Formation, based or abundant well-preserved palynomorphs. The total thickness is 400m and the contact with the underlying Awaynat Wanin Formation is disconformable. Gundobin divided the sequence at the type locality into two members distinguished by thicker cyclothems and an increased claystone content in the upper member. The lower member contains 35 to 40 cycles ranging in thickness from 3m to 15m; the upper member comprises nine cycles ranging in thickness from 25m to 35m. According to Whitbread and Kelling the lower part represents a fluvially-dominated deltaic sequence whilst the upper part indicates a more wave dominated environment with hummocky cross-stratification suggesting tempestite facies towards the top.[164]

The Marar Formation outcrops extensively (but patchily) in the area west of the Al Qarqaf Arch and has been mapped at Hamadat Tanghirt, Hasi Anjiwal and Wadi Irawan. In these areas the upper 25m of the Marar

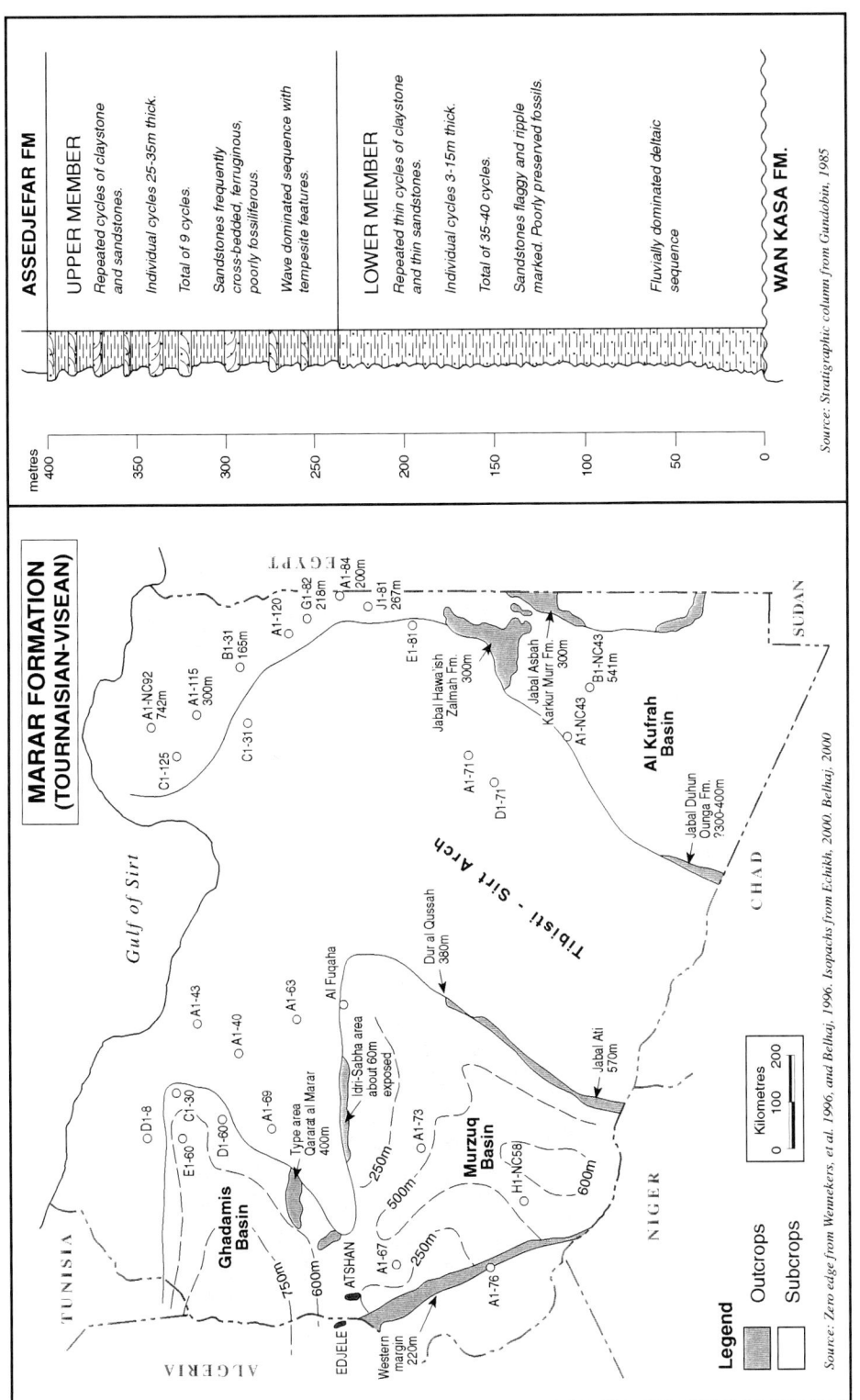

ASSEDJEFAR FM

UPPER MEMBER

Repeated cycles of claystone and sandstones.

Individual cycles 25-35m thick.

Total of 9 cycles.

Sandstones frequently cross-bedded, ferruginous, poorly fossiliferous.

Wave dominated sequence with tempesite features.

LOWER MEMBER

Repeated thin cycles of claystone and thin sandstones.

Individual cycles 3-15m thick.

Total of 35-40 cycles.

Sandstones flaggy and ripple marked. Poorly preserved fossils.

Fluvially dominated deltaic sequence

WAN KASA FM.

Source: Stratigraphic column from Gundobin, 1985

MARAR FORMATION (TOURNAISIAN-VISEAN)

Source: Zero edge from Wennekers, et al. 1996, and Belhaj, 1996. Isopachs from Echikh, 2000. Belhaj, 2000

Legend

Outcrops

Subcrops

Figure 3.23   Marar Formation, Distribution and Stratigraphic Column

The Marar Formation represents a fluvial/deltaic sequence characterised by repeated cycles of claystones and thin cross-bedded sandstones. In many areas (but not at the type location) the formation is capped by an horizon of stromatolitic algae called the *Collenia* Beds. The formation contains abundant plant fossils including the trunks of large lycopods, and some poorly preserved marine fossils. It is present in both the Murzuq and Ghadamis Basins, and in the Al Kufrah Basin the equivalents are the Ounga, Zalmah and Karkur Murr Formations. In Algeria the formation contains the 'D' reservoir units in the Edjeleh area, and in Libya the same units contain hydrocarbons on the Atshan field.

Formation is characterised by the presence of several horizons full of *Collenia*, a stromatolitic alga which forms spectacular mounds up to 2m in diameter. The surface of these algal biostromes is domed and undulating and the stromatolites are often coloured purple, red or brown. This horizon is widespread and was studied by Freulon in Algeria who named it the *Collenia* Beds. The unit contains fossils which allow it to be dated as Visean in age.[165]

On the western flank of the Murzuq Basin the Marar Formation has been mapped from Tikiumit to the border with Niger. In this area the Marar Formation lies unconformably on Wan Kasa Formation in the south, and on progressively older formations to the north indicating significant pre-Carboniferous erosion on the flank of the subsurface Al Qarqaf Ridge. It averages 220m in thickness and is capped by *Collenia* Beds about 20m thick. Sand content increases significantly towards the south. In the Al Awaynat area the Marar Formation contains a rich fossil flora (including ferruginised logs) which indicates a Visean age. The middle part of the formation contains a diverse fauna of brachiopods and pelecypods.[166]

The Marar Formation is present in the subsurface of the Murzuq Basin, where its cyclic nature is evident, but it thins dramatically in wells C1, E1 and F1-NC 58 on an intra-basin ridge. The *Collenia* Beds are present in wells in the northwestern part of the basin, but not in wells to the southeast. The Marar Formation is also present in wells A1-64, A1-73 and A1-NC 58 where it has a thickness of 120m, 225m and 435m respectively. In concession NC 115 the thickness ranges from 154 to 226m and the formation comprises a fluvial channel, delta front and prodelta complex.[167]

On the south side of the Al Qarqaf Arch the Marar Formation has been mapped in the Idri/Sabha/Al Fuqaha area where incomplete sections are exposed which show a significant increase in carbonate content. In the Sabha area

brecciated dolomites and limestone coquinas occur within the Marar sequence and contain rich faunas of brachiopods, crinoids and gastropods of Visean age. The exposed thickness of the Marar Formation is about 60m.[168]

In the Dur al Qussah area the Marar Formation is represented by a lower shaly unit with a basal conglomerate containing bone fragments, followed by typical Marar cyclic units of shales and sandstones containing plant remains. The total section is 380m thick. Further south the sequence thickens to 570m and becomes progressively more continental towards the border with Niger.[169]

In the Kufrah Basin the equivalent of the Marar Formation in the Jabal Hawa'ish area is the Zalmah (Dalma) Formation which was first described by Vittimberga and Cardello from the northeast Kufrah Basin. They named the same sequence at Jabal Duhun the Ounga Formation, and at Jabal Arknu the Karkur Murr Formation. The formation comprises 500m of continental sandstones, conglomerates and siltstones containing plant remains similar to those in the Murzuq Basin. A comparable section is present in the outcrops north of Jabal Arknu where the thickness reaches 300m. These outcrops have been studied by Turner who interpreted the Zalmah Formation in this area as mainly braided-stream deposits with tidal channels, passing upwards into marine shoreface sequences full of ripple cross laminations. In the southwest, at Jabal Duhun, a marine intercalation has been found within the Zalmah Formation containing brachiopods, pelecypods and bryozoa of Visean age, which represents the furthest extent of the marine lower Carboniferous transgression in this area. The formation is absent in the A1-NC 43 well where post-Carboniferous rocks unconformably overlie the Binem Formation. In the B1-NC 43 well it has a thickness of 541m and comprises a dominantly shaly lower part overlain by alternating cycles of sandstones and shales containing a rich and varied microflora. Palynological analysis shows that there is no hiatus between the latest Famennian palynozone of the Binem Formation and the earliest Tournaisian palynozone of the Zalmah Formation. The upper part of the Zalmah Formation in the B1 well contains early Visean palynofloras.[170]

The Marar Formation has been encountered in many wells in the Ghadamis Basin. Studies by Massa et al. indicated that the lowest palynozone of the Tournaisian is missing in the central Ghadamis Basin, and this is confirmed by studies on wells in Agip's concession NC 40, indicating a hiatus between the Awaynat Wanin Formation and the Marar Formation in this area. Belhaj presented an isopach map of the Marar Formation which showed a maximum thickness of 1060m in the Ghadamis area. The formation comprises a basal conglomerate full of fish fragments, molluscs, and crinoids, followed by cyclic sediments similar to those found at outcrop. Sandstones however make up only about 20 per cent of the total sequence. In the Jifarah Basin the Marar Formation exhibits marginal swamp features.[171]

In the Zamzam Depression the Marar Formation is present in the E1-60, C1-30 and D1-60 wells reaching a thickness of 670m in the centre of the depression (Figure 3.23). It comprises numerous shallow-marine to deltaic cycles which unconformably overlie the Awaynat Wanin Formation. The uppermost palynozone of the Strunian is absent, which confirms the hiatus noted above in the Ghadamis Basin. The three palynozones indicative of the Tournaisian and early Visean are all present. No major hydrocarbon reservoirs have been found in the Marar Formation in the Ghadamis Basin, but at Atshan, on the southern margin of the basin, oil and gas have been found in a number of thin sandstones low in the Marar section.[172]

In Cyrenaica, Carboniferous rocks are patchily distributed as a result of both non-deposition and subsequent erosion. Shallow marine and deltaic rocks equivalent to the upper part of the Marar Formation have been found in several wells. The most complete sequence is present in well A1-NC 92, where 1000m of sandstones, siltstones and shales have yielded both early and late Visean palynomorphs. Only late Visean assemblages have been found in wells further south such as A1-115, A1-37, B1-31, G1-82, A1-84 and J1-81, where the thickness is reduced to about 330m.[173]

In the Illizi Basin of Algeria the equivalent of the Marar Formation is the Issandjel Formation composed of repeated cycles of shales, claystones and thin sandstones, which includes several thin reservoirs, the 'D' reservoirs, which contain minor oil and gas at Edjeleh, Zarzaitine, El Adeb Larache and Tiguentourine.[174]

3.6.2 Assedjefar Formation

The Assedjefar Formation was defined in 1952 by Lelubre for a sequence of deltaic and shallow marine rocks conformably overlying the Marar Formation in the Hamadat Tanghirt area of west Libya. A type locality was subsequently designated by Collomb (Figure 3.24), 80km west of Awaynat Wanin. In the type area there is a rapid transition from a dominantly sandy facies in the east, comprising coarse-grained, cross-bedded sandstones with fossil wood, to a more shaly facies in the west. In both areas the clastic sequence is overlain by marls and thin limestones. In remapping the area for the Geological Map of Libya, Berendeyev recognised at least four incomplete and truncated cycles of shaly limestones, siltstones, and cross-bedded sandstones with occasional conglomeratic intervals. The total thickness is 120m. Berendeyev distinguished two members separated by a minor unconformity marked by a

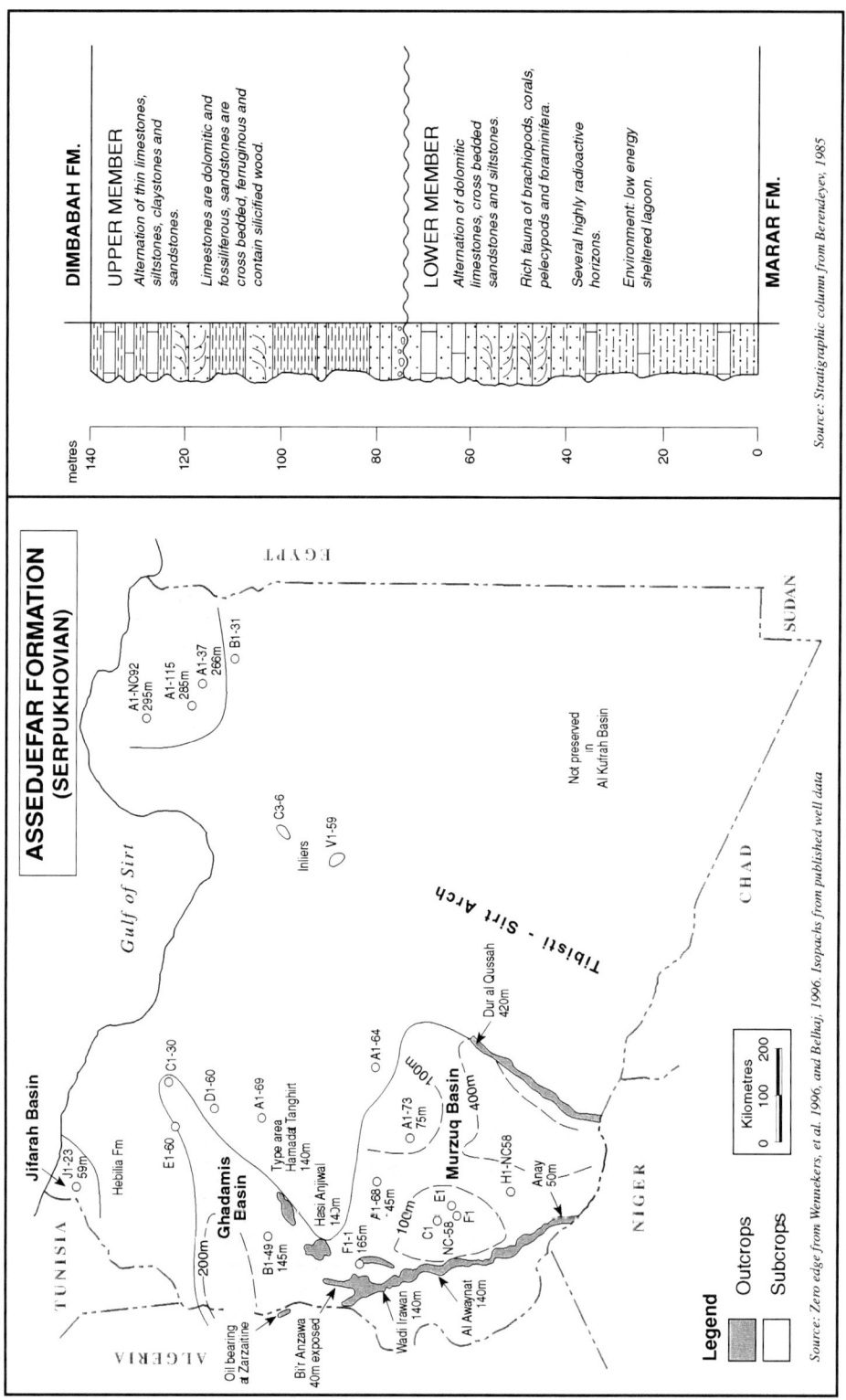

DIMBABAH FM.

UPPER MEMBER

Alternation of thin limestones, siltstones, claystones and sandstones.

Limestones are dolomitic and fossiliferous, sandstones are cross bedded, ferruginous and contain silicified wood.

LOWER MEMBER

Alternation of dolomitic limestones, cross bedded sandstones and siltstones.

Rich fauna of brachiopods, corals, pelecypods and foraminifera.

Several highly radioactive horizons.

Environment: low energy sheltered lagoon.

MARAR FM.

Source: Stratigraphic column from Berendeyev, 1985

ASSEDJEFAR FORMATION (SERPUKHOVIAN)

EGYPT

A1-NC92 295m
A1-115 285m
A1-37 266m
B1-31

C3-6
Inliers
V1-59

Not preserved in Al Kufrah Basin

SUDAN

Tibisti - Sirt Arch

CHAD

Dur al Qussah 420m

Gulf of Sirt

Jifarah Basin
J1-23 59m
Hebilia Fm

TUNISIA
ALGERIA

200m

Oil bearing a Zarzaitine
Bir Anzawa 40m exposed

C1-30
D1-60
E1-60
A1-69
Type area Hamadat Tanghirt 140m
B1-49 145m
Hasi Anjwal 142m
F1-1 165m
A1-68 45m
C1 E1 NC-58 F1

Ghadamis Basin

Wadi Irawan 140m
Al Awaynat 140m

NIGER

A1-64
A1-73 75m
Murzuq Basin 400m
100m
H1-NC58
Anay 50m

Legend

Outcrops
Subcrops

Kilometres
0 100 200

Source: Zero edge from Wennekers, et al. 1996, and Belhaj, 1996. Isopachs from published well data

Figure 3.24   Assedjefar Formation, Distribution and Stratigraphic Column

The Assedjefar Formation represents a low-energy lagoonal to shallow shelf environment in which dolomitic limestones, siltstones and ferruginous and gypsiferous sandstones were deposited. The sequence contains a rich marine fauna plus abundant silicified wood. A type section was designated in the Hamadat Tanghirt area, and the formation outcrops extensively along the western margin of the Murzuq Basin. It is present in the subsurface Murzuq and Ghadamis Basins, but no equivalents have been preserved in the Al Kufrah Basin. In northern Libya equivalents of the Assedjefar Formation are present in the Jifarah Basin, in two small inliers on the Sirt Arch, and in northeastern Cyrenaica. The formation contains a few poor reservoir sands which contain oil at the Zarzaitine field in Algeria.

conglomerate. The Assedjefar Formation contains a rich faunas of brachipods, pelecypods corals and foraminifera. The lower member of Berendeyev contains faunas indicative of the Visean while the upper member contains a Serpukhovian fauna. Previous authors attributed the lower member to the Serpukhovian and the upper member to the Bashkirian. Vachard and Massa considered the Assedjefar Formation to be limited to the Serpukhovian.[175]

The formation has been mapped around the Al Qarqaf Arch in the areas of Qararat al Marar, Bi'r Anzawa and Hasi Anjiwal. These outcrops confirm the cyclic nature of the Assedjefar Formation and the rapid facies changes. Deposition occurred in a low energy sheltered lagoonal environment with variable amounts of terrigenous input. At Hasi Anjiwal seven cycles are recognisable with a total thickness of 140m. Certain horizons have anomalously high radioactive values, due to the presence of uranium derived as volcanic dust. The formation is not exposed on the southern flank of the Al Qarqaf Arch in the Idri, Sabha and Al Fuqaha areas.[176]

On the western flank of the Murzuq Basin the Assedjefar Formation is mappable from Tikiumit to South Anay. The thickness ranges from 140m in the Hasi Anjiwal area to 50 m at Anay reflecting a gradual change from a clastic section in the north to a more calcareous facies in the south. The fining and thickening upwards cycles suggest a regressive sequence compared to the transgressive character of the Marar Formation. One horizon in the middle part of the formation is famous for its cannon-ball sandstone concretions which reach up to 1.5m in diameter, which were called by the French 'gres a champignons'.[177]

In the subsurface of the Murzuq Basin the Assedjefar Formation is composed predominantly of dark-coloured shales and fine-grained argillaceous sandstones, with oolitic limestones in the western part of the basin. The formation thins over the arch as exemplified by the C1, E1 and F1-NC 58 wells. Palynological studies on the Braspetro wells gave no evidence of sediments younger than Visean age. The formation is also present in wells F1-1, A1-68 and A1-73 where the thickness is 165m, 145m and 75m respectively. It is absent in the A1-64 well.[178]

The formation thickens to 420m in the Dur al Qussah Trough where the lithology comprises alternating shaly marls, thin sandstones and limestones with a rich fauna of Visean and Serpukhovian age. The formation continues southwards towards the border with Niger.[179]

In the Al Kufrah Basin the Zalmah Formation (equivalent to the Marar Formation) is overlain by Nubian Sandstones in the outcrops of Jabal Awaynat, Jabal Zalmah and Jabal Duhun, and in the B1-NC 43 well by continental sandstones of late Permian age. It is clear that the Kufrah Basin was emergent during the middle and late Carboniferous period.[180]

The Assedjefar Formation is present in the subsurface of the southern Ghadamis Basin and has been studied in wells A1-49 and B1-49. In this area it is divisible into a lower sandy member and an upper dominantly calcareous member with a total thickness of 135m. In the Zamzam Depression a small remnant of Assedjefar Formation is preserved around well C1-30 where a sequence of micaceous mudstones, fine-grained sandstones and microcrystalline dolomites has been dated as late Visean-early Serpukhovian. Further east in the Sirt Basin Wennekers et al. recorded quartzites on both the Nasser and Bilhizan fields which contain palynomorphs of Visean to Serpukhovian age – presumably in small remnant inliers. In the Jifarah Basin the J1-23 well contains a 187m section of limestones, dolomites and sandstones, containing foram-inifera, which Massa, Termier and Termier assigned to the Hebilia Formation. Originally regarded as partly equivalent to the Assedjefar Formation this unit is now believed to be younger.[181]

In Cyrenaica there appears to be a major hiatus between the equivalent of the Marar Formation and rocks of late Carboniferous age, except in the northeast where Serpukhovian siltstones and limestones have been recorded in wells A1-NC 92, A1-115, A1-37, A1-19 and C1-33. It is likely that much of the Cyrenaican Platform was emergent during this time with sedimentation restricted to the northeastern periphery.[182]

In the Illizi Basin of Algeria the equivalents of the Assedjefar Formation, immediately overlying the *Collenia* beds, are the Oued Assekaifaf Formation which contains abundant fossils of early Serpukhovian age, and the Oued Oubaranat Formation which contains foraminifera of late Serpukhovian age. These units contain a number of minor reservoir sands known as the 'B' reservoir units which are well developed on the Tihemboka Uplift, and contain gas at the Ohanet and In Akamil fields and oil at Zarzaitine.[183]

3.6.3 Dimbabah Formation

The Dimbabah Formation represents the last of the Carboniferous marine formations preserved in Libya. It was originally defined by Lelubre for a series of mainly carbonate rocks overlying the Assedjefar Formation in the Hamadat Tanghirt area (Figure 3.25). Collomb provided a more detailed description but did not designate a type section. In the Hamadat Tanghirt area the Dimbabah Formation comprises a lower unit of gypsiferous claystone, minor siltstones and thin-bedded argillaceous limestones containing the stomatolitic alga *Collenia* near the base. The algal horizon, which is distinguishable from that in the Marar Formation by its softness and lighter colour, is patchily developed but very widespread and has been traced as far as Jadu in Chad. The upper unit comprises dolomites and dolomitic limestones interbedded with claystones and

siltstones. The total thickness of the formation in the Hamadat Tanghirt area is about 65m. The environment of deposition is interpreted as a shallow basin, subject to occasional periods of evaporation, with limited input of clastic rocks. In his re-examination of the area, Berendeyev did not mention the *Collenia* occurrence, but reported a rich fauna of brachiopods, echinoderms and gastopods which give a Bashkirian age for the lower unit and a Moscovian to Gzelian age for the upper unit. The Gzelian age has not been confirmed from any other location in Libya. If this determination is reliable it represents a considerable expansion of the age of the formation, which was not previously believed to extend above the Moscovian. Vachard and Massa and Mergl and Massa regarded the Dimbabah Formation as restricted to the Bashkirian-Moscovian.[184]

West of the Al Qarqaf Arch the Dimbabah Formation outcrops in the Qararat al Marar, Hasi Anjiwal and Bi'r Anzawa areas where it averages 120m in thickness. Here the *Collenia* horizon is present, and the claystones contain significant amounts of gypsum and halite. A band of oolitic sandstone containing analcite occurs in the middle of the formation, normally only a few metres thick, but reaching almost 30m at Al Awaynat. Analcite is a zeolite which, in its bedded form, is characteristic of alkaline lakes. Other authors prefer a volcanic explanation for this mineral. The analcite horizon contains 13.5ppm of thorium. The thickness of the Dimbabah Formation averages 120m in this area, but the upper horizons have been removed by post-Carboniferous erosion at Anay where only the basal 30m is preserved. The Dimbabah Formation in this area contains a rich fauna of foraminifera, conodonts, corals and brachiopods which prove a Bashkirian-Moscovian age.[185]

Outcrops of the Dimbabah Formation can be traced southwards along the western margin of the Murzuq Basin from Tikiumit to the border

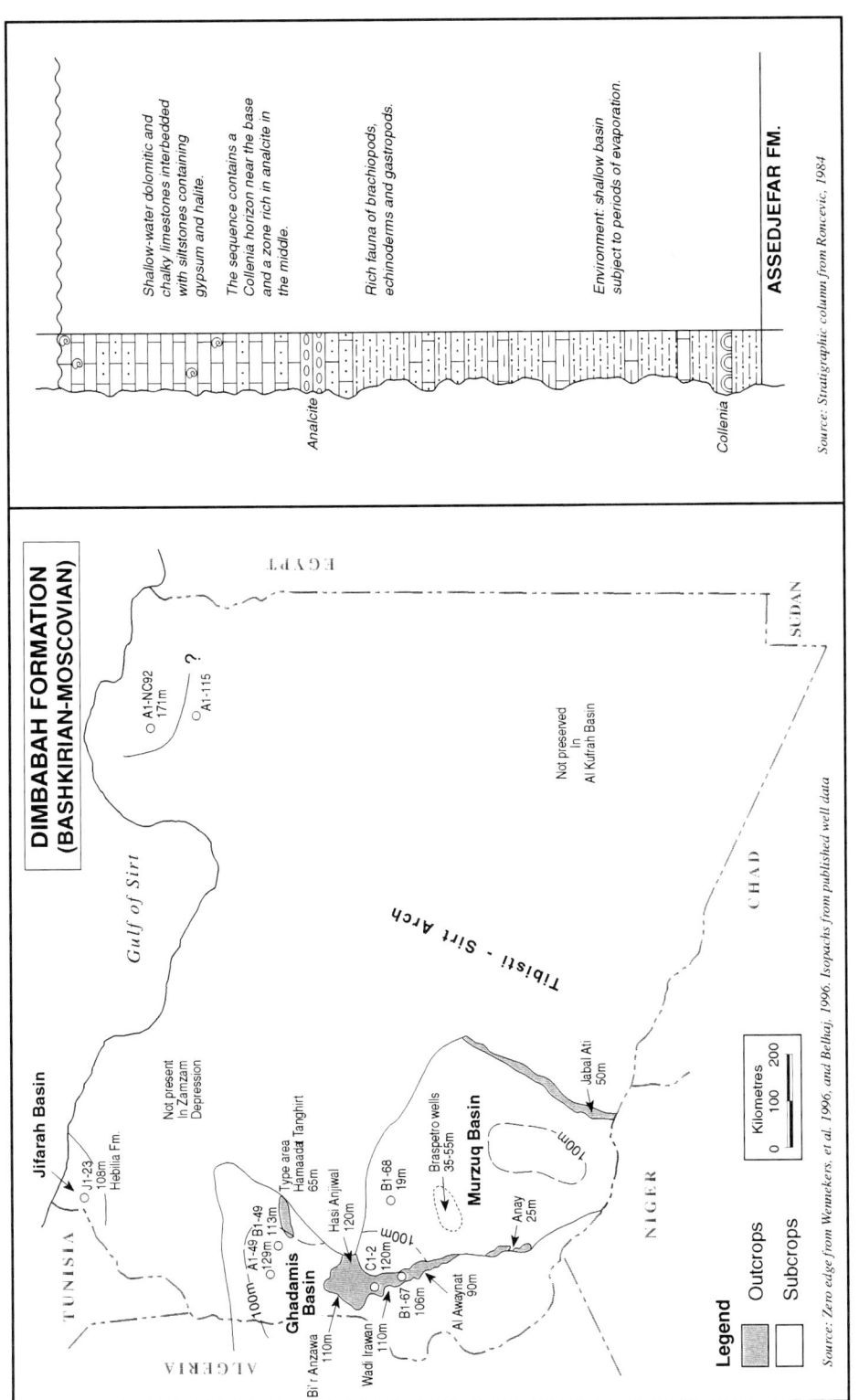

Figure 3.25   Dimbabah Formation, Distribution and Stratigraphic Column

The Dimbabah Formation represents the last of the Palaeozoic marine deposits preserved in Libya before the late Carboniferous uplift and regression. The formation comprises chalky, dolomitic limestones and gypsiferous siltstones which were deposited in a shallow marine basin subject to occasional periods of evaporation. The type area of the Dimbabah Formation was defined in the Hamadat Tanghirt area and the formation is present in the Murzuq Basin and in the southern Ghadamis Basin. It has not been preserved in the Al Kufrah Basin. In northern Libya two areas of late Carboniferous marine limestones are preserved in northeastern Cyrenaica and in the Jifarah Basin which are equivalent to the Dimbabah Formation.

with Niger. The formation thins from 110m at Wadi Irawan to 25m at Anay due to post-Carboniferous erosion. In this area the formation maintains its dominantly calcareous aspect with the *Collenia* horizon marking the base, and with occasional lumachelles of pelecypods and broken shell fragments. Clastic input was limited to fine-grained reddish-brown siltstones and rare microconglomerates. The sequence is characteristic of a shallow evaporitic basin. Dolomites are common, and lenses of celestite are also present within the carbonates.[186]

In the subsurface of the Murzuq Basin the Dimbabah Formation has been encountered in many wells. It thins from 120m in the C2-1 well to 106m in B1-67 to less than 60m in the Braspetro wells, and to only 19m in B1-68. Well data suggest a gradual transition from open-marine shallow shelf conditions in the north to progressively more evaporitic conditions in the south. Palynological studies on well samples (where the upper part has been removed by erosion) give a Bashkirian to Moscovian age.[187]

On the eastern flank of the Murzuq Basin Klitzsch reported outcrops of rhythmical alternations of marls, limestones and dolomites, containing brachiopods and nautiloids of Serpukhovian to Moscovian age, which reach a thickness of 50m in the Jabal Ati area (Figure 3.25). The formation has been traced into eastern Niger where it becomes progressively more evaporitic, but where *Collenia* is still present in association with gypsum deposits. The Dimbabah Formation is not present in the Al Kufrah Basin.[188]

In the subsurface of the Ghadamis Basin the Dimbabah Formation is represented by microcrystalline limestones and dolomitic marls containing brachiopods and fusulinid foraminifera which indicate a Bashkirian to early Moscovian age. The thickness is 113m in the B1-49 well and 129m in A1-49. The formation is not preserved in the Zamzam

Depression. In the Jifarah Basin the upper part of the Hebilia Formation in well J1-23, comprising 108m of foraminiferal limestones and crystalline dolomites was originally thought to be equivalent to the Dimbabah Formation, but it is now clear that the Hebilia Formation unconformably overlies the Dimbabah Formation. In Cyrenaica there is no evidence of the presence of rocks of Bashkirian age, but 170m of Moscovian oolitic limestones, shaly limestones and sandstones have been penetrated in well A1-NC 92 which may be equivalent to the Dimbabah Formation of western Libya. It is likely that most of Cyrenaica was emergent during that time.[189]

In the Illizi Basin of Algeria the equivalent of the Dimbabah Formation is the El Adeb Larache Formation with a dominant lithology of marls and bioclastic limestones with gypsum, containing a foraminiferal fauna which gives a Bashkirian-Moscovian age. The formation reaches 80m in thickness near the Libyan border and thickens westwards. The insignificant 'A' reservoir occurs near the base of the formation.[190]

3.6.4 Tiguentourine Formation

The Tiguentourine Formation represents the final regressive event of Carboniferous sedimentation prior to the culmination of the Hercynian orogeny. By Kasimovian/Gzelian times a major marine regression had taken place and much of Libyan territory had become emergent. Sediments deposited during this time began with shallow-marine sequences in the Murzuq Basin but were rapidly replaced by continental, fluvial and lagoonal deposits. They form part of a thick sequence of continental rocks deposited on the north African Platform between the late Carboniferous uplift and the mid-Cretaceous marine transgression. These rocks occur widely in Algeria where they were first lumped into a broad group called the 'Continental Post-

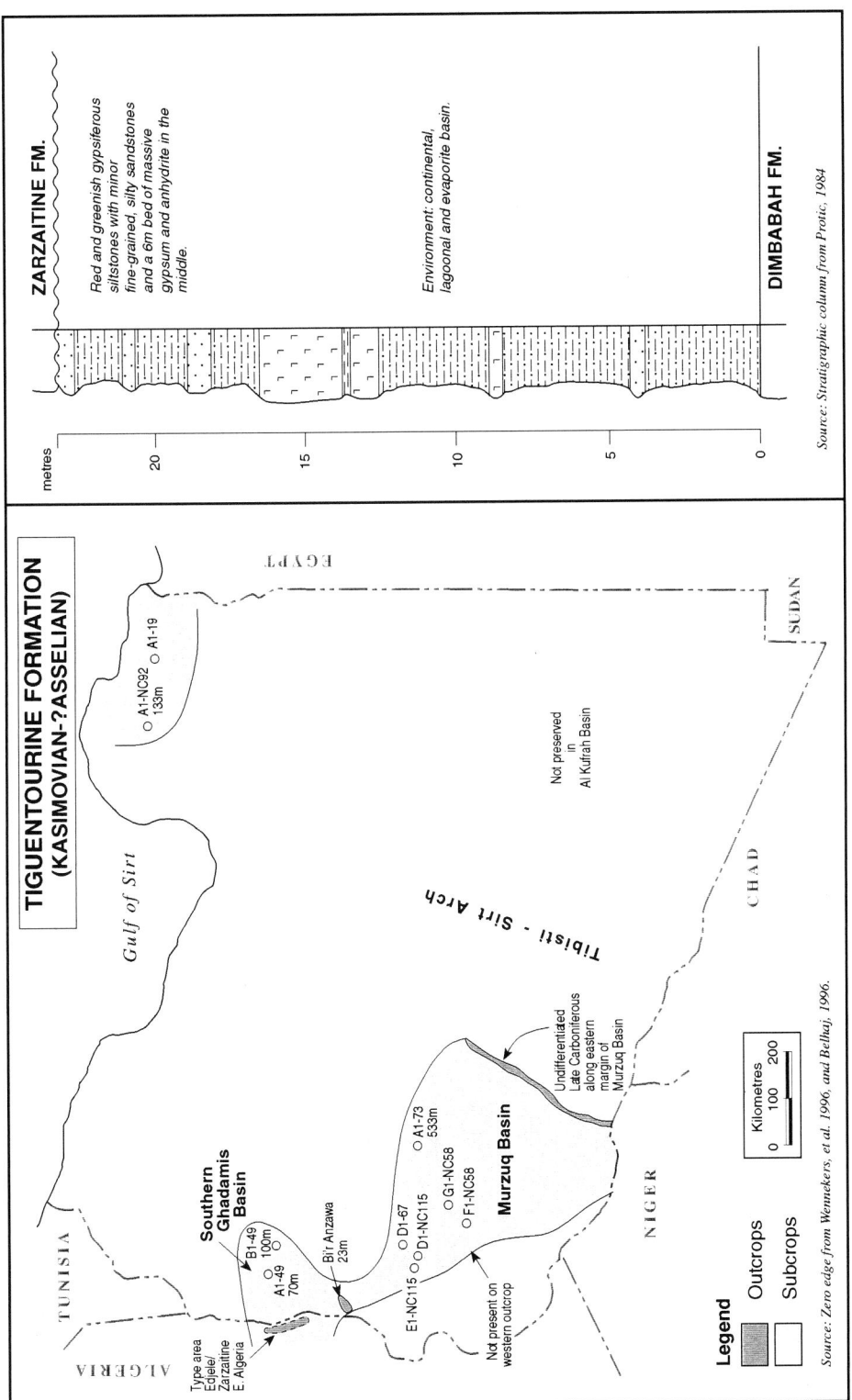

Figure 3.26 Tiguentourine Formation, Distribution and Stratigraphic Column

The Tiguentourine Formation was defined in the Edjele area of eastern Algeria. It outcrops in Libyan territory in the Bi'r Anzawa area, and possibly on the eastern flank of the Murzuq Basin, although these outcrops have not been precisely dated. It represents the final regressive event before the Hercynian orogeny. In the Murzuq Basin it begins with a marine phase, but is quickly replaced by the continental fluvial, lagoonal and evaporite basin conditions which are characteristic of the formation. It is present over much of the Murzuq and southern Ghadamis Basins. Equivalent rocks are present in northeastern Cyrenaica. On palynological evidence the age of the formation is Kasimovian – ?Asselian.

Tassilien' by Kilian, and were later subdivided by de Lapparent and Lelubre into three formations: the Tiguentourine, Zarzaitine and Taouratine Formations. These rocks are well developed on the Tihemboka Arch around Edjele on the Algerian-Libyan border.[191]

The Tiguentourine Formation, in the type area of eastern Algeria, comprises 80m of reddish dolomitic claystones with gypsum, interbedded with thin limestones and dolomites containing ostracods, and minor sandstones (Figure 3.26). It has been mapped in the Bi'r Anzawa area of Libya where 23m of red and greenish gypsiferous siltstones are exposed, with a 6m bed of massive gypsum and anhydrite in the middle. The age was given by Protic as late-Carboniferous, but correlation with wells in the northwestern Murzuq Basin, which have been dated palynologically, suggests that it extends into the early Permian.[192]

According to the surveyors of the Geological Map of Libya the Tiguentourine Formation is not present at outcrop along the western margin of the Murzuq Basin where the Dimbabah Formation is overlain directly by the Triassic Zarzaitine Formation. However, in the subsurface of the northern Murzuq Basin it has been identified in wells D1-67, E1 and D1-NC 115, and A1-73, reaching a thickness of 533m in the latter well. Further south the Tiguentourine Formation is present in the Braspetro wells where a lower marine sequence has been dated as late-Carboniferous and the upper more continental section gives an early Permian age. In the subsurface the formation extends southwards into the Jadu Trough of Niger. On the eastern flank of the basin thick continental sandstones are present but their age has not been established.[193]

No rocks equivalent to the Tiguentourine Formation have been found either at outcrop or in wells in the Al Kufrah Basin. In northeast Libya 120m of silty shales with thin sandstones and coal laminations, capped by sandy limestones in wells A1-NC 92 and A1-19, are probably equivalent to the Tiguentourine Formation of west Libya. These rocks have been dated as Gzelian to Asselian on the basis of palynology.[194]

In the Ghadamis Basin the Tiguentourine Formation is present in the subsurface comprising red-brown dolomitic shales, shaly dolomites and anhydrites. The thickness is 70m in well B1-49 and 100m in well A1-49, and the age has been determined by palynology as Gzelian to Asselian.[195]

The Tiguentourine Formation in the Illizi Basin comprises continental and lacustrine sediments with a unit of gypsum in the middle. The thickness ranges from 200m to 280m. The age is assumed to be late Carboniferous, but on the basis of recent palynological work in Libya probably extends into the early Permian.[196]

3.7 Permian

The Tiguentourine Formation extends into the Asselian stage of early Permian age. Thereafter the record is very fragmentary due to the effects of the Hercynian orogeny. There is evidence of continued marine sedimentation in Cyrenaica at least until Ufimian time, but virtually all the rest of Libya became emergent, major new tectonic elements were established and a continental regime was established over most of Libya which continued until the mid-Cretaceous. In the Murzuq Basin estimates of the amount of uplift during the orogeny range between 300 and 2000m. During this period the exposed Palaeozoic and Precambrian deposits were subjected to extensive erosion, and thick deposits of continental alluvial, fluvial and lacustrine sediments accumulated in the newly formed interior basins. During the late Permian the proto-Tethyan shoreline became established close to the crest of the Jabal Nafusah Uplift, north of which marine late Permian rocks are present in wells in the Jifarah Basin (Figure 3.27).

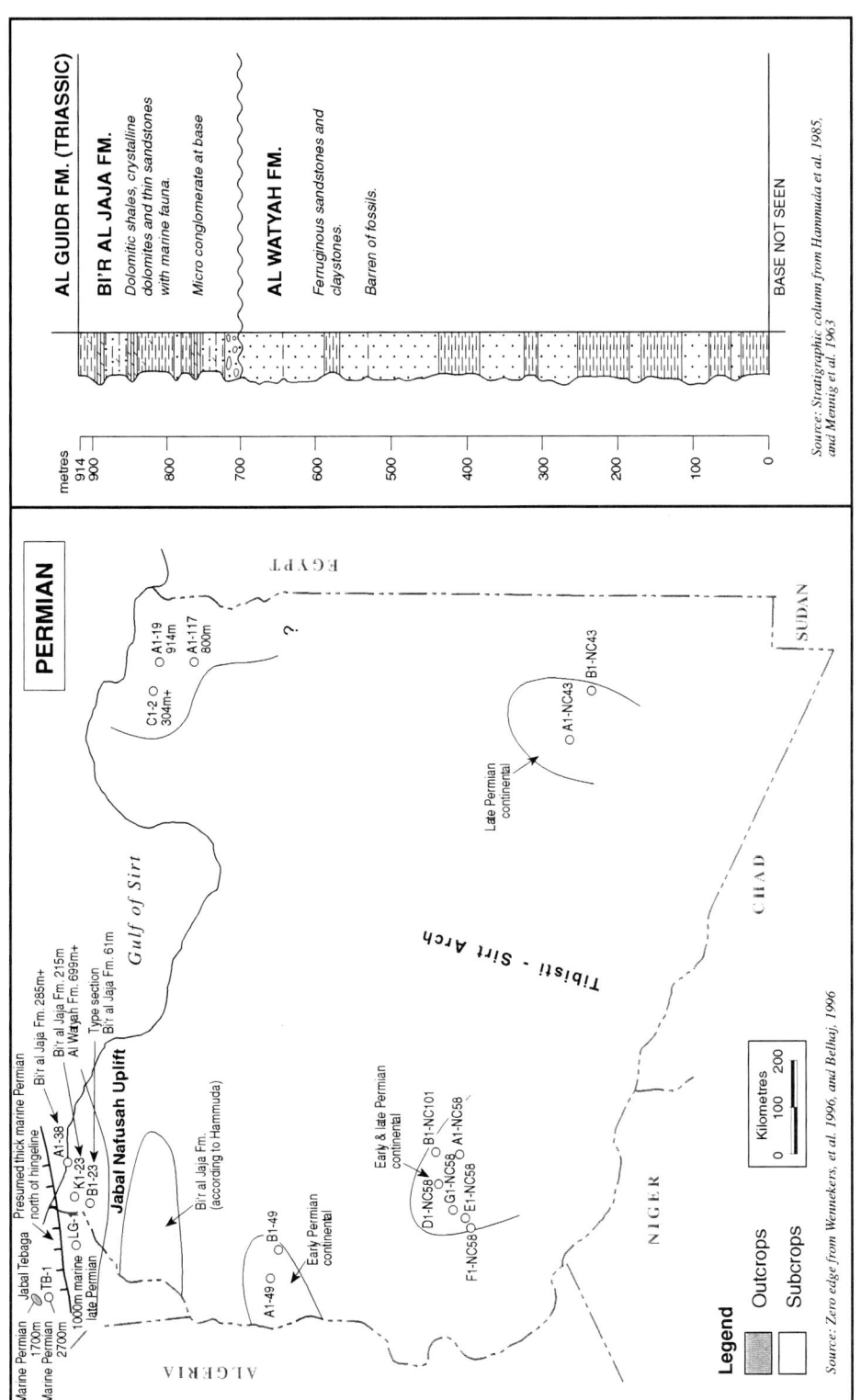

Figure 3.27   Permian, Distribution and Stratigraphic Column

Palynological data proves that the Tiguentourine Formation extends into the early Permian in the Murzuq, southern Ghadamis Basins, and northeastern Cyrenaica. Late Permian continental rocks are present in the Murzuq and Al Kufrah Basins. Thick marine late Permian carbonates are exposed at Jabal Tebaga in Tunisia and are presumed to extend into the western offshore area of Libya. These rocks thin southwards both in Tunisia and Libya. They are present in wells in the Jifarah Basin of northwest Libya and pinch out against the Jabal Nafusah Uplift. They may also be present in the northern Ghadamis Basin. The stratigraphic column is based on the K1-23 well, but in the absence of published data the Al Watyah lithology has been generalised.

No outcrops of Permian marine rocks are known onshore in Libya, but a unique inlier at Jabal Tebaga, 25km northwest of Medenine in Tunisia, exposes 1700m of upper Permian marine sediments, and a further 2700m were penetrated in the nearby Tebaga well. This sequence thins rapidly southwards to less than 1000m in the Kasbah Leguine and Kirchaou wells. Details of the Permo-Carboniferous section in wells on the Kirchaou Permit in southern Tunisia have been provided by Benzarti. The KLF-1 well found 1540m of late Permian carbonates beneath a cover of Mesozoic rocks, unconformably underlain by thin early Permian clastics. The KJD-1 well penetrated 986m of uppermost Permian clastics underlain by 2052m of late Permian dolomites and limestones, and towards the Libyan border the BMT-1 well drilled early Triassic (Scythian) marine rocks, conformably underlain by 555m of late Permian dolomites and evaporites, 190m of early Permian dolomites, and 900m of Carboniferous limestones. The biostratigraphy and palaeoenvironment of the Permo-Triassic sequence of southern Tunisia has also been reviewed by Kilani-Mazraoudi, et al. In Cyrenaica marine Permian rocks are restricted to the early Permian. The problem of the Carboniferous-Permian boundary in northeast Libya was discussed by Brugman, et al. They took the view that it is not possible to precisely define the Carboniferous-Permian boundary in northeast Libya since it has not been possible to separate the Gzelian from the Asselian. In general terms it occurs within the Tiguentourine Formation and its equivalents.

In the offshore, Permian rocks are too deep to have been penetrated by drilling but seismic data show that thin Permian rocks can be interpreted in the southern Sabratah Basin and in the Misratah Basin but pinch out before reaching the Jarrafa Arch and are not present further north on the Pelagian Block. Further east, in the Sirt Embayment seismic evidence suggests that thin Permian rocks extend to the foot of the lower Sirt Slope.[197]

3.7.1 Al Watyah Formation

In Libyan territory Permian rocks have been encountered in wells in the Jifarah Basin north of the Jabal Nafusah Uplift. The name Al Watyah Formation was proposed for a sequence of claystones and sandstones with ferruginous cement occurring in the K1-23 well beneath the Bi'r al Jaja Formation of known late Permian age. The thickness is 699m, with the base not reached. The formation is unfossiliferous and its attribution to the early Permian is based solely on stratigraphic position. The formation has been interpreted as marginal marine to fluvial, and shows delta-front, tidal-channel and distributary-mouth bar characteristics (Figure 3.27).

In Cyrenaica early Permian marine rocks have been found overlying the Tiguentourine Formation apparently without interruption. Well A1-19 shows a sequence of 510m of limestones, sandy shales and dolomites which have been dated as Sakmarian. These rocks represent a marginal marine environment which becomes progressively more deltaic upwards. An overlying sequence of continental to deltaic aspect is found in wells A1-19, A1-117 and C1-2. This unit, which has a thickness of 400m, has been dated as Sakmarian to Ufimian. The distribution of these rocks is not precisely known, but a NNW-SSE half-graben was shown by Keeley and Massoud in this area during the early Permian which they named the Tehenu Basin. These formations have not been named or formally described, but they are obviously of some importance, and a formal description is urgently needed. They are unconformably overlain by rocks of middle Triassic age.[198]

3.7.2 Bi'r al Jaja Formation

Unconformably overlying the Al Watyah Formation in the K1-23 well is a fully marine deposit of limestones, dolomites, clays and fine-grained sandstones, comparable to the sequences in the Kasbah Leguine wells in Tunisia. This unit has been named the Bi'r al Jaja Formation. The formation contains foraminifera and a rich palynological assemblage which confirm an age extending from Tatarian to Scythian. The thickness in well K1-23 is 215m. Hammuda et al. formally designated a type section for the Bi'r al Jaja Formation in well B1-23 where it is only 61m thick. They also designated two hypostratotypes: wells K1-23, and A1-38 near Az Zawiyah. Evidence from Tunisia suggests that the formation may thicken dramatically to the north. Southwards the formation pinches out on the flank of the Jabal Nafusah Uplift, but Hammuda also reported its presence in the northern Ghadamis Basin (Figure 3.27).[199]

Elsewhere in Libya late Permian rocks have been identified in several areas, but all in continental facies. In the Murzuq Basin palynological investigation has been carried out on several wells. Wennekers et al. showed a widespread area of Permian continental rocks reaching over 300m in thickness in the centre of the basin (attributed by Sikander to the Tiguentourine Formation), and Meister showed progressive thickening of the Permian section from west to east, reaching 600m in well B1-NC 101. The E1-NC 58 well has also been examined palynologically, but much of the section is barren and the results were inconclusive. Wennekers et al. also recorded the occurrence of a small Permian inlier on the Amal field in the Sirt Basin.[200]

In the Al Kufrah Basin Carboniferous rocks are unconformably overlain by 'Continental Post Tassilien' rocks. Palynological investigation of wells A1 and B1-NC 43 showed that the lower part of this sequence in both wells can be dated as late Permian (109m in A1-NC 43 and 315m in B1), and it is likely that this is also the case in other parts of the basin.[201]

In northeast Libya no evidence has been found of Permian rocks younger than Ufimian age, and it is assumed that the region must have been emergent during this time.[202]

Bellini and Massa provided no evidence for the presence of Permian rocks in the Ghadamis Basin, however as noted above Hammuda showed the presence of Bi'r al Jaja equivalents as far south as latitude 30°30'N. Otherwise it is likely that the only Permian rocks in northwest Libya (south of the Jabal Nafusah Arch) are those previously mentioned as forming the uppermost unit of the Tiguentourine Formation. Similarly in eastern Algeria Permian age rocks appear to be limited to undifferentiated continental rocks, interbedded with andesite flows, at the base of the Continental Post Tassilien.[203]

Chapter 4

STRATIGRAPHY
PART TWO: MESOZOIC

MESOZOIC

The collision of western Gondwana with Laurasia, and the late Palaeozoic Hercynian orogeny produced profound consequences in Libya (and incidentally the most important mass extinction event in the Phanerozic record). A series of new tectonic elements was formed with a dominantly southwest-northeast trend, chief of which was the Tibisti-Sirt Arch. This broad uplift, 400km wide and 900km long, extends from the Tibisti Mountains to the Gulf of Sirt. From the known thickness of Palaeozoic sediments in the Ghadamis and Murzuq Basins the Arch must have been uplifted by at least 3000m. Other similar but smaller uplifts were produced: the Ennedi-Al Awaynat Uplift in the southeast, the Al Qarqaf Arch in west Libya, and the Nafusah Uplift in the northwest. Mesozoic outcrops and the intervening arches are shown on Figure 4.1. The effect of these events was threefold: the Palaeozoic depocentres were separated into the Al Kufrah, Murzuq and Ghadamis basins, widespread and prolonged erosion occurred on the emergent arches, and major marine regression took place over most of Libya, with a shoreline becoming established in the vicinity of the Jabal Nafusah and the northern margin of the Jabal al Akhdar. This new regime was established by early Permian times, and persisted until the marine transgression which began in early Cretaceous. Continental sediments were deposited over much of Libyan territory, with marine rocks being confined to the northernmost areas, and the offshore. The stratigraphic nomenclature of the Mesozoic rocks of Libya is shown in Figure 4.2.

4.1 TRIASSIC

Triassic rocks in Libya occur in four main domains: the predominantly marine rocks of the Nafusah escarpment and the offshore, the eastern extension of the Algerian Triassic Basin into the Ghadamis area, the grabens in the subsurface of eastern Libya, and the continental rocks of the interior. Rubino, et al. divided the Triassic sequence of the Nafusah escarpment into four sequences separated by unconformities. Sequence 1 equates to the Al Guidr/ Kurrush Formations, sequence 2 to the Al Aziziyah, sequence 3 to the Abu Shaybah Formation and sequence 4 to the Abu Ghaylan carbonates. In the offshore the Triassic is mostly too deep to have been penetrated by wells but seismic data shows thick Triassic sediments to the north of the Sabratah-Cyrenaica Fault. Triassic rocks are well developed in the Sabratah Basin and evaporites are present in the late Triassic in the western part of the basin. Over 1000m of Triassic sediments have been interpreted on the fault terrace north of well E1-34, and the sequence thickens gradually northwards, reaching 1500m in the Misratah Basin and 3000m on the Medina Bank. In the Sirt Embayment the Triassic is thinner, averaging no more than 500m on the upper Sirt Slope, and probably less than 200m on the Cyrenaica Ridge. Three distict facies associations have been identified in onshore Cyrenaica: non-marine sandstones of Anisian-Ladinian age

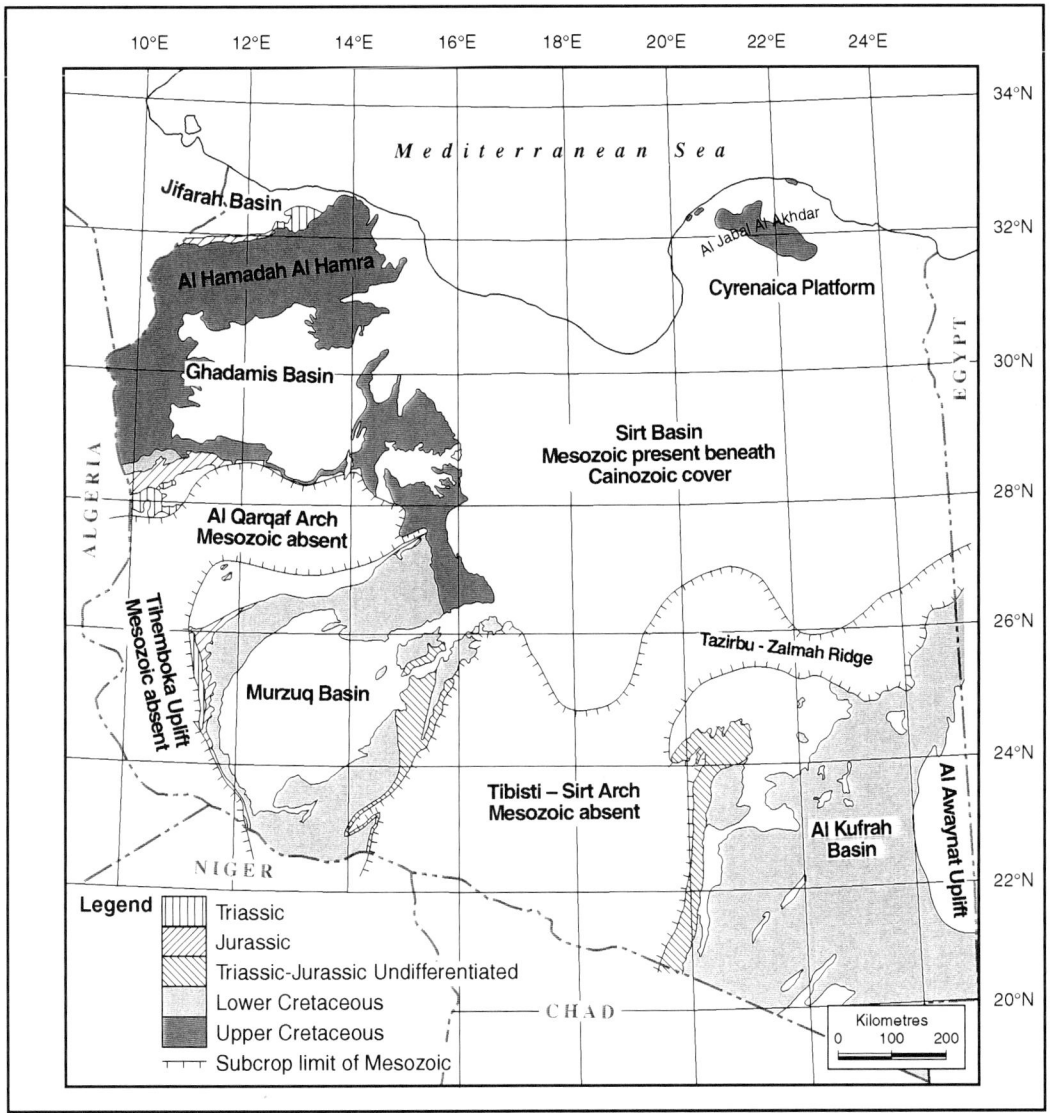

Sources: Geological Map of Libya, scale 1:1,000,000, 1985, Wennekers, et al. 1996

Figure 4.1   Libya: Mesozoic Outcrops and Subcrop Limit

Mesozoic rocks are present over much of Libya, with the exception of the principal Hercynian uplifts. Rifting began in the Triassic and a series of horsts and grabens developed on the Sirt Arch, some of which contain Triassic and Jurassic sediments. During this period marine rocks were confined to the northern margin of Libya and the interior was dominated by continental deposition. The early Cretaceous is also dominated by the presence of widespread continental sandstone including the Nubian and Messak Formations. The rift phase ended with the collapse of the Sirt Arch in the early Cretaceous and the flooding of northern Libya by a marine transgression. The collapsed Sirt Arch was progressively flooded during the late Cretaceous, and by the end of the Maastrichtian only a few small islands remained.

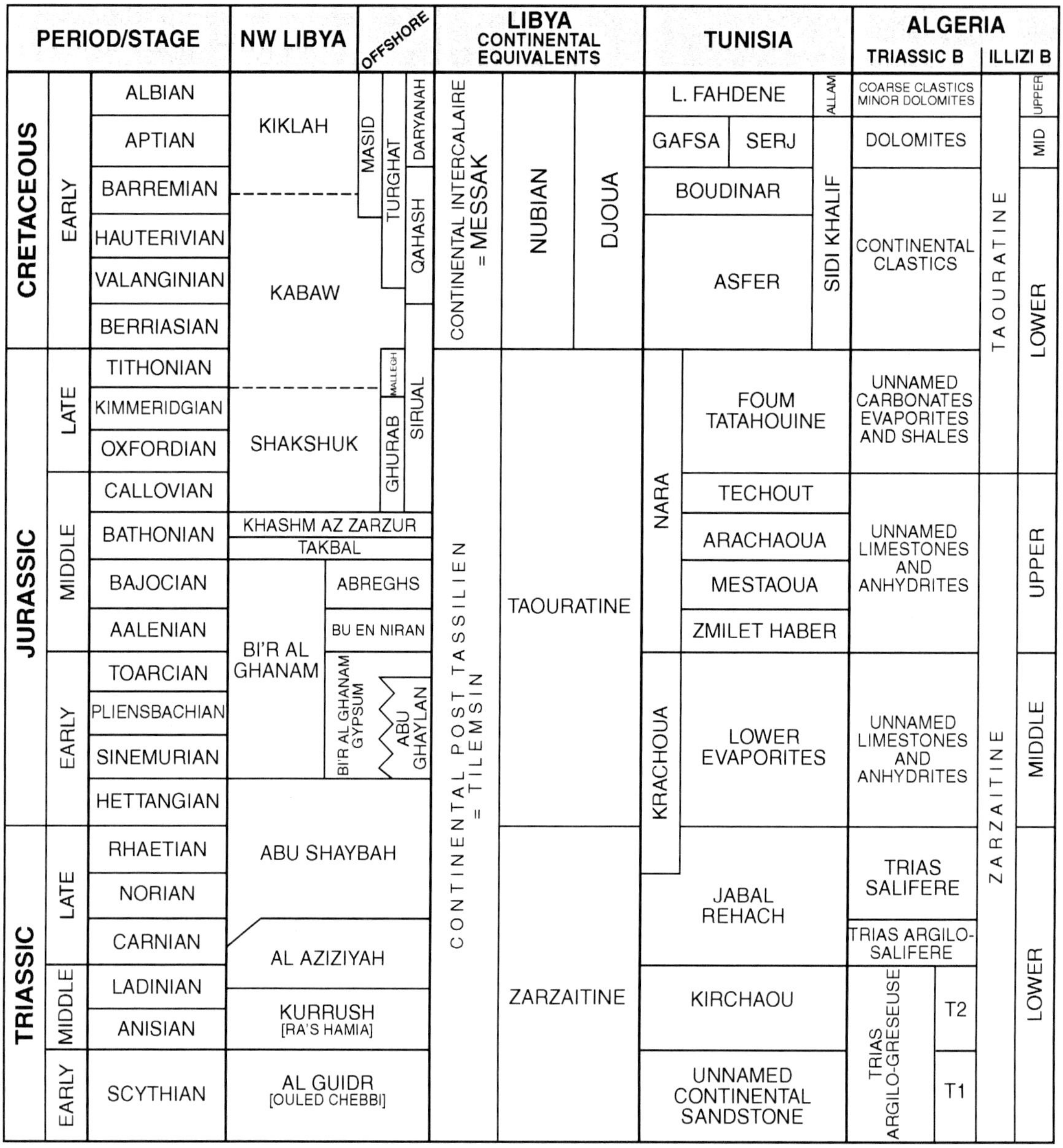

Figure 4.2 Lithostratigraphic Nomenclature for the Triassic, Jurassic and Lower Cretaceous Rocks of Libya and Adjacent Areas

Source: See accompanying text

This chart shows the current lithostratigraphic nomenclature for the Mesozoic of Libya (except for the Upper Cretaceous which is shown on Figure 4.10). The oblique line represents differing views on the age of the Abu Shaybah Formation. No attempt has been made to show unconformities or gaps in the section on this chart.

in southern Cyrenaica, lacustrine and lagoonal deposits of Anisian age in eastern Cyrenaica, and shallow, shelfal carbonates of Anisian to Rhaetian age in northern Cyrenaica. No deposits of early Scythian or Norian age have been found in this area.[1]

4.1.1 Al Guidr (Ouled Chebbi) Formation

The oldest Triassic rocks in northwest Libya were described from wells in the Jifarah Basin, and informally named the Ouled Chebbi Formation by Mennig et al. in 1963. It is known only in the subsurface, although it outcrops in Tunisia near Medenine. The formation was mentioned only incidentally by Mennig and was not formally described or defined. Subsequently Hammuda et al. gave a formal description, designated the B1-23 well as the type locality, and renamed the unit the Al Guidr Formation. The redefined formation comprises 244m of fine- to medium-grained lignitic sandstones with minor grey, green and reddish micaceous shales. The sandstones contain rare dolomite bands and are pyritic towards the base. The formation is probably unconformable with the underlying late Permian Bi'r al Jaja Formation since there is a distinct change in dip direction between the two formations on the dipmeter log in well A1-38. The environment of deposition appears to be continental to marginally marine in well A1-38. It is overlain conformably by the middle Triassic Kurrush Formation. The formation is unfossiliferous and its assignment to the early Triassic is based solely on stratigraphic position, although Rubino, et al. indicated a definite early Scythian to late Ladinian in Tunisia. In well K1-23 the formation reaches a thickness of 866m, becoming coarser towards the top. It is also present in wells A1-23, F1-23, G1-23, J1-23, M1-23 and A1-131. (Figure 4.3).[2]

Belhaj reported the presence of marginally marine sandstones and mudstones in wells D1-8, B2-34, A1-8 and A1-34 which he assigned to the Al Guidr Formation and which he implied contain Scythian palynomorphs. Details however were not provided. This is unfortunate since it would represent the first positive dating of the Al Guidr Formation. At present the formation has not been recognised outside the Jifarah/Zamzam area, but it is probably present at depth in the offshore Sabratah Basin, and it extends westwards into Tunisia where it is represented by 800m of unnamed massive red sandstones. No rocks of Scythian age have been found in Cyrenaica.

In the Triassic Basin of Algeria the Triassic sequence is divisible into a lower clastic section and an upper evaporite section. The lower part of the clastic sequence, which contains the T1 reservoir, has been dated palynologically as early Triassic. This unit, which represents a major braided-stream complex, extends along the eastern flank of the El Biod Arch in Algeria and passes into a fluvial-deltaic system south of the Dahar Uplift in Tunisia. It enters Libyan territory north of Ghadamis, and is present in the subsurface in former concession 90. This unit is an approximate age equivalent of the Al Guidr Formation.[3]

4.1.2 Kurrush (Ra's Hamia) Formation

The term Kurrush Formation was introduced by Desio in 1960 for a series of reddish, fine-grained, micaceous sandstones, silty shales and dolomitic limestones exposed in several small domal inliers north of Gharyan. In the same year Burollet named this same unit the Ra's Hamia Formation, but Desio's name has priority. Recent remapping by IRC confirmed that the formation outcrops in a small area north of Gharyan. About 40m is exposed but the base is not seen. The upper contact with the Al Aziziyah Formation is gradational. The formation contains a rich fauna of pelecypods, echinoderm fragments, gastropods, foramin-

148

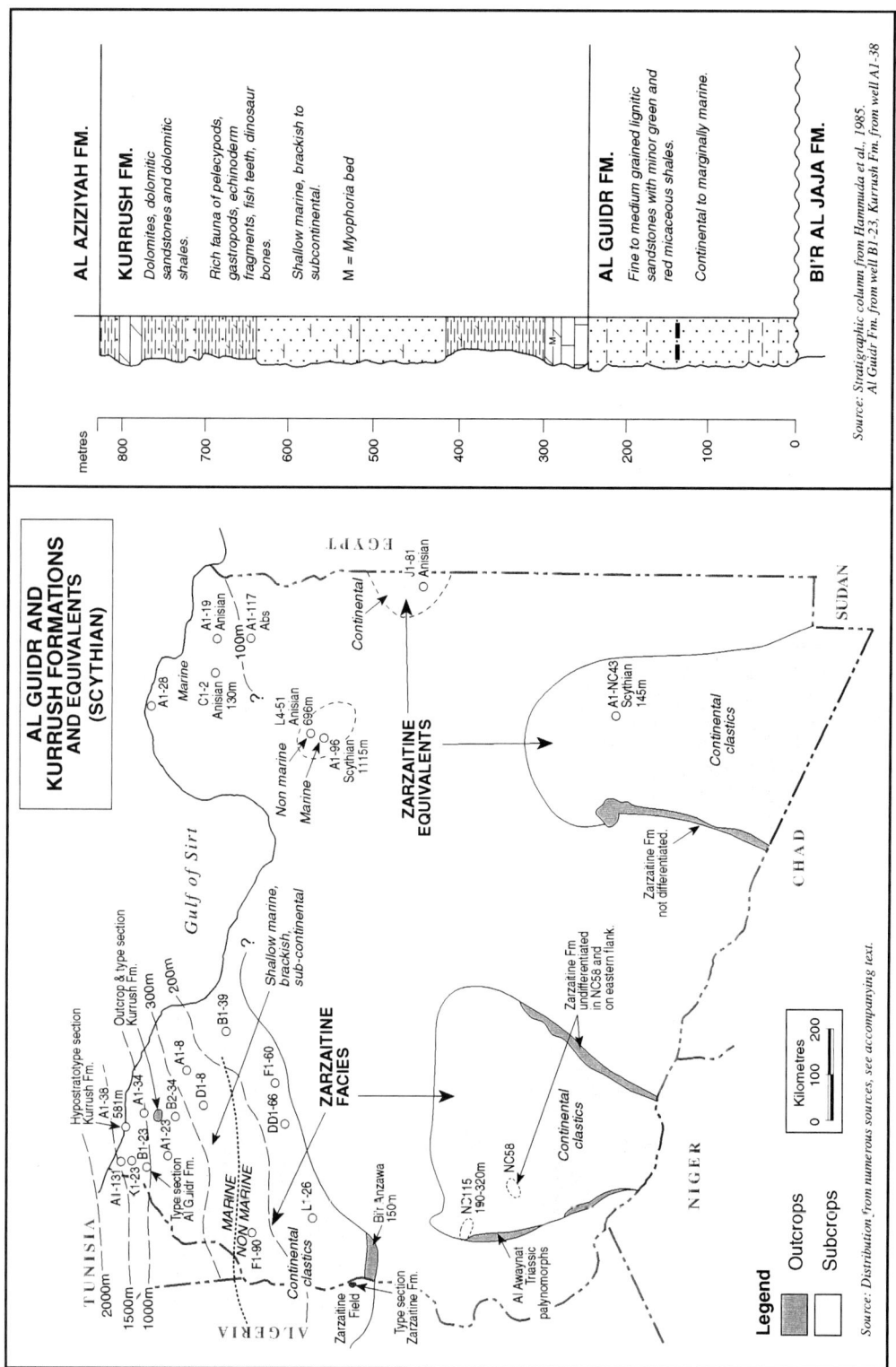

Figure 4.3   Al Guidr and Kurrush Formations: Distribution, Equivalents and Stratigraphic Column

The Al Guidr and Kurrush Formations are confined to northwestern Libya and adjacent areas of Tunisia and Algeria. These rocks form minor oil reservoirs in the area of former concession 90. Further south continental equivalents are represented by the lower part of the Zarzaitine Formation. In Cyrenaica marine, lagoonal and continental equivalents have been identified, and in the Al Kufrah Basin early Triassic continental equivalents are present.

ifera, fish teeth, and bone fragments of the sauropod *Nothosaurus*, which allow it to be dated as Anisian/Ladinian (Figure 4.3).[4]

The formation is present in the subsurface of the Jifarah Basin. In well A1-38, where the thickness is 581m, a hypostratotype was designated by Hammuda et al. The basal contact with the Al Guidr Formation appears to be gradational. The maximum known thickness of 993m has been found in well A1-131, and the formation occurs in many of the Jifarah Basin wells. A zone rich in *Myophoria* characterises the basal 75m in the subsurface. A sedimentological study by Mennig et al. showed that the formation ranges from subcontinental in the south, through brackish, to a zone with marine incursions in the north. The formation thickens rapidly northwards, and presumably becomes very thick in the Sabratah Basin, but has not been penetrated in this area.

The Kurrush Formation has a wide distribution in the subsurface of the Ghadamis Basin where it unconformably overlies the truncated Palaeozoic formations which were folded, uplifted and eroded during the Hercynian orogeny. It averages 250m in thickness in the northern part of the basin, but thins southwards and pinches out south of wells DD1-66 and F1-60. The formation is an excellent aquifer in this area. The eastern limit of the Kurrush Formation was mapped by Belhaj. He showed the formation to be present in wells E1-60, D1-8, A1-8, B2-34 and A1-34, but absent in all wells to the southeast. In southern Tunisia the Kirchaou Sandstone which has been dated as Anisian-Ladinian in age is an equivalent of the Kurrush Formation, and in central Tunisia it is equivalent to part of the Jabal Rheouis Formation.[5]

In Algeria the T2 reservoir of the Trias Argilo-greseuse has been dated palynologically as mid-Triassic, although it may extend into the late Triassic. It represents a fluvial sequence and floodplain shales with crevasse splays and palaeosols overlain by sabkha evaporites. It extends into Libyan territory north of Ghadamis and is present in some of the old concession 90 wells where it represents a modest oil reservoir. It is regarded as correlative with the Kurrush Formation of the Jifarah Basin.

In Cyrenaica sediments of middle Triassic age have been identified in three distinct facies. A non-marine sequence of reddish sandstones with thin shales and coals, about 120m thick, is present in well J1-81a, and similar sequences have been found in wells I1c-81a and A1a-84. This sequence contains palynomorphs which were originally attributed to the Norian-Rhaetian, but have subsequently been reassigned to the Anisian-Ladinian. In the Maragh Graben lacustrine and lagoonal shales and sandstones ranging from 425m to 570m in thickness have been dated as latest Scythian to Anisian, and further north restricted marginal marine carbonates and clastics have been found in wells A1a-117, A1-19, C1-2 and A2-2. These littoral sediments, which reach a thickness of 400m, have been dated as Ladinian-Carnian in age, and are taken to represent the southern margin of the Tethys Ocean in this area.[6]

4.1.3 Al Aziziyah Formation

The Al Aziziyah Formation was first described by Parona as the Azizia Limestone from a quarry outcrop near Gharyan. The Gharyan outcrops were later studied by Christie who established a type section, and formalised the unit as the Al Aziziyah Formation. The formation has undergone numerous revisions since 1955. The area was remapped by El Hinnawy and Cheshitev for the Geological Map of Libya in 1975. The formation was redefined and a new type section was selected. The base was taken at the first appearance of carbonates overlying the Kurrush Formation. The upper boundary was defined at the unconformity with the overlying continental sandstones of the Abu

Shaybah Formation. As redefined at the type locality the Al Aziziyah Formation comprises 160m of dark grey, hard, dolomitic limestones with occasional marls, claystones and chert bands. The top is marked by a thin unit with phosphate bands. The formation contains an abundant pelecypod fauna which indicates a Ladinian/Carnian age. The formation also contains characteristic algal laminae. The sedimentology of the formation was examined by Assereto and Benelli who recognised six lithofacies which collectively suggest a low-energy, inner-neritic, subtidal to intertidal environment. A subsequent review by Fatmi et al. in 1980 proposed the subdivision of the formation into two members.[7]

Outcrops of the formation are restricted to the area north of Gharyan, but it occurs in the subsurface throughout the Jifarah Basin and extends into Tunisia. The distribution and stratigraphy of the Al Aziziyah Formation is shown on Figure 4.4. Hammuda et al. designated a hypostratotype section in well A1-38 to illustrate the considerable thickening towards the north. In the well the Al Aziziyah Formation has a thickness of 706m. The lithology is essentially the same as at outcrop. Banerjee reported that the formation is also present in the subsurface of the Ghadamis Basin. The equivalent formations in Tunisia appear to be the lower part of the Jabal Rehach Dolomite in the south and the Jabal Rheouis Formation in central Tunisia.[8]

In the subsurface of the Ghadamis Basin the Aziziyah Formation can be traced across the northern part of the basin from the Tunisian border in well K1-90 to well D1-70 and eastwards to F1-60, but it cannot be differentiated to the south of this line. Palynological studies on wells southeast of Tripoli, reported by Belhaj, show the presence of fine-grained sandstones and mudstones in wells D1-8, B2-34 and A1-34 containing Ladinian palynomorphs, overlain by dolomites and anhydrites containing Carnian palynomorphs. Belhaj also showed the presence of undifferentiated Triassic clastics in the western Sirt Basin in wells A1-40, A1-3, A1-17, but without palynological control.

In Cyrenaica limestones of restricted marginal-marine origin, ranging from 250m to 400m in thickness, have been dated as Ladinian-Carnian in wells A1-19, C1-2, A1-117 and A2-2. A more open-marine facies containing Rhaetian-Hettangian assemblages has been found in well A1-28 with a thickness of 890m. These rocks are equivalent to the Al Aziziyah Formation of northwest Libya. No evidence has been found for the presence of the Norian in Cyrenaica, and the area was probably emergent during this time.[9]

In Algeria the clastic sequences of the early and middle Triassic are replaced by a sequence called the Trias argilo-salifere which has been dated as Carnian. This unit probably extends into the area of former concession 90 and current concession NC 100, where it forms a seal for the underlying sandstone reservoirs.[10]

4.1.4 Abu Shaybah Formation

At outcrop in the Gharyan area, the Al Aziziyah Formation is unconformably overlain by 125m of red fluvial sandstones and conglomerates which were named the Bu Sceba Group by Christie. Several revisions were made before the area was remapped for the Geological Map of Libya by El Hinnawy and Cheshitev. As currently defined the formation comprises vari-coloured fine to coarse-grained cross-bedded sandstones with a few thin dolomite and calcareous horizons towards the top. The lower contact with the Abu Shaybah Formation is distinctly unconformable, but the upper contact with the Jurassic Abu Ghaylan Formation is conformable. The formation outcrops at the foot of the escarpment north of Gharyan, and in three small inliers south of Ra's Ajdir. The formation is poorly fossiliferous with only a few undiag-

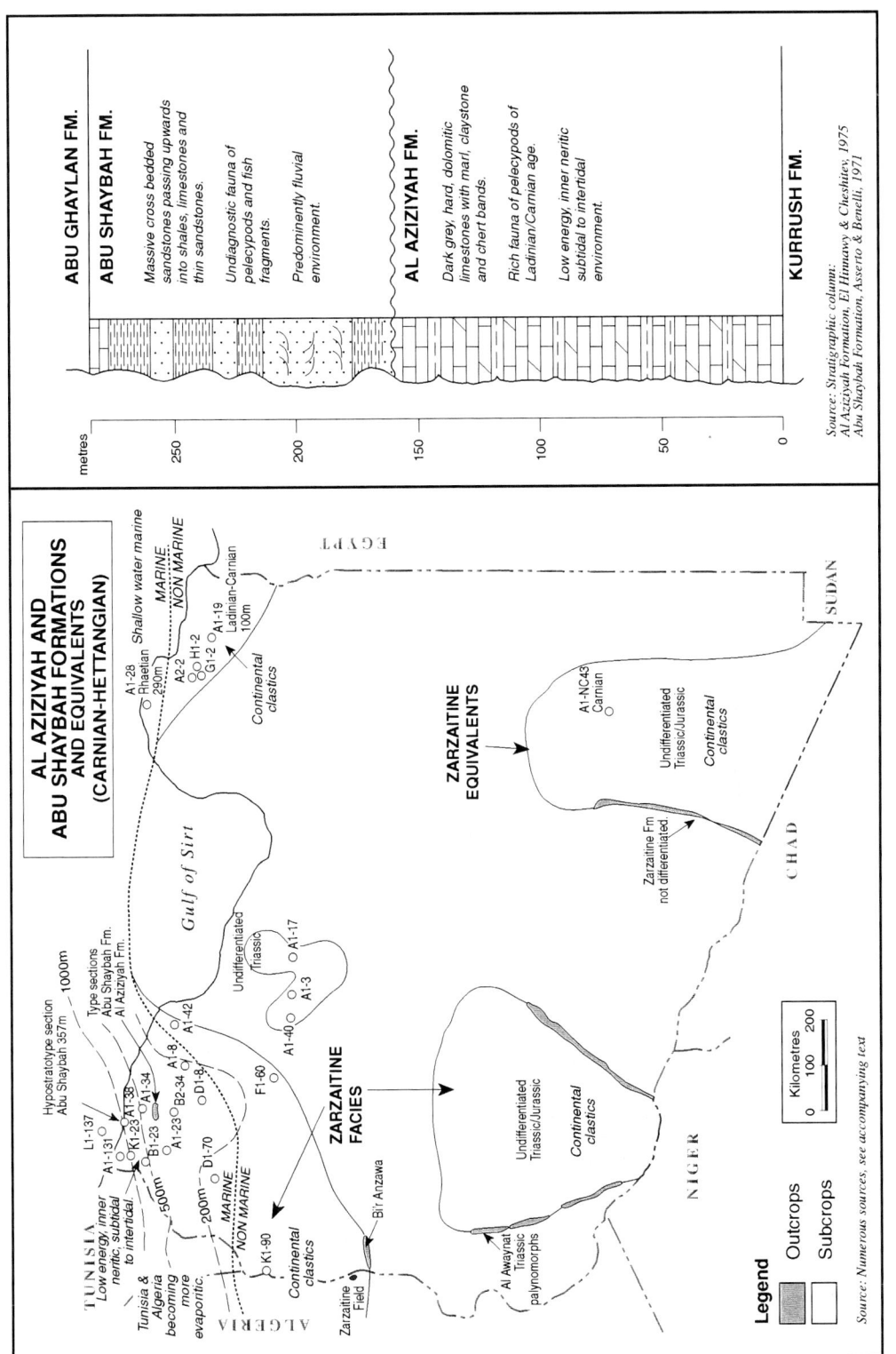

Figure 4.4   Al Aziziyah and Abu Shaybah Formations: Distribution, Equivalents, and Stratigraphic Column

The Al Aziziyah and Abu Shaybah Formations are confined to northwest Libya and adjacent areas of Tunisia and Algeria, although marine, brackish and continental equivalents have been found in Cyrenaica (see text). Further south continental equivalents are present in the southern Ghadamis and Murzuq basins, and in the Al Kufrah Basin.

nostic pelecypods and fish fragments being found in the carbonates near the top. The attribution to the late Triassic-early Jurassic is based solely on stratigraphic position. El Hinnawy and Cheshitev favoured a late Carnian age, Fatmi et al. believed the Norian to be absent (as in Cyrenaica) and assigned the Abu Shaybah Formation to the Rhaetian-Hettangian, whilst Hammuda et al. suggested a Norian to Hettangian age. The sedimentology was examined by Assereto and Benelli who recognised four lithofacies indicative of a fluvial environment. Palaeocurrent measurements indicate a west to northwest direction of sediment transport.[11]

Hammuda et al. described a hypostratotype in well A1-38. Here the formation has a thickness of 357m and is composed of silty and sandy shales, interbedded with clean porous sandstones which may represent point bars. The formation is over 200m thick in wells C2-23, D1-23 and B1-23. In the Sabratah Basin, the L1-137 well penetrated 56m into the Abu Shaybah Formation.[12]

The Abu Shaybah Formation can be traced across the northern part of the Ghadamis Basin, but it is difficult to distinguish from the Al Aziziyah Formation. In the area south and east of Tripoli palynological evidence shows both the Norian and Rhaetian intervals to be absent. In wells B2-34 and A1-42 dolomites and anhydrites have been dated as Hettangian which suggests an increasing lagoonal/marine influence in this area. Westwards into Tunisia the Abu Shaybah Formation progressively interfingers with the basal Lower Evaporites until in southern Tunisia the interval is entirely evaporitic. In central Tunisia the Abu Shaybah formation is equivalent to the upper part of the Jabal Rheouis Formation which comprises alternating beds of limestones, shales, siltstones and gypsum. In the Gulf of Gabes a thick series of halites occur and piercement salt-domes are present in the area north of Ra's Ajdir. The salt

horizon is present throughout the Tunisian offshore and extends into the Sabratah Basin of Libya.[13]

In Algeria the upper Triassic is represented by massive halites of the Trias Salifere, which form an effective seal for the underlying Lower Triassic sandstone reservoirs. These evaporites extend into the westernmost part of Libya in the area of former concession 90.[14]

In the eastern Sirt Basin the Triassic palynological assemblages at Jakharrah and well L4-51 are described in section 4.1.5. Two other fields, As Sarah and Tuama, produce oil from pre-Upper Cretaceous reservoirs which are regarded as late Triassic in age based on palynological dating of the overlying black claystones. In Cyrenaica shallow-water limestones, containing Rhaetian palynomorphs were found in well A1-28. The thickness of the limestones is 290m.[15]

The distribution and stratigraphy of the Abu Shaybah Formation and its equivalents is shown on Figure 4.4.

TRIASSIC CONTINENTAL EQUIVALENTS

In the Murzuq Basin the 'Continental Post-Tassilien' has been subdivided into three formations: the Tiguentourine Formation of late Carboniferous-early Permian age, the Zarzaitine Formation of Triassic age, and the Taouratine Formation of Jurassic age.[16]

4.1.5 Zarzaitine Formation

The Zarzaitine Formation was first described from the Zarzaitine field area of eastern Algeria. In this area the formation was divided into three informal members by Lefranc: a lower unit of multi-coloured claystones and fine-grained sandstones, containing Triassic dinosaur remains, a middle unit of mottled claystones, argillaceous limestones and dolomites, with thin bands of gypsum and anhydrite, and an upper unit of friable, argillaceous sandstones, claystones, dolomites and gravels. The total thickness is about 400m. In Algeria the age is considered to extend from Triassic to mid-Jurassic. The Jurassic attribution is based on the presence of late Jurassic plants in the overlying basal Taouratine Formation. It outcrops in Libyan territory on the southern flank of the Ghadamis Basin in the Bi'r Anzawa, Hasi Anjiwal, Wadi ain Armas and Hamadat Tanghirt areas. In these areas it comprises 150m of alternating bands of gypsiferous and ferruginous claystones, dolomitic marls, micaceous siltstones, quartzose sandstones and gravels, with a distictive brown colour. The threefold division was not adopted during the recent mapping by IRC.[17]

The formation outcrops along the western margin of the Murzuq Basin from Wadi Irawan southwards to the border with Niger. (Figures 4.3 and 4.4). The results of this mapping programme were summarised by Grubic et al. and Jakovljevic. In this area the formation unconformably overlies the Dimbabah Formation. It maintains its dominantly siltstone character and the thickness ranges from 140m at Wadi Irawan to 80m at Anay. The upper contact is erosional and is marked by a lateritic horizon, 3-4m thick. The formation contains fresh-water pelecypods, large fossil logs up to 5m in length, and rare reptile bones. The environment of deposition includes braided-stream, alluvial-plain and lacustrine facies. Measurements of cross-bedding direction indicate a dominant transport direction towards the southwest. Palynomorphs recovered from outcrops in the Al Awaynat area confirm a definite Triassic age.[18]

The Zarzaitine Formation is present in the wells of the northern Murzuq Basin where, in concession NC 115, it ranges in thickness from 190 to 320m, but it has not been differentiated in the Braspetro wells further south, nor on the eastern flank of the basin where thick

'Continental Post-Tassilien' rocks occur, but there is no reason to assume that it does not form part of this sequence.[19]

In the Al Kufrah Basin little information is available on the age of the 'Continental Post-Tassilien' sequence. Outcrops of undifferentiated continental rocks have been mapped on the western flank of the basin, but not on the eastern side. However in well A1-NC 43 palynological studies revealed a definite Triassic age for part of this sequence. The interval 540m to 685m, consisting of continental siltstones and sandstones, contains a Scythian assemblage. Higher in the same well a second palynological assemblage, 289m thick, contains a Carnian assemblage. The areal extent of the Zarzaitine Formation is shown on Figures 4.3 and 4.4.[20]

In recent years major new information has been obtained from wells in the northeastern Sirt Basin which suggests that the rifting phase in the Sirt Basin began as early as the Triassic. A palynological study of the pre-Upper Cretaceous ('PUC A') section in well A1-96 (Jakharrah field), a unit formerly equated with the Amal Formation, yielded a rich palynomorph assemblage of late Scythian to Middle Triassic age. This unit consists of organic-rich shales and sandstones with some oil-source potential, deposited in conditions alternating between oxic and anoxic. The thickness reaches 440m at Jakharrah. Thusu suggested that this sequence was deposited in an incipient rift, marking the beginning of the syn-rift phase in the Sirt Basin.[21]

Studies on the Amal Formation in well L4-51 revealed a similar picture. In this well a rich assemblage of miospores and pollen of Anisian age was found. In the same sequence a fossil fish was discovered which was identified as the Triassic species *Cleithrolepis major*. The sediments were interpreted as representing a lacustrine to lagoonal environment, on the floor of a shallow graben. This unit represents a potential petroleum source rock in the Maragh Graben. The thickness in well L4-51 is 825m. This discovery has major implications, not only in the dramatic revision of the age of the Amal Formation, but also in providing evidence of the earliest stages of rifting in the Sirt Basin. Re-examination of well J1-81A in the Qattara Graben close to the Egyptian border also showed the presence of a Middle Triassic assemblage which had previously been regarded as Jurassic. The environment of deposition in this case is also continental.[22]

PERMO-TRIASSIC IGNEOUS ROCKS

Evidence is beginning to accumulate, as discussed above, that the rifting phase in the Sirt Basin began in the Triassic, and the Triassic sediments found in shallow grabens in the northeast Sirt Basin provide support for this view. In addition, the North African margin experienced a marked increase in igneous activity during the late Permian and Triassic. Wilson and Guiraud demonstrated that mid- to late Permian basalts occur in an offshore well in the Gulf of Sirt, mid-Permian granites and gran-odiorites have been penetrated on the Waddan horst, and basalts and microsyenite sills of Scythian age have been encountered on the Amal field. In the Western Desert of Egypt dolerite sills of Triassic age have been reported. These occurrences are indicative of crustal instability and the initiation of rifting during this period.[23]

4.2   JURASSIC

As in the Triassic, marine Jurassic rocks, deposited on the southern margin of the Tethys Ocean, are largely confined to the northern areas of Libya, whilst the interior is dominated by continental rocks which form part of the 'Continental Post-Tassilien sequence'. Jurassic rocks were encountered in the A1-38 well in the Jifarah Basin and in one or two wells in the Sabratah Basin, but in general it is too deep to have been reached by most wells in the offshore. However, seismic data show that Jurassic rocks have a wide distribution in the offshore north of the Sabratah-Cyrenaica Fault. Jurassic rocks reach 700m in thickness in the H1-137 area, and north of well K1-137 they contains thick evaporites, which form salt domes in the western Sabratah Basin. The Jurassic is thin in the Misratah Basin and on the Jarrafa Platform but thickens to the north, reaching 1200m on the Medina Platform. In the Sirt Embayment Jurassic rocks are present beneath the mid-Cretaceous unconformity in the Sirt Trough and extend as far as the Ionian abyssal plain where middle Jurassic rocks overlie oceanic crust.[24]

4.2.1   Abu Ghaylan Formation

The carbonate overlying the Abu Shaybah Formation in the Gharyan area was named the Bu Gheilan Limestone by Christie. It outcrops at Abu Ghaylan, north of Gharyan, and for about 20km along the escarpment. The relation-ship of the Abu Ghaylan Formation with the Bi'r al Ghanam Formation has been the subject of some dispute. The Abu Ghaylan Formation pinches out abruptly west of Wadi Zaggut, and appears to pass laterally into the Bi'r al Ghanam evaporites. Some authors, such as Christie, favoured a lateral facies change, and this is supported by apparent interfingering of the two formations at Kaf Tahishmat, but others, such as Desio, et al., Burollet and Magnier, preferred an erosional explanation. Syndepositional folding within the Abu Ghaylan Formation was reported by Gray, which suggests that slumping probably occurred whilst the sediments were still uncon-solidated. The stratigraphy of the Abu Ghaylan Formation is shown on Figure 4.5.[25]

At the type locality, 60m of dolomitic limestone are exposed, interbedded with marls,

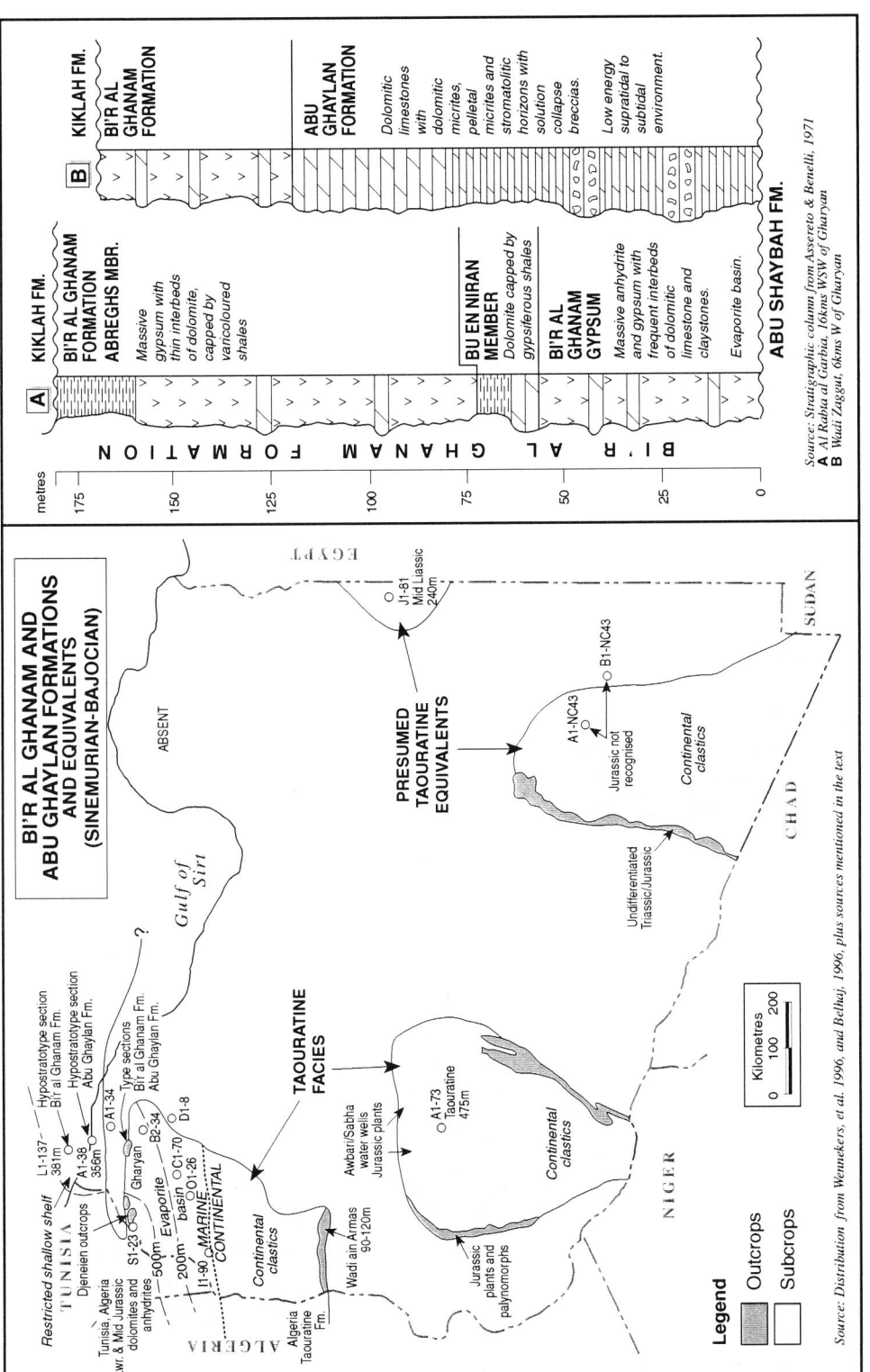

Figure 4.5   Bi'r al Ghanam and Abu Ghaylan Formations: Distribution, Equivalents and Stratigraphic Column

The Abu Ghaylan Formation has a limited distribution in the Gharyan area, west of which it passes laterally into the Bi'r al Ghanam Formation. Both formations are present in the subsurface of the Jifarah and Sabratah Basins, and the Bi'r al Ghanam Formation also extends into the Ghadamis Basin. South of Ghadamis continental equivalents belonging to the Taouratine Formation are present in the southern Ghadamis and Murzuq Basins. Equivalents may also be present in the Al Kufrah Basin, as part of the undifferentiated Continental Post Tassilien.

thin sandstones and breccias. The contact with the underlying Triassic Abu Shaybah Formation is abrupt, but the upper contact is variable. Where it is overlain by Bi'r al Ghanam evaporites the contact is conformable, but where it is overlain by the Kiklah Formation the contact is unconformable. Assereto and Benelli studied the sedimentology and recognised a number of low-energy supratidal to subtidal facies including laminated dolomitic micrites, pelletal micrites, stromatolite horizons and solution-collapse breccias. The fossil content is poor and undiagnostic. The age was assumed by Christie to be early Jurassic.

The outcrops were re-examined by El Hinnawy and Cheshitev, who concluded that the Abu Ghaylan Formation was affected by local uplift and erosion. They showed the formation as a time equivalent of the lower unit of the Bi'r al Ghanam Formation, and believed that the age was probably Norian-Rhaetian. Subsequently new evidence has come to light which clearly shows the presence of both the Abu Ghaylan and Kiklah Formations east of the Ra's al Tahunah High but their absence on it, implying an erosive limit of the formation in this area.[26]

Although the Abu Ghaylan Formation is only sparsely exposed at outcrop, it is present in the subsurface of the Jifarah Basin and Hammuda et al. designated a hypostratotype in the A1-38 well where the formation has a thickness of 356m, showing a dramatic increase from the Gharyan outcrops. The formation comprises crystalline dolomites with relict pellet and oolite structures, interbedded with minor shales and anhydrites. Similar lithologies have been encountered in the offshore L1-137 well.[27]

Thin dolomitic limestones have been found in the Ghadamis Basin in well B1-70 which may be attributable to the Abu Ghaylan Formation, but the unit is impersistent. Further east dolomites and anhydrites in wells B2-34 and A1-42 contain a Hettangian palynological assemblage which Belhaj attributed to the Bi'r

al Ghanam Formation. The Hettangian age however suggests that they should be attributed to the Abu Shaybah Formation. There is no evidence of Sinemurian to Bajocian rocks in the Zamzam Depression or in the Sirt Basin.

In Tunisia it is equivalent to part of the Lower Evaporites, and in Algeria dolomitic limestones and anhydrites dated as Liassic are probably age equivalents of the Abu Ghaylan Formation. In Cyrenaica, the Rhaetian marine carbonates in well A1-28 have already been mentioned. This sequence is overlain by 600m of shallow-marine carbonates containing Liassic palynomorphs which are age equivalent to the Abu Ghaylan Formation.[28]

4.2.2 Bi'r al Ghanam Formation

In the area west of the Wadi Zaggut on the escarpment west of Gharyan the Abu Shaybah Formation is overlain by a thick evaporite sequence first described and named the Bir al Ghnem Group by de Lapparent. Christie subsequently showed that it appears to be the lateral equivalent of the Abu Ghaylan Formation in the Gharyan area.[29]

Numerous revisions have been made to the description and definition of the formation prior to the re-survey for the Geological Map of Libya in 1975. The current view is that the formation can be divided into three members in the type area, from bottom to top the Bi'r al Ghanam Gypsum, the Bu en Niran and the Abreghs Gypsum members. In this area the formation is about 370m in thickness. A stratigraphic column is shown on Figure 4.5. The lower member comprises 270m of massive gypsum and anhydrite with frequent interbeds of dolomitic limestone and claystone. The Bu en Niran Member, which reaches a thickness of 20m, consists of a basal limestone with an overlying sequence of red and green gypsiferous shales. The top member, the Abreghs Gysum, has a lower 60m interval of gypsum and an upper 20m section of varicoloured shales. In the type area the formation rests on the Abu Shaybah Formation, but further east it is underlain by the Abu Ghaylan Formation. It is overlain in the west by the Takbal Formation and in the east by the Kiklah Formation. West of the type area the formation is exposed in several small inliers on the Ra's Ajdir sheet, where the thickness increases to over 750m. Eastwards it pinches out rapidly and appears to pass laterally into the Abu Ghaylan Formation (Figure 4.7). A poor fauna of pelecypods has been reported but it is not age diagnostic. The formation was deposited in an enclosed, subsiding, evaporitic basin. Petrographic studies have revealed a variety of environments of deposition ranging from supratidal through lagoonal to marginal sea pools. The evaporites were originally deposited as anhydrite and the gypsum, which predominates at outcrop, is of secondary origin. The age of the formation, purely on the basis of stratigraphic position, is assumed to be Sinemurian to Bajocian. The reserves of gypsum-anhydrite in this area have been assessed as 79.7 billion metric tons.[30]

The formation is present in the subsurface in the Jifarah Basin and Hammuda et al. designated a hypostratotype in the L1-137 offshore well with a thickness of 382m. The threefold division is not recognisable at this location. The lithology comprises anhydrite with interbeds of dolomite and limestone. It lies conformably on the Abu Shaybah Formation and is overlain unconformably by the Kiklah Formation. It is also present in the A1-38 well. In the Tunisian offshore it is present in several wells north of Ra's Ajdir. Evaporites attributable to the Bi'r al Ghanam Formation are present in the northern Ghadamis Basin. They thin rapidly southwards from a maximum of 750m in well S1-23 to 150m in well O1-26. They can be traced westwards to well I1-90 near the Tunisian border and eastwards as far as well C1-70. In the Triassic Basin of Algeria anhydrites, halite,

marls and limestones of Liassic and early Dogger age are the time equivalents of the Bi'r al Ghanam Formation. In southern Cyrenaica it is difficult to recognise the extent of Liassic deposition, as most wells reveal a red-bed continental facies which is mostly unfossiliferous. Well J1-81 contains middle Liassic palynomorphs, but this appears to be an isolated occurrence.[31]

4.2.3 Takbal Formation

The 'Tocbal Limestone' was first described by Desio, et al. as a 38m limestone unit exposed on the escarpment near Kiklah. The formation outcrops along the escarpment east and west of Kiklah. It was remapped by El Hinnawy and Cheshitev who described a 22m section comprising varicoloured shales, thin marly limestones, dolomites and occasional beds of gypsum near the base. The formation contains a fauna of pelecypods, echinoid fragments, fish scales and shark's teeth which indicate a Bathonian age. The basal contact is conformable with the Bi'r al Ghanam Formation but the upper contact is clearly unconformable with the Kiklah Formation. Further west, towards Yifran, it is comformably overlain by the Khashm az Zarzur Formation (Figure 4.7). It has a limited distribution. Outcrops extend into the Ra's Ajdir sheet area and the formation is present in the subsurface in wells I2-23, S1-23 and P1-23, but it has not been recognised in the offshore. In the Ghadamis Basin a thin limestone overlying the Bi'r al Ghanam Formation may represent the Takbal Formation, but it it is only locally developed. In Tunisia it is probably time-equivalent to the Arachoua Formation. A stratigraphic column is shown on Figure 4.6.

The Takbal Formation does not appear to be present in the L1-137 well, offshore, where the Bi'r al Ghanam Formation is overlain by equivalents of the Khashm az Zarzur Formation.[32]

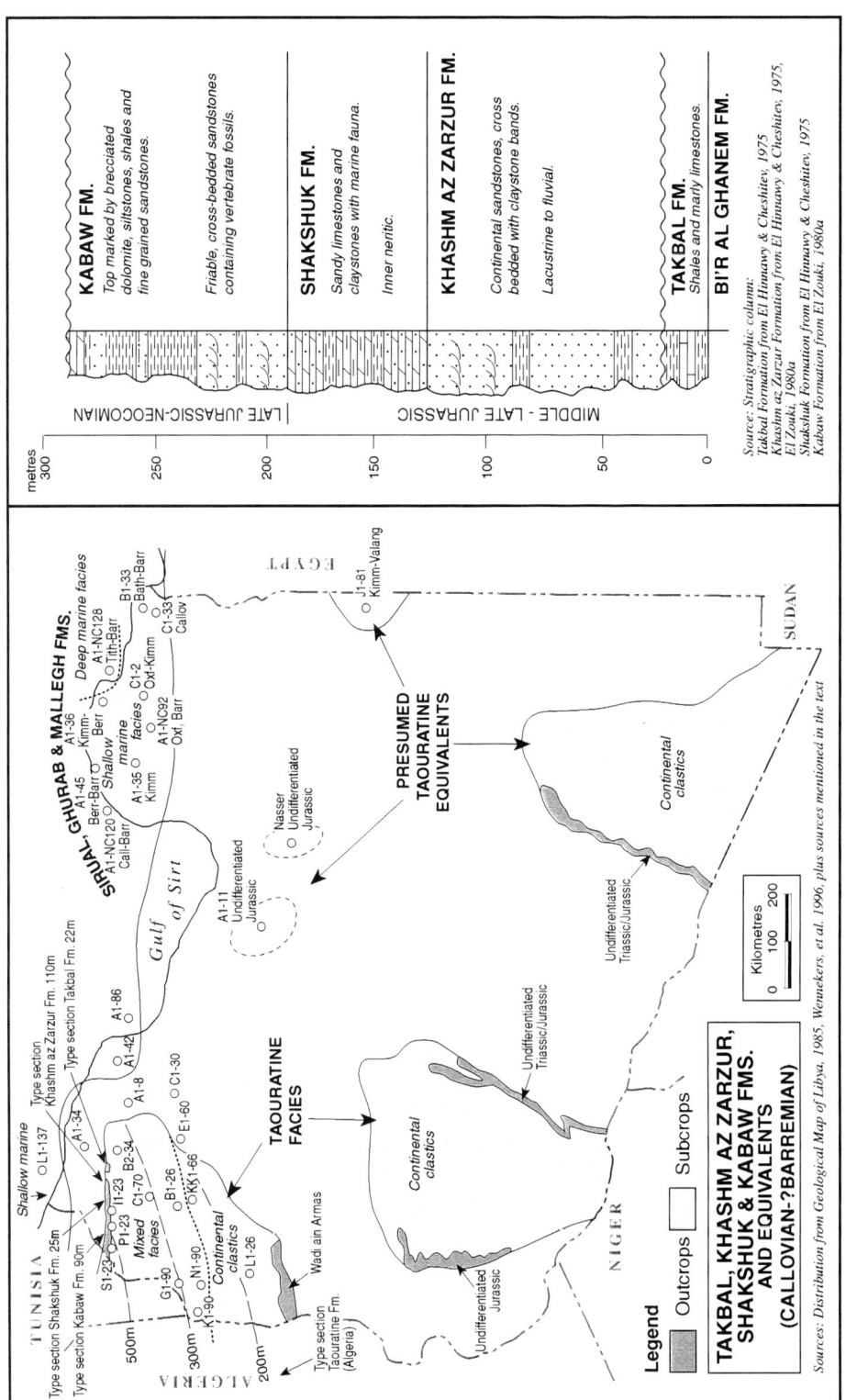

Figure 4.6   Takbal, Khashm az Zarzur, Shakshuk and Kabaw Formations: Distribution, Equivalents and Stratigraphic Column

The Takbal Formation has a limited distribution in the Gharyan area and in the subsurface nearby. The overlying Khashm az Zarzur, Shakshuk and Kabaw Formation are exposed on the Jabal Nafusah escarpment and are present in the northern Hammadah al Hamra. To the north equivalents of these formations are found in the Sabratah Basin and in the Misratah area. Southwards these formations pass into the upper part of the continental Taouratine Formation. Equivalents may also be present in the Al Kufrah Basin as part of the undifferentiated Continental Post Tassilien sequence. In northern Cyrenaica marine rocks of the Sirual, Ghurab and Mallegh Formations are partial age equivalents of the Jabal Nafusah succession.

4.2.4 Khashm az Zarzur Formation

There has been much controversy concerning the section between the Takbal Formation and the Sidi as Sid Formation. Basically this section comprises a predominantly continental sequence which can be divided into a number of distinctive units. Writers before 1975 recognised six lithostratigraphic subdivisions which appeared to be progressively truncated to the east of Gharyan by a major unconformity at the base of the uppermost unit, the Kiklah Formation. However El Hinnawy and Cheshitev, whilst remapping this area for the Geological Map of Libya, took the view that the major unconformity was located at the base of the Sidi as Sid Formation and that all the continental units below could be grouped together as members of a redefined Kiklah Formation. Subsequent work has shown that this view is not tenable. It has been demonstrated that the Khashm az Zarzur, Shakshuk and Kabaw formations are separated from the Kiklah Formation by a major unconformity, and that a further unconformity separates the Kiklah from the Sidi as Sid Formation. These relationships are shown on Figure 4.7. Consequently the concept of a Kiklah Formation extending down into the Jurassic and including the Shakshuk and Khashm az Zarzur formations must be abandoned.[33]

The Khashm az Zarzur Formation was named by El Hinnawy and Cheshitev for the two units overlying the Takbal Formation which were formerly known as the Giosc Shale and the Chameau Mort Sandstone. It is best developed in the area near Yifran where it is 110m thick. Further west near Jadu a 70m section is present with the base not visible. It continues to outcrop westwards to the Tunisian border. It is composed of two 'members', a lower lacustrine claystone with traces of salt and gypsum, with fresh water pelecypods, plant remains and fossil wood (formerly the Giosc Shale). The upper unit (formerly known as the Chameau Mort Sandstone) is a continental sandstone with marked cross-bedding and occasional claystone bands, which has been interpreted as a subaerial deposit close to a shoreline. SEM analysis of sand grains confirms an aeolian origin. Plant fossils have been recovered from the formation but are not age diagnostic. However palynological analysis of samples from outcrop near Yifran yielded a Bathonian to Tithonian age. On the basis of regional correlation perhaps the upper part of this sequence should be attributed to the Shakshuk Formation. Both members of the Khashm az Zarzur Formation can be traced in the subsurface of the Ghadamis Basin where they thin southwards. The formation is well developed in the area between wells S1-23 and B1-26 and can be traced as far as the K1-90 well near Ghadamis. South of Tripoli, well B2-34 drilled a mudstone sequence with interbedded dolomites and sandstones, 150m thick, which contains a palynological assemblage of Bathonian age which may be equivalent to the Khashm az Zarzur Formation. Further south the formation passes laterally into the Taouratine Formation (Figure 4.6).[34]

In the Sabratah Basin Hammuda et al. described a 156m section in the L1-137 well composed of sandstones, pyritic shale and argillaceous dolomites which, on the basis of stratigraphic position is probably equivalent to the Khasm az Zarzur Formation.[35]

In Cyrenaica rocks equivalent in age to the Khashm az Zarzur Formation are present in well B1-33 where oolitic limestones are interbedded with shales which underlie a carbonate sequence of definite Callovian age.[36]

4.2.5 Shakshuk Formation

The Shakshuk Formation was originally defined by Burollet from near Jadu where it comprises alternating limestones and claystones with occasional sandy horizons. The thickness

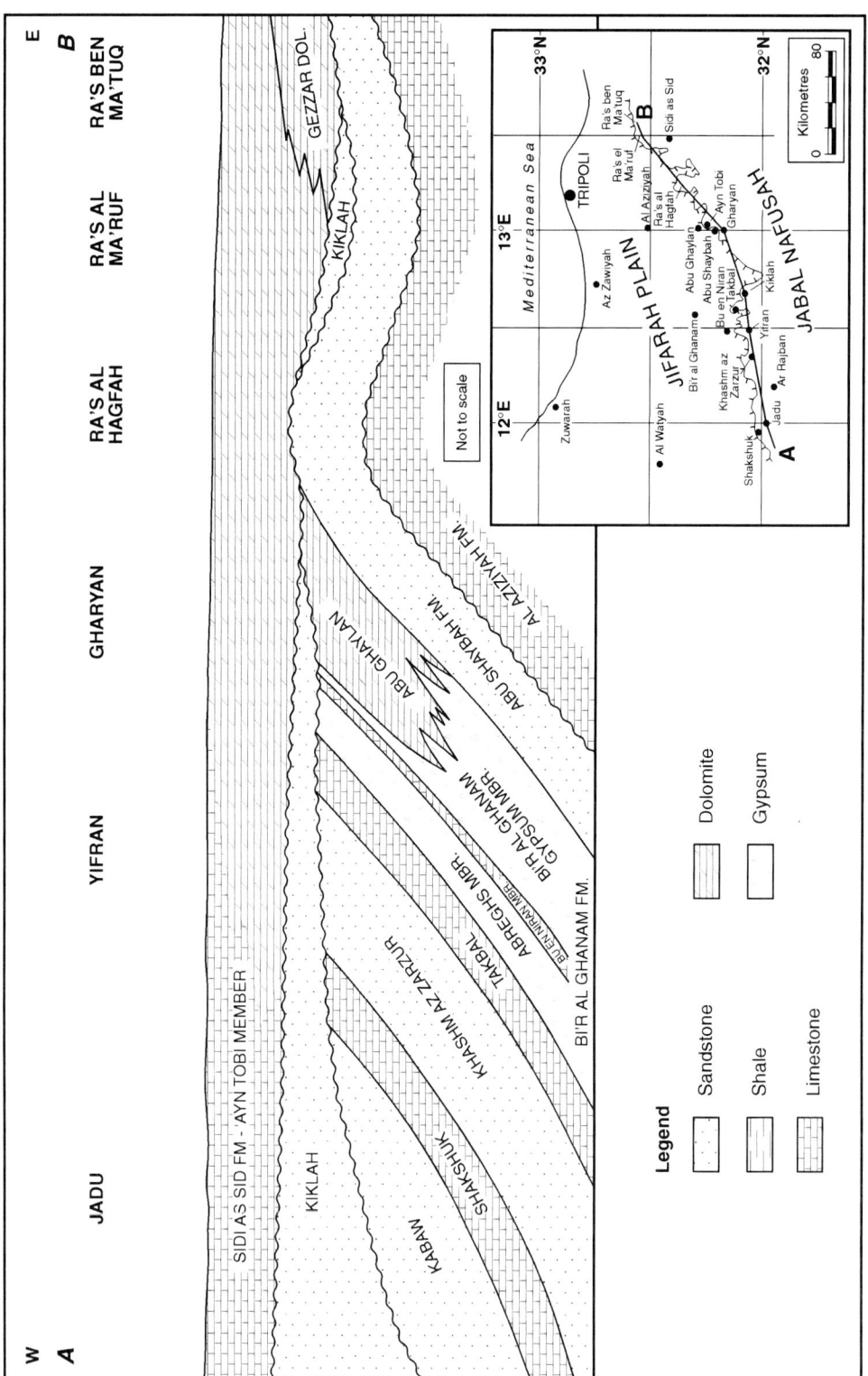

Figure 4.7 Geological Cross-Section to show Mesozoic Stratigraphic Relationships on the Jabal Nafusah Escarpment

Sources: Fatmi and Sbeta, 1991, El Zouki, 1980, Assereto and Benelli, 1971, Hammuda, 1971

Recent studies confirm the presence of a major unconformity at the base of the Kiklah Formation. Further unconformities have been identified at the base of the Sidi as Sid Formation and at the base of the Abu Shaybah Formation. A small remnant of Kiklah Formation has been found east of the Al Hagfah anticline, at Ra's al Ma'ruf. The index map shows the location of the section.

at the type locality is 25m. It contains a rich fauna of pelecypods, gastropods, brachiopods, foraminifera, echinoids, and plant debris, which give an age variously interpreted as Bathonian to Kimmeridgian, but the fauna from the Nalut area suggests a Bathonian-Callovian age. In the area east of Tripoli the A1-86 and A1-42 wells both penetrated continental sandstones, mudstones and coals with abundant miospores of Callovian to Kimmeridgian age which are assumed to be partly equivalent to the Shakshuk Formation. In the Ghadamis Basin the Shakshuk Formation can be traced as far south as wells K1-90 and KK1-66, and as far east as well C1-70. (Figure 4.6).[37]

Offshore, in the L1-137 well, Hammuda et al. described a section of crystalline dolomites and micritic limestones which is probably equivalent to the Shakshuk Formation. The limestones contain the foraminiferal genus *Kurnubia*, which indicates a late Jurassic age. In the well this sequence is 93m thick.[38]

In Cyrenaica marine carbonates equivalent in age to the Shakshuk Member are present in wells B1-33, C1-33, C1-2, A1-NC 92, A1-35 and A1-36 in paralic to transitional facies.[39]

The sequences in the two deep offshore wells in Cyrenaica, A1-NC 120 and A1-NC 128, have been described by Duronio, et al. He introduced five new formation names; the Sirual, Ghurab, and Mallegh formations of late Jurassic age, and the Qahash and Daryanah formations of early Cretaceous age. Unfortunately only the Daryanah Formation was formally defined. This is regrettable since there are inconsistencies in the ages and descriptions of the four informal 'formations'. In the A1-NC 120 well the lower part of the Sirual 'Formation' is equivalent in age to the Shakshuk Formation (late Jurassic). It comprises low-energy, shallow-water carbonates deposited in a dominantly open marine environment. The dominant lithology is oolitic and pelletal limestones and solution breccias with a fauna of planktonic foraminifera,

bryozoa, pelecypods and echinoderms. The sequence has been significantly affected by diagenesis and recrystallization. The faunal assemblages prove the presence of the Callovian, Oxfordian, and Kimmeridgian stages.

In well A1-NC 128 the setting is different. Here the Callovian to Kimmeridgian sequence is dominated by brown silty marls and limestones and contains several levels of brecciated material which include reefal and shallow-shelf debris which has been redeposited by gravity flow. This unit was named the Ghurab Formation' by Duronio. It contains a mostly planktonic fauna and is interpreted as a deep sea deposit with abundant evidence of gravity flows.[40]

4.2.6 Kabaw Formation

The Kabaw Formation was defined by Burollet in the Jadu area. It consists of poorly consolidated, cross-bedded sandstones, becoming finer grained and more silty towards the top. Thin cherty, brecciated dolomites are present in the Nalut area. The total thickness at Jadu is 90m, but the formation thickens westwards towards Nalut and southern Tunisia. It contains lignite, wood fragments, and a rich assortment of teeth, bones and reptilian remains. Hammuda reported that in Tunisia the lower part contains the index fossil *Pseudo-cyclammina jaccardi* which marks the Jurassic-Cretaceous boundary in Algeria. The reptilian remains include a dinosaur vertebra, sharks teeth, fish scales and teeth, crocodile teeth, and turtle shells which suggest a Wealden age. SEM analysis of sand grains from the Kabaw Formation shows evidence of both fluvial and alluvial characteristics. A paly-nological study in the Nalut area shows that the lower part contains late Jurassic spores and pollen. It seems likely that the Kabaw Formation ranges from late Jurassic to Neocomian in age. The Kabaw Formation is present in wells in the Ghadamis Basin, but is difficult to differentiate from the Kiklah Formation. It has a thickness of about 50m in the area of wells L1-26, N1-90, I1-90 and G1-90. In the subsurface south of Misratah, palynological evidence demonstrates a hiatus in which Tithonian to Hauterivian age rocks are absent. Equivalents of the Kabaw Formation are evidently absent in this area. (Figure 4.6).[41]

Overlying the *Kurnubia*-bearing limestones in offshore well L1-137 is a section of fine-grained dolomitic sandstones, siltstones and shales, which is assumed to equate to the combined Kabaw/Kiklah formations of the Jabal Nafusah. This sequence is overlain by volcanic rocks of early Cretaceous age. The total thickness of this unit in the well is 110m.[42]

In Cyrenaica age-equivalents of the Jurassic part of the Kabaw Formation are present in the A1-36, B1-33, and A1-45 wells, mainly as open-marine, shallow-shelf limestones and paralic claystones.[43]

Offshore Cyrenaica late Jurassic rocks are found in both the A1-NC 120 and A1-NC 128 wells. In A1-NC 120 the open marine carbonates of Kimmeridgian age are overlain by dolomitic mudstones with minor dolomite stringers, and thin coals. This sequence is interpreted as a very low-energy, restricted environment with a high rate of sedimentation. A fauna including *Clypeina jurassica, Kurnubia palastiniensis* and *Salpingoporella annulata* suggests a Kimmeridgian-Tithonian age. This unit is overlain by a distinctive stylobreccia composed of bioclastic and skeletal limestones, which has been greatly affected by pressure solution. The unit was probably deposited on a shallow carbonate shelf subject to episodes of reworking, resedimentation a diagenesis. The fauna indicates a Berriasian age. Duronio et al. assigned all of these units to the upper part of the Sirual 'Formation'.[44]

In A1-NC 128, the gravity flow deposits of the Kimmeridgian are overlain by brown marls and mudstones with intervals of coarse-grained sandstones and conglomerates which probably also originated as gravity flows. The fauna include *Calpionella alpina*, which indicates a Tithonian age. Duronio named this unit the Mallegh 'Formation'. On the basis of the descriptions provided by Duronio it is difficult to distinguish any significant difference between the Ghurab and Mallegh 'formations'. The Tithonian sequence in well A1-36 is comparable with A1-NC 128, where it is also difficult to distinguish late Jurassic and Tithonian sediments. However, both wells at Tithonian level show a deep-water depositional environment, characterised by the deep-water calpionellid *Calpionella alpina*.[45]

JURASSIC CONTINENTAL EQUIVALENTS

The 'Continental Post-Tassilien' Group of Kilian is divided into three formations of which the Taouratine is the youngest. It overlies the Triassic Zarzaitine Formation and is overlain by the 'Continental Intercalaire' of Kilian which is usually taken to equate with the Nubian Sandstone.[46]

4.2.7 Taouratine Formation

The Taouratine Formation was first described by de Lapparent and Lelubre from the Taouratine area in the eastern Illizi Basin of Algeria. In this area it can be subdivided into three units, a lower member of alternating mottled sandy claystones and ferruginous sandstones which contains plants and vertebrate remains indicative of late Jurassic age, a middle member of sandy dolomitic limestones and marls of mid-Jurassic to early Cretaceous age, and an upper unit of claystones, friable sandstones and occasional limestones which have been dated as Vraconian (late Albian-early Cenomanian). The thickness is about 350m.

(Figure 4.2). In Libya the Taouratine Formation is normally restricted to the continental Jurassic, and therefore does not correspond to the Algerian nomenclature. The Taouratine Formation extends into the subsurface of the western Ghadamis Basin and is present from Wadi ain Armas northwards, eventually interfingering with marine rocks towards Nalut. The distribution of the formation is shown on Figures 4.5 and 4.6.[47]

The Taouratine Formation outcrops in the Wadi ain Armas and Hamadat Tanghirt areas on the south flank of the Ghadamis Basin. In this region two units are recognised, a lower member of red claystones and thin ferruginous sandstones which contains numerous fern and conifer remains, which allow this unit to be dated as mid-Jurassic, and an upper member of cross-bedded sandstones, conglomerates and claystones which also contains a rich Jurassic palaeoflora. The fossil flora is similar to that found in the Kiklah Formation in northern Libya. The total thickness ranges between 90m and 120m in thickness. These sediments were deposited in a lacustrine, swamp or flood plain environment.[48]

The formation has also been mapped on the western flank of the Murzuq Basin as far as Anay, south of which it is overstepped by the Messak Formation It outcrops at the base of the great Messak escarpment. In this area the Taouratine Formation unconformably overlies the Zarzaitine Formation, with a thick lateritic zone marking the junction which suggests a long period of non-deposition. The sequence comprises numerous stacked channel-fill deposits of a braided-river system, generally 2m to 3m thick, and 10m to 15m wide with coarse-grained sandstones and lenses of conglomerate marking the axis of the active river channels. The formation contains silicified plant remains and fossil logs, in one case up to 29m in length. The plants and palynological assemblages from this area confirm a Jurassic age with no evidence of Cretaceous forms.[49]

The Taouratine Formation is present in wells in the northern Murzuq Basin where a thickness of 400m was reported in well L1-NC 115, but it has not been differentiated in the Braspetro wells of concession NC 58, nor on the eastern margin of the basin where undifferentiated 'Continental Post-Tassilien' rocks outcrop from the Dur al Qussah area to the border with Niger. However evidence from boreholes in the Sabha-Awbari area show the presence of claystones, siltstones and coals containing Jurassic plant remains and microfossil assemblages, and a large area of Taouratine outcrop is shown on the 1985 Geological Map of Libya in the Sabah, Samnu, Az Zighan area. The Taouratine Formation is reported to be present in the A1-73 well near Murzuq with a thickness of 475m, and Wennekers et al. reported the presence of 1000m of Jurassic section in the central Murzuk Basin, based on proprietary data. Similarly in the Al Kufrah Basin thick undifferentiated 'Continental Post-Tassilien' rocks outcrop on the western flank of the basin, and it is likely that some of these rocks are equivalent to the Taouratine Formation of western Libya, but palynological investigation of samples from the A1 and B1-NC 43 wells has not yielded any firm evidence of the presence of Jurassic continental deposits. These sequences are almost always barren of microfossils.[50]

In the Sirt Basin isolated occurrences of rocks dated as Jurassic have been found in wells on the Zaltan and Al Hufrah fields, presumably as isolated inliers on the Palaeozoic surface.[51]

In southern Cyrenaica several wells penetrate continental rocks of Jurassic age. Wells in concession 65 and well J1-81a contain red continental sandstones and shales which contain two Jurassic assemblages, the first of middle Liassic age and the second of late Jurassic to early Neocomian age which can therefore be regarded as Taouratine Formation equivalents. The area further north, where marine rocks are found, has already been discussed.[52]

4.3 LOWER CRETACEOUS

Rocks of early Cretaceous age in Libya can be assigned to several depositional provinces: the Kabaw and Kiklah facies of northwest Libya, the marine facies of the northwestern offshore, the Messak facies of the Murzuq Basin, the Nubian facies of the Al Kufrah and Sirt Basins, and the marine rocks of the northeastern offshore. These rocks mark the end of the syn-rift phase, which was terminated by a marine transgression beginning tentatively in the Neocomian-Aptian (with the Variegated Shale of the Nubian) and culminating with a major flooding event in the Cenomanian. These rocks contain some of the best oil reservoirs in Libya, and contain enormous reserves of oil particularly in the southeastern Sirt Basin.

Kilian separated the distinctive group of continental rocks lying between the 'Continental Post-Tassilien' and the marine Cretaceous under the name 'Continental Intercalaire'. These rocks have been variously referred to as Kiklah Formation, Messak Formation, Nubian Sandstone and Djoua Group. None of these names is entirely satisfactory, but they have been adopted in various parts of Libya. The Kabaw Formation and its equivalents, which range in age from late Jurassic to Barremian, have already been discussed.[53]

NORTHWESTERN LIBYA

4.3.1 Kiklah Formation

As noted above the expansion of the Kiklah Formation by El Hinnawy and Cheshitev to include the Kabaw, Shakshuk and Khashm az Zarzur Formations is no longer tenable. The Kiklah Formation must now be restricted to the sequence lying above the Kabaw Formation and beneath the Sidi as Sid Formation. Consequently El Hinnawy and Cheshitev's Ar Rajban Member is a junior synonym of the Kiklah Formation and must be abandoned.[54]

The Kiklah Formation was first described by Christie from the Gharyan area. It is present on the Jabal Nafusah escarpment for almost 400km from the Tunisian border to east of Gharyan. It overlies the early Cretaceous, Jurassic and Triassic sediments of the Jabal Nafusah with marked unconformity, and in turn is unconformably overlain by the Sidi as Sid Formation. In the Gharyan area the Kiklah Formation overlies the Abu Ghaylan Formation. Further west at Yifran it overlies the Khashm az Zarzur Formation and at Jadu it overlies the Kabaw Formation. The thickness varies from 60m at Gharyan, 30m at Yifran, 50m at Jadu and thickens westward towards the Tunisian border. It has been demonstrated that both the Kiklah and Abu Ghaylan formations extend further east than was previously thought. The age of the Kiklah Formation has not been established with any certainty, but a palynological study at Nalut has yielded a Barremian-Aptian assemblage in the lower part, and a late Aptian to Albian age assemblage in the upper part. Tekbali assigned an early Albian to Vraconian age to the Kiklah Formation based on angiosperm and gymnosperm palynomorphs, without disclosing the locality of the samples. He interpreted a wet, tropical environment for the formation which contrasts with a drier environment in neighbouring areas of Gondwana. A stratigraphic column for the Kiklah Formation is shown on Figure 4.8.[55]

In the Gharyan area the Kiklah Formation is composed of impersistent beds of mudstone, sandstone and conglomerate which were probably deposited in a braided-river system carrying high sediment loads, with frequent migration of the principal channels. SEM studies show pitting and breakage characters on the sand grains which are diagnostic of a fluvial environment. In the Yifran area the dominant lithology is micaceous mudstone which has been interpreted as a flood plain deposit. Further west, from Jadu to the Tunisian border the lower part of the formation is dominated by coarse, cross-bedded sandstones with pebbles lining the principal scour channels. The upper part is largely made up of siltstones and mudstones containing gypsum. This sedimentological pattern is interpreted as a long meandering river system with the mudstones representing overbank deposits. On the western Jabal Nafusah escarpment Burollet described two further units lying between the Kiklah Formation and the Sidi as Sid Formation. These are the Jadu (Giado) Dolomite and the Wazzan (Uazzan) Sands. These are local units which are now included within the Kiklah Formation. The Jadu Dolomite can be traced southwards into the Ghadamis Basin where it has been recorded as far south as former concession 66.[56]

The formation is present over much of the Ghadamis Basin from well K1-90 on the Tunisian border to L1-26, and DD1-66, and it thickens eastwards reaching 300m in well B1-69. In the Misratah area it has been found in wells B2-34, A1-8, A1-42, C1-39 and A1-86, where it is represented by non-marine to marginally marine sandstones and dolomites which contain miospores of Barremian-Aptian age.[57]

In the offshore Sabratah Basin, Hammuda et al. designated the offshore L1-137 well as a hypostratotype for the Kiklah Formation. Unfortunately this designation was made using the El Hinnawy and Cheshitev definition of the Kiklah Formation which is no longer valid. In fact only part of the section described by Hammuda is attributable to the Kiklah Formation as currently defined. The well contains a 110m section of dolomitic sandstone which is believed to be equivalent to both the Kiklah and Kabaw formations.[58]

In Tunisia the Kiklah Formation is equivalent to the Boudinar, Gafsa and the lower part of the Fahdene Formations.[59]

OFFSHORE NORTHWESTERN LIBYA

4.3.2 Turghat Formation

The extension of the Kiklah Formation into the offshore has already been mentioned. Its extension into the Sabratah Basin is uncertain since it has not been penetrated north of L1-137. On the northern flank of the Sabratah Basin the early Cretaceous is represented by a thick series of shallow-water carbonates which have been named the Turghat Formation, with a type section defined in the A1-NC35a well. The formation is composed of a variety of shallow-water limestones and dolomites, ranging from pure crystalline dolomite through fossiliferous calcarenites to argillaceous limestones and calcareous shales. In the type well it is capped by volcanic rocks. The incomplete thickness penetrated in the well is 1026m with the base not reached. The formation has been encountered only in the one well, but it probably extends westwards into concession 137 and into Tunisian waters where it is the likely equivalent of the Gafsa Formation. The formation contains a rich fauna of large foraminifera with a total absence of planktonic foraminifera. This underlines the shallow water nature of the formation which probably represents deposition in a sheltered embayment or back-reef environment. The foraminifera give an early Aptian age for the lower part and an Apto-Albian age for the upper part.[60]

4.3.3 Masid Formation

The Masid Formation represents deeper-water, more open-marine conditions than in the Turghat Formation. A type section was defined in the B1-NC35a well located close to the Zohra Graben fault zone, south of the Jarrafa Arch. The formation is composed of soft grey shales with minor thin limestone bands. The shales are usually silty or sandy, but at the base are black

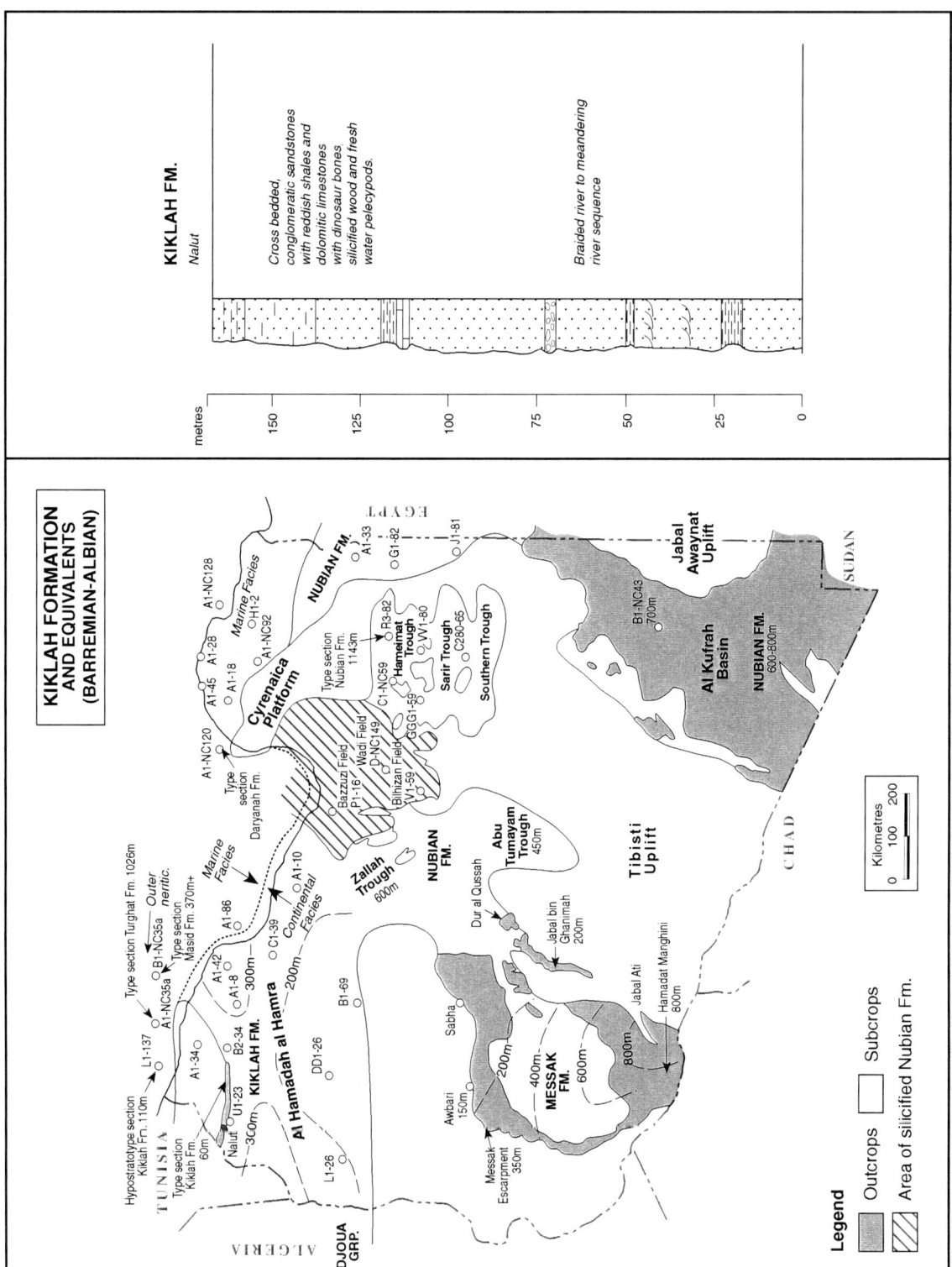

KIKLAH FM.

Nalut

Cross bedded, conglomeratic sandstones with reddish shales and dolomitic limestones with dinosaur bones silicified wood and fresh water pelecypods.

Braided river to meandering river sequence

metres

150

125

100

75

50

25

0

Source: Stratigraphic column: Kiklah Formation from Novovic, 1977

KIKLAH FORMATION AND EQUIVALENTS (BARREMIAN-ALBIAN)

Sources: Distribution from Geological Map of Libya, 1985, Wennekers, et al., 1996, El Hawat, et al., 1996, Ibrahim, 1991, Thusu, et al., 1988, Bonnefous, 1972

Legend

Outcrops

Subcrops

Area of silicified Nubian Fm.

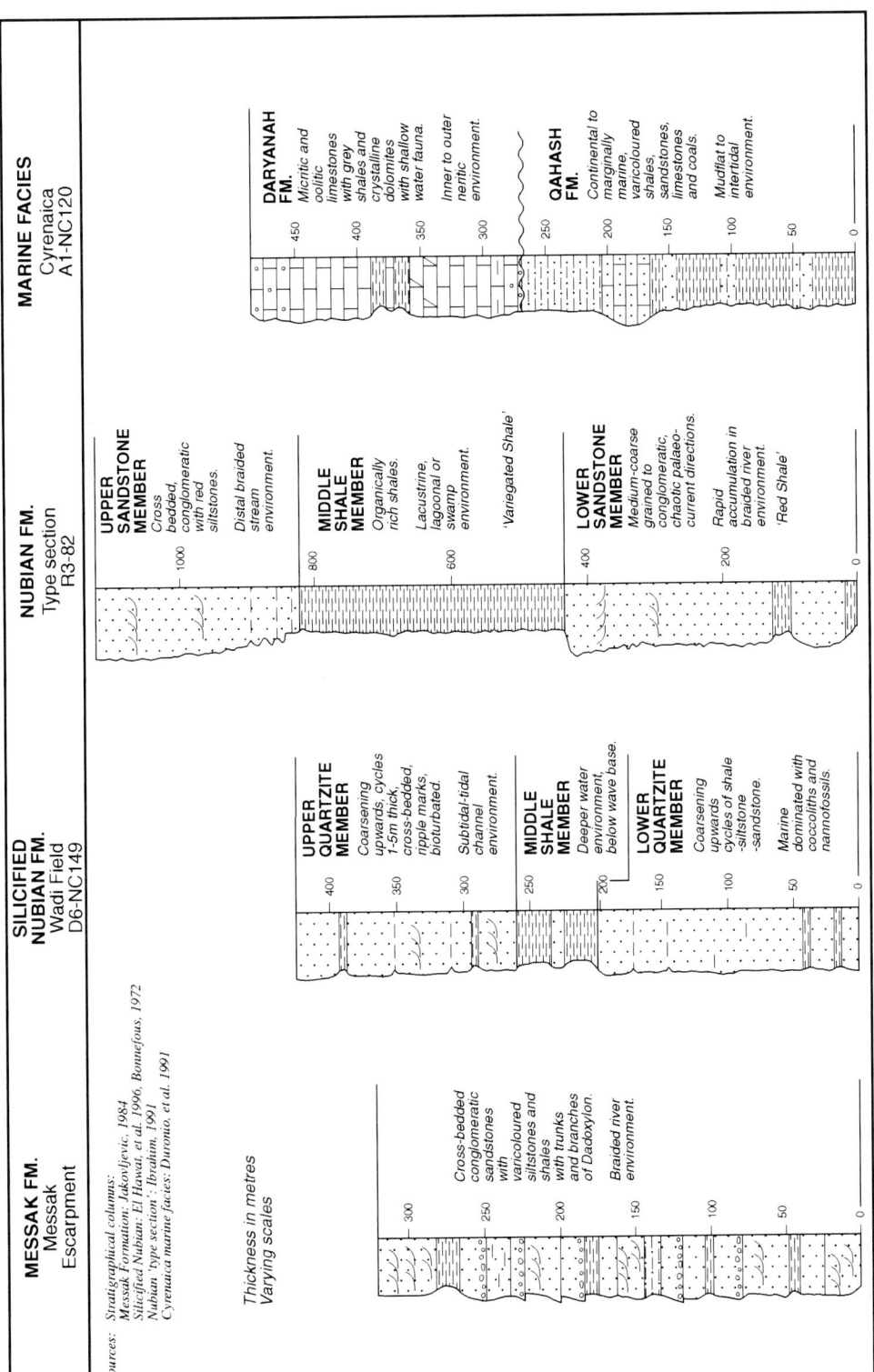

Figure 4.8 Kiklah Formation: Distribution, Equivalents and Stratigraphic Columns

The Kiklah Formation overlies the Kabaw Formation and older deposits in the Jabal Nafusah with marked unconformity. It is present over much of the Hamadah al Hamra and extends into the offshore Sabratah Basin. In the Murzuq Basin the early Cretaceous is represented by the Messak Formation which forms a prominent escarpment along its western outcrop. The Nubian Formation outcrops extensively in the Kufrah Basin, and is well developed in the subsurface of the eastern Sirt Basin where it is a prolific oil reservoir. In the central Sirt Basin the Nubian Formation is heavily silicified. Marine early Cretaceous equivalents are present in northern Cyrenaica (Qahash and Daryanah Formations), and in the eastern Sabratah Basin (Turghat and Masid Formations).

and indurated. The limestones are argillaceous. It has an unconformable contact with the overlying Alalgah Formation; the base was not reached in the well. The incomplete thickness in the well is 370m. The formation contains a rich fauna of planktonic and benthonic foraminifera, echinoderms, ostracods and radiolaria. The environment of deposition is probably outer neritic. The fauna suggests a late Albian-early Cenomanian (Vraconian) age. This formation is also believed to have been penetrated in well F1-NC 35a. It is probably equivalent to the upper part of the Turghat Formation in the area further west, and in Tunisia it probably equates with the upper part of the Gafsa Formation and the lower part of the Fahdene Formation.[61]

LOWER CRETACEOUS CONTINENTAL EQUIVALENTS

4.3.4 Messak Formation

The Messak Formation was first defined by Klitzsch from the Jabal Messak escarpment south of Awbari. It comprises 330m of continental sandstones and shales, frequently conglomeratic, often cross-bedded, and occasionally ferruginous. The base is unconformable with the underlying 'Continental Post-Tassilien' (called the Tilemsin Formation by Klitzsch in this area), and is unconformably overlain by marine Cretaceous rocks. Sparse plant remains in the formation were previously referred to the Wealden or Upper Jurassic. However a recent study of borehole samples from the Sabha area has yielded a rich spore assemblage which gives a precise late Berriasian age for these sediments in this area. Near the type area a palynological assemblage has been found which suggests a range from late Jurassic to early Cretaceous.[62]

The formation has a wide distribution around the margin of the Murzuq Basin, and has been remapped by IRC, along the western and northern peripheries (Figure 4.8). The formation

reaches 350m in thickness on the western margin where it is composed of massive cross-bedded sandstones, often microconglomeratic, interbedded with varicoloured sandy siltstones containing trunks and branches of *Dadoxylon*. These rocks were deposited in a braided-river environment. Northwards the formation thins to about 150m at Awbari where it has been split into two members, a lower Jarmah Member comprising 90m of cross-bedded sandstones, medium- to coarse-grained, and often conglomeratic, interbedded with varicoloured siltstones and silty claystones, and an upper Awbari Member composed of 55m of massive cross-bedded, pebbly sandstones which forms the top of the Messak escarpment in the Awbari-Jarmah area. These units correspond to the Germa beds and Ubari Sandstone of previous authors. They have been mapped eastwards as far as Tmassah. In the Jarmah area underground aqueducts called foggaras have been dug into the escarpment to provide water for extensive agricultural schemes possibly dating from the time of the Garamantes.[63]

The eastern Murzuq outcrops were examined and measured by Goudarzi. In the northern part of the Dur al Qussah area the Messak Formation is composed almost entirely of silicified and well cemented conglomeratic boulders of Palaeozoic origin. These boulder beds pass laterally into gravels and conglomerates in which the size of the blocks gradually decreases. On the eastern flank of the Murzuq Basin the Messak Formation forms the rugged ridge of the Jabal bin Ghanimah. The thickness of the formation increases from 100m at Dur al Qussah to 200m at Jabal bin Ghanimah. In this area plant remains have been found which indicate a Wealden age. Further south the Messak Formation forms extensive outcrops on the southern flank of the Murzuq Basin in the Jabal Ati and Hamadat Manghini areas, where it reaches 800m in thickness. The outcrops continue southwards into Niger. The Messak

Formation outcrops disappear beneath the Murzuq Sand Sea but the formation is present in the wells drilled within the sand sea area. It reaches a thickness of over 800m in the Braspetro wells in concession NC 58, in the centre of the basin.[64]

The environment of deposition was studied by Lorenz and de Castro who demonstrated extensive tracts of overloaded braided-streams flanked by flood-plain siltstones and claystones. Evidence of humid, tropical, lateritic weathering has also been deduced.[65]

4.3.5 Nubian (Sarir) 'Formation'

The term Nubian Sandstone was introduced into Libya by Desio in 1935. The name originated from the Nile valley where it was used by Russeger in 1837 as a general term for the widespread non-marine sandstones which characterise the upper Nile valley. In southwestern Egypt the Nubian 'Formation' has been described by Klitzsch. In Libya the terms Nubian Sandstone, Nubian Formation or Nubian Group have been used by different authors in different senses for either all or part of the non-marine sandstones underlying the marine Cretaceous. In view of these conflicting opinions many authors have recommended abandoning the term Nubian altogether, and it has not been used by the surveyors of the Geological Map of Libya. But, as Barr and Weegar pointed out, the name is still commonly used by petroleum geologists in Libya, particularly in the southeastern Sirt Basin where it is a prolific oil producing horizon. It can perhaps be accepted as a broad informal subdivision which clearly does not satisfy the Code of Stratigraphic Nomenclature, but which nevertheless continues to have some value. In the eastern Sirt Basin some authors have preferred to use terms such as Sarir Sandstone and Kalanshiyu (Calanscio) Sandstone to indicate late Jurassic to early Cretaceous continental

clastics specific to this area. In this account Nubian 'Formation' is used in the sense of the non-marine sequences lying between the Taouratine Formation and the marine Cretaceous of Cenomanian age. Essentially therefore it is equivalent to the Messak Formation and the Kabaw/Kiklah Formations, and is mainly of early Cretaceous age.[66]

The term Djoua Group was introduced by Burollet for rocks in the Fort Flatters area of Algeria corresponding to the 'Continental Intercalaire' of Kilian. They comprise two subdivisions, a lower bone bed, and an upper bed of gypsiferous shale, which reach a combined thickness of 200m. The sequence contains fragments of fish, reptiles and pelecypods which indicate an early Cretaceous age. These rocks are approximate equivalents of the Kiklah Formation, Messak Formation and Nubian Sandstones in other parts of Libya (Figure 4.8).[67]

In the Al Kufrah Basin Nubian Sandstones outcrop across the entire surface of the basin over an area of some 240,000km^2. (Figure 4.8). They consist of fine- to coarse-grained continental sandstones, frequently cross-bedded, varying in colour from yellow to red to purple and are generally poorly cemented. Breccias and conglomerates are common. The B1-NC 43 well penetrated 700m of barren sandstones which are assigned to the Nubian Formation but no age determination was possible, and well A1-NC 34 penetrated 600m with porosities of 21-30%. South of Jabal al Awaynat Burollet divided the Nubian into a lower sandstone unit which he believed to be Jurassic in age, a middle Chieun Limestone of supposed Wealden age, and an upper sandstone of early Cretaceous age.[68]

The Nubian 'Formation' is widely distributed in the Sirt Basin. It is particularly well known in the southeastern part of the basin where it is a prolific oil reservoir at As Sarir, Messlah, and many other fields. The 'formation' comprises fluvial and lacustrine sandstones ranging from very fine- to coarse-grained, quartzitic, poorly-sorted, often with a clay matrix. A basal conglomerate is present in some places. Some of the coarser sandstones have excellent porosity. The sandstones are frequently cross-bedded and often show ripple-marks. Carbonaceous debris is found within the shaly intervals and occasional lignite beds are found. Fossils are very rare, but charophytes of Aptian age have been found along with brackish-water ostracods in the Variegated Shale 'Member' of the Nubian 'Formation' of concession 82. In general, the Nubian Formation' unconformably overlies basement, Palaeozoic, or Triassic/Jurassic graben-fill sediments, and is overlain by the transgressive marine Cretaceous.[69]

An informal Nubian 'type section' was illustrated by Ibrahim from the R3-82 well where the 'formation' has a thickness of 1143m. In the type well it is divisible into three units, a lower sandstone of 470m, a middle variegated shale of 445m and an upper sandstone of 228m. A similar subdivision was made by Rossi et al. in a study of 84 wells in the Hameimat Trough. They attempted a correlation between the sequences found in the Hameimat Trough with global sea level changes and concluded that the lower sand could be assigned to the Tithonian-Valanginian, the middle shale to the Hauterivian-Barremian and the upper sand to the Aptian-Albian. This three-fold subdivision is widespread, but local pinchouts and onlaps frequently mask the typical subdivision. (Figure 4.8). Ambrose suggested that serious errors have been made in the correlation of the Sarir Sandstone. Using Agoco data he showed that two shale units are present and that the full sequence from top to bottom is: Upper Sarir Sandstone, Variegated Shale, Middle Sarir Sandstone, Red Shale and Lower Sarir Sandstone. This throws into question many of the correlations previously made, and has major implications for the distribution of petroleum reserves within the various members of the Sarir

Sandstone. According to Ambrose the Lower Sarir Sandstone, which rests on basement, represents early rift deposition and is typified by alluvial fans, conglomerates and braided-plain deposits. The Red Shale was deposited in a playa-sabkha environment. The Middle Sandstone, which is one of the main reservoir units in the eastern Sirt Basin, represents a fluvial system deposited by river systems flowing towards the centre of the Hameimat Trough. The Variegated Shale, which reaches a thickness of 500m, contains palaeontological evidence of an Aptian-Albian age. It represents a transition to lacustrine conditions, and is a significant petroleum source rock. The Upper Sarir Sandstone is another major reservoir rock in the southeastern Sirt Basin. It unconformably overlies the Variegated Shale and represents a range of distal alluvial-plain, deltaic coastal-plain, tidal-flat and marsh environments.

The Nubian 'Formation' in the eastern Sirt Basin area represents syn-rift deposition. According to Ibrahim, floral and palynological determinations in the southeastern Sirt Basin yield a Neocomian to Aptian age for the 'formation', although it may extend into the late Jurassic in the Siwa, Hameimat and Najah Troughs. A palynological study yielded early Cretaceous miospores from the Nubian Formation in well C280-65, a Barremian-Aptian age in well C1-NC 59 in the Hameimat Trough, and a Jurassic age in well C2-NC 59.[70]

The sedimentology of the Nubian 'Formation' in the marginal sub-basins of the southeastern Sirt Basin was studied by El Hawat. He demonstrated that the lower sandstone member accumulated rapidly, close to the basement source area in a braided-river environment. The chaotic palaeocurrent directions may indicate contemporaneous tectonism. The middle sandstone member, found in the As Sarir area, shows two differing facies, one representing a marginal marine association and the other a fluvial facies association. The variegated shale member, present in the Hameimat Trough, shows clear evidence of deposition under low-energy conditions in a lacustrine, lagoonal or swamp environment. The shales contain a high organic content which suggests deposition under anoxic conditions. The shales contain charophytes, ostracods and dinoflagellate cysts of probable Barremian-Aptian age. The upper sandstone in the Sarir area represents a distal braided-stream environment close to the shoreline. In the Hameimat Trough this unit shows alluvial-plain and meander-belt features at some distance from the shoreline. In another study Abdulghader identified four basic lithofacies within the Nubian 'Formation' of the eastern Sirt Basin; a meandering river facies with point bars, levées and overbank deposits, a relatively high-energy alluvial-plain association with low-sinuosity braided-streams, a swamp facies, and a relatively deep-water lacustrine facies.[71]

Until recently the presence of Nubian sediments in the deep troughs of the Sirt Basin was speculative. It was generally assumed that the rift grabens which developed on the Sirt Arch were floored with basement quartzites. But in recent years more and more evidence has come to light to suggest that Nubian Sandstones are present in the rift grabens. It has been shown above that rifting began as early as the Triassic, so it is not unreasonable to suppose that Nubian Sandstones were also deposited in these rift grabens, especially following the intensification of rifting which began in the Neocomian. As early as the 1970's both Hea and Bonnefous pointed to highly silicified sequences in the central Sirt Basin which they believed were of Nubian age. Bonnefous examined two wells in detail, D2-104A on the Wadi field in the Maradah Trough, and V1-59 on the Bilhizan field on the Al Bayda Platform adjacent to the Maradah Trough. In both wells he found evidence of early Cretaceous (Berriasian-Barremian) nannofossils and coccoliths in the

lower, marginally-marine quartzites overlying the Cambro-Ordovician. The massive quartzitic sands higher in the section are Upper Cretaceous in age. Since then much new data has become available. Wennekers et al. provided data on six wells which have been reassessed as containing early Cretaceous microfaunas. Bonnefous' work on the Wadi field has been rechecked and verified, and El Hawat provided data on the P1-16 well on the Bazzuzi field adjacent to the Maradah Trough. El Hawat concluded that Nubian sediments are present in all the deep troughs in the centre of the Sirt Basin (Figure 4.8). The presence of nannofossils and coccoliths shows that the Nubian succession in the deep troughs is at least partly marine. The troughs connected northwards with the Tethys Ocean and it is believed that marine incursion into the troughs began at least as early as the Neocomian, immediately following the initial foundering of the Sirt Arch. The incursion extended for several hundred kilometers in the Maradah Trough, and extended onto the margin of the Al Bayda Platform, and onto the Manzilah Ridge. Subsequent diagenesis has had a profound effect on these sediments, particularly in the form of quartz overgrowths, converting them into indurated quartzites with minimal residual porosity.[72]

In the western Sirt Basin Wennekers et al. showed the presence of unsilicified Nubian rocks in both the Zallah and Abu Tumayam troughs, associated with possible lacustrine shales which may represent local source rocks for this area. The thickness of the Nubian rocks in the Zallah Trough is 600m, and in the Abu Tumayam Trough 450m. These rocks exhibit both marine and non-marine characteristics.[73]

In Cyrenaica Nubian rocks have been found in many wells, although definitive age dates are relatively rare. Three palynological assemblages have been identified which prove an early Cretaceous age for these rocks. A late Neocomian to early Barremian assemblage has been found in wells VV1-80 and GGG1-59, a late Barremian to Aptian assemblage in wells R3-82 and R8-82, and an Aptian-Albian assemblage in well J1-81A.[74]

NORTHEASTERN LIBYA

In northeast Libya the early Cretaceous was dominated by the Jabal al Akhdar Trough, a narrow, east-west, fault-bounded trough lying between the stable Cyrenaica Platform to the south and the Al Marj ridge to the north. The trough was subsequently inverted to form the Jabal al Akhdar Uplift (Figure 4.9). The trough can be traced from Al Abyar to Bamba, and sequences of Lower Cretaceous rocks over 2000m thick have been proved in wells A1-18, B1-36, A1-36 and offshore well A1-NC 128. The thickness reaches 2750m in well A1-18 and 3300m in A1-NC 128, compared with only 250 to 300m on the Cyrenaica Platform, and 900m on the Al Marj ridge.

The only stratigraphic names so far assigned to the Lower Cretaceous rocks of northern Cyrenaica are the informal names introduced by Duronio et al. The problems associated with the Upper Jurassic lithostratigraphic names of Duronio et al. have already been mentioned. There are similar problems with the Lower Cretaceous names. Duronio showed the Qahash 'Formation' as being present in both the A1-NC 120 and A1-NC 128 offshore wells. The sequence in A1-NC 120 on the Al Marj Ridge is represented by continental to marginally marine sediments of Valanginian to Barremian age, whilst A1-NC 128 is located in the offshore Jabal al Akhdar Trough and is represented by deep-water, gravity-flow deposits of Berriasian to Hauterivian age. Despite the totally different facies and environment of deposition Duronio et al. assigned both of these sequences to the Qahash 'Formation'. These problems urgently need to be addressed.

4.3.6 Qahash 'Formation'

The Qahash 'Formation' has not been formally defined. The term was introduced by Duronio et al. for subsurface sequences encountered in the A1-NC 120 and A1-NC 128 wells. In well A1-NC 120 the stylobreccia of the Sirual 'Formation' is overlain by a predominantly shaly sequence grading upwards into shaly continental sandstones with frequent coaly intervals. It contains rare non-diagnostic fossils, but on the basis of stratigraphic position it is believed to be Valanginian to Barremian in age. The thickness in the well is probably less than 200m (Figure 4.8). In well A1-NC 128, on the other hand, the deep-water conditions of the late Jurassic continued into the Berriasian, Valanginian and Hauterivian stages and the sediments deposited during this time are dominated by mass-gravity transport. Material derived from reefs and shallow carbonate platforms was carried into deep water by turbidity flows and gravity slides, producing a mixed assemblage of rock types within a deep water mudstone matrix. The thickness of the formation in this well is probably more than 1500m.

The thick sequences found in the Jabal al Akhdar Trough are not exclusively bathyal in character. Other wells show a variety of facies from open-marine to the shallow-water sandstones and shales with gypsum found in well B1-18. It seems that deep-water conditions are largely confined to the eastern part of the trough.[75]

4.3.7 Daryanah Formation

The Qahash 'Formation' is overlain in the offshore Cyrenaica wells by the Daryanah Formation which has been formally defined. Duronio et al. selected the section in well A1-NC 120 as the type section. Here the formation comprises 238m of shallow water

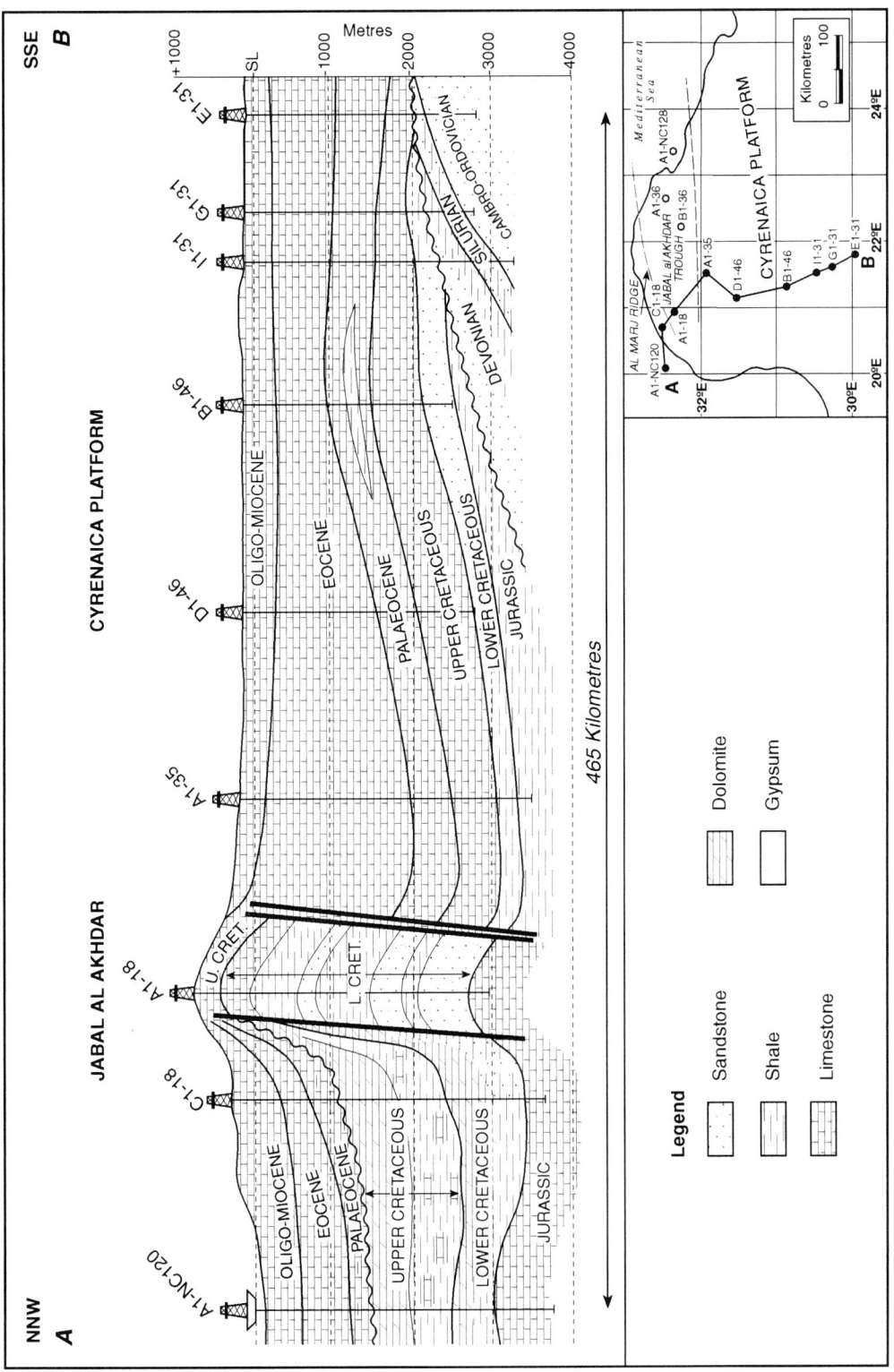

Figure 4.9 Geological Cross-Section through the Jabal al Akhdar Uplift

Sources: Boote, et al., 1998, El Hawat & Shelmani, 1993, Pallas, 1980

The section shows the dramatic thickening of the Lower Cretaceous section in the Jabal al Akhdar Trough, and its subsequent inversion to form the Jabal al Akhdar Uplift. The trough continues eastwards as far as the offshore A1-NC128 well.

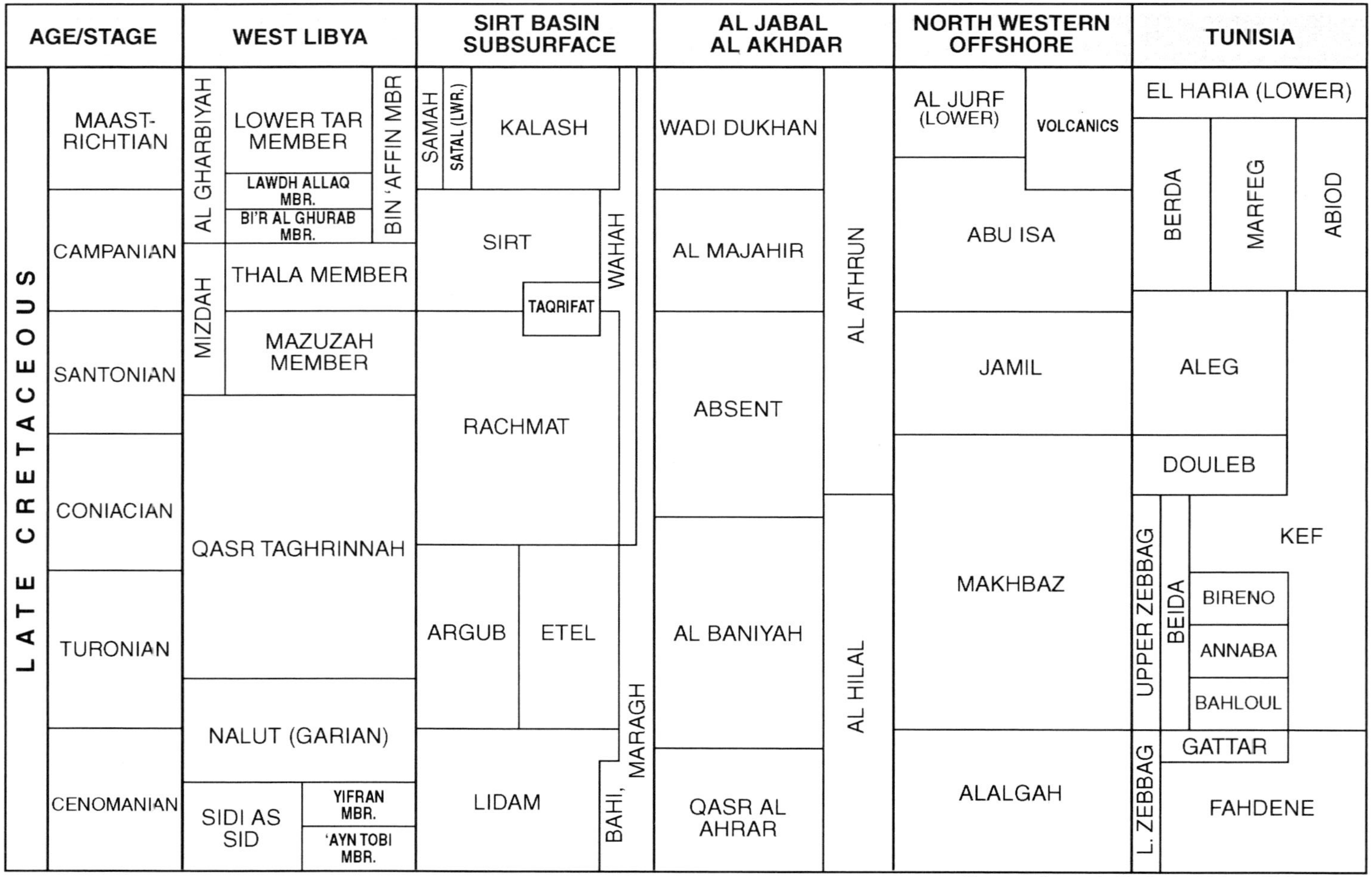
AGE/STAGE
WEST LIBYA
SIRT BASIN SUBSURFACE
AL JABAL AL AKHDAR
NORTH WESTERN OFFSHORE
TUNISIA
LATE CRETACEOUS
MAAST-RICHTIAN
CAMPANIAN
SANTONIAN
CONIACIAN
TURONIAN
CENOMANIAN
AL GHARBIYAH
LOWER TAR MEMBER
LAWDH ALLAQ MBR.
BI'R AL GHURAB MBR.
BIN 'AFFIN MBR
MIZDAH
THALA MEMBER
MAZUZAH MEMBER
QASR TAGHRINNAH
NALUT (GARIAN)
SIDI AS SID
YIFRAN MBR.
'AYN TOBI MBR.
SAMAH
SATAL (LWR.)
KALASH
SIRT
WAHAH
TAQRIFAT
RACHMAT
ARGUB
ETEL
MARAGH
BAHI,
LIDAM
WADI DUKHAN
AL MAJAHIR
ABSENT
AL BANIYAH
QASR AL AHRAR
AL ATHRUN
AL HILAL
AL JURF (LOWER)
VOLCANICS
ABU ISA
JAMIL
MAKHBAZ
ALALGAH
EL HARIA (LOWER)
BERDA
MARFEG
ABIOD
ALEG
DOULEB
KEF
UPPER ZEBBAG
BEIDA
BIRENO
ANNABA
BAHLOUL
GATTAR
L. ZEBBAG
FAHDENE

Source: General: Geological Map of Libya, 1985
West Libya: Megerisi and Mamgain 1980a,
Nairn and Salaj, 1991
Sirt Basin: Barr and Weegar, 1972
Al Jabal al Akhdar: El Hawat and Shelmani, 1993
Northwestern offshore: Hammuda et al., 1985
Tunisia: Fournie, 1978

Figure 4.10 Lithostratigraphic Nomenclature for the Upper Cretaceous Rocks of Libya and Equivalents

This chart shows the current lithostratigraphic nomenclature for the Upper Cretaceous rocks of Libya and is based on sources shown, with some recent additions, and updated transliteration of Arabic names.

micritic and oolitic limestones and dolomites with dark grey shales and crystalline dolomites in the middle part. It contains large foraminifera, gastropods, echinoderms and pelecypods which indicate a Barremian-Aptian-Albian age, suggesting an age equivalence with the Kiklah Formation of northwest Libya.(Figure 4.8). In well A1-NC 128 the Daryanah Formation shows a gradual shallowing of the basin. The lower shales and marls were deposited in deep water, with occasional turbidite flows. The middle shales, carbonates and calcareous sandstones were deposited in an outer shelf environment, whilst the upper siltstones and limestones represent a transition from outer neritic to inner neritic.[76]

Similar rocks have been found in several onshore wells in northern Cyrenaica including C1-33, B1-36, A1-18, F1-2, A1-28, A1-117, B1-33 and C1-18. Palynological investigation of these wells showed an unbroken succession from late Hauterivian to late Aptian age, covering a range of environments from outer to middle shelf in the northern wells to basin margin (supratidal) in the southern wells. These rocks can be regarded as nearshore equivalents of the Daryanah Formation.[77]

4.4   UPPER CRETACEOUS

Continental conditions dominated the interior of Libya from the Permian to the mid-Cretaceous. As pointed out above, evidence is accumulating that the rift phase began in the Triassic, and culminated in the early Cretaceous with the collapse of the northern Sirt Arch. The rift phase ended with marine incursion from the Tethys Ocean. As proved by the presence of nannofossils and coccoliths, the incursion into the deepest rift troughs began in the early Cretaceous. This marks the beginning of the post-rift phase of basin development. The initial incursion of the early Cretaceous was followed by a major flooding event in the Cenomanian

which extended marine conditions over most of the Sirt Basin, except for the major platform areas which remained as large islands or peninsulas. Palaeontological evidence proves that, despite several interruptions, the general trend during the late Cretaceous was for the platform areas to be progressively covered by marine sediments, so that by the end of the Cretaceous the troughs had been almost entirely filled, and only a few very small islands remained above sea level. It is likely that continental sediments continued to be deposited futher south, and Wennekers et al. suggested that up to 300m of undifferentiated continental clastics may be present in the Murzuq Basin beneath the recent dune cover, although no rocks of this age have yet been positively identified either at outcrop or in the subsurface. The late Cretaceous rocks of Libya have received great attention from petroleum geologists and a great many formation names have been defined. The current stratigraphic nomenclature is shown on Figure 4.10. In view of this greater amount of detail the late Cretaceous formations will be reviewed by region, starting in the west and progressing eastwards.[78]

NORTHWESTERN LIBYA

4.4.1   Sidi as Sid Formation

The Sidi as Sid Formation was named and described by El Hinnawy and Cheshitev for a marine sequence in the Gharyan area overlying the Kiklah Formation in the west and the Abu Shaybah Formation in the east. The formation comprises two members, a lower carbonate unit named the 'Ayn Tobi Member, and an upper marlstone unit, the Yifran Member.

A type section for the 'Ayn Tobi Member was designated by El Hinnawy and Cheshitev on the Gharyan Dome, as a replacement for the previous composite type section selected by

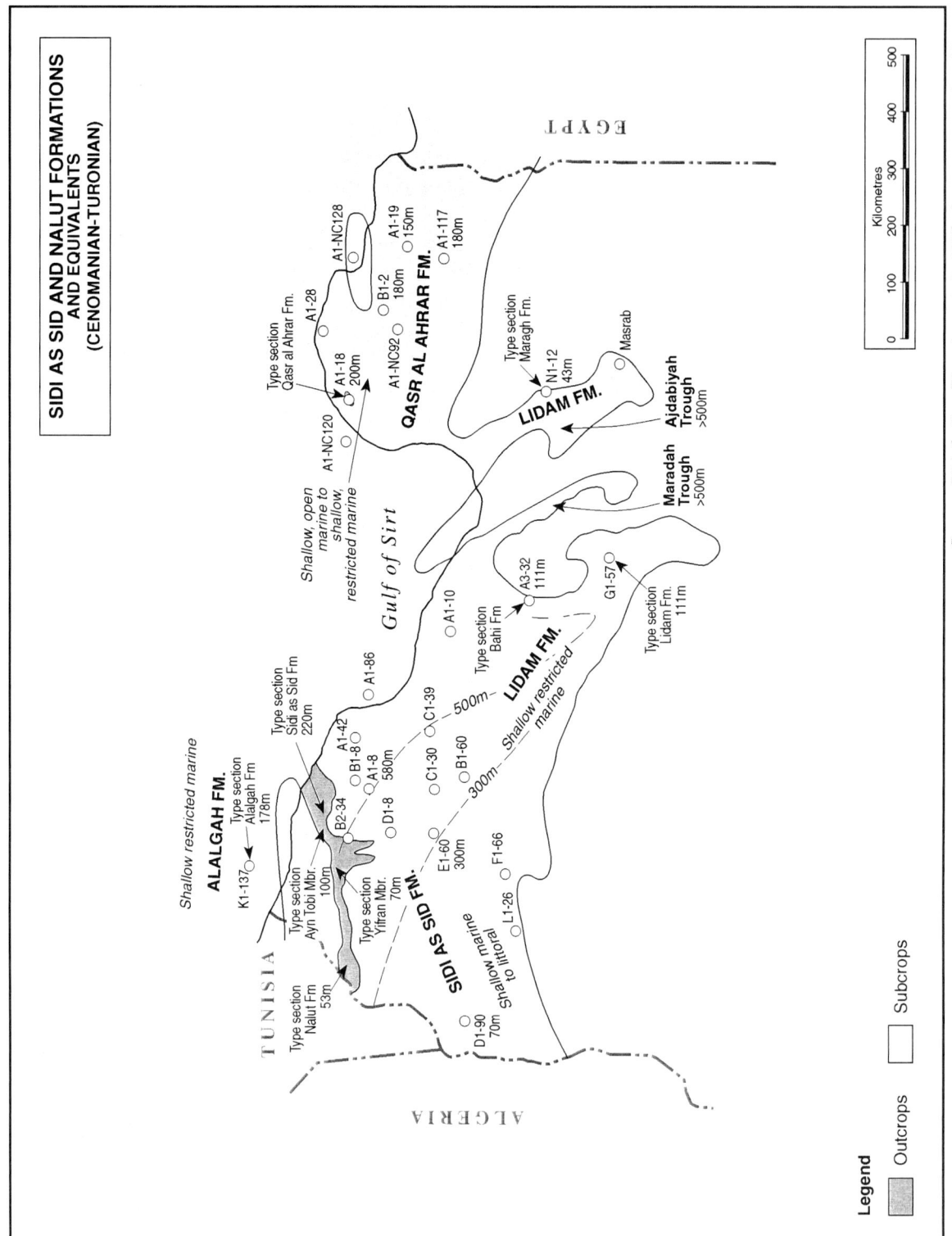

SIDI AS SID AND NALUT FORMATIONS
AND EQUIVALENTS
(CENOMANIAN-TURONIAN)

Source: Distribution: Wennekers, et al., 1996, plus sources mentioned in text

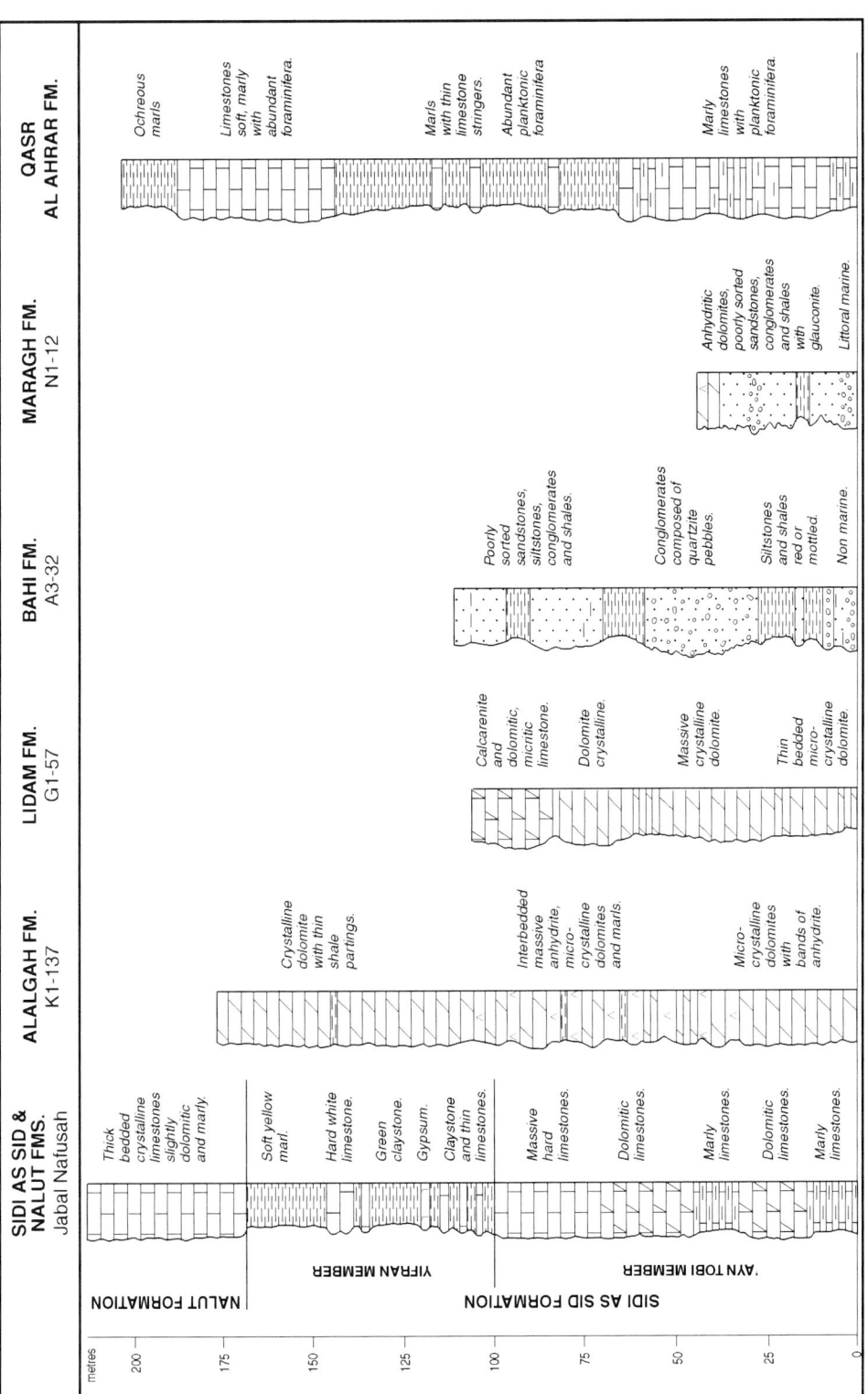

**Figure 4.11 Sidi as Sid and Nalut Formations:
Distribution, Equivalents and Stratigraphic Columns**

*Sources: Sidi as Sid Formation: El Hinnawy & Cheshitev, 1975
Nalut Formation: Novovic, 1977
Atalgah Formation: Hammuda, et al., 1985
Lidam Formation: Barr and Weegar, 1972
Qasr al Ahrar Formation: Klen, 1974, El Hawat and
Shelmani, 1993, Duronio, et al., 1991*

The Cenomanian transgression produced a major flooding event which established marine conditions over much of northern Libya. The sediments are principally shallow water carbonates, evaporites and marls. This sequence is represented in the Jabal Nafusah and Al Hamadah al Hamra area by the Sidi as Sid and Nalut Formations, in the northwestern offshore by the Alalgah Formation, in the Sirt Basin by the Lidam Formation (with the Bahi and Maragh Formations as local clastic wedges), and in Cyrenaica by the Qasr al Ahrar Formation.

Christie. At this location the member comprises 100m of grey and yellow crystalline limestones and dolomites with thin marlstone interbeds. The base is marked by a thin sandstone horizon which contains quartz pebbles derived from the underlying Kiklah Formation. At the type locality the middle of the limestone is marked by a distictive horizon full of rudistids and other pelecypods. The lower contact is unconformable, the upper contact with the Yifran Member is gradational. The member contains a rich fauna of pelecypods, echinoderms, ostracods and algae which indicate a Cenomanian age. (Figure 4.11). An exhaustive study of the diagenesis of the Ayn Tobi dolomites has been conducted by El Bakai. He recognised five diagenetic types of dolomite reflecting successive stages of the diagenetic process.[79]

El Hinnawy and Cheshitev also defined a type section for the Yifran Member, close to the village of Yifran. This unit is composed of 70m of soft, thinly bedded alternations of marly limestone, claystones, marlstones and bedded gypsum. The gypsum horizon develops westwards, but is not present in the outcrops to the east. The fauna is generally poor, but some fragmentary pelecypods have been found along with fish teeth, echinoid debris, foraminifera and ostracods. The age is almost certainly Cenomanian. The environment of deposition ranges from lagoonal in the west through littoral to low-energy neritic in the east. (Figure 4.11).[80]

The formation outcrops along the entire length of the Jabal Nafusah escarpment from Al Khums to the Tunisian border, but the two-fold subdivision is not maintained. Westward the sequence thins to 65m at Nalut and becomes progressively more marly. Eastwards the formation thickens to 380m at Al Khums and becomes more calcareous and dolomitic. The Gezzar Dolomite and Scersciara Member of Desio are now included within the Sidi as Sid Formation.[81]

The formation is present beneath younger rocks in the Al Hamadah al Hamra where the lithology is predominantly anhydrite ranging to dolomite and dolomitic limestone. It can be traced as far south as wells L1-26, F1-66, and B1-60, but it is overstepped and does not outcrop around the southern margin of the basin. (Figure 4.11). It is present in several wells in the subsurface north of the Al Qarqaf Arch, including E1-60, D1-8, B2-34, A1-8, A1-42, C1-39, A1-86 and A1-10. It is composed of unconsolidated sands and crystalline dolomite, reaching a thickness of 64m in offshore well A1-86. There is no evidence that the Cenomanian transgression extended south of the Al Qarqaf Arch.[82]

4.4.2 Nalut Formation

In the western Jabal Nafusah the Sidi as Sid Formation is overlain by 30 to 40m of dolomitic limestone named the Nalut Limestone by Zaccagna. The equivalent section near Gharyan was named the Garian Limestone, but the former name has priority and has been adopted by the surveyors of the Geological Map of Libya. A type section was defined by El Hinnawy and Cheshitev east of Gharyan where 53m of hard, crystalline, dolomitic limestone overlie the Yifran Member of the Sidi as Sid Formation. The formation contains bands of chert and concretions. It contains rare, badly-preserved pelecypods, which suggest a late Cenomanian to Turonian age. It is traceable to Nalut and the Tunisian border, and eastwards it outcrops extensively south of Kiklah and Gharyan and to the east of Tarhunah. (Figure 4.11). In this area it is truncated by erosion to a thickness of about 15m. It extends beneath the Al Hamadah al Hamra, where it has been encountered in most of the wells, but does not outcrop on the southern margin. In the Misratah area it has been recorded in wells B1-8, A1-8 and C1-30. A study of the dolomitization of the Nalut Formation based on outcrop samples showed that the process was attributable to the circulation of magnesium-rich waters during the early stages of burial prior to complete compaction and lithification of the rock.[83]

4.4.3 Qasr Taghrinnah Formation

In the Gharyan area the Nalut Formation is overlain by a predominantly marly sequence which was first described by Christie and is now called the Qasr Taghrinnah Formation. El Hinnawy and Cheshitev selected a type section to the southeast of the village of Qasr Taghrinnah where 71m of soft yellow marls interbedded with hard crystalline to marly limestones are exposed. The formation is richly fossiliferous with a fauna of gastropods, pelecypods, echinoids, brachiopods, ostracods and benthonic foraminifera which indicate a Turonian to early Santonian age. The formation outcrops over a wide area on the Al Hamadah al Hamra plateau from the Tunisian border to Wadi Ka'am south of Al Khums, and southwards to Bani Walid. (Figure 4.12). These outcrops show rapid lithological variation. In the west near Nalut the formation is composed predominantly of marl with a conspicuous marly limestone development in the middle, whereas further east at Mizdah the lower half of the formation is composed of massive gypsum with thin partings of marl and limestone, and an upper unit of chalky limestone. At Bani Walid the formation is composed entirely of chalky limestone. The thickness varies from 250m in the east to 60m in the west.[84]

The Qasr Taghrinnah Formation outcrops in the Ghadamis/Bi'r Amasin/Wadi Tanarut area where the upper 44m are exposed. In this area the lithology is dominantly marly to sandy limestone with thin beds of marl, gypsum and claystone. The fossil fauna in this area clearly demonstrates that the formation extends into the Santonian. The formation is present beneath a thin cover in the Al Hamadah al Hamra and has been reported in wells N1-90, A1-66, and A1-60. East of Bani Walid it passes into dolomitic limestones of the Argub Formation which have been dated as Turonian in wells A1-43 and A1-25.[85]

Further south the Qasr Taghrinnah Formation outcrops on the southern margin of the Al Hamadah al Hamra in the area of Wadi ain Armas/Hamadat Tanghirt, where the formation shows a much more littoral character. (Figure 4.12). It comprises 50m of alternating flaggy sandstones, massive cross-bedded sandstones and claystones with virtually no carbonate content. This clearly indicates proximity to a shoreline, to the extent that in the concession 49 wells this sequence was confused with the Nubian Sandstone.[86]

4.4.4 Mizdah Formation

Much of the surface of the Al Hamadah al Hamra plateau is occupied by outcrops of the Mizdah Formation. The term was introduced by Burollet for a series of massive dolomites containing rudists, capped by dolomitized limestones and shales which outcrop in the Mizdah area. The formation was subsequently subdivided by Jordi and Lonfat. Following remapping of the Mizdah area, the formation was redefined to include two members, a lower Mazuzah Member and an upper Thala Member.[87]

The type area of the Mazuzah Member is located west of Mizdah where it is composed of 30m of thick-bedded, crystalline limestone and dolomite, with some intervals of marly and chalky limestone. Fossils are common in the lower part of the formation including pelecypods – especially *Inoceramus*, ammonites, ostracods and foraminifera, which indicate a Santonian age. The sequence is interpreted as a shallow-water, high-energy

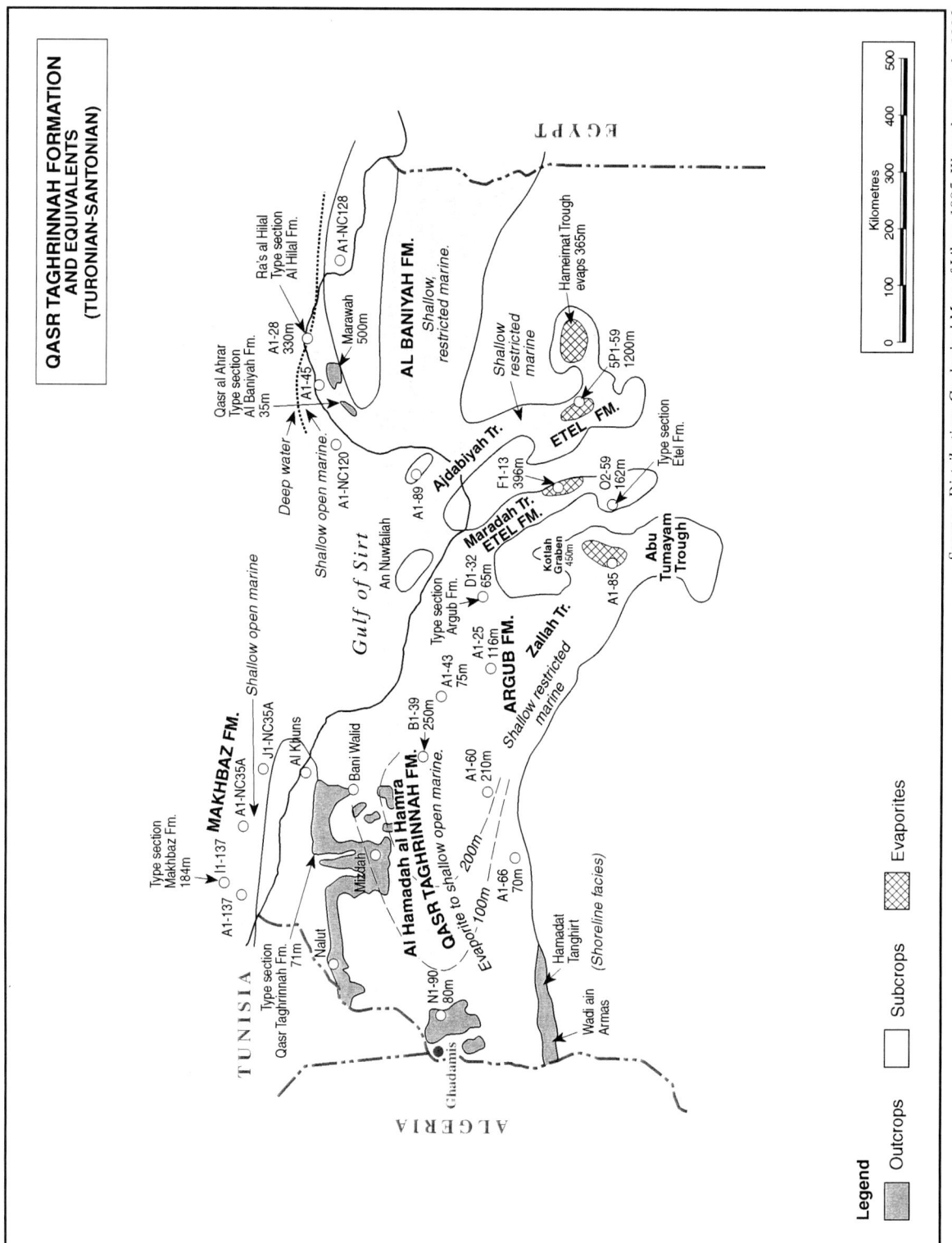

QASR TAGHRINNAH FORMATION AND EQUIVALENTS (TURONIAN-SANTONIAN)

Source: Distribution: Geological Map of Libya, 1985, Wennekers, et al., 1996

Legend

Outcrops

Subcrops

Evaporites

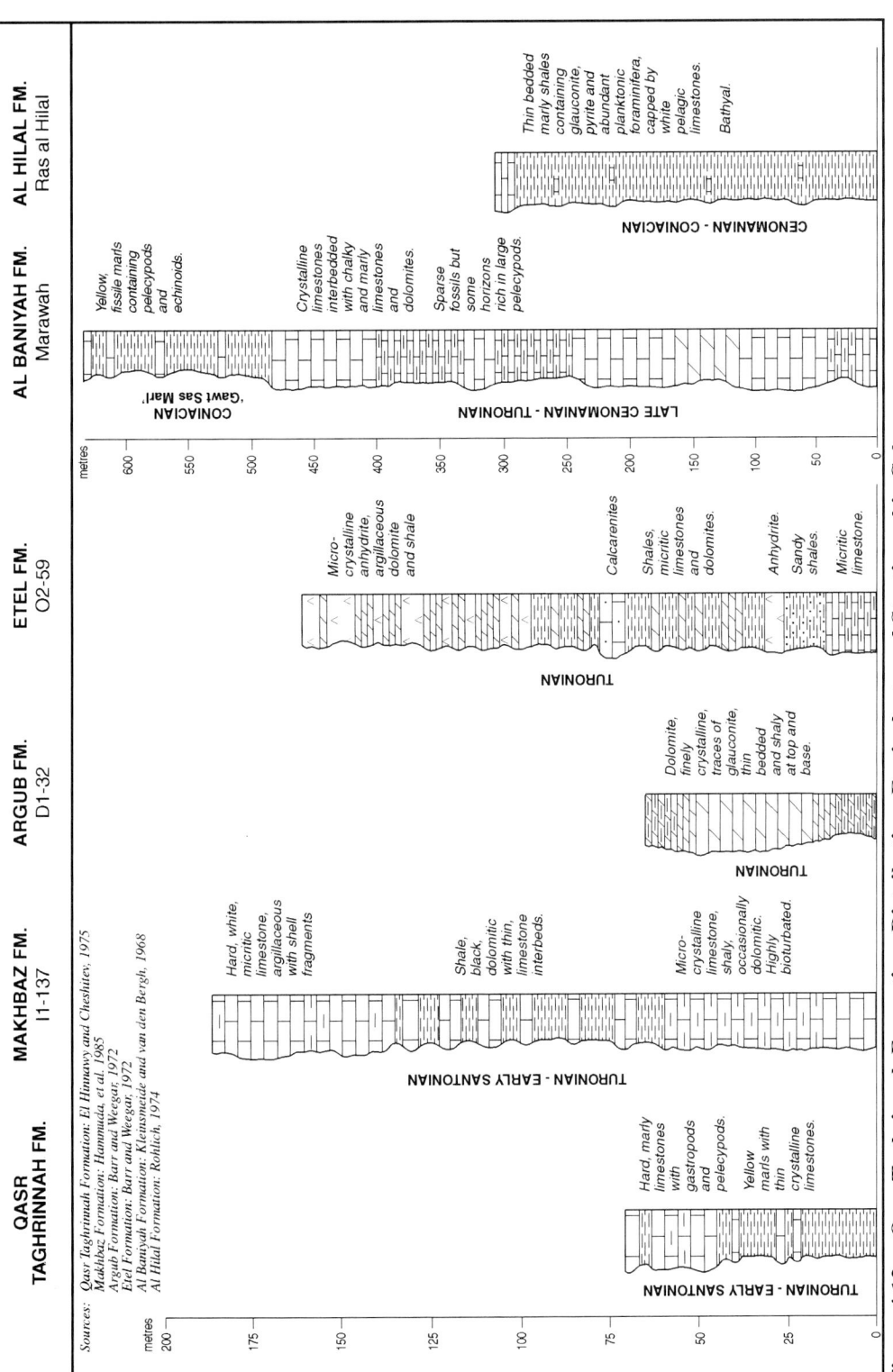

Figure 4.12   Qasr Taghrinnah Formation: Distribution, Equivalents and Stratigraphic Columns

The Turonian-Coniacian section is represented by generally shallow water carbonates, passing into evaporites towards the south and east. The Qasr Taghrinnah Formation outcrops extensively on both the northern and southern margins of the Hamadah al Hamra. In the northern Sirt Basin the Argub Formation is a dolomitic unit, which interfingers with the Etel Formation further south which is predominantly an evaporite sequence. Offshore northwest Libya the Makhbaz Formation represents a shallow shelf environment. In the Jabal al Akhdar the Al Baniyah Formation is exposed in two small inliers which exhibit a similar shallow shelf setting. The Al Hilal Formation in northern Cyrenaica represents deeper water sedimentation. Note the scale change for the Al Baniyah and Al Hilal stratigraphic columns.

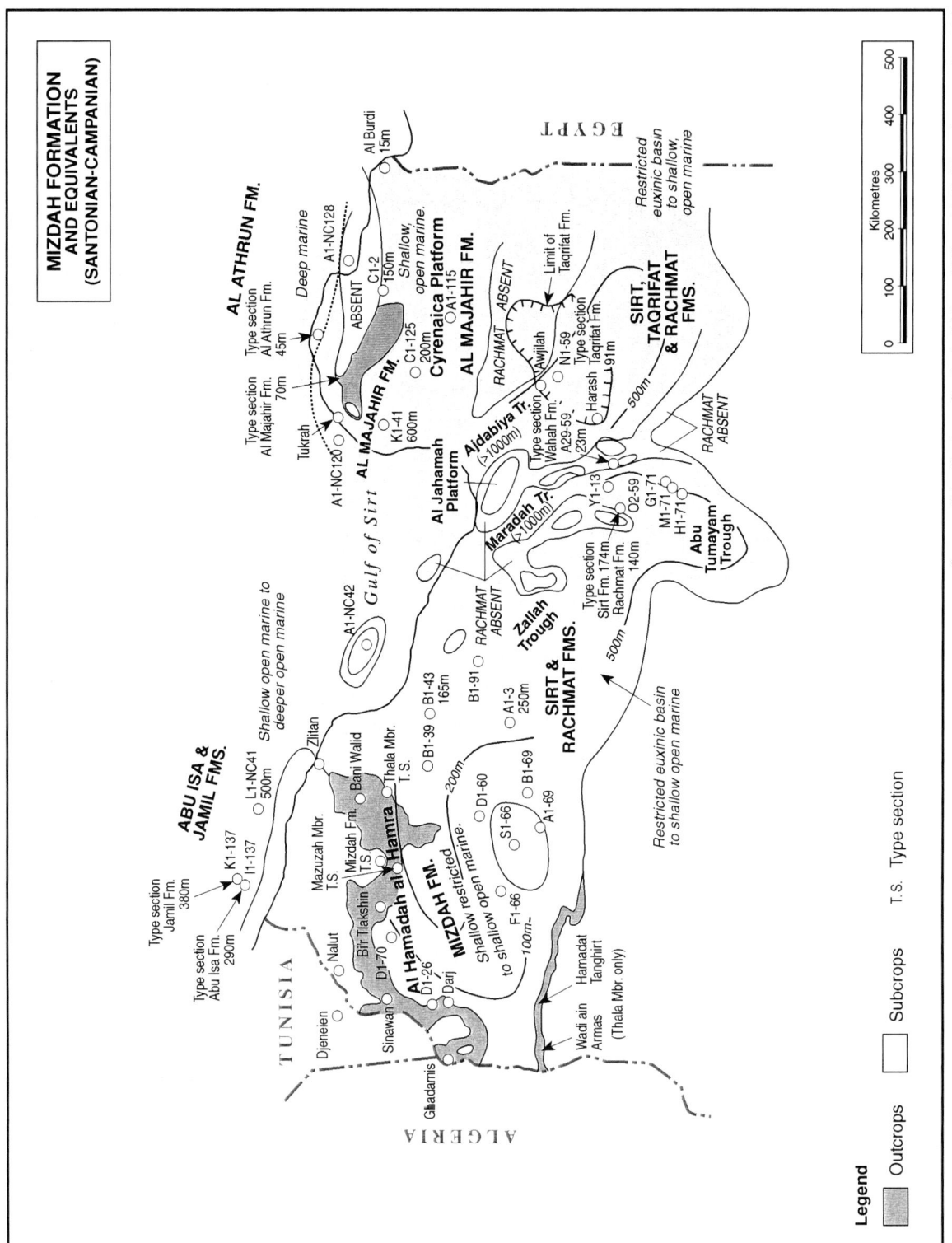

Source: Distribution: Geological Map of Libya, 1985, Wennekers, et al., 1996

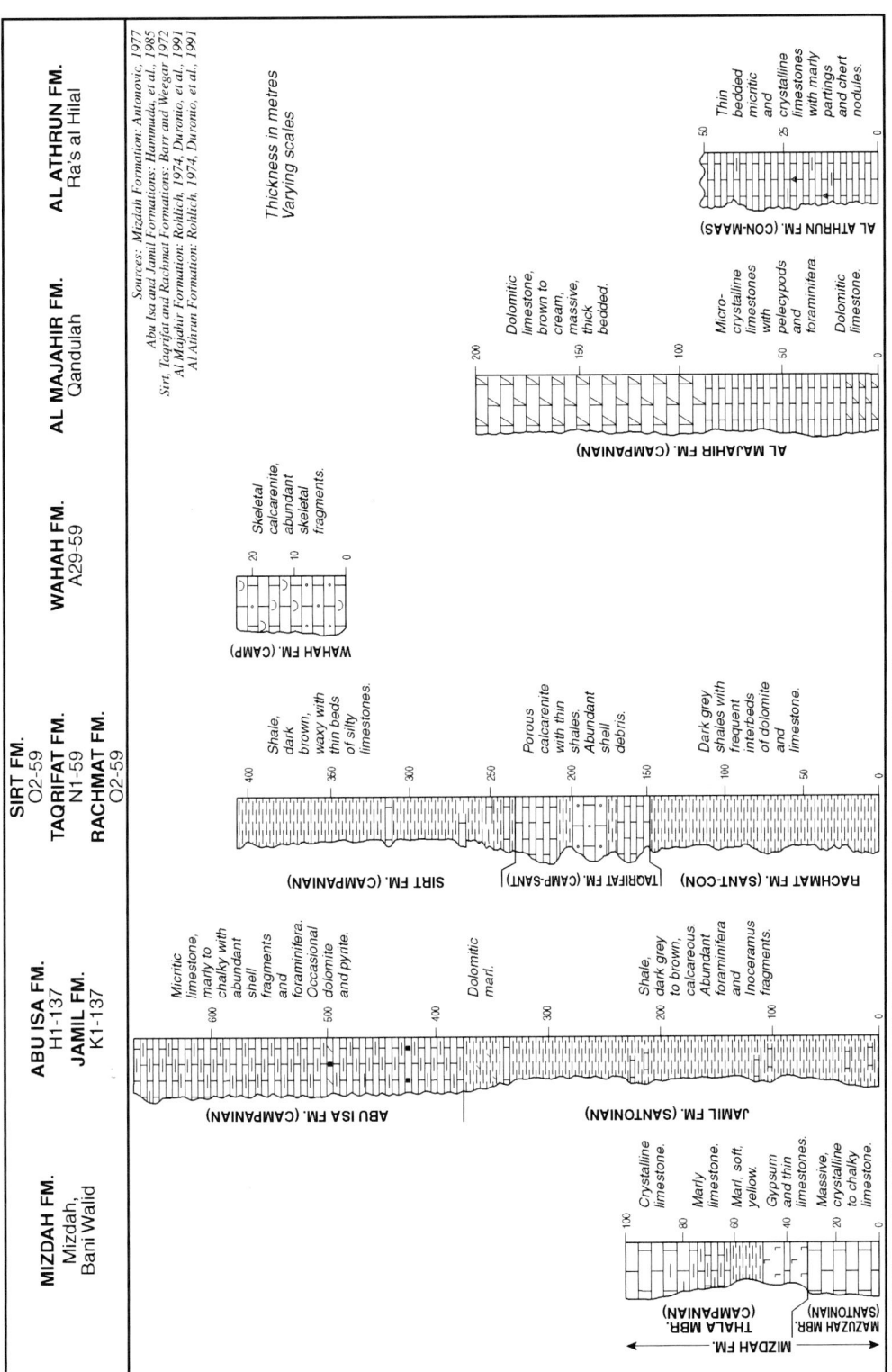

Figure 4.13   Mizdah Formation: Distribution, Equivalents and Stratigraphic Columns

Marine transgression continued during the Santonian-Campanian with the progressive onlap of the Az Zahrah-Al Hufrah, Al Bayda, Zaltan and Amal Platforms. By the end of the Campanian only a few scattered islands remained. The Mizdah Formation and its equivalents in the Sirt Basin represent shallow water carbonates and shales for the most part. Deep water conditions are found only in the Al Athrun Formation offshore northern Cyrenaica. This sequence includes the Sirt Formation which is the principal source rock in the Sirt Basin. The Wahah Formation, which is a major oil bearing formation, represents a shoreline deposit fringing the remaining islands and peninsulas in the Campanian sea. Note the varying vertical scales for the stratigraphic columns.

deposit with periodic connection to the open sea. The inoceramid fauna has been studied in some detail. The fauna is highly distinctive, and significantly different from that of the Lower Tar Member. The assemblage contains some 'southern' species, which gives support for the existence of a trans-Sahara seaway at this time.[88]

The Mazuzah Member has a wide distribution from the Tunisian border almost to the coast at Zlitan. (Figure 4.13). On the Tunisian border the thickness varies from 5m to 15m and the lithology is dominantly sandy dolomite. There is a gradual thickening eastwards to 25m at Bani Walid, with a transition from sandy dolomite in the west to marly limestone with chert in the east. There are small isolated outcrops of the Mazuzah Member at Ghadamis and Darj. On the southern rim of the Al Hamadah al Hamra plateau the Mazuzah Member is overstepped by the Thala Member and does not outcrop in this area.[89]

The overlying Thala Member is best exposed in the Wadi Suf Ajjin near Bani Walid, although no type section has been defined. In this area the Thala Member comprises 72m of massive gypsum, marl, and marly limestone. The limestones contain a rich fauna of pelecypods, gastropods, and ostracods which suggest a Campanian age for this unit. It shows a transition from a lagoonal to a shallow neritic environment of deposition. The Thala Member outcrops in a wide arc from Ghadamis, Sinawan, Bi'r Tlakshin, south of Mizdah to east of Bani Walid. (Figure 4.13). These outcrops show a transition from lagoonal evaporites and shallow-water limestones in the southern outcrop areas (Bi'r Amasin, Ghadamis, Mizdah) to marls and thin limestones further north (Djeneien, Nalut, Bani Walid). A further outcrop belt occurs on the southern margin of the Al Hamadah al Hamra plateau in the Wadi ain Armas/Hamadat Tanghirt area. The lithology in this area comprises sandy limestones, shales and stringers of gypsum about 35m thick, suggesting nearshore to lagoonal conditions reminiscent of the underlying Qasr Taghrinnah Formation. The Mizdah Formation has been penetrated by many wells on the Hamadah al Hamra where it averages about 150m in thickness.[90]

In the subsurface north of the Jabal as Sawda a sequence of crystalline and chalky limestones, claystones and anhydrites has been found in wells B1-69, A1-3 and B1-91 containing a Santonian microfossil assemblage, with an overlying Campanian assemblage, which suggests that the Mizdah Formation extends into this region, but it is absent in an area north of the Al Qarqaf Arch in wells A1-69 and S1-66.[91]

4.4.5 Al Gharbiyah Formation

Studies in the Qarayrat al Gharbiyah area southeast of Mizdah have proved the presence of carbonates containing late Campanian microfossils in rocks which had previously been considered as Lower Tar Member of the Zimam Formation. Detailed work in this area prompted Nairn and Salaj to propose a revision to the late Cretaceous stratigraphy of northwest Libya. Between the Thala Member of the Mizdah Formation and the Upper Tar Member of the Zimam Formation they recognised three sedimentary sequences, two of which had not been previously recognised. They proposed a new formation, the Al Gharbiyah Formation made up of three members, of which the upper unit is the Lower Tar Member, previously included within the Zimam Formation (Figure 4.10).[92]

As redefined, the Al Gharbiyah Formation comprises a succession of generally shallow-water carbonates, divisible into three sequences, a lower Bi'r al Ghurab Member, middle Lawdh Allaq Member, and an upper Lower Tar Member. (Figure 4.14). However, in addition to these three members it is necessary to include a fourth, the Bin 'Affin Member, which is a distinctive littoral facies of the Al Gharbiyah Formation. The Al Gharbiyah Formation outcrops over large areas of Al Hamadah al Hamra. To the south of Jabal as Sawda it is represented by a sandy facies which was named the Bin 'Affin Member by Fürst. It represents a shoreline facies marking the southern limit of the Maastrichtian sea in the area to the east of the Al Qarqaf Arch. To the northeast of Tmassah its southern limit is concealed beneath the volcanics of the Al Haruj al Aswad. The type section of the Bin 'Affin Member is located in this area where it is made up of coarse-grained sandstones and conglomerates which reach a maximum thickness of 100m, but are more commonly 30 to 35m thick. The pebbles are largely derived from the underlying Nubian Sandstone. Iron and manganese are common constituents of the sandstone, and salt and gypsum are found in some areas. The calcareous sandstones contain a poorly preserved fauna which includes *Omphalocyclus macroporus*, an index fossil of the Maastrichtian. The coarse-grained sandstone facies continues northwards as far as the Jabal as Sawda, north of which it passes into the marlstone and shallow-water limestone facies of the Al Gharbiyah Formation. A sedimentological study of the Bin 'Affin Member by El Haddad, et al. described three lithofacies all indicative of a fluvial environment. The Bin 'Affin Member contains widespread copper and uranium mineralisation extending over almost the entire outcrop belt. The minerals were probably derived from the underlying Precambrian granites, granodiorites and diorites (Figure 4.14).[93]

Further north the Al Gharbiyah Formation is represented by three carbonate sequences which have been given the status of members. The Bi'r al Ghurab Member, at the type locality southeast of Mizdah, unconformably overlies the Thala Member. The unconformity is marked by a ferruginous crust and a basal conglomerate. The member comprises 10 to 15m of gypsiferous clays, dolomitic limestones, marls and skeletal coquinoid limestones which are capped by a layer of phosphatic nodules. The fauna includes planktonic foraminifera which indicate a late Campanian age. The areal extent of this unit has yet to be defined.[94]

The middle cycle, the Lawdh Allaq Member, is defined by a type section southwest of Mizdah. The member is composed of 55m of sandy limestones, shelly, coquinoid limestones, and gypsiferous marls. The member contains an adundant fauna including *Inoceramus* and other pelecypods, *Orbitoides media*, and planktonic foraminifera of late Campanian and early Maastrichtian age. This unit too is capped by a layer of phosphate pebbles. Further micropalaeontological work is required before the extent of this member can be determined. A pelagic facies of this member was named the Bi'r az Zamilah Member by Nairn and Salaj. It is developed in the area north of Al Gharbiyah and is composed of 25m of chalk and micritic limestone overlain by marls, but has not yet been fully described.[95]

The uppermost member of the Al Gharbiyah Formation is the Lower Tar Member which was originally included in the Zimam Formation. The Zimam Formation was defined by Jordi and Lonfat from a type section in the southeastern part of the Hamadah al Hamra at Wadi Zimam near Suknah. The section at Wadi Zimam is over 190m thick and bridges the boundary between the Cretaceous and Palaeocene. They divided the formation into two members, the Tar Marl Member (145m) and the Had Limestone Member (45m). They also divided the Tar Marl into a lower unit of Maastrichtian age, and an upper unit of Danian age, with a shell bed, the Socna Mollusc Bed, at the base of the Upper Tar Marl. The age of the Lower Tar Marl has been established as Maastrichtian on the basis of foraminifera from both surface and subsurface samples. According to Nairn and Salaj, it was

186

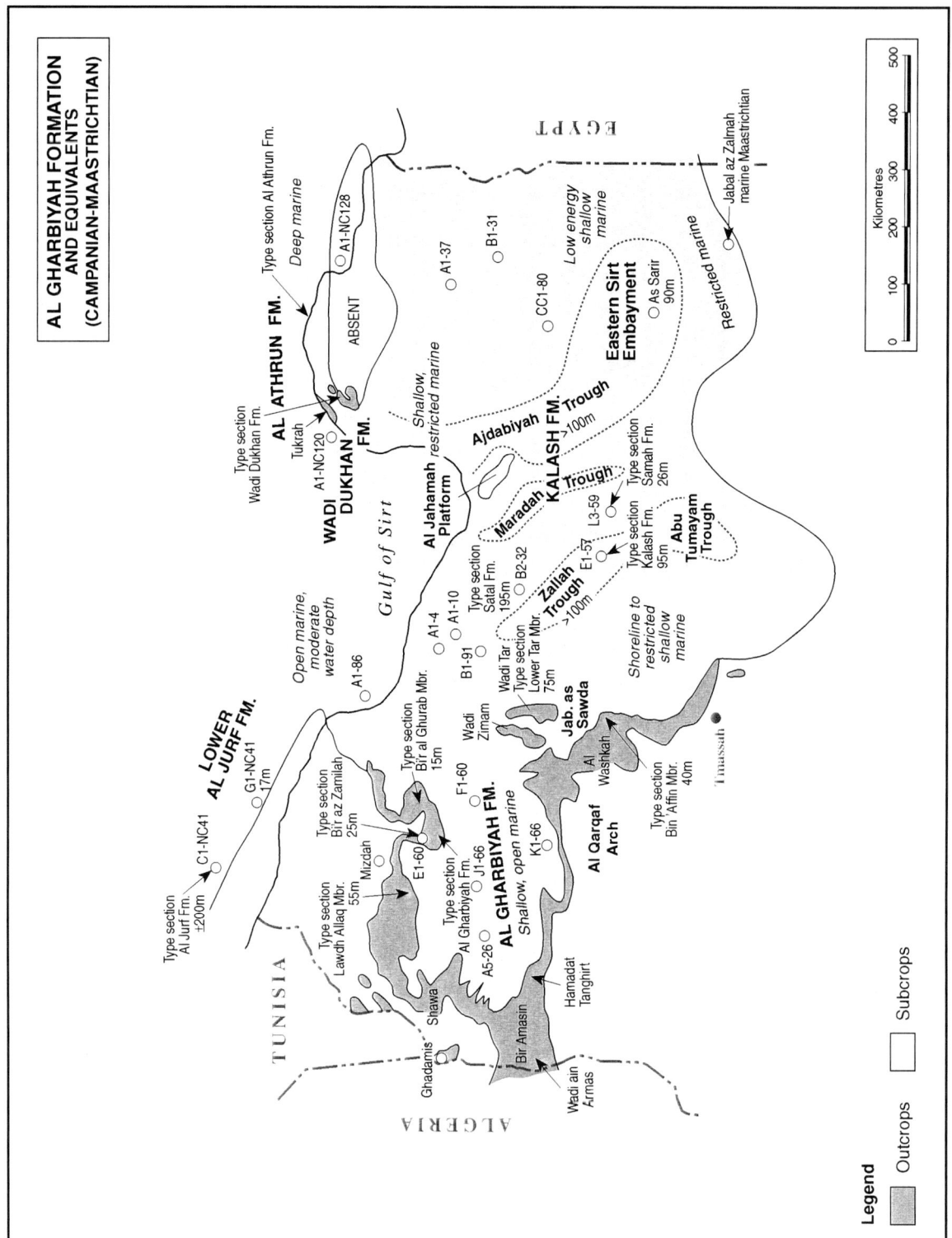

Source: Distribution: Geological map of Libya, 1985, Wennekers, et al., 1996

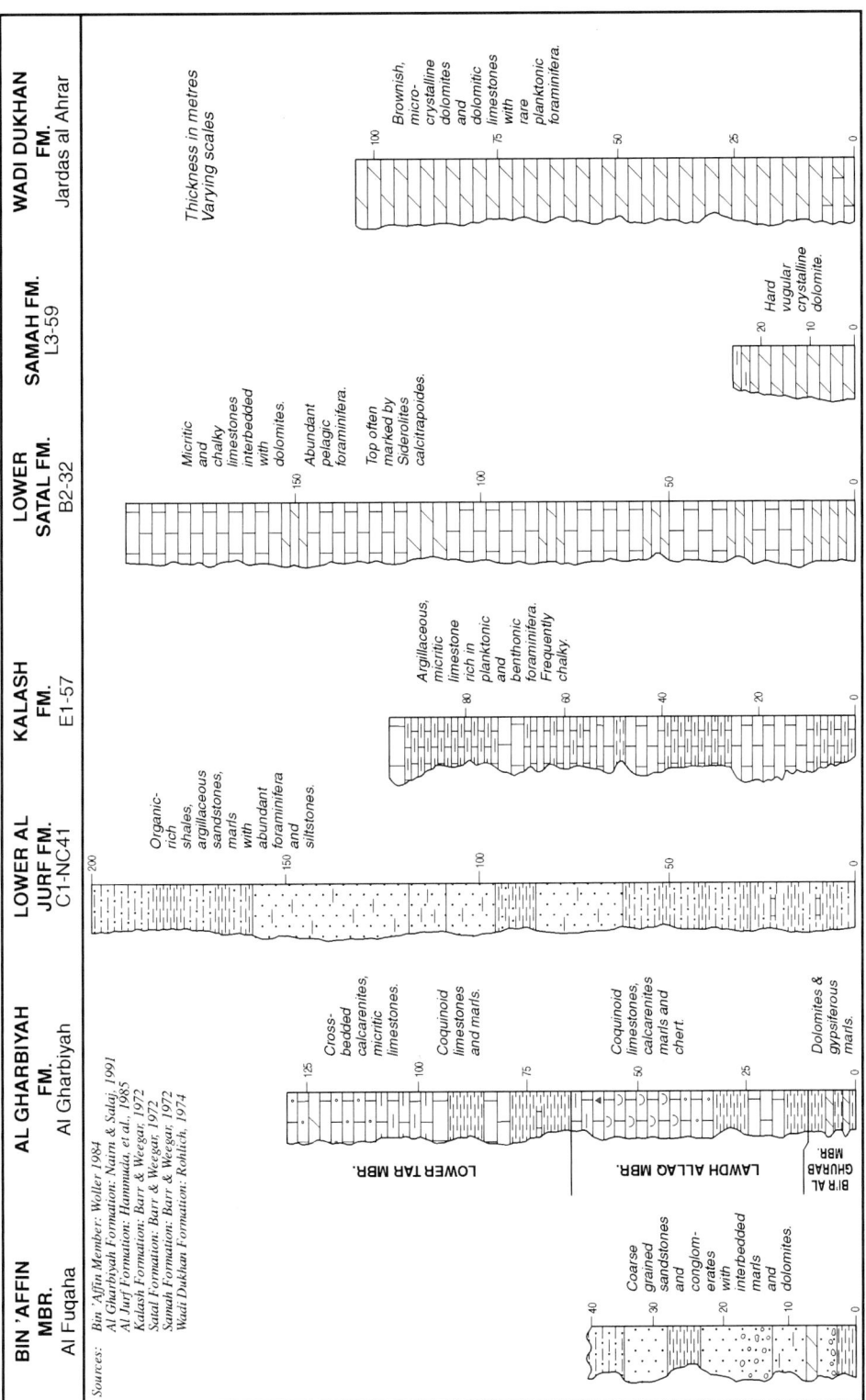

Figure 4.14   Al Gharbiyah Formation: Distribution, Equivalents and Stratigraphic Columns

The Upper Cretaceous transgression reached its maximum extent in the Maastrichtian covering virtually the whole of northern Libya, except for the northern Al Hammadah al Hamra which was emergent during this period, and the Jabal al Akhdar Ridge. The sediments deposited during the Maastrichtian are mostly shallow water carbonates (Al Gharbiyah Formation), shoreline clastics (Bin 'Affin Member), chalky and micritic limestones (Kalash Formation) and nearshore restricted carbonates (Satal, Samah and Wadi Dukhan Formations). Deeper water deposits are found offshore.

not realised at the time that the Wadi Tar section was incomplete. As redefined the Lower Tar Member is now included in the Al Gharbiyah Formation, and the Zimam Formation was redefined to include only the Palaeocene Upper Tar and Had Limestone members.[96]

At Wadi Tar the Lower Tar Member comprises 75m of calcareous, slightly gypsiferous shales and fossiliferous marlstones, with rare thin stringers of compact reddish limestone. The sequence contains pelecypods and a rich microfauna of planktonic foraminifera, plus the index fossil *Omphalocyclus macroporus*, which defines its Maastrichtian age. Subsurface data show that in some areas the Lower Tar Member can reach a thickness of 240m.[97]

On the northern flank of the Al Qarqaf Arch the Lower Tar Member rests directly on Hasawnah Formation with a conglomerate marking the base. The member comprises interbedded marlstones and sandy limestones with a lumachelle of pelecypod shells in the middle, known as the *Overwegi* Beds. The lumachelle also contains *Omphalocylus macroporus*. The total thickness of the Lower Tar Member in this area is about 50m. The lithology varies rapidly and further west it is made up of cavernous dolomite overlain by calcarenites and cross-bedded sandstones.[98]

Following the outcrops northwards the Lower Tar Member passes from a mixed lithology of thin-bedded limestones and sandstones at Bi'r Amasin to a much more calcareous facies at Shawa where shallow water limestones and marls contain a rich fauna of oysters and other pelecypods. The species of *Inoceramus* found in the Lower Tar Member in this area are quite distinct from those found in the Mazuzah Member of the Mizdah Formation.[99]

The Lower Tar Member outcrops extensively along the northern Hamadah al Hamra. The thickness increases to over 150m and the lithology is dominated by marls, gypsiferous clays and shallow-water limestones with coquinas. The presence of extensive karstification and caliche deposits at the top of the Lower Tar Member in the Wadi Zamzam area has been interpreted as evidence for emergence and weathering in this area prior to the deposition of the Upper Tar Member. In the subsurface the Lower Tar Member is present in many wells on the Hamadah al Hamra where the average thickness is around 120m. Further east wells such as A1-86, A1-4, A1-10 and B1-91, prove the passage of the Lower Tar Member into the northwestern Sirt Basin.[100]

In summary, the extensive outcrops of the Lower Tar Member and its clastic equivalent demonstrate three characteristic facies: coarse clastic shoreline deposits in the southeast (Bin 'Affin Member), an inner shelf, clastic/carbonate facies in the southwest, and a shallow-water carbonate facies in the north and east.

NORTHWESTERN OFFSHORE

4.4.6 Alalgah Formation

In the southern Sabratah Basin the Kiklah Formation is overlain by a sequence of dolomites and anhydrites which have been named the Alalgah Formation by Hammuda. A type section was designated in the K1-137 well where the formation is made up of 178m of brown, crystalline, sparry dolomite with thin black shale partings, interbedded with massive nodular anhydrites. In the type well the formation is underlain by volcanic rocks (which makes this a strange choice for a type section), but in other places it is underlain by Kiklah, Turghat or Masid formation. It is overlain by the more open marine Makhbaz Formation. The formation was deposited in a shallow lagoon which was frequently isolated from the open ocean and subjected to evaporation.

The formation has been found in the south of the Sabratah Basin in concessions 137, NC 41 and NC 35a. (Figure 4.11). It is unfossiliferous, but on the basis of stratigraphic position it is equivalent to the Sidi as Sid Formation of the Jabal Nafusah. In Tunisia the Alalgah Formation passes laterally into the carbonates and evaporites of the lower Zebbag Formation, and the marls and shales of the Fahdene Formation.[101]

4.4.7 Makhbaz Formation

In the offshore southern Sabratah Basin the Alalgah Formation is overlain by a series of micritic to finely-crystalline limestones, sometimes dolomitic, with thin intercalations of calcareous shales. Hammuda named this unit the Makhbaz Formation and defined a type section in the I1-137 well close to the Tunisian border. The thickness in the type well is 184m. It has been penetrated by many wells on the southern margin of the basin, as far east as well J1-NC 35a in the Misratah Basin. (Figure 4.12). No diagnostic fauna has been reported, but on stratigraphic grounds it has been assigned to the Turonian-Coniacian. It is believed to equate approximately with the Qasr Tigrinnah Formation of northwest Libya and with the Upper Zebbag and Douleb carbonates of Tunisia, which contain both reservoir and source rocks.[102]

4.4.8 Jamil Formation

In the Sabratah Basin the carbonates of the Makhbaz Formation are overlain by a thick shale sequence named the Jamil Formation by Hammuda. The formation is defined by a type section in well K1-137 on the southern flank of the basin. In the type well the Jamil Formation is composed of 380m of calcareous shales, with thin stringers of crystalline to dolomitic limestone, resembling the Rachmat Formation of the Sirt Basin. The shales contain a pelagic microfauna including *Globotruncana concavata*, an index fossil of the Santonian, plus fragments of the pelecypod *Inoceramus*. It was probably deposited in an outer neritic environment. The formation is known from the southern and central Sabratah Basin with a maximum recorded thickness of 500m in well L1-NC 41. (Figure 4.13). The Jamil Formation correlates approximately with the Aleg shales and upper part of the Kef Formation in Tunisia, the Mazuzah Member of the Mizdah Formation in northwest Libya, and the Rachmat Formation of the Sirt Basin.[103]

4.4.9 Abu Isa Formation

In the Sabratah Basin the Jamil Formation is succeeded by a thick formation of micritic and chalky limestones which were named the Bu Isa Formation by Hammuda (National Atlas transliteration: Abu Isa). The type section was defined in the H1-137 well on the southern margin of the Sabratah Basin. The well shows a sequence of 290m of uniform, hard, micritic to microcrystalline limestones, passing into chalky to marly limestones, with a thin dolomite horizon in the middle. The formation contains shell fragments and planktonic foraminifera, which give a Campanian age. It was deposited in a quiet environment in an outer neritic, shelfal setting. It has conformable contacts with both the underlying Jamil Formation and the overlying Al Jurf Formation. It extends northwards to at least the C1-NC 41 well in the centre of the basin. (Figure 4.13). Westwards it can be traced into the area between the Elyssa and Tanit wells in Tunisian waters. The Abu Isa Formation is the approximate equivalent of the Berda, Marfeg and Abiod carbonates in Tunisia, the Thala Member of the Mizdah Formation in northwest Libya, and the Sirt Formation in the Sirt Basin.[104]

4.4.10 Al Jurf Formation (Lower part)

The Al Jurf Formation was defined in the C1-NC 41 well in the centre of the Sabratah Basin by Hammuda for a sequence of shales, marls, sandstones and limestones lying between the Farwah Formation above and the Abu Isa Formation below. It spans the Cretaceous-Palaeocene boundary, and foraminiferal evidence suggests that the age ranges from mid-Maastrichtian to latest Palaeocene. The planktonic microfaunas of the Maastrichtian are particularly rich. The total thickness in the type well is 414m, but only part of this is attributable to the Cretaceous. The formation was deposited in an outer shelf setting. The shales are organically rich and have minor source potential. In some areas volcanic rocks are present in this part of the section. Southwards from the type well the formation thins towards the basin margin and in well G1-NC 41 only 17m of Al Jurf Formation is present. (Figure 4.14). The lower part of the Al Jurf formation correlates with the lower El Haria shales in Tunisia, and in Libya it is equivalent to the Lower Tar Member, and to the Kalash Formation of the Sirt Basin.[105]

SIRT BASIN SUBSURFACE

4.4.11 Bahi Formation

The Bahi Formation represents a distinctive facies which is only present in the subsurface of the western Sirt Basin. It is a significant hydrocarbon reservoir in the Sirt Basin. The formation was first defined by Barr and Weegar who designated a type section in well A3-32 in the Bahi field. In the type well the formation overlies quartzites and is composed of 111m of poorly-sorted sandstones, siltstones, conglomerates and shale, with a basal conglomerate containing rounded quartzite pebbles. (Figure 4.11). The formation is unfossiliferous,

and is assumed to be non-marine. However the top few metres contain glauconite and probably marks a transition to a marine environment. At the type locality it is overlain by marine Cenomanian shales, but in other places it is time-transgressive and may be as young as Maastrichtian in some areas. It represents a fluvial environment which was ultimately overwhelmed by the late Cretaceous marine transgression, but at differing times depending on the local palaeogeography. Thicknesses up to 122m have been encountered. It is frequently overlain by the Lidam Formation, but sometimes by younger Cretaceous units. The formation is known from the northern Az Zahrah-Al Hufrah Platform and neighbouring troughs, and is patchily developed in other areas such as the Al Bayda and Zaltan Platforms, and in the offshore area north of Al Aqaylah. Evidence from the As Sarir field shows the presence of ?Albian to Cenomanian fresh-water algae and sporomorphs in a sand/shale unit known locally as the Busat 'Formation'. This unit is probably equivalent to the Bahi Formation of the western Sirt Basin. Its palynomorph content suggests a non-marine environment of deposition.[106]

The petrography and diagenesis of the Bahi Formation have been studied by Sghair. He categorised the Bahi Formation as a sub-arkosic quartz arenite, and he has recognised a sequence of diagenetic events involving periods of quartz and carbonate cementation, with subsequent partial dissolution of the carbonate cement. He concluded that the sediments were derived from a mature, low relief cratonic interior by erosion of existing and recycled quartz sandstones.[107]

4.4.12 Maragh Formation

In the eastern Sirt Basin the equivalent of the Bahi Formation was named the Maragh Formation by Barr and Weegar. The type section was defined in the N1-12 well on the Amal field. In the type well the formation comprises a basal conglomerate with quartzitic and volcanic pebbles in a clay matrix containing glauconite. (Figure 4.11). The conglomerate is overlain by a poorly-sorted, friable sandstone, partly con- glomeratic, and containing glauconite, hematite, and dolomite, and is capped by sandy, crystalline dolomite with traces of anhydrite. The thickness in the type well is 43m. The Maragh Formation, as defined by Barr and Weegar, differs from the Bahi Formation in its marine character. According to them it represents a littoral deposit formed by the erosion of local highs with very limited trans- portation distances. The formation thickens on the flanks of local highs and reaches over 150m on the flank of the Amal structure. The Maragh Formation is a significant oil reservoir in the eastern Sirt Basin. It was originally regarded as unfossiliferous, and was assumed to be strongly diachronous, but more recent work has revealed the presence of several age-diagnostic paly- nomorphs. A study on the palynology of the Maragh Formation in the A1-96 well gave a Cenomanian to early Turonian age, and a marginal marine environment of deposition.[108]

A study by Sghair and El Alami however indicates that essentially the Bahi and Maragh formations are similar. They demonstrated a passage from marine to non-marine environment and recognised four distinctive facies: shallow marine, swamp, palaeosol, and alluvial fan. The complete transition is seen from the A1-96 well (alluvial fan) to the B5-96 well (entirely marine). According to this study the Maragh Formation is confined to the eastern Sirt Basin, and its age is probably no younger than Santonian.[109]

The Maragh Formation has also been studied in the Masrab area of the eastern Sirt Basin. In this area it has been divided into a lower sandstone member and an upper shale/evaporite member. The sandstone represents a marine transgressive unit, whilst the upper member represents a supra-tidal sabkha environment rich in algal material. It is overlain by Lidam carbonates and in this area appears to be of Cenomanian age.[110]

4.4.13 Lidam Formation

In the subsurface of the Sirt Basin, the first distinctive marine formation overlying the Nubian Sandstone was named the Lidam Formation by Barr and Weegar. A type section was chosen in the G1-57 well in the southern Zallah Trough. In this area the Lidam Formation is characterised by a hard, brownish sucrosic dolomite, and pelletal calcarenite. The basal unit is frequently sandy, reflecting derivation from the underlying Nubian Sandstone. Fossils are rare due to dolomitisation, but fragments of pelecypods, echinoids and algae have been found, and ostracods and large foraminifera are relatively common. The large foraminifera, par- ticularly *Ovalveolina ovum*, prove a Cenomanian age for this formation. In the type well the thickness is 111m, but according to Wennekers et al. thicknesses up to 600m have been recorded in the trough areas. The formation is underlain by quartzites, Nubian Sandstone, Bahi or Maragh formations, depending on location. The environment of deposition is lagoonal to intertidal. As the first unit of the Cenomanian marine transgression, the distribution of the Lidam Formation gives a good indication of the palaeogeography of the Sirt Basin at this period. Baird, et al. published a map showing that the formation is present in the Zallah, Abu Tumayam, Maradah, Ajdabiya, and Hameimat Troughs, but is not present on the major platform areas such as the Az Zahrah- Hufrah, Zaltan and Amal platforms. (Figure 4.11). The Lidam Formation is a significant oil reservoir in several areas, particu- larly in the western Sirt Basin. There is a gradual passage from a dominantly dolomite lithology in the north to a mainly limestone facies in the south. The formation is generally overlain by shales or carbonates of Cenomanian-Turonian age.[111]

The diagenetic history of the Lidam Formation in the northwest Sirt Basin was studied by El Bakai. He concluded that the formation has been greatly affected by diagenesis which began shortly after deposition with the burrowing activities of algae and fungi, and ended with extensive dolomitization which occurred after lithification and compaction.[112]

In the eastern Sirt Basin the Lidam Formation is present in the Masrab area. Here crystalline dolomites are interbedded with bioclastic and pelletal limestones representing a passage from sabkha deposits to subtidal lagoonal limestones. The Lidam dolomite forms a significant oil reservoir in this area, and an isopach of the formation in the Jalu area was published by Mansour and Magairhy.[113]

4.4.14 Argub Formation

In the western Sirt Basin the subsurface equivalent of the Nalut Formation is the Argub Formation, established by Barr and Weegar in the D1-32 well, north of the Bahi field. At this location it comprises 65m of hard, crystalline dolomite, with thin stringers of shale. The dolomite is slightly glauconitic and sometimes vugular. South of the Bahi field the dolomite passes into limestone. The formation does not contain diagnostic fossils but on the basis of stratigraphic position is believed to be Turonian in age, although Wennekers et al. suggested, on the basis of unpublished data, that it may be younger. It has a conformable relationship with both the underlying Lidam Formation and the overlying Rachmat Formation. The formation is present over much of the northwestern Sirt Basin, except for the crestal areas of the platforms, and thickens considerably into the trough areas. (Figure 4.12).[114]

4.4.15 Etel Formation

On the Al Bayda Platform the Lidam Formation is overlain by a sequence of thin-bedded dolomites, anhydrites, shales and siltstones which Barr and Weegar named the Etel Formation. The formation contains thick evaporites in several of the trough areas. The dolomites are often brown, very finely crystalline or with a sucrosic texture. The type section was defined in the O2-59 well, on the southeast margin of the Al Bayda Platform. Here the formation has a thickness of 163m. It has a sharp contact with the underlying Lidam Formation, and in other areas it rests unconformably on the Bahi Formation or on pre-Upper Cretaceous rocks. The contact with the overlying Rachmat Formation is gradational. It has a widespread distribution in the south-central Sirt Basin, but is absent over the Az Zahra-Al Hufrah, Zaltan and Amal platforms. (Figure 4.12). Offshore it is absent on the An Nufaliyah High and in the area of well A1-89. In the trough areas the thickness increases dramatically. In the southern Ajdabiya Trough well 5P1-59 proved a thickness of almost 1200m, in the Kotlah Graben the thickness reaches 450m, and in the Maradah Trough a thickness of 396m was reported in the F1-13 well, and the thickness almost certainly increases further north. Thick evaporite deposits have been encountered in four of the principal troughs: in the Kotlah Graben/southern Zallah Trough (wells A1-59F, A1-85), in the southern Maradah Trough (Wadi field), in the southern Ajdabiya Trough (5P1-59), and in the Hameimat Trough. An isopach map of the Hameimat salt deposits was presented by Mansour and Magairhy which shows a maximum thickness of 365m. The formation was deposited in a sabkha/lagoonal environment with intervals of very shallow nearshore marine incursions. El Alami, et al. published a palaeogeographic map of the Etel Formation showing a mainly evaporitic tidal-flat facies, over most of the Sirt Basin, except for the emergent platforms, with open marine carbonates to the north and west. The formation is assigned a Turonian age on the basis of stratigraphic position, but again Wennekers et al. suggested a younger age. At As Sarir marginal marine sands containing Turonian dinoflagellate cysts may represent littoral equivalents of the Etel Formation. Evidence is accumulating that the Turonian represents a 'greenhouse' period during which atmospheric CO_2 levels and ocean temperatures were much higher than at present. Sea levels were very high and anoxic conditions became widely established. This has become known as the Turonian anoxic event.[115]

4.4.16 Rachmat Formation

In the subsurface of the Sirt Basin the Argub and Etel formations are overlain by a thick shale section named the Rachmat Formation by Barr and Weegar. They defined a type section in the O2-59 well on the Al Bayda Platform. At this location the Rachmat Fomation is composed of 140m of dark grey shales with frequent interbeds of dolomite and limestone. (Figure 4.13). The shales are fissile, and usually slightly calcareous, and contain glauconite and pyrite. The dolomites and limestones are dense, finely crystalline and occasionally anhydritic. In the type well the formation overlies the Etel Formation but in other places it overlies older Cretaceous formations, or rests directly on quartzites. The formation is widespread over much of the Sirt Basin except for the major platform areas such as the Az Zahrah-Al Hufrah, Zaltan and Amal platforms. This is well illustrated on the Awjilah field where the Rachmat Formation is present low on the flank of the structure, but not on the crest. In this area it is composed mostly of fine-grained argillaceous sandstone and silty shale. It is thickest in the troughs and reaches 610m in the Hameimat Trough south of the Awjilah structure, and more

than 365m in the Maradah Trough close to the Wadi field. Wennekers et al. reported thicknesses exceeding 900m. On the As Sarir field anoxic marine shales containing Turonian to early Campanian palynomorphs are probably equivalent to the Rachmat Formation. In the Maradah Trough the formation contains high pressure gas within thin limestone stringers which caused a spectacular blowout in well Y1-13. The formation contains foraminifera and ostracods which indicate a Coniacian-Santonian age. It is therefore partially equivalent to the Qasr Taghrinnah Formation of the Al Hamadah al Hamra area. The junction with the overlying Sirt Formation is often marked by a zone containing phosphate pellets.[116]

4.4.17 Taqrifat Formation

In the eastern Sirt Basin a distinctive limestone horizon is present between the Rachmat and Sirt formations which was informally called the Rakb Carbonates by Williams, formally defined as the Tagrifet Limestone by Barr and Weegar, and is now properly called the Taqrifat Formation. It is unclear why Barr and Weegar chose this name since the Taqrifat oasis is located in the Zallah Trough some 420km west of the type well. The type section was defined in the N1-59 well, located on the Amal Shelf adjacent to the southern margin of the Awjilah field. In the type well the formation comprises 91m of porous calcarenite, with interbeds of calcareous shale. The basal member contains quartz grains, and higher in the section shell-debris and pellets occur within the limestone. The limestones contain rudistids, *Inoceramus* and other pelecypods, gastropods, bryozoa, algae, ostracods and foraminifera, indicative of very shallow-water deposition (Figure 4.13). In the Hameimat Trough it is characterised by the presence of *Inoceramus* prisms and *Rotalia cayeuxi*. The age of the formation is late

Santonian to early Campanian. The Taqrifat Formation is restricted to the eastern Sirt Basin, and has been mapped between the Awjilah area and the Harash field. An isopach map of the Taqrifat Formation in the area east of Jalu, where the formation is oil bearing, was published by Mansour and Magairhy. The maximum recorded thickness is 335m. The formation is a significant oil reservoir on the Awjilah and Nafurah fields and in the Latif area. The upper part of the Taqrifat Formation comprises dark-brown argillaceous micritic limestone containing planktonic foraminifera. It passes laterally into Sirt Formation shales, and probably has some local oil-source potential. The lithology and diagenesis of the formation have been described in some detail by Williams. He believed that the Taqrifat Formation represents a temporary shallowing of the Upper Cretaceous seas in this area, with the upper part indicating a return to low-energy, open-marine conditions. It is probably equivalent to the lower part of the Sirt Formation in other areas.[117]

4.4.18 Sirt Formation

In the subsurface of the Sirt Basin the Rachmat Formation is overlain by a thick shale formation which was named the Sirte Shale by Barr and Weegar. They selected a type section in well O2-59 on the edge of the Al Bayda Platform. In this well the Sirt Formation is composed of 174m of dark brown, waxy shales. The shales are subfissile to splintery and contain an abundant fauna of planktonic foraminifera which indicate a Campanian to early Maastrichtian age. The shales are carbonaceous and calcareous and contain thin stringers of shaly limestone. They were deposited in a restricted euxinic basin and they represent the principal oil source rocks of the Sirt Basin. The shales contain zones with very high radioactive values, clearly visible on radioactive wireline logs, which coincide with the zones of maximum organic richness. The lower boundary with the Rachmat Formation is often marked by a bed of phosphate pellets, and a change from the grey, poorly calcareous shales of the Rachmat Formation to the waxy brown calcareous shales of the Sirt Formation. The formation is usually overlain by the Kalash Formation, or in some areas by the Wahah Formation.

The formation is present throughout the Sirt Basin, except on the few islands which still existed in the Campanian sea. (Figure 4.13). By this time marine conditions covered all but the highest points of the pre-transgression surface. As in the Etel and Rachmat formations the thickness increases markedly from the platforms into the troughs, although the proportionate amount of thickening is less than in the underlying formations. Baird et al. reported thicknesses of 760m in the Zallah Trough, 300m in the Maradah Trough and 600m in the Ajdabiya Trough. At the As Sarir field about 75m of anoxic, marginal-marine shales containing abundant Campanian marine palynomorphs represent the Sirt Formation. In the Hameimat Trough the Sirt Formation is characterised by the planktonic foraminifera *Gansserina gansseri* and *Globotruncana linneiana* which indicate a late Campanian-Maastrichtian age in this area. In the southern Maradah Trough Hammuda described the presence of sand-bodies within the Sirt Formation in wells G1-, H1-, and M1-71, which represent potential oil reservoirs. They form large nearshore sandspits and bars connected to the Dayfah-Al Wahah Ridge, and were formed during the Campanian marine transgression by the erosion of clastic sediments from the ridge.[118]

4.4.19 Kalash Formation

Over much of the Sirt Basin the Sirt Formation is overlain by a dense micritic limestone which was named the Kalash Limestone by Barr and Weegar. They selected a type section in well E1-57 in the southern Zallah Trough near to the Al Kotlah Graben. At this location the formation is composed of 95m of argillaceous micritic limestone rich in both planktonic and benthonic foraminifera, and rare shell debris. It represents deposition in a neritic, low-energy environment, in a shallow sea with several small islands. Adjacent to some of the islands the limestone becomes sandy, and in the eastern Sirt Basin it is predominently chalky. (Figure 4.14). The age based on both surface and subsurface data has been established as Maastrichtian and earliest Danian. The contact with the underlying Sirt Formation is gradational, but in some areas it is underlain by locally developed formations such as the Samah Dolomite or the Wahah Limestone. It is unconformably overlain by the Al Hagfah Formation or by the Lower Sabil carbonates. The thickness varies from a few metres on the platforms to 150m in the trough areas. The Kalash Formation represents the maximum extent of the Cretaceous marine transgression which covered most of the Sirt Basin, except for a very few remaining islands and peninsulas. At As Sarir it is represented by 90m of dark grey marine shale containing very high levels of amorphous organic material and dinoflagellate cysts of Maastrichtian age. Its southeastern limit has recently been extended to Jabal az Zalmah on the northern margin of the Al Kufrah Basin. In this area Lüning, et al. discovered 25m of evaporites, shales and dolomites which contain poorly preserved planktonic foraminifera which indicate a Campanian-Maastrichtian age. Analysis of strontium isotope ratios from an unaltered echinoid shell yieded an absolute age of 65.7 Ma, which confirms a late Maastrichtian age for this deposit.

The Kalash Formation forms a poor oil reservoir in some areas including the B1-10 area on the Waddan Platform, but its effectiveness is limited by very low permeability. The planktonic foraminiferal fauna confirms a Maastrichtian age for the formation, although recent palaeontological evidence shows that in some areas, such as the Jabal field on the Zaltan Platform, it extends into the lowermost Danian. This was also confirmed in well A1-NC 29A on the Az Zahrah-Al Hufrah Platform where the Kalash Formation contains *Eoglobigerina minitula*, which is an index fossil of the lowermost Danian. It is equivalent to the Lower Tar Member of the Al Gharbiyah Formation, and is at least a partial equivalent of the Wahah and Lower Satal Formations.[119]

4.4.20 Samah Formation

On the Al Bayda Platform the Kalash Formation is underlain by a distinctive dolomitic unit which was named the Samah Dolomite by Barr and Weegar. The type section was defined in the L3-59 well on the Samah field. In the type well the formation comprises 26m of coarsely crystalline, vugular dolomite, heavily fractured and containing thin shale partings. The formation rests directly on quartzites, and a thin conglomerate of quartzite pebbles marks the base of the formation. It is overlain by Kalash Formation. The distribution of the formation is extremely limited, and is confined to the area of the Al Bayda Platform immediately adjacent to the Samah field. (Figure 4.14). It forms the principal oil reservoir on the Samah field. The maximum recorded thickness is about 90m. The formation contains a poorly preserved fauna of pelecypod fragments, including rudists, and foraminifera which suggest a late Cretaceous age. It seems likely that the formation was deposited in very shallow clear seas during the late stages of the Cretaceous marine transgression. It may represent a local, shallow water facies of the Kalash Formation.[120]

4.4.21 Satal Formation (Lower Member)

On the crest of the Az Zahrah-Al Hufrah Platform the Sirt Formation is overlain by a thick series of micritic and chalky limestones and dolomites which were named the Satal Formation by Barr and Weegar. A type section was chosen in the B2-32 well on the Az Zahrah West field. Like the Al Jurf Formation offshore the Satal Formation bridges the Cretaceous-Palaeocene boundary. It is divided into lower and upper members which correspond to the Maastrichtian and Danian portions respectively. In the type well the Lower Satal Member is composed of 195m of micritic limestones, frequently becoming chalky, interbedded with thin bands of dolomite. The Satal Formation shows a progressive shallowing of the sea. Pelagic foraminifera dominate the lower part of the Lower Satal Member whilst benthonic foraminifera dominate the upper part. The Upper Satal Member is composed largely of dolomites with traces of anhydrite. The Lower Satal Member represents a shallow-water equivalent of the Kalash Formation, and the transition from thick Lower Satal porous carbonate to thin, tight, Kalash Limestone takes place within a few kilometres. The passage into the Upper Satal Member is marked by an abrupt change from micritic limestone to dolomite. The presence of *Siderolites calcitrapoides, Orbitoides* and *Omphalocyclus* proves a Maastrichtian age for the Lower Satal Member. It is restricted to the higher parts of the Az Zahrah-Al Hufrah Platform. (Figure 4.14). Its transgressive character is evident from the fact that in several wells it directly overlies quartzites. The Satal Formation forms a major oil reservoir on the Bahi and Az Zahrah fields, but most of the oil is reservoired in the Upper Satal Member which has much better reservoir characteristics.[121]

4.4.22 Wahah Formation

The Waha Limestone (National Atlas of Libya transliteration Al Wahah) was first defined by Barr and Weegar from the A29-59 well on the Wahah field, where it forms a major oil reservoir. In the type well it is composed of 23m of sandy, skeletal limestone containing abundant shell fragments, algae and benthonic foraminifera. (Figure 4.13). The sequence overlies granite and is overlain by Kalash Formation. The sand content tends to decrease upwards. The shell debris includes rudistid and echinoid fragments and the large Maastrichtian foraminifera *Omphalocyclus macroporus, Siderolites calcitrapoides* and *Orbitoides media.* The Wahah Formation was regarded by Barr and Weegar as a shoreline facies of the Kalash Formation, and is characteristically developed as an apron around the small islands which remained on the platform areas in the Maastrichtian sea. On the platform the thickness of the Wahah Formation ranges between zero and 110m. Rapid facies changes occur within the formation. Frequently the clastic content increases to such an extent that the lithology becomes a calcareous sandstone rather than a sandy limestone. Laterally the formation passes rapidly into open marine facies. Shorewards the formation wedges out and the crestal part of the island or peninsula is frequently 'bald'. The formation is a major hydrocarbon reservoir on the Zaltan Platform and nearby platform areas.[122]

The Wahah Formation on the Ar Raqubah field has been described by Broughton. He demonstrated the lateral passage from open marine to nearshore facies, wedge-out on the flank of the structure, and the bald crest. He also showed the complex relationships between facies zones, reservoir quality, faulting and water encroachment. In another study El Ghoul highlighted the difficulties of identifying the Wahah Formation from seismic data on the

Zaltan Platform, and the need for seismic attribute and velocity studies to effectively pinpoint the pinchout margins and facies changes. The relationship of the Wahah Formation to the Sirt and Kalash formations on the Zaltan Platform was studied by Jones. He demonstrated from core studies that in the area around the Nasser field the Kalash Formation is separated from the Wahah Formation by an unconformity. On the basis of foraminiferal content he regarded the Wahah Formation as age equivalent to the upper part of the Sirt Formation. He also illustrated the sequence in the YYY1-6 well which is located in the Maradah Trough close to the Zaltan Platform boundary fault. In this well, beneath typical Wahah Limestone, a 550m sand-wedge is present which interfingers basinward with Sirt Formation shales. Similar sand-wedges are known in other areas. These sand-wedges apparently represent large volumes of clastic material eroded from the platform margin and dumped into the adjacent troughs, presumably as slumps or debris flows. Jones showed the YYY1-6 sand-wedge as Wahah equivalent, but admitted that no palaeontological confirmation of Maastrichtian age was available.[123]

NORTHEASTERN LIBYA

4.4.23 Qasr al Ahrar Formation

In northern Cyrenaica Upper Cretaceous rocks are attributable to either a platform facies (in the south and west), or a basinal facies (in the north and east), with some interfingering in the transition zone. The oldest Cretaceous rocks are exposed in a small inlier at Qasr al Ahrar (formerly Jardas al Abid), near the site of the first wildcat well in Libya, A1-18, where a deeply eroded anticline exposes rocks of Cenomanian age. The geology of this inlier was described by Kleinsmeide and van den Berg, Pietersz, and Barr. The oldest sequence exposed

in the inlier was named the Jardas al Abid Formation, but following renaming of the village is now called the Qasr al Ahrar Formation. It was re-examined during the remapping for the Geological Map of Libya. The outcrop shows the top of a sequence of green and yellow ochreous marls with thin marly limestones. Only about 10m is exposed at outcrop, but a further 200m was penetrated in the A1-18 well. A description was given by Hawat and Shelmani. The formation contains a rich fauna of planktonic foraminifera, pelecypods and ostracods which indicate a late Cenomanian age. The formation has been penetrated in several wells in northern Cyrenaica, and in offshore well A1-NC 120 the Qasr al Ahrar Formation unconformably overlies the Albian Daryanah Formation. (Figure 4.11). Here the formation represents a regressive cycle, passing from open-marine conditions to restricted lagoonal conditions at the top. Nearshore marine limestones containing early Cenomanian palynomorphs have also been reported from wells A1-28, A1-19, A1a-117, A1-NC 92 and B1-2.[124]

4.4.24 Al Baniyah Formation

In the Qasr al Ahrar inlier of Cyrenaica and in another inlier near Marawah the Qasr al Ahrar Formation is overlain by a limestone named by Kleinsmeide and van den Berg the Benia Limestone, and later renamed the Al Baniyah Formation. A type section close to Qasr al Ahrar was selected by Klen, but at this locality only 35m of section is present due to pre-Campanian erosion. Further east the thickness reaches 300m, and locally may attain 600m. The formation is composed of microcrystalline limestones containing rudists and large gastropods, followed by yellow dolomitic limestones, and capped by fossiliferous marls, formerly known as the Ghawt Sas (Got Sas) Marl Member. At Marawah the limestone is

500m thick and the marls about 140m. (Figure 4.12). The formation contains a rich fauna of echinoids, pelecypods, gastropods and foraminifera, which indicate a late Cenomanian to Coniacian age. The environment of deposition was predominantly open-marine, inner to outer neritic. It is equivalent to the major part of the deep-water Al Hilal Formation to the northeast. The formation was affected by uplift, folding and erosion prior to the deposition of the overlying Al Majahir Formation. This episode was studied in detail by Rohlich, et al. who demonstrated by palaeontological analysis that the unconformity can be dated as middle to late Santonian. In offshore well A1-NC 120 the Al Baniyah Formation is well developed, and is overlain by Al Hilal Formation. In the well the Santonian unconformity can be identified at the base of the Al Majahir Formation. El Hawat claimed that the unconformity is traceable in the Sirt Basin and offshore.[125]

4.4.25 Al Majahir Formation

In northern Cyrenaica the Al Baniyah Formation is unconformably overlain by the Al Majahir Formation. The hiatus represented by the unconformity was marked by a period of folding, uplift and erosion and Coniacian and early Santonian platform sediments appear to be absent. The formation was first described by Rohlich based on a type section southeast of Qandulah. Here 70m of limestones and dolomites unconformably overlie the Al Baniyah Formation. A basal zone of chalky limestone is overlain by 12m of dolomite or dolomitic limestone, which in turn is overlain by a thick sequence of microcrystalline limestones containing planktonic foraminifera, the large foraminifera *Orbitolites* and *Rhapydionina* and fragments of pelecypods, gastropods and echinoids. The fossil assemblage suggests a Campanian age. It grades upwards

into the Wadi Dukhan Formation. The Al Majahir Formation is equivalent to the Al Feitah Limestone of previous authors.[126]

The formation outcrops over a large area in Al Jabal al Akhdar, but unfortunately was misidentified during the remapping of the area for the Geological Map of Libya. On the Zawiyat Masus and Bi'r Hakim sheets it was confused with the Eocene-Oligocene Al Khowaymat Formation. Subsequent studies by Megerisi and Mamgain demonstrated that the lower unit of the Al Khowaymat Formation is the same as the upper part of the Al Majahir Formation and is Cretaceous in age. The Al Khowaymat Formation is therefore compromised and they recommended that the name should be suppressed. The 1:1,000,000 geological map published by IRC in 1985 shows the revised extent of the Al Majahir Formation. As revised it is evident that the Al Majahir Formation outcrops over some 4,000km² in the Jabal al Akhdar as a sequence of dolomitic limestones with a Campanian fauna. (Figure 4.13). The outcrop thickness rarely exceeds 70m but it is known to reach a thickness of 200m in the subsurface. On the coast near Al Burdi, Schiettecatte described a small exposure of 15m of chalky limestone containing Campanian foraminifera, which suggests that the formation extends in the subsurface at least as far as the Egyptian border. Westwards the formation has been recognised in the A1-NC 120 offshore well northwest of Binghazi. The well penetrated dolomitic, chalky and skeletal limestones with mudstones, with a foraminiferal fauna of Campanian age. It is not present in the A1-NC 128 well drilled near Bamba, where a major hiatus is recognised in which Eocene Darnah Formation rests directly on Albian Daryanah Formation.[127]

On the coast near Tukrah four small inliers expose Cretaceous limestones which Klen named the Tukrah Formation. Unfortunately the name Tocra Limestone had already been used by Desio in a different sense, for Eocene limestones exposed at the foot of the escarpment southeast of Tukrah. The outcrops described by Klen are small and only expose a few metres of compact limestone with chert nodules. They contain a foraminiferal assemblage of Campanian age, and are almost certainly equivalent to the Al Majahir Formation. There is no need for a separate formation name for these limestones, particularly one which has been previously used in a different sense.[128]

4.4.26 Wadi Dukhan Formation

The Wadi Dukhan Formation was established by Pietersz for a thick series of dolomites and dolomitic limestones in the area west of Qasr al Ahrar which he believed to be of Maastrichtian to lowermost Cainozoic age. Subsequently Klen restricted the definition to include only the sequence lying between the Al Majahir Formation and the Al 'Uwaliyah Formation. As redefined the sequence comprises 50m to 150m of brownish microcrystalline dolomites which rest conformably on the Al Majahir Formation. The contact with the overlying Al 'Uwayliah Formation is sharp but apparently conformable. Exposures are limited to the Qasr al Ahrar and Al 'Uwayliah areas, and a small outcrop near Jardas al Jarari. The fossil content is very sparse but Rohlich and Klen reported the presence of rare Maastrichtian planktonic foraminifera with an absence of Danian species. The extent of this formation is poorly known. (Figure 4.14). It is not present in offshore well A1-NC 120 where the Maastrichtian is represented by deep-water facies of the Al Athrun Formation, and it is absent in the A1-NC 128 well due to erosion.[129]

4.4.27 Al Hilal Formation

A shale/limestone sequence, which contains a complete succession of foraminiferal zones from early Cenomanian to Maastrichtian, is exposed at Marsa al Hilal on the northern coast of Cyrenaica. It was first recognised by Barr and Hammuda, who named the lower part the Hilal Shale and the upper part the Al Atrun Limestone. The name was formalised as the Al Hilal Formation by Rohlich. This formation represents the basinal equivalent of the Qasr al Ahrar and Al Baniyah Formations on the platform to the southwest. At Marsa al Hilal the Al Hilal formation is made up of thin-bedded, marly shales, containing glauconite, pyrite and a rich fauna of planktonic foraminifera. (Figure 4.12). The depositional setting is regarded as bathyal, and the sequence appears to be condensed. The upper few metres contain white pelagic limestone interbeds which show a transition into the overlying Al Athrun Formation. The base is not seen at outcrop but nearby well data shows a thickness of 330m. The age, based on foraminifera and dinocysts, has been variously assessed as ranging from early Cenomanian to mid-Coniacian (Barr), Albian to Coniacian (Banerjee) and Coniacian to Santonian (El Mehdawi). A study based on calcareous nannofossils suggested that it is limited to the Santonian. The limited exposure at Marsa al Hilal is probably due to the presence of a northeast-southwest anticline which reaches the shoreline in this area. The upper part of the formation is present in the A1-NC 120 well, where it onlaps onto the Al Baniyah Formation, and is probably also present in the A1-45 well, further east, where open marine shales and limestones of early Cenomanian age have been encountered. It is not present in the A1-NC 128 well near Bamba due to erosion.[130]

4.4.28 Al Athrun Formation

The upper part of the Cretaceous sequence exposed on the coast east of Marsa al Hilal was named the Atrun Limestone by Barr and Hammuda, and the Al Athrun Formation by

Rohlich. At the type section in Wadi al Athrun 45m of thin-bedded micritic and microcrystalline limestones, with marly partings are exposed. The outcrops show several zones of contorted bedding suggesting slumping whilst the sediments were still soft, perhaps caused by tectonic movements connected with the Santonian unconformity, visible to the southwest. The contact with the underlying Al Hilal formation is conformable, whilst that with the overlying Apollonia Formation is represented by a rugged erosion surface and is unconformable. The formation is rich in microfossils and contains assemblages representative of each of the planktonic foraminiferal zones from the late Coniacian to the late Maastrichtian. On the basis of calcareous nannofossils however, the formation is Campanian in age. The formation also contains a late Cretaceous coccolith assemblage which was described by Hay. The abundant planktonic fauna and paucity of nearshore fossils suggests that the Al Athrun Formation represents an outer-shelf to bathyal environment of deposition. The outcrop area of the formation is limited to the coastal section near Wadi al Athrun, but the formation has been encountered in several wells including offshore well A1-NC 120 where soft chalk and marls containing Maastrichtian planktonic foraminifera were referred to the Al Athrun Formation. (Figure 4.13).[131]

Chapter 5

STRATIGRAPHY
PART THREE: CAINOZOIC

The Cretaceous-Tertiary boundary in Libya is not marked by any apparent disruption, and no evidence has yet been found of the iridium layer which marks the boundary in many parts of the world. The Upper Tar Member of the Danian Zimam Formation shows very little change in facies from the Lower Tar Member of the Maastrictian Al Gharbiyah Formation, and to the northwest both the Al Jurf Formation of the Libyan offshore and the El Haria Formation of Tunisia bridge the boundary without apparent interruption. On the Az Zahrah-Al Hufrah Platform micropalaeontological evidence showed the Cretaceous-Palaeocene boundary to be located within (but near the top) of the Kalash Formation.

The rift grabens in the Sirt Basin, which played such a conspicuous part in late Cretaceous sedimentation, had been largely infilled by the end of the Maastrichtian, and the topography of the intervening horsts had been levelled to the extent that only very small islands remained.

Nevertheless gentle tectonic forces during Palaeocene time resulted in the gradual tilting of northernwestern Libya towards the east. The Hamadah al Hamra became emergent by the end of the Palaeocene whilst subsidence continued in the Sirt Basin. This regional tilting became one of the most persistent features of Cainozoic deposition in Libya.

Cainozoic outcrops are widespread in Libya, and show predominently shallow-water marine lithologies (Figure 5.1). These same rocks contain major oil and gas accumulations in the subsurface in a variety of different facies,

including reefs, bioherms and nearshore sands. They contain excellent shale and evaporite seals, but source rock quality is generally poor. Most of the potential Cainozoic source rocks are lean, and they are usually immature.

5.1 PALAEOCENE

The Palaeocene sequences in Libya are dominated for the most part by shallow-water carbonates deposited in restricted shallow-shelf to open-marine environments. The carbonate platforms usually have a clearly definable ramp margin with more argillaceous sediments in the deeper-water areas basinward of the platform margin. Reefal developments, with corals, algae and bryozoa as the frame-building elements, exhibit a wide variety of forms in the Sirt Basin. Pinnacle reefs located basinward of the platform margin are present at Intisar, As Sahabi and Al Mheirigah, whereas the Zaltan complex shows evidence of an atoll-type of morphology. Patch reefs and bioherms are also present. Palaeocene reservoirs contain about 30% of the oil reserves of the Sirt Basin.

There is growing evidence, based on ostracod studies, to suggest that during the late Palaeocene a short-lived marine link (the trans-Sahara seaway) existed between north Africa and Nigeria. The vast majority of species found in late Palaeocene rocks in the Sirt Basin are identical with those of Nigeria. The same studies showed that the seaway lay through Algeria west of the Hoggar Massif via the Gao Strait and Illemeden Basin into the Niger valley. The connection however was brief and did not persist into the Eocene.[1]

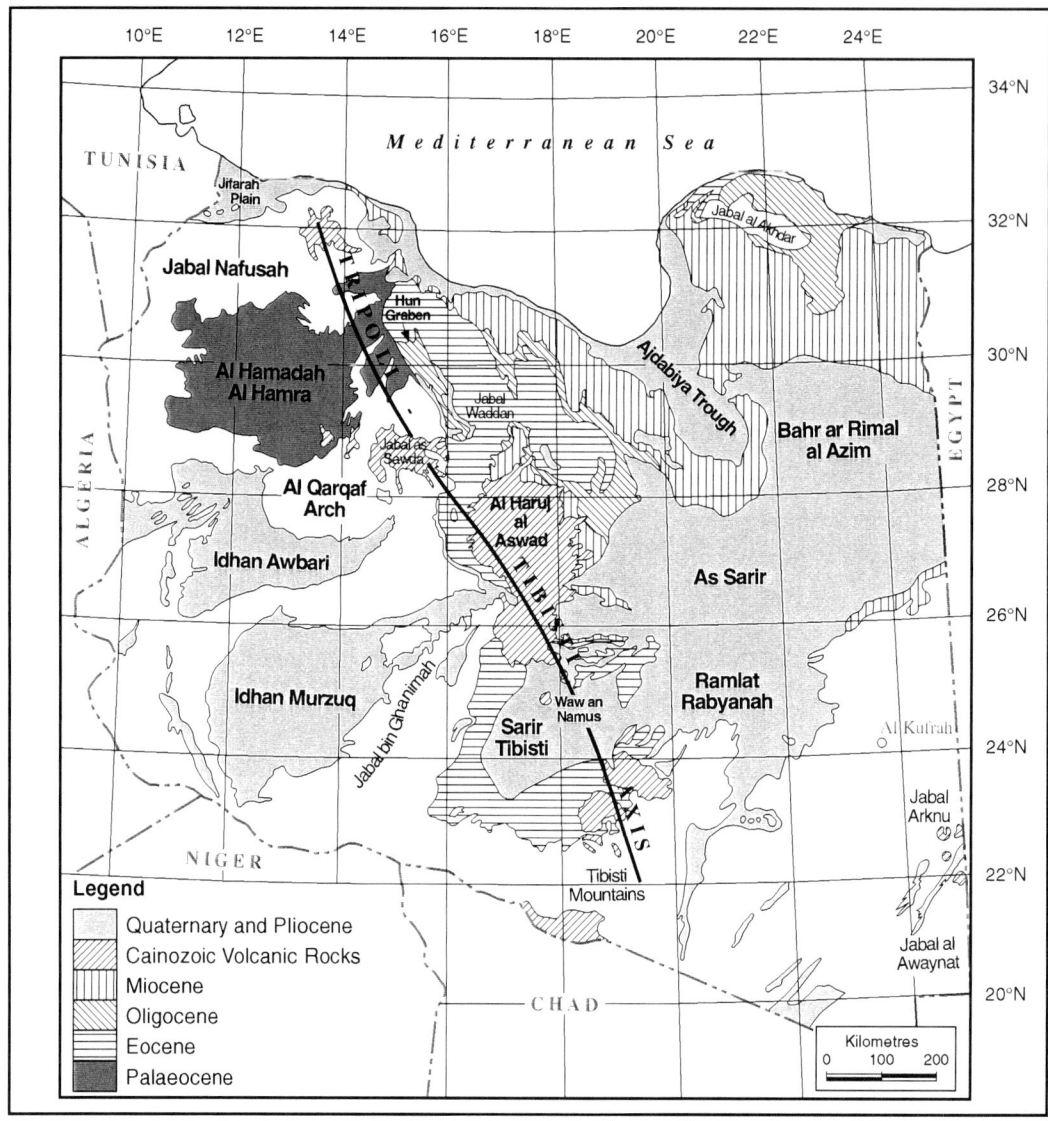

Source: Geological Map of Libya, 1:1,000,000, 1985

Figure 5.1   Libya: Cainozoic Outcrops

The Cainozoic outcrops in Libya illustrate the gradual withdrawal of marine deposition from western Libya during the Palaeocene, followed by a marine transgression which extended to the foot of the Tibisti Mountains during the Eocene. Oligocene sediments indicate a period of regression followed by tectonic instability during the Miocene. Eocene granites were intruded into the Jabal al Awaynat complex in the form of stocks and ring intrusions, and volcanic activity developed along the Tripoli-Tibisti axis, culminating in the late Miocene, with the production of extensive basaltic lava flows. The Messinian evaporitic event is well documented both offshore and onshore.

The stratigraphy of the Palaeocene will be reviewed in four areas: the west Libyan outcrops, the northwestern offshore, the subsurface of the Sirt Basin and Cyrenaica. Lithostratigraphic nomenclature for the Palaeogene succession is shown on Figure 5.2.

WEST LIBYA

5.1.1 Zimam Formation

The Zimam Formation was first described by Jordi and Lonfat from the Wadi Tar area northwest of Suknah, as discussed in chapter 4 (Figure 5.3 and 5.4). It was subsequently revised by Nairn and Salaj who restricted its use to the upper two members, the Upper Tar Member and the Had Member. The type section of the Upper Tar Member was defined at the entrance to Wadi Tar (Figure 5.3) where it is composed of 67m of fossiliferous brown marls, shales and mudstones with thin stringers of shaly limestone. The base is marked by a conspicuous shell bed, known as the Suknah Mollusc Bed which contains pelecypods, particularly *Venericardia* sp., *Cardita libyica*, gastropods, and foraminifera. The foraminifera include reworked Maastrichtian forms, but the presence of *Globoconusa daubgergensis* proves a Danian age for this unit. Where the Suknah Mollusc Bed is absent the highest appearance of *Omphalocyclus macroporus* is taken to indicate the top of the underlying Lower Tar Member.

The Upper Tar Member is succeeded conformably by the Had Member which was also defined by Jordi and Lonfat from the same type section as the Upper Tar Member (Figure 5.4). This unit comprises 50m of hard, compact dolomites, dolomitic limestones, chalk and marlstones which represent an inner protected subtidal to supratidal environment. The formation contains a sparse fauna of algae, gastropods, echinoids, foraminifera and nautiloids which indicate a Danian (Montian) age.[2]

The Zimam Formation outcrops over an area of 40,000km[2] in Al Hamadah al Hamra, with the hard siliceous limestones of the Had Member forming the flat featureless surface of much of the plateau. The Upper Tar Member (Figure 5.3) thins from 65m in the north at Bani Walid, to 10m on the southern outcrop near Awaynat Wanin. The lithology varies from calcarenites and marls in the south to calcilutites and dolomitic marls in the north. The unit contains an abundant fauna and the Suknah Mollusc Bed is usually recognisable at the base. The Had Member (Figure 5.4) has a fairly constant thickess in the range 10 to 25m, and the lithology varies from calcarenite to siliceous crystalline limestone with chert nodules. Dolomitic limestones are sometimes present and several dolomitic breccia horizons have been reported. These lithologies suggest deposition in a relatively shallow-water, low-energy environment.[3]

Outcrops of both members can be traced southwards towards Tmassah until they disappear beneath the lavas of Al Haruj al Aswad. In this area the Upper Tar Member consists of about 8 to 15m of yellow dolomite and marlstone, and the Had Member comprises 6m of slightly siliceous, microcrystalline dolomite. Gohrbandt suggested that the Zimam Formation extended south of Al Haruj al Aswad towards Waw al Kabir, but these rocks are regarded as Eocene by the compilers of the 1985 Geological Map of Libya.[4]

5.1.2 Shurfah Formation

The Shurfah Formation, which conformably overlies the Zimam Formation in the eastern Hamadah al Hamra, was established by Jordi and Lonfat for a series of marls, chalks and limestones in the Wadi Tar area, northwest of Suknah (Figure 5.5). Three members have been identified, the Abu Ra's Member at the base, overlain by the Qaltah and Ammur Members.

The Abu Ra's Member was defined in the Wadi Tar as a 32m sequence of well-bedded, platy marlstones, and chalky limestones with occasional dolomitic limestone beds (Figure 5.5). The chalky limestones are often brecciated and conglomeratic. The unit contains

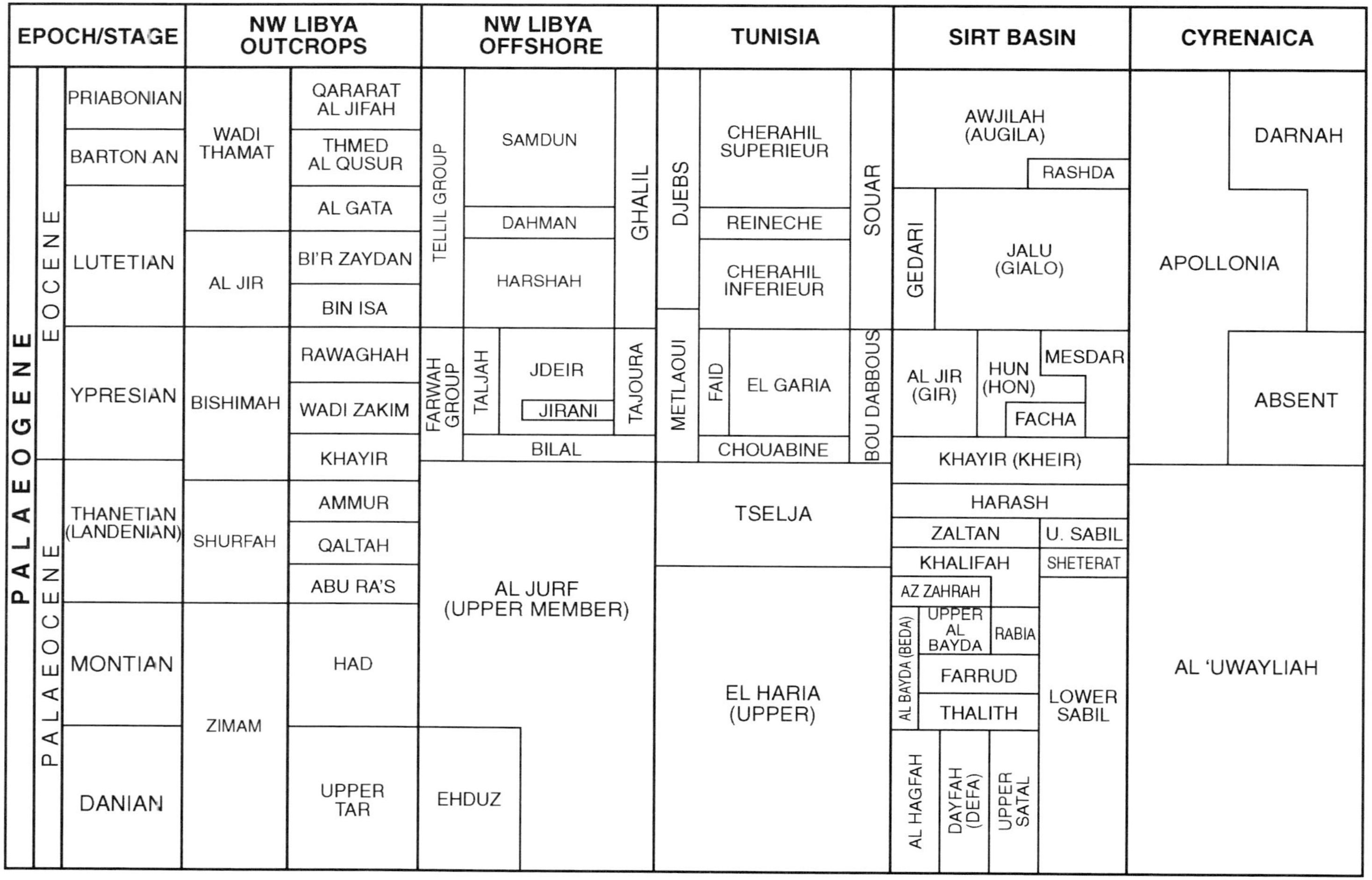

Figure 5.2   Lithostratigraphic Nomenclature for the Palaeogene Rocks of Libya and Adjacent Areas

EPOCH/STAGE		NW LIBYA OUTCROPS	NW LIBYA OFFSHORE	TUNISIA	SIRT BASIN	CYRENAICA
PALAEOGENE **EOCENE**	PRIABONIAN	WADI THAMAT (QARARAT AL JIFAH, THMED AL QUSUR)	SAMDUN / GHALIL	CHERAHIL SUPERIEUR / DJEBS / SOUAR	AWJILAH (AUGILA) / RASHDA	DARNAH
	BARTONIAN	AL JIR (AL GATA, BI'R ZAYDAN, BIN ISA)	DAHMAN, HARSHAH / TELLIL GROUP / TAJOURA	REINECHE, CHERAHIL INFERIEUR / METLAOUI / FAID	GEDARI / JALU (GIALO) / HUN (HON) / MESDAR / FACHA / AL JIR (GIR)	APOLLONIA
	LUTETIAN					
	YPRESIAN	BISHIMAH (RAWAGHAH, WADI ZAKIM, KHAYIR)	FARWAH GROUP / JDEIR / JIRANI / BILAL / TALAH / BOU DABBOUS	EL GARIA / CHOUABINE	KHAYIR (KHEIR) / HARASH	ABSENT
PALAEOCENE	THANETIAN (LANDENIAN)	SHURFAH (AMMUR, QALTAH, ABU RA'S)	AL JURF (UPPER MEMBER)	TSELJA	ZALTAN, U. SABIL, SHETERAT / KHALIFAH / AZ ZAHRAH, UPPER AL BAYDA / AL BAYDA (BEDA), RABIA / LOWER SABIL	AL 'UWAYLIAH
	MONTIAN	ZIMAM (HAD)		EL HARIA (UPPER)	DAYFAH (DEFA) / FARRUD / THALITH / UPPER SATAL / AL HAGFAH	
	DANIAN	UPPER TAR	EHDUZ			

Sources: see text

This table shows the current lithostratigraphic nomenclature for the Palaeogene sediments of Libya with updated transliteration of Arabic names. There are some obvious problems. The Al Jir Formation of the surface outcrops is Lutetian in age, whereas the Al Jir (Gir) Formation as defined in the subsurface is Ypresian in age, despite obvious lithological similarities. This problem is discussed in the text. No attempt has been made to show unconformities or gaps in the section on this chart.

a fauna of pelecypods, bryozoa, dasycladacean and codiacian algae and foraminifera which although not age diagnostic, suggests a Thanetian age.

The Qaltah (formerly Guelta) Member was originally described by Burollet from the western edge of the Waddan Platform (Figure 5.5). Jordi and Lonfat included the unit as the middle member of their Shurfah formation and described a representative section from Wadi Tar. Here the sequence is 50m thick and is made up of blocky chalk and soft chalky limestone with lenses of silicified limestone and thin intervals of dolomite and marl. This lithology suggests a shallow-marine environment of deposition ranging from brackish and littoral to inner neritic. The fauna includes abundant foraminifera and ostracods but macrofossils are not common. The assemblage indicates a Thanetian age.

The upper member, originally named the *Operculinoides* Limestone by Burollet and the *Operculina* Limestone by Jordi and Lonfat, was renamed the Ammur Member by Shakoor and Shagroni with a type locality on the eastern scarp of the Hun Graben (Figure 5.5). The sequence comprises 10 to 12m of cavernous dolomitic limestone rich in the large foraminifer *Operculina canalifera*, deposited in a nearshore littoral environment. The unit contains a rich foraminiferal fauna which has been studied in some detail, plus calcareous algae and some fragmentary macrofossils. The age of the faunal assemblage is Thanetian.[5]

The Shurfah Formation outcrops extensively on the eastern Hamadah al Hamra, in small patches along the eastern scarp of the Hun Graben, and along an arcuate belt from Jabal as Sawda to Al Haruj al Aswad (Figure 5.5). The last outcrop in this chain is an inlier in the lavas of Al Haruj al Aswad. The Abu Ra's Member varies in thickness from 5m at Bani Walid to 36m at Qasr ash Shwarif. The lithology is predominantly chalky, dolomitic marl, and solution brecciation is common. Calcarenites and calcilutites become common in the Qasr ash Shwayrif and Al Washkah areas. The Qaltah Member ranges in thickness from 8m in the south to 55m in the Hun Graben. The dominant lithology is dolomitic chalky limestone, passing into chalky marl to the north and into argillaceous dolomite in the south. The Ammur Member thickens from northwest to southeast and reaches its maximum thickness of 31m in the Tmassah area (Figure 5.5). The lithology is mainly skeletal limestones and argillaceous dolomites, passing into calcarenites in the Al Washkah area. Gohrbandt claimed that typical Shurfah Formation sediments extend to the foothills of the Tibisti Massif, but these sediments are regarded as Eocene by the compilers of the 1985 Geological Map of Libya. Further east, towards Jabal Nuqay, Fürst described a 70m sequence of unfossiliferous shales and shaly sandstones, overlying basement granite, which he tentatively assigned to the Palaeocene, but the age attribution is based solely on their position beneath Eocene claystones and evaporites.[6]

NORTHWEST LIBYA OFFSHORE

5.1.3 Al Jurf Formation (Upper part)

The Maastrictian part of the Al Jurf Formation was described in Chapter 4. Sedimentation continued without interruption into the Danian with a lithology of silty marls, black shales and shaly sandstones (Figures 5.3). Well C1-NC 41 in the Sabrathah Basin was selected as the type well by Hammuda et al. The total thickness of the formation in the well is 414m, but part of this is Maastrictian in age and attributable to the lower part of the Al Jurf Formation. The faunal richness decreases in the upper part of the formation, but evidence from well D2-NC 41 and elsewhere proves a continuous sequence of planktonic index fossils from low in the Danian to the top of the

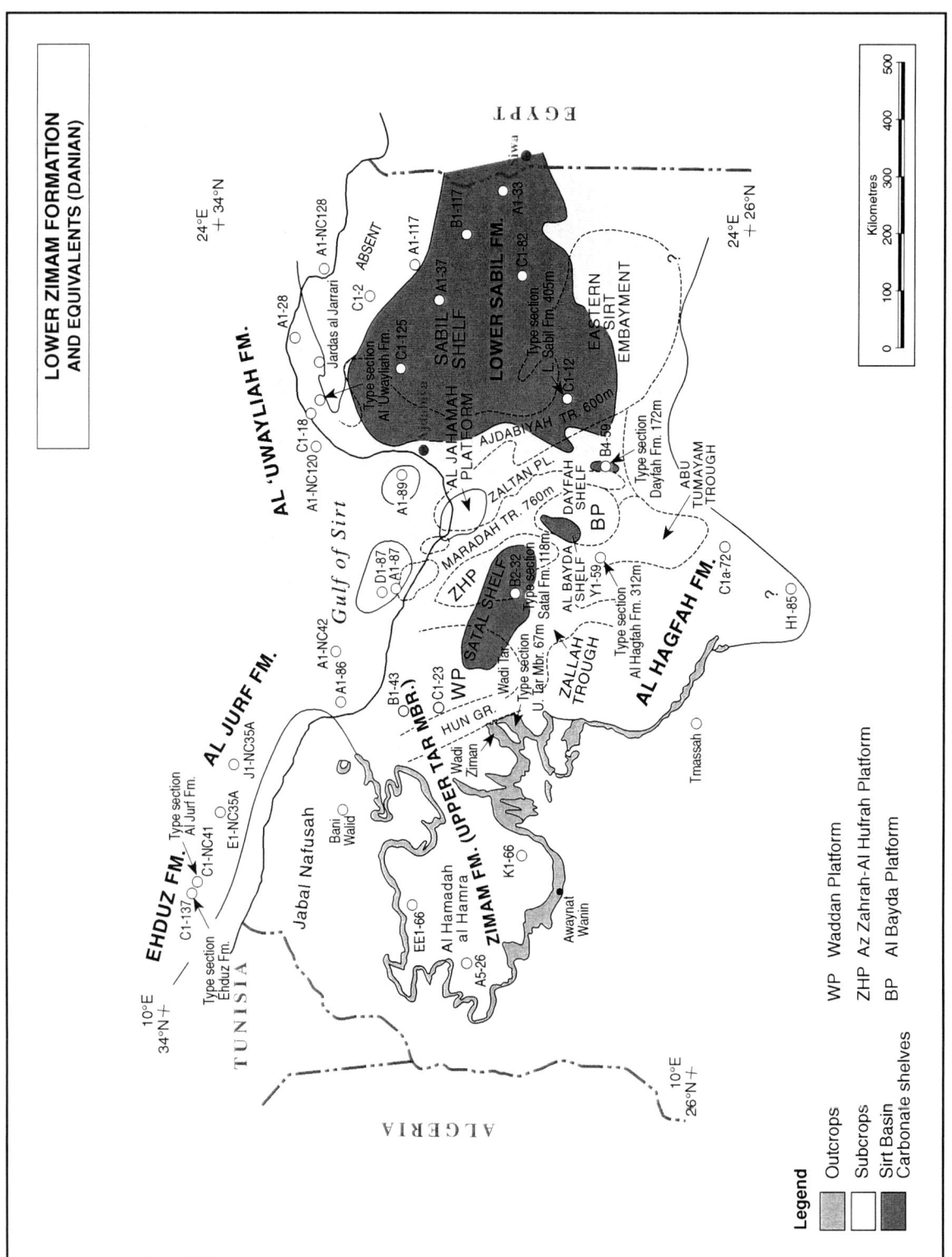

LOWER ZIMAM FORMATION
AND EQUIVALENTS (DANIAN)

Sources: Distribution: Geological Map of Libya, 1985, Bezan, 1996

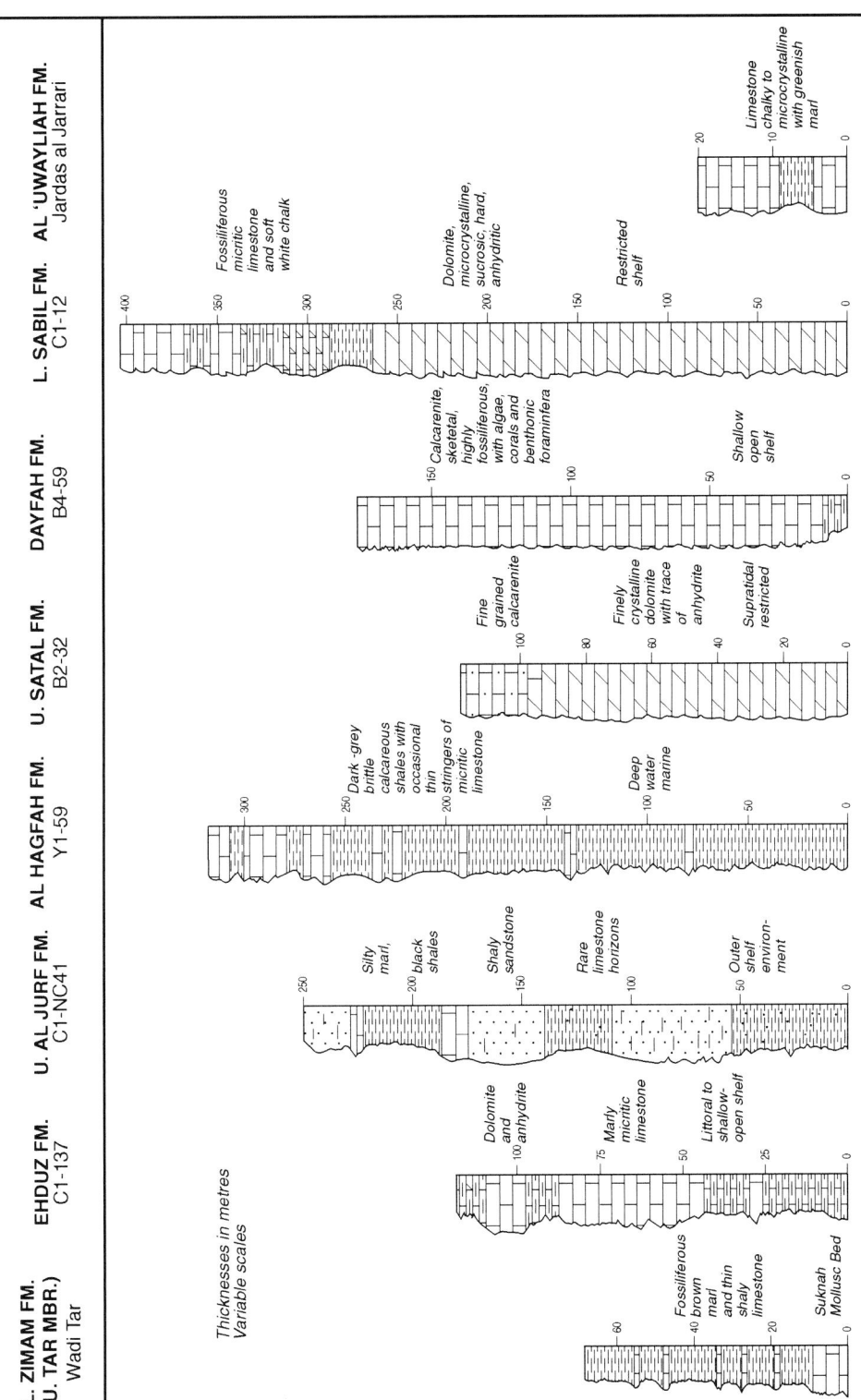

**Figure 5.3   Lower Zimam Formation (Upper Tar Member):
Distribution, Equivalents and Stratigraphic Columns**

By the beginning of the Palaeocene the rift-troughs of the Sirt Basin had been largely infilled and Palaeocene sedimentation was dominated by a number of carbonate shelves separated by deeper-water areas. Regional tilting led to a gradual withdrawal of the sea from west Libya. Four carbonate shelves dominated the early Palaeocene: Satal, Al Bayda, Dayfah and Sabil, with deeper-water areas in the Zallah, Maradah, and Ajdabiya Troughs. Open marine conditions existed in the northwestern offshore and in northern Cyrenaica.

208

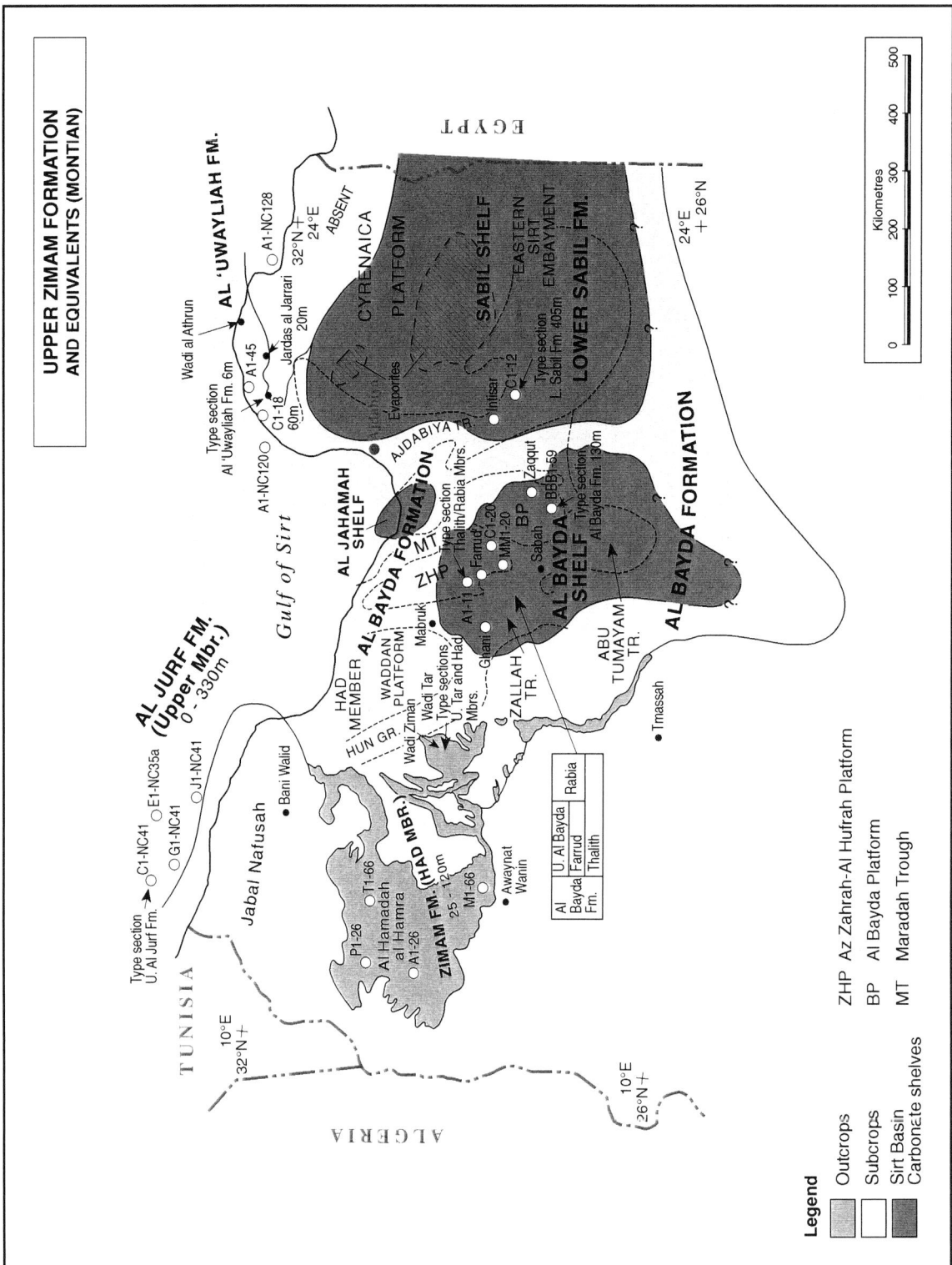

Source: Distribution: Geological Map of Libya, 1985, Bezan, 1996, Wennekers, et al., 1996

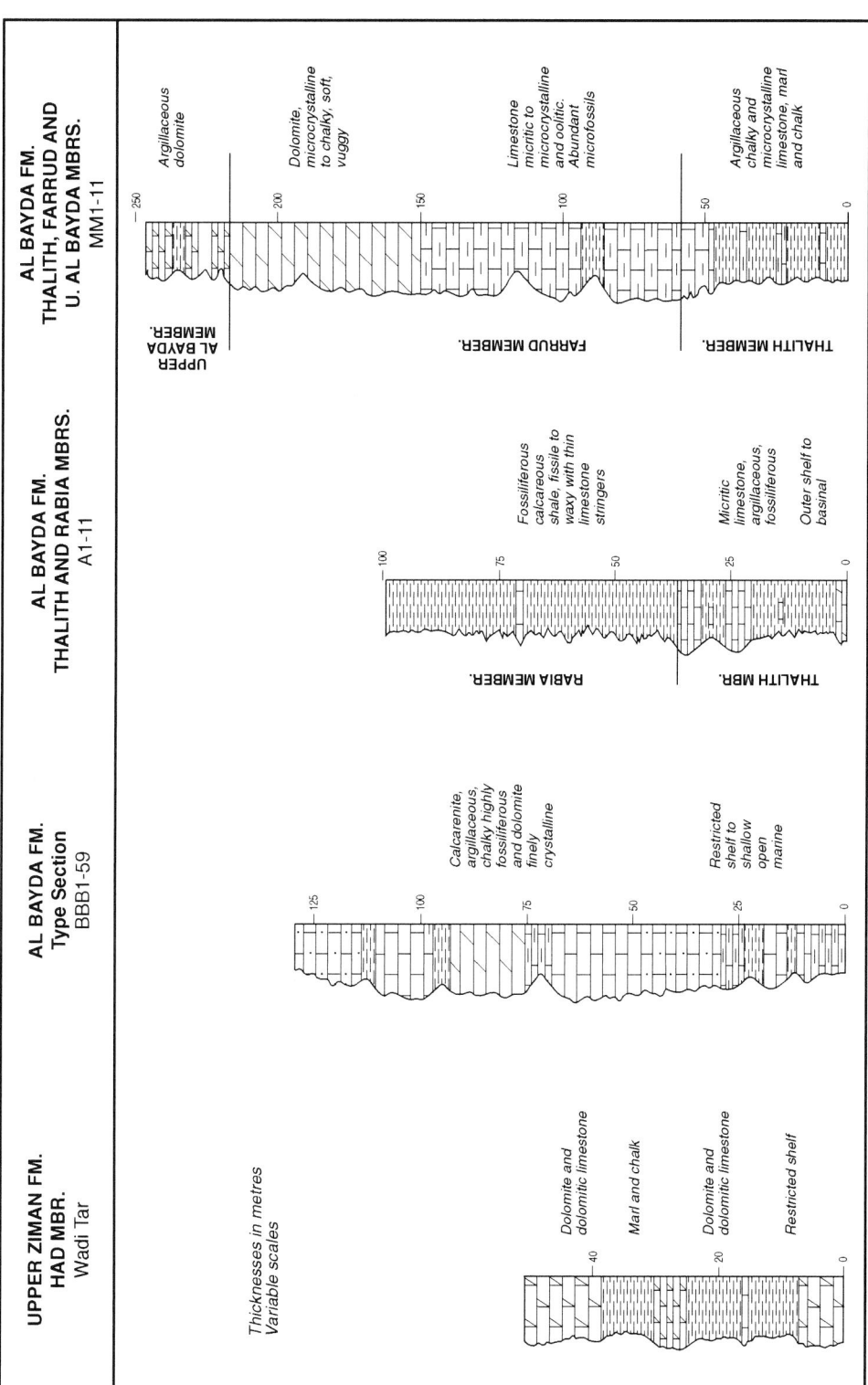

Figure 5.4   Upper Zimam Formation (Had Member):
Distribution, Equivalents and Stratigraphic Columns

The Had Member, which outcrops over a large area on the Al Hamadah al Hamra, is a shallow-water carbonate deposited on a rising shelf. In the Sirt Basin carbonate shelves spread over much of the basin, attracting shallow-water limestone, dolomite and anhydrite, whilst marls and shales were deposited in the deeper-water areas. In northern Libya deposition of the Al 'Uwaliah and Al Jurf Formations continued. The type section for the Al Bayda Formation was not well chosen. A much more representative section is found in the MM1-11 well.

Thanetian (Figures 5.3, 5.4 and 5.5). The black shales represent a potential source rock for the overlying Farwah Group reservoirs, although the Lower Eocene shales are a more likely candidate. In the type well the Al Jurf Formation is overlain by carbonates of the Farwah Group. The lithology of the formation suggests deposition in open marine conditions in an outer shelf environment. It thins towards the basin margin to the south, and only 17m is present in well G1-NC 41 (Figure 5.4). The upper Al Jurf Formation is age equivalent to both the Zimam and Shurfah Formations of northwest Libya, and to the upper part of the El Haria Formation plus the Tselja Formation in Tunisia.[7]

5.1.4 Ehduz Formation

The Ehduz Formation was established by Hammuda et al. for an open marine carbonate unit lying between the Al Jurf Formation and the Bilal Formation (Figure 5.3). A type section was defined in well C1-137, in the western Sabratah Basin where 115m of marly, micritic limestone is capped by a zone of microcrystalline dolomite. This sequence is interpreted as a littoral to shallow open shelf environment of deposition. The Ehduz Formation contains a distictive fauna of planktonic foraminifera which includes *Subbotina triloculinoides*, an index fossil of Danian age (Figure 5.3). It appears that the Ehduz Formation represents a nearshore equivalent of part of the Al Jurf Formation. The formation has also been reported from well F1-NC 41. It is equivalent to the Upper Tar Member of the Zimam Formation of Al Hamadah al Hamra area, and to part of the El Haria Formation in Tunisia.[8]

SIRT BASIN

The Palaeocene is an epoch of great economic importance in the Sirt Basin since it contains over 10 billion barrels of original oil reserves. It has received correspondingly detailed attention from stratigraphers and sedimentologists, and there is an extensive literature on the Palaeocene stratigraphy of Libya.

5.1.5 Al Hagfah Formation

Barr and Weegar introduced the term Hagfa Shale for a sequence of calcareous shales lying between the Kalash Formation and the Al Bayda Formation (Figure 5.3). They selected well Y1-59 in the Al Kotlah Graben as the type section. At this location the formation comprises 312m of dark-grey, brittle, calcareous shales with occasional very thin stringers of micritic limestone. The limestone intervals become more common towards the top. Both top and bottom contacts are conformable. The fauna of the Al Hagfah Formation is dominated by planktonic foraminifera which suggests a deep-water marine environment of deposition. The assemblage includes several Danian index fossils. The formation has an extensive distribution in the Sirt Basin, controlled by the palaeogeography of the early Palaeocene, which is significantly different from that of the late Cretaceous. By Danian times the rift troughs of the Cretaceous had been infilled, and a much more gentle topography of carbonate shelves separated by shallow basins had developed (Figure 5.3). The Al Hagfah Formation is characteristic of the shallow basins. It is present throughout the central Sirt Basin as a deeper-water equivalent of the shelfal carbonates. An example of this transition is provided on the Al Bayda Platform where the Al Hagfah Formation passes laterally into shallow-water carbonates on the Amin Ridge (Figure 5.3). A similar situation is present on the Zaltan Platform where the Nasser carbonate build-up passes laterally into Al Hagfah shales (Figure 8.16). An isopach map of the Al Hagfah Formation was published by Bezan which shows major depocentres in the

Ajdabiya Trough (600m), Maradah Trough (760m), south of the Satal Bank (450m) and south of the Sabil Bank (150m). The formation has minor source rock potential, although the total organic carbon level rarely exceeds 1%.[9]

5.1.6 Dayfah (Defa) Formation

The term Defa Limestone was introduced by Barr and Weegar for a section of skeletal limestones and calcarenites developed on the Dayfah field, overlying the Wahah Formation (Figure 5.3). As mentioned above, the formation is a shallow-water equivalent of the Al Hagfah Formation. The type section was selected in the B4-59 well. In this well 172m of massive, well-bedded limestones are present, which show a variety of lithofacies ranging from algal and coralline calcarenites to pelletal, chalky and micritic facies. The dasycladacean algae frequently form algal mats which develop a characteristic fenestral fabric. The formation contains an abundant fauna of corals, algae, molluscs and foraminifera which are not age diagnostic, but on the basis of stratigraphic position the formation is assigned to the Danian. Bezan, et al. however believed that the formation extends into the Montian in the Dayfah field discovery well B1-59, on the basis that it is overlain by the Upper Al Bayda Member. The formation was deposited in extremely shallow water, but the extent of the Dayfah Shelf is extremely limited, occurring over an area of only 19x10km on the southern Zaltan Platform (Figure 5.3). It is interpreted as a small carbonate shoal within an area of deeper-water conditions. A similar, even smaller shoal developed at Nasser (Zaltan) further north and formed the foundation on which the Zaltan carbonate complex developed in the late Palaeocene. The formation is highly porous and represents an excellent reservoir. It forms the principal reservoir on the Dayfah field, one of the giant fields of the Sirt Basin.[10]

5.1.7 Satal Formation (Upper Member)

The lower part of the Satal Formation, which is Maastrichtian in age, was discussed in Chapter 4. In the area of the type well, B2-32, on the Az Zahrah West field, the upper member of the Satal Formation comprises 118m of massive crystalline dolomite with traces of anhydrite and fine-grained calcarenite (Figure 5.3). The member contains a fauna dominated by foraminifera, algae and molluscs, which give a Danian age for this unit. The contact between the lower member and the upper member is marked by an increase in dolomite content in the upper unit and a decrease in the chalk component. The upper-member was deposited on a shallow carbonate shelf analogous to the Dayfah and Sabil shelves. The extent of the Upper Satal carbonate shelf has been determined from well data by Bezan. He showed the shelf extending over an area of 12,000km[2] from the Waddan Uplift to the Manzilah Ridge, with a smaller, isolated bank on the Al Bayda platform (Figure 5.3). It is present over the crestal part of the Az Zahrah-Al Hufrah Platform where it has a thickness of 130m with excellent porosity and permeability. In this area it shows a transition from marginal to lagoonal and tidal flat facies. Laterally it passes rapidly into Al Hagfah Shales. To the north the Al Jahamah Platform was emergent during the early Palaeocene and the offshore wells A1-89, E1-87 and D1-87 also show a total absence of early Palaeocene sediments (Figure 5.3). The Upper Satal Member forms the principal reservoir on several fields including the Az Zahrah-Al Hufrah, Ali, Almas and Arbab fields.[11]

5.1.8 Sabil Formation (Lower Member)

In the eastern Sirt Basin the Al Hagfah Formation passes into a calcareous facies which was named the Sabil Formation by Barr and

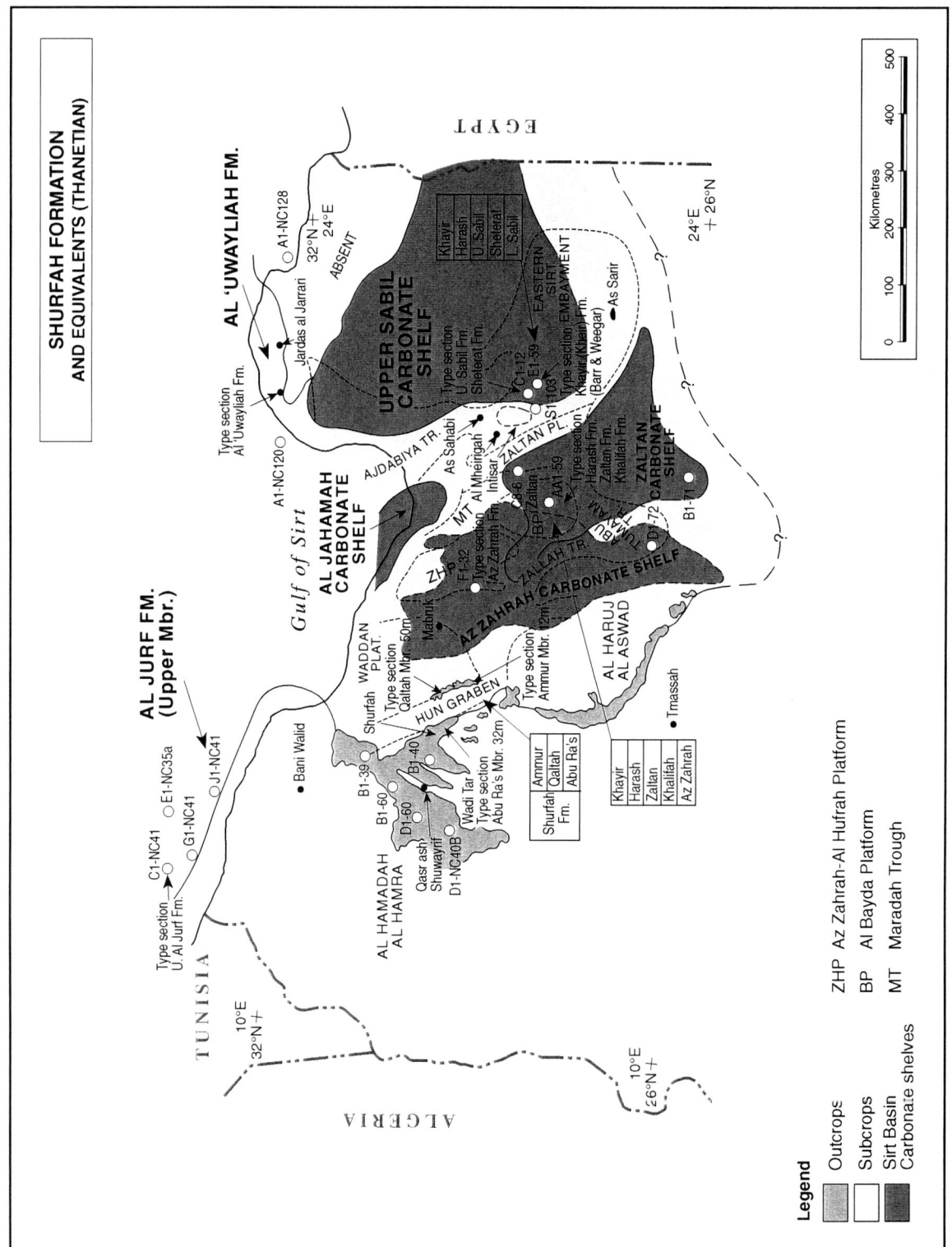

Sources: Distribution: Geological Map of Libya, 1985, Carbonate Shelves: Bezan, 1996

213

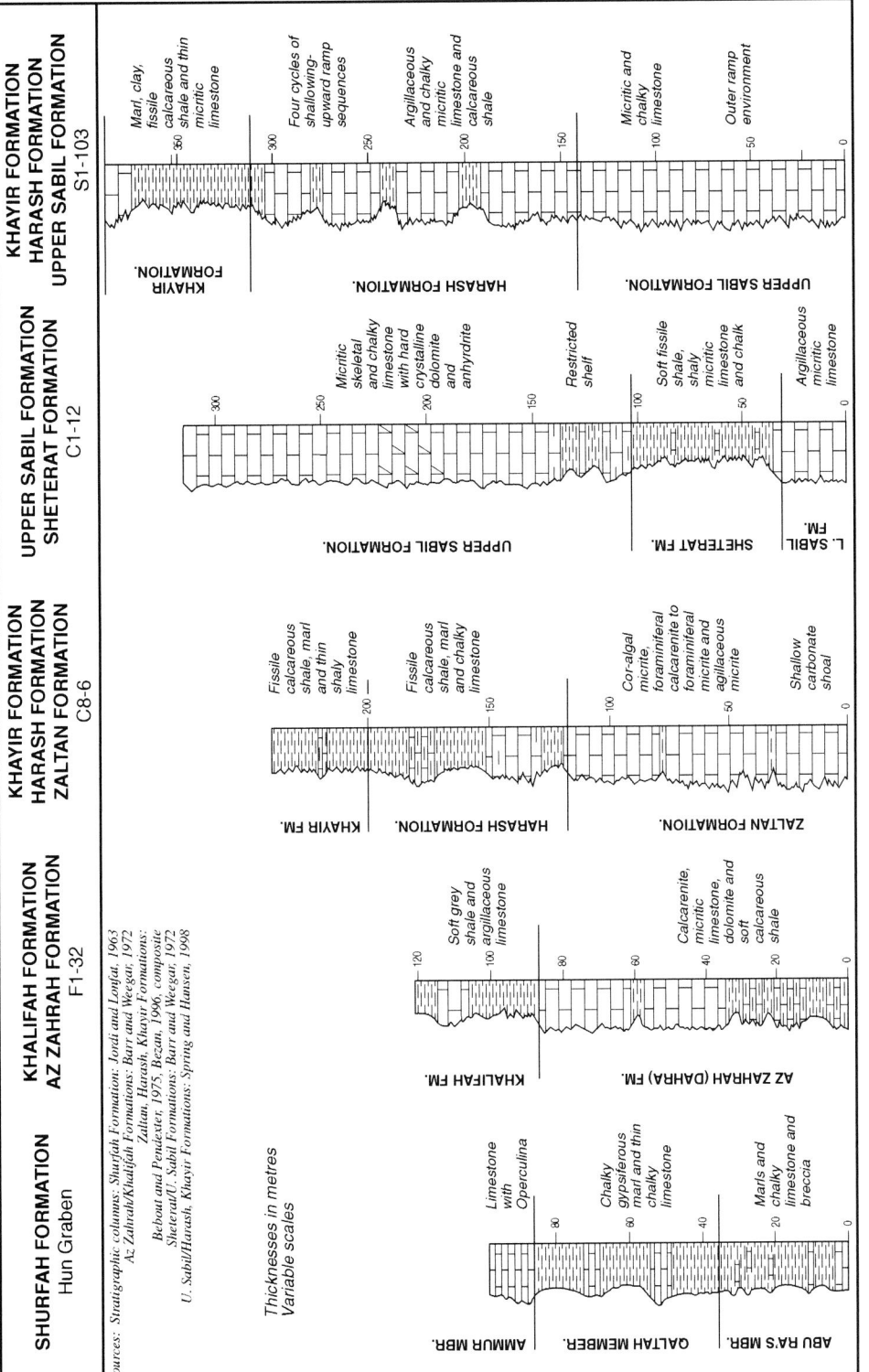

Figure 5.5   Shurfah Formation: Distribution, Equivalents and Stratigraphic Columns

The Shurfah Formation is defined from type sections on both sides of the Hun Graben. It outcrops on the eastern Hamadah al Hamra, and southwards as far as Al Haruj al Aswad. In the subsurface, shallow-water carbonates were deposited on extensive shelves separated by deeper-water areas in which shales and lime muds were deposited. The Zaltan shoal complex is located on the northern edge of the Zaltan carbonate shelf, and the Mabruk and Az Zahrah-Al Hufrah fields are situated on the northern margin of the Az Zahrah carbonate shelf. The Intisar, Al Mheirigah and As Sahabi reefs are situated within a re-entrant between the Zaltan and Sabil carbonate shelves.

Weegar (Figure 5.3 and 5.4). They defined a type section in well C1-12 near the Farigh field in the southern Ajdabiya Trough, and split the formation into upper and lower members separated in the type well by a unit which they called the Sheterat Formation. The Sabil Shelf is much larger than the Satal and Al Bayda shelves, extending over an area of 140,000km².

The lower part of the Sabil Formation in the type well is composed of fossiliferous microcrystalline limestone and sucrosic or crystalline dolomite (Figure 5.3). Traces of anhydrite occur with the dolomite, and anhydrites become a significant component on the Sabil shelf during the Montian. The limestone is occasionally chalky or micritic. In the type well the unit reaches a thickness of 405m. It is present over much of the eastern Sirt Basin, thickening to 610m in the area west of Intisar. Towards the east the formation becomes progressively more dolomitic. Its extent has been documented by Bezan. He showed the formation to be present over much of northeastern Libya from the Mediterranean coast near Ajdabiya, southwards to include the As Sahabi, Al Mheirigah and Intisar areas, then eastwards, including most of the eastern Sirt Basin to the Egyptian border near Siwa (Figures 5.3 and 5.4). To the south of the Sabil Shelf the formation passes into deeper-water facies of the Al Hagfah Formation. The fauna is not age diagnostic but the Lower Sabil Member is assigned to the Danian and Montian on the basis of its stratigraphic position overlying the Kalash Formation. The Lower Sabil was deposited in a quiet, low-energy, shallow-water environment with occasional periods of emergence. The character of the shelf margin for the Lower Sabil has not been positively established, but by analogy with the Upper Sabil it may represent a barrier-reef type margin.[12]

5.1.9 Al Bayda (Beda) Formation

Barr and Weegar introduced the name Al Bayda (formerly Beda) Formation for the carbonate sequence overlying the Al Hagfah Formation (Figure 5.2). They chose the BBB1-59 well on the eastern margin of the Al Bayda Platform as the type well. In fact, during the Montian, the Al Bayda carbonate shelf extended from the Abu Tumayam area in the south to the Bahi field in the north (Figure 5.4). They assigned a middle Palaeocene (Montian) age to the formation based on the foraminiferal fauna. The Montian stage has since been abandoned as a valid stage name as it has been shown to represent a facies variant of the Danian. In Libyan usage however it is convenient to regard the Montian as equivalent to the late Danian. In the type well the Al Bayda Formation comprises 130m of skeletal, oolitic and micritic limestones with minor dolomite, which often shows a fenestral fabric. The limestones are often highly fossiliferous, including molluscs, echinoids and corals, plus dasycladacean algae and foraminifera. The formation is widespread over the western part of the Sirt Basin and Barr and Weegar introduced a number of subdivisions relevant to specific areas (Figure 5.4). In the area of the Al Hufrah field Barr and Weegar divided the formation into two members, a lower carbonate unit named the Thalith Member, and an upper shale unit named the Rabia Member. The A1-11 well on the Al Hufrah field was selected as the type well for both members. On the Manzilah Ridge the entire Montian section is represented by carbonates. Indeed, the Upper Satal, Al Bayda and Az Zahrah Formations, form an unbroken carbonate sequence in this area.[13]

The Thalith Member (Figure 5.4) is composed of argillaceous, microcrystalline limestone interbedded with fissile calcareous shale and chalky marl. In the type well it is 36m thick. It extends as far south as the Al Kotlah

Graben and Al Bayda Platform. The distribution of the member was shown by Bezan, et al. The Thalith Member reaches a thickness of over 213m in wells FFF1-11 and 4G1-11 near the Ghani field. North of Raqubah the thickness ranges between 60m in BB1-20 to 36m in the type well, A1-11. The environment of deposition ranges from lagoonal to open shelf.[14]

The Rabia Member (Figure 5.4) is a sequence of calcareous shale and soft mudstone with thin argillaceous limestones. The thickness in well A1-11 is 65m. It is limited in extent to the area around the Az Zahrah-Al Hufrah fields. In the southwestern part of the Sirt Basin Barr and Weegar suggested that the Rabia Member can be subdivided into a lower and upper unit, but they provided no details of this subdivision.[15]

The main problem with Barr and Weegar's definition of the Al Bayda Formation is that it does not adequately represent the typical form of the formation in the area where it is best developed. Bezan showed that the carbonate bank on which the Satal Formation was deposited expanded southwards during the Montian to occupy the entire western Sirt Basin south of the Az Zahrah field. The sections in wells C1-20, west of Ar Raqubah field, and MM1-11 slightly further west, were used by Bezan to illustrate a more typical development. In these wells the Thalith Member is overlain by carbonates, the lower of which has been named the Lower Al Bayda, Farrud or Meem Member, and the upper is referred to as the Upper Al Bayda or Awra (Ora) Member (Figure 5.2). This three-fold subdivision is maintained in the Zallah Trough, as illustrated by well GGG1-11 south of the Ghani field.

Further south, in the area of the Sabah field, a dolomitic zone is developed in the middle of the Al Bayda Formation. This unit forms one of the principal exploration targets in the area. It has been studied by Garea who showed the presence of three shallowing-upwards cycles, which pass from a lagoonal to a supratidal environment and end with a thin evaporite unit. The effect of the subsequent diagenesis has been to generally enhance reservoir quality.

The Farrud Member is a skeletal, oolitic calcarenite passing to dolomite, with excellent reservoir characteristics, which is productive in several fields including Ghani (Figure 5.4). At Ghani, according to Abushagur and LeMone, the Farrud Member is composed of 45m of intertidal dolomites which were deposited in tropical conditions in shallow water with moderate wave and current activity. This unit reaches a maximum thickness of 186m in well H1-47 in the Al Kotlah Graben, but a thickness of 90m is more common. The palaeogeography of the Al Farrud Member on the Bayda Platform has been reconstructed by Sinha and Mriheel, who showed shelfal carbonates passing southwards into oolite shoal and calcarenite facies. It is also a major exploration target in the Abu Tumayam Basin where it is composed of porous calcarenite, skeletal, oolitic and micritic limestone and rare dolomite, deposited in a shoreface environment. The average thickness in this area is 90m. An isopach map of this unit was published by Bezan, et al.[16]

The Upper Al Bayda Member represents a restricted shelfal dolomitic facies, which also has good reservoir characteristics. The thickness varies from a maximum of 100m in well CC1-47 to a minimum of 12m in well MMM1-11. This unit is oil productive on several fields on the Al Bayda Platform where it is composed of shelfal limestone, oolites and calcarenites. To the east of the carbonate bank, in the Maradah Trough, deeper-water conditions prevailed but occasional low-stand episodes allowed carbonates to extend out into the basin, as on the Wadi field, where a thin Lower Al Bayda carbonate is found sandwiched between two deeper-water shale sections. This unit is referred to informally as the Zaqqut or M-59 'Member', and contains oil at both Zaqqut and Wadi.

In the eastern Sirt Basin Sabil carbonates are equivalent to the Al Bayda Formation, and in the north carbonates were also deposited on the Al Jahamah Platform. According to Bezan, the Al Bayda Formation reaches a maximum thickness of 506m in the F1-57 well in the Zallah Trough with an average of 180m on the platform areas.[17]

5.1.10 Khalifah Formation

The Khalifah Formation was established by Barr and Weegar for a predominently shaly unit lying between the Al Bayda and Zaltan Formations (Figure 5.2). The AA1-59 well, west of Zaqqut on the Al Bayda Platform, was selected as the type well (Figure 5.4). Here the formation can be divided into a lower shale unit and an upper more calcareous unit. The lower unit, 100m thick, is composed of fissile, slightly calcareous shale with thin limestone beds, and the upper unit, 60m thick, is made up of argillaceous micritic limestone, interbedded with calcareous shale. On the platform crestal areas the lower unit passes into a shallow-water carbonate facies, the Az Zahrah (Dahra) Formation (Figure 5.5). Barr and Weegar added the comment that in areas such as the Al Wahah field the Al Bayda Formation is represented by a shale section "which is then included within the lower Khalifah Formation". Whilst this is a legitimate application of the rules of stratigraphic nomenclature it is nevertheless confusing since the shaly facies of the Al Bayda Formation had already been designated the Rabia Member. The Rabia Member is Montian in age, whereas the Khalifah Formation is early Thanetian on the basis of its planktonic foraminifera fauna. Of course, where Khalifah shales directly overlie Rabia shales it is virtually impossible to separate them, which is presumably what Barr and Weegar had in mind.[18]

During the early Thanetian the area of the western carbonate platform decreased in size due to a marine transgression, and sedimentation on the Sabil shelf assumed a more open marine aspect. By the time of deposition of the Upper Khalifah Member open marine conditions had become established over most of the Sirt Basin, except for the Al Jahamah Platform and part of the Sabil Platform. It was during this interval that the Sheterat Formation was deposited.

The Khalifah Formation extends over much of the Sirt Basin (Figure 5.5). On the Az Zahrah-Al Hufrah Platform the Khalifah Shale is well developed forming a seal for the underlying Az Zahrah carbonate reservoir. The formation ranges in thickness from 30m north of the Az Zahrah Shelf to 60m in the Maradah Trough. Further east it reaches 120m in the Ajdabiyah Trough, decreasing to 30m in the area south of the Sabil Shelf. On the Zaltan Platform, in the area of the Zaltan field, the Upper Khalifah Member includes a thin but distictive limestone horizon which is known locally as the Cra Limestone and is composed of foraminiferal micrite and *Discocyclina* micrite. This unit marks the beginning of the shallowing phase which produced the overlying Zaltan Formation.[19]

5.1.11 Az Zahrah (Dahra) Formation

As mentioned above, Barr and Weegar defined the Az Zahrah Formation as the lateral shallow-water equivalent of the lower Khalifah Formation (Figure 5.2). The type section was defined in the F1-32 well, the discovery well of the Az Zahrah East field (Figure 5.5). In this well 95m of calcarenite, micritic limestone, dolomite and calcareous shale overlie Rabia Member shales (Montian) and underlie Upper Khalifah shales (Thanetian). This limestone unit is often referred to in oil company reports as the Mabruk Limestone since it forms the principal reservoir on the Mabruk field. It is also the main producing unit on the Az Zahrah-Al Hufrah fields. The formation is best developed on the Az Zahrah-Al Hufrah Platform, but is also present in a more shaly facies in the eastern Zallah Trough and, according to Bezan, passes beneath the Al Haruj al Aswad volcanics as far south as well D1-72 in the Abu Tumayam Basin (Figure 5.5). It is not present over most of the Al Bayda Platform where it is represented by Khalifah Formation shales. Az Zahrah equivalent carbonates are present on the Al Jahamah Platform but cannot be separated from the Upper Al Bayda carbonates. Similarly on the Sabil Platform the Lower Sabil early Thanetian carbonates cannot be distinguished from the 'Montian' carbonates. The thickness of the Az Zahrah Formation varies from a maximum of 120m in the Zallah Trough area to less than 100m in the Az Zahrah-Al Hufrah area.[20]

5.1.12 Sheterat Formation

Reference was made above to the marine trangression during early Thanetian times which led to a temporary deepening of the sea on the Sabil Platform. This episode gave rise to the Sheterat Formation which was first described by Barr and Weegar from the C1-12 well which is located on the Sabil Shelf west of the Al Farigh field (Figure 5.5). In this well the Sheterat Formation is represented by 44m of soft, fissile shale, shaly micritic limestone and chalk. The formation thins towards the Amal field and is absent east of the Jalu field, but it thickens west of the type well towards the Ajdabiya Trough where thicknesses in excess of 760m have been recorded. The Sheterat Formation separates the Lower from the Upper Sabil carbonates. Where the Sheterat Formation is absent it is impossible to separate these two units. The Sheterat Formation is equivalent to the Khalifah shales of the western Sirt Basin, and represents a short-lived flooding event.[21]

5.1.13   Zaltan Formation

Barr and Weegar established the Zaltan Formation (originally Zelten Limestone) for a thick carbonate sequence overlying the Khalifah Formation in the central Sirt Basin. They designated the AA1-59 well, on the Al Bayda Platform, as the type well (Figure 5.5). This is a curious choice since the formation is much better developed on the Zaltan Platform, particularly on the Nasser (Zaltan) field. In the type well the formation comprises 102m of shaly, chalky micritic limestone with some calcarenite and thin stringers of shale, becoming biohermal in places. Barr and Weegar remarked that in other places the formation becomes highly fossiliferous, vuggy and dolomitic which is much more typical of its character on the Nasser field (Figure 5.5). No diagnostic fauna was reported by Barr and Weegar, but subsequent studies have shown the presence of *Globorotalia pseudomenardii* and *Morozovella velascoensis*, index fossils of mid- to late Thanetian age.[22]

Detailed analysis of the Zaltan Formation on the Nasser field was conducted by Bebout and Pendexter who recognised fifteen carbonate facies ranging from coralgal micrite and foraminiferal calcarenite to foraminiferal micrite and argillaceous micrite. Porosity reaches 40% in the coralgal-micrite facies, attributable to the grain-supported nature of the unit, subsequent leaching and lack of compaction. In the area of the Nasser field the formation is about 115m thick. It represents a carbonate shoal with the different facies indicating subtle changes of water-depth, current direction and clastic supply. At the time of maximum regression the shoal was emergent and subjected to fresh-water leaching. Finally a regressive phase covered the area with foraminiferal micrite facies and development of the shoal came to an end. Bebout and Pendexter nowhere mention the word reef, but subsequent studies have shown that the Zaltan stucture began to develop early in the Danian, and there is evidence of atoll and lagoonal conditions during the early Thanetian, which established the foundation on which the late Thanetian shoal accumulated. The Zaltan shoal contains over 2 billion barrels of original oil reserves.[23]

The palaeogeography of the basin at this time has been mapped by Bezan. The main Zaltan shelf extended westwards from the Nasser field across the southern part of the Az Zahrah-Al Hufrah Platform, Al Bayda Platform and as far south as well B1-71, and along the southern margin of the Sirt Basin to a point south of the Sarir field (Figure 5.5). It should be pointed out however that on the Az Zahrah-Al Hufrah Platform the development of high-energy, shallow-water carbonates is limited to a small area west of the Manzilah Ridge. Over the rest of the platform the Zaltan Formation is represented by argillaceous and micritic limestones. The extension of the shelf in the area south of the Al Bayda Platfrom was confirmed by Sinha and Mriheel. In the northern Sirt Basin smaller carbonate shelves developed on the Waddan Platform and on the Al Jahamah/An Nuwfaliyah Platforms. In the east the Sabil Shelf represented a supratidal environment on which the Upper Sabil carbonates were deposited. In other areas, such as the Zallah, Maradah and Ajdabiya Troughs deeper-water carbonate muds were deposited. The thickness of the Zaltan Formation on the shelves averages 100-120m and in the deeper-water facies reaches 140m in the Zallah and Maradah Troughs, and over 150m in the Ajdabiya Trough.[24]

5.1.14 Sabil Formation (Upper Member)

On the Sabil Shelf the Sheterat Formation is overlain by limestones and dolomites which Barr and Weegar named the Upper Sabil Carbonates (Figure 5.2). The type section was described from the same well as the Lower Sabil Carbonates, C1-12, located close to the Al Farigh field (Figure 5.5). In this well the member comprises 240m of micritic, skeletal and chalky limestone, and hard crystalline dolomite, with minor anhydrite. Dolomite and anhydrite become more abundant in the area east of the Amal field. The formation represents deposition on a shallow, sometimes restricted shelf. The thickness averages about 250m, but thickens to over 300m in the area west of Amal. In the Intisar area the thickness is about 120m and further north, in the Ajdabiya Trough, it passes into a deeper-water facies.[25]

The Upper Sabil Member is oil bearing at As Sahabi and Intisar where pinnacle reefs are developed. The Intisar reefs have been studied by Terry and Williams and more recently by Brady et al. and Gumati. Well data shows that the Upper Sabil Shelf has an abrupt margin, which Brady regarded as a barrier reef. In the Intisar area the shelf margin forms a re-entrant about 25km long by 20km wide in which the Upper Sabil is drastically reduced in thickness and which is largely filled with shale. The pinnacle reefs developed within the re-entrant, starting as foraminiferal mounds on the surface of the Upper Sabil carbonate and developing into coral reefs completely surrounded by shale. Several facies have been identified, of which the most conspicuous are the lower algal-foraminiferal unit and an upper coral reef unit. About twenty pinnacle reefs have been found, most of which are oil bearing. The Intisar 'A' reef is 365m thick and about 6km in diameter and contains original oil reserves of 1.2 billion barrels. Porosity in the coral reef facies is 20 to 25%. The coral reef unit, which forms a cap to the pinnacle, contains coral colonies in growth position which are surrounded by a micitic matrix. The original aragonite of the corals has been removed by solution leaving extensive mouldic porosity.[26]

5.1.15   Harash Formation

The term Harash Formation was introduced by Barr and Weegar for a shale and limestone sequence overlying the Zaltan Formation, and originally formed part of the Jabal Zaltan Group (Figure 5.2). The type section was defined in the AA1-59 well on the Al Bayda Platform (Figure 5.5). In the type well the formation is composed of 88m of calcareous fissile shales, and soft chalky limestones with common *Operculina*, bryozoa and algae. On the Nasser field this formation includes two thin oil-bearing limestones. It is overlain by the Khayir Member of late Thanetian-early Ypresian age, and is therefore regarded as late Thanetian in age.[27]

The formation reaches its thickest development in the Zallah Trough where 120m are present. It thins to 30m on the Az Zahrah-Al Hufrah and Zaltan Platforms, 60m in the Maradah and Ajdabiya Troughs and 30m in the southeast Sirt Basin. The carbonate shelf facies continues southwards from the Al Bayda Platform onto the Southern Shelf. It is absent over most of the Sabil Shelf, probably having been removed by later erosion. In the area around Intisar it forms the seal for the Intisar reefs, and it is difficult to distinguish from the overlying shales of the Khayir Formation.[28]

The Harash Formation was studied in detail in the Intisar area by Spring and Hansen who demonstrated that the formation infills the irregular surface of the Upper Sabil shelf, and envelopes the Intisar reefs in the Intisar re-entrant. They showed that on the eastern margin of the re-entrant the Upper Sabil Shelf shows a steep, rimmed carbonate platform margin, whilst on the southern flank it shows a gentle carbonate ramp geometry. Five shallowing-upward carbonate sequences were recognised within the Harash Formation in the Intisar embayment, reflecting cyclic sea level changes, which can be correlated to the sea level curves published by Haq et al.[29]

5.1.16   Khayir (Kheir) Formation

The Khayir Formation (Kheir) was originally named by Burollet as a member of his Jabal Waddan Formation for a series of marls overlying the Ammur Member at outcrop in Wadi Ammur. The name was adopted by Barr and Weegar and raised to the status of formation, based on a subsurface type section in well E1-59, the Jalu field discovery well (Figure 5.5). Subsequent work on the outcrops led to Burollet's Khayir Member being incorporated into the Bishimah Formation. This has produced the anomaly of having a Khayir Member defined by a surface type section in western Libya, and a Khayir Formation defined by a subsurface type section in eastern Libya. They are approximate age equivalents, but the nomenclatural confusion needs to be addressed. The Khayir Member is discussed further below, in the section on Eocene stratigraphy.[30]

The Khayir Formation, as defined by Barr and Weegar (Figure 5.2), comprises 84m of marl, clay, fissile calcareous shale and thin stringers of micritic limestone. In the type well it directly overlies the Upper Sabil Member and is overlain by Eocene carbonates. The formation contains index fossils from both the late Thanetian and early Ypresian, so it clearly crosses the Palaeocene-Eocene boundary. This demonstrates a difference between the Khayir Formation and the Khayir Member since the latter contains no Palaeocene fossils and is entirely Ypresian in age. The Khayir Formation has a wide distribution in the Sirt Basin (Figure 5.5). It averages 60m in thickness in the southeastern Sirt Basin, and 90m in the Ajdabiya Trough, but is absent on the Sabil Shelf to the north of the Amal field due to subsequent erosion. Bezan provided evidence of the presence of redeposited Palaeocene material in several wells in the Ajdabiya Trough. In the central Sirt Basin it has a thickness of 60m on the Al Jahamah and Zaltan Platforms, where it is locally known as the Meghil Formation. On the Southern Shelf, south of the Dayfah field, a dolomite facies is developed within the Khayir Formation. In the western Sirt Basin it averages 60m on the Az Zahrah-Al Hufrah Platform and 90m in the Zallah Trough. The Khayir Formation represents an effective seal for the underlying carbonates within the Harash, Zaltan and Upper Sabil Formations over much of the basin.[31]

NORTHEAST LIBYA

5.1.17 Al 'Uwayliah Formation

In Al Jabal al Akhdar Palaeocene outcrops have been found in only a few scattered localities. The first, discovered by Barr in 1968, is in a small quarry 6km east of Al 'Uwayliah which was subsequently designated the type section for the Al 'Uwayliah Formation (Figure 5.2). This outcrop exposes only 6m of white chalk and greenish marl, in a tectonically disturbed area, in which neither the upper nor lower contacts are visible. The formation contains a good assemblage of planktonic foraminifera spanning the complete Thanetian stage (Figure 5.5). The age of the fauna has been confirmed in a detailed study by Eliagoubi. Nannofossil evidence also confirms that at the type locality the age is latest Palaeocene. Further east, at Jardas al Jarrari, 20m of chalky to microcrystalline limestone and greenish marls are exposed containing foraminifera of Danian age, overlying the dolomites of the Wadi Dukhan Formation (Figure 5.3). Another Palaeocene exposure has been found at Wadi al Athrun where a thin bed containing Thanetian foraminifera is present beneath the Apollonia Formation, which also contains reworked fragments of limestone from the Al Athrun Formation. In the subsurface it has been encountered in the C1-18 well (over 60m thick), in the offshore A1-NC 120 well, and in wells close to the Egyptian border. In well A1-NC 120 the age has been determined as Thanetian, possibly extending into the Montian, but Danian sediments appear to be missing at this location. From the meagre evidence available a tentative picture of Palaeocene geology can be reconstructed. It appears that the contact between the Cretaceous Wadi Dukhan Formation and the Palaeocene Al 'Uwaliyah Formation is conformable, but that it represents an abrupt change from the restricted environment of the Wadi Dukhan dolomites to the open marine conditions of the Al 'Uwaliyah Formation. It is likely that Danian deposition extended over much of northern Cyrenaica, and continued through the Montian and Thanetian, but early Eocene tectonic activity, named by Barr and Berggren the Cyrenaican orogeny, uplifted and folded the rocks of the Jabal al Akhdar, and much of the Palaeocene was removed by erosion. In the Ash Shulaydimah Trough the Palaeocene is largely represented by dolomites which formed as secondary replacement of limestone. It has been speculated that the large amounts of magnesium required for this process were derived from partially evaporated lagoons around the periphery of the embayment. Three stages have been identified in the dolomitization process.[32]

5.2   EOCENE

The pattern of sedimentation established during the Palaeocene continued into the Eocene, but was affected by significant tectonic events both in the western Sirt Basin and in Cyrenaica, reflecting the closure of the Tethys Ocean. The principal axis of deposition continued to shift to the east, and the whole of the Hamadah al Hamra became emergent. Eocene rocks outcrop extensively at Jabal Waddan, on the Az Zahrah-Al Hufrah Platform, in the Zallah Trough, and to the west of Al Haruj al Aswad. Further south there are extensive outcrops of Eocene littoral rocks which extend as far as the foothills of the Tibisti Massif. Eocene rocks are present in the subsurface throughout the Sirt Basin, where they form major hydrocarbon reservoirs. In Cyrenaica Eocene rocks outcrop on the northern flank of the Jabal al Akhdar, and continuous outcrops are present along the coast from Tulmaythah to Darnah. In the Sirt Basin the dominent lithologies are shallow-water carbonates, dolomites and evaporites. The ring intrusions of Jabal Awaynat have been radio-metrically dated as Eocene, and at Gharyan the earliest plateau lavas are also Ypresian in age, with the later phonolite domes yielding a Bartonian age. The Eocene is reviewed under four regional headings: western outcrops, north-western offshore, Sirt Basin subsurface, and northeastern Libya.[33]

WESTERN OUTCROPS

The outcrops north of Al Haruj al Aswad have been studied in detail, particularly during the mapping campaign for the Geological Map of Libya. A summary of this work are presented below.

The outcrops to the south of Al Haruj al Aswad, by contrast, have received only the most cursory attention. Eocene rocks outcrop over much of the Tibisti Arch, except where covered by Tertiary/Quaternary lavas and recent desert sands. Little has been published on these extensive outcrops. Desio briefly described gypsiferous shales and limestones in the Waw al Kabir area which contain pelecypods of Ypresian age. From the Jabal Nuqay area Fürst described a sequence which contains claystones and gypsum of possible Ypresian age, followed by fossiliferous claystones which contain foraminifera, echinoids and molluscs of Lutetian age. The most detailed work in this area was done by Savage and Wight who studied the Eocene outcrops in the Ash Sharit and Dur at

Talah areas, southeast of Al Haruj al Aswad (Figure 5.6). They showed that a full Eocene sequence is present in this area with a thickness of 210m, comprising a lower unit of gypsiferous shale containing pelecypods of Ypresian age, a middle claystone unit with a marine fauna of Lutetian age, and an evaporite unit of Bartonian/Priabonian age. A rich fauna of vertebrates, including whales, turtles, sirenians, crocodiles and fish has been recovered from the base of the evaporite unit. Patchy outcrops of these beds extend for 300km from Ash Sharit towards Jabal Nuqay.[34]

5.2.1 Bishimah Formation

The Bishimah Formation was first defined by Jordi and Lonfat for a series of marls, chalks and evaporites of Ypresian age, lying between the Shurfah and Al Jir Formations in the Wadi Zimam area on the western flank of the Hun Graben (Figure 5.6). Following the mapping for the Geological Map of Libya the formation has been divided into three members, the Khayir, Wadi Zakim and Rawaghah Members (Figure 5.2).

The nomenclatural confusion of the Khayir Member was mentioned in section 5.1.16. The Khayir Member at outcrop was first described by Burollet, and was later incorporated by Jordi and Lonfat into their Bishimah Formation. Barr and Weegar adopted the name for a subsurface formation in the eastern Sirt Basin which contains foraminifera of both Thanetian and Ypresian age. The type section of the Khayir Member was established by Jordi and Lonfat in the Wadi Khayir on the western flank of the Hun Graben (Figure 5.6). At this location 32m of chalky and gypsiferous marl with thin dolomitic limestones are exposed, with a poor fauna of echinoids, gastropods, foraminifera including *Operculina*, and dasycladacean algae. The fauna is dominently Ypresian in character, but the member may extend into the late Palaeocene, as

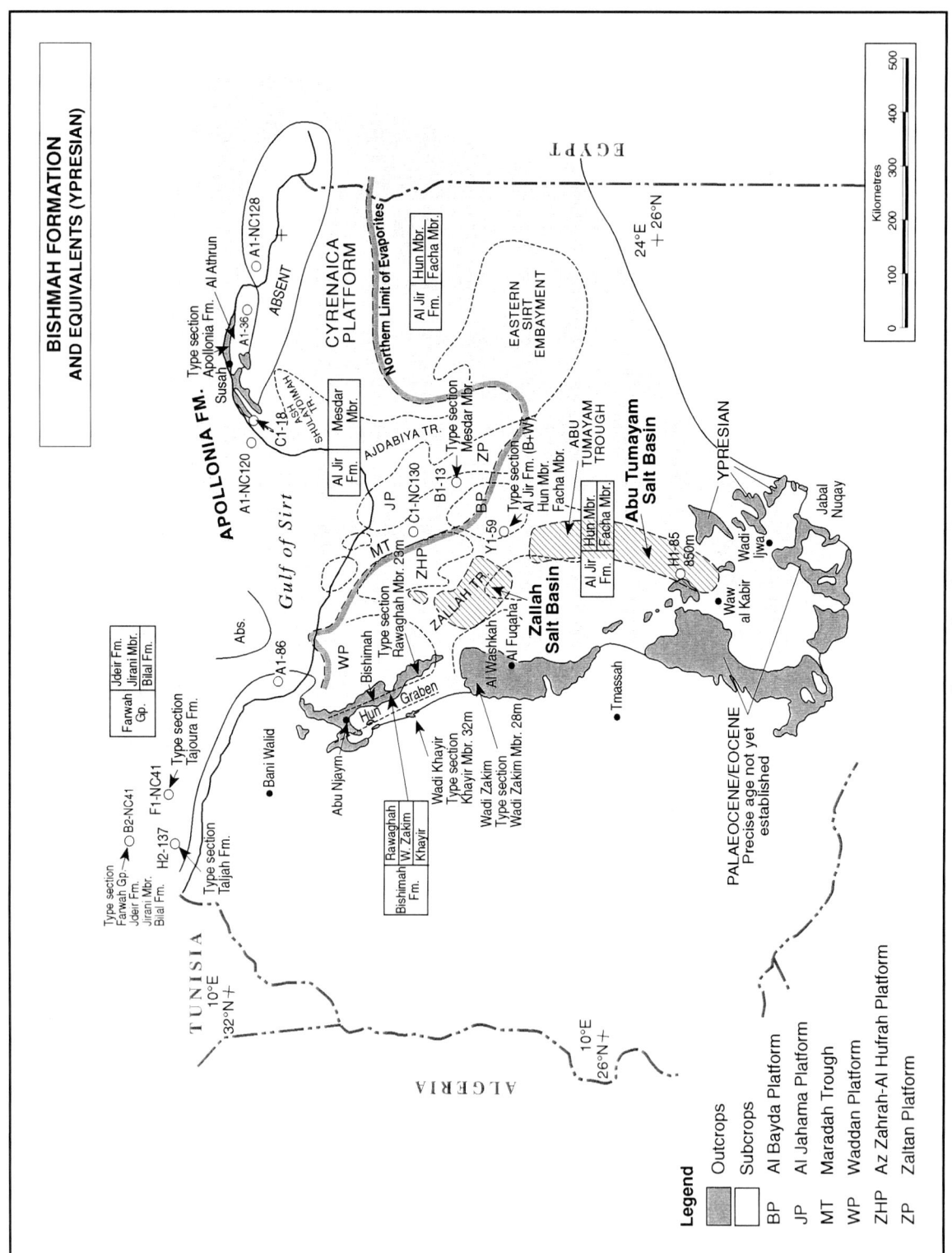

BISHMAH FORMATION AND EQUIVALENTS (YPRESIAN)

Sources: Distribution: Geological Map of Libya, 1985, Abugares, 1996, Wennekers, et al., 1996

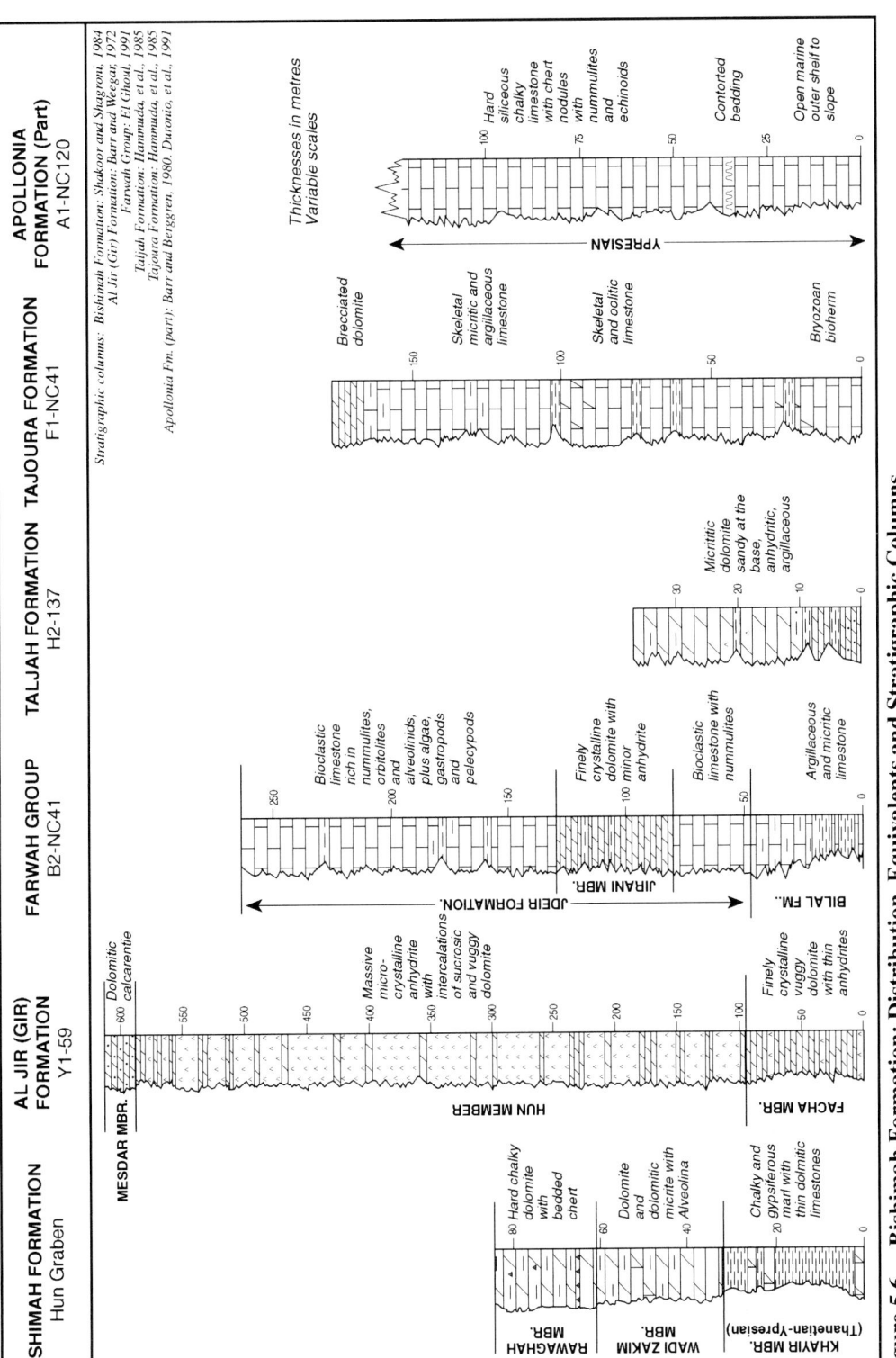

Figure 5.6   Bishimah Formation: Distribution, Equivalents and Stratigraphic Columns

The Bishimah Formation is defined from outcrops around the Hun Graben. In the subsurface it is age-equivalent to the Al Jir Formation as defined by Barr and Weegar, which contains thick evaporites in the western Sirt Basin, and passes into an open shelf environment further north. Offshore the Farwah Group contains the Jdeir Formation which is the main reservoir on the Bouri field. The Taljah and Tajoura Formations are nearshore equivalents. In Cyrenaica the Apollonia Formation is mainly Lutetian in age, but sedimentation began in the early Ypresian in the A1-NC 120 well. The Facha Member is a significant oil reservoir in the western Sirt Basin.

it does in the subsurface of the Sirt Basin. The member was deposited in a neritic to littoral environment, becoming brackish in some areas.[35]

The Wadi Zakim Member, previously known as the *Flosculina* Limestone, was first described by Burollet and subsequently revised by Fürst. The unit was renamed by Woller and a type section was described from Wadi Zakim, 40km south of Hun (Figure 5.6). At the type locality 28m of dolomite, dolomitic micrite and chalky limestone are exposed, with a 4m interval crowded with several species of *Alveolina*. It also contains planktonic foraminifera which confirm a Ypresian age. The member was deposited in shallow marine conditions in a neritic to littoral setting.[36]

The Rawaghah Member forms the upper unit of the Bishimah Formation. It was first described by Burollet from an outcrop on the Jabal Waddan where 23m of chalky dolomite overlie the Wadi Zakim Member. The dolomite is hard and massive and contains numerous chert nodules and lenses of silicified dolomite. It is fossiliferous and occasionally brecciated. The unit contains a fauna of shallow-water fossils including pelecypods and abundant foraminifera of Ypresian age, and was deposited in a lagoonal to littoral environment. The low Mg content of the Rawaghah dolomites suggests that the original composition was probably aragonite, which was dolomitized by seepage of concentrated brine from the overlying Al Jir evaporites.[37]

The Bishimah Formation has a wide distribution from Bani Walid in the north to Tmassah in the south (Figure 5.6). Where fully developed, in the Abu Njaym, Hun and Al Washkah areas, the thickness averages 100m. In the northern area at Bani Walid and Al Qaddahiyah it is not possible to differentiate the three members and the formation is composed of 30 to 35m of thick-bedded marly and cherty limestones. The Khayir Member reaches a maximum thickness of 32m in the Abu Njaym and Hun areas where it is made up of marl with thin dolomitic limestones. Futher south it passes into dolomitic limestone with traces of gypsum, and is reduced in thickness to about 12m. The Wadi Zakim Member reaches 50m in thickness at Hun where it is composed of dolomitic micrite and marly dolomite. It is truncated by erosion southwards, and towards Al Haruj al Aswad it is reduced to 2m of chalky dolomite. The Rawaghah Member is best developed in the Abu Njaym and Al Washkah areas where it reaches a thickness of 55m. It is characterised by the presence of chalk and chalky dolomite with masses of tabular chert. It is truncated by erosion in the Al Fuqaha and Al Haruj al Aswad areas.

The extensive outcrops in the area between Waw al Kabir and Jabal Nuqay have been mentioned above. These exposures have not been studied in detail and their precise age has not yet been determined. Wight regarded them as predominently Palaeocene and Lower Eocene in age, but the Geological Map of Libya shows them as undifferentiated Eocene. Outcrops fringing the volcanic rocks north of Wadi Ijwa however consisting of 65m of gypsiferous shales and limestones with pelecypods have been dated as Ypresian by Fürst and for the sake of convenience all these exposures are shown on figure 5.6. Further palaeontological control is required in this area.[38]

5.2.2 Al Jir Formation (sensu IRC)

The term Al Jir Formation is fraught with problems. It was first introduced by Burollet as the Gir Gypsum for a series of rocks overlying the Bishimah Formation, and is named after the Jabal al Jir, southeast of Waddan. The formation was inadequately defined and no type section was proposed. Barr and Weegar regarded Burollet's name as invalid, and they adopted the name Gir Formation (new transliteration Al Jir) for a series of rocks overlying the Khayir

Formation in the subsurface of the Sirt Basin. They regarded Burollet's Gir Gypsum as "probably equivalent to at least the upper portion of the Gir Formation". In 1977, during the mapping for the Geological Map of Libya Mijalkovic renamed Burollet's Gir Gypsum as the Al Jir Formation and described a composite section from the Al Qaddahiyah area, 250km north of the Jabal al Jir, where no significant evaporites are present (Figure 5.8). Finally, during mapping of the Hun area, Shakoor and Shagroni described a section in the Wadi Faras near Waddan, 20km north of the Jabal al Jir, where gysum and anhydrite are both present. There are obvious similarities between the sequence exposed at Wadi Faras and the subsurface Al Jir Formation of Barr and Weegar, and most authors have regarded them as correlative. However the fact remains that at outcrop the Al Jir Formation is Lutetian in age and is underlain by 75m of Ypresian Bishimah Formation, whilst the Al Jir Formation of Barr and Weegar is Ypresian in age and is underlain by Thanetian-earliest Ypresian Khayir Formation. There is an urgent need to resolve these problems.[39]

Burollet included two members within the Al Jir Formation, and this arrangement has been retained by the surveyors of the Geological Map of Libya. The lower unit, the Bin Isa Member, was named by Burollet from a locality north of Abu Njaym (Figure 5.8). Here the Bin Isa Member comprises 24m of hard, compact chalk and chalky limestone containing chert nodules and thin stringers of gypsum. The member thins and pinches out towards the southwest. The upper unit was named the *Orbitolites* Limestone by Burollet, but this unit had previously been named the Bir Ziden Formation by Desio, and it is now known as the Bi'r Zaydan Member. It too is named after a well in the area north of Abu Njaym (Figure 5.8), and is composed of 27m of compact white limestones containing *Orbitolites complanatus* and calcareous algae.

The two-fold subdivision of the Al Jir Formation is only recognisable over a small area north of Abu Njaym. Elsewhere the Al Jir Formation is undivided.[40]

The Al Jir Formation outcrops over much of the Jabal Waddan, and on the western part of the Az Zahrah-Al Hufrah Platform (Figure 5.8). The average thickness is about 60m. The formation shows a complex inter-relationship of evaporitic and shallow-shelf facies, with the evaporitic facies being best developed in the southern part of the Jabal Waddan, from where Burollet first named the Gir Gypsum. Nearby, at Wadi Faras, near Jabal al Jir, the formation comprises 58m of anhydrite, dolomite, and gypsum with thin siliceous limestones and marls. Further north on the Jabal Waddan the formation passes into a shallow, restricted-marine environment dominated by marly and chalky limestones, cherty dolomites and thin bands of gypsum and anhydrite. On the Az Zahrah-Al Hufrah Platform the outcrops are all in an open-marine facies with fossiliferous, chalky and siliceous limestones with a rich fauna of pelecypods, bryozoa, echinoids, foraminifera and calcareous algae. The foraminifera give a clear Lutetian age for this formation.

In the far south a few isolated outcrops at Ash Sharit and Bi'r al Ma'ruf near the southeastern limit of Al Haruj al Aswad have been dated as Lutetian, and can be assigned to the Al Jir Formation. These outcrops were described by Fürst and comprise about 75m of claystones and siltstones containing both marine and non-marine fossils.[41]

5.2.3 Wadi Thamat Formation

The Wadi Thamat Formation was introduced by Desio, and subsequently modified by Burollet, for the late Eocene sequence overlying the Al Jir Formation and underlying the Oligocene Ma'zul Ninah Formation (Figure 5.2). A type section was designated in

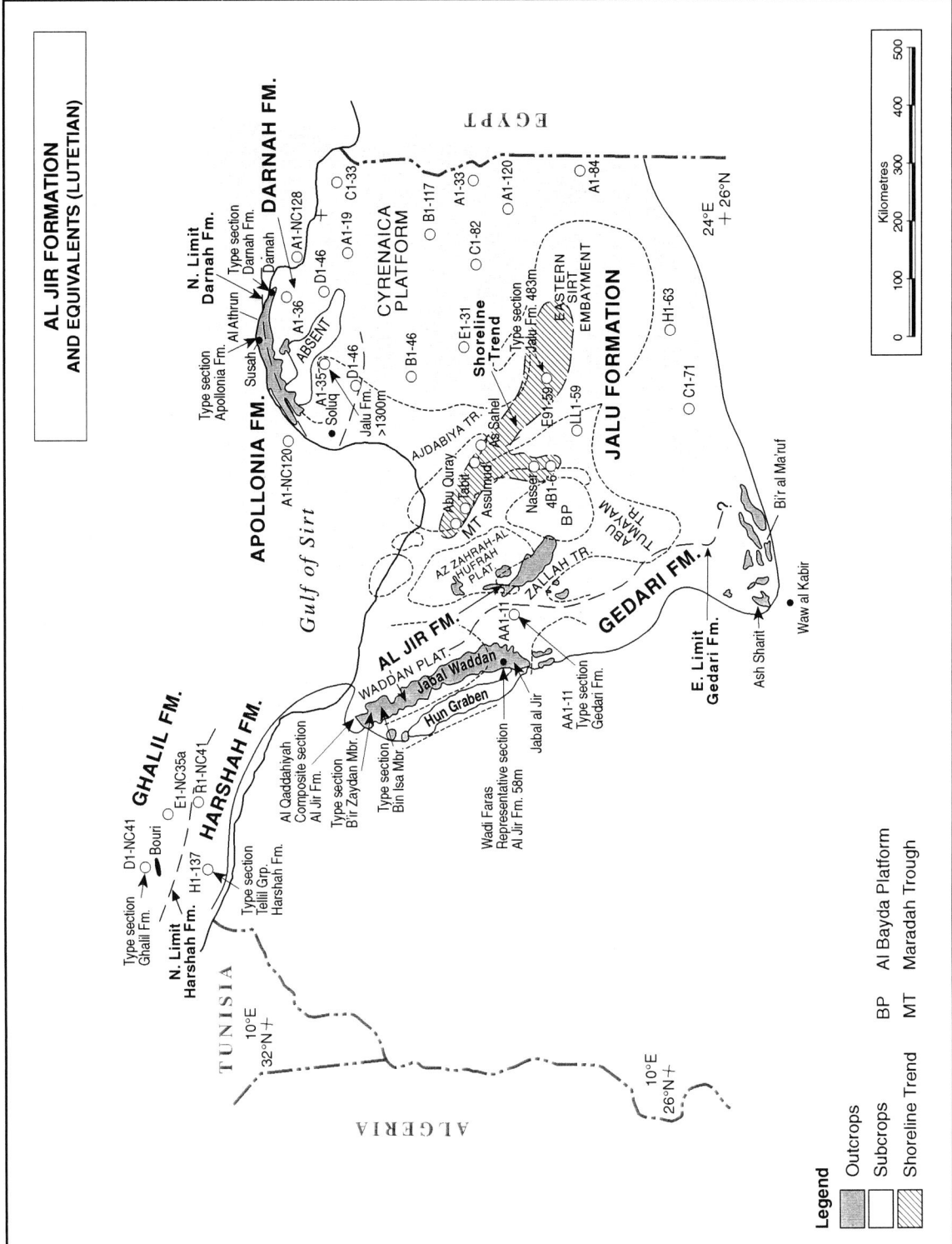

AL JIR FORMATION
AND EQUIVALENTS (LUTETIAN)

Sources: Distribution: Geological Map of Libya, 1985, Wennekers, et al., 1996

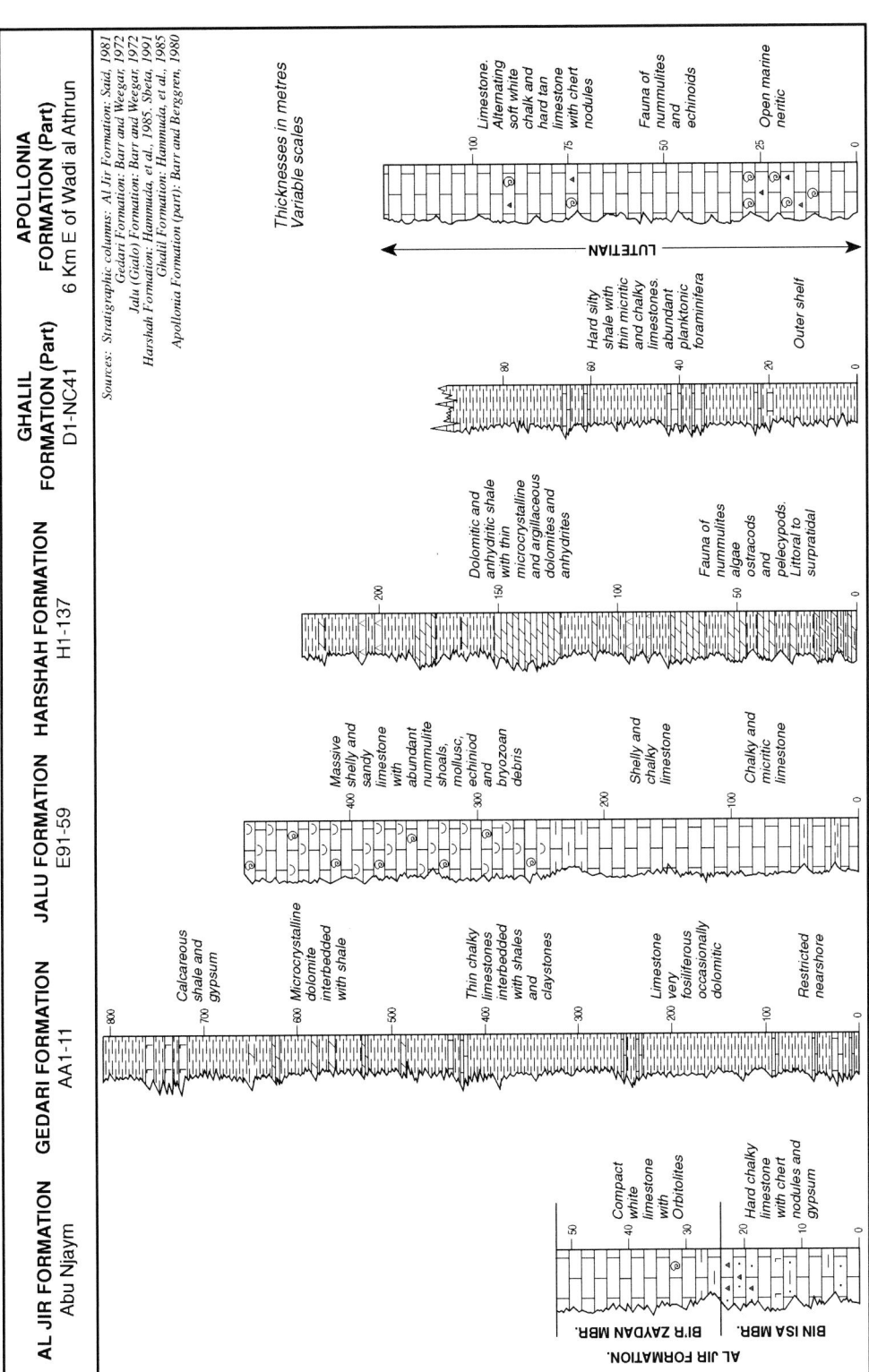

Figure 5.8   Al Jir Formation: Distribution, Equivalents and Stratigraphic Columns

Al Jir Formation was defined from the outcrops of the Jabal Waddan. It is not age equivalent of the subsurface Al Jir (Gir) Formation as defined by Barr and Weegar. In the Zallah Trough, and as far south as Ash Sharit, it is equivalent to the Gedari Formation, and in the Sirt Basin it is equivalent to the Jalu (Gialo) Formation. Offshore northwest Libya the Harshah Formation represents a shallow water facies which passes basinward into the outer shelf facies of the Ghalil Formation. In Cyrenaica the Apollonia Formation passes shorewards into the littoral Darnah Formation. The Jalu Formation is a major petroleum reservoir in the Sirt Basin.

the Wadi Thamat on the northern edge of the outcrop (Figure 5.9). The type section was not well chosen since the formation develops a much more distinctive character further south. Subsequent work in the Zallah area allowed the formation to be subdivided into three members, from bottom to top the Al Gata, Thmed al Qusur and Qararat al Jifah Members (Figure 5.2), and this scheme was adopted during the mapping for the Geological Map of Libya.[42]

The Al Gata Member was named by Goudarzi for a series of marls, dolomitic limestones and micritic limestones containing oyster-rich coquinas, on the eastern flank of the Jabal Waddan (Figure 5.9). The thickness at the type locality is about 75m. The member thins to the north and thickens to the south. The Al Gata Member contains a fauna of pelecypods, gastropods, echinoids and foraminifera which suggest a late Lutetian age for this unit. It was deposited in a neritic, shallow-marine environment.[43]

The Thmed al Qusur Member was named by Burollet from a locality in the Zallah Trough for a distinctive unit of white chalk and chalky limestone with chert nodules, about 15m thick (Figure 5.9). It contains moulds of the large gastropod *Rostellaria*, plus a poor fauna which is not age diagnostic. On the basis of stratigraphic position the member is probably Priabonian in age. It represents a littoral to lagoonal depositional setting.[44]

The uppermost member, named the Qararat al Jifah Member by Goudarzi, is based on a composite type section in the Zallah Trough (Figure 5.9). It is composed of 128m of fossiliferous and coquinoid limestone, dolomitic limestone with traces of gypsum, and greenish marl. The limestone is richly fossiliferous, particularly with oysters and nummulites, and also contains some vertebrate remains. The presence of *Nummulites fabiani* gives a late Lutetian age for the Qararat al Jifah Member.[45]

The Wadi Thamat Formation outcrops over the eastern Jabal Waddan, the Zallah Trough, the Az Zahrah-Al Hufrah Platform and the Al Bayda Platform (Figure 5.9). The three-fold subdivision is present across the entire outcrop, except in the Abu Njaym area. The average thickness is 150m. The most distictive feature is the white chalky limestone of the Thmed al Qusur Member, which is often highly porous and cavernous. It has a thickness of 12-16m and contains a fauna of gastropods and pelecypods. Occasionally it passes into a saccharoidal dolomite. The Al Gata Member is less distinctive. It has a highly variable lithology, from marls, clays and bedded gypsum at Zallah to fossiliferous limestones, coquinas and chalky limestones at Abu Na'im. In the Wadi bu ash Shaykh area it contains numerous quartz geodes measuring up to 20cm in diameter. The thickness reaches 80m at Abu Na'im. The most extensive outcrops are formed by the thick upper member, the Qararat al Jifah Member. This unit is frequently over 100m thick and reaches 140m near Zallah. It is composed of alternating beds of calcarenite, shelly limestones, coquinas full of *Exogyra*, marls, dolomites and gypsum. It contains a fauna of molluscs, nummulites, gastropods, and in the Maradah area, vertebrate bones. Overall the foraminiferal fauna gives a late Lutetian age for the Wadi Thamat Formation. The area around the Al Hufrah field was studied in detail by Anketell and Kumati who regarded the Eocene stratigraphy in this area as sufficiently distinct to warrant the establishment of a new set of formations and members, only one of which has been formally described. The new units reflect local differences in the Al Hufrah area.[46]

NORTHWEST LIBYA OFFSHORE

Farwah Group

In the northwestern offshore of Libya the Eocene was divided by Hammuda et al. into two groups, a lower Farwah Group, which includes the Bilal, Taljah and Jdeir Formations and the Jirani Member, and an upper Tellil Group which includes the Harshah, Dahman and Samdun Formations. Comprehensive reviews of both groups were subsequently presented by Sbeta. He showed that the Farwah Group was deposited on an eroded surface ranging from Palaeocene in the Bouri area to progressively older rocks further south, ending with pre-Santonian volcanics in wells G1-137 and H1-137. The stratigraphy of the Farwah Group was later revised by El Ghoul.[47]

5.2.4 Bilal Formation

The Bilal Formation was established by Hammuda, et al. for a series of shallow-water sediments overlying the Al Jurf Formation in the Sabratah Basin. A type section was selected in the B2-NC 41 well where Hammuda et al. described a 135m section composed of argillaceous, micritic and skeletal limestones with traces of chert, and an upper bed of dolomite (Figure 5.6). Subsequently El Ghoul revised the type section. He showed that the lower 44m were lithologically more akin to the underlying Palaeocene section, and contained a Thanetian fauna, and that the upper 45m is attributable to the Jdeir Formation. The revised Bilal Formation type section is therefore reduced to a thickness of 46m and the lower marly section and the upper carbonate units are now excluded. The formation contains a rich fauna of large foraminifera which includes *Orbitolites, Fasciolites*, and *Nummulites*, and the rotalid *Lockhartia*, indicating deposition on an open shallow shelf. The planktonic foraminiferal fauna of the revised formation now indicates an exclusively Ypresian age.

The Bilal Formation has a wide distribution in the Sabratah Basin with a variety of different lithologies (Figure 5.6). A map showing its distribution and facies (although before the El Ghoul revision) was published by Sbeta. Most typically it is represented by argillaceous micritic limestones rich in planktonic foraminifera with rare phosphate nodules. The type section is not a good illustration of this facies. It reaches a maximum thickness of 213m in well H1-NC 41, but thins towards the south before pinching out south of the K1-137 and J1-137 wells. A shoreline facies of coarse sandstones and conglomerates is developed in the G1-137 and H1-137 wells. Further east the Bilal Formation is present in concession NC 35a, where the average thickness is 70m. Microfossils of mid-Ypresian age have been found at the top of the Bilal Formation in well F1-NC 35a. Hammuda et al. equated the Bilal Formation with the Taljah, Tajoura and Hallab Formations in the Sabratah Basin, but El Ghoul demonstrated that the Hallab Formation is Bartonian in age and recommended the suppression of this name. More generally the Bilal Formation is equivalent to the lower part of the Bishimah Formation of the Hamadah al Hamra, to the Khayir Formation and possibly the Facha Member of the Sirt Basin, and to the Chouabine Formation of Tunisia, which also contains abundant phosphatic nodules.[48]

5.2.5 Jdeir Formation

The Jdeir Formation was established by Hammuda et al. for a skeletal carbonate, rich in nummulites, which forms the principal hydrocarbon reservoir in the Sabratah Basin (Figure 5.2). A type section was selected in well B2-NC 41 on the Bouri field (Figure 5.6). In this well the Jdeir Formation is composed of two units, a lower bioclastic limestone rich in nummulites, *Orbitolites* and alveolinids, and an upper unit of fossiliferous bioclastic and micritic limestone containing numerous nummulites, calcareous algae, gastropods, pelecypods and ostracods. The nummulites are so numerous that they form highly porous nummulite shoals. The total thickness in the type section is 131m. The type section was subsequently revised by El Ghoul, on the basis that the Jirani Member forms a tongue within the Jdeir Formation, and that the 45m below the Jirani Member should be included within the Jdeir Formation rather than the Bilal Formation. The revised thickness of the formation in the type well is 214m, including the Jirani Member (Figure 5.6). The formation contains a rich fauna of nummulites and other benthonic foraminifera which suggest a late Ypresian age for this formation. It was deposited on a shallow ramp margin, in water depths probably between 30m and 60m. A review of the Jdeir Formation on the Bouri field was published by Bernasconi, et al. The crest of the Bouri structure is occupied by a nummulitic bank, with two other less extensive nummulitic banks below, formed during temporary marine regressions. The initial site of the shoal was determined by a swell caused by an underlying Jurassic salt wall. The Jirani Dolomite and the Jdeir Formation nummultic banks represent excellent hydrocarbon reservoirs with porosity values between 10 and 25%, and averaging 16%.

The Jdeir Formation is present over a wide area in the Sabratah Basin (Figure 5.7). Sbeta published an isopach map of the Jdeir Formation showing a maximum thickness of over 200m in the E1-NC 41 and L1-NC 35a wells. It thins towards the southwest to only 37m in well F1-137, and passes into the Taljah Formation south of well P1-NC 41. Further east it averages 180m on concession NC 35a, towards the shelf edge. It is not present on the Jarrafa Arch in wells F1-NC 35a and B1-NC 35a, nor in well M1-NC 41 and an arcuate belt extending towards well L1-137. In general the facies changes from a shelf-edge environment in the Bouri area through a middle shelf facies in the C1-NC 41 area to the sabkha facies of the Taljah Formation in the region of the A1-137 well. The mid-shelf unit consists of pelletal and bioclastic limestones with traces of dolomite, and a reduced abundance of nummulites, and was deposited in a low-energy nearshore to restricted environment. The shelf-edge unit contains banks of nummulites, dasycladacean algae and echinoderm fragments. Present day *Nummulites* (and *Operculina*) are known to thrive in a middle to lower ramp environment in water depths of around 50m, and Eocene forms are assumed to have occupied a similar ecological setting. A study of the nummulitic banks on the Hasdrubal field in Tunisia interpreted the environment of deposition as a storm-dominated mid-ramp setting.[49]

Hammuda et al. introduced the Jirani Dolomite as a new stratigraphic term for a unit underlying the Jdeir Formation without defining its status (Figure 5.6). A type section was chosen in the B2-NC 41 well, the same well as the Jdeir Formation type section. The Jirani Dolomite was defined as a 38m sequence of finely-crystalline dolomite, with minor anhydrite, which is often argillaceous. Locally it passes into dolomitic limestone with a fauna of shallow-water foraminifera, ostracods and algae. It represents a high stand during which restricted and emergent conditions prevailed. The sedimentology of the Jirani Dolomite has been studied by Giaj-Via, et al. They recognised a cyclic sequence of lagoonal and sabkha deposits with excellent porosity in the diagenetic cap. It has limited distribution mainly to the south of the Bouri field, westwards to B1-137, eastwards to E1-NC 35a and southwards to H1-NC 41. It pinches out before reaching well H1-NC 35a (Figure 5.7). El Ghoul assigned this unit as a member within the Jdeir Formation.

The Jdeir Formation is equivalent to the Wadi Zakim and Rawaghah Members onshore and to the Hun and Mesdar Members of the Sirt Basin. In Tunisia it is equivalent to the carbonates of the Metlaoui Formation. The Metlaoui Formation was deposited on a ramp, and passes notheastwards into the basinal

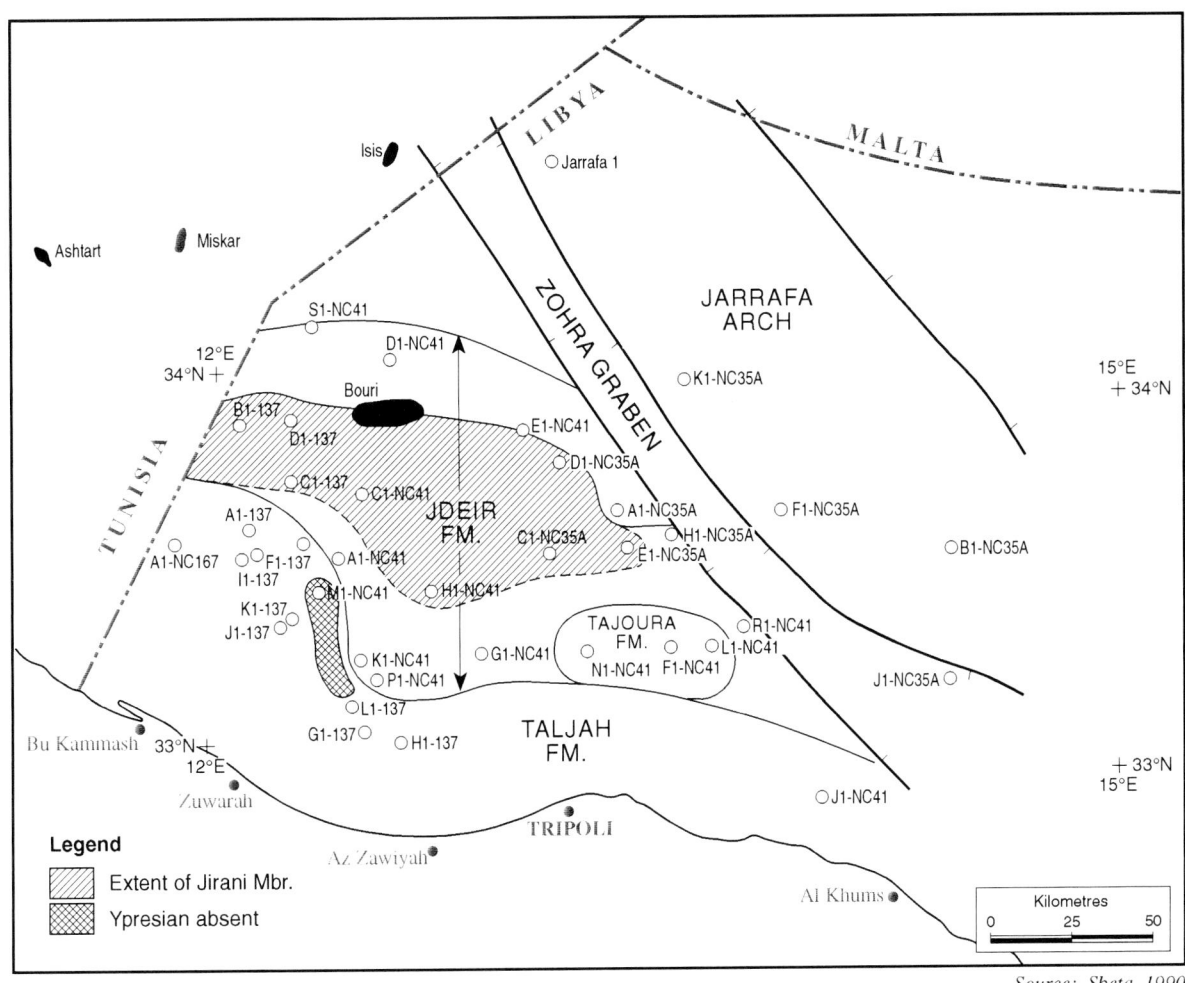

Source: Sbeta, 1990

Figure 5.7   Jdeir Formation, Offshore: Distribution and Equivalents

This map provides details which it was not possible to show on Figure 5.6. The Jdeir Formation, of Ypresian age, is the offshore equivalent of the Bishimah Formation. Three main facies are developed. The Jdeir Formation, which is the principal reservoir in the Sabratah Basin, is a nummulitic ramp sequence. It includes a dolomite interval named the Jirani Member. The Taljah Formation represents a nearshore sequence of micitic dolomite and anhydrite. The Tajoura Formation represents a localised bryozoan-oolite shoal. Columnar sections are shown on Figure 5.6.

sediments of the Bou Dabbous Formation. Several facies zones are recognisable on the ramp: sabkha (Faid Formation), restricted lagoon (Ain Merhotta Formation), and high energy nummulite shoal (El Garia Formation). The El Garia Formation is the Tunisian equivalent of the Jdeir Formation. Excellent outcrops of the nummulitic trend are present on the Kesra Plateau, northwest of Kairouan in Tunisia where a full range of facies is present from back-bank to forebank and storm deposits. The outcrops are closely analogous to the nummulitic trend in the Sabratah Basin.[50]

5.2.6 Taljah Formation

The Taljah Formation (Figure 5.2) represents a nearshore dolomitic equivalent of the Jdeir formation. It was first defined by Hammuda et al. in the H2-137 well on the southern flank of the Sabratah Basin (Figure 5.6). In the type well it comprises 37m of micritic dolomite, sandy at the base, anhydritic in the middle and argillaceous towards the top. In the type section the formation is unfossiliferous, but since it lies between the Bilal Formation below and the Harshah Formation above, it is assumed to be a shoreward equivalent of the Jdeir Formation. The contact with the Bilal Formation is unconformable. The formation is limited in extent and is confined to the southern margin of the Sabratah Basin in the vicinity of wells H1 and H2-137 and G1-137, where it represents a southern facies of the Farwah Group (Figure 5.7). A map showing the distribution of the Taljah Formation was presented by Sbeta. It is equivalent to the Faid Formation in Tunisia, which has a similar restricted and evaporitic character.[51]

5.2.7   Tajoura Formation

In a small area around well F1-NC 41 in the southeastern part of the Sabratah Basin the equivalent of the Jdeir Formation is represented by shallow-water carbonates which form a bryozoan bioherm. The Tajoura Formation was established by Hammuda et al. with a type section in the F1-NC 41 well (Figure 5.6). The lower 30m of the formation form the bioherm which is full of bryozoan fragments. The middle part of the formation is composed of skeletal and oolitic limestone, with bryozoa, ostracods, pelecypods, echinoids, algae and foraminifera. The upper part is predominantly dolomitic with some brecciated zones, interbedded with marl and shale. The total thickness is 175m (Figure 5.6). The lower contact is conformable

with the underlying Bilal Formation, and the upper contact is unconformable with the overlying Harshah Formation.

The Tajoura Formation has a very limited distribution, and is confined to the region encompassed by wells F1, F2-NC 41, N1-NC 41 and L1-NC 41 (Figure 5.7). A map showing its distribution was presented by Sbeta. Despite the biohermal character of the formation, no hydrocarbons have yet been found within it.

The fauna of the Tajoura Formation is not age diagnostic, but on the basis of stratigraphic position it is assumed to be Ypresian in age. It passes northwards into the Jdeir Formation.[52]

Tellil Group

The Tellil Group represents the younger of the two groups established by Hammuda et al. for the Eocene of the northwestern offshore. It includes four formations of Lutetian to Priabonian age. A comprehensive review of the Tellil Group was published by Sbeta.[53]

5.2.8   Harshah Formation

The Harshah Formation was established by Hammuda et al. for a series of dolomitic limestones, marls, dolomites and anhydrites found in the H1-137 well on the southern margin of the Sabratah Basin, which was designated the type well (Figure 5.8). The sequence lies between the Taljah Formation below and the distinctive Dahman Limestone above. The formation comprises 230m of alternating carbonates, marls and evaporites, deposited in a littoral to supra-tidal environment. The marls occasionally contain lignite, and anhydrite is common.

The Harshah Formation is present over the entire southern Sabratah Basin. Its thickness varies rapidly from 361m in well F1-NC 41 to 143m in well G1-NC 41, but averages about 230m (Figure 5.8). It pinches out southwards before reaching the present-day coastline, and it passes into the deeper-water shales of the Ghalil Formation to the north, before reaching the Bouri area. Westwards the formation passes into the evaporites of the Djebs Formation in Tunisian waters. The Harshah Formation was deposited on a ramp, and five progressively deeper-water facies zones have been recognised, ranging from a supra-tidal belt of dolomitic and anhydritic shales to skeletal micrites rich in nummulites, molluscs and echinoids. The formation contains good quality potential reservoir rocks, although no hydrocarbon accumulations have yet been encountered. The fauna of nummulites, algae, ostracods and pelecypods, is not age diagnostic, but on stratigraphic position it can be assigned to the Lutetian. The Harshah Formation is age equivalent of the Al Jir Formation of west Libya. Onshore in Tunisia it equates to the shelly limestones of the lower Cherahil Formation.[54]

5.2.9 Dahman Formation

In the H1-137 well the Harshah Formation is capped by a distinctive chalky and nummulitic limestone which Hammuda et al. called the Dahman Limestone (Figure 5.2). It was subsequently elevated to the status of a formation by Sbeta. It has a thickness of only 23m in the type well but thickens to 64m in well K1-137 (Figure 5.9). It has been likened to the carbonate shoals which occur in the Jalu Formation in the Sirt Basin which form important hydrocarbon reservoirs. It is best developed in the K1/J1-137 area and thins rapidly to the east where it rarely exceeds 25m in thickness. Northwards it passes into the shales of the Ghalil Formation. In Tunisia it is probably equivalent to the Reineche Member of the Cherahil Formation, which has similar characteristics. The Reineche Member is known to be mid-Lutetian in age.[55]

5.2.10 Samdun Formation

The Samdun Formation was named by Hammuda et al. for a sequence of soft greenish shales and clays with minor siltstones, dolomites and anhydrites (Figure 5.2). It was deposited in a shallow, nearshore environment with considerable clastic input. It is a shallow-water equivalent of the Ghalil Formation. The type section is in the H1-137 well, on the southern margin of the Sabratah Basin (Figure 5.9). In the type well the Samdun Formation is composed of 337m of alternating shales, clays, and thin stringers of dolomite and anhydrite. The shales are frequently silty and the dolomites are microcrystalline. It is not easy to differentiate the Samdun Formation from the Harshah Formation where the Dahman Limestone is not present. The formation ranges in thickness from 369m in H1-NC 41 to 160m in well I1-137, and pinches out before reaching the present-day coastline. Northwards it passes into the deeper-water shales of the Ghalil Formation (Figure 5.9). The fauna is not age diagnostic but the formation is capped by the *Nummulites vascus* marker bed which marks the base of the Oligocene, so the Samdun Formation is assigned a late Lutetian to Priabonian age. In Tunisia the Samdun Formation is equivalent to the upper part of the Cherahil Formation.[56]

5.2.11 Ghalil Formation

The deeper-water equivalent of the Harshah, Dahman and Samdun Formations is the Ghalil Formation, named by Hammuda et al. from a type section in the D1-NC 41 well, located in the centre of the Sabratah Basin, north of the Bouri field (Figure 5.2). Here the Ghalil Formation is composed of 184m of hard silty shale grading to soft silty marl. Thin bands of micritic and chalky limestone are present, particularly in the lower part. The formation was deposited in relatively deep water in an outer

234

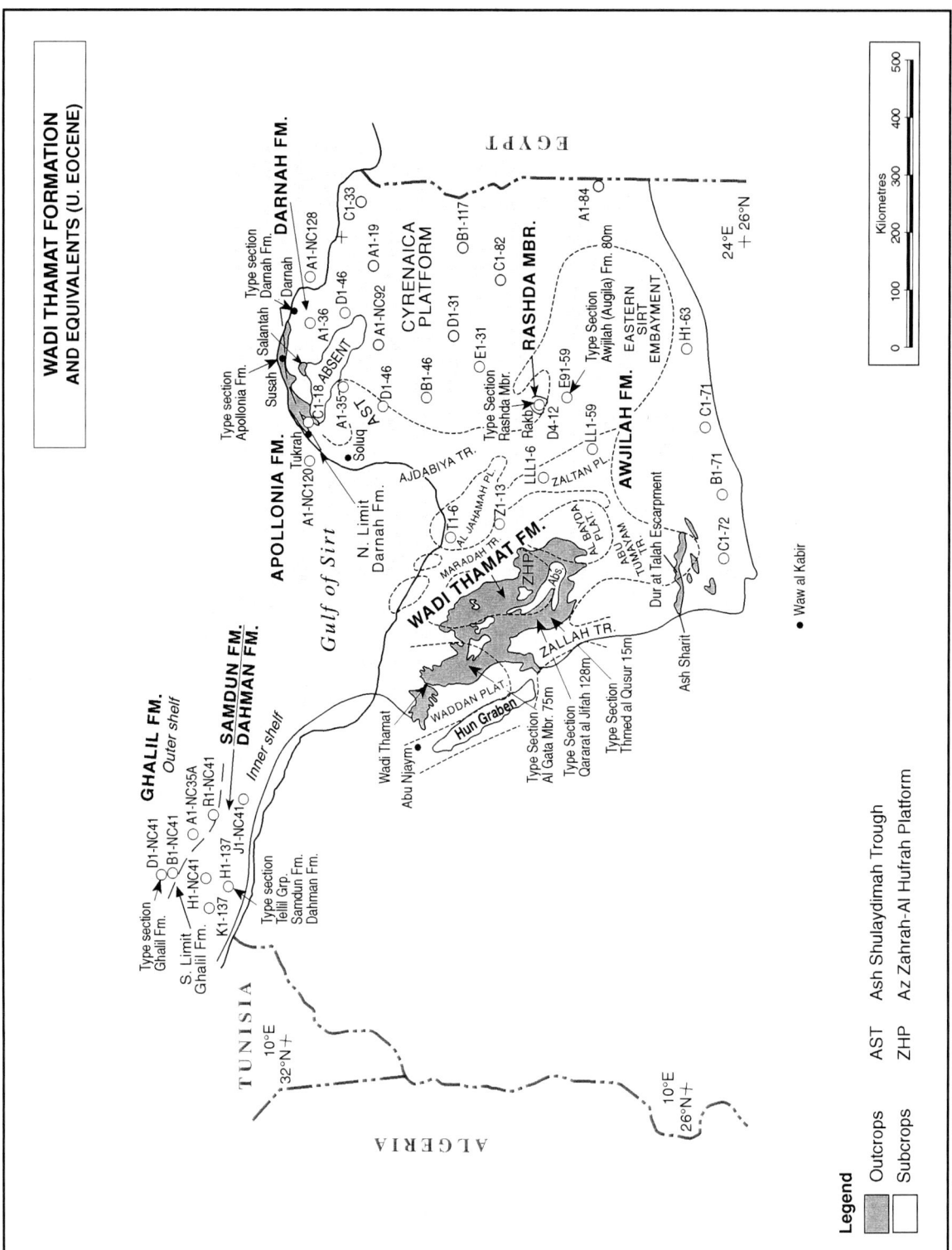

Sources: Distribution: Geological Map of Libya, 1985, Wight, 1980, Wennekers, et al., 1996

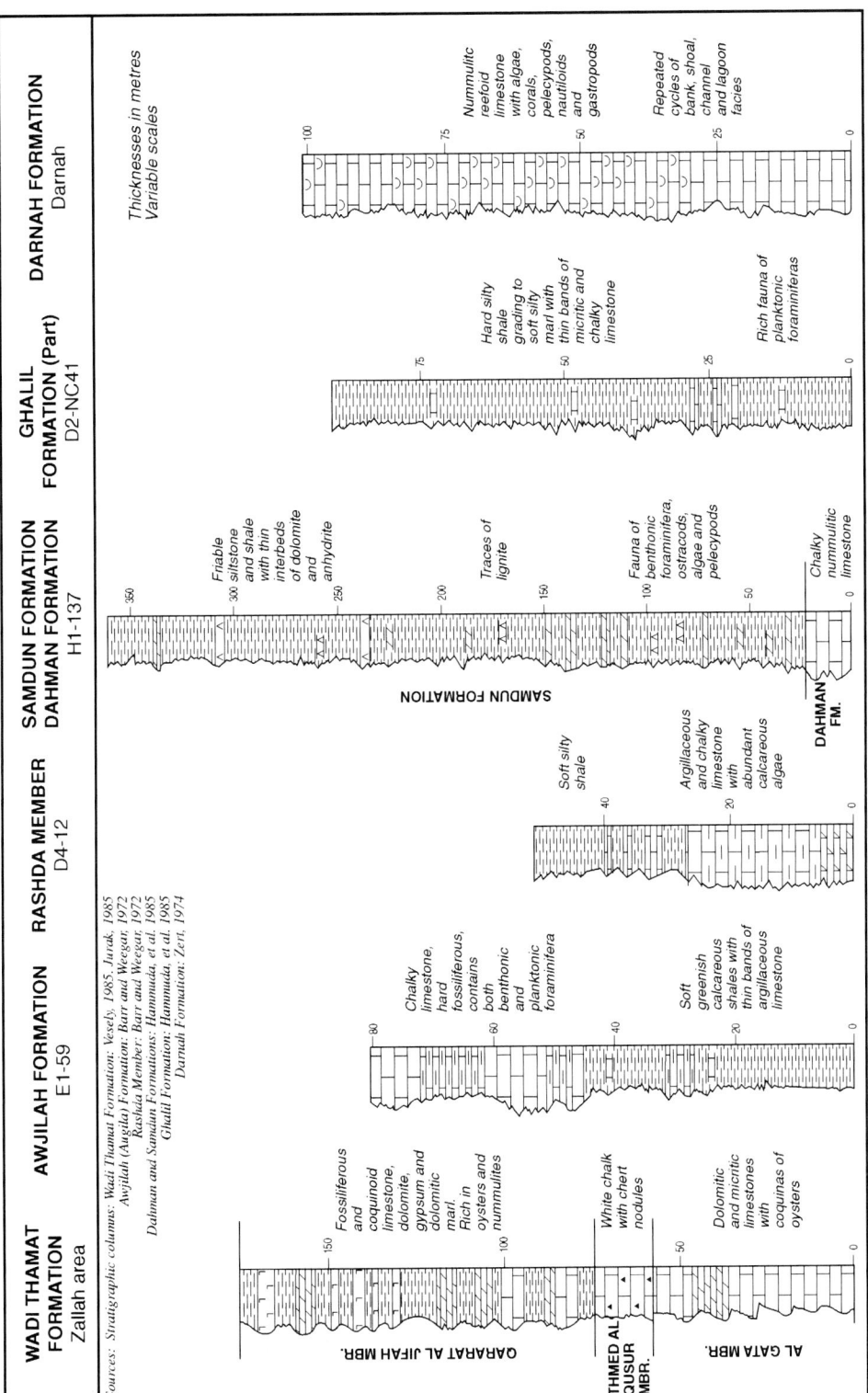

Figure 5.9   Wadi Thamat Formation: Distribution, Equivalents and Stratigraphic Columns

The Wadi Thamat Formation was described from the extensive outcrops in the Zallah Trough and Az Zahrah-Al Hufrah Platform. In the Dur at Talah outcrops a rich fauna of vertebrates, including whales and crocodiles, has been recovered. In the Sirt Basin shales and micrites of the Awjilah Formation are age equivalents of the Wadi Thamat Formation. The northwestern offshore reveals a nearshore facies (Dahman and Samdun Formations) and an outer shelf facies (Ghalil Formation), and in Cyrenaica nummulitic shoals are common in the Darnah Formation, whilst the Apollonia Formation represents more open marine conditions.

shelf setting. It has an extensive distribution in the central part of the Sabratah Basin. The thickness ranges from 190m at Bouri, 116m at E1-NC 41, 85m at C1-NC 35a and 72m at A1-NC 35a (Figure 5.9). The formation contains a rich fauna of planktonic foraminifera including several zonal markers which span the entire interval from the basal Lutetian to the top Priabonian. There is palaeontological evidence to suggest that the formation may extend into the Ypresian, although it is underlain by Jdeir carbonates in the type well. The planktonic fauna confirms the outer shelf environment of deposition. The shales may have some source rock potential where thermally mature. The Ghalil Formation corresponds to the Souar Formation in Tunisia, which was deposited in a similar environment.[57]

SIRT BASIN

5.2.12 Al Jir (Gir) Formation (sensu Barr and Weegar)

The confusion concerning the definition of the Al Jir Formation has been discussed above (section 5.2.2). Barr and Weegar unwisely adopted a discredited name, originally introduced by Burollet, and applied it to a subsurface sequence overlying the Khayir Formation in the Sirt Basin. It has generally been assumed that the Al Jir Formation as defined by IRC at outcrop equates with the Al Jir Formation as defined in the subsurface. The problem is that at outcrop the Al Jir Formation overlies the Bishimah Formation and is of definite Lutetian age whereas the subsurface formation overlies the Khayir Formation and is of definite Ypresian age.

As defined by Barr and Weegar (Figure 5.2) the Al Jir Formation comprises three members, a lower Facha Member, middle Hun (Hon) Member and upper Mesdar Member. The Y1-59 well, near the Khalifah field on the Al Bayda

Platform, was selected as the type section for the Facha and Hun Members (Figure 5.6). At this location the Facha Member comprises 95m of finely crystalline vuggy dolomite with thin interbeds of micritic limestone, and anhydrite. The Hun Member consists of 492m of massive, microcrystalline anhydrite with intercalations of sucrosic and often vuggy dolomite. The type section of the Mesdar Member was selected in a different well, the B1-13 well in the Maradah Trough west of the Meghil field. (Figure 5.6). At this location the member occupies the entire interval between the Khayir Formation and the Jalu Formation. It is composed of 550m of massive bioclastic and micritic limestone, sometimes cherty, with minor intercalations of dolomite and shale. The limestone is fossiliferous and contains lumachelles of nummulites.[58]

The Al Jir Formation has been the subject of several studies. The biostratigraphy of the Al Jir Formation was examined by Abugares. Based on both planktonic and nannofossil assemblages he demonstrated that the Al Jir Formation in the Sirt Basin is of mid-Ypresian age, and that the contact with the overlying Jalu Formation is unconformable. He presented isopach and facies maps of the Al Jir Formation which show that over the Palaeocene carbonate shelves the formation averages about 400m in thickness, whilst in the Zallah, Abu Tumayam and Maradah Troughs the thickness exceeds 900m. He also showed that the three members of the Al Jir Formation represent facies zones. The Facha Member represents a restricted shelf facies which is present around the margin of the Sirt Basin and rarely exceeds 180m in thickness. The Hun Member exhibits an evaporitic facies which is best developed in the Abu Tumayam Basin (Figure 5.6). In well H1-85 it occupies the entire interval of the Al Jir Formation, and has a thickness of 850m. In this area the Hun Member contains abundant halite, which gives way to anhydrite further north. Both the Facha and Hun Members reappear on the Amal Platform in the eastern part of the basin. The centre of the basin is dominated by the massive carbonates of the Mesdar Member which represent an open marine facies. The Mesdar Member is present over a wide area of the Maradah Trough, Zaltan Platform and Ajdabiya Trough, with a thickness which varies from 300m on the Zaltan Platform to 825m in well C1-NC 130 in the Maradah Trough (Figure 5.6). The porous dolomites of the Facha Member are important oil reservoirs, particularly in the Zallah Trough where they are sealed by evaporites of the Hun Member. Twelve fields have been discovered to date.[59]

The sedimentology of the Facha Member in the Zallah Trough was studied by Elag. The wells in this area show an upward progression from open marine carbonates with planktonic foraminifera through bioclastic limestones with benthonic foraminifera to argillaceous dolomite and finally to porous micro-crystalline dolomite The upper unit provides the best reservoir rock.[60]

The diagenesis of the Facha Member in the Zallah Trough was studied by Lashab and West who concluded that the dolomitization was caused by reflux seepage, in which concentrated brines seeped into the Facha carbonates from the overlying evaporites of the Hun Member.[61]

5.2.13 Jalu Formation

The term Gialo Limestone was introduced by Barr and Weegar for a thick series of carbonates overlying the Al Jir Formation on the Jalu field (Figure 5.2). A type section was designated in the E91-59 well (Figure 5.8). The formation comprises massive shelly, sandy and chalky limestones rich in nummulites, with mollusc, echinoid and bryozoan debris. At least five species of *Nummulites* are present indicating a Lutetian age, possibly extending into earliest Bartonian. The nummulites often occur in banks several metres thick. At the type well the formation is 483m thick, and it has a widespread

distribution in the Sirt Basin, with an average thickness of about 500m. The Jalu Formation thickens dramatically into the Ash Shulaydimah Trough, just south of the Jabal al Akhdar Uplift, and reaches a thickness of 1300m in the A1-35 well (Figure 5.8). The Jalu Formation is the principal reservoir in the Jalu field and is oil bearing in several neighbouring pools. The shoreline trend continues to the northwest and the formation is gas-productive at As Sahel and Assumud and contains gas and oil at Tabit and Abu Quray. It also contains oil further south on the Zaltan Platform, but reservoir quality is generally poor in this area. The Jalu Formation extends as far west as the Al Bayda Platform, beyond which it passes into the shaly dolomites of the Gedari Formation. It is age equivalent to the Al Jir Formation (sensu IRC) at outcrop on the Jabal Waddan, and to part of the Tellil Group offshore.[62]

5.2.14 Gedari Formation

In the western Sirt Basin the Al Jir Formation in the subsurface is overlain by a thick series of shales, marls, dolomites, rare gypsum, and chalky limestones which Barr and Weegar assigned to the Gedari Formation (Figure 5.2). The name is derived from the Wadi al Gedari near the Az Zahrah field, but the type section was selected in the AA1-11 well in the Zallah Trough, near the Ghani field (Figure 5.8). The formation is 802m thick in the type well. According to Barr and Weegar the formation outcrops in the western part of concession 11. The Geological Map of Libya shows outcrops of the Al Gata Member and the Al Jir Formation in this area, which are age equivalents of the Gedari Formation. The Gedari Formation contains an abundant fauna of echinoids, bryozoa, pelecypods, gastropods and foraminifera, mostly of Lutetian age, and indicative of shallow-water deposition close to a shoreline. The Gedari Formation can be regarded as a nearshore equivalent of the more open-marine Jalu Formation. To the east it passes gradually into the Jalu Formation. It is restricted to the western margin of the Sirt Basin, particularly the Zallah Trough and Abu Tumayam Basin.[63]

5.2.15 Awjilah (Augila) Formation

In the eastern Sirt Basin the Jalu Formation is overlain by a shaly sequence which passes upwards into argillaceous limestone. This formation was named the Awjilah (Augila) Formation by Barr and Weegar (Figure 5.2) with a type section in the E1-59 well, the discovery well of the Jalu field. At this location the formation is composed of 80m of calcareous shales and chalky limestone containing both benthonic and planktonic foraminifera, both of which indicate a Bartonian-Priabonian age (Figure 5.9). The Awjilah Formation was deposited in shallow-water open-sea conditions. The lower unit is dominated by soft, greenish shales with thin limestone bands, whilst the upper unit consists of hard, sandy limestone, which is frequently argillaceous and glauconitic. In some areas a thin, friable quartz sandstone, which can reach 30m in thickness, occurs between the two members. The Awjilah Formation occurs over the east and central Sirt Basin as far west as the Al Bayda Platform, where the formation outcrops. On the Cyrenaica Platform the Awjilah Formation passes into an open-marine carbonate facies, and reaches a thickness of 200m in the B1-46 well (Figure 5.9).

In the vicinity of the Rakb D field the lower member of the Awjilah Formation passes into a fossiliferous micritic limestone facies which Barr and Weegar termed the Rashda Member. They chose a type section in the D4-12 well on the Rakb field where 51m of argillaceous and chalky limestones overlie the Jalu Formation. The limestone is dolomitic at the base and contains abundant calcareous algae. It is overlain by limestones forming the upper member of the Awjilah Formation. The Rashda Member has a very limited distribution in the Rakb field area, where it forms a productive oil reservoir (Figure 5.9).[64]

NORTHEAST LIBYA

5.2.16 Apollonia Formation

The tectonism which uplifted, folded and eroded the pre-Eocene rocks of the Jabal al Akhdar ceased during the early Eocene, and a new transgression began which progressively onlapped the eroded terrain of the Jabal al Akhdar. The crestal area was not covered until middle Eocene times. The open-marine facies to the north is represented by the Apollonia Formation whilst the shallower-water nummulitic facies further south is represented by the Darnah Formation (Figure 5.2).

The Eocene rocks of northern Cyrenaica were first studied by Gregory in 1911 who named three units, a lower Apollonia Limestone, a middle Darnah (Derna) Limestone and an upper Salantah (Slonta) Limestone. He described the Apollonia Limestone from the Susah area, but at this locality only the upper 30m of the formation is exposed, consisting of finely crystalline to chalky limestone. The sequence was subsequently examined by Pietersz, who established the Apollonia Formation, and described outcrops in the hills south of Susah where 75m of hard, siliceous and chalky limestones are exposed, containing chert nodules, and a fauna of nummulites and echinoids which indicate a Lutetian age (Figure 5.6). Limestones of the Darnah Formation overlie the formation, but the base is not exposed. Subsequent mapping has shown that the formation outcrops along the coast almost continuously from Tukrah to Darnah, but it wedges out rapidly to the south where it interfingers with the Darnah Formation (Figures 5.6, 5.8 and 5.9). The Apollonia Formation is mostly Lutetian in age, but at Wadi Athrun, where the lower unconformable contact with the Al Athrun Formation can be seen, the lower 33m has been dated as Ypresian. By contrast, the section at Wadi al Qal'ah near Al Hilal, which also exposes the contact with the underlying Al Athrun Formation, the formation is wholly Lutetian in age (Figure 5.8). Further east along the coast 122m of Apollonia Formation are exposed which is all Lutetian in age. Other exposures reported by Klen and El Khoudary show that further west the formation extends into the late Bartonian, and at Wadi Bakur near Tukrah the entire sequence is Priabonian in age (Figure 5.9). This sequence was originally named Tocra Limestones by Desio who correctly identified it as Eocene on the basis of its nummulite fauna. More recent studies based on nannofossils show that it extends from late Eocene to basal Oligocene in age. Unfortunately, it was subsequently confused with the underlying Maastrichtian sequence which is now assigned to the Wadi Dukhan Formation. The thickness of the Apollonia Formation is known to reach 380m southeast of Al Athrun, and in the subsurface it increases to 870m in the C1-18 well. In the subsurface of the Ash Shulaydimah Trough the upper Palaeocene-lower Eocene seqeuence is locally known as the Antlat Formation. It comprises argillaceous microcrystalline limestones with a fauna of planktonic foraminifera. This formation has modest potential source rock characteristics in some areas. In the A1-NC 120 well the Apollonia Formation is represented by open-marine, deep-platform and slope argillaceous micrites. The basal contact with the Al Uwayliah Formation is conformable and palaeontological analysis shows a complete sequence from early Ypresian to late Priabonian (Figure 5.6). In the east the Apollonia Formation is over 300m thick in the

A1-36 well. The Apollonia Formation is demonstrably diachronous. The formation represents deposition in an open-marine setting, and Rohlich suggested a turbidite environment for some of the exposures, based on contorted-bedding features, erosion channels and repeated cycles of fining-upwards carbonate sequences. In some areas, such as Wadi al Athrun, the Apollonia Formation shows colour banding which has been shown to relate to the more advanced stage of diagenesis reached by the white chalky limestones compared to that of the hard, tan limestones.[65]

5.2.17 Darnah Formation

Gregory showed the Apollonia Limestone to be overlain by the Darnah (Derna), and Salantah (Slonta) Limestones. The Salantah Limestone has long proved difficult to separate from the Darnah Limestone and it is now incorporated within the Darnah Formation. It is now generally accepted that the Apollonia Formation is the lateral equivalent of at least part of the Darnah Formation (Figure 5.2).

Gregory described the Darnah Limestone from the coastal exposures at Darnah where 100m of nummulitic and reefal limestones overlie the microcrystalline limestones of the Apollonia Formation. A type section was later defined in this area by Zert (Figure 5.9). The reefal limestones contain corals, algae and nummulites of Bartonian-Priabonian age. Subsequent work has shown the Darnah Formation to have a wide distribution in the northern Jabal al Akhdar and southwards around the Qasr al Ahrar and Jardas al Jarrari inliers, from where the Salantah (Slonta) Limestone was first described. In this area it is represented by massive crystalline and chalky limestones with occasional marl partings. It is also present in the subsurface in wells A1-36 and A2-2. The thickness increases westwards to 270m at Al Bayda, but wedges out rapidly to the south

(Figure 5.9). The formation contains nummulites, which sometimes form shoals similar to those in the Jalu Formation, and a rich shallow-water fauna of corals, pelecypods, nautiloids and gastropods. The nummulites give a Lutetian to Priabonian age for this formation. The lateral passage from Darnah Formation to Apollonia Formation can be seen in the area between Susah and Wadi al Athrun. Offshore the A1-NC 120 well shows the presence of only the Apollonia Formation, but in well A1-NC 128, near Bamba, shallow-water carbonates of the Darnah Formation are present, unconformably overlying Albian limestones. The lower part shows a nummulitic-bank facies, which is overlain by open-marine platform carbonates. The sequence covers the interval from early Lutetian to late Priabonian. It is likely that marine conditions spread over the entire Jabal al Akhdar by Lutetian times, but towards the end of the Eocene there was a shallowing of the sea followed by a regression, which left the entire area emergent. El Hawat and Shelmani examined the Darnah Formation in detail and have been able to recognise several cycles and shoal/channel sub-facies particularly within the upper part of the formation.[66]

5.3   OLIGOCENE

The Oligocene in Libya marks a period of regression in which the shoreline migrated northwards. There is evidence that minor oscillations occurred in the mid-Oligocene followed by a minor transgression in the late Oligocene. Oligocene outcrops are widespread in a belt extending from the Hun Graben to the area east of Al Haruj al Aswad, and Oligocene rocks are present throughout the Sirt Basin. Significant outcrops of Oligocene rocks are also present in the Jabal al Akhdar. The lithostratigraphic nomenclature for the Palaeogene is shown on Figure 5.10.

WESTERN OUTCROPS

The youngest rocks in the Hun Graben are of Oligocene age, and outcrops cover an area of 8000km^2 between Abu Njaym and Waddan. Further east outcrops occur along the eastern margin of Az Zahrah-Al Hufrah Platform from the An Nuwfaliah area to the Al Bayda oilfield. A third area of outcrops is present east of Al Haruj al Aswad, with the southernmost outliers extending as far as the Dur at Talah escarpment (Figure 5.11). The northern areas have been mapped for the Geological Map of Libya and are described below.

In addition to the lagoonal and nearshore rocks of northern Libya continental rocks of Oligocene age are present in the Zallah Trough and around the Al Qarqaf Arch. Near Zallah fluvial-channel sands and conglomerates contain abundant fossil wood and vertebrate bones. Towards the Hun Graben these rocks pass laterally into the Ma'zul Ninah Formation. A number of isolated pockets of laminated micritic limestone have been found on the Jabal Hasawnah which contain a rich fauna including mammals (hyracoids), frogs and fish. The fauna indicates an Oligocene age. These sediments were assigned to the Tarab Formation by Jurak (Figure 5.11), and were deposited in a lacustrine environment.[67]

Extensive outcrops of Oligocene continental and transitional beds outcrop over an area of 180,000km^2 east of Al Haruj al Aswad. These rocks have been studied in the Abu Na'im area where they comprise 40m of cross-bedded sandstones, sandy siltstones, gravels and gypsiferous claystones, which contain roots and trunks of broad-leaved trees. Further south at Dur at Talah about 30m of micritic limestones and gypsiferous siltstones, overlain by 20m of massive cross-bedded and ripple-marked sandstones are found. These were named the Idam Unit and Sarir Unit by Wight. They represent lacustrine and fluvial deposits and the

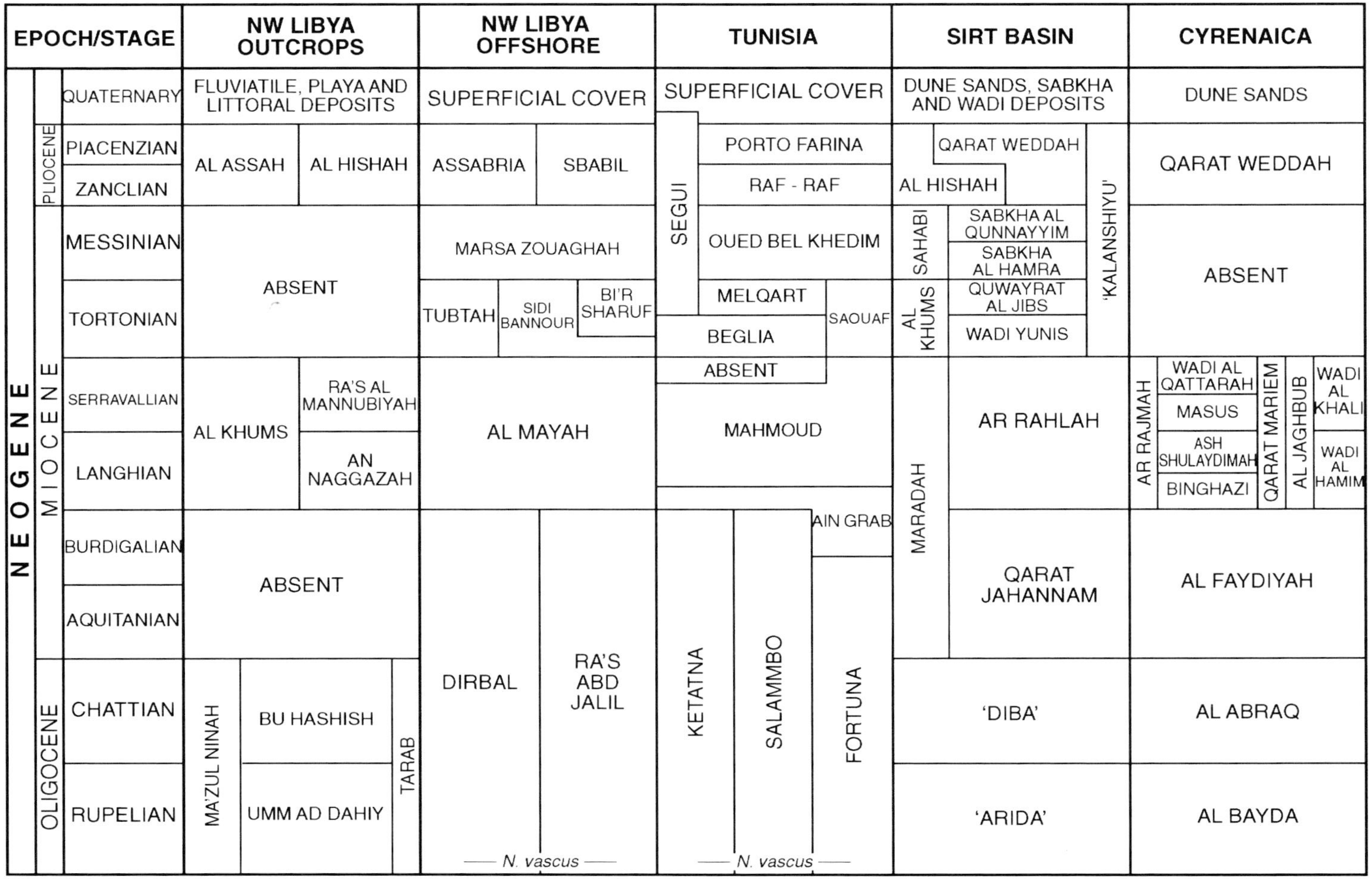

Figure 5.10 showing the lithostratigraphic nomenclature table for the Neogene Rocks of Libya and Adjacent Areas.

EPOCH/STAGE		NW LIBYA OUTCROPS	NW LIBYA OFFSHORE	TUNISIA	SIRT BASIN	CYRENAICA
NEOGENE	PLIOCENE — QUATERNARY	FLUVIATILE, PLAYA AND LITTORAL DEPOSITS	SUPERFICIAL COVER	SUPERFICIAL COVER	DUNE SANDS, SABKHA AND WADI DEPOSITS	DUNE SANDS
	PIACENZIAN	AL ASSAH / AL HISHAH	ASSABRIA / SBABIL	PORTO FARINA / RAF-RAF	QARAT WEDDAH / AL HISHAH; SABKHA AL QUNNAYYIM, SABKHA AL HAMRA, QUWAYRAT AL JIBS, WADI YUNIS; 'KALANSHIYU'	QARAT WEDDAH
	ZANCLIAN			OUED BEL KHEDIM		ABSENT
	MESSINIAN	ABSENT	MARSA ZOUAGHAH (TUBTAH, SIDI BANNOUR, BI'R SHARUF)	MELQART / BEGLIA / SEGUI / SAOUAF / ABSENT	SAHABI / AL KHUMS	AR RAJMAH (WADI AL QATTARAH, MASUS, ASH SHULAYDIMAH, BINGHAZI) / QARAT MARIEM / AL JAGHBUB (WADI AL KHALI, WADI AL HAMIM)
	TORTONIAN		AL MAYAH	MAHMOUD	MARADAH	
	SERRAVALLIAN	RA'S AL MANNUBIYAH			AR RAHLAH	AL FAYDIYAH
	LANGHIAN	AN NAGGAZAH / AL KHUMS				
	BURDIGALIAN	TARAB	RA'S ABD JALIL / DIRBAL	AIN GRAB / FORTUNA / SALAMMBO / KETATNA	QARAT JAHANNAM	
	AQUITANIAN	ABSENT				
OLIGOCENE	CHATTIAN	BU HASHISH / MA'ZUL NINAH			'DIBA'	AL ABRAQ
	RUPELIAN	UMM AD DAHIY			'ARIDA'	AL BAYDA

Source: see text

Figure 5.10   Lithostratigraphic Nomenclature for the Neogene Rocks of Libya and Adjacent Areas

This table shows the current lithostratigraphic nomenclature for the Neogene rocks of Libya, with updated transliteration of Arabic names. One area of concern is the relationship between the Maradah Formation and the Al Khums Formation which is illustrated in more detail on Figure 5.14. No attempt has been made to show unconformities on this chart.

242

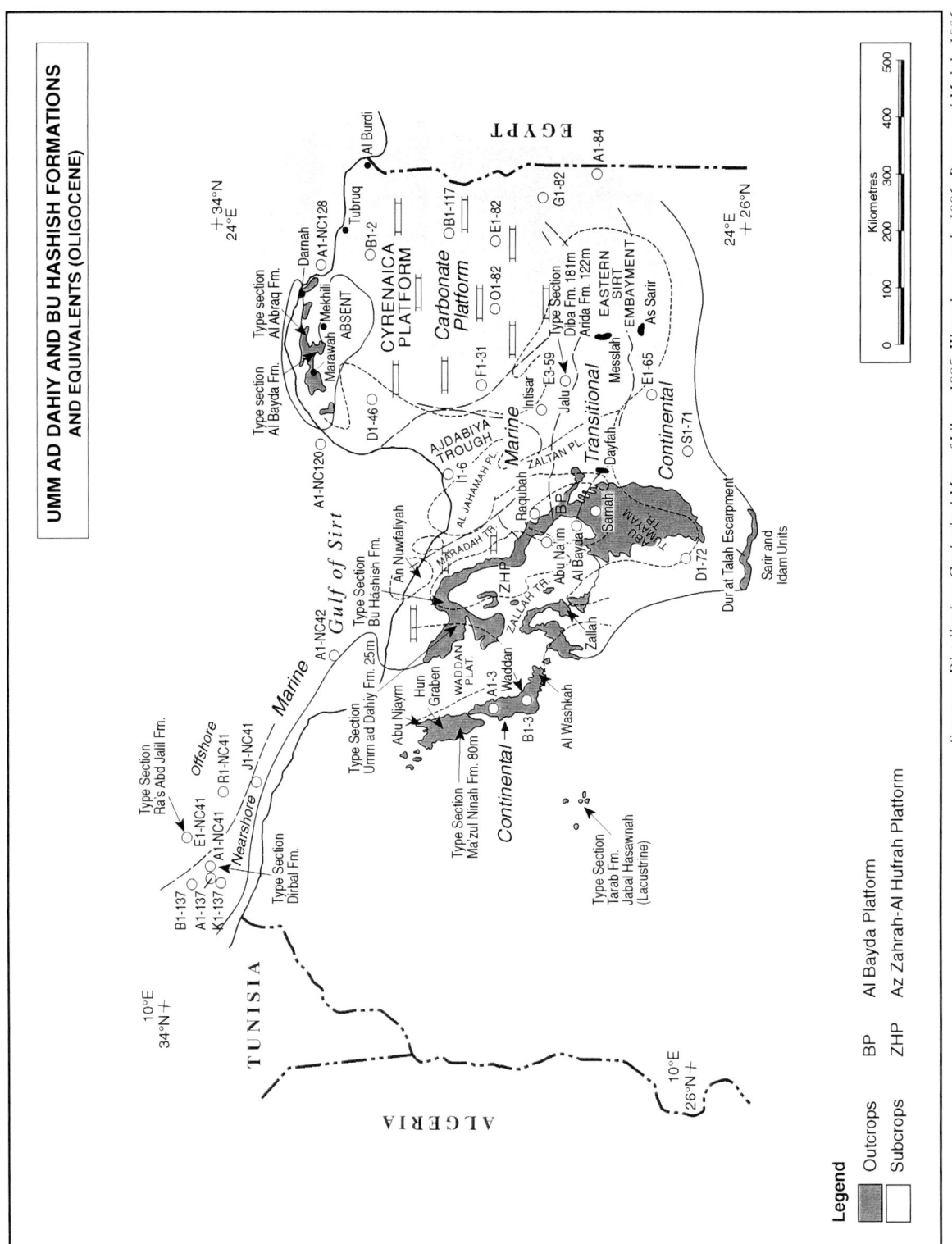

Sources: *Distribution: Geological Map of Libya, 1985, Wennekers, et al., 1996, Bezan and Malak, 1996*

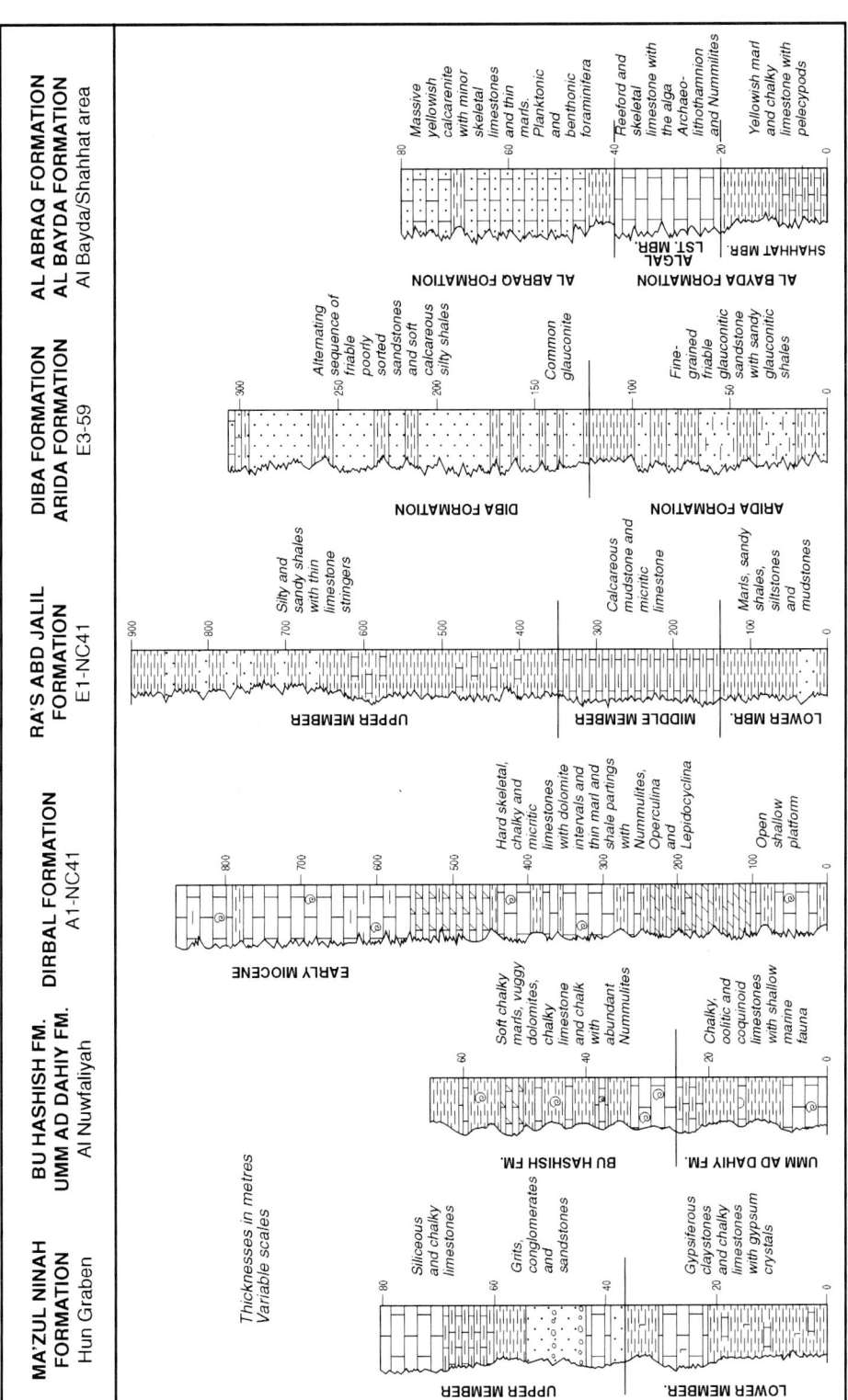

**Figure 5.11 Umm ad Dahiy and Bu Hashish Formations:
Distribution, Equivalents and Stratigraphic Columns**

*Sources: Stratigraphic columns: Ma'zul Ninah Formation: Shakoor & Shagroni, 1984
Umm ad Dahiy and Bu Hashish Formations: Srivastava, et al., 1980
Dirbal and Ra's Abd Jalil Formations: Hammuda, et al., 1985
Arida and Diba Formations: Barr and Weegar, 1972
Al Bayda and Al Abraq Formations: Rohlich, 1974*

The Umm ad Dahiy and Bu Hashish Formations represent the marine Oligocene succession which is exposed on the western flank of the Sirt Basin, whilst the Ma'zul Ninah Formation represents continental equivalents in the Hun Graben. Offshore an open shallow shelf facies is present in the Dirbal Formation and a deeper water facies in the Ra's Abd Jalil Formation. In the Sirt Basin the Arida and Diba Formations reflect nearshore marine clastic sedimentation, and in Al Jabal al Akhdar the Al Bayda and Al Abraq Formations represent a variety of nearshore carbonate facies.

Idam Unit contains a remarkable fauna including rodents, hyracoids, proboscideans, crocodiles, chelonians, fish and a snake, all consistent with an early Oligocene age (Figure 5.11). These rocks also contain silicified trees and the fruits of water-lilies.[68]

5.3.1 Ma'zul Ninah Formation

Extensive outcrops of Oligocene rocks are preserved in the Hun Graben where they represent the latest basin-fill of the graben. The Ma'zul Ninah Formation was named by Jordi and Lonfat from a hill in the western Hun Graben, but was not described until the remapping for the Geological Map of Libya in 1981/4 (Figure 5.10). Shakoor and Shagroni established a type section near Ra's Ma'zul where the formation is 80m thick with the base not exposed. They divided the formation into upper and lower members. The lower member comprises 35m of gypsiferous claystones with soft chalky limestones containing gypsum crystals. The upper member, which is 45m thick, contains friable grits, conglomerates and sandstones, overlain by gypsiferous marls and chalky limestones which become siliceous at the top (Figure 5.11). The base of the formation is visible in the northern part of the graben where it variously overlies the Wadi Zakim, Khayir or Ammur Members, which indicates a period of erosion prior to the deposition of the Ma'zul Ninah Formation. The formation has a limited distribution within the Hun Graben. It varies in thickness from 25 to 80m, but in the A1-3 well it reaches 120m and in the B1-3 well, 200m.

The formation is characterised by the presence of gypsiferous claystones and conglomerates, and bedded gypsum is found in the Al Washkah area (Figure 5.11). The conglomerates are composed mostly of calcareous pebbles containing Eocene and late Cretaceous microfaunas. Fossils are rare, but charophytes, ostracods and gastropods have been found in the

area west of Zallah. The formation was deposited in a restricted lagoon with periodic marine influxes. South of Hun the Ma'zul Ninah Formation is overlain by lavas of the Jabal as Sawda which have been radiometrically dated as Middle Miocene in age. On the basis of its fauna and stratigraphic position the Ma'zul Ninah Formation is interpreted as Oligocene in age. The area south of the Hun Graben is deformed by a complex series of drape folds which have been interpreted as a response to relay growth faulting in the underlying Wadi Thamat Formation.[69]

5.3.2 Umm ad Dahiy Formation

Further to the east marine Oligocene rocks are exposed on the Az Zahrah-Al Hufrah Platform (Figure 5.11). A sequence of limestones and dolomites overlying the Qararat al Jifah Member was named the Dor el Abd Formation by Burollet after an outcrop northwest of the Bahi field which has since been renamed Dur Umm ad Dahiy (Figure 5.10). At this locality 25m of chalky, oolitic and coquinoid limestones are exposed which contain a distinctive fauna of *Operculinoides*, *Operculina*, bryozoa, algae and ostracods. Nummulites are conspicuously absent. Dolomitic marls and clays are interbedded with the limestones. The formation extends over a distance of 350km from An Nuwfaliyah to Abu Na'im (Figure 5.11). The thickness remains relatively constant and rarely exceeds 40m. The limestones pass gradually into dolomites towards the south and ultimately into gypsiferous claystones and siltstones. This transition indicates a progressively more restricted environment of deposition towards the south. The age of the formation is bracketed as Rupelian by the presence of age diagnostic species of *Nummulites* in both the underlying and overlying formations. The Umm ad Dahiy Formation contains patch reefs in the area north of the Bahi field and near Ar Raqubah field. The mounds occur within a fairly narrow band, but extend over tens of kilometers in length. Individual mounds reach about 10m in thickness. The main framebuilders are scleractinian corals.[70]

5.3.3 Bu Hashish Formation

The sequence of green shales and vugular dolomites overlying the Umm ad Dahiy Formation was named the Greir bu Hascisc Formation by Burollet, from a locality on the northern margin of the Az Zahrah-Al Hufrah Platform. Recent mapping has demonstrated that the formation can be traced to beyond the Ar Raqubah field (Figure 5.11). In the type area it comprises 40m of soft chalky marls, vuggy dolomites, chalky limestones and chalk. At the base is a marker bed containing *Nummulites fichteli*. The nummulites are very abundant and often form coquinas. In addition pelecypods, gastropods and echinoids are common, plus the large foraminifer *Myogypsina complanata*. This fauna dates the formation as Chattian in age. The Bu Hashish Formation marks the end of Oligocene marine deposition in northwest Libya, and it is unconformably overlain by the Al Khums Formation of Middle Miocene age. The formation thins southwards. In the Maradah area it is about 35m thick and is composed of calcareous and gypsiferous sandstone, sandy dolomite and bedded gypsum. At Abu Na'im it is reduced to a thickness of 15m where it is composed of dolomitic sandstone, dolomite with abundant nummulites, and traces of gypsum. The formation was deposited in an inner neritic to lagoonal environment.[71]

NORTHWEST LIBYA OFFSHORE

5.3.4 Dirbal Formation

The Dirbal Formation was established by Hammuda, et al. for a thick sequence of carbonates overlying the *Nummulites vascus* marker bed in the Sabratah Basin (Figure 5.10). A type section was designated in the A1-NC 41 well in the southern part of the basin. In the type well the formation has a thickness of 873m, with the *N.vascus* bed at the base. It consists of hard, skeletal, chalky and micritic limestones with minor dolomite intervals, and thin marly and shaly partings (Figure 5.11). It is unconformably overlain by the middle Miocene Al Mayah Formation. Biostromes of bryozoans, algae and large foraminifera occur at several levels which frequently contain large dissolution cavities. The dominant foraminifera are *Nummulites, Operculina* and *Lepidocyclina*. On the basis of the fauna the age of the formation ranges from early Oligocene to early Miocene. The Dirbal Formation can be divided into three major sequences separated by unconformities reflecting marine regressions. Karst features and greatly enhanced vugular porosity at these unconformities reflects periods of exposure and weathering. The formation is limited to the southern part of the Sabratah Basin and has been penetrated in wells A1-137, B1-137, K1-137, A1-NC 41 and J1-NC 41 (Figure 5.11). It passes laterally into the deeper-water facies of the Ra's Abd Jalil Formation. The bryozoan/algal biostromes are characteristic of the Dirbal Formation. Although they rarely extend for more than a few kilometers, they are numerous and occur at several different horizons. Oil was found in this facies in the J1-137 well. The Dirbal Formation is the equivalent of the Ketatna Formation in the Tunisian offshore which also ranges in age from early Oligocene to early Miocene. The Ketatna Formation is overlain by shales of middle Miocene age.[72]

5.3.5 Ra's Abd Jalil Formation

The deeper-water equivalent of the Dirbal Formation (Figure 5.10) in the Sabratah Basin is the Ra's Abd Jalil Formation, named by

Hammuda, et al. from a type section in the E1-NC 41 well, located east of the Bouri field. In the type well the formation has a thickness of 902m, and it conformably overlies the Ghalil Formation. It is composed of three members, a lower sequence of marls, sandy shales, siltstones and mudstones, a middle unit of calcareous mudstone to argillaceous micritic limestone and marl, and an upper unit of predominently silty and sandy shales with thin limestones. The formation is present in the deeper parts of the Sabratah Basin as far east as well R1-NC 41 and as far south as the F1-137, I1-137, M1-NC 41 area (Figure 5.11). It represents an outer neritic environment of deposition. The fauna is dominated by planktonic foraminifera, which show a complete sequence of zones from Rupelian to Aquitanian. It is overlain, without any apparent break, by the Al Mayah Formation of Langhian-Serravallian age. A similar outer neritic facies is present in Tunisian waters where it has been named the Salammbo Formation.[73]

SIRT BASIN

5.3.6 Arida Formation

The Arida Formation (Figure 5.10) was established by Barr and Weegar for a sequence of clastic sediments unconformably overlying the Awjilah Formation in the eastern Sirt Basin. They selected the E3-59 well on the Jalu field as the type well. At this location the Arida Formation is 122m thick, and comprises a lower sandstone unit divisible into three or four cycles of fine-grained, friable, glauconitic sands separated by thin shales, overlain by an upper unit of silty and glauconitic shale (Figure 5.11). The full assemblage represents a nearshore to littoral environment of deposition. It is overlain by non-marine sands of the Diba Formation. The formation is poorly fossiliferous, but the upper shale unit contains foraminifera which indicate an Oligocene age. The lower sandstone unit is a major oil reservoir on the Jalu field where it is locally known as the Chadra Sandstone. An isopach map of this unit in the Jalu area was published by Mansour and Magairhy where the Chadra Sandstone attains a maximum thickness of 150m.[74]

5.3.7 Diba Formation

The Arida Formation (Figure 5.10) is overlain in the E3-59 well by an alternating sequence of sandstones and shales which Barr and Weegar named the Diba Formation. The sands are poorly sorted and extremely friable. Glauconite is a common constituent, but fossils are rare. The shales are calcareous, soft and frequently silty. Thin stringers of sandy limestone occur near the top of the formation. The thickness in the type well is 181m. According to Barr and Weegar the formation has a broad distribution over the south-central Sirt Basin (Figure 5.11), but they pointed out that where the Arida shale member was not developed it became difficult to separate the Diba from the Arida Formation.[75]

The two-fold separation of the Oligocene rocks by Barr and Weegar in the eastern Sirt Basin has not stood the test of time. It may have local application in the Jalu area, but it is not applicable on a regional scale. Benfield and Wright elected to divide the Oligocene into a lower marine member and an upper non-marine member. Their work showed that the Oligocene as a whole thickens from about 250m in the Messlah field area towards the northwest. At Intisar the thickness is 490m and in the Ajdabiya Trough it reaches more than 2000m. There is also a local thickening in the Sarir area where the Oligocene is about 370m thick. The non-marine member is confined to the area south of Intisar and Jalu, and the maximum recorded thickness of this unit is 400m near the Dayfah field (Figure 5.11). The marine unit thickens towards the north. They were able to demonstrate a substantial regression during the Oligocene, when the shoreline migrated northwards from Dayfah to Jalu, with the establishment of fluvial conditions south of this line. This was followed by a less dramatic transgression during the late Oligocene, which established a shoreline extending from Al Jabal field to As Sarir field (Figure 5.11). Dolomites become a dominant constituent of the Oligocene in the area between Jalu and Messlah.[76]

In a more recent study, Bezan and Malak refined the sedimentological model for the Oligocene. They demonstrated the presence of three depositional environments (Figure 5.11). In the south they recognised a non-marine clastic facies in which braided fluvial channels pass laterally into meandering channels with overbank deposits, and ultimately to a deltaic facies. Further north, in a belt extending from the Al Bayda field to the Messlah field, they recognised a transitional facies in which marine and non-marine deposits are intermixed. This suggests that a number of transgressions and regressions occurred in this area during the Oligocene. Estuarine channels, tidal flats, offshore bars, and barrier islands have all been recognised within this complex, intercalated with sequences which are clearly fluvial. North of the Jalu field, in a broad belt extending from the Cyrenaica Platform to the outcrops on the Az Zahrah-Al Hufrah Platform, a marine environment was present. Shallow-water limestones, dolomites and some evaporites characterise the Cyrenaica Platform and Amal Shelf, which pass abruptly into deeper-water marls and mudstones in the Ajdabiya Trough. Well evidence shows that Oligocene mudstones reach a thickness of 2280m in the area of the present-day coastline. West of the Ajdabiya Trough the sequence thins to less than 300m and is composed mainly of open-shelf carbonates in an open shallow-shelf setting. In the Maragh Graben wells on the As Sarah and Jakharrah fields reveal a complex pattern of transitional deposition showing examples of tidally-influenced to wave-dominated deltaic and estuarine conditions. The Oligocene sequence in this area is about 250m thick and contains discrete sandstone members which have been assigned to the informal Fellaga and Chadra Members. As an interesting aside, two meteorite impact craters have been found northeast of Kufrah in an area of Nubian Sand outcrop. The craters, which appear to have been formed simultaneously by a double impact have diameters of 5.1 and 2.8km. They have been indirectly dated by fission-track analysis of Libyan desert glass found nearby as late Oligocene in age.[77]

NORTHEAST LIBYA

5.3.8 Al Bayda Formation

The Oligocene sequence of the Jabal al Akhdar has been revised several times since Gregory first assigned the entire sequence to the Cyrene Limestone. It has subsequently become evident that the Oligocene rests unconformably on eroded and weathered Eocene rocks, and that there are two major sequences within the Oligocene of Cyrenaica, separated by an unconformity. The lower sequence was named the Al Bayda Formation by Rohlich, and the upper sequence the Al Abraq Formation (Figure 5.10). The choice of the name Al Bayda Formation poses a problem since it duplicates that of the Al Bayda (Beda) Formation of Palaeocene age in the Sirt Basin, and even if the name Beda Formation is preferred there is still the potential for confusion between the Al Bayda Formation of the Jabal Akhdar and the Al Bayda Platform in the Sirt Basin.[78]

The Al Bayda Formation of Rohlich consists of two members, a lower Shahhat Member and an upper Algal Limestone Member. The type section of the formation was defined just south of Al Bayda where both members are present (Figure 5.11). At this locality the Shahhat Member is composed of 20m of yellowish marl and chalky limestone, containing pelecypod shells still in growth position. The Algal Limestone Member is 20m thick and is made up of thick-bedded reefoid and skeletal limestone full of the calcareous alga *Archaeolitho-thamnion*. It also contains foraminifera which indicate a Rupelian age. The Shahhat Member is only present north of Salantah, and the Algal Limestone Member overlaps it and extends further to the south. In the Darnah area only the Algal Limestone Member is present. The formation reaches a maximum thickness of 70m near Marawah (Figure 5.11). There is some evidence of deformation between the two members which may indicate tectonic activity during this period. This was followed by a major regression which corresponds to the global fall in sea level during the mid-Oligocene. Offshore in the A1-NC 120 well there is an unconformity between the Apollonia Formation and the Al Bayda Formation, but no obvious breaks in sedimentation within the Oligocene. The Al Bayda Formation is represented by open shallow-platform limestones containing *Nummulites vascus, N. fichteli, Lepidocyclina*, pelecypods and bryozoa. A similar sequence with a similar fauna is present in the A1-NC 128 well.[79]

5.3.9 Al Abraq Formation

An unconformity separates the Al Abraq Formation (Figure 5.10) from the Al Bayda Formation and represents a major regression in the mid Oligocene. The Labrak Calcarenite Member was established by Kleinsmiede and van den Berg but was subsequently redefined as a formation by Rohlich on the basis of the unconformity separating it from the underlying Al Bayda Formation. A type section was defined close to the village of Al Abraq. At this

location 36m of massive yellowish calcarenites are exposed, unconformably overlying the Al Bayda Formation. The calcarenites are interbedded with skeletal limestones and thin marl beds, and dolomitic limestones are present in other areas (Figure 5.11). The formation has an average thickness of 60-80m but reaches 130m near Darnah. The transgressive nature of the formation can be seen in the western part of the Jabal al Akhdar where it progressively overlies formations as old as the Campanian Al Majahir Formation. The Al Abraq Formation represents a wide range of depositional environments from open-shelf to brackish. It contains a rich fauna including planktonic foraminifera which suggest a late Rupelian to early Chattian age, plus *Nummulites fichteli, N. vascus*, and other large foraminifera. The Al Abraq Formation has a patchy distribution. Eroded remnants are preserved over much of the northern Jabal al Akhdar, but it is absent along a ridge extending from Marsa al Hilal southwards to Mekhili (Figure 5.11). It has been reported as far west as Soluq and as far east as Tubruq and Al Burdi (Bardia), but these outcrops are questionable and probably belong to the Miocene Al Faydiyah Formation. It is present in the A1-NC 120 offshore well where it consists of fossiliferous open marine limestones containing *Operculina*, bryozoa and algae, but it is absent in the A1-NC 128 well.[80]

5.4 MIOCENE

Miocene outcrops cover 180,000km^2 of northern Libya, amounting to more than 10% of the total area of the country. The principal outcrops are a coastal belt extending from Al Khums to Ra's Lanuf, the Sirt Basin embayment from the coast to beyond the Sarir oilfield, and the entire Cyrenaican Platform, north of the Amal oilfield. For the most part these rocks represent fluvial, littoral and nearshore deposits forming part of a regressive sequence which culminated in the Messinian low-stand, and the incision of the Sahabi channel which has cut 200m below present-day sea level. These rocks have been extensively studied particularly by Magnier, Selley, El Hawat, Benfield and Wright, and the surveyors of the Geological Map of Libya, but despite these studies the stratigraphic nomenclature of the Miocene in Libya is still in a state of some confusion. Early Miocene rocks are not found at outcrop west of An Nuwfaliah. Mid-Miocene rocks occur from Al Khums to the Egyptian border, and late Miocene rocks are confined to the Ajdabiya Trough and offshore. The extensive volcanic outcrops of the Tibisti Massif, Al Haruj al Aswad, Jabal as Sawda and Gharyan are mostly late Miocene to recent in age. Radiometric dating has shown that the Jabal as Sawda lavas range from Langhian to Tortonian in age, whilst the later volcanics at Gharyan date from the Serravallian. The Al Haruj al Aswad lavas are the youngest, giving absolute ages of 8 Ma to 0.4 Ma.[81]

The stratigraphy will be reviewed for the coastal plain, the northwestern offshore, Sirt Basin and Cyrenaica.

COASTAL PLAIN

5.4.1 Maradah Formation

Early Miocene rocks are not present along the coastal plain west of An Nuwfaliyah and between Al Nuwfaliyah and Al Khums middle Miocene sediments rest unconformably on progressively older rocks. To the east of An Nuwfaliyah, however, early Miocene rocks are present. The basal Miocene unconformity becomes less marked, although it is still recognisable at Maradah where rocks of Aquitanian age unconformably overlie the Bu Hashish Formation, with a conglomerate or breccia at the base. The rocks overlying the Oligocene in the Maradah area were named the Marada Series by Desio, which was subsequently formalised as

the Maradah Formation (Figures 5.10 and 5.14). These beds are well exposed in the cliffs flanking the Maradah oasis, and the sedimentology of the Maradah Formation was described in detail by Selley and by El Hawat. They interpreted a fluvial setting, with estuaries draining into lagoons, and fringed by offshore bars and barrier beaches. During the mapping for the Geological Map of Libya the Maradah Formation was divided by Mastera into two members, a lower Qarat Jahannam Member and an upper Ar Rahlah Member (Figure 5.12). The Qarat Jahannam Member type section is located 40km west of Maradah. It comprises 123m of stacked clastic sequences of cross-bedded fluvial sandstones, grading into siltstones and silty claystones with traces of gypsum. The sequence is cut by numerous erosion channels. The thickness of the unit reaches 150m on the southern margin of the basin, but thins towards both the northeast and northwest. Fossils are generally rare but occasional bands of echinoids, marking brief marine incursions, indicate an Aquitanian to Burdigalian age.[82]

The type section for the Ar Rahlah Member was defined from outcrops to the east of Maradah (Figure 5.12). This member marks an increasing marine influence, and most of the sequence was deposited in a littoral setting. At the type locality 120m of calcarenites, calcareous sandstones and sandy limestones overlie the Qarat Jahannam Member. The lithology is very varied and includes oolites, dolomites and rare gypsum. The limestones contain pelecypods, corals, echinoids, bryozoa, algae and foraminifera, frequently in coquinoid mounds. Subsurface evidence shows that the thickness of the formation increases rapidly into the centre of the basin reaching over 800m northeast of the Hutaybah gas field. The fauna suggests a Burdigalian to Serravallian age, which implies that the Maradah Formation is distinctly diachronous in this area. At the northern end of the outcrop the Qarat Jahannam

250

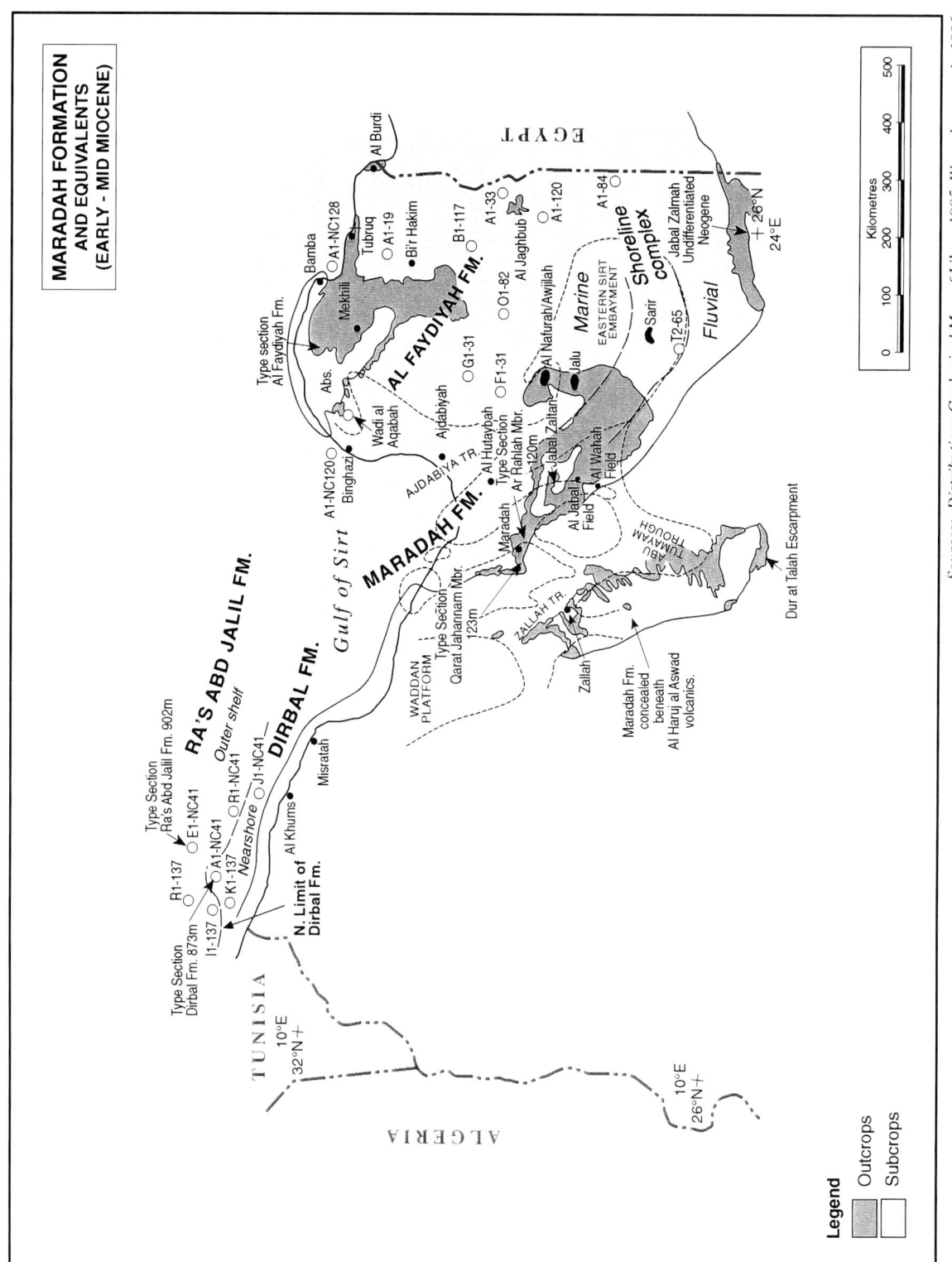

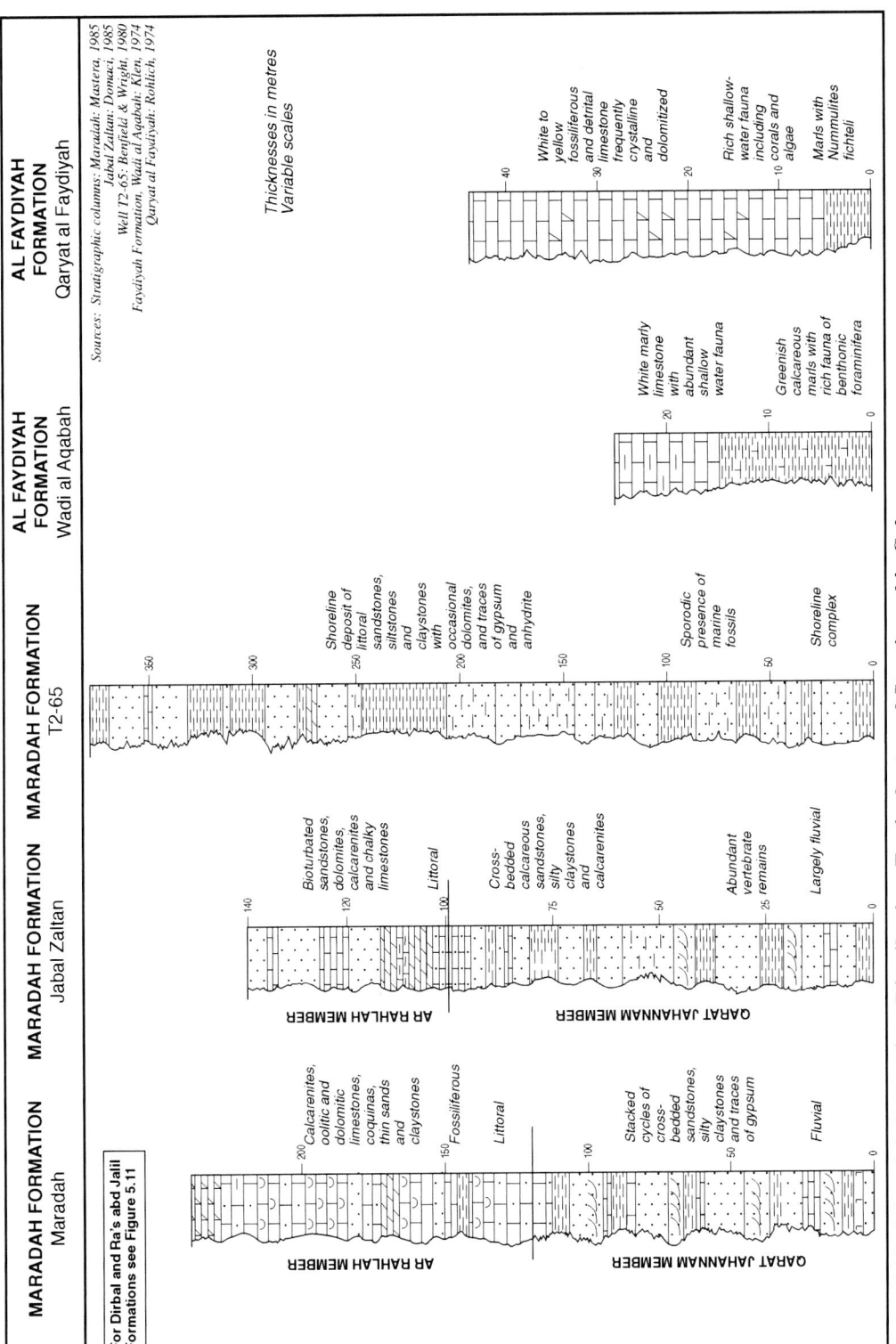

Figure 5.12 Maradah Formation: Distribution, Equivalents and Stratigraphic Columns

The Maradah Formation extends into the Serravallian in the type area, but in other areas it appears to be confined to the early Miocene. At Maradah it represents a fluvial/estuarine complex which passes northwards into a marine facies. The large outlier partly concealed beneath the Haruj al Aswad lavas is totally continental in character. Offshore northwest Libya it equates to the upper part of the shallow-marine Dirbal Formation and the upper part of the deeper-water Ra's Abd Jalil Formation. In Cyrenaica the open shallow-platform limestones of the Al Faydiyah Formation are correlative with the Maradah Formation

252

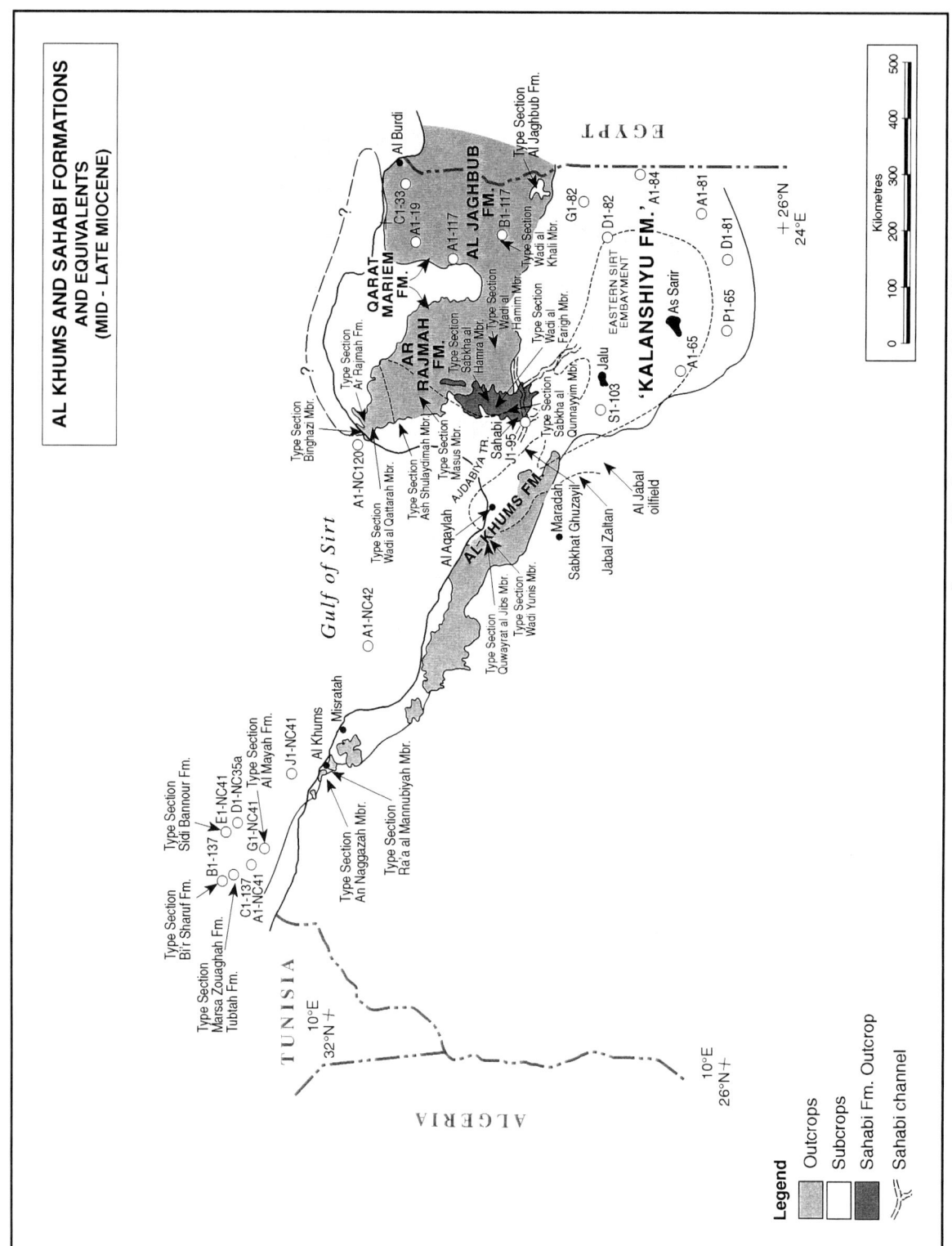

AL KHUMS AND SAHABI FORMATIONS
AND EQUIVALENTS
(MID - LATE MIOCENE)

Sources: Distribution: Geological Map of Libya, 1985, Benfield and Wright, 1980

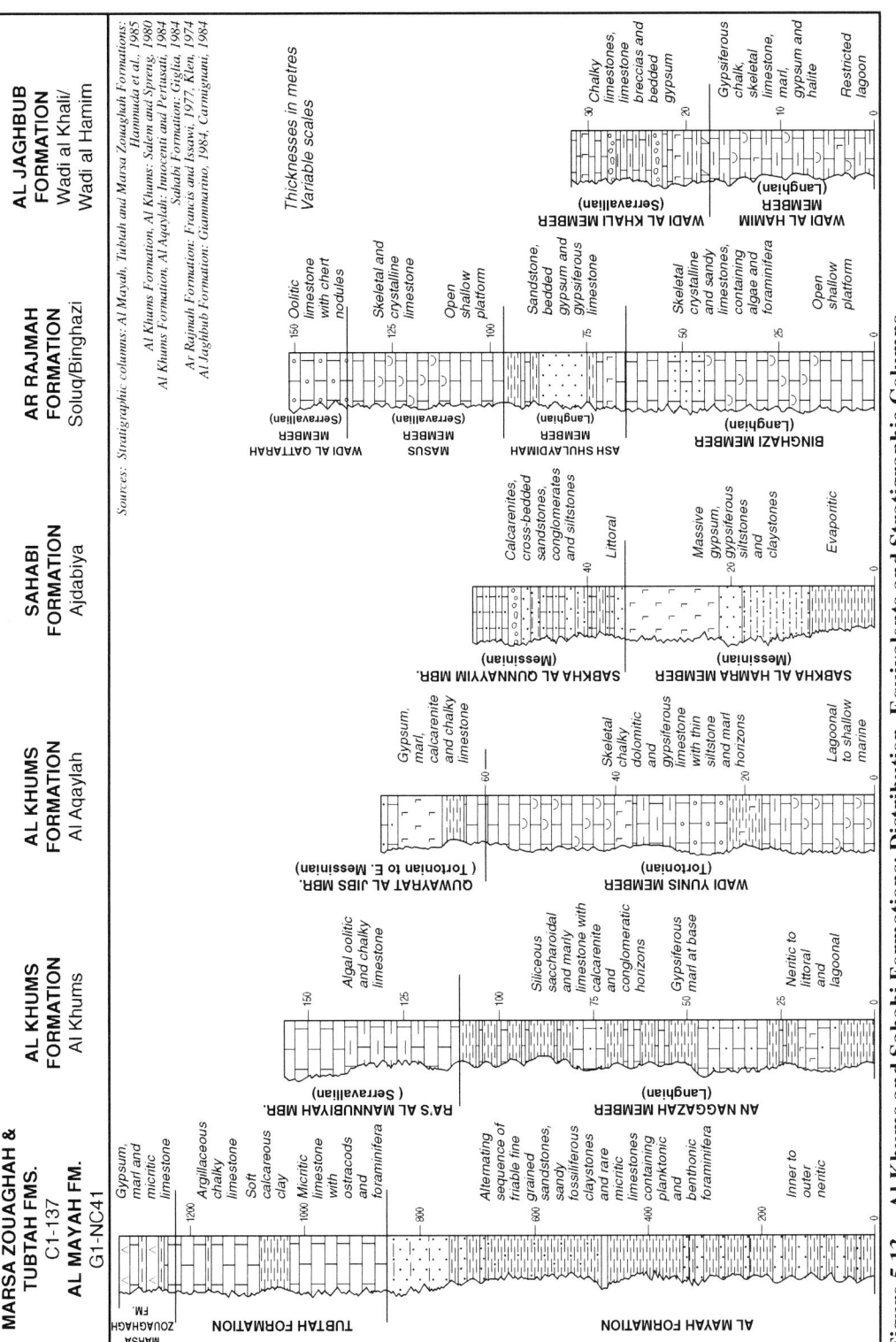

Figure 5.13 Al Khums and Sahabi Formations: Distribution, Equivalents and Stratigraphic Columns

Al Khums Formation outcrops along the coast from Al Khums to Al Aqaylah and inland to Sabkhat Ghuzayil. It is middle Miocene in age in the west, but late Miocene in the east. It exhibits a variety of nearshore and lagoonal facies. In the Ajdabiya Trough it is overlain by the late Miocene Sahabi Formation of lagoonal and evaporitic facies. Offshore the middle and late Miocene are represented by clastics, carbonates and evaporites of the Al Mayah, Tubtah, and Marsa Zouaghah Formations. In Cyrenaica only the Middle Miocene is present, represented by carbonates of the Ar Rajmah and Al Jaghbub Formations. Both the Sahabi and Marsa Zouaghah Formations show evidence of the Messinian salinity crisis. The Sahabi area is famous for its vertebrate faunas which have been found in both late-Miocene and early-Pliocene deposits.

Member passes into marine facies and becomes indistinguishable from the overlying Ar Rahlah Member.[83]

The Maradah Formation is not present on the Az Zahrah-Al Hufrah and Al Bayda Platforms, except for a few erosional remnants, but it outcrops extensively in the Zallah and Abu Tumayam Troughs, and presumably is present beneath the Al Haruj al Aswad volcanics. This suggests that subsidence of the Zallah and Abu Tumayam Troughs continued as late as mid-Miocene times. In these areas the two-fold subdivision is maintained, and the character of the two members remains distinct. Outcrops continue along the eastern margin of Al Haruj al Aswad, as far south as the Dur at Talah escarpment.[84]

The main outcrop of the formation continues east of Maradah to Jabal Zaltan, with a broad arc of outcrops extending around the southern margin of the Ajdabiya Trough (Figure 5.12). Important vertebrate remains have been recovered from the Jabal Zaltan outcrops which confirm a Burdigalian age. These faunas are very rich and include aquatic types such as sirenians and crocodiles (some up to 8m in length), and land mammals such as deinotheres and pigs, mastodons, rhinoceroses and giraffes. Root casts are also common in the fluviatile facies.[85]

5.4.2 Al Khums Formation

Evidence from the Maradah area makes it clear that the Al Khums Formation overlies the Maradah Formation (Figures 5.10 and 5.14). West of An Nuwfaliyah the Maradah Formation is not present and the Al Khums Formation rests unconformably on progressively older rocks until near Al Khums it overlies the Sidi as Sid Formation of Cenomanian age. There is considerable relief on the pre-Miocene surface. Rocks of Miocene age were recognised in this area by Floridia in 1939, but were not studied in detail until the 1960's. The Al Khums Formation was established by Mann for a sequence of about 70m of shallow-marine limestones exposed near Al Khums (Figure 5.13). At the type locality the Al Khums Formation consists of a basal unit of gypsiferous marls with pelecypods, followed by siliceous, saccharoidal and marly limestones and calcareous clays. A calcarenite horizon overlain by a conglomerate of disintegrated limestone pebbles marks the middle of the formation. The upper part contains fine-grained algal and oolitic limestones and contains a marine fauna of foraminifera, gastropods, pelecypods and ostracods, which collectively suggest a Langhian age. There is no evidence of early Miocene deposition in this area. The lithofacies suggest a number of local transgressions and regressions and the environment of deposition varies from neritic to littoral and lagoonal. The area around Al Khums was subsequently studied by Salem and Spreng. They recognised two members within the formation, a lower An Naggazah Member and an upper Ra's al Mannubiyah Member (Figure 5.13). The lower unit is characterised by a basal sand and reefal development along the former shoreline. The upper unit contains a limestone conglomerate overlain by chalky limestones. The base of the upper unit contains abundant specimens of the alveolinid foraminifer *Borelis melo*. They recognised two transgressive phases separated by a regressive cycle, and they concurred with a mid-Miocene age for the formation. Foraminiferal studies from outcrops near Al Khums suggest a Middle Miocene age, but ostracod faunas from the same area indicate a predominently Tortonian age.[86]

Outcrops of the Al Khums Formation can be traced along the coastal plain eastwards from Al Khums. The thickness averages 50m, but the lithology is quite variable. At Misratah the dominant lithology is fossiliferous and microcrystalline limestones with a fauna including some planktonic foraminifera. Reefoid mounds

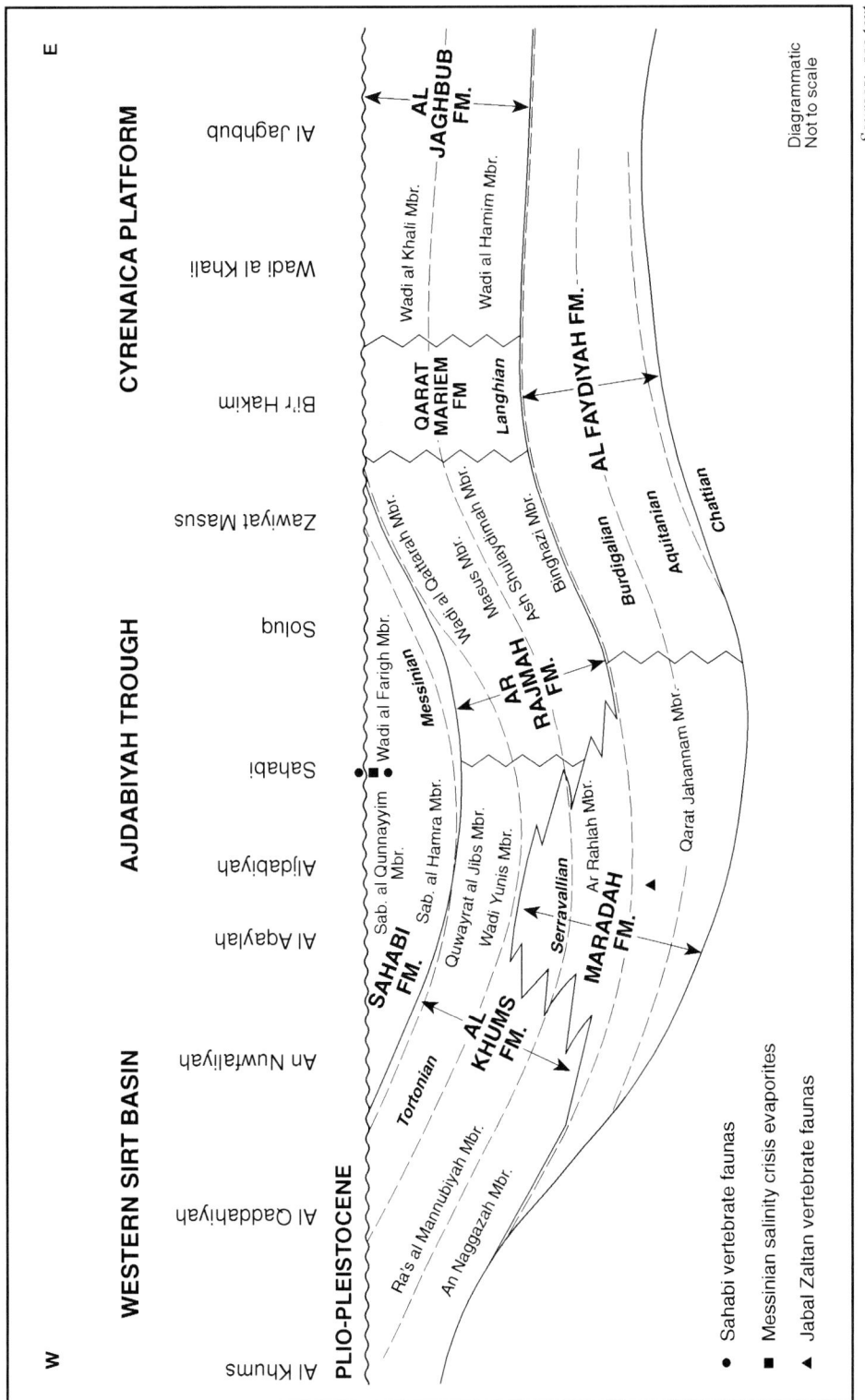

Figure 5.14 The Miocene in Libya: Stratigraphic Terminology and Relationships

The stratigraphic nomenclature of the Miocene in Libya is complex. This diagram shows the relationships between the formations and members which have been described onshore. The most striking feature is the apparent diachronous development of the Maradah Formation in the Al Aqaylah/Ajbabiya area. More work is needed to determine whether this is a real event or an artifact.

are common, with the alga *Lithothamnion* as the main reef-building organism. At Bani Walid the formation contains gypsiferous marls and marly limestones with coquinas of pelecypod shells. At Al Qaddahiyah and Abu Njaym the dominant lithology is chalky limestone which contains abundant marine fossils, including both planktonic and benthonic foraminifera, and reefoid mounds similar to those at Misratah. At An Nuwfaliyah the lithology ranges from dolomitic and crystalline limestones to limestones with thick interbeds of gypsum. These sequences exhibit a range of environments from lagoonal to open marine, and the age appears to be mid-Miocene.

There are a few isolated outliers of Al Khums Formation in the Wadi bu ash Shaykh and Zallah areas, but no outcrops have been found on the eastern flank of Al Haruj al Aswad.[87]

East of An Nuwfaliyah the Al Khums Formation overlies the Maradah Formation with a distinct unconformity. In the Al Aqaylah area the Al Khums Formation has been subdivided into two members, a lower Wadi Yunis Member and an upper Quwayrat al Jibs Member (Figure 5.13). The lower unit is represented by 60m of skeletal, chalky, dolomitic and gypsiferous limestones, with occasional siltstones, marls and bedded gypsum. The sequence contains a poor fauna of foraminifera and ostracods which are not age definitive, but suggest a likely Tortonian age. The member was deposited in lagoonal conditions with marine incursions. The Quwayrat al Jibs Member comprises about 15m of marls, bedded gypsum, calcarenites and sandstones, which contain a fauna of ostracods and foraminifera which indicate a Tortonian to early Messinian age. It is clear that the Al Khums Formation in this area includes rocks of a much younger age than in the Al Khums area where the upper section has been removed by erosion. It is also evident that, as currently defined, the Al Khums Formation is strongly diachronous in the Al Aqaylah/Ajbabiya area. The stratigraphic relationships between the Maradah and Al Khums formations are shown of Figure 5.14.[88]

In the Maradah area only the Wadi Yunis Member is preserved, and is composed of 30m of oolitic calcarenites passing northwestwards into dolomitic calcarenites with gypsum cement. The unit contains fragmentary remains of algae, foraminifera, pelecypods and echinoids, indicating frequent marine incursions. Similarly in the outcrops further east around Hutaybah only the lower member is present, showing about 25m of fluvial sands with vertebrate bones, and littoral calcarenites and dolomites containing corals and foraminifera. The age of these rocks is Tortonian to early Messinian. Outcrops reappear on the eastern flank of the Ajdabiya Trough where they are assigned to the Ar Rajmah Formation (section 5.4.10).[89]

5.4.3 Sahabi Formation

In the area southeast of Ajdabiya the middle Miocene sequence is overlain by about 55m of sandy limestones, claystones and evaporites which have been assigned to the Sahabi Formation (Figures 5.10 and 5.14). The formation was first named by Desio from outcrops around the fort of Sahabi, in the area famous for its vertebrate remains. Subsequently the formation was redefined to include only the late Miocene rocks of this area, whereas Desio also included the Pliocene rocks. Similarly, in their study of the Neogene sediments of the eastern Sirt Basin, Benfield and Wright established the Kalanshiyu (Calanscio) Formation for the late Miocene and Pliocene deposits of the area (Figure 5.13). Unfortunately both Desio's Sahabi Series and Benfield and Wright's Kalanshiyu Formation span the important Messinian event which marks the Miocene-Pliocene boundary. For this reason Giglia and others redefined the Sahabi

Formation to include only the late Miocene sequence, and referred the Pliocene sequences to the Qarat Weddah Formation.[90]

The Sahabi Formation, as revised, was subdivided into two members, a lower Sabkha al Hamra Member and an upper Sabkha al Qunnayyim Member. (Figure 5.13) The lower member unconformably overlies limestones of middle Miocene age, and is about 35m thick. It shows two different facies, a transgressive facies of calcarenites, friable sandstones and gypsiferous siltstones, and an upper regressive facies of gypsiferous claystones and siltstones and thick beds of selenitic gypsum up to 8m thick. The member contains a rich fauna which indicates a late Tortonian to early Messinian age. The overlying Sabkha al Qunnayyim Member, which reaches a thickness of 20m, represents a transgressive marine interval which overlaps the older member. It is composed of cross-bedded sandstones and sandy limestones, and represents a littoral environment of deposition. The age of the Sabkha al Qunnayyim Member is Messinian (Figure 5.14). In the area north of Sahabi the Sabkha al Qunnayyim Member passes into a high-energy sandbar facies. This facies has been called the Wadi al Farigh Member, but has only local distribution. The Sahabi Formation can be traced southwards as far as the Amal oilfield. In this area the two members can still be recognised but the sequence is considerably more sandy. The Sabkha al Qunnayyim Member contains vertebrate bones and shark's teeth.[91]

The celebrated vertebrate fauna of Sahabi, which was discovered by Italian soldiers in the late 1920's, has been studied in detail, particularly by an international group of specialists during the period 1975 to 1987. It is clear however that the stratigraphic terminology used by the group does not conform to the scheme developed by the IRC (Industrial Research Centre) geologists for the Geological Map of Libya. The Sahabi Formation as defined by IRC is Tortonian to Messinian in age and includes the evaporite sequence which correlates with the Messinian salinity crisis of the Mediterranean basin. The Sahabi study group, on the other hand, use the term Sahabi Formation for the richly fossiliferous horizons overlying the evaporite unit which are all Pliocene in age. The majority of the astonishing vertebrate fossils come from the Pliocene beds. The pre-evaporite beds at Sahabi, which form the floor of the Sabkha al Qunnayyim, contain patch reefs and a fauna of corals, pelecypods, gastropods and echinoderms. They also contain vertebrate remains, particularly sirenians and cetaceans. This unit represents the upper part of the Sabkha al Qunnayyim Member. It is overlain by an evaporite sequence up to 25m thick made up of gypsum with an admixture of sand and clay. This corresponds to the Messinian salinity crisis event. The fauna of the Pliocene beds is referred to in a later section. The Messinian crisis is discussed under the heading Marsa Zouaghah Formation.[92]

One other feature remains to be mentioned. In a contribution to the Deep Sea Drilling Project Barr and Walker described the Sahabi channel. The channel was discovered by seismic and magnetic surveys and has been penetrated by three exploration wells. It represents a deeply incised fluvial valley which reaches 480m below present sea level. Its course can be followed over a distance of 200km (including tributaries), and it ranges in width from less than 1km to more than 9km (Figure 5.13). The channel was penetrated by well J1-95, northwest of the Amal field, and is cut into sediments of late Miocene age. The channel is filled with sands of probable Pliocene age. Barr and Walker suggested that the channel was incised during the late Messinian, which lends support to the theory of an abrupt drop in sea level of at least 500m during the Messinian salinity event. A similar channel, which is 570m below present sea level, has been found beneath the Nile delta at Cairo.[93]

An outcrop of Sahabi Formation has been mapped on the western flank of the Ajdabiyah Trough, extending for 190km along the coast from An Nuwfaliyah to south of Al Aqaylah.[94]

NORTHWEST LIBYA OFFSHORE

5.4.4 Al Mayah Formation

As discussed above both the Dirbal and Ra's Abd Jalil Formations cross the Oligocene-Miocene boundary without any evidence of a break in sedimentation, and they probably extend in age to the top of the Burdigalian. They are overlain by a series of marine clastic rocks which Hammuda et al. named the Al Mayah Formation (Figure 5.10). The type section is defined in the G1-NC 41 well located on the southern flank of the Sabratah Basin (Figure 5.13). In the type well the formation is composed of 856m of interbedded shales and sandstones, with a distinctive 13m micritic limestone band which contains abundant fossil fragments. The sandstones are usually very fine-grained, and frequently argillaceous. The shales are calcareous and fossiliferous. The contact with the Dirbal Formation is conformable and indeed gradational, but the contact with the overlying Sidi Bannour Formation is sharp and distinct. The formation is present over much of the Sabratah Basin, but gradually thins to the west and north. In A1-NC 41 it is 634m thick, but in F1-137 only 419m thick. Further east in the C1-NC 35a well it is represented by 1090m of inner to outer neritic marly limestones, but it passes into thick bathyal shale facies in the region of A1-NC 35a (Figure 5.13). The great thicknesses encountered on the Pelagian Shelf give an indication of the rapid subsidence of this area during the Miocene. In Tunisia it is equivalent to the Mahmoud Formation, which has a similar lithology and fauna to the Al Mayah Formation. Both formations contain molluscs, bryozoa, gastropods and planktonic foraminifera which give a middle to late Miocene age. The formation therefore is age equivalent of the Al Khums Formation.[95]

5.4.5 Sidi Bannour Formation

Overlying the Al Mayah Formation in the Bouri area is a sequence of sandy claystones, mudstones and sandstones. Hammuda et al. named this unit the Sidi Bannour Formation and designated the E1-NC 41 well as the type section (Figure 5.10). In the type well the formation is 441m thick and it is overlain by the Pliocene Marsa Zouaghah Formation. The lithology represents deposition on an open shallow platform and the fauna comprises benthonic foraminifera, ostracods, gastropods, bryozoa, algae and echinoderms. The fauna is not age diagnostic but the formation is assumed to be late Miocene, most probably Tortonian. It has a similar age and lithology to the Saouaf Formation in Tunisia.[96]

5.4.6 Bi'r Sharuf Formation

The Bi'r Sharuf Formation (Figure 5.10) was defined by Hammuda et al. as a series of micritic and pelletal limestones, marls, and silty sandstones, with traces of gypsum towards the top, which overlie the claystones and sandstones of the Sidi Bannour Formation in the B1-137 well, close to the Tunisian median line. The thickness in the type well is 353m. The fauna of this formation has not been published, but it is believed to represent an open shallow shelf to restricted environment. It is overlain by the gypsiferous beds of the Marsa Zouaghah Formation and is therefore asssumed to be Tortonian in age.[97]

5.4.7 Tubtah Formation

Hammuda et al. established the Tubtah Formation (Figure 5.10) for a carbonate sequence overlying the Al Mayah Formation in the western part of the Sabratah Basin. The type section was defined in the C1-137 well, not far from the type well of the Bi'r Sharuf Formation. In this well 359m of micritic limestones and calcareous claystones overlie the Al Mayah Formation with an abrupt but apparently conformable contact (Figure 5.13). The limestones contain a meagre fauna of ostracods and benthonic foraminifera which is not age diagnostic. On the basis of stratigraphic position the formation is assumed to be Tortonian in age. It represents deposition on an inner shallow platform.

The differences between the Sidi Bannour, Bi'r Sharuf and Tubtah Formations are very slight. They are all Tortonian in age. They were all deposited in shallow shelf settings with similar faunas, and they have similar thicknesses. The Sidi Bannour Formation is predominantly a claystone unit whilst the Bi'r Sharuf and Tubtah Formations are dominated by micritic limestones, but the differences are insufficient to justify three separate formations, particularly since the type wells of the Bi'r Sharuf and Tubtah Formations are only 20km apart. On a regional scale it is clear that the dominant lithology of this unit is shallow-water, low-energy limestones with occasional pelletal and oolitic intervals, which compares closely to the Melqart Formation in Tunisia. On this basis the unit can be traced from the Tunisian median line at B1-137, through the C1-137, I1-137, F1-137 A1-NC 41, M1-NC 41, C1-NC 35a wells, and probably as far east as J1-NC 41. There is evidence however that east of E1-NC 41 the formation passes into a bathyal facies in wells D1-NC 35a and E1-NC 35a (Figure 5.13).[98]

5.4.8 Marsa Zouaghah Formation

The Marsa Zouaghah Formation (Figure 5.10) is the youngest of the Miocene formations in the Libyan offshore area. It is an evaporitic and carbonate unit which marks a peripheral manifestation of the Messinian salinity crisis, during which huge volumes of evaporites were produced in the deep Mediterranean basins. The formation was established by Hammuda et al. for the evaporitic sequence found in well C1-137, southwest of Bouri (Figure 5.13). In this well 93m of interbedded gypsum, marls, dolomites and micritic limestones overlie the carbonates of the Tubtah Formation with apparent conformity. The upper contact is unconformable with the silty sands of the Pliocene Assabria Formation. The limestones contain poor faunas of benthonic foraminifera, ostracods and gastropods. Detailed biostratigraphic work in other wells proves that the formation is Messinian in age. The formation has been studied by Mriheel and Alhnaish who published isopach, lithofacies and structure maps. These show that the facies passes from a marginal sabkha close to the present coastline to an open lagoon in the Bouri area. Further east the formation passes into a hypersaline lagoon facies in the NC 35a area. A patch of aeolian and fluvial clays and sandstone extends northwards from Tripoli to the edge of the hypersaline lagoon. The thickness reaches 105m in the Bouri area and 150m in well J1-NC 35a in the Misratah Trough. In Tunisia it is equivalent to the Oued Bel Khedim Formation which is composed of marls and bedded gypsum.[99]

A special study of the Messinian succession in well B1-NC 35a was conducted by van Hinte et al. with the purpose of establishing the relationship between the micropalaeontological record and the Messinian salinity crisis. They were able to show that the marl below the anhydrite is late Miocene in age and the marl above the anhydrite is Pliocene. The evaporite zone is barren of fauna. They established that in the B1-NC 35a well the evaporite zone spans a period of 0.8Ma. Estimates of palaeobathymetry show that the Messinian marls accumulated in a water depth of about 400 to 500m, the anhydrite in less than 30m, and the Pliocene marls in about 600m water depth. They concluded that sea level must have undergone a dramatic fall during the late Messinian, but recovered very rapidly in the early Pliocene. This is consistent with the deep channels excavated into the bedrock both onshore and on the continental shelf, and with the dessicated deep basin model of Hsu to explain the evaporites in the central Mediterranean basins.[100]

There is an extensive literature on the Messinian crisis. Evidence from the Oued Bel Khedim Formation in Tunisia was presented by Wiman, and more general discussions of the Messinian 'event' are given by Hsu and Sonnenfeld.[101]

SIRT BASIN

The Neogene succession in the southeastern Sirt Basin was studied by Benfield and Wright. They assigned the Lower and Middle Miocene sequences to the Maradah Formation, and the post-Middle Miocene section to a new formation which they named the Kalanshiyu (Calanscio) Formation. They demonstrated that the Maradah Formation can be traced well beyond the Sarir field, and they selected the T2-65 well as a representative section. In this well the Maradah Formation is 500m thick (Figure 5.12). They were able to demonstrate a passage from fluvial sedimentation in the south, through a shoreline complex extending north-westwards from As Sarir, to a marine environment at Jalu, Awjilah and An Nafurah. North of Jalu the thickness increases to over 850m into the Ajdabiya Trough. They recognised that in this area the Maradah Formation extends from Aquitanian to Serravallian in age and a mid-Miocene age was confirmed by the presence of *Borelis melo* in well C1-95. The continental deposits on the northern flank of Jabal az Zalmah, shown on the Geological Map of Libya as indifferentiated Neogene, may also be attributable to the Maradah Formation (Figure 5.12).[102]

The Maradah Formation is overlain by late Miocene and Pliocene rocks which Benfield and Wright named the Kalanshiyu (Calanscio) Formation. This formation is equivalent to the Al Khums, Sahabi and Qarat Weddah Formations of the Ajdabiya/Sahabi area, and is of questionable value since it pays no regard to the critical Messinian event which marks the top of the Miocene. The Kalanshiyu Formation was differentiated from the Maradah Formation by a change to a more fluvial environment in which carbonates are largely absent. This transition is illustrated in well S1-103 at Intisar where the change to fluvial deposition is very abrupt. The formation was laid down on an alluvial plain with meandering fluvial channels and occasional overbank deposits represented by thin claystone horizons, and shallow ephemeral lakes. Benfield and Wright chose the S1-103 well as representative of the Kalanshiyu Formation. The well contains 190m of fluvial sands which includes Pliocene sands which later authors have assigned to the Qaret Weddah Formation. The Kalanshiyu 'Formation' has a wide distribution in the southeast Sirt Basin, extending from south of As Sarir to the Ajdabiya Trough (Figure 5.13). The Kalanshiyu Formation exceeds 200m in thickness in the Hameimat Trough east of Jalu, but averages only 150m in the Intisar area. Figure 5.14 demonstrates the problems with Miocene stratigraphic nomenclature in Libya. In particular major confusion has been caused by the inconsistent definition of the Maradah and Al Khums Formations in different areas, and by the plethora of members which have been erected. It is difficult to believe that the situation shown on Figure 5.14 represents a realistic picture of stratigraphic relationships within the Miocene, and a thorough revision is urgently needed.[103]

CYRENAICA

Early and middle Miocene outcrops cover much of Cyrenaica, and these rocks have been extensively studied. From a petroleum point of view they have only marginal interest because of their surface or near-surface occurrence, but for the sake of completeness they will be reviewed briefly.

5.4.9 Al Faydiyah Formation

The sequence of marls and limestones overlying the Al Abraq Formation in the central Jabal al Akhdar was named the Faidia Formation by Pietersz. He recognised a lower marl member and an upper limestone member. A type section was selected near the village of Qaryat al Faydiyah, south of Shahhat. The formation was redefined during mapping for the Geological Map of Libya, but some differences of opinion have been expressed, both regarding the extent, and the age of the formation.

As redefined by Rohlich (Figure 5.10) in the type area the Al Faydiyah Formation rests unconformably on the Al Abraq Formation. The lower 4m are composed of marls, claystones and micritic limestones which contain *Nummulites fichteli*, which indicates a late Oligocene age, but the areal extent of the basal unit is limited. The upper member, 40m thick, consists of yellow, detrital limestones, which are frequently crystalline and dolomitized (Figure 5.12). The formation contains a rich fauna of pelecypods, gastropods, echinoids, algae, corals and foraminifera which indicate an early Miocene age.[104]

Eastwards in the Bamba area the lower part of the formation contains gypsum interbedded with claystones which suggests a local lagoonal facies in this area. The coastal outcrop extends to Tubruq where the facies shows a return to the yellow detrital limestones characteristic of the type locality. The thickness at Tubruq is 60m.

An isolated exposure at Al Burdi close to the Egyptian border shows 58m of yellow crystalline limestones, with abundant *Operculina*, pelecypods and echinoids, with minor marls and claystones (Figure 5.12). The formation is clearly transgressive and oversteps older rocks towards the south. Near Mekhili it rests on carbonates of the Al Majahir Formation of Campanian age. The thickness of the formation varies from a few metres to 150m in the central Jabal al Akhdar. The formation is present in the A1-NC 120 well, offshore Binghazi, where shallow-water carbonates containing *Miogypsina* and *Operculina* conformably overlie the Al Abraq Formation. It is also present in the A1-NC 128 well, offshore Bamba, where a similar facies is present, but here it rests unconformably on the middle Oligocene Al Bayda Formation.[105]

South of the Jabal Akhdar the Al Faydiyah Formation outcrops along the southern flank of the jabal, with a large lobe extending southwards in the area west of Bi'r Hakim (Figure 5.12), although there has been some disagreement on correlation in this area. The basal marl is conspicuous in the west but passes into carbonates in the east. The thickness in this area varies from zero, on the Cretaceous core of the Jabal al Akhdar, to 30-40m. A further isolated inlier of Al Faydiyah Formation is present at the Al Jaghbub oasis. Further south it passes beneath the middle Miocene rocks of the Ar Rajmah Formation.[106]

The foraminiferal fauna was studied by Eliagoubi from boreholes in the Darnah area and he reported planktonic foraminiferal assemblages extending into the middle Miocene. This is a unique observation. No middle Miocene forms have been recorded from any other locality where the Al Faydiyah Formation is exposed. El Hawat and Shelmani believed that the *Nummulites fichteli* found at the type locality are probably reworked, so it is likely that the Al Faydiyah Formation extends from Aquitanian to possibly Langhian in age (Figure 5.10).[107]

5.4.10 Ar Rajmah Formation

Desio introduced the name Regima Limestone for a transgressive carbonate sequence in the Binghazi area which progressively overlies rocks ranging from the Al Faydiyah Formation to the Eocene Darnah Formation. The formation represents the youngest Miocene sequence in Cyrenaica, and outcrops cover an area of 60,000km². It was subsequently redefined as the Ar Rajmah Formation. Many local names have been used, but during the remapping of the area for the Geological Map of Libya, four main members were recognised, from bottom to top the Binghazi Member, Ash Shulaydimah (Sceleidima) Member, Masus Member and Wadi al Qattarah Member (Figure 5.10).[108]

The Binghazi Member (Figure 5.13) comprises about 30m of crystalline, fossiliferous limestones, green fossiliferous clays and soft algal limestones, which are exposed on the plain around Binghazi and as far east as Bi'r Hakim. The member contains a rich fauna of both planktonic and benthonic foraminifera, including *Borelis melo*, which gives a Langhian age for this unit. It also contains the calcareous algae *Lithothamnion* and *Halimeda*. The member was deposited in an open shallow-platform to littoral environment. In the A1-NC 120 well a similar sequence is present with gypsum and gypsiferous limestones at the top which may equate with the overlying Ash Shulaydimah Member.[109]

The Ash Shulaydimah (formerly Sceleidima) Member (Figure 5.13) is characterised by porous gypsiferous limestone, gypsiferous clays and bedded gypsum, and sandstone with thin limestone bands. The unit is best exposed on the Ash Shulaydimah escarpment where it reaches a thickness of 30m. It thins eastwards and appears to pinch out in the Zawiyat Masus area. It represents a sabkha deposit with fluctuating marine and evaporitic episodes. The marine intervals contain a diverse pelecypod fauna indicative of a mid-Miocene, probably Langhian age.[110]

The Masus Member (Figure 5.13), which forms a broad plateau in the Soluq-Zawiyat Masus-Bi'r Hakim area, is composed of 40m of hard crystalline limestone, with a dolomitic and oolitic horizon at the base. The limestones contain a rich foraminiferal fauna including *Borelis melo*, and the member was deposited on an open shallow shelf.[111]

The uppermost member of the Ar Rajmah Formation is the Wadi al Qattarah Member (Figure 5.13), which was defined in the Wadi al Qattarah, southeast of Ar Rajmah. At this location the member is composed of 15m of soft oolitic limestone with chert nodules in the upper part. Further south tidal flat and salt marsh environments are found, separated from the open marine facies by a barrier beach. This member is only found in the Binghazi-Al Bayda area. The unit contains a fauna of benthonic foraminifera of probable Serravallian age, and was deposited in a high-energy littoral environment.[112]

The Ar Rajmah Formation outcrops over much of the western Cyrenaican Platform, and is present in the A1-NC 120 well (Figure 5.13). Eastwards it passes laterally into the Al Jaghbub Formation, and westwards into the Al Khums Formation. Late Miocene rocks were not deposited on the Cyrenaican Platform, and are not present in the A1-NC 120 well, but are present in the Ajdabiya Trough where the Ar Rajmah Formation is overlain by the Sahabi Formation (Figure 5.14).

5.4.11 Al Jaghbub Formation

The Al Jaghbub Formation (Figure 5.10) was introduced by Desio for a series of limestones and marls and thin sands overlying rocks of early Miocene age at the Al Jaghbub oasis and over a large area of the eastern Cyrenaican Platform. Westwards it passes laterally into the Ar Rajmah Formation (Figures 5.13 and 5.14). At Al Jaghbub the formation is 120m thick and is composed of hard, white to yellow fossiliferous limestones, soft clays, marls and thin sandstones. It outcrops extensively around Al Burdi and on the sea cliffs south of the town the thickness reaches 140m. In this area the limestone contains abundant echinoids, oysters, gastropods and bryozoa, and biohermal reefs are common.[113]

Near Bi'r Hakim a transitional facies between the Al Jaghbub and Ar Rajmah Formations is developed. This has been called the Qarat Mariem (originally Qarat Meriam) Formation, but has a very localised occurrence and probably does not merit formation status (Figure 5.14). It is a highly fossiliferous reefal limestone, about 14m thick, deposited on an open shallow platform. It contains abundant fossils including several which prove a Langhian-Serravallian age. This unit is only found in the Bi'r Hakim and Zawiyat Masus area.[114]

The Al Jaghbub Formation extends as far west as Wadi al Hamim beyond which it passes laterally into the Ar Rajmah Formation. In this area two members have been recognised, a lower Wadi al Hamim Member, 20m thick, characterised by gypsiferous chalk and calcarenite with interbeds of marl, gypsum and halite, and an upper Wadi al Khali Member, 14m thick, which is composed of chalky limestones, limestone breccia and bedded gypsum (Figures 5.10 and 5.14). The Wadi al Khali Member can be traced as far east as Al Muffawaz and Al Jaghbub. The Wadi al Hamim Member has been dated as Langhian, and the Wadi al Khali Member is Serravallian in age.[115]

5.5 PLIOCENE and QUATERNARY

Pliocene rocks are present at Al Assah on the Tunisian border (Al Assah Formation), at Al Qaddahiyah, south of Misratah (Al Hishah Formation), in the Ajdabiyah Trough, along the southern flank of the Cyrenaica Platform between Sahabi and Al Jaghbub (Qarat Weddah Formation), and offshore.

The Al Assah Formation (Figure 5.10) is an evaporitic deposit of massive gypsum which reaches over 200m in thickness in the C1-31 well. It is predominently Pliocene in age but may extend into the Villafranchian. Offshore in the Sabratah Basin 122m of marine nearshore sands have been described from well B1-NC 41 by Hammuda et al. as the Assabria Formation (Figure 5.10). It contains a fauna of benthonic foraminifera, gastropods, bryozoans, pelecypods and ostracods. It unconformably overlies the Marsa Zouaghah Formation and is assumed to be early Pliocene in age. A deeper-water unit overlies the Assabria Formation in well F1-NC 41, which Hammuda et al. named the Sbabil Formation. This formation, which is 570m thick in the well, is composed of sandy fossiliferous marls, which contain planktonic foraminifera of Zanclian age. The formation is developed on the upper continental slope, but is not present on the Medina Bank. It may equate to the Raf-Raf Formation in Tunisia which has a similar lithology.[116]

The Al Hishah Formation (Figure 5.10) is a deltaic to estuarine deposit of calcarenite and sandstone with gypsum, which reaches a thickness of 20m in the Al Qaddahiyah area. Similar deposits have been found at Al Aqaylah, Maradah, Bi'r Zaltan and Sabkhat Ghuzayil which have also been referred to the Al Hishah Formation. The age is Pliocene, but may extend into the early Pleistocene.[117]

The Qarat Weddah Formation (Figure 5.10) was defined at the Al Jaghbub oasis, and can be traced westwards to the Al Aqaylah area. The formation covers a range of environments, but is dominated by thick aeolian sands, with intervals of lacustrine clays and marls with traces of gypsum. However, silty clays containing planktonic foraminifera of early Pliocene age were penetrated in the A1-NC 120 well, offshore Binghazi, indicating an outer shelf setting in this location. The thickness of the Qarat Weddah Formation reaches 60m at Al Aqaylah, but thins towards the east. It clearly overlies the Al Hishah Formation where both formations are present, but at Ajdabiya and Wadi al Hamim only the Qarat Weddah Formation is present. The age is probably Pliocene to early Pleistocene. As discussed above. the Qarat Weddah Formation was included within Benfield and Wright's Kalanshiyu (Calanscio Formation). The southern limit of the formation beneath the Quaternary cover is difficult to determine, but it may extend as far as the As Sarir area.

Both the Al Hishah and Qarat Weddah Formations are younger than the Sahabi channel which was discussed above in the section on the Sahabi Formation. Vertebrate fossils at Sahabi have been recovered from late Miocene beds below the Messinian evaporites, but the bulk of the finds have been made in the post-Messinian beds, which have been dated as early to mid-Pliocene. These are usually assigned to the Qarat Weddah Formation. An astonishing range of vertebrates has been recovered from these beds including shark teeth, sirenian skeletons, toothed whales, turtles, crocodiles, birds and land vertebrates. The latter group includes snakes, birds, monkeys, carnivores, probiscids, equids, pigs, anthracotheres, hippopotamus,

giraffe, bovids, and three probable hominid fragments. In the Murzuq Basin Plio-Pleistocene lacustrine deposits, dated by ostracods, probably equate to the Al Hishah and Qarat Weddah Formations.[118]

Post-Pliocene sediments cover almost one third of the land surface of Libya. Alluvial fans of Pleistocene age are found in front of the Jabal Nafusah escarpment, with playa deposits at the foot of the slope, and cemented littoral sands along the coastline. Quaternary deposits include extensive sabkha deposits south of Misratah and in the Ajdabiya region, and in the interior a large Pleistocene lake developed in the Al Kufrah area, and extensive fresh-water limestones are found in the Murzuq Basin, but by far the most spectacular and extensive Quaternary deposits are the sand seas of the interior. Sand seas cover enormous areas, particularly in eastern Libya where the Great Sand Sea extends for 400km from Al Jaghbub to Jabal Zalmah. The Rabiyanah, Tibisti, Murzuq and Awbari sand seas cover equally extensive areas in south and west Libya. In most of these areas linear dunes are dominant, frequently reaching heights of 200m and extending for tens of kilometers. These dune fields present formidable obstacles to geological and geophysical prospectors. The classic studies on the morphology and movement of dune sand by Bagnold in the 1930s were conducted in the Great Sand Sea of eastern Libya. The mountainous areas of Jabal Awaynat, the Tibisti Mountains and the Akakus Mountains provide some of the most dramatic desert scenery in North Africa.[119]

Chapter 6

STRUCTURE

An outline of the structural development of north Africa in plate tectonic terms was given in Chapter 2. This chapter takes a closer look at the structural provinces of Libya in order to provide a framework for understanding the relationship between stratigraphy, structural development, hydrocarbon generation and hydrocarbon accumulation. The chapter will review the main structural provinces of Libya, and highlight the timing of the main tectonic events.

The broad outlines of the structure of Libya were established by Italian geologists in the 1920s and 1930s, but it was the arrival of the oil companies in the late 1950s which allowed the broad outlines to be converted into a detailed picture. Oil companies introduced gravity, magnetic and seismic techniques and over 9000 wells have been drilled, which have allowed a detailed picture of the subsurface structure to be compiled. Seismic acquisition and processing methods have become progressively more powerful and sophisticated, and specific techniques have been developed to resolve a wide range of problems from the velocity effects of dune sands to identifying stratigraphic traps. The increasing use of 3D seismic has greatly improved the reliability of fault mapping, and the application of seismic attribute analysis has allowed a much better definition of facies changes in the subsurface. Several generalised structure maps of Libya have been published. Goudarzi and Smith, Mikbel, and Anketell produced basement structure maps, Goudarzi and Smith produced a base Mesozoic map, and Wennekers, et al. produced a top Cretaceous structure map of the Sirt Basin.[1]

The African plate at the present time can be subdivided into either two or three sub-plates. Burke and Dewey favoured two sub-plates, a west African and an east African, but later authors recognised that the east African sub-plate is itself divisible into two along the line of the Central African shear and the south Sudan rifts. The Mediterranean is an amalgam of several microplates created during the opening and closing of the Tethys Ocean. Many of the sub-plates are separated by megashears, which have exerted a major impact on the tectonic history of the region. In Libya the junction between the west African and eastern African sub-plates is represented by a sinistral megashear which extends from the Tibisti Mountains along the western margin of the Sirt Basin, and perhaps continues as the eastern margin of the Pelagian Block (Figure 6.21). The northern margin of the African plate is represented by the Atlas Fold belt and the northern border of the Pelagian Block. The block foundered during the Miocene as a result of wrenching associated with the emplacement of the Calabrian arc. The southern border of the Pelagian Block is marked by the South Atlas Shear Zone which extends from Algeria through Tripoli to Cyrenaica as the Sabratah-Cyrenaica Fault Zone (Figure 6.21). These megashears propagated several important associated structures which have had a major influence on the tectonic history of Libya, and which will be discussed in this chapter. (Figures 2.7, 6.1 and 6.21).[2]

The structure of Libya will be analysed region by region approximately from south to north (Figure 6.1). Similarities and differences will be drawn between these regions, and an overall synthesis will be presented.

6.1 SOUTHERN LIBYA

6.1.1 Basement

The core of southern Libya is formed by Precambrian basement massifs in the form of a letter 'w', with the eastern prong represented by Jabal Awaynat, the middle prong by the Tibisti Massif and the western prong by the Tihemboka Arch (Figure 6.1). Jabal Awaynat and the eastern Tibisti Massif form part of the Archaean and Proterozoic Nile Craton, the Tihemboka Arch forms part of the Neoproterozoic East Saharan craton, and the western Tibisti Massif forms part of the Pan African remobilized belt. These terranes were emplaced during the late Precambrian Pan-African orogeny, accompanied by intense deformation, shearing and metamorphism. The main phase of the Pan-African orogeny was completed by about 540 Ma, but post-orogenic magmatism continued until about 440 Ma during which time granites and syenites were intruded. This final phase was accompanied by an extensional regime which produced a series of gentle north-south to northwest-southeast swells and troughs, including the Tihemboka, Tripoli-Tibisti, Haruj and Kalanshiyu-Awaynat Uplifts (Figure 2.6).[3]

Southern Libya Evolution

Despite patches of Cambro-Ordovician continental clastics preserved in grabens on the Tibisti Massif at Uzu and Bardai, it is likely that the exposed Precambrian terranes were being actively eroded during the Cambro-Ordovician and formed the principal sediment source for the thick clastic apron which was deposited over much of the passive northern margin of Gondwana. Palaeocurrent measurements lend some support to this concept. In the northeastern Tibisti Massif Selley recorded northeasterly palaeocurrent directions whilst the dominant direction in the Kufrah Basin is northwesterly to northerly. During the late Ordovician the Precambrian terranes were occupied by continental ice sheets, and outwash fans, glacio-marine diamictites and dropstones from these ice sheets are preserved in the Ordovician sediments of the Murzuq and Illizi Basins.[4]

By Silurian times erosion had considerably reduced relief on the Tibisti Massif and associated uplifts, and the Silurian transgression covered the Tihemboka, Tripoli-Tibisti, Haruj and Awaynat Uplifts, leaving perhaps only the Tibisti Massif emergent. Mid-Devonian tectonism in the Murzuq Basin is taken to represent the initial stage of the collision between Gondwana and Laurasia. Major uplift occurred and erosion removed Silurian and early Devonian sediments from a large part of the basin. The area was peneplained prior to the deposition of the Awaynat Wanin Formation, and it is likely that Upper Devonian and Carboniferous rocks blanketed the entire area. Tectonism increased in intensity during the late Carboniferous culminating in the Hercynian orogeny, which by the end of the Carboniferous had produced a new series of structural elements with an east-west to southeast-northwest orientation (Figure 2.7). The Kalanshiyu Trough was inverted to form the Tibisti-Sirt Uplift which remained a positive element until the mid-Cretaceous. By the end of the Carboniferous all of southern Libya was emergent and was subjected to prolonged erosion and continental deposition.[5]

Basement was probably exhumed at Jabal Awaynat, on the Tihemboka Arch and over much of the Tibisti-Sirt Uplift following the Hercynian orogeny. The subcrop pattern beneath the post-Hercynian deposits is shown on Figure 3.3. The situation in southern Libya remained essentially unchanged until the early

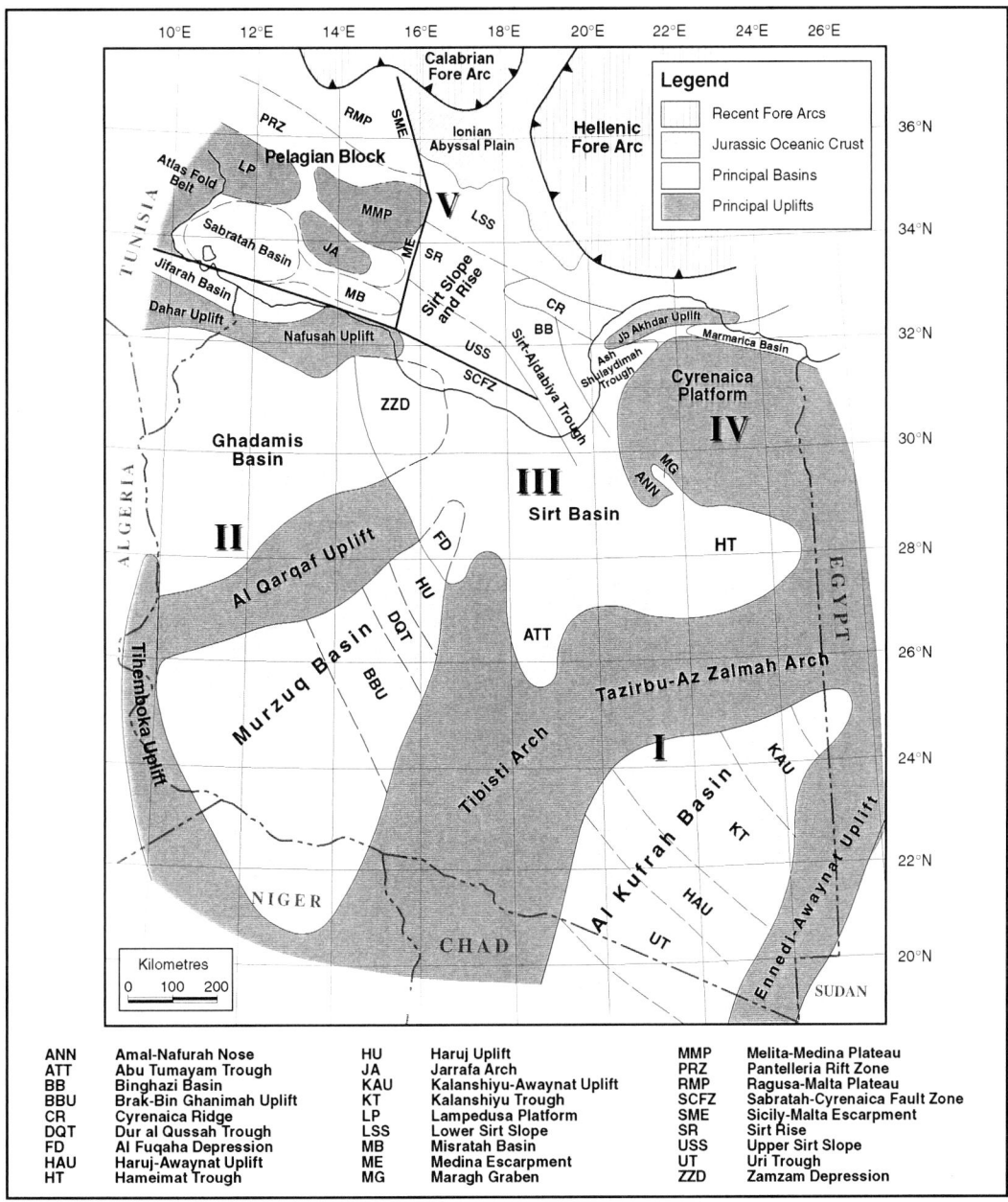

Figure 6.1   Libya, Principal Tectonic Elements

Libya can be divided into five main structural provinces: I - Southern Libya, including the Precambrian Basement terrains and Al Kufrah Basin. II - Western Libya, including the Murzuq Basin, Al Qarqaf Arch, Ghadamis Basin, Nafusah Arch and Jifarah Basin. III - Sirt Basin, including the Abu Tumayam Trough and the Hameimat Trough. IV - Cyrenaica, including the Cyrenaica Platform and the Jabal al Akhdar Uplift. V - Offshore, including the Sabratah Basin, Jarrafa Arch and Melita-Medina Platform. Further north, in the central Mediterranean, the offshore area makes contact with the Calabrian and Hellenic Fore Arcs.

Cretaceous when crustal extension in the Tethys Ocean resulted in the collapse of the Sirt Arch to form the Sirt Basin. Tethys began to close during the mid-Cretaceous due to a change in the direction of motion of the African plate relative to Europe, and the more rapid movement of northeast Africa compared with northwest Africa, has been attributed to the operation of two plates in Africa during the Cretaceous. The line of junction between the two plates extends from the Benue Trough in Nigeria to the Tibisti Massif. The collapse of the Sirt Arch produced an embayment which extended to the foot of the Tibisti Massif. This embayment was progressively flooded by successive transgressions, reaching a maximum flooding level during the Eocene.[6]

The closure of Tethys led to magmatic activity during the Cainozoic. Ring intrusions were injected into the Jabal Awaynat basement complex during the Eocene and extensive volcanic centres developed along the line of the old Tripoli-Tibisti Uplift from the Tibisti Massif to Gharyan (Figure 5.1). This line may also be connected with the plate boundary between the west African and east African plates. Volcanic activity continued in the Tibisti Mountains in Chad until recent times.[7]

6.1.2 Al Kufrah Basin

Al Kufrah Basin covers an area of 500,000km[2] in southeastern Libya, Chad, Sudan and Egypt. In Chad it is known as the Erdis Basin, in Sudan as the Mourdi Basin and in Egypt as the Dakhla Basin. It is surrounded by basement highs: Jabal Nuqay and Jabal Duhun to the west, the Borkou Massif and the Ennedi Mountains to the south in Chad, Jabal Awaynat and Jabal Azbah to the east and Jabal az Zalmah to the north. The distance from Jabal Zalmah to the Ennedi Mountains is 1000km (Figure 6.2). The basin is floored by basement rocks similar to those exposed at Jabal Awaynat which belong to the Nile craton suite. These rocks have been penetrated in both the A1 and B1-NC 43 wells. The total basin fill reaches 3500m in the northern part of the basin. Reviews of the Kufrah Basin have been published by Bellini and Massa, Bellini et al. (including a detailed aeromagnetic study), and Lüning et al. (with particular reference to the Silurian source rock potential).[8]

The present-day structure of the Kufrah Basin shows a northeast-southwest trend, largely as a result of Hercynian tectonics (Figure 6.2). An almost complete sequence of Palaeozoic rocks is preserved from Cambrian to late Carboniferous, overlain by 'Continental Post Tassilien' rocks of Permian, Triassic and Cretaceous age. The total basin fill reaches 4500m and two sub-basins have been recognised (Figure 7.2). The northern sub-basin, just south of the Kufrah oasis was evaluated by the two deep Agip wells, A1 and B1-NC 43. The southern sub-basin, close to the border with Chad, has not been drilled. These sub-basins may reflect the early Palaeozoic Kalanshiyu and Uri Troughs.[9]

The aeromagnetic survey published by Bellini, et al. showed several intrusive bodies with a northeast-southwest alignment, but the age of these intrusions has not been established. The basement is heavily block faulted with a strike-slip component which Bellini, et al. believed to be related to a basement megashear (Figure 6.2). The dominent fault pattern also has a northeast-southwest trend and fault blocks tilted towards the southeast have been interpreted from air photographs at outcrop, and from seismic data in the subsurface. The fragmentary depth-to-basement contours shown on Figure 6.2 have been interpreted from the magnetic data of Bellini et al.[10]

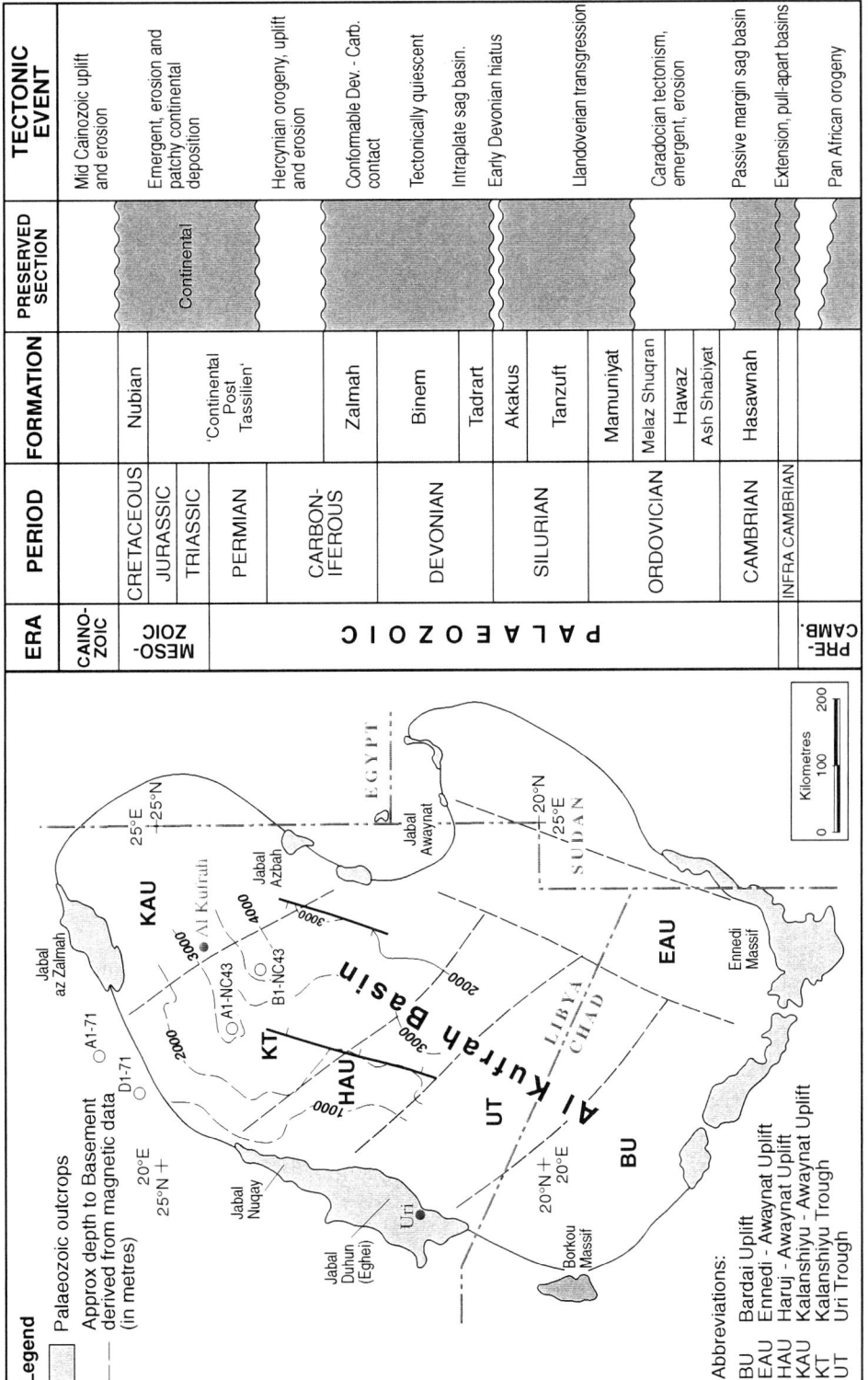

ERA		PERIOD	FORMATION	PRESERVED SECTION	TECTONIC EVENT
CAINO-ZOIC					Mid Cainozoic uplift and erosion
MESO-ZOIC		CRETACEOUS	Nubian	Continental	Emergent, erosion and patchy continental deposition
		JURASSIC	'Continental Post Tassilien'		
		TRIASSIC			
PALAEOZOIC		PERMIAN			Hercynian orogeny, uplift and erosion
		CARBON-IFEROUS	Zalmah		Conformable Dev. - Carb. contact
		DEVONIAN	Binem		Tectonically quiescent
			Tadrart		Intraplate sag basin.
			Akakus		Early Devonian hiatus
		SILURIAN	Tanzuft		Llandoverian transgression
		ORDOVICIAN	Mamuniyat		Caradocian tectonism, emergent, erosion
			Melaz Shuqran		
			Hawaz		
		CAMBRIAN	Ash Shabiyat		Passive margin sag basin
			Hasawnah		Extension, pull-apart basins
PRE-CAMB.	INFRA CAMBRIAN				Pan African orogeny

Sources: Luning, et al., 1999. Bellini, et al., 1991. Klitzsch, 1971.
Contours derived from magnetic data of Bellini, et al., 1991

Figure 6.2   Al Kufrah Basin, Tectonic Elements and History

Al Kufrah Basin extends from Libya into neighbouring Chad, Sudan and Egypt. It is a Palaeozoic basin and its periphery is ringed by Palaeozoic outcrops. The current NE-SW trend was imposed during the Hercynian orogeny, but the basement structure shows indications of the late Pan-African NW-SE trend.

Al Kufrah Basin Evolution

Following the Pan-African orogeny the Kufrah Basin shows evidence of extensional tectonics during the Infracambrian to early Cambrian with the development of pull-apart basins formed as a result of shearing along the Pannotian suture (Figure 2.3). During the early Palaeozoic the Kufrah Basin formed part of the passive margin of Gondwana on which transgressive fluvial and deltaic sediments were deposited in a ramp setting. As discussed above it is likely that the Tibisti Massif was emergent during the Cambro-Ordovician. Indeed it has probably remained emergent throughout the entire Phanerozoic. The Hasawnah Formation is very thick in the Kufrah Basin but the Ash Shabiyat, Hawaz and Melaz Shuqran Formations are absent. This hiatus reflects uplift and erosion during the Caradocian and is associated with known tectonism of this age in Algeria and Morocco. This episode was followed by the deposition of a thick glacio-marine sequence in the late Ordovician. Palynological evidence is lacking to prove the nature of the Ordovician-Silurian contact, but Grignani claimed that deposition of the Tanzuft Formation began in this area during the late Ordovician. In other areas the basal Silurian is represented by black shales of Rhuddanian age, but so far no evidence of black shales has been found in the Kufrah Basin. The facies of the Silurian sediments found in the two deep Agip wells is indicative of a shallow marine siltstones and sandstone environment, and outcrop samples from the periphery of the basin show a similar picture.[11]

There is some dispute about the amount of tectonic activity in the Kufrah Basin during the mid-Palaeozoic (Figure 6.2). Bellini and Massa invoked tectonic events with a northwesterly trend to explain the Silurian transgression and thickness variations between the basin and the surrounding uplifts, and a second tectonic cycle beginning in the late Devonian-early Carboniferous with a northeasterly trend. This conforms to Klitzsch's concept of two principal tectonic episodes during the Palaeozoic. Lüning et al. questioned the tectonic trend lines of Klitzsch as schematic and ambiguous, (a view shared by Selley), and they found no evidence of significant tectonism in the Kufrah Basin during the period from the Infracambrian extensional phase until the late Carboniferous (despite the absence of almost the entire Ordovician section). There is no palynological evidence for the presence of late Silurian or Lochkovian rocks, and deposition in the Devonian began in the Pragian and continued until the Visean. Lüning, et al. maintained that all of the observed unconformities, thickness changes, transgressions and regressions could be accounted for by eustatic sea level changes. They regarded the Palaeozoic in the Kufrah Basin as representing a tectonically quiescent period during which gentle subsidence produced an intraplate sag on the passive margin of Gondwana. This is in marked contrast to the Murzuq Basin in which was tectonism was very active particularly during the Devonian.[12]

The tectonic effects caused by the collision between Gondwana and Laurasia spread from northwest Africa into Libya during the Carboniferous as a compressional event which produced the Tibisti-Sirt Arch and the Ennedi-Awaynat Uplift, and obliterated large sections of the Tripoli-Tibisti Uplift, Haruj Uplift and Kalanshiyu Trough (Figure 2.7). The Kufrah Basin was isolated from the northern margin of Gondwana, and a continental regime was established which persisted until the Cretaceous. Marine sedimentation in the Kufrah Basin ended in the late Devonian, and there is an apparent hiatus in the continental rocks of the Carboniferous, with no evidence of any deposits of Serpukhovian to Gzelian age.

The tensional regime which led to the collapse of the Sirt Arch and the formation of the Sirt Basin in the Cretaceous also affected the Kufrah Basin. Major subsidence occurred accompanied by rifting, and thick Nubian Sandstones were deposited, particularly in the southern part of the basin. Evidence that marine conditions reached the Jabal Zalmah area during the Cretaceous has been provided by Lüning et al. who reported the presence of marine carbonates, shales and evaporites from a small outlier on the northern flank of the jabal (Figure 4.14). The basin was uplifted during the late Eocene and has been emergent ever since.[13]

6.2   WESTERN LIBYA

6.2.1 Murzuq Basin

The Murzuq Basin, like the Kufrah Basin, developed as an intracratonic sag on the passive margin of Gondwana during the Palaeozoic. It extends southwards into Niger where it is known as the Jadu Basin, and is surrounded by uplifts; Al Qarqaf Arch to the north, Tihemboka Uplift to the west, Jadu Plateau to the south and the Northwest Tibisti Ridge to the east. It covers an area of 320,000km^2, and contains a total basin-fill of up to 4000m of Palaeozoic and Mesozoic sediments (Figure 6.3).[14]

The present-day structure of the Murzuq Basin can only be determined by subsurface methods since the entire basin centre is covered by the Murzuq Sand Sea. A review of the structure of the basin was presented by Echikh and Sola. Previously a basement structure map of the northern part of the basin was published by Mikbel, and a generalised structure map on the top Ordovician by Meister, et al. showed a single basin depocentre at a subsea depth of −2000m located close to the A1-NC 58 well. It is now evident that this is a gross oversimplification. Pierobon produced a cross-section illustrating the importance of the Brak-Bin Ghanimah Uplift and the Tiririne High. Echikh and Sola showed seven principal tectonic

elements. From west to east these are the Tihemboka Arch, Al Awaynat Trough, Tiririne High, Awbari Trough, Idhan Depression, Brak-Bin Ghanimah Uplift and Dur al Qussah Trough (Figure 6.3). A map of the basin is shown on Figure 6.3 and a cross-section through the basin on Figure 6.4. The depth contours to the top Ordovician on the map are from Echikh and Sola.[15]

Tihemboka Arch

The Murzuq Basin is bounded to the west by the Tihemboka Arch, which has been a positive feature since early Palaeozoic times. It roughly follows the line of the Libyan-Algerian border. It affected sedimentation throughout the Palaeozoic and has been intermittently reactivated since then. Uplift during the Devonian shifted the main depocentre of the Murzuq Basin from the Al Awaynat Trough to the Idhan Depression, and reactivation during the Hercynian orogeny separated the Murzuk Basin from the Illizi Basin. Precambrian basement is exposed in the south in the Adrar in Yahia area, and further north the N1 to N4-1 wells west of Atshan proved a total Palaeozoic thickness of only 1200m (Figure 6.6). The arch plunges northwards and is last seen in the Zarzaitine-In Amenas area of Algeria.[16]

Al Awaynat Trough

In the area east of Al Awaynat a narrow trough is preserved in the re-entrant between the Tihemboka Arch and the Tiririne High. It contains a significant Lower Devonian section which is not present on the Tiririne High to the east. Southwards it passes into the Idhan Depression, and to the north it is cut by the Tumarolin wrench fault.[17]

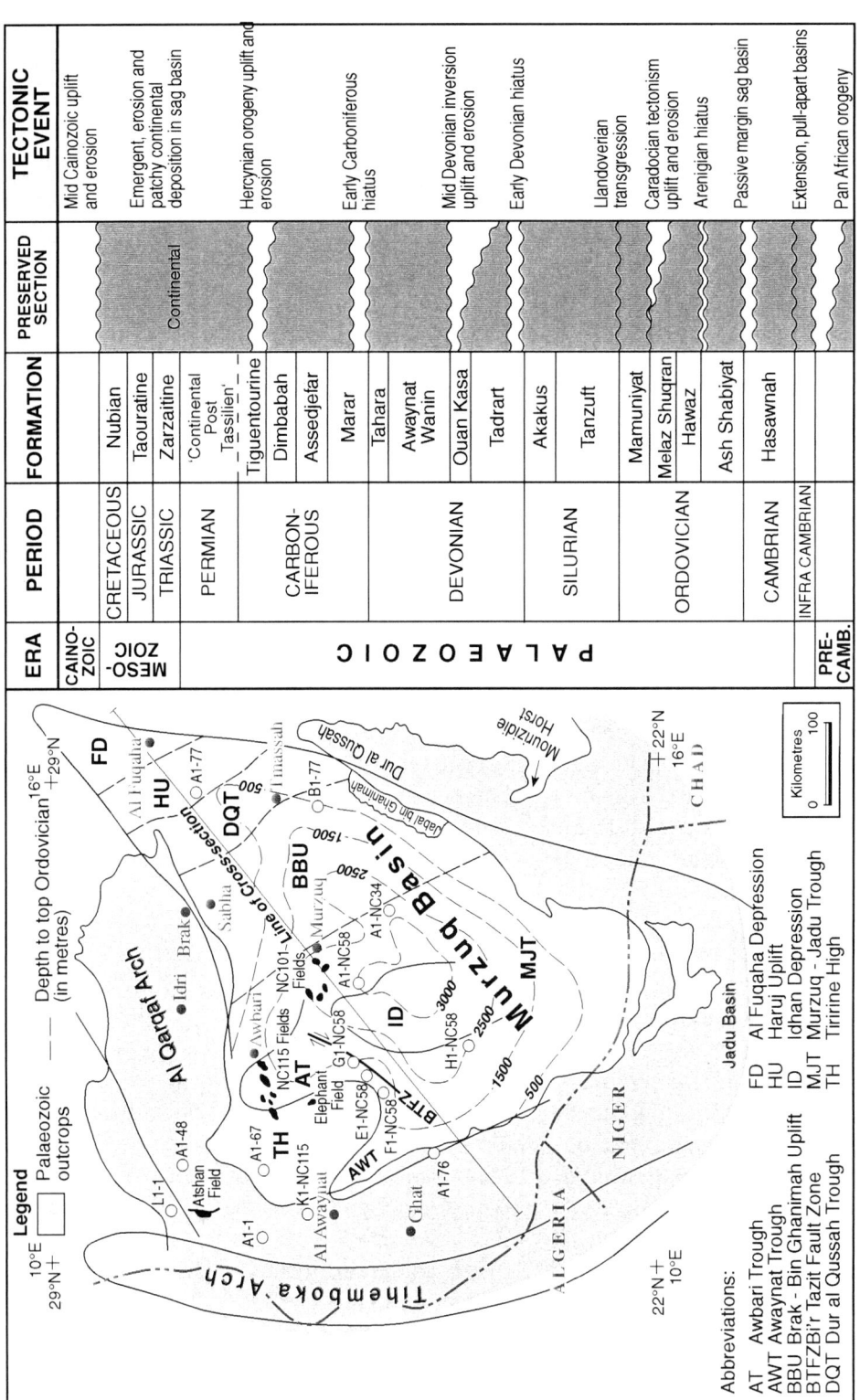

Sources: *Klitzsch, 1966, Pierobon, 1991, Meister et al., 1991, Sutcliffe, et al., 2000, Echikh, 2000, Echikh, 2000*
Contours modified from Echikh and Sola, 2000

Figure 6.3 Murzuq Basin and Al Qarqaf Arch: Tectonic Elements and History

The Murzuq Basin is a Palaeozoic sag basin, which underwent a major inversion during the mid-Devonian. The current SW–NE trend was imposed during the Hercynian orogeny, but the basement structure retains indications of the earlier Pan-African trend. The Awaynat Trough and Idhan Depression are early Devonian depocentres, whilst the Tirrine High and Brak-Bin Ghanimah Uplift reflect the mid-Devonian uplift and erosion.

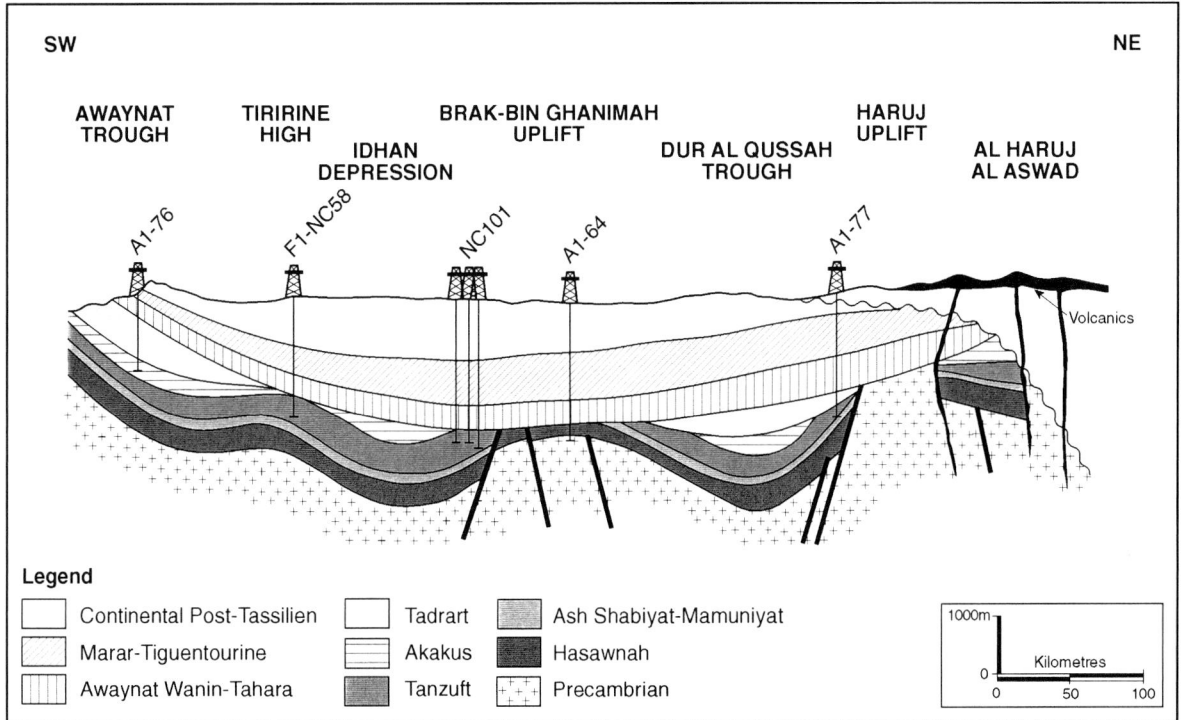

Sources: Echikh, 2000, Sutcliffe, et al., 2000, Pierobon, 1991, Klitzsch, 1971

Figure 6.4 Murzuq Basin, Diagrammatic Cross-Section

This section across the Murzuq Basin shows the importance of the Brak-Bin Ghanimah and Haruj Uplifts during the early Palaeozoic, and the regional mid-Devonian tectonic event which folded, uplifted and peneplained the entire area. Only a few small inliers of Tadrart Formation are preserved beneath the unconformity. The section also shows the possibility that a remnant of the Kalanshiyu Trough may be preserved beneath the Al Haruj al Aswad volcanic centre.

Tiririne High

The Tiririne High is bounded to the west by the north-south Tumarolin wrench fault, which according to Echikh and Sola, shows a dextral displacement in Carboniferous formations near the Tihemboka Arch of up to 20km. En echelon folds flank the fault, especially in the Atshan area. Further south the Tiririne High is flanked by the Al Awaynat Trough. Two further dextral wrench faults, which can be traced over a distance of 150km, have been mapped in a highly disturbed area in the area of wells C1, F1 and G1-NC 58 (Figure 6.3). On seismic sections these faults exhibit compressional flower structures. The Hawaz and Melaz Shuqran Formations are thin or missing on the Tiririne High suggesting Caradocian erosion in this area. The area was reactivated during the early Carboniferous when both the Marar and Assedjefar Formations were removed by erosion.[18]

Awbari Trough

The area between the Tirirene High and the Brak-Bin Ghanimah Uplift is occupied by the Awbari Trough, commercially important because it contains the NC 115 and NC 174 oilfields. In the eastern part of the trough reverse faults with a northwest orientation perhaps represent reactivation of Pan-African fault zones. The Ordovician sequence thins on the upthrown side of the reverse faults and in many cases roll-over is present on the overriding fault block. The trough subsided during the mid-Devonian, and formed the main axis of the basin during the later Palaeozoic.[19]

Idhan Depression

The Idhan Depression marks the main depocentre of the basin from mid-Devonian times when it subsided rapidly. It is separated from the Awbari Trough only by the disturbed zone of the Bi'r Tazit wrench fault (Figure 6.3). The NC 101 fields lie around the northern rim of the depression. In many cases the structures are formed by drape of Silurian shales over relict Mamuniyat glacial topography. Typically these structures have a vertical relief of up to 70m. Other structures include Hercynian anticlines, some of which are faulted, with closures up to 40m, and anticlines affecting the Devonian and younger horizons, but not the Ordovician and Silurian.[20]

Brak-Bin Ghanimah Uplift

The Brak-Bin Ghanimah Uplift, sometimes known as the Traghan High, forms part of the Tripoli-Tibisti Uplift of Klitzsch. This feature originated, like the Tihemboka Arch, during the early Palaeozoic and remained active until the mid-Devonian. The Hawaz and Melaz Shuqran Formations are thin or absent in this area, due to Caradocian erosion, and the Lower Devonian and Silurian section is missing due to mid-Devonian erosion so that Mamuniyat Formation underlies Awaynat Wanin Formation. Evidence from wells such as C1-NC 101 close to the Tanzuft edge show no indication of littoral facies suggesting that the edge is erosional rather than depositional (Figure 8.4). The uplift

subsequently became inactive and thereafter the area formed part of the greater Murzuq Basin.[21]

Dur al Qussah Trough

As originally defined by Klitzsch the Dur al Qussah Trough forms a counterpart to the adjacent Tripoli-Tibisti Uplift (Figure 2.6). It trends northwest-southeast and was active during the early Palaeozoic when thick sequences of Cambro-Ordovician, Silurian and Lower Devonian rocks were deposited. Like the Brak-Bin Ghanimah Uplift it became inactive following the mid-Devonian tectonism and became part of the greater Murzuq Basin during late Devonian and Carboniferous times. Echikh and Sola mentioned the presence of numerous northeast-southwest oriented structures in this area, affected by a group of wrench faults visible on the Devonian outcrops. In the Mourizidie area the eastern margin of the trough is faulted, with Devonian outcrops on the downthrown side juxtaposed against Precambrian basement to the east, suggesting a vertical displacement of around 1000m. A trough in the Al Fuqaha area in which thick Silurian shales were found in a water well may represent a preserved fragment of the Proterozoic – early Palaeozoic Kalanshiyu Trough.[22]

Murzuq Basin evolution

Evidence from sedimentological studies has confirmed that the Al Qarqaf Arch is a Hercynian feature, and that during the early Palaeozoic the basin was open to the north. Palaeocurrent measurements by Selley demonstrated that the Murzuq Basin was not finally closed off to the north until the Mesozoic. The tectonic history of the Murzuq Basin can be divided into three phases. During the early Palaeozoic the Murzuq Basin was represented by several northwest-southeast troughs and uplifts which were open to the north. After the mid-Devonian tectonism a greater Murzuq Basin became established, still open to the north, and after the Hercynian orogeny the northward opening was closed by the formation of the Al Qarqaf Arch. Two major periods of subsidence can be identified, during the Permian and late Cretaceous-Palaeocene, followed by periods of uplift in the late Permian and late Cainozoic. The magnitude of the Permian event is difficult to assess since the evidence has largely been removed as a result of erosion.[23]

The western part of the basin is floored by Precambrian basement belonging to the East Saharan craton, which is exposed at Anay and in the Hoggar Mountains of Algeria. The eastern part of the basin is floored by metamorphic rocks of the Pan-African remobilised belt which are exposed in the Tibesti Mountains, in several inliers on the Qarqaf Arch and on the western flank of Al Haruj al Aswad. Pan-African rocks also form basement in the Illizi Basin of Algeria (Figure 3.1). On the eastern and southern flanks of the basin sporadic occurrences of continental sandstones of Infracambrian to Cambrian age suggest the presence of possible pull-apart basins as described in the Kufrah Basin. It has been suggested, on the basis of Landsat data, that basement in the Murzuq Basin shows evidence of strike-slip faulting formed as a result of intraplate stresses, which exerted a controlling influence on later tectonic trends.[24]

The absence of Hasawnah Formation in well O1-1 on the flank of the Tihemboka Arch suggests that this area may have been emergent at this time. However, the early Palaeozoic succession in general is much more complete in the Murzuq Basin than in the Kufrah Basin. An almost complete sequence of late Cambrian and Ordovician rocks is present which can be divided into three transgressive megasequences: Hasawnah-Ash Shabiyat, Hawaz-Melaz Shuqran, and Mamouniyat. The Caradocian tectonism, which is clearly defined in Morocco and Algeria, is reflected in a widespread erosional unconformity at the base of the Mamuniyat Formation. Outcrop data suggest that Cambro-Ordovician rocks thin over the Tripoli-Tibisti Uplift and Haruj Uplifts (Figure 2.7), which were evidently positive features during the Cambro-Ordovician. Sequences of this age however are well developed in the Dur al Qussah Trough, which can be regarded as a depression between the two uplifts.[25]

In contrast to the Kufrah Basin, the Silurian shows an excellent development of the Rhuddanian hot shale, which accumulated in hollows and glacially produced channels on the Ordovician surface (Figure 7.5) prior to flooding of the basin by open marine conditions. Silurian outcrops on the eastern flank of the Murzuq Basin show a similar picture to the Cambro-Ordovician; thin or absent on the uplifts and thick in the troughs. On the Brak-Bin Ghanimah Uplift the Silurian is absent, with the erosional edge located just east of the NC 101 oilfields (Figure 8.4). The Silurian transgression was terminated by the development of prograding deltas in the late Silurian, represented in the Murzuq Basin by the Akakus Formation. Late Silurian and Lochkovian rocks are not present in the basin and their absence marks a hiatus at the Silurian-Devonian boundary.

The Devonian in the Murzuq Basin was greatly affected by mid-Devonian inversion and erosion (Figure 6.4). Pierobon and Meister et al. demonstrated that the Lower Devonian is absent in the centre of the basin, and Sutcliffe, et al. mapped the extent of the inversion. Echikh and Sola published a subcrop map showing the eroded sequences underlying the mid-Devonian unconformity, and the transgression of the overlying middle and upper Devonian rocks (Figures 3.39 and 3.20). The inversion took place in stages, beginning with an unconformity at the base of the Eifelian, reaching a climax with a regional unconformity at the base of the

Givetian and closing with a less widespread unconformity at the base of the Frasnian. Regional data suggests that the Tihemboka Arch was also emergent during much of the Devonian. Middle and late Devonian rocks were deposited on the eroded surface left after the mid-Devonian peneplanation. The Tirirene High remained emergent for much of the middle and late-Devonian and was only finally covered during the Famennian. Similarly the Brak-Bin Ghanimah Uplift was covered during the Frasnian. Upper Devonian rocks outcrop on the southern flank of the Qarqaf Arch, where they contain commercial deposits of sedimentary ironstone. The mid-Devonian tectonism was a turning point in the development of the Murzuq Basin. It marked the end of the early Palaeozoic tectonic influence and the beginning of the Hercynian tectonic regime (Figure 6.4).[26]

During the Carboniferous marine conditions persisted for much longer than in the Kufrah Basin, continuing until Moscovian times. The Hercynian orogeny reached its climax during the late Carboniferous and produced major tectonic changes in west Libya. The Qarqaf Arch was uplifted and subjected to severe erosion. It has remained emergent to the present day. The uplift effectively separated the Murzuq Basin from the Ghadamis Basin, although a narrow saddle remained to the west of the Arch. From the Permian to the mid-Cretaceous protracted erosion occurred in the Murzuq Basin and enormous thicknesses of continental Zarzaitine, Taouratine and Nubian sands accumulated which reach 1500m in the basin centre. The eroded rim of the upper part of this sequence, locally known as the Messak Sandstone, forms a dramatic escarpment west of Awbari and around the western margin of the basin.

The events in Tethys during the Cretaceous had little impact on the Murzuq Basin, and in contrast to the Kufrah Basin there is no evidence of any significant rifting or warping during the Cretaceous. Gentle regional uplift and tilting during the Cainozoic ensured that the basin remained emergent during this entire period.[27]

6.2.2 Al Qarqaf Arch

Al Qarqaf Arch is a conspicuous WSW-ENE structural feature separating the Murzuq Basin from the Ghadamis Basin. It covers an area of 25,000km^2 at outcrop and its core is formed by the Jabal Hasawnah, rising to an elevation of 800m. The present-day structure of the arch shows a major uplift (but not a horst) which exposes Palaeozoic and Precambrian rocks at an elevation of +800m, compared with a depth to basement of –5180m in the Ghadamis Basin to the north. Well evidence shows that the Arch plunges in the subsurface both to the southwest and northeast. The arch is onlapped by Ordovician rocks on the western plunging nose and by late Devonian rocks on the southern flank. The northern flank is concealed beneath Cretaceous cover. Outcrop and photogeological studies show a fault and fracture pattern with a dominent WSW-ESE trend.

Five small inliers of Precambrian basement are present on Jabal Hasawnah. These comprise anorogenic granites and low grade metasediments, which have been dated as Pan-African (551 to 476 Ma). Both the granites and metasediments retain evidence of early Proterozoic relict ages. These rocks form part of the Pan-African remobilised belt which is widely exposed in the western Tibisti Mountains.[28]

The Upper Cambrian Hasawnah Formation is exceptionally well exposed on the Jabal Hasawnah, and has been described in chapter 3. The overlying Ordovician section has been difficult to elucidate due to rapid facies changes, local onlap and extensive channelling. Both the Melaz Shuqran and Mamuniyat Formations show evidence of periglacial activity. It has been suggested that in this area, by the time of deposition of the Mamuniyat Formation the ice sheet was in retreat, but other evidence suggests that deeply incised valleys in the Mamuniyat Formation on the western Qarqaf Arch were excavated by glacial action.[29]

As mentioned above, palaeocurrent evidence suggests that Al Qarqaf Arch was not a positive feature during the early Palaeozoic, and that the Murzuq Basin was open to the north. The situation on Al Qarqaf Arch is complicated by the mid-Devonian inversion in the northern Murzuq Basin which was discussed in the previous section, and by the subsequent removal of much evidence during the erosion following the Hercynian Unconformity. It is likely that the northwest-southeast Pan-African trend determined depositional patterns during the early Palaeozoic, and that this episode was terminated by the mid-Devonian inversion in the Murzuq Basin. Subsequent Upper Devonian and Carboniferous sediments were laid down in a broad, shallow basin which probably extended over Al Qarqaf Arch and perhaps far to the east. (Carboniferous inliers have been found in the Sirt Basin). This phase was terminated by the Hercynian orogeny which created the Al Qarqaf Arch, from which almost the entire Palaeozoic section has since been eroded. The arch has continued to influence sedimentation to the present day. Subcrop pinch-out edges on the flanks of the arch reflect this complex evolution.[30]

6.2.3 Ghadamis Basin

The Ghadamis Basin, like the Al Kufrah and Murzuq Basins, is a large intracratonic sag basin developed on the passive northern margin of Gondwana. It covers an area of 350,000km^2, with the basin centre located in Algeria. The Libyan portion represents the eastern flank of the basin, rising towards the Tripoli-Tibisti Uplift, with a small sub-basin, the Zamzam Depression extending towards Misratah. The basin contains up to 6000m of basin fill in Algeria, but not more than 5200m in Libya. The basin is bounded in the west by the Amguid-El Biod Uplift in Algeria, to the south by the Hoggar Massif in Algeria and the Qarqaf Arch in Libya, and to the north by the Dahar-Nafusah Uplift. To the east the basin wedges out beneath the western part of the Sirt Basin (Figure 6.5).

The most conspicuous feature of the basin is the Hercynian unconformity which truncates the Palaeozoic succession and is overlain by a Mesozoic basin (the Hamadah Basin in Libya) with a markedly different basin configuration to that of the Palaeozoic. The tectonic history of the basin has some similarities with the Murzuq Basin, but some important differences, due to its location closer to the continental margin and to the Tethys Ocean. A map of the basin is shown on Figure 6.5 and a cross-section of the Palaeozoic Basin on Figure 6.6, and a cross-section of the overlying Mesozoic Basin on Figure 6.7. The depth contours to the top Ordovician shown on Figure 6.5 are from Echikh.[31]

Ghadamis Basin Evolution

The early Palaeozoic history of the basin was controlled by the northwest-southeast Pan-African tectonic trend, although little evidence of this has come to light due to the thickess of the overlying sediments. The basin narrows southwards, confined between the Tripoli-Tibisti and Tihemboka Uplifts, into the Murzuq Basin as discussed above. To the east the Palaeozoic section pinches-out against the Tripoli-Tibisti Uplift, but to the west the Tihemboka Uplift did not extend further north than the Edjeleh field, and the basin widened into a broad depocentre extending into Tunisia and Algeria west of the Tihemboka Uplift. Basement in much of this area is formed by Pharusian accreted terranes, but further south, and particularly in the Illizi Basin of Algeria, it is represented by rocks of the Pan-African remobilised belt.[32]

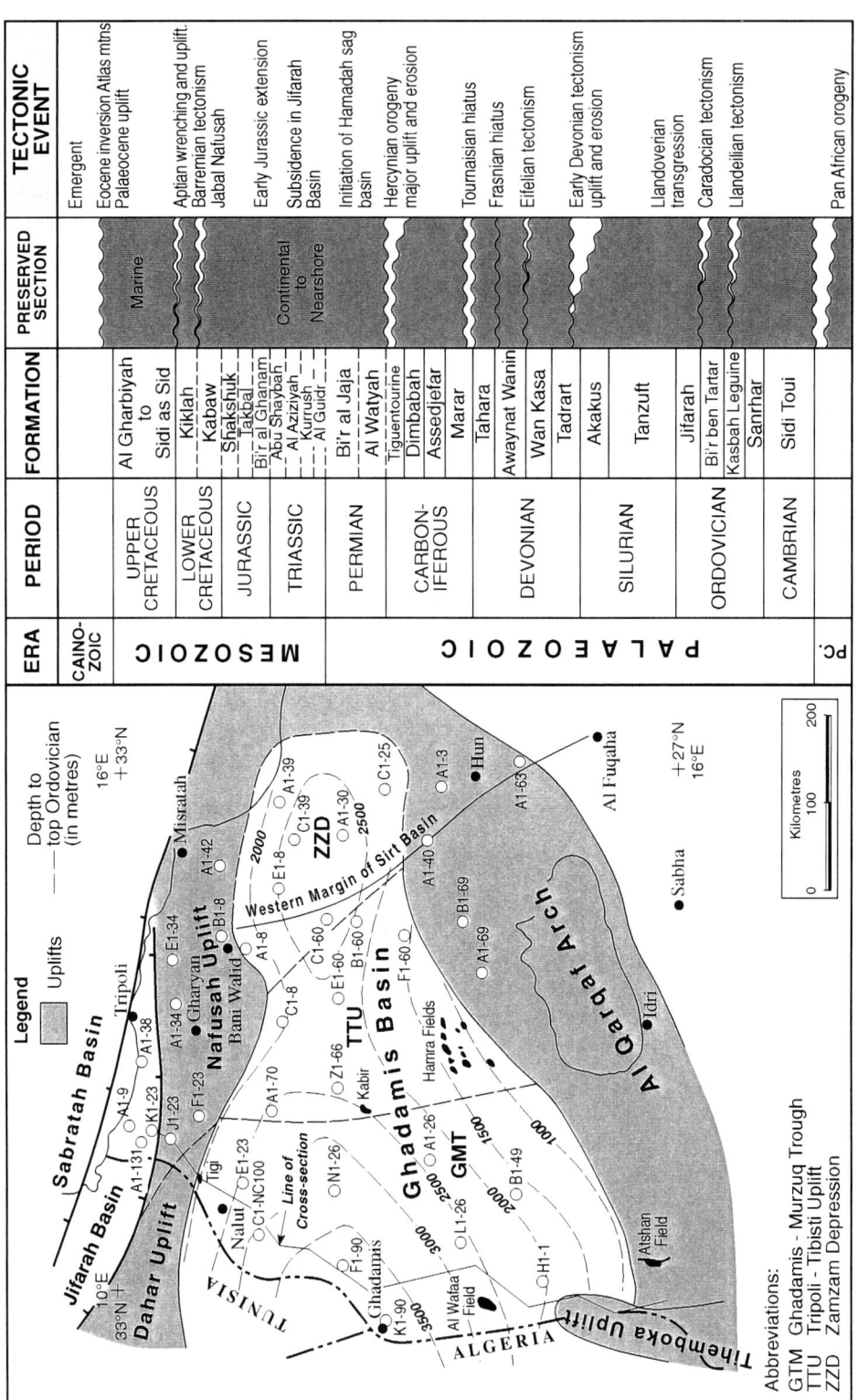

The table within the figure:

ERA	PERIOD	FORMATION	PRESERVED SECTION	TECTONIC EVENT
CAINO-ZOIC				Emergent
MESOZOIC	UPPER CRETACEOUS	Al Gharbiyah to Sidi as Sid	Marine	Eocene inversion Atlas mtns Palaeocene uplift
	LOWER CRETACEOUS	Kiklah / Kabaw / Shakshuk / Takbal		Aptian wrenching and uplift. Barremian tectonism Jabal Nafusah
	JURASSIC	Bi'r al Ghanam		Early Jurassic extension
	TRIASSIC	Abu Shaybah / Al Azziyah / Kurrush / Al Guidr	Continental to Nearshore	Subsidence in Jifarah Basin
PALAEOZOIC	PERMIAN	Bi'r al Jaja / Al Watyah		Initiation of Hamadah sag basin
	CARBON-IFEROUS	Tiguentourine / Dimbabah / Assedjefar / Marar		Hercynian orogeny major uplift and erosion
	DEVONIAN	Tahara		Tournaisian hiatus
		Awaynat Wanin / Wan Kasa / Tadrart		Frasnian hiatus
				Eifelian tectonism
		Akakus		Early Devonian tectonism uplift and erosion
	SILURIAN	Tanzuft		Llandoverian transgression
	ORDOVICIAN	Jifarah / Bi'r ben Tartar / Kasbah Leguine / Sanrhar		Caradocian tectonism
				Llandeilian tectonism
PC.	CAMBRIAN	Sidi Toui		Pan African orogeny

Sources: Echikh, 1998, Boote, et al., 1998, Belhaj, 1996, Kruseman and Floegel, 1980, Klitzsch, 1971
Contours from Echikh, 1998

Figure 6.5   Ghadamis Basin and Adjacent Areas, Tectonic Elements and History

The Ghadamis Basin is a Palaeozoic sag basin, which was deformed and eroded during the Hercynian orogeny. It is overlain by the Hamadah Basin, a Mesozoic post-orogenic sag basin. The Nafusah Uplift is a Hercynian feature which was reactivated during the late Cainozoic. The Jifarah Basin began to subside during the early Mesozoic, and major subsidence occurred in the Miocene. The Zamzam Depression represents the eastern prolongation of the Ghadamis Basin beneath the western margin of the Sirt Basin. The Pan-African basement trend of the Tripoli-Tibisti Arch is barely perceptible, and much less significant than in the Murzuq Basin.

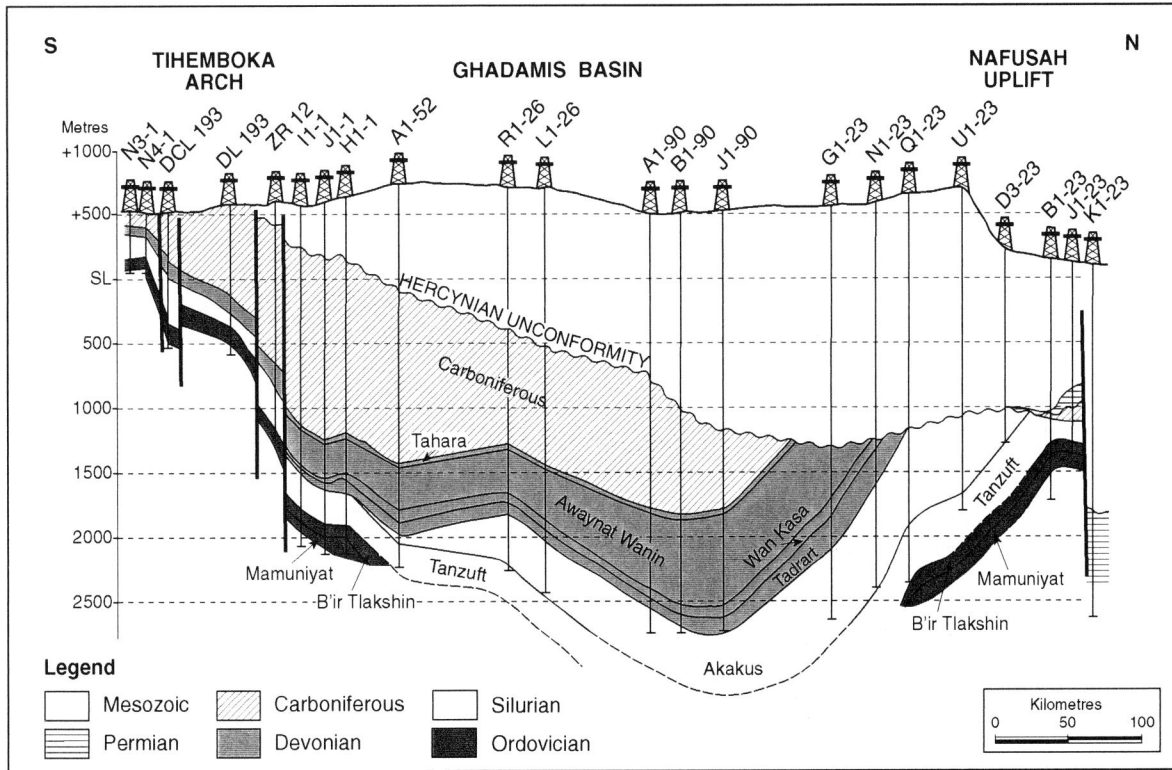

Figure 6.6   Ghadamis Basin, N-S Structural Section

This cross-section illustrates the Palaeozoic succession in the Ghadamis Basin, and the effect of the Hercynian unconformity. Truncation of the Palaeozoic is particularly marked on the southern flank of the Jabal Nafusah Uplift. Further north a thick Permo-Triassic section is present in the Jifarah and Sabratah Basins.

The final pulses of Pan-African tectonism continued into the Ordovician. During Llandeilian times, uplift and erosion occurred on the Tihemboka Arch and the Ahara Uplift in Algeria, and during the Caradocian, folding, faulting, uplift and erosion occurred which removed much of the early and middle Ordovician section on the Dahar Uplift in Tunisia. It was onto this irregular surface that the late Ordovician ice-sheet encroached from the south. Volcanic rocks were produced during this final tectonic phase which have been penetrated in several wells in the eastern Illizi Basin.[33]

Tanzuft shales were deposited throughout the basin, but Rhuddanian radioactive shales are confined to the early depocentres where euxinic conditions existed prior to the basin-wide flooding event. Early Devonian tectonic activity led to uplift and erosion of the Akakus Formation over parts the basin. This is evident from well evidence in the Illizi Basin where only the Lower Akakus Sandstone is preserved, and similar truncation has been recorded in Libya. Folding of this age is evident on the Kabir trend (Figure 6.5) in Libya where early Devonian sands unconformably overlie folded Akakus sands. The Eifelian unconformity at the base of the

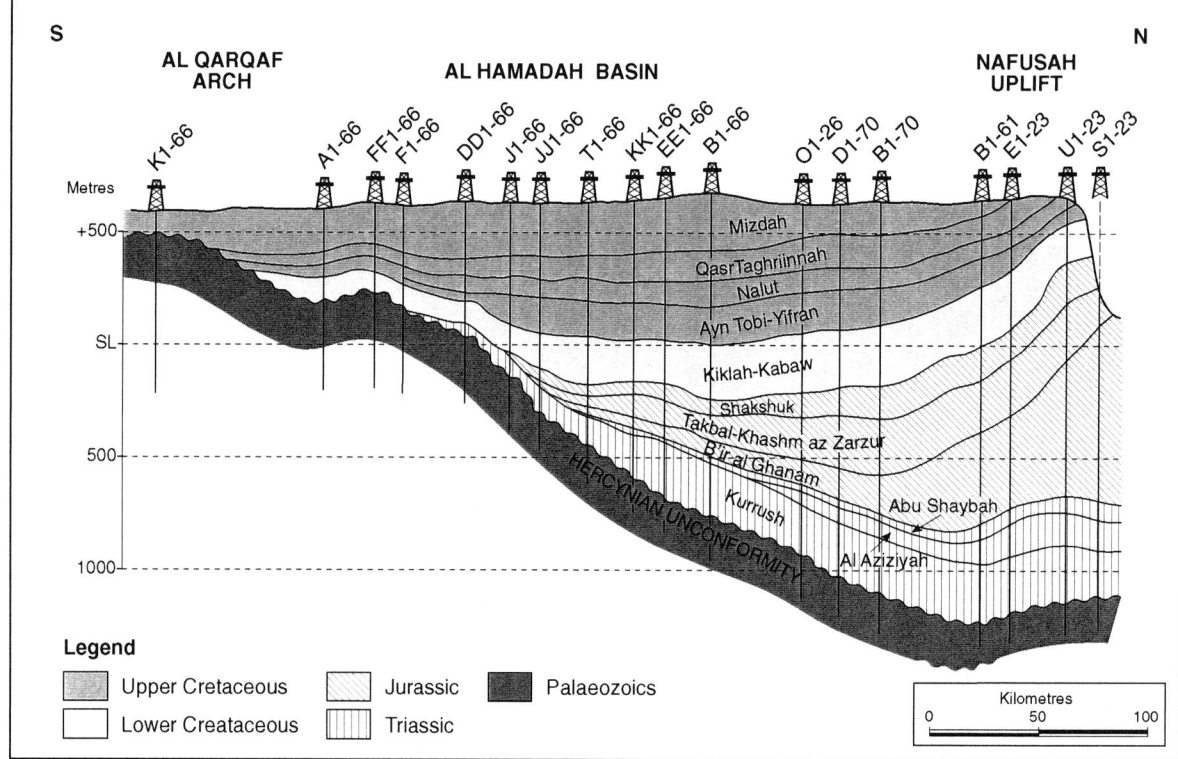

Source: Montgomery, 1994, Sinha, 1980

Figure 6.7   Al Hamadah Basin, N-S Structural Section

The Hamadah Basin is the Mesozoic Basin which overlies the Palaeozoic Ghadamis Basin. The basin contains a thick succession of Mesozoic rocks which were uplifted along the reactivated Jabal Nafusah Uplift during the mid-Tertiary, producing the Jurassic and Triassic outcrops along the Jabal Nafusah escarpment. The section also shows the extent of the mid-Cretaceous transgression across the entire basin.

Awaynat Wanin Formation which was noted in the Murzuq Basin is also present in the southern Ghadamis Basin, and a further unconformity has been recorded at the base of the Frasnian which is associated with the radioactive unit known locally as the Cues Limestone which forms a petroleum source rock in the northern part of the basin.[34]

Over most of the Ghadamis Basin rocks of Tournaisian age are absent and in the Illizi Basin there is evidence of thickness variations and erosion on the northwest flank of the Tihemboka Uplift which indicate continued tectonic activity during the early

Carboniferous. The Hercynian orogeny reached its peak during the late Carboniferous and major new tectonic elements were formed, including the Qarqaf and Nafusah Uplifts in Libya, the Dahar Arch in Tunisia and the Talemzane and El Biod Arches in Algeria. The entire area was uplifted and subjected to intense erosion during the Permian which left the basin surrounded by highs which, in the case of the Nafusah and Qarqaf Arches, were eroded to their Cambro-Ordovician roots. A subcrop pattern of progressively younger rocks can be traced into the centre of the basin where a complete section up to late

Carboniferous is preserved. Within the Libyan part of the basin structural trends with a northeast-southwest alignment were formed during the Hercynian upheaval (Figure 6.5).[35]

Thick successions of Mesozoic continental deposits, including Triassic sandstones and evaporites, were deposited in the post-Hercynian sag basin (known as the Triassic Basin in Algeria and the Hamadah Basin in Libya) in which the depocentre was located much further north than during the Palaeozoic (Figure 6.7). The basin was affected by an extensional episode during the early Jurassic which produced block faulting in eastern Algeria. Jurassic transgressive sequences are followed by an early Cretaceous regression which was terminated by deformation during the Aptian related to the opening of the Mesogean spreading axis in the southern Tethys. Wrenching and uplift of Aptian age has been reported from the Illizi Basin and the Tihemboka Uplift was reactivated, faulted and uplifted.[36]

A major marine transgression followed the Aptian deformation in which late Cretaceous and Palaeocene sediments spread as far south as the Qarqaf Arch, but thereafter the area was gently uplifted and tilted and has been emergent since the mid-Palaeocene. In Algeria and Tunisia the Atlas Basin was inverted in the early Eocene as the western Tethys closed, culminating in thrusting and nappe formation as the Kabylie terrane was welded onto the African plate during the Miocene. This fold belt forms the northern margin of the Ghadamis Basin in Algeria.[37]

6.2.4 Nafusah Uplift

The Nafusah Uplift (Figure 6.5) is a major east-west ridge which separates the Ghadamis Basin from the Jifarah Basin. It extends for 400km from Misratah to the Tunisian border, and continues in Tunisia as the Dahar Uplift

and in Algeria as the Talemzane Arch. It is bounded to the north by the Jifarah Fault and the Jifarah Basin and to the south by the Ghadamis Basin. The southern margin is also partially faulted. During the early Palaeozoic the area of the Nafusah Uplift formed part of the Ghadamis Basin and a thick sequence of Palaeozoic rocks was deposited. Nevertheless there is evidence that the Tripoli-Tibisti Uplift exerted an influence on sedimentation during the early Palaeozoic, since the thickness of the Ordovician is reduced in the area of wells C1-8 and F1-60 (Figure 6.5), and the presence of microconglomerates overlying the Mamuniyat Formation in wells E1-23 and A1-61 implies that glacio-marine influences extended this far north.[38]

The Hercynian orogeny inverted the northern Ghadamis Basin into a prominent east-west arch which was extensively eroded during the Permian, exposing the Precambrian core of Pharusian rocks in the area of wells A1-34 and D1-24 (Figure 6.5), and granite was encountered in water-well 19894/72, 50km south of Tripoli. Cambro-Ordovician rocks subcrop the Mesozoic over much of the rest of the arch. Near the Tunisian border Silurian shales are preserved in a saddle over the crest of the arch in the region of well J1-23 (Figure 3.3). On the southern flank successively younger rocks subcrop the Hercynian unconformity in the Ghadamis Basin.[39]

During the Permian the Nafusah Uplift formed a barrier between the marine sequences of the Tethys Ocean to the north and the Continental Post Tassilien rocks to the south. By Triassic times the arch had been greatly reduced in elevation and Triassic, Jurassic and early Cretaceous sediments were deposited over the arch and into the Mesozoic Hamadah Basin to the south. The arch was covered by the Cenomanian transgression and deposition of marine rocks continued until Eocene times.[40]

The Nafusah Arch was reactivated during the mid-Tertiary in response to the closing of Tethys. It was subjected to uplift accompanied by wrench faulting on the Jifarah Fault (Figure 6.5). Dextral strike-slip faulting can be seen at several locations along the line of the fault. Eocene tectonism also led to the production of basaltic sills and flows near Gharyan, and the volcanic activity has continued until recent times. A major escarpment formed along the line of the Jifarah Fault and Jurassic and Triassic rocks

outcrop both on the escarpment and on the Jifarah Plain.[41]

6.2.5 Jifarah Basin

The Jifarah Basin (Figures 6.5 and 6.8) is bounded by the Jifarah Fault and Nafusah Uplift to the south and the Sabratah-Cyrenaica fault system and Sabratah Basin to the north, and represents a downfaulted terrace on the unstable continental margin. It underlies the Jifarah Plain and extends westwards into Tunisia north of the

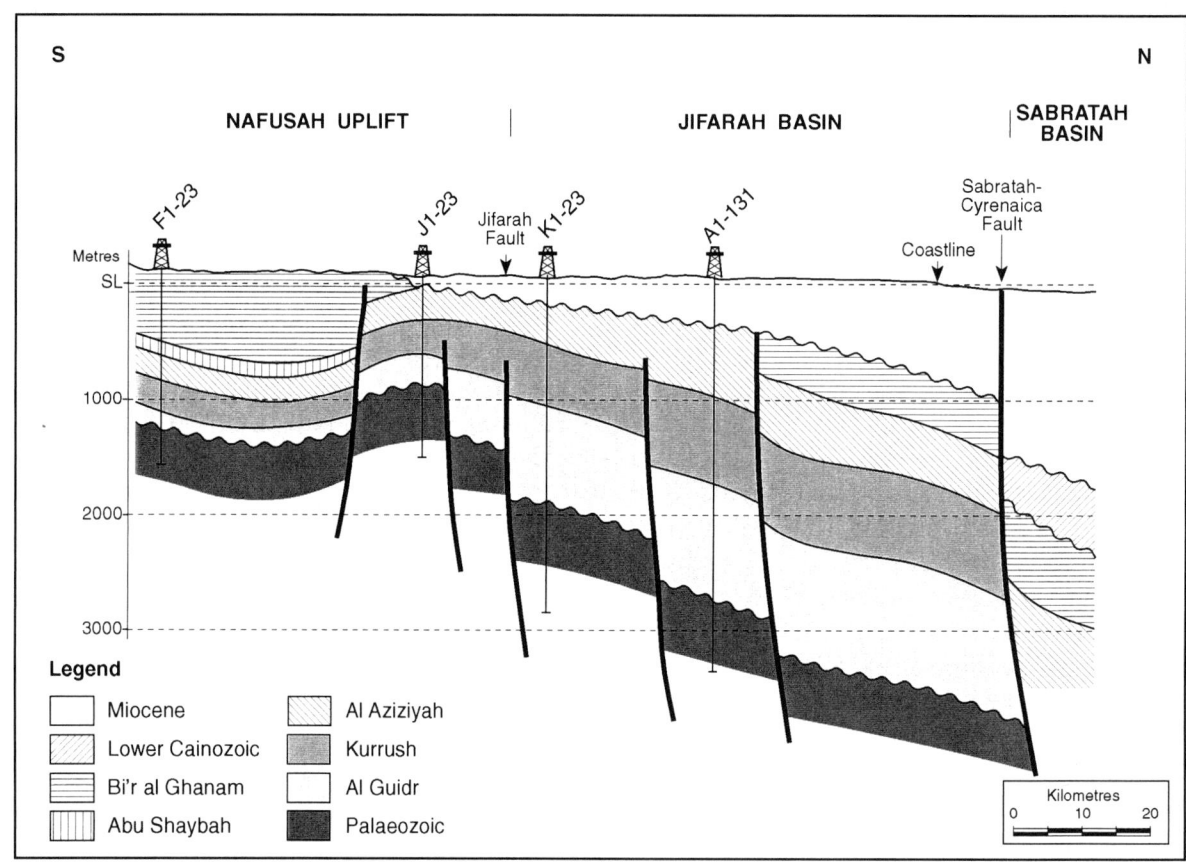

Source: Kruseman and Floegel, 1980

Figure 6.8   Jifarah Basin, Diagrammatic Cross-Section

The Jifarah Basin is a complex wrench-faulted area forming a terrace between the Nafusah Uplift and the Sabratah Basin. The section shows the effect of syn-depositional faulting during the Triassic, and the dramatic thickening of the Triassic formations towards the north. Most of the Mesozoic section has been removed by subsequent erosion, and during the Miocene the basin subsided as a result of instability on the Pelagian Block to the north.

Dahar Uplift covering an area of 15,000km², more than half of which is in Tunisia.[42]

The basin has been penetrated by numerous wells (mostly water wells) which reveal a stratigraphy extending from Palaeozoic to Jurassic in age, overlain by a thick Miocene cover. Pre-Permian rocks, encountered on the northern flank of the Nafusah Arch in wells F1-23, M1-23, S1-23 in Libya, and DB 1 in Tunisia, are presumed to continue northwards, forming basement in both the Jifarah and Sabratah Basins. North of the Jifarah Fault a thick Permian sequence has been found in wells in Tunisia, and in wells K1-23, A1-131 and A1-38 in Libya. This sequence ends abruptly at the Jifarah Fault. The Jifarah Fault is part of the Sabratah-Cyrenaica wrench zone, which marks the boundary between the relatively stable shelf to the south and the unstable continental margin to the north. The Mesozoic basin-fill of the Jifarah Basin was affected by syn-depositional faulting during the Triassic and by shearing during the Neocomian which reflect major tectonic activity along the southern margin of Tethys (Figure 2.9). The structure of the Jifarah Basin was studied by Anketell and Ghellali. They demonstrated that the dominent fault directions in the Mesozoic section are east-west and NNW-ESE with an en echelon arrangement which they interpreted as riedel shears and imbricate fan splays formed as a result of strike-slip movement on the South Atlas-Jifarah dislocation. Riedel shears are slip surfaces formed during the early stages of shearing. They developed a model which suggested an analogy between the imbricate fan splays visible in the Wadi Ghan area south of Al Aziziyah with the strikingly similar fault trends in the Sirt Basin, and attributed both to strike-slip faulting associated with underlying basement dislocations.[43]

Sedimentation continued until the Palaeocene, but the basin was caught up in the Eocene tectonism which reactivated the Nafusah Arch and introduced a new generation of faulting in the Jifarah Basin. Extensive pre-Miocene erosion removed much of the Mesozoic sediments from the basin, and the pre-Miocene subcrop shows a complex pattern. Well A1-38 shows Miocene rocks resting on Lower Jurassic Bi'r al Ghanam Formation, whereas water wells west of well A1-131 show Miocene rocks resting on Middle Triassic Kurrush Formation. By comparison, on the Nafusah escarpment south of the Jifarah Fault, almost the entire early Mesozoic sequence is preserved. Since the beginning of the Miocene the basin has been subsiding towards the north.[44]

6.3   SIRT BASIN

A vast amount of data is available for the Sirt Basin as a result of oil exploration activities extending over more than forty years. Thousands of wells have been drilled, gravity, magnetic and seismic data have been gathered, and as a result the basin is far better known than any other area in Libya, although it can fairly be claimed that the deep troughs are still imperfectly known. Studies on the subsidence history of the Sirt basin have been published by Gumati and Kanes, Gumati and Nairn, van der Meer and Cloetingh and Baird, et al. and the results of a gravity study of the Sirt Basin were presented by Suleiman, et al.[45]

The Sirt Basin covers an area of 600,000km² in central Libya (Figure 6.9) and contains a basin-fill which reaches a thickness of 7500m. It opens onto the Sirt Slope and Sirt Rise to the north and gradually wedges out towards the basin margin to the east, south and west. The basin is characterised by a basin-fill which is entirely Mesozoic and Cainozoic in age, and by the presence of a series of platforms and deep troughs. In the northwest the dominant structural trend is northwest-southest, in the eastern part of the basin it is east-west, and in the southwest it is NNE-WSW. This has led to

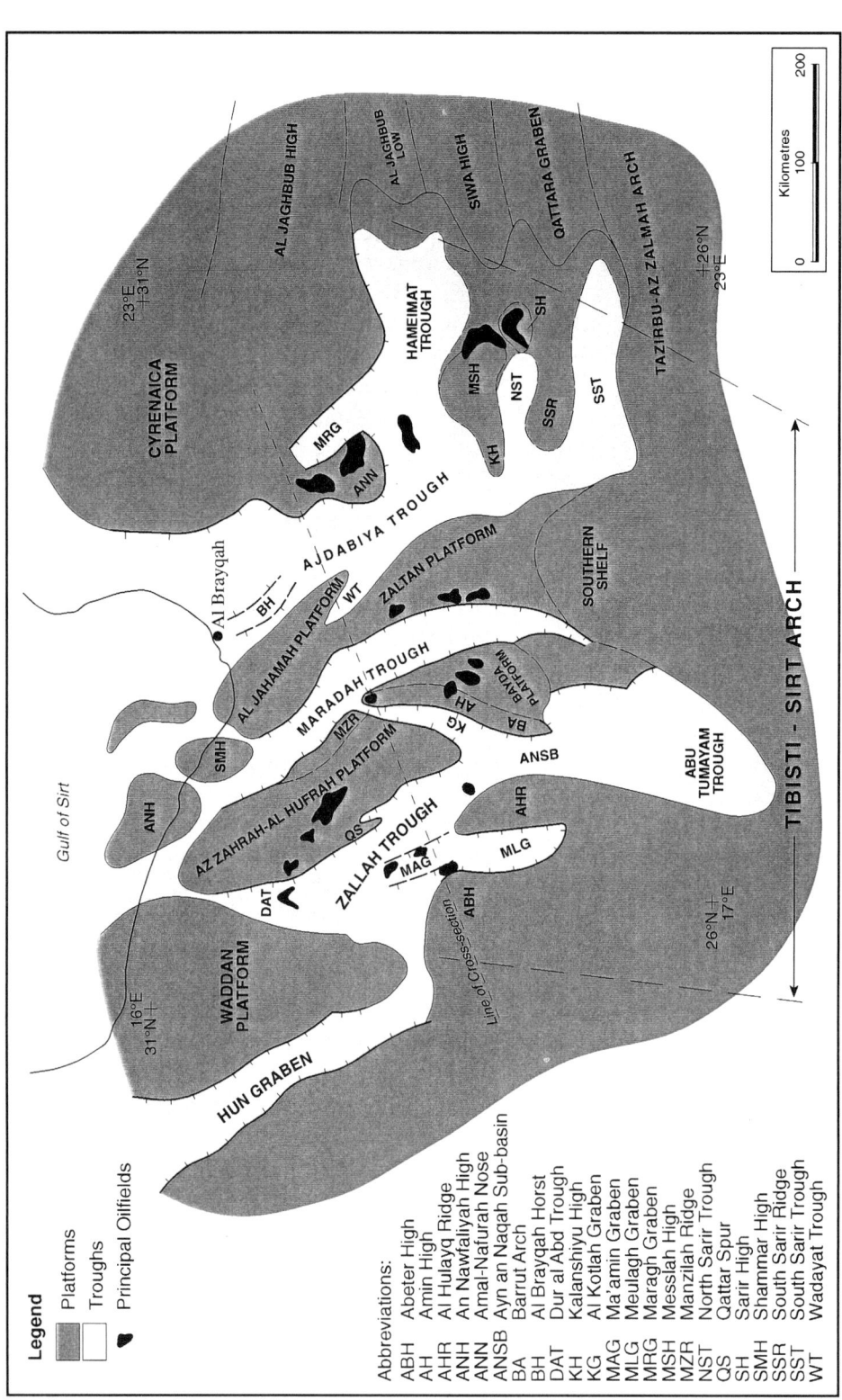

Figure 6.9   Sirt Basin, Tectonic Elements

Rifting began during the Triassic and increased in intensity until the Cretaceous. The Sirt Arch collapsed during the Aptian to form the Sirt Basin, which is characterized by a series of horsts and grabens which follow the trend of the pre-Aptian rifts. Rapid subsidence in the grabens outpaced deposition during the Santonian to late-Campanian, but subsidence then ceased and during the Maastrichtian the grabens were infilled. Sag basins developed over the site of the troughs during the early Cainozoic, which subsequently coalesced to form one large basin centred on the Ajdabiya Trough where subsidence has continued to the present day.

Sources: *General: Goudarzi, 1980, Mikbel, 1977, 1979, 1981, Wennekers, et al., 1996, Baird, et al., 1996
Anketell, 1996. Specific: El Arnauti and Shelmani,
1988, Ibrahim, 1991, Schroter, 1996, Knyit, et al., 1996
Sinha and Mriheel, 1996, Johnson and Nicoud, 1996
Roohi, 1996 a,b, Sgair and El Alami, 1996, Ambrose, 2000*

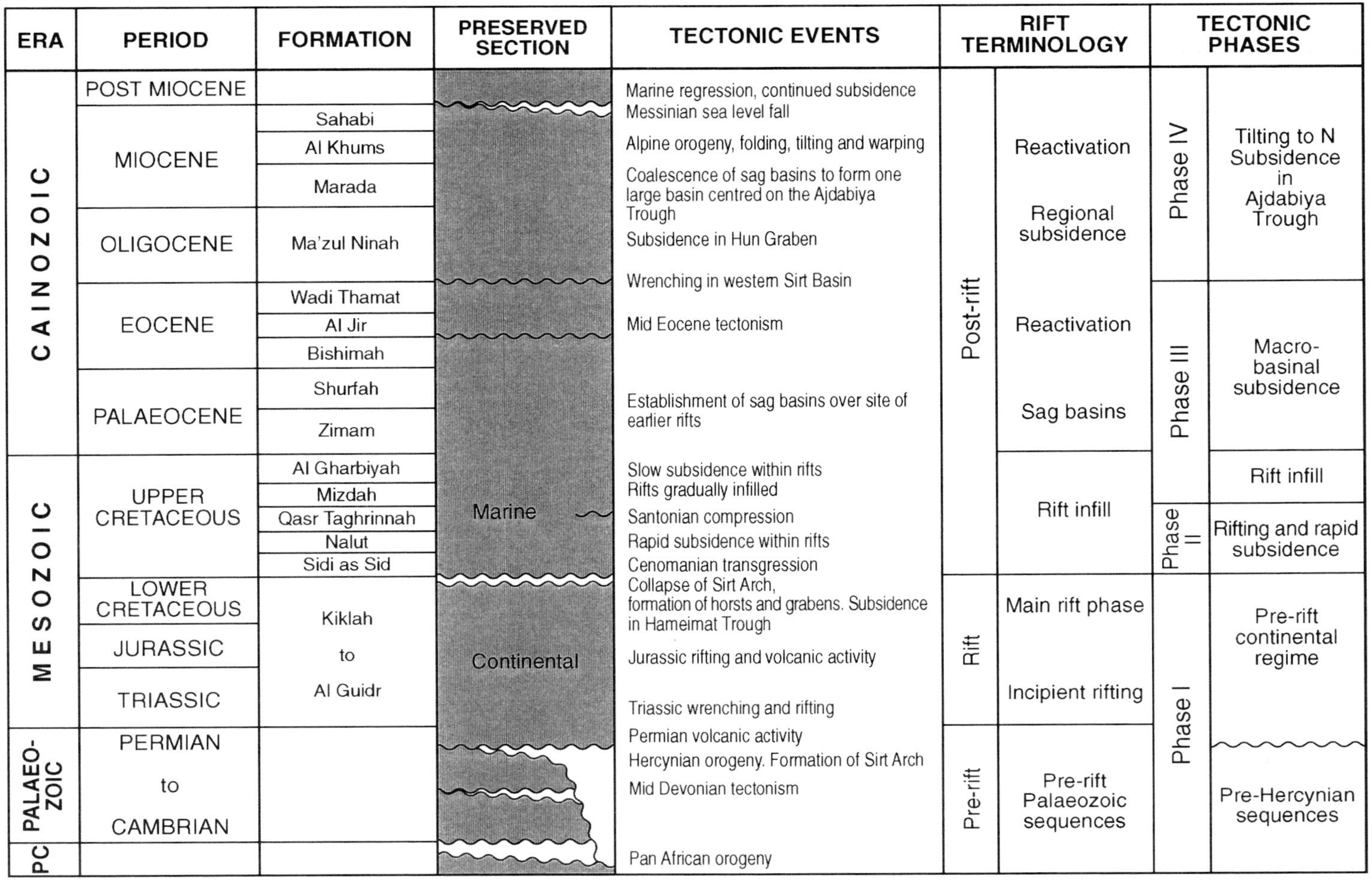

ERA	PERIOD	FORMATION	PRESERVED SECTION	TECTONIC EVENTS	RIFT TERMINOLOGY		TECTONIC PHASES	
CAINOZOIC	POST MIOCENE		Marine	Marine regression, continued subsidence; Messinian sea level fall	Post-rift	Reactivation	Phase IV	Tilting to N Subsidence in Ajdabiya Trough
	MIOCENE	Sahabi / Al Khums / Marada		Alpine orogeny, folding, tilting and warping; Coalescence of sag basins to form one large basin centred on the Ajdabiya Trough		Regional subsidence		Macro-basinal subsidence
	OLIGOCENE	Ma'zul Ninah		Subsidence in Hun Graben			Phase III	
	EOCENE	Wadi Thamat / Al Jir / Bishimah		Wrenching in western Sirt Basin; Mid Eocene tectonism		Reactivation		
	PALAEOCENE	Shurfah / Zimam		Establishment of sag basins over site of earlier rifts		Sag basins		Rift infill
MESOZOIC	UPPER CRETACEOUS	Al Gharbiyah / Mizdah / Qasr Taghrinnah / Nalut / Sidi as Sid	Continental	Slow subsidence within rifts; Rifts gradually infilled; Santonian compression; Rapid subsidence within rifts; Cenomanian transgression; Collapse of Sirt Arch, formation of horsts and grabens. Subsidence in Hameimat Trough		Rift infill	Phase II	Rifting and rapid subsidence
	LOWER CRETACEOUS	Kiklah to Al Guidr		Jurassic rifting and volcanic activity	Rift	Main rift phase		
	JURASSIC			Triassic wrenching and rifting		Incipient rifting	Phase I	Pre-rift continental regime
	TRIASSIC			Permian volcanic activity; Hercynian orogeny. Formation of Sirt Arch	Pre-rift	Pre-rift Palaeozoic sequences		
PALAEOZOIC	PERMIAN to CAMBRIAN			Mid Devonian tectonism				
PC				Pan African orogeny				Pre-Hercynian sequences

Sources: Numerous sources, refer to text. The Rift terminology is from Gras and Thusu, 1998, and the Tectonic phases from Baird, et al., 1996

Figure 6.10   Sirt Basin, Tectonic History

In terms of continental margin rifting the rift phase extended from the Triassic to the mid-Cretaceous and was terminated by the Cenomanian marine transgression. There is no doubt however that maximum subsidence in the grabens occurred during the Cenomanian to late Campanian period. This has led to some confusion of terminology as illustrated on the chart, although the main tectonic events are not in dispute.

the suggestion that the Sirt Basin may be related to a triple junction in the underlying plate, but as will be shown, the timing of tectonic activity in the eastern arm is not consistent with a triple-junction model. The Sirt Basin is bordered to the northeast by the Cyrenaica Platform, to the southeast by the Tazirbu-Az Zalmah Arch, to the south by the Southern Shelf, and to the west by the western Tibisti spur and the Al Qarqaf Arch (Figures 6.1 and 6.9). An east-west cross-section of the basin is shown on Figure 6.11.

The origin of the Sirt Basin has been attributed to different causes by different authors, and the main concepts were reviewed in chapter 2. It is likely that the Sirt Basin owes its origin to a combination of causes. The timing of the various tectonic episodes is not yet fully resolved, particularly in respect of the rift-phase. Baird, et al. presented a very detailed account of what might be termed the traditional view, that rifting did not begin before the mid-Cretaceous, and was the dominant force during the Cenomanian to late Campanian. It was followed by a period of rift-infill extending to the end of the Maastrictian, after which a large sag basin developed over much of the Sirt Basin. The final phase, starting in the Oligocene, was dominated by tilting towards the northeast and continued subsidence in the Ajdabiya Trough (Figure 6.10). Other authors, particularly Finetti for the offshore, and Gras and Thusu for the onshore, presented evidence to suggest that incipient rifting began as early as the Triassic, and claimed that the syn-rift phase extended only from the Triassic to the early Cretaceous, being terminated by the thermal sag which gave rise to the Cenomanian marine transgression. Wennekers, et al. showed rifting continuing to the present time. This is partly a question of semantics. The sequence of events is not in doubt. The evidence for incipient rifting in the Triassic is now firmly established, both in the eastern Sirt Basin and offshore, but rifting was intermittent and the rifts were probably shallow.

Anketell demonstrated that rifting was certainly active in the Hameimat Trough during the Neocomian, and in the western Sirt Basin during the Aptian. Finetti, and Gras and Thusu are correct in terms of marginal basin terminology to end the syn-rift phase with the Cenomanian transgression, and to refer the post-Albian events to the post-rift phase, but in practical terms the rift troughs in the Sirt Basin experienced their maximum subsidence during the late Cretaceous, and later reactivation is evident in some of the troughs. Most of the subsidence in the Hun Graben took place during the Oligocene, and subsidence is still taking place in the Ajdabiya Trough. The two different point of view are shown on Figure 6.10. The following review will attempt to highlight the principal features and relate them to probable causes.

The nature of the faults which control the Sirt Basin grabens is important, particularly in respect of oil migration. The asymmetry of several of the grabens in the western Sirt Basin has led to the suggestion that the bounding faults are listric faults which sole-out beneath the Mesozoic, with the master-fault and associated synthetic faults on the eastern basin margins, and the antithetic faults forming the western margins. This is a plausible suggestion, but still remains to be demonstrated in kinetic terms. Another point, which also has a bearing on oil migration, is that post-Maastrichtian sag over the earlier grabens may be more a function of compaction than of continued subsidence. In any event, the drape-folds overlying the graben margins form excellent potential oil traps.[46]

6.3.1 Sirt Basin Evolution

Basement in the Sirt Basin has been penetrated in a number of deep oil wells and in general comprises Pharusian accreted oceanic terranes north of latitude 27°N, whilst to the south of this line it is composed of Pan-African

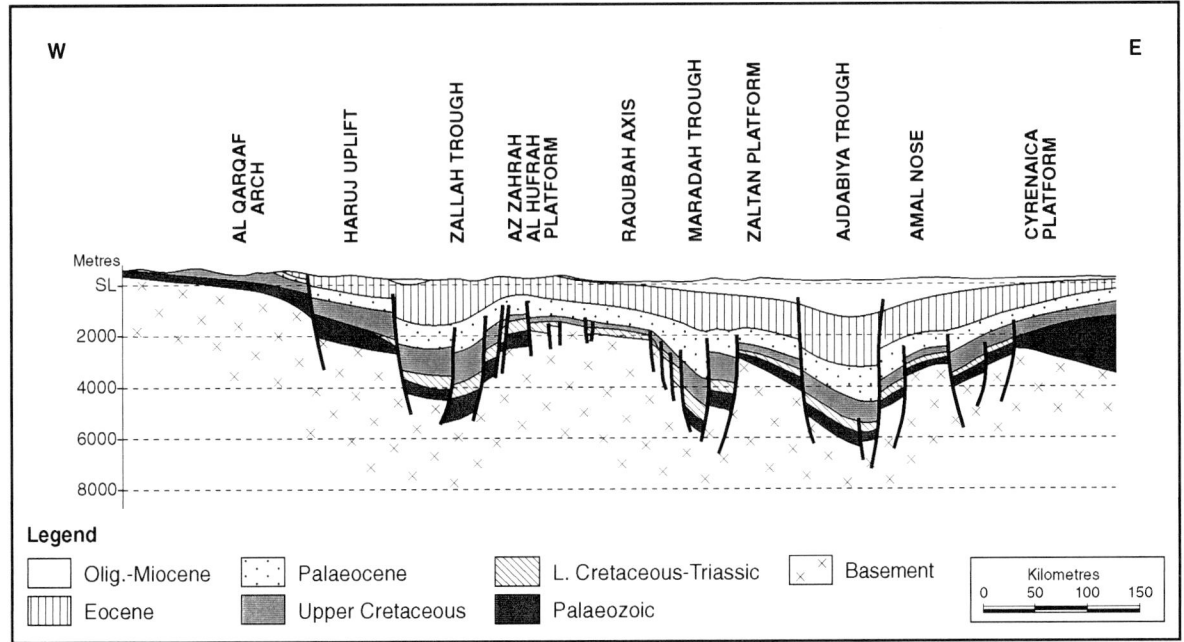

Sources: Bonnefous, 1972, Massa and Delort, 1974, Parsons et al., 1980
Peterson, 1985, Gumati, et al., 1996

Figure 6.11   Sirt Basin, Diagrammatic Cross-Section

This east-west section across the Sirt Basin shows the effect of the fragmentation of the Sirt Arch into a series of horsts and grabens during the mid-Cretaceous. The grabens were largely infilled by the end of the Cretaceous and were replaced by broad sag basins during the early Cainozoic. Subsidence during the late Cainozoic was largely confined to the Ajdabiya Trough.

remobilised continental terranes (Figure 3.1). Both of these tectonic provinces were assembled during the Pan-African orogeny. Following the orogeny early Palaeozoic sediments were deposited in the Kalanshiyu Trough, which probably became inactive after the mid-Devonian. Mid-Devonian uplift, erosion and peneplanation was probably followed by Middle and Upper Devonian and Carboniferous deposition over all of central Libya. The Kalanshiyu Trough, along with other early Palaeozoic tectonic elements, was inverted during the Hercynian orogeny to form the Sirt Arch (Figure 2.7), extending northeastwards from Chad towards Cyrenaica. The arch was faulted and deformed and subjected to intense erosion which ultimately removed the entire Palaeozoic succession over the crest of the arch.

The only traces of the Kalanshiyu Trough now remaining are the lower Palaeozoic sequences preserved in the northern depocentre of the Kufrah Basin, in the Abu Tumayam Trough, in the complex area beneath the Al Haruj al Aswad volcanic centre, in the Al Fuqaha Depression, where the Tanzuft Shale is exceptionally thick, and in the Zamzam Depression, where well evidence proves a thick lower Palaeozoic succession.[47]

Whilst the Tibisti-Sirt Arch was being unroofed in the Permian, volcanic activity was active on the Waddan Platform and at Amal, and dextral wrenching was taking place along the suture between Gondwana and Laurasia. Pangaea began to break-up during the Triassic and a spreading axis developed in the Proto-Tethys. Rifting began on the Tibisti-Sirt Arch accompanied by volcanic activity. Triassic clastics and lacustrine clays have been found in the Sirt Basin which may represent the earliest deposits in the incipient rifts. Opening of the central Atlantic during the Jurassic induced an apparent eastern motion of Africa relative to Europe and sinistral shearing has been reported from Tunisia and from the Jifarah Basin. The earliest oceanic crust in the central Tethys belt has been dated as mid-Jurassic. On the Sirt Arch rifting and volcanic activity continued through the Jurassic and into the early Cretaceous accompanied by the accumulation of extensive areas of continental sands. It has been suggested that at this time the Sirt Arch was located over a fixed-mantle hot-spot which thinned and weakened the overlying crust.[48]

During the early Cretaceous the Apulian plate separated from north Africa, creating a tensional regime on the plate margin. It also increased instability in the stretched and rifted Tibisti-Sirt Arch. Subsidence and pull-apart occurred in the Hameimat and Sarir Troughs, and the Messlah and South Sarir horsts were formed during the Neocomian. The Jaghbub Trough on the Cyrenaica Platform originated at the same time. All of these features have an east-west alignment. Nubian sandstones accumulated in many areas of Libya, particularly in the rift basins and structural depressions. Early Cretaceous granites and volcanic rocks have been reported from wells in the Sirt Basin. During the Aptian the sea-floor spreading axis in the central Tethys switched to the Mesogean location adjacent to the African continental margin. According to Anketell's tectonic model this induced a system of WNW-ESE major wrench zones, which were the primary controls on the geometry of the Sirt Basin.[49]

The continued tensional regime led to the eventual collapse of the Tibisti-Sirt Arch into a fragmented assemblage of platforms and troughs during the Aptian, and flooding of the collapsed area by shallow epicontinental seas of the Tethys Ocean during the Cenomanian. Burke and Dewey proposed a two-plate model for Africa during the Cretaceous with the boundary between the two plates roughly corresponding to the western margin of the Sirt Basin, adding further to the instability of the area. It seems highly likely, on the basis of the data presented above, that a number of different causes were involved in the collapse and fragmentation of the Sirt Arch. Gras and Thusu recognised pre-rift (Palaeozoic) and syn-rift (Triassic to early Cretaceous) phases, and a post-rift phase characterised by graben fill, basin sag and subsidence (late Cretaceous to recent), as shown on Figure 6.10.[50]

The South Atlas-Jifarah Shear Zone, discussed in the previous section, may also have been associated with the collapse of the Sirt Arch. Antetell suggested a general model to relate the tectonics of the Sabratah and Sirt Basins with that of Cyrenaica. He recognised the South Atlas-Jifarah Fracture Zone as a dextral shear zone which marks the southern margin of the Pelagian Platform. To the east, in the Sirt Basin, the fracture zone feathers into a series of northwest-southeast splays which governed the location of the Sirt Basin troughs, and which can be regarded as pull-apart grabens. The NNE-SSW trending Abu Tumayam Trough is believed to have formed at the same time with its trend determined by the structural grain of the underlying basement (Figure 6.21).[51]

The post-rift period is marked by a marine transgression in the Cenomanian which flooded the troughs during a period of rapid subsidence, representing post-rift sag. The organic-rich Sirt Shale was deposited during the Campanian after which continued subsidence led to the gradual onlap of the platforms during the late Cretaceous (Figure 6.12). The late Cretaceous was generally a period of extension, with some block faulting, except for a brief compressional phase during the Santonian. Subsidence of the

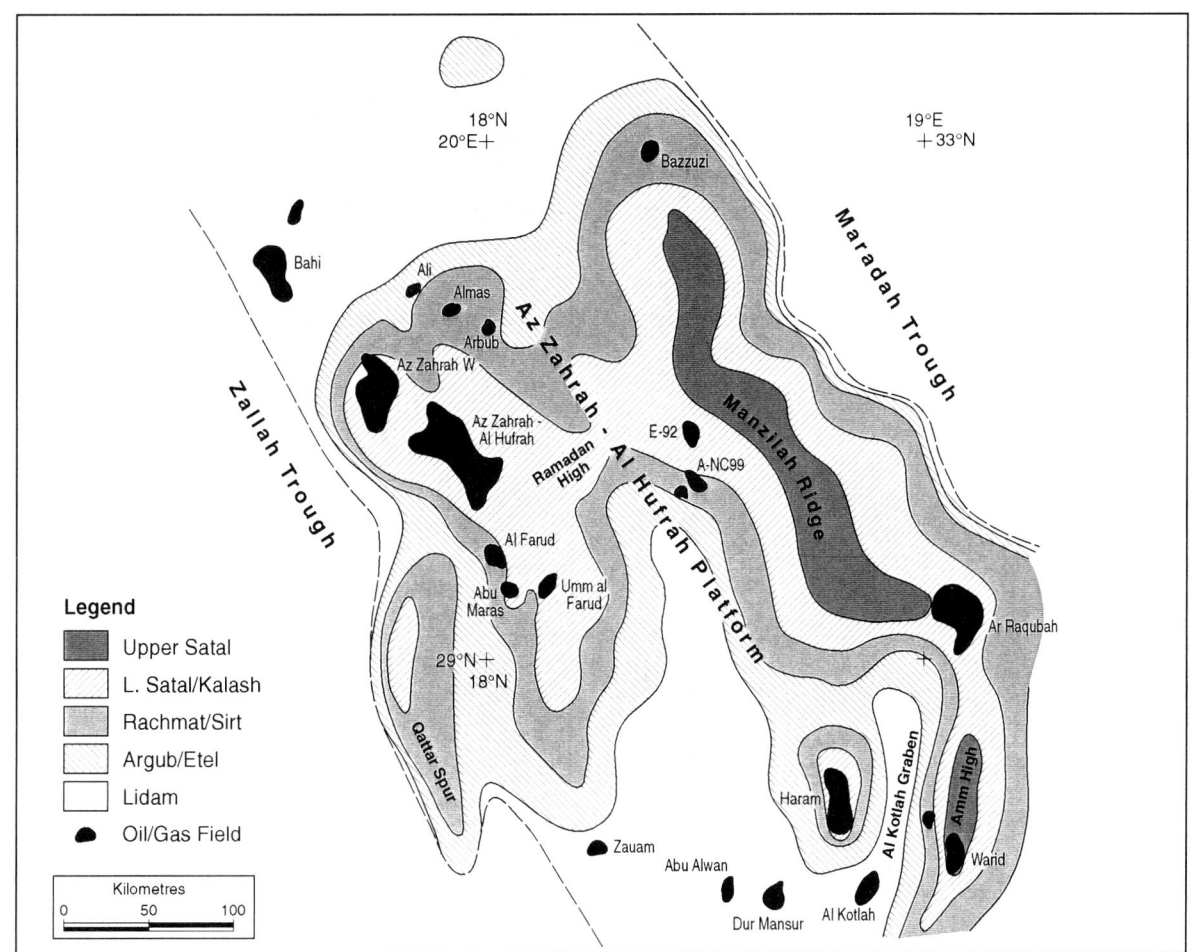

Source: *Roohi, 1996b*

Figure 6.12   Sirt Basin, Az Zahrah-Al Hufrah Platform. Onlap of Pre-Upper Cretaceous Unconformity

Az Zahrah-Al Hufrah Platform was onlapped progressively from the Cenomanian to the Danian with the Manzilah Ridge and the Amin High being the last areas to be covered. This map has important implications for the distribution of reservoirs and seals on the platform during the late Cretaceous.

Kotlah Graben and the break-up of the Jahamah Platform into three distinct blocks occurred during the Campanian. By the end of the Cretaceous only a few small islands remained in the Maastichtian sea. During the Palaeocene carbonate platforms developed separated by deeper-water depressions in which pinnacle reefs flourished. The absence of Maastrichtian and Danian rocks from the northern Al Bayda Platform suggests that this area remained emergent. On the southern flank of the Jabal al Akhdar Uplift rapid subsidence during the Ypresian produced the Ash Shulaydimah Trough. Tilting of the Az Zahrah-Al Hufrah Platform towards the north, and an unconformity at the base of the Lutetian on the Jahamah Platform testify to the early effects of Alpine tectonism. During the late Eocene the western margin of the Az Zahrah-Al Hufrah platform was affected by sinistral strike-slip faulting, and

Upper Eocene rocks rest unconformably on faulted Lower and Middle Eocene rocks. This brief late Eocene compressive episode is anomalous in an otherwise tensional regime. The crestal area of the platform was probably emergent during the late Eocene. Growth faults on the western margin of the Ajdabiyah Trough indicate subsidence of over 700m during the late Eocene. During the Oligocene the Hun Graben subsided, and in the Ajdabiya Trough a further 500m of subsidence can be inferred from well evidence. Volcanic rocks were erupted from Eocene to recent times along the line of the Tripoli-Tibisti Uplift from the Tibisti Massif to Gharyan. In the Miocene much of the Sirt Basin was tilted towards the northeast, and great thicknesses of sediment accumulated in the Ajdabiya Trough. Gentle folding occurred in the southern Sirt Basin. The Messinian event, during which sea level fell by over 500m at the end of the Miocene, was discussed in chapter 4. A summary of the tectonic history of the Sirt Basin is given on Figure 6.10.[52]

6.3.2 Structural Components of the Sirt Basin

Within the Sirt Basin about a dozen distinct tectonic subdivisions can be distinguished, which will be reviewed from west to east.

Hun Graben

The Hun Graben extends for 300km from Wadi Zamzam to Suknah and averages 40km in width. It is bounded to the west by the Haruj Uplift, and to the east by the Waddan Platform (Figure 6.9). Northwards it plunges into a sag overlying the Zamzam Depression and southwards it is deflected eastwards by the subsurface nose of the Qarqaf Uplift. Surface mapping and Landsat data suggests that it links with the Zallah Trough by means of a zone of complex relay growth faulting to the south of

the Waddan Uplift. The dominent structural grain, reflected by numerous faults along the graben margin, is NNW-SSE. Depth to the top Cretaceous in the centre of the trough is about 760m and to the base Mesozoic 1500m near Abu Njaym in the north but only 800m in the southern part of the trough. Depth to basement varies from 2300m to 3000m with its surface dipping towards the northeast. Displacement on the boundary faults reaches 2000m in the northeast, but only about 500m on the western margin. The Hercynian subcrop is represented by Cambro-Ordovician rocks in the south, and by Silurian and Devonian rocks to the north. The graben-fill comprises rocks from early Cretaceous to Oligocene in age. Rifting may have begun with the collapse of the Sirt Arch, but the most active phase of rifting and subsidence was during the Oligocene-Miocene interval, perhaps associated with the nearby volcanic activity on the Tripoli-Tibisti axis. Baird et al. regarded the Hun Graben as a hinge rift, rather than a crestal rift, in contrast to the Zallah, Maradah and Ajdabiya Troughs.[53]

Waddan Platform

The Waddan Platform is a major platform area covering 16,000km^2 located between the Hun Graben and the Dur al Abd Trough (Figure 6.9). The northern limit is ill-defined as it plunges towards a saddle separating it from the subsurface plunge of the Nafusah Uplift. The platform terminates southwards against the northeastern extension of the Qarqaf Arch. Basement is at a depth of only 1500m west of the Zallah Trough, but the platform is tilted towards the northeast, and depth to basement adjacent to the Dur al Abd Trough is 3000m. As in the Hun Graben, the Hercynian subcrop is formed by Cambro-Ordovician rocks in the south and Silurian and Devonian rocks further north. The Mesozoic and Cainozoic cover, 500m thick in the south and 1500m thick in the northeast, comprises possible Nubian rocks in the structural lows, and post-rift marine rocks which progressively onlapped the platform during the late Cretaceous, followed by Cainozoic rocks up to late Eocene in age. The dominant fault direction is NNW-SSE with a conjugate set of ENE-WSW faults on the northern margin of the platform. The northeastern margin of the platform is extensively faulted, due to dextral wrenching on an unnamed WSW-ESE basement fault postulated by Anketell. The northeasterly tilt and warping is a late feature dating from the early Miocene.[54]

Zallah Trough

The Zallah Trough is an irregular faulted graben which connects northwards with the Dur al Abd Trough and southwards with the Abu Tumayam Basin. Its western margin is formed by the Waddan Platform in the north and the Abeter High further south (Figure 6.9). The Al Hulayq Ridge projects into the basin from the south, and west of the spur a branch of the Zallah Trough, the Meulagh Graben, passes beneath the Al Haruj al Aswad volcanic centre. The eastern side of the trough is formed by the Az Zahrah-Al Hufrah Platform, with the Qattar Spur (well H1-11) forming a small projection into the trough. Further south the Zallah Trough connects with the Al Kotlah Graben and Ayn an Naqah sub-basin. This is an extremely complex area in terms of structural relationships. The connection between the Hun Graben and the Zallah Trough south of the Waddan Platform is complicated by wrench tectonics, and the situation beneath the Haruj al Aswad volcanics is difficult to establish with any reliability. Klitzsch envisioned the preservation of a remnant of the Kalanshiyu Trough Palaeozoic rocks beneath the Al Haruj al Aswad volcanics, but this has yet to be substantiated (Figure 6.4). The Zallah Trough is best regarded as a composite basin linking several diverse structural elements. Depth to the top Cretaceous in the centre of the basin is 2250m and to the base Mesozoic about 4800m. The Upper Cretaceous is 2000m thick compared with only 300m on the adjacent platforms. The pre-Hercynian subcrop is composed entirely of Cambro-Ordovician rocks. The trough is filled with syn-rift sediments of unsilicified Nubian Sandstones and possibly older rocks which reach 750m in thickness, and post-rift sediments of late Cretaceous to Miocene age. The Sirt Shale has a thickness of 760m in the Zallah Trough, and both the Turonian and Eocene sequences contain thick evaporite deposits.[55]

Detailed structural studies have been conducted by Schröter in the northern Zallah Trough and by Knytl et al., in the southern area. Schröter defined the Central Zallah Trough and the Ar Ramlah syncline in which the Ghani and Adh Dhi'b oilfields are located, and the Ma'amir Graben in which the Hakim and Fidda fields are located (Figure 6.9). To the north the Facha Graben flanks the Az Zahrah-Al Hufrah Platform, and in the south a narrow trough (the Ayn an Naqah sub-basin) connects with the Abu Tumayam Trough. Schröter divided the tectonic history of the basin into six phases. Nubian sands were deposited during the rift phase, followed by marine transgression and subsidence during the late Cretaceous. A regressive cycle in the Palaeocene gave way to a period of downwarping during the Eocene, uplift and compression in the Oligocene and early Miocene, and regional tilting to the east from the mid-Miocene. Schröter assigned the entire sequence from the Nubian to the base Eocene to the syn-rift phase, but thermal subsidence in the Sirt Basin began in the mid-Cretaceous, and led to the Cenomanian transgression which should be regarded as the beginning of the post-rift stage. In fact, subsidence within the trough has continued to the present day and the eastern boundary fault reaches the surface, close to well H1-11.[56]

Knytl et al. described the structure of the southern part of the Zallah Trough on the western flank of the Al Hulayq Ridge in the area of the Al Abraq oil discovery. They demonstrated an extremely complex zone of drag folds within the Palaeocene and Eocene sections. They explained these features in terms of two dextral wrench faults, one marking the western boundary of the Al Hulayq Spur and the other a conjugate splay fault. The combination drag folds and wrench faults have produced spectacular flower structures which form traps for several small oil pools in the area. The wrench faults have been dated as post-Oligocene in age.[57]

Dur al Abd Trough

The Zallah Trough narrows northwards and becomes shallower to form the Dur al Abd Trough which separates the Az Zahrah-Al Hufrah Platform from the Waddan Platform in the area of the Mabruk field. The trough is difficult to follow north of the Mabruk field where it loses its identity in the faulted northeast flank of the Waddan Platform (Figure 6.9). Several authors show the trough continuing northwards to the coast near Sultan, and this is supported by a significant re-entrant in the Miocene outcrop in this area. The Mabruk field lies within the Dur al Abd Trough with the Facha field on its faulted western flank. The trough is only 50km wide in the Mabruk area but broadens out further south. It extends 175km northwards from the Facha field. Cambro-Ordovician quartzites subcrop beneath the Hercynian unconformity. In the Mabruk area depth to the top Cretaceous is 1500m, to the base marine Cretaceous 2100m, and to the Hercynian unconformity 2300m. As in the Zallah Trough, Nubian sands are present, and

rifting was active during the early Cretaceous, if not earlier. The oldest marine Cretaceous is represented by Lidam Formation dolomites of Cenomanian age, although the northern part of the Mabruk field was not covered until Turonian times, indicating a pre-Upper Cretaceous high in this area. The Mabruk field is reservoired in carbonates of Palaeocene age. The trough is extensively faulted by numerous normal faults with a predominently NNW-SSE trend, and the trough margins are fault bounded, with displacements of about 250m at Lidam level. Graben fill was completed by the end of the Cretaceous but subsidence continued during the Cainozoic forming a structural sag over the Cretaceous graben. The trough was reactivated during the late Eocene when it was tilted towards the east and deformed. The youngest rocks preserved in the trough are early Miocene in age.[58]

Abu Tumayam Trough

The Zallah Trough narrows to the east of the Al Hulayq Spur and connects with the Al Kotlah Graben in the region of the Dur/Mansour and Al Kuf oilfields. The connection is further constricted by the Barrut Arch (wells D1-85, A1-72 and I1-72) to form the Ayn an Naqah sub-basin, before broadening into the Abu Tumayam Trough further south. The Abu Tumayam Trough, sometimes known as the Dur at Talah Trough, is the most southerly of the Sirt Basin troughs. It is bounded to the west by the Al Hulayq Ridge, but further south its margin is masked beneath the Al Haruj al Aswad volcanics. Evidence from wells F1-85, C1-85 and E1-85 however suggests that it pinches out westwards against the old Haruj Uplift. To the south it thins against the Al Abraq slope between wells E1-72 and D1-72, to the east it thins against the Southern Shelf, and to the northeast it has a faulted margin with the Al Bayda Platform (Figure 6.9). The trough covers an area of about 8500km^2. It had long been assumed that the Abu Tumayam Trough connects to the Maradah Trough over a structural saddle to the south of the Al Bayda Platform, but a study by Sinha and Mriheel provided evidence to show that most of this area is occupied by a broad horst, and that any connection is likely to be very shallow.[59]

Unlike the Zallah Trough Cambro-Ordovician rocks are not present, and syn-rift rocks rest directly on Precambrian basement. The depth to the top Cretaceous in the centre of the basin is 3600m and to the Hercynian unconformity about 4800m. The syn-rift section comprises unsilicified Nubian sands, and the post-rift section ranges from late Cretaceous (Lidam Formation) to early Miocene in age, and the western margin is covered by Neogene lavas. Evaporites are present within the Upper Cretaceous Etel Formation (Figure 4.12), and in the Eocene Al Jir Formation (Figure 5.6). Hydrocarbons have been found in Palaeocene carbonates at Ayn an Naqah and Barrut. Johnson and Nicoud recognised six phases of structural development for the Abu Tumayam Trough: a pre-rift sequence, a structurally active syn-rift sequence extending to the mid-Cretaceous, followed by a graben-fill sequence in the late Cretaceous. The Cainozoic is represented by a structurally inactive period in the Palaeocene dominated by basin sag, and a reactivated period of faulting and graben-fill during the early Eocene. The final phase from mid-Eocene to present is marked by basin subsidence and tilting. The formation of the Barrut Arch belongs to the final phase of tectonic development. The overall orientation of the trough from NNE-SSW has been attributed to the influence of the structural grain of the underlying Tibisti-Sirt Uplift.[60]

Az Zahrah-Al Hufrah Platform

The Az Zahrah-Al Hufrah Platform is located between the Dur al Abd Trough and the Maradah Trough. The southern boundary is taken at the Al Kotlah Graben which separates it from the Al Bayda Platform. The platform dips towards the northeast and its northern boundary becomes indistinct in an area affected by wrench-faulting close to the present coastline. The platform occupies an area of 40,000km^2 (Figures 6.9, 6.10).

Depth to the top-Cretaceous ranges from 900m on the western margin near the Az Zahrah field to 1600m further north, and the thickness of the Upper Cretaceous is about 300m. Depth to the base-Mesozoic ranges from 1300m in the south to 2000m in the north. Anketell and Kumati described the western platform boundary fault in the Al Hufrah region. They demonstrated a complex series of en echelon faults indicating a sinistral strike-slip movement, with associated riedel shears forming small-scale horst and graben structures. The fault lies close to the assumed junction between the basement west and east African plates, which were active during the Cretaceous. The platform margin faults were reactivated as sinistral wrench faults in the Eocene in response to the more rapid movement of the east African plate in relation to the west African plate.[61]

The sub-Hercynian surface of the Az Zahrah-Al Hufrah Platform is largely composed of quartzites. These were formally assigned to the Al Hufrah Formation of supposed Cambro-Ordovician age. Wennekers et al. demonstrated that the Al Hufrah Formation at the type locality can no longer be regarded as Cambro-Ordovician in age. However the P1-16 well on the Manzilah Ridge has yielded acritarchs of late Cambrian age. On the Qattar Spur the H1-11 well penetrated granite beneath the marine Cretaceous. Unsilicified Nubian sands are preserved in some of the palaeotopographic depressions, for example in wells BBB1-11 and KK1-11. Roohi published a map showing the progressive onlap of the late Cretaceous formations onto the platform (Figure 6.12). During the Cenomanian a large island extended from the Az Zahrah field to the Ar Raqubah field. Lidam Formation dolomites are present both north and south of the island in the area of the Bahi field, and in the Dur-Mansur, Abu Alwan and Zauam areas. Onlap continued during the Santonian-Campanian period, flooding the areas of the Bazzuzi Ali, Almas and Arbab fields. In the Maastrichtian the transgression spread over the Az Zahrah, Al Hufrah, Umm al Furud, Haram, and E-92 areas, and the Qattar Ridge, but the Manzilah Ridge, which forms the eastern edge of the platform, and the Ar Raqubah field were not onlapped until Danian times. This onlap pattern allows several discrete tectonic elements to be identified, which controlled deposition during the late Cretaceous-early Palaeocene. Emergent areas included the Qattar, Az Zahrah-Al Hufrah, Ramadan, Haram, and Manzilah Ridges, which formed a large irregular-shaped island, surrounded by marine conditions to both north and south. At the present time the Manzilah Ridge is 600m structurally lower than the Az Zahrah field which demonstrates that considerable ENE tilt has been imposed on the Az Zahrah-Al Hufrah Platform since Danian times. It is evident that most of this tilt was imposed during the late Eocene-Oligocene tectonic disturbance, and was accompanied by extensive faulting and fracturing. This late movement has had a controlling effect on the migration of hydrocarbons on the platform. Roohi recognised three principal fault trends; NNE-SSW which he associated with the pre-rift section, NW-SE syn-sedimentary faults associated with the Cretaceous, and a NNW-SSE group of faults developed during the late Eocene tectonic phase. Fault displacement on the platform rarely exceeds 60m, but the platform boundary faults have displacements of 800 to 1000m.[62]

Al Bayda Platform

A narrow graben called the Al Kotlah Graben or the Jarad Trough separates the Az Zahrah-Al Hufrah Platform from the Al Bayda Platform (Figure 6.9). The graben has a NNE-SSW trend and is aligned with the western margin of the Abu Tumayam Trough. The trough was formed as part of the Sirt Basin rift system and contains a syn-rift fill of Nubian sands, and a post-rift graben-fill of Cenomanian to Maastrichtian age (including Turonian evaporites), overlain by a sag basin containing Palaeocene to Oligocene rocks. The graben contains the Al Kotlah field, and the Campanian Sirt Shales have a very high total organic carbon content in this area. Depth to the base Tertiary is about 1700m and the thickness of the Upper Cretaceous reaches 3000m in the south. Depth to the Hercynian unconformity varies from about 2200 to 3900m. The graben, on the model proposed by Anketell, would represent an accomodation feature within the polygonal mesh produced within the dextral wrench system which he envisaged as being active in this area during the Aptian (Figure 6.21).[63]

The Al Bayda Platform has a NNE-SSW structural grain, more similar to that of the Al Kotlah Graben and Abu Tumayam Trough than the Az Zahrah-Al Hufrah Platform. Cambro-Ordovician rocks subcrop the Hercynian unconformity on the northern part of the platform, and these rocks pinch-out onto Precambrian basement in the south. Depth to the top Cretaceous is 1300m to 1800m and the Upper Cretaceous is about 300m thick. Depth to the Hercynian unconformity is 1800 to 2300m. To the west the platform is bounded by the Zallah Trough north of the Barrut Arch, and by the Abu Tumayam Trough to the south. The southern limit is less clear. Sinha and Mriheel have made a strong case for suggesting that the Al Bayda Platform does not end abruptly along a faulted margin, but merges southwards across a shallow saddle with the Southern Shelf. To the east the platform has a faulted margin with the Maradah Trough (Figure 6.9). The platform covers an area of approximately 3200km[2].[64]

The elevated ridge along the margin of the

Al Kotlah Graben where the Warid and Al Bayda fields are located, has been named the Amin High (Figure 6.9). This ridge may connect to the Manzilah Ridge to the north, and continues southwestwards as the Barrut Arch, separating the Zallah Trough from the Abu Tumayam Trough. The platform had significant relief at the time of the Cenomanian transgression and the palaeotopography was progressively onlapped during the late Cretaceous. The area southwest of Samah and around the Tibisti (Al Awra/Ora) field were onlapped during the Cenomanian, and the Samah, Balat and Bilhizan areas during the Turonian to Santonian. The eastern platform margin, from Zaqqut to the A-NC 107 field, remained emergent until Campanian-Maastrictian times, and the Amin High, flanking the Al Kotlah Graben was not fully onlapped until the Danian. The marine Cretaceous is overlain by Palaeocene carbonates which contain most of the oil found on the Al Bayda Platform, which in turn are overlain by Eocene and Oligocene rocks. The platform was warped and tilted during the late Eocene-early Miocene with the crest now centred on the Al Bayda field and dipping eastwards towards the Maradah Trough. The area south of the Al Bayda field has a southerly dip towards the Abu Tumayam Trough.[65]

Sinha and Mriheel conducted a detailed study on the area south of the Al Bayda Platform and concluded that this area is occupied by a horst which links the Al Bayda Platform to the Southern Shelf. The horst has been penetrated by ten wells on concession 71 which demonstrate a shelfal location during the Palaeocene. The situation in the Upper Cretaceous is more equivocal, since well evidence suggests a gradual transition from nearshore facies on the Dayfah Ridge to deeper-water facies to the west. The horst is probably faulted along its eastern and western margins.[66]

The Southern Shelf (Figure 6.9) forms a salient projecting into the Sirt Basin from the south. It represents the crest of the Hercynian Tibisti-Sirt Arch, with Precambrian basement subcropping beneath the Hercynian unconformity. The unconformity is covered by less than 1250m of Mesozoic and Cainozoic sediments.

Maradah Trough

The Maradah Trough, sometimes called the Al Hagfah Trough, is a deep fault-bounded graben extending for almost 400km from west of the Dayfah field to the coast at As Sidrah. The width varies from 10 to 40km. It is bounded on the west by the Az Zahrah-Al Hufrah Platform and the Al Bayda Platform and on the east by the Al Jahamah Platform and the Zaltan Platform (Figure 6.9). The deepest part of the graben is adjacent to the Al Jahamah Platform and it shallows to both north and south. The northern extremity is constricted between the Az Zahrah-Al Hufrah Platform and the An Nuwfaliyah High, but passes over a structural saddle onto the offshore Upper Sirt Slope. The southern extremity pinches out between the Zaltan Platform and the 'South Bayda Horst' mentioned in the previous section.[67]

Depth to the top Cretaceous varies from 3000m in the centre of the basin to 2000m in the north and south. Upper Cretaceous sediments are 1800m thick in the basin centre compared with less than 150m on the adjacent Zaltan Platform. Depth to the Hercynian unconformity ranges from 4600m in the centre to 3500m on the margin. Cambro-Ordovician rocks subcrop the Hercynian unconformity. Rifting probably began in the Triassic although no evidence has yet been found of Triassic or Jurassic syn-rift sediments. The oldest rocks dated with certainty are quartzites containing marine fossils of early Cretaceous age, first reported by Bonnefous from the Wadi field, and subsequently confirmed in other wells. El Hawat et al. developed a model suggesting that the deep troughs were flooded by early Cretaceous seas which deposited marine sandstones which were later silicified, whilst non-marine Nubian sediments were deposited elsewhere. The Nubian quartzites are overlain by Cenomanian dolomites of the Lidam Formation and a full sequence of late Cretaceous sediments, including evaporites in the Turonian and thick organic-rich shales in the Campanian. The thickness of marine Cretaceous sediments in the Maradah Trough reaches 1800m. During the Campanian in the Dayfah area, clastic sediments eroded from the emergent platform margins were re-deposited as huge sandbars and sandspits on the seaward margin of the platform. Further north thick sand wedges have been found on the down-thrown side of the platform margin fault which are believed to be of Maastrichtian age. The Cainozoic sediments in the Maradah Trough, ranging from Palaeocene to Miocene in age, are generally thicker than on the adjacent platforms and show a deeper-water character, implying a long period of continuous subsidence over the site of the Cretaceous grabens. Late Eocene deformation reactivated earlier faults and produced an asymmetrical basin profile with the eastern margin becoming deeper than the western side.[68]

El Batroukh and Zentani presented a gravity map of the northern Maradah Trough which indicates a total sedimentary column of 3600 to 4600m overlying basement. The faulted trough margins are very clear with the faults having a preferred NW-SE and N-S orientation. They identified several structures within the trough which they interpreted as deep horsts which may be comparable with the Wadi field further south. There is some suggestion that the line of the Al Kotlah Graben may continue across the Maradah Trough and may account for the re-entrant between the Zaltan and Al Jahamah Platforms in the area of the B1-94 well. This may reflect an underlying Hercynian fault zone associated with the western margin of the Tibisti-Sirt Arch.[69]

Zaltan and Al Jahamah Platforms

The Zaltan and Jahamah Platforms, and their northward extensions, the Shammar High and the Al Nuwfaliyah High, form a series of linked platforms extending from the Dayfah field into the offshore north of An Nuwfaliyah (Figure 6.9). To the west these elements are flanked by the Maradah Trough and to the east by the Ajdabiya Trough. Southwards the Zaltan Platform merges with the Southern Shelf. A structural saddle in the region of the Attahadi gas field separates the Zaltan and Al Jahamah Platforms. Similar saddles separate the Shammar and An Nuwfaliyah Highs. The western margin is defined by a platform-margin fault complex with a vertical displacement reaching 1000m in places. The eastern margin is less clearly defined, and the throw on the boundary fault is much less (Figures 6.9, 6.11). Depth to the top Cretaceous is 1650m at the Dayfah field to 2650m at Hutaybah. The thickness of the Upper Cretaceous is less than 200m on the western margin thickening to 600m on the eastern flank. Corresponding depths to the Hercynian unconformity are 1900m and 2800m. A north-south cross-section of the Zaltan Platform is shown on Figure 8.16.[70]

Cambro-Ordovician quartzites subcrop the Hercynian unconformity over much of the Zaltan Platform, as confirmed by palynological analysis of the quartzites at Attahadi which gives a late Cambrian age. In the extreme south the quartzites pinch-out onto Precambrian basement. Wennekers et al. demonstrated however that not all the quartzites are of Cambro-Ordovician age, and they reported Lower Carboniferous rocks in well C3-6 beneath the Zaltan field. Both the Hutaybah and Dayfah fields are underlain by granite. The Hercynian unconformity had considerable relief. Patches of Nubian or Bahi sands are present in structural lows, but in general the marine Cretaceous section is very thin. Lidam

Formation is present in the trough between the Zaltan and Al Jahamah Platforms and low on the eastern flank of the platform, but the western crest of the platform was not onlapped until Maastrichtian times, and even then a few small islands remained. The Cainozoic section is over 1800m thick over much of the platform. The Palaeocene carbonates include the shoal complexes of Dayfah and Nasser which developed on residual highs, and now form major oilfields. The carbonates are overlain by Eocene and Oligocene rocks which have been tilted to the northeast and slightly deformed. Miocene littoral deposits complete the Cainozoic succession.[71]

The Zaltan Platform is separated from the Al Jahamah Platform by a structural saddle in which the Attahadi gas field is located, which marks the confluence of two negative structural trends. On the one hand a faulted trough passes northeastwards (to the west of Hutaybah) on trend with the Al Kotlah Graben, and on the other, the Wadayat Trough extends southeastwards linking to the Ajdabiya Trough (Figure 6.9). Anketell regarded the line of the Wadayat Trough as the site of an underlying dextral shear which also determined the southern boundary of the Shammar and An Nuwfaliyah Highs. The Wadayat Trough is flanked to the north by the Assumud Ridge which forms the eastern part of the Al Jahamah Platform. The platform extends from the As Sahel field in the east to the coast at Ra's Lanuf. To the northwest the platform continues as the Shammar and An Nuwfaliyah Highs beyond which it passes onto the Sirt Rise. The most significant difference between the Jahamah Platform and the Zaltan Platform is that the Jahamah Platform was emergent during the Danian. Indeed on the northern flank of the An Nuwfaliyah High the total Palaeocene section is less than 30m thick. The Cainozoic succession is completed by Eocene carbonates and marine Oligocene and Miocene sediments, which show

evidence of Eocene-Miocene reactivation accompanied by tilting and gentle folding. The southwestern margin of the Al Jahamah Platform is faulted facing both the Maradah Trough and the Wadayat Trough with fault displacements of 800-900m. The northern flank facing the Ajdabiya Trough is also faulted with displacements of 400-500m.[72]

Ajdabiya Trough

The Ajdabiya Trough is the deepest trough in the Sirt Basin, containing 8000m of post-Hercynian sediments. The Ajdabiya Trough is the main depocentre for Oligocene and Miocene sediments in the Sirt Basin with 1500m of Oligocene and 2100m of Miocene sediments in the centre of the trough, and active subsidence has continued to the present day. Upper Cretaceous rocks reach a thickness of 2400m in the southern Ajdabiya Trough compared with 600m on the adjacent platforms. Because of its depth little is known about the northern part of the trough. The Ajdabiya Trough covers an area of 22,500km^2 and extends from the coast at Al Brayqah to the Kalanshiyu High in the south. South of the Amal/Nafurah Nose it connects with the Hameimat Trough and with the North and South Sarir Troughs. It is flanked by the Al Jahamah and Zaltan Platforms to the west, the Cyrenaica Platform to the east and the Hameimat Basin to the southeast. Northwards it opens onto the offshore Upper Sirt Slope (Figure 6.9).[73]

Cambro-Ordovician rocks subcrop the Hercynian unconformity over most of the trough with Precambrian basement subcropping in the south. Evidence from the eastern Sirt Embayment shows the presence of Triassic and Jurassic rocks forming the oldest part of the syn-rift sequence, and the same situation may be present in other parts of the Ajdabiya Trough. The main syn-rift deposition occurred in the early Cretaceous when the Nubian (Sarir)

Sandstone accumulated in rift troughs and topographic lows on the irregular pre-Cretaceous surface. Nubian Sands are known as far west as the Harash field, but to the north they pass into a quartzitic facies, which is probably typical of the northern Ajdabiyah Trough. The Nubian may pass into a marine facies in the northern part of the trough. The post-rift sequence comprises Upper Cretaceous to Late Miocene rocks which reach a thickness of 7600m in the basin centre. Thick evaporites were deposited during the Turonian in the southern Ajdabiya Trough (Figure 4.12). In the Palaeocene, pinnacle reefs developed around the margin of the trough, and many of these reefs are oil bearing, particularly at Intisar, As Sahabi and Al Mheirigah. Most of the post-Cretaceous rocks in the centre of the Ajdabiya Trough are argillaceous in character due to their deposition in a rapidly subsiding environment. The rapid subsidence during the Oligocene and Miocene was accompanied by deformation and faulting.[74]

Little is known about the deep structure of the Ajdabiya Trough. Parsons, et al. described it as a half-graben, and Wennekers, et al. referred to a deep, intra-trough horst feature identified on both gravity and seismic data which they called the Al Brayqah Horst (Figure 6.9). This feature was shown as being capped by Palaeocene rocks and covered by Oligo-Miocene shales. The trough margins are better known. The southwestern margin is a gentle ramp margin with small down-to-the-basin faults. The Wadayat re-entrant, south of the Assumud Ridge, is an assymetrical trough with a ramp on the southern side and a major faulted margin on the northern side. The northern flank of the Jahamah Platform is poorly known but is believed to be a steep, fault-assisted margin. The eastern flank of the trough is more complex. The northeastern margin abuts against the Cyrenaica Platform and the Amal/Nafurah Nose, with a series of terraces which may represent relay-ramp faulting. A north-south faulted threshold gives access to the eastern Sirt Embayment, whilst the Ajdabiya Trough continues southwards to pinch out between the Kalanshiyu and Chadar fields (Figure 8.17).[75]

Eastern Sirt Embayment (Hameimat and Sarir Troughs)

The structure of the eastern arm of the Sirt Basin is complex, but can be subdivided into five main components which have a predominently east-west orientation: the Hameimat (Mar, Metem) Trough, Messlah-Kalanshiyu (Wasat) High, North Sarir Trough, South Sarir Ridge, and South Sarir Trough (Figure 6.9). These features are related to pre-Mesozoic structural trends. The Hameimat Trough is on trend with the Al Jaghbub Trough, an Hercynian graben, penetrated by wells G1-82 and A1-120, in which the depth to the Hercynian unconformity is 1850m. The Messlah High is on trend with the Siwa High, revealed in well A1a-84, in which the Hercynian unconformity is at a depth of only 950m, and the South Sarir Trough corresponds to the Qattara Graben, in which the J1-81a well was drilled, where the Hercynian unconformity is at a depth of 2450m. It is evident that the later structural trends were largely determined by the configuration of the Hercynian unconformity. The northern boundary of the embayment is heavily faulted and is related to the underlying South Cyrenaica Fault of Anketell which is a dextral wrench fault. The eastern margin pinches out gently onto Palaeozoic rocks near the Egyptian border. The southern margin of the embayment, which may be related to the 'southern fault zone' of Anketell, abuts the Tazirbu-Az Zalmah High on which the E1-81 well was drilled, where the depth to the unconformity is only 330m. To the west the embayment connects to the Ajdabiya Trough over a complex series of faults and faulted ramps.[76]

The Hercynian unconformity in the eastern Sirt Embayment is subcropped largely by

Precambrian basement, except along the northern margin where Palaeozoic rocks are present. Incipient rifting probably began in the Triassic, and Triassic fossils have been found in the Amal Formation, formerly regarded as Cambro-Ordovician in age in the Maragh Graben. The Amal Formation is present on the Amal Nose, in the Maragh Graben and in the northern Hameimat Trough, and it is likely that Triassic rocks form the earliest syn-rift sediments in these areas. Continental Jurassic rocks have been identified in wells in concession 65 near Sarir, in well C2-NC 59 in the Hameimat Trough, and J1-81a on the eastern margin of the trough. The Lower Cretaceous Nubian (Sarir) Sandstone is widespread in the eastern Sirt Embayment and forms the principal oil reservoir of the area. The sandstone was deposited in the principal rifts and troughs of the eastern Sirt Embayment and abrupt thickness variations are characteristic of the formation (Figure 8.19). Nubian Sandstones are absent from the intervening horsts. The Nubian sequence is truncated by a sharp unconformity which has removed the upper part of the formation from many areas, including the Sarir oilfield. This has been designated the intra-Aptian unconformity. It clearly indicates a late period of faulting, uplift and erosion prior to the Cenomanian marine transgression. Post-rift thermal subsidence allowed the area to be invaded by the Cenomanian sea which deposited shales and carbonates on the peneplained surface. Rapid subsidence in the Hameimat Trough during the Turonian-Campanian produced over 1200m of sediment in this area compared with only 300m in the Sarir area. Thick Turonian evaporites were deposited in the Hameimat Trough, contemporary with those in the southern Ajdabiyah Trough (Figure 4.12). The post-Turonian section in the eastern Sirt Basin is relatively constant in thickness and little disturbed by tectonic activity. This is in marked contrast to the western Sirt Basin where

this section was greatly affected by tectonic events. The Upper Cretaceous section is overlain by a full succession of Cainozoic sediments up to mid-Miocene in age.[77]

The structure of the Hercynian unconformity shows that the eastern Sirt Basin is asymmetric and is much deeper in the Hameimat Trough than further south. Depth to the unconformity in the Hameimat Trough is 4500m whereas in the North Sarir Trough it is 3300m and in the South Sarir Trough only 2750m. On the Messlah High depth to the unconformity is 2700m. The highs all correspond to basement horsts which were probably never covered by Nubian sediments. The bounding faults of the horsts are almost invariably oriented east-west. The Nubian sequence reaches a thickness of 1182m in the Hameimat Trough, but is reduced to 440m at Sarir. The faults marking the northern margin of the basin have a displacement of 1200m, whilst the Messlah Horst has a relief of about 500m. The eastern and southern flanks of the basin are ramp margins, with some small down-to-the-basin faults on the southern flank. The western margin plunges over a series of fault terraces into the southern Ajdabiya Trough.[78]

Anketell regarded the east-west oriented Messlah High and South Sarir Ridge as belonging to earliest phase of tectonic activity in the formation of the Sirt Basin. He interpreted the area as a pull-apart basin formed during the Neocomian, prior to the dextral wrenching in the western Sirt Basin. In this model the eastern Sirt Embayment was formed by north-south extensional stress on the northern margin of the African plate, during the separation of the Apulian plate from the African margin (Figure 2.10).[79]

6.4   CYRENAICA

Cyrenaica occupies an area of 160,000km^2 extending from Binghazi to the Egyptian border and from the northern coastline to the northern margin of the Hameimat Trough. Two major tectonic provinces dominate the area, the Cyrenaica Platform in the south and the Jabal al Akhdar Uplift and fold belt in the north. In addition the Maragh Graben, Ash Shulaydimah Trough and Marmarica Basins form significant re-entrants on the margin of the Cyrenaica Platform (Figure 6.13). The stratigraphy and structure of Cyrenaica differ in several important respects from the rest of onshore Libya. The area owes its tectonic evolution to a different tectonic regime from that which controlled the development of the Sirt Basin. Whereas the Sirt Basin represents a pull-apart, dilational basin within a dextral wrench setting, the Cyrenaican Platform was formed within a splay wedge, and the Jabal al Akhdar Uplift represents a dextral contractional duplex.[80]

6.4.1 Structual Evolution of Cyrenaica

Basement rocks in Cyrenaica, penetrated in well C1-82, are composed exclusively of Phasusian accreted terranes. Red beds containing late Proterozoic acritarchs have been found in three wells, and this sequence probably reaches a thickness of 750m in well D1-31.

The deposition of the Palaeozoic rocks was influenced by the Kalanshiyu-Awaynat Uplift (Figure 2.6), and almost all the Palaeozoic formations thicken to the north and northeast. No evidence of Cambrian rocks has been found in Cyrenaica and the Precambrian rocks are unconformably overlain by rocks of Caradocian and Ashgillian age which show some similarities with the glacio-marine deposits of the same age in Morocco. Palynological evidence shows no major break at the Ordovician-Silurian boundary and the Silurian exhibits a typical lower shaly unit and upper sandy unit of Llandoverian, Wenlockian and possibly Ludlovian age. The Silurian-Devonian junction is unconformable and Cyrenaica is interpreted as being emergent during the Pridolian,

Lochkovian and Pragian stages. A complete sequence of Eifelian to Famennian age rocks is present, thickening from south to north, and is best developed in well C1-125, but another hiatus during the Tournaisian possibly reflects the initial contact between Gondwana and Laurasia. An almost complete Carboniferous sequence was deposited, again thickening towards the north, which extends from early Visean to Gzelian age in well A1-NC 92, although the Bashkirian and Kasimovian stages may not be present.[81]

The Hercynian orogeny culminated during the late Carboniferous-early Permian and the Sirt Arch was formed at this time (Figure 2.7). Southern Cyrenaica was uplifted, folded and subjected to erosion which removed part of the Carboniferous, particularly along the southern rim of the Cyrenaica Platform. Progressively more Palaeozoic section is preserved on the Cyrenaica Platform towards the north, and in northern Cyrenaica deposition appears to be uninterrupted across the Carboniferous-Permian boundary. Well A1-19 shows a passage from Gzelian to Asselian rocks, overlain by Sakmarian to Ufimian limestones and shales. This predominently marine Permian section is comparable to that found in Tunisia. No late Permian or early Triassic rocks are known from Cyrenaica and this suggests a period of emergence, but marine middle and late Triassic rocks are present in wells A1-19, C1-2 and A1-28, and continental Triassic rocks are known from the Maragh Graben.[82]

In northeast Cyrenaica marine Jurassic rocks of middle and late Jurassic age are present in wells A1-NC 120, A1-36, A1-NC 128, and B1-33, and early Jurassic rocks may be present below the total depth of the wells. These rocks pass southwards into continental facies. During the mid- to late Jurassic, subsidence began in the Jabal al Akhdar Trough, and Upper Jurassic rocks thicken significantly into the trough. Marine sedimentation continued during the

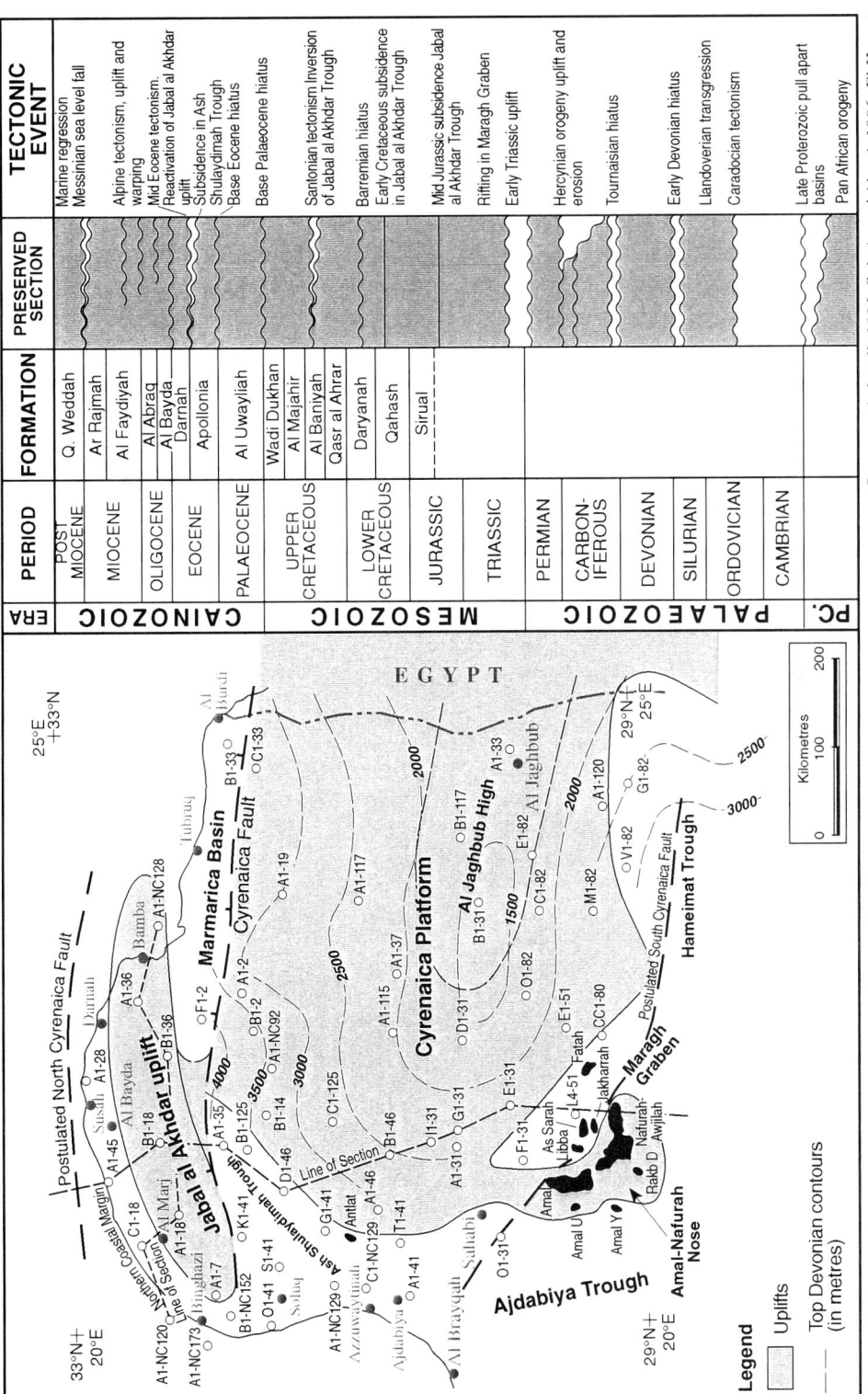

ERA		PERIOD	FORMATION	PRESERVED SECTION	TECTONIC EVENT
CAINOZOIC		POST MIOCENE	Q. Weddah		Marine regression / Messinian sea level fall
		MIOCENE	Ar Rajmah		Alpine tectonism, uplift and warping
		OLIGOCENE	Al Faydiyah		
		EOCENE	Al Abraq / Al Bayda / Darnah		Mid Eocene tectonism. / Reactivation of Jabal al Akhdar uplift / Subsidence in Ash Shulaydimah Trough
		PALAEOCENE	Apollonia		Base Eocene hiatus / Base Palaeocene hiatus
MESOZOIC		UPPER CRETACEOUS	Al Uwayliah		
			Wadi Dukhan		Santonian tectonism Inversion of Jabal al Akhdar Trough
			Al Majahir		
		LOWER CRETACEOUS	Al Baniyah		Barremian hiatus
			Qasr al Ahrar		Early Cretaceous subsidence in Jabal al Akhdar Trough
			Daryanah		
		JURASSIC	Qahash		Mid Jurassic subsidence Jabal al Akhdar Trough
			Sirual		Rifting in Maragh Graben
		TRIASSIC			Early Triassic uplift
PALAEOZOIC		PERMIAN			Hercynian orogeny uplift and erosion
		CARBON-IFEROUS			Tournaisian hiatus
		DEVONIAN			Early Devonian hiatus
		SILURIAN			Llandoverian transgression
		ORDOVICIAN			Caradocian tectonism
PC.		CAMBRIAN			Late Proterozoic pull apart basins / Pan African orogeny

Sources: Anketell, 1996, Sghair and Alami, 1996, El Hawat and Shelmani, 1993, Duronio, et al., 1991, Thusu, et al., 1988, Ghori, 1991, El Arnauti and Shelmani, 1985, 1988, Pallas, 1980, Rohlich, 1978

Figure 6.13   Cyrenaica, Tectonic Elements and History

Cyrenaica forms the bridge between the stable African margin to the south and the Tethyan open marine domain to the north. Permian to mid-Cretaceous open marine rocks are only found in northern Cyrenaica, whilst the area to the south contains mostly continental and shallow shelf equivalents. The Jabal al Akhdar developed as a trough during the late Jurassic and early Cretaceous and was inverted in two stages, during the Santonian and mid-Eocene.

early Cretaceous, and Berriasian and Valanginian carbonates are present in well B1-33 in the Marmarica Basin. An unconformity at the base of the Barremian is recognisable over a large area of northern Cyrenaica. Rapid subsidence in the Jabal al Akhdar Trough led to the deposition of 2300m of Lower Cretaceous rocks in well A1-36 and 3300m in well A1-NC 128, compared with only about 600m to both north and south (Figures 6.13, 6.14 and 6.15). The trough was oriented east-west and extended from well A1-7 to offshore well A1-NC 128.[83]

The collapse of the Sirt Arch and the accompanying thermal subsidence led to the Cenomanian transgression during which marine conditions were established over northern Cyrenaica and in the Hameimat Trough, but the Cyrenaica Platform was not covered until Coniacian times. Tectonic events in the Santonian, which Barr and Berggren named the Cyrenaican orogeny, folded and inverted part of the Jabal al Akhdar Trough into an uplifted horst, the Jabal al Akhdar Uplift, which influenced sedimentation during the late

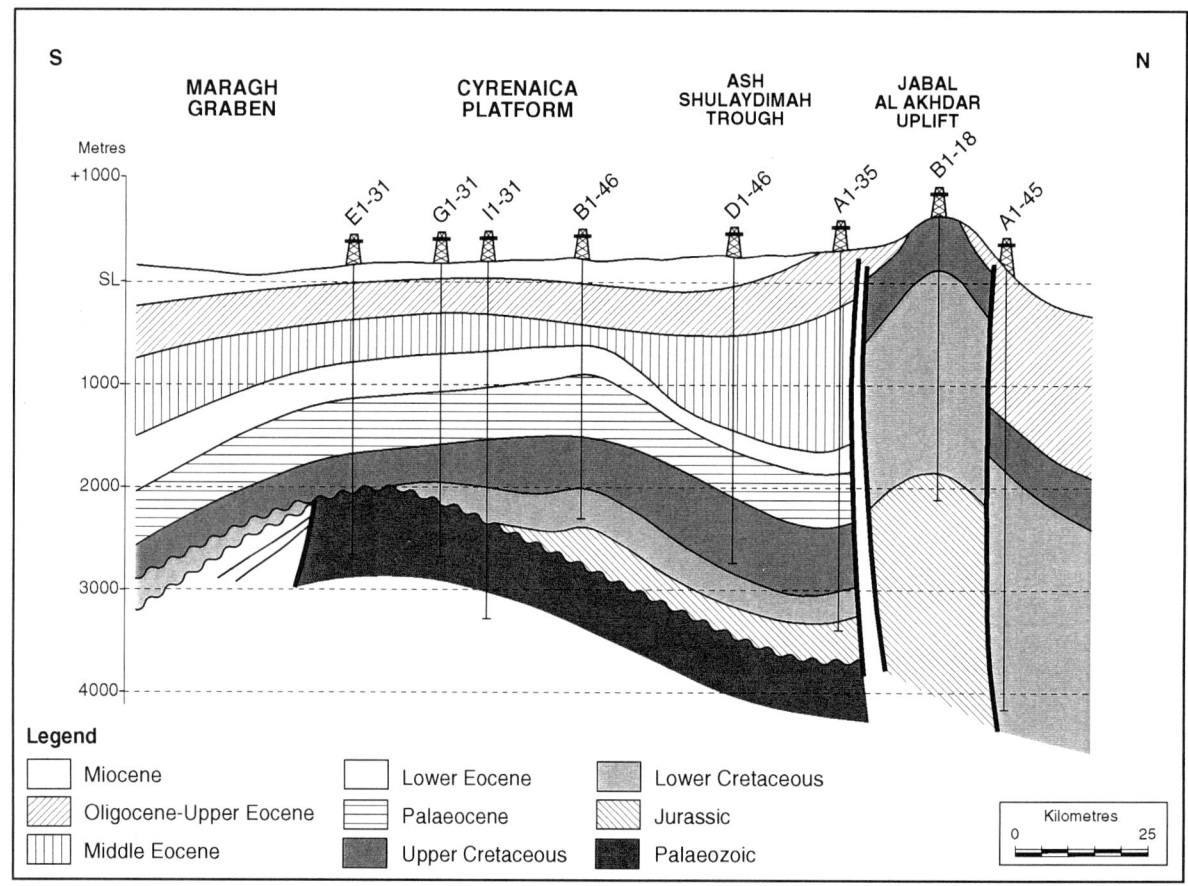

Sources: Pallas, 1980, Ghori, 1991, El Hawat and Shelmani, 1993

Figure 6.14   Cyrenaica Platform, Diagrammatic Cross-Section

This section shows the stable Cyrenaica Platform passing northwards into the unstable area of the Ash Shulaydimah Trough which subsided during the middle Eocene, and the Jabal al Akhdar Uplift. The Jabal al Akhdar Trough developed during the late Jurassic and early Cretaceous, and was inverted during the Santonian, and further uplifted during the Eocene.

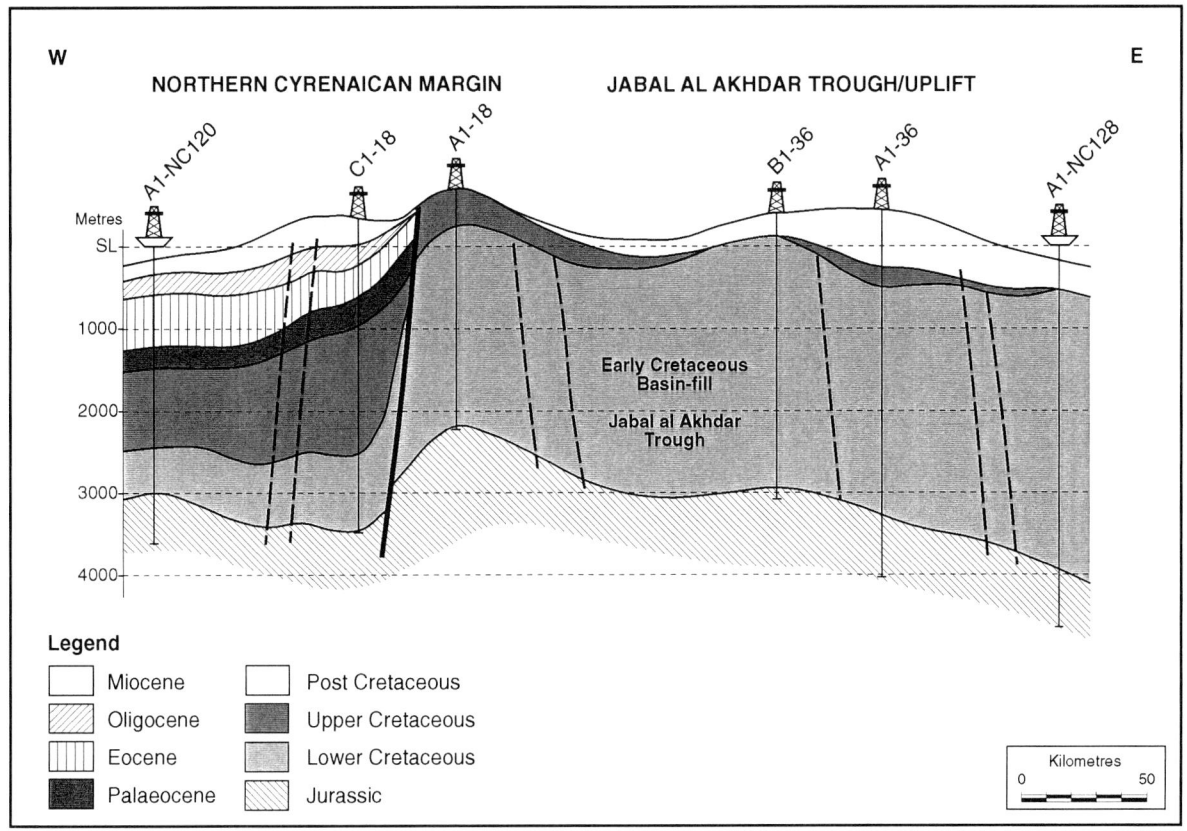

Sources: Modified from El Hawat and Shelmani, 1993, Ghori, 1991, Duronio, et al., 1991

Figure 6.15   Jabal al Akhdar Uplift, Diagrammatic Cross-Section

The Jabal al Akhdar Trough began to form during the late Jurassic and contains over 2500m of early Cretaceous rocks compared with only about 700m to north and south. This longitudinal section along the axis of the trough shows the effect of the inversion, which occurred in two stages, in the Santonian and mid-Eocene. The mid-Eocene reactivation is most pronounced in the west, compared with the A1-36 and A1-NC 128 areas to the east.

Cretaceous (Figures 6.14 and 6.15). Over a large part of the uplift Santonian to Maastrichtian rocks are absent and in other areas a Santonian unconformity reflects the tectonic event. On the northern flank of the uplift deep-water marine conditions were present, as proved by the unbroken sequence of late Cretaceous planktonic foraminiferal zones at Ra's al Hilal. By contrast, south of the uplift, a shallow carbonate shelf became established over the Cyrenaica Platform during the late Cretaceous and carbonate deposition persisted in this area until Eocene times. During the Eocene,

extensive subsidence in the Ash Shulaydimah Trough to the south of the Jabal al Akhdar Uplift indicates a further period of instability. Over 1700m of middle Eocene carbonates and evaporites are present in well A1-35 compared with an average of only 900m on the Cyrenaican Platform. During the mid- to late Eocene the western Jabal al Akhdar Uplift was reactivated and further elevated (Figures 6.14 and 6.15). However, as discussed below, well evidence shows that the mid-Eocene reactivation did not affect the same area as the Santonian uplift. In well A1-NC 128 marine deposition resumed during the late Eocene, and open marine conditions extended along the northern flank of the uplift, whilst the western part of the uplift, around well A1-18, became emergent and has remained so ever since.[84]

Continuing instability is marked by several unconformities and oscillating shorelines within the Oligocene-Miocene succession. During this period the carbonate sedimentation, which had dominated the area since the late Cretaceous, was replaced by a dominently clastic succession, with the Jabal al Akhdar Uplift remaining as an emergent ridge. Late Miocene rocks are not present over most of Cyrenaica, representing a period of regression, which ended with the Messinian salinity crisis and the excavation of the As Sahabi channel. Gentle arching during the Pliocene and withdrawal of the sea to the present shoreline led to erosion and the unroofing of a large area of Cretaceous rocks on the northern margin of the Cyrenaica Platform.[85]

6.4.2 Structural Components of Cyrenaica

Cyrenaica Platform

The Cyrenaica Platform occupies the major part of Cyrenaica, being bordered to the north by the Jabal al Akhdar Uplift and to the south by the Maragh Graben and Hameimat Trough. Eastwards the platform dips gradually into the Western Desert Basin of Egypt, and northeastwards over a faulted margin into the Marmarica Basin (Figure 6.13). The top of the basement is highest along the southern margin of the platform where it was found at 1830m in well M1-82. It plunges into a trough in the region of A1-115 at an estimated depth of about 4850m before rising slightly towards the Cyrenaica Fault in the north. At top Devonian level (depth contours shown on Figure 6.13) the platform culminates on the Al Jaghbub High at a subsea depth of 1525m (well B1-31), and dips northwards to a depth of 3950m in well B1-2. Southwards it plunges into the Al Jaghbub Trough (Figure 6.9) and reaches a depth of 2600m in well G1-82. The top Carboniferous shows a similar picture. Carboniferous rocks subcrop the Hercynian unconformity over much of the platform, including the Al Jaghbub High. At top Cretaceous level the platform has a gentle WSW dip from a depth of 350m in well C1-33 to 1750m in well A1-31. A stable shallow-carbonate shelf became established during the late Cretaceous which persisted until the Eocene. In terms of regional tectonics Anketell interpreted the Cyrenaica Platform as representing a splay wedge bordered to the north by the Cyrenaica Fault and to the south by the South Cyrenaica Fault (Figure 6.21). These are both basement faults with a dextral wrench component, and mark respectively the southern limit of the Jabal Akhdar foldbelt and the northern limit of the Hameimat Trough. The western margin of the platform is formed by ramps and terraces on the edge of the Ajdabiya Trough.[86]

Jabal al Akhdar Uplift

The Jabal al Akhdar Uplift was formed by the inversion of the Jabal al Akhdar Trough during the Santonian, with reactivation during the Eocene, and it extends east-west from Qasr al Ahrar towards Darnah. The uplift is a dramatic faulted arch with steeply dipping faulted margins, exposing Cretaceous rocks in the core of the fold (Figures 6.14 and 6.15). Inversion took place in two stages, during the Santonian and mid-Eocene. However, as mentioned above, the Eocene reactivation did not affect the same area as the Santonian uplift. The location of the trough can be deduced from well evidence. Abnormally thick Lower Cretaceous sediments are found in wells A1-18, B1-18, A1-45, A1-28, B1-36, A1-36 and

A1-NC 128. Inversion during the Santonian led to emergence and erosion and Santonian to Maastrichtian rocks are absent in all these wells. Reactivation of the uplift during the early and middle Eocene affected the area of wells A1-18, B1-18, B1-36 and A1-28 where Cainozoic rocks are absent or very thin. Well A1-45 has a thick Eocene sequence and was evidently located north of the Eocene uplift. Wells A1-36 and A1-NC 128 also contain a substantial mid-Eocene to early Miocene section, and this area too was not involved in the Eocene uplift. In summary, it appears that the Santonian inversion affected most of the area of the Jabal al Akhdar Trough, but the Eocene reactivation affected only the western part of the former trough. The uplift was produced by wrench faulting and compression within what Anketell has termed the Al Jabal al Akhdar Duplex, located between the Cyrenaica and the North Cyrenaica Faults. Local NE-SW tectonic features, such as the Tulmitah Horst and the Al Marj Graben are consistent with the dextral wrench model.[87]

Marmarica Basin

There is some confusion about the definition of the Marmarica Basin. El Arnauti and Shelmani described the basin as containing thick Triassic, Jurassic and Lower Cretaceous marine to paralic sediments, situated between the Jabal al Akhdar Uplift and the Cyrenaica Platform. They mentioned the A1-36 well as being typical of the basin, and made specific reference to late Cretaceous and Tertiary subsidence relative to the rising Jabal al Akhdar Uplift. Published well evidence however appears to contradict this view. The key wells A1-36, B1-36 and A1-NC 128, as outlined above, all show very thick lower Cretaceous sequences, and were caught up in the Jabal al Akhdar inversion during the Santonian, eroded during the late Cretaceous and early Cainozoic and were probably not submerged again until the middle Eocene. These

wells are located on the Santonian Uplift, not to the south of it. However well B1-33, drilled much further east near Al Burdi, shows a continuous marine sequence from Jurassic to Cainozoic with no evidence of a Santonian unconformity, and this well would probably be a better candidate for a 'type' Marmarica Basin well. It seems preferable to limit the term Marmarica Basin to the marine sediments wedged between the Cyrenaica Platform and the Jabal al Akhdar Uplift (Figure 6.13). This however would exclude the three wells mentioned above.

The basin forms a fault-bounded re-entrant between the Jabal al Akhdar Uplift to the north and the Cyrenaica Platform to the south, with the faults downthrowing into the basin. According to El Arnauti and Shelmani, the faults were active during the Jurassic and early Cretaceous. This area is located south of the Jabal al Akhdar Trough and probably formed part of the northern margin of the Cyrenaica Platform, open to the Tethys Ocean. The southern margin of the basin coincides with the line of the Cyrenaica Fault, and the basin can be interpreted as the trailing edge of the Jabal al Akhdar Duplex. The western limit of the basin is formed by a saddle which separates it from the Ash Shulaydimah Basin to the west. Anketell and some other authors preferred to use the term Marmarica Basin in the wider context of the tilted and down-faulted eastern part of the Cyrenaican Platform. The area of the A1-36 and A1-NC 128 wells, which formed part of the Jabal al Akhdar Trough, and which was inverted during the Santonian they referred to as the Darnah Trough.[88]

Northern Coastal Margin

The northern coastal margin is the area north of the Jabal al Akhdar Uplift, which was exposed to open-marine conditions for much of the Mesozoic. The continental shelf in northern

Cyrenaica is very narrow and the slope plunges rapidly into deep water. Seismic data suggest that a full sequence of Mesozoic marine rocks is present offshore, but only 50km north of the coastline the sedimentary wedge comes into contact with the thrust belt of the Hellenic fore-arc. During the early Cretaceous the northern margin remained relatively stable compared with the Jabal al Akhdar Trough to the south, and Lower Cretaceous sediments thin from the trough onto the stable northern margin, as seen in wells A1-NC 120 and C1-18.

Well A1-45 on the coast near Susah contains about 1500m of Eocene rocks and was not apparently affected by the Eocene reactivation, but well A1-28 further east has only a very thin Eocene cover and was involved in the Eocene uplift. The deep-water, uninterrupted Cretaceous section found at Ra's al Hilal, close to well A1-28, is difficult to explain in relation to the well evidence. It appears that the northern margin sequence found in wells A1-NC 120 and C1-18 passes around the northern margin of the Jabal al Akhdar Uplift, and connects to the Marmarica Basin, which has a similar Mesozoic sequence (Figure 6.13). Anketell referred to the area of the C1-18 and A1-NC 120 wells as the Binghazi Trough. Northwards the sedimentary wedge plunges rapidly into deep water over a series of faults and terraces.[89]

Al Jaghbub High

The Al Jaghbub High is an east-west Hercynian ridge forming the southeastern part of the Cyrenaica Platform, which was revealed by wells, B1-31, E1-82 and A1-33 (Figure 6.13). The horst was uplifted during the Hercynian orogeny and extensively eroded. At the present time Devonian rocks reach their highest elevation of 1500m subsea in well B1-31, whilst the Carboniferous culmination is further east in well A1-33 at a depth of 1112m. The high was peneplained following the

Hercynian orogeny and had no effect on post-Palaeozoic deposition.[90]

Maragh Graben

The Maragh Graben, sometimes known as the Khatt Graben, is a re-entrant between the Amal Platform and the Cyrenaican Platform, opening to the southeast into the Hameimat Trough (Figure 6.13). It was originally believed to be floored by thick Amal Formation sandstones of Cambro-Ordovician age, but the discovery of Triassic fossils in the Amal Formation in wells L4-51 and A1-96 now suggests that the formation is at least in part of Triassic age and that incipient rifting began during the Triassic. The Amal Formation is at least 300m thick in the graben. Subsidence within the graben permitted the accumulation of thick alluvial fans of the Maragh Formation which are well developed on the Jakharrah and As Sarah fields, and a thick upper Cretaceous section is present starting with Rachmat Formation of Santonian-Coniacian age. Subsidence ceased at the end of the Cretaceous and the subsequent history of the trough was similar to that of the adjacent Amal Nose and Cyrenaica Platform.[91]

Amal/Nafurah Nose

The Amal/Nafurah Nose forms the south-western margin of the Cyrenaica Platform (Figure 6.13). It is developed over a basement high and granite was encountered at a depth of 2900m in well B8-12. The granite is covered by thin Amal Formation clastics, now known to be at least in part Triassic in age. Following the collapse of the Sirt Arch the ridge remained emergent during the Cenomanian to Coniacian period and the Taqrifat Formation, of Santonian age, is the first Upper Cretaceous unit to cover the crest of the ridge. The area remained as a shallow carbonate shelf from the Santonian to the late Eocene. The nose is tilted to the west, and plunges over a faulted margin into the Ajdabiya Trough. It is separated from the Maragh Graben by a major fault which was active during the Cretaceous, and was reactivated during the late Eocene. The large Amal, An Nafurah/Awjilah and Ar Rakb oilfields are located on the nose.[92]

Ash Shulaydimah Trough

The western margin of the Cyrenaican Platform is marked by a series of downfaulted terraces and ramps. Further north in the re-entrant between the Cyrenaica Platform and the Jabal al Akhdar Uplift a basin is present which has been given many different names – Ash Shulaydimah (formerly Sceleidima), Soluq, Antlat, Ajdabiya, Hamama. The name Ash Shulaydimah is preferred by most recent authors (Figures 6.13, 6.14). The defining feature of the trough is the abnormally thick Middle Eocene section which developed during the period of instability which led to the reactivation of the Jabal al Akhdar Uplift. The trough forms a counterpart of the Marmarica Basin, from which it is separated by a structural saddle. Interestingly both troughs lie on the line of a postulated late Miocene wrench zone contemporary with the Medina and Sirt Wrenches. Unfortunately pre-Upper Cretaceous rocks have not been reached in the Antlat area. The oldest rocks found to date are siltstones of Santonian age (in well U1-41), and the incomplete Upper Cretaceous section reaches a thickness of 2500m. The overlying Palaeocene-Eocene section is 1750m thick in well G1-41. The post-Eocene section simply thickens westwards into the Ajdabiya Trough. The trough narrows eastwards adjacent to the Jabal al Akhdar Uplift where it has been called the Hamama Trough. Well A1-35, which penetrated to the Jurassic shows an extremely thick development of middle Eocene carbonates (Figure 6.11). The

Palaeocene-Eocene section in the well reaches 1900m in thickness, of which 1350m are assigned to the middle Eocene. Oil has been found in the Eocene section in the Antlat area. This basin can be regarded as a downfaulted block wedged between the Jabal al Akhdar Uplift to the north and the Cyrenaican Platform to the south. The northern bounding fault is coincident with Anketell's Cyrenaica basement fault, and the southern bounding fault is parallel to the Jabal al Akhdar fold axis.[93]

6.5   OFFSHORE

The offshore area of Libya is divisible into three broad zones, the Pelagian Block in the west, downfaulted during the Miocene tectonism, the Sirt Embayment, extending from Misratah to the Ajdabiya Trough, and the Cyrenaican Margin (Figure 6.16). Bathymetric contours show the distinctive character of the three zones. The Pelagian Block is mostly in water depths of less than 500m, the Sirt Slope and Rise deepens steadily to the abyssal plain at a depth of 3500m. The Cyrenaican margin drops precipitously to a water depth of 3000m within 30km of the northern coastline. Tectonic development of the offshore can be assigned to four main phases. Continental rifting occurred during the late Triassic in the Sabratah Basin, the Binghazi Basin and the area from Malta to north of the Cyrenaica Ridge. During the mid-Jurassic the rift phase developed into crustal extension with the formation of oceanic crust on the Ionian abyssal plain. Extension continued and during the late Cretaceous the Sabratah Basin developed as a pull-apart trough. The Alpine orogeny produced intense rifting in the Pelagian and Ionian Seas, resulting in the formation of the Pantelleria Rift Zone, and subsidence in both the Sabratah and Binghazi Basins.

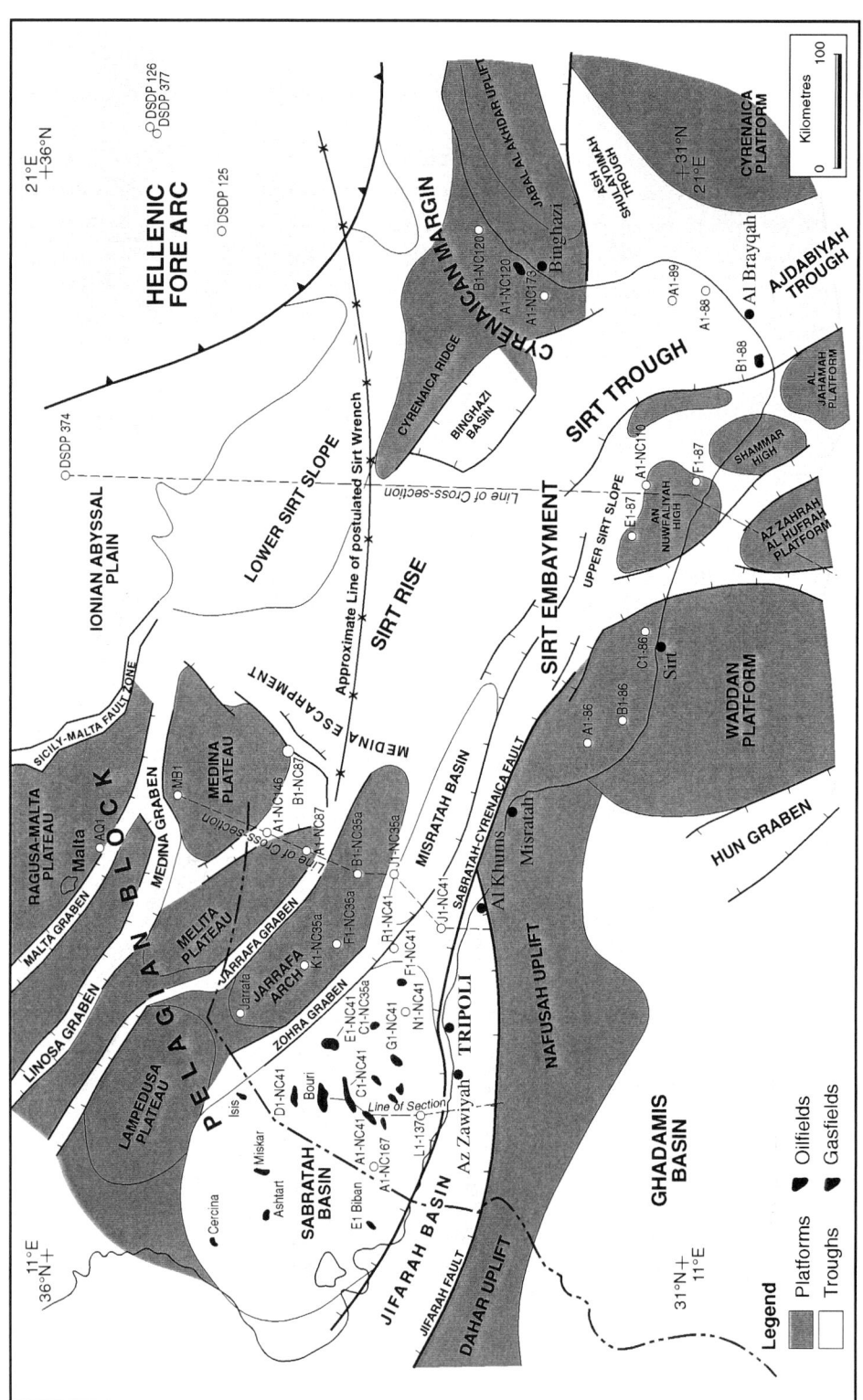

Figure 6.16   Libyan Offshore, Tectonic Elements

Sources: Finetti, 1982, 1985, Del Ben and Finetti, 1991
Burollet, 1978, Hammuda et al. 1985, Hammuda et al. 1991
Sheta, 1990, 1991, Jongsma, 1991, El Ghoul, 1991
Anketell, 1996, Smith and Kurki, 1996

The Libyan offshore is divisible into three areas, the Pelagian Block, the Sirt Embayment and the Cyrenaican margin. The area is dominated by Alpine tectonics, particularly wrench faulting associated with the emplacement of the Calabrian and Hellenic Arcs. The entire area subsided during the Miocene, and the grabens and faulted margin of the Pelagian Block are all of Miocene/Pliocene age. Evidence from the Ionian abyssal plain shows Middle Jurassic and later rocks overlying Jurassic oceanic crust The thrust sheets of the Hellenic Arc approach within 50km of the Cyrenaican coastline.

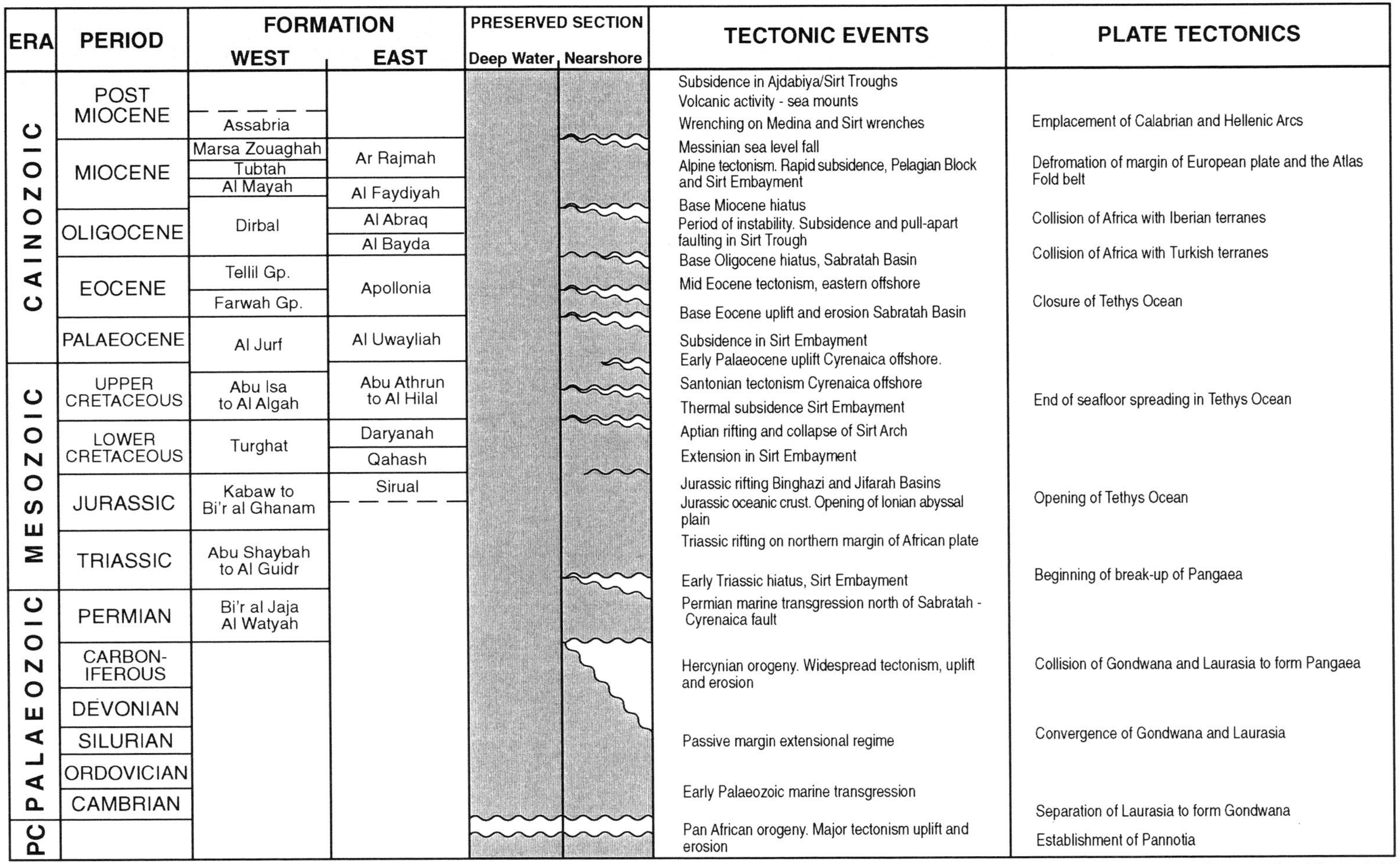

ERA	PERIOD	FORMATION WEST	FORMATION EAST	PRESERVED SECTION (Deep Water, Nearshore)	TECTONIC EVENTS	PLATE TECTONICS
CAINOZOIC	POST MIOCENE	Assabria			Subsidence in Ajdabiya/Sirt Troughs. Volcanic activity - sea mounts. Wrenching on Medina and Sirt wrenches	Emplacement of Calabrian and Hellenic Arcs
CAINOZOIC	MIOCENE	Marsa Zouaghah; Tubtah; Al Mayah	Ar Rajmah; Al Faydiyah		Messinian sea level fall. Alpine tectonism. Rapid subsidence, Pelagian Block and Sirt Embayment. Base Miocene hiatus	Deformation of margin of European plate and the Atlas Fold belt
CAINOZOIC	OLIGOCENE	Dirbal	Al Abraq; Al Bayda		Period of instability. Subsidence and pull-apart faulting in Sirt Trough. Base Oligocene hiatus, Sabratah Basin	Collision of Africa with Iberian terranes
CAINOZOIC	EOCENE	Telil Gp.; Farwah Gp.	Apollonia		Mid Eocene tectonism, eastern offshore. Base Eocene uplift and erosion Sabratah Basin	Collision of Africa with Turkish terranes
CAINOZOIC	PALAEOCENE	Al Jurf	Al Uwayliah		Subsidence in Sirt Embayment. Early Palaeocene uplift Cyrenaica offshore.	Closure of Tethys Ocean
MESOZOIC	UPPER CRETACEOUS	Abu Isa to Al Algah	Abu Athrun to Al Hilal		Santonian tectonism Cyrenaica offshore. Thermal subsidence Sirt Embayment	
MESOZOIC	LOWER CRETACEOUS	Turghat	Daryanah; Qahash		Aptian rifting and collapse of Sirt Arch. Extension in Sirt Embayment	End of seafloor spreading in Tethys Ocean
MESOZOIC	JURASSIC	Kabaw to Bi'r al Ghanam	Sirual		Jurassic rifting Binghazi and Jifarah Basins. Jurassic oceanic crust. Opening of Ionian abyssal plain	Opening of Tethys Ocean
MESOZOIC	TRIASSIC	Abu Shaybah to Al Guidr			Triassic rifting on northern margin of African plate	
PALAEOZOIC	PERMIAN	Bi'r al Jaja Al Watyah			Early Triassic hiatus, Sirt Embayment. Permian marine transgression north of Sabratah - Cyrenaica fault	Beginning of break-up of Pangaea
PALAEOZOIC	CARBONIFEROUS				Hercynian orogeny. Widespread tectonism, uplift and erosion	Collision of Gondwana and Laurasia to form Pangaea
PALAEOZOIC	DEVONIAN				Passive margin extensional regime	Convergence of Gondwana and Laurasia
PALAEOZOIC	SILURIAN				Early Palaeozoic marine transgression	
PALAEOZOIC	ORDOVICIAN					
PC	CAMBRIAN				Pan African orogeny. Major tectonism uplift and erosion	Separation of Laurasia to form Gondwana. Establishment of Pannotia

Source: See Figure 6.16

Figure 6.17 Libyan Offshore, Tectonic History

The geology and tectonics of the Libyan offshore reflect very clearly the plate tectonic history of north Africa. The Palaeozoic is dominated by the northward movement of Gondwana culminating in the Hercynian orogeny. The Mesozoic reflects the break up of Pangaea and the opening of the Tethys Ocean, and the Cainozoic reflects the closure of Tethys culminating in the Alpine orogeny. The significant differences between the western and eastern offshore areas are largely due to Alpine tectonism particularly the wrenching caused by the emplacement of the Calabrian Arc.

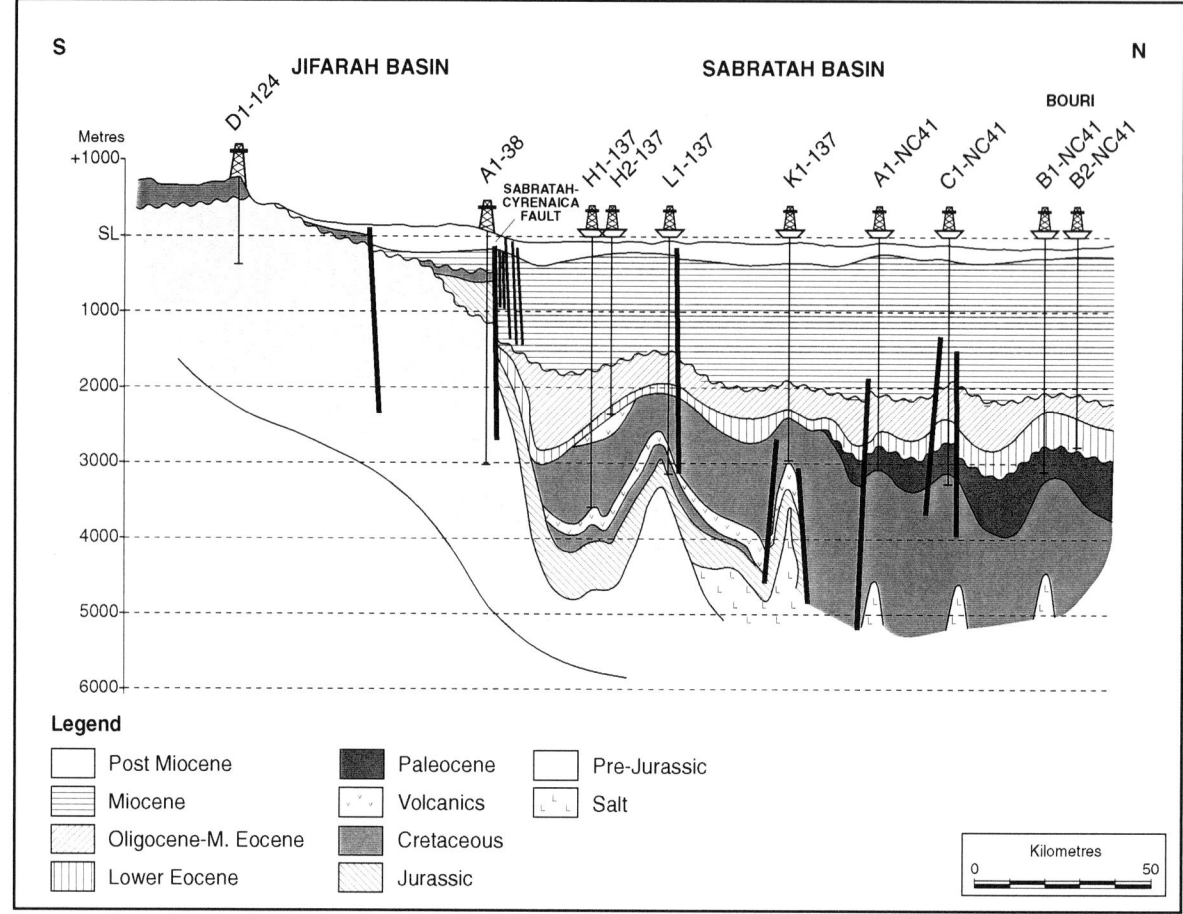

Source: Redrawn from Sbeta, 1991

Figure 6.18   Sabratah Basin, Diagrammatic Cross-Section

This section across the Sabratah Basin shows the importance of the Sabratah-Cyrenaica Fault which was at its most active during the Miocene. Over 2000m of Miocene and post-Miocene rocks are present over most of the basin and the entire Pelagian Block subsided during this period. The late Triassic-early Jurassic salt has also had a major impact on structure, and in the western part of the basin most of the structures are salt related. There are also significant volcanic horizons particularly within the Cretaceous.

6.5.1 Pelagian Block

The western offshore region of Libya forms part of the Pelagian Block which represents the northern margin of the African Plate where it impinges against the Calabrian Fore-Arc. The block has been severely affected by Miocene tectonics caused by the eastward movement of the Calabrian Fore-Arc. This movement produced major wrenching on the Pelagian Block, particularly on the Medina and Sirt Wrenches, and the effects can be seen in the pull-apart grabens in the Pantelleria Rift Zone. From north to south the Pelagian block can be divided into the following tectonic elements: the Ragusa-Malta Plateau, the Pantelleria Rift Zone, Lampedusa-Melita-Medina Plateau, and the Sabratah Basin (Figure 6.16). The Pelagian Block is terminated to the east by the Medina Escarpment, a dramatic post-Miocene fault scarp which probably follows the line of a basement fracture zone (Figures 6.16, 6.18). Since only the two southernmost elements occur within Libyan territory, the following review will concentrate on these areas.[94]

Sabratah Basin

The Sabratah Basin (Figures 6.16, 6.17 and 6.18) is a pull-apart basin within a dextral shear zone. Its southern margin coincides with the Sabratah-Cyrenaica Fault Zone, a basement shear which connects westwards to the South Atlas Fracture in Algeria. The orientation of the shear zone may have been controlled by the Hercynian Nafusah Uplift which has a similar trend. Depth to the base Cainozoic in the basin ranges from 2800m on the southern flank to 5300m in the basin centre. The Jifarah Basin is located on a terrace to the south of the Sabratah-Cyrenaica Fault Zone, where the depth to the base Cainozoic is only about 500m. This gives some indication of the throw on the basin margin fault (Figure 6.18). The Sabratah Basin passes westwards into Tunisian waters and gradually pinches out towards the Atlas fold belt. The northern limit of the basin is formed by the Lampedusa Plateau in the west, and the Melita-Medina Plateau to the east.[95]

The oldest rocks encountered in the the area are late Triassic in age in well L1-137, located in the southern Sabratah Basin. These rocks were deposited in a rift basin, and it is likely that they are underlain by older Triassic and Permian rocks as seen in the Jifarah Basin and in Tunisia. The Bi'r al Ghanam Formation in well L1-137 has a strongly evaporitic character, and these rocks pass into thick salt deposits in the western Sabratah Basin. In this area the late Triassic-early and middle Jurassic evaporites show diapiric features in the form of salt domes, swells and walls (Figure 6.18). Most of the hydrocarbon discoveries in the western Sabratah Basin are located on salt-related structures.

Upper Jurassic and early Cretaceous sediments are overlain by widespread volcanic rocks which are probably of early Cretaceous age. Marine rocks of Neocomian to Albian age are represented by the Turghat Formation, and extensive carbonate shelves were developed over much of the basin. Shallow-water to restricted conditions became established during the late Cretaceous, although deeper water shales are present in the centre of the basin suggesting the presence of a pull-apart trough in this area. Organic shales comparable to the Sirt Shales of the Sirt Basin have not been found in the Sabratah Basin. The nearest equivalent is the Jamil Formation of Santonian age, which has rather modest TOC values.[96]

Palaeocene rocks are only preserved in the northern part of the basin where the contact with the Cretaceous appears to be uninterrupted. A major basal Eocene unconformity reflects a tectonic episode which resulted in tilting and temporary emergence of the southern part of the basin. Pre-Eocene rocks are progressively truncated to the south, so that in the L1-137 well Eocene rocks rest on Santonian/Campanian rocks. A widespread marine transgression during the early Eocene led to the deposition of the Farwah Group, which contains the principal oil reservoirs of the Libyan offshore, and the regressive Tellil Group which is composed of nearshore and restricted shallow-shelf lithologies. Alpine rifting led to subsidence in the centre of the basin where deeper-water deposits accumulated. An unconformity separates the late Eocene from the Oligocene, which is represented by carbonates and deeper-water shales in the basin centre. The eastward displacement of the Calabrian Arc during the Miocene produced great instability and rapid subsidence on the Pelagian Block. Over 2000m of Miocene sediments were deposited on the block, and the wrenching produced pull-apart grabens not only in the Pantelleria Rift Zone but also further south in the Jarrafa Graben and the

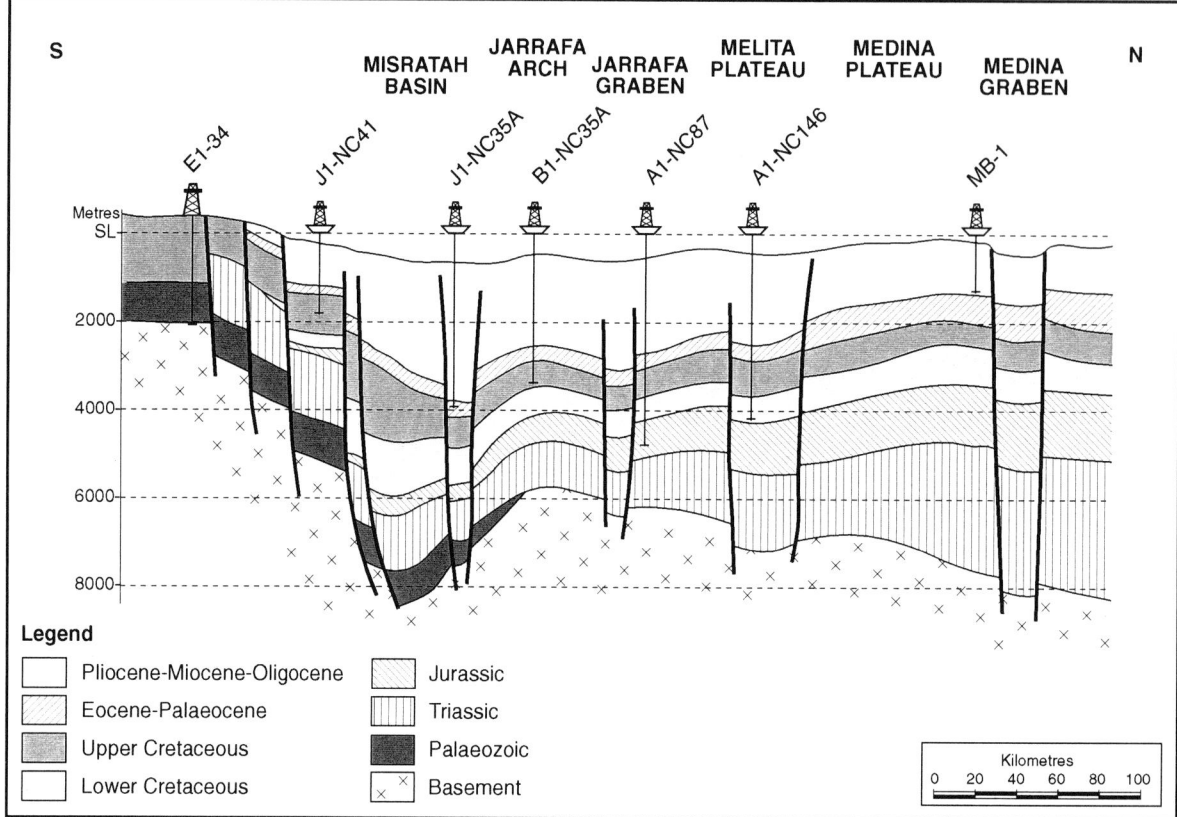

Source: Modified from Del Ben and Finetti, 1991

Figure 6.19   Libyan Western Offshore, Diagrammatic Cross-Section

This section crosses the easten part of the Pelagian Block, and shows the relation of the Misratah Basin to the Jarrafa Arch and Medina Plateau to the north. The Palaeozoic section wedges out and does not extend north of the Misratah Basin, whereas by contrast the Triassic and Jurassic section gradually thickens, reaching 6000m beneath Malta. The Pelagian block subsided during the Miocene and 4000m of Neogene section is present in the Misratah Basin.

graben south of the Jarrafa Arch (Figure 6.16). Late Miocene and Pliocene rocks are preserved in the basin and evidence of the Messinian salinity event has been described from well B1-NC 35a, where 10m of anhydrite marks the top of the Messinian. A summary of the tectonic history of the northwestern offshore is shown on Figure 6.17.[97]

Jarrafa Arch

In Tunisian waters the Sabratah Basin is bounded to the north by the Lampedusa Plateau.

In the Libyan area the situation is more complex since the Jarrafa Arch is intercalated between the basin and the Melita-Medina Plateau. The Arch, on which the Jarrafa, B1-NC 35a and F1-NC 35a wells were drilled, is flanked by two narrow pull-apart grabens, the Jarrafa and Zohra Grabens which were formed as part of the Sirt Wrench during the early Pliocene (Figures 6.16, 6.17). The Eocene Farwah nummulitic carbonates, which form the principal hydrocarbon reservoir in the Sabratah Basin, are not present on the Jarrafa Arch, where thin Eocene shales rest directly on Cretaceous carbonates. Depth to the base Eocene on the arch is 2400m in well F1-NC 35a, compared with 2900m in the Sabratah Basin in well C1-NC 35a.[98]

Melita Medina Plateau

Little is known about the Melita Medina Plateau (Figure 6.16). Details of the A1-NC 146, A1-NC 87 and B1-NC 87 wells have not been released, but geophysical studies by Del Ben and Finetti suggested that the platform is tilted towards the south with a raised southern margin where the B1-NC 35a well was drilled. The base Tertiary rises from a depth of 3500m on the Melita Plateau to 2000m on the Medina Plateau with the two plateaux separated by a NW-SE pull-apart graben. The grabens are all late features dating from the Miocene rifting associated with the emplacement of the Calabrian Arc (Figures 6.16, 6.17). Using geological and seismic data from Malta and Tunisia, Bishop and Debono predicted the presence of a thick Cretaceous carbonate plateau, possibly with a wedge of Palaeocene/Eocene carbonates on the western part of the plateau, overlain by a thick Oligocene-Miocene-Pliocene sequence.[99]

Misratah Basin

The Misratah Basin is an asymmetric half-graben which links the Sabratah Basin to the Sirt Embayment. It was formed during the Miocene rifting episode as an eastward extension of the Sabratah Basin which also subsided during this period (Figures 6.16, 6.17, 6.19). According to the geophysical studies of Del Ben and Finetti the depth to the top Cretaceous in the graben is 4000m compared with 1400m in well J1-NC 41, drilled on a terrace on the south flank of the graben, and 2900m in well B1-NC 35a, which was drilled on the raised rim of the adjacent Jarrafa Arch

(Figure 6.19). Depth to the base Mesozoic in the graben varies from 6000m in the north to 7800m in the south. Displacement on the southern boundary fault (the Sabratah-Cyrenaica Fault Zone) is 3500m at top Palaeozoic level.[100]

6.5.2 Sirt Embayment

To the east of the Medina Escarpment the Sirt Embayment represents the area where the Sirt Basin opens onto the Sirt Slope and Sirt Rise (Figure 6.16). The detachment of the Apulian plate in the early Cretaceous introduced great instability along the north African plate margin which led to crustal stretching and ultimately to the collapse of the Sirt Arch in the Aptian and to the establishment of the Sirt Basin. Almost all of the offshore wells in this area have been drilled on the An Nuwfaliyah and Al Jahamah Platforms. These wells reveal a truncated Palaeozoic sequence in which Silurian rocks subcrop the Hercynian unconformity. Wells on the offshore part of the Jahamah Platform reached the Cambro-Ordovician. There is no evidence of Permian or Triassic rocks in the nearshore wells, although these rocks can be interpreted on seismic data to the north of the Sabratah-Cyrenaica Fault Zone. Early Cretaceous Nubian sandstones are present in the nearshore wells B1-88 and D1-88. An unconformity in the Senonian may reflect the Santonian tectonism in Cyrenaica during which the Jabal al Akhdar Trough was inverted (Figure 6.17). Palaeocene sediments in the Sirt Embayment mostly comprise outer-shelf black mudstones and chert. On the An Nuwfaliyah High and in well A1-89 Danian sediments are absent. (Figure 5.3) It has been suggested that nummulitic mounds could be present in the Eocene on the seaward extension of the Jahamah Platform. Seismic data shows evidence of the early/middle Eocene unconformity which is well documented in the Jabal al Akhdar. Late Eocene tectonism produced pull-apart basins

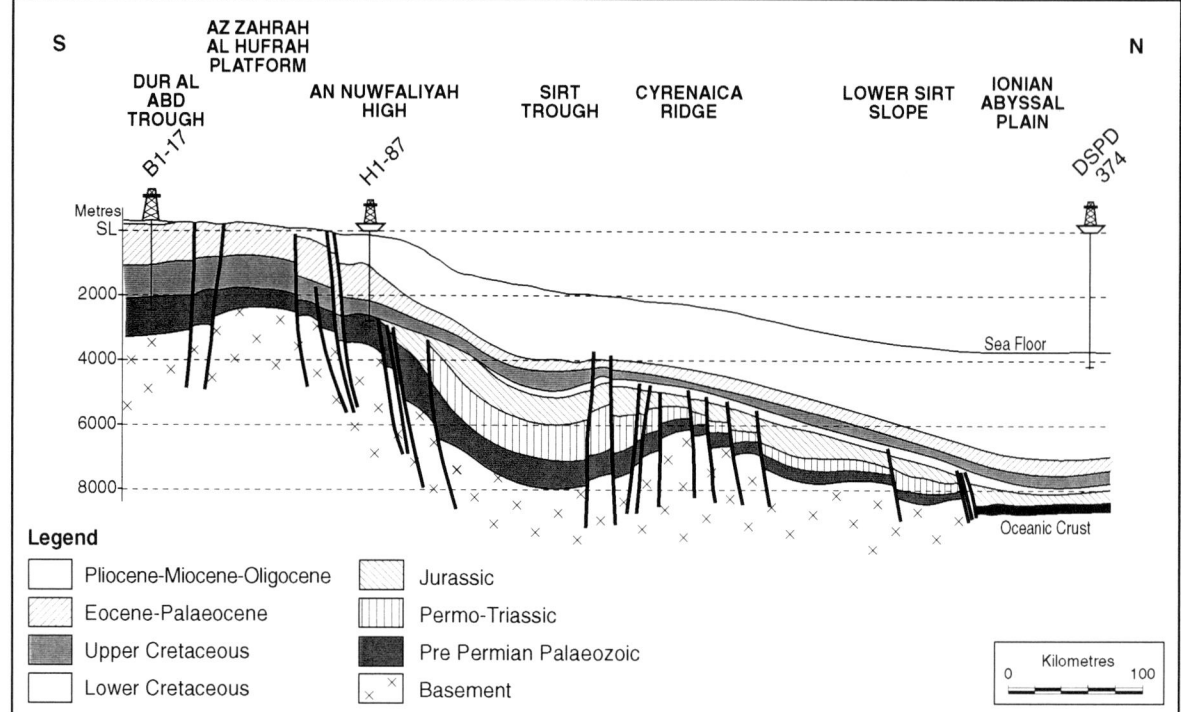

Source: Redrawn from Del Ben and Finetti, 1991

Figure 6.20   Libyan Eastern Offshore, Diagrammatic Cross-Section

In contrast to western Libya the area of the Sirt Embayment shows a typical slope and rise configuration, interrupted only by the Sirt Trough and the Cyrenaica Ridge. The total sediment thickness is much less than on the Pelagian Block. The Triassic-Jurassic sequence nowhere exceeds 1500m in thickness, or the Neogene more than 3000m. Palaeozoic rocks extend to the edge of the Ionian abyssal plain, beyond which Jurassic and later sediments overlie Middle Jurassic oceanic crust.

which are reflected in the variable thickness of Oligocene sediments in the Sirt Embayment. Rapid subsidence of the Ajdabiya Trough during the Miocene led to the accumulation of 2000m of clastic sediments in the offshore extension of the trough. A series of deep canyons cross the continental shelf at the top of the Miocene, represent the Messinian sea-level fall. The channels have been excavated to a depth of at least 350m.[101]

Only one well has been drilled to the north of the Sabratah-Cyrenaica Fault, but seismic data shows that this area marks the junction between the northern margin of the African plate, the Calabrian arc to the northwest and the Hellenic arc to the northeast. Finetti and Del Ben attributed the origin of the Sirt Embayment to the detachment of the southern tip of the Apulia Plate from north Africa during the Aptian. On the Sirt continental rise (Figure 6.20), north of the Sirt-Cyrenaica Fault, the section thickens rapidly by the addition of Permian, Triassic and Jurassic sediments, but on the Ionian abyssal plain near the site of DSDP hole 374 seismic data shows that middle to late Jurassic rocks overlie Jurassic oceanic basement (Figures 6.16, 6.20). The pre-Upper Cretaceous section is extensively faulted, reflecting the Aptian tectonism and break up of the Sirt Arch. The late Cretaceous to Oligocene sequence maintains a fairly constant thickness of about 1000m out to the foot of the Sirt Rise. The Miocene-Recent section however thickens to over 3000m on the abyssal plain. Messinian salt can be traced from the Ionian abyssal plain onto the imbricated thrust sheets of the Hellenic fore-arc. In the deep Mediterranean there is evidence of rifting in the middle to late Triassic on the lower Sirt Rise, which culminated during the middle Jurassic with the opening of the Ionian abyssal plain. A further tensional phase is evident in the Senonian, which correlates with tectonic events in Cyrenaica. Major tectonic upheaval during the Miocene-Pliocene, associated with the emplacement of the Calabrian arc, resulted in subsidence in the Sirt Trough, the seaward continuation of the Ajdabiya Trough, and the eruption of submarine volcanoes which are represented by sea-mounts on the present day seafloor.[102]

6.5.3 Northern Cyrenaica Offshore

In a very real sense Cyrenaica represents the northern margin of the African plate. The continental margin plunges very steeply into deep water and the thrust sheets of the Hellenic fore arc lie only 50km offshore. Seismic evidence suggests that a full sequence of marine Mesozoic rocks is preserved offshore, although the oldest rocks penetrated in offshore wells are Callovian in age. The Hellenic fore-arc curves away to the northwest and the sedimentary wedge plunges into a deep basin at the foot of the continental rise. From the upper continental slope near Binghazi the Cyrenaica Ridge can be traced northwestwards for a distance of 200km (Figures 6.16, 6.20). Depth to basement on the ridge is about 6000m, compared with 8000m to both north and south. South of the ridge the Binghazi Basin represents the deepest part of the offshore Ajdabiya Trough where the base Mesozoic is at a depth of around 7000m. It has been suggested that the Binghazi Basin represents a pull-apart basin initiated during the late Triassic rifting and reactivated during the Miocene rifting phase which also produced the Medina and Sirt Wrenches (Figure 6.16). In this area seismic data shows that the Cainozoic sequence thickens towards the abyssal plain whilst the Mesozoic section thins. Triassic rocks may not be present on the Ionian abyssal plain and in this area Jurassic rocks probably overlie basement.[103]

Well evidence for onshore northern Cyrenaica was discussed above under the section on the northern coastal margin. In the offshore of western Cyrenaica well A1-NC 120 showed a full marine succession from mid-Jurassic to mid-Miocene in age interrupted only by a brief period of emergence during the early and middle Palaeocene. Further south well A1-89 penetrated Lower Cretaceous marine carbonates, which demonstrates that early Cretaceous marine conditions extend around the western margin of the Cyrenaican Platform. Unfortunately well A1-88 did not reach the Cretaceous due to its location in the Ajdabiya Trough, but wells B1-88 and D1-88 further south show that marine conditions did not reach this area. The Santonian unconformity which represents an important tectonic episode in the Jabal al Akhdar is found in the wells offshore western Cyrenaica. In the Palaeocene carbonates form the dominant lithology and it has been speculated that reefal conditions may be present on the shelf margin. During the Eocene carbonates prograded from the east into the Ajdabiya Trough. The Eocene shelf edge can be traced on seismic data from the Ajdabiya region northwards to a location offshore Binghazi. The early-mid Eocene unconformity is visible on seismic data and the very thick Oligocene clastics found in well A1-89 may have accumulated within a pull-apart graben formed during the late Eocene tectonism.[104]

6.6 Structural Synthesis

This section attempts to draw together the threads from the five structural provinces just described in an effort to identify the similarities and differences between them, and to identify the main tectonic events which have shaped the structure of Libya.

Palaeozoic

The Palaeozoic tectonic history can be reconstructed from data obtained from the Kufrah, Murzuq and Ghadamis Basins, and fragmentary data is available from wells in Cyrenaica. The Pan-African orogeny established the basement terranes of Libya in their present disposition during the late Proterozoic, leaving an irregular topography dominated by mountains in the south and an irregular peneplain further north. Extension during the Infracambrian, caused by shearing on the Pannotian suture, produced block faulting and pull-apart basins in both the Kufrah and Murzuq Basins, and in Cyrenaica where Infracambrian red beds have been found. The final phase of the Pan-African orogeny imposed a NW-SE trend over much of Libya which influenced deposition during the Cambro-Ordovician. The Hasawnah Formation is thick in the Kufrah and Murzuq Basins but is absent on the Tihemboka Uplift, and the entire Cambro-Ordovician section thins onto the Tripoli-Tibisti and Haruj Uplifts, but is thick in the Dur al Qussah Trough. The Ordovician was deposited in a ramp setting on the passive margin of Gondwana. However tectonism during the Caradocian uplifted and removed the Ash Shabiyat, Hawaz and Melaz Shuqran Formations from much of the Kufrah Basin and resulted in a dramatic fall in sea level prior to the deposition of the Mamuniyat Formation in the Murzuq Basin. Further north Caradocian tectonism produced uplift on the Dahar Uplift, and associated volcanism. The

suggestion that these events can be ascribed purely to eustatic changes is difficult to support. In Cyrenaica Cambrian rocks have not been found.

The short-lived late Ordovician glaciation involved a continental ice sheet which extended into southern Libya with marine conditions further north. Evidence of glacial conditions has been found in the Murzuq and Kufrah Basins and on the Al Qarqaf Uplift. Rocks of Caradocian-Ashgillian age which show glacial characteristics have also been found in Cyrenaica. Melting of the ice sheets led to a major marine transgression during the early Silurian during which shales were deposited over much of western Libya. These shales contain rich source rocks in the Murzuq and Ghadamis Basins but are much more silty in the Kufrah Basin. Several episodes of uplift and erosion during the Devonian attest to the earliest stages of the Hercynian orogeny, and significantly these events are more evident in the west than in the east. In the early Devonian uplift removed the Akakus Formation from large parts of the Ghadamis Basin including the Kabir trend. A base Devonian unconformity is also present in Cyrenaica. A major mid-Devonian tectonic inversion in the Murzuq Basin, where much of the centre of the basin was uplifted eroded and peneplained prior to deposition of the middle Devonian, is mirrored by unconformities in the Eifelian and Frasnian in the Ghadamis Basin. In the Kufrah Basin evidence for mid-Devonian tectonism is less obvious, and in Cyrenaica the Devonian section is substantially complete.

The effects of the Hercynian orogeny became more evident during the Carboniferous. Tournaisian rocks are absent over much of Libya and despite numerous oscillations regression became the dominant feature throughout the Carboniferous. Continental conditions became established in the Kufrah Basin during the mid-Devonian, in the Murzuq

Basin during the Moscovian and in the Ghadamis Basin and Cyrenaica during the latest Carboniferous. The Hercynian orogeny culminated during the late Carboniferous with the formation of the Nafusah Uplift, Al Qarqaf Arch, Tibisti-Sirt Arch and part of northern Cyrenaica. Most of Libya became emergent and erosion during the Permian stripped several of the new uplifts down to their Precambrian roots. Permian continental deposits are present in the Kufrah Basin and probably also in the Murzuq and Ghadamis Basins. However north of the Sabratah-Cyrenaica fault there is an apparently uninterrupted passage from the Carboniferous into marine Permian rocks, and these sequences can be traced on seismic data into very deep water in the Sirt Embayment. Marine Permian rocks are also present in the Jifarah Basin

Mesozoic

The Permian-Triassic junction in the Jifarah and Sabratah Basins is probably unconformable, and in Cyrenaica is clearly unconformable since the entire late Permian and early Triassic is missing. The Triassic marks the beginning of the break-up of Pangaea in which shearing and rifting occurred along the northern margin of the African plate. Mid- to late Triassic rifting is evident in the Sabratah Basin, the Binghazi Basin and the deep offshore, and there is also evidence of rifting in the Sirt Basin. Marine Triassic rocks are present north of the Sabratah-Cyrenaica fault and extend as far as the abyssal plain. Onshore post-tectonic subsidence produced sag basins which contain thick continental deposits of Permian, Triassic, Jurassic and early Cretaceous age. In west Libya the Hamadah Basin is a sag basin which developed over the site of the Palaeozoic Ghadamis Basin. In the Sirt Basin Triassic continental sandstones have been found in the Maragh Trough and they are likely to be present in many other areas. In the Jifarah and Sabratah

Basins evaporites accumulated during the late Triassic-early Jurassic, and these rocks show evidence of subsequent flowage.

By the mid-Jurassic continued crustal extension had led to the formation of oceanic crust on the Ionian abyssal plain, which continued to form until the late Cretaceous. At DSDP site 374 middle Jurassic sediments overlie oceanic crust (Figure 6.20). Seismic data shows that on the deep ocean floor Jurassic sediments overlap Triassic rocks. There is evidence of sinistral shearing in the Jifarah Basin, and onshore Jurassic block-faulting has been reported in the Hamadah Basin, and rifting probably continued on the Sirt Arch. Marine Jurassic rocks are present north of the Sabratah-Cyrenaica fault, and further south continental rocks continued to accumulate in sag basins and incipient rifts. In the late Jurassic the Jabal Akhdar Trough began to subside and a thick sequence of upper Jurassic rocks is preserved in the trough.

The Jabal Akhdar Trough continued to subside during the early Cretaceous, while open-marine conditions formed the southern margin to the Tethys Ocean to the north and in the Sirt Embayment. Further west, major carbonate platforms became established on the Pelagian Block particularly on the Melita-Medina Plateau and Jarrafa Platform. Onshore, Nubian Sandstones accumulated in subsiding basins and rifts in the Kufrah, Murzuq, and Hamadah Basins, on the Sirt Arch and in the Hameimat Trough. An unconformity at the base of the Barremian in Cyrenaica, subsidence during the Neocomian in the Hameimat Basin and faulting and shearing in the Jifarah Basin during the Neocomian are early indicators of increasing instability in the region.

The detachment of the Apulian plate from the north African margin and the establishment of the Mesogean axis of seafloor spreading during the early Cretaceous produced wrenching along the line of the Sabratah-Cyrenaica fault, and associated NW-SE imbricate fan faults in the Jifarah Basin, which Anketell used as an analogy to explain the NW-SE orientation of the horsts and grabens in the western Sirt Basin. Anketell's structural model is discussed below. The stresses produced by the opening of the Mesogean spreading axis led to the collapse of the Sirt Arch probably during the Aptian, to produce a complex arrangement of horsts and grabens in the Sirt Basin. Aptian tectonism has been reported from the Hamadah Basin where it is accompanied by a marine regression, and is also evident in the deep offshore in the Sirt Embayment where widespread faulting of Aptian age is visible on seismic data. In terms of classical continental margin terminology this tectonic culmination represents the end of the rift phase, although subsidence within the rifts continued for much longer (Figure 6.11).

The collapse of the Sirt Arch was followed by thermal subsidence which led to a widespread marine transgression in the Cenomanian which extended to the Al Qarqaf Arch in the west and the Jabal az Zalmah in the east. The grabens of the Sirt Basin were flooded, and active subsidence continued in the Zallah, Maradah, and Ajdabiya Troughs. During the Campanian-Santonian subsidence outpaced deposition and deep-water conditions became established in the grabens in which organic-rich shales were deposited, which form the principal oil source rocks of the Sirt Basin. In Cyrenaica compression during the Santonian led to the inversion of the Jabal Akhdar Trough into an arcuate uplift, and to a widespread unconformity which extends into the offshore Sirt Embayment. By the Maastrichtian subsidence had ceased in the Sirt Basin grabens and the seas gradually onlapped the remaining topographic highs, leaving only a few small islands still emergent.

Cainozoic

The Palaeocene marked the beginning of a new tectonic phase in the Sirt Basin, characterised by the development of broad sag basins, which were only partially coincident with the earlier rift grabens, separated by carbonate platforms. Southern Libya remained emergent and tilting in the Hamadah Basin led to the emergence of this area by the end of the Palaeocene. Offshore northern Cyrenaica uplift produced a brief hiatus during the Palaeocene as seen in the A1-NC 120 well. Reefs developed in the Ajdabiya Trough during the Palaeocene and reefal conditions may be present offshore Cyrenaica, but in the Sirt Embayment deep-water shales accumulated during this period.

The Eocene marks a period of increasing tectonism caused by the closing of the Tethys Ocean. Uplift on the southern margin of the Sabratah Basin caused erosion of the Palaeocene from this part of the basin, after which a carbonate platform became established. Similar carbonate platforms are present in the Sirt Basin and perhaps in the offshore Sirt Embayment. In Cyrenaica rapid subsidence occurred in the Ash Shulaydimah Trough during the middle Eocene, and a middle Eocene unconformity is visible on seismic data in the offshore Sirt Embayment. This was followed in the late Eocene by reactivation and further elevation of the Jabal Akhdar Uplift, and by the creation of pull-apart basins offshore. Volcanism in the Eocene resulted in the ring intrusions of the Jabal Awaynat and volcanic flows on the Jabal as Sawda. In west Libya the Nafusah Arch and Jifarah Basin were uplifted and deformed by wrench faulting.

A change from dominantly carbonate to clastic deposition occurred during the Oligocene. Tectonism continued and in the Sabratah Basin the Eocene-Oligocene contact is unconformable, and it was during this period that the main subsidence in the Hun Graben took place, accompanied by regional tilting to

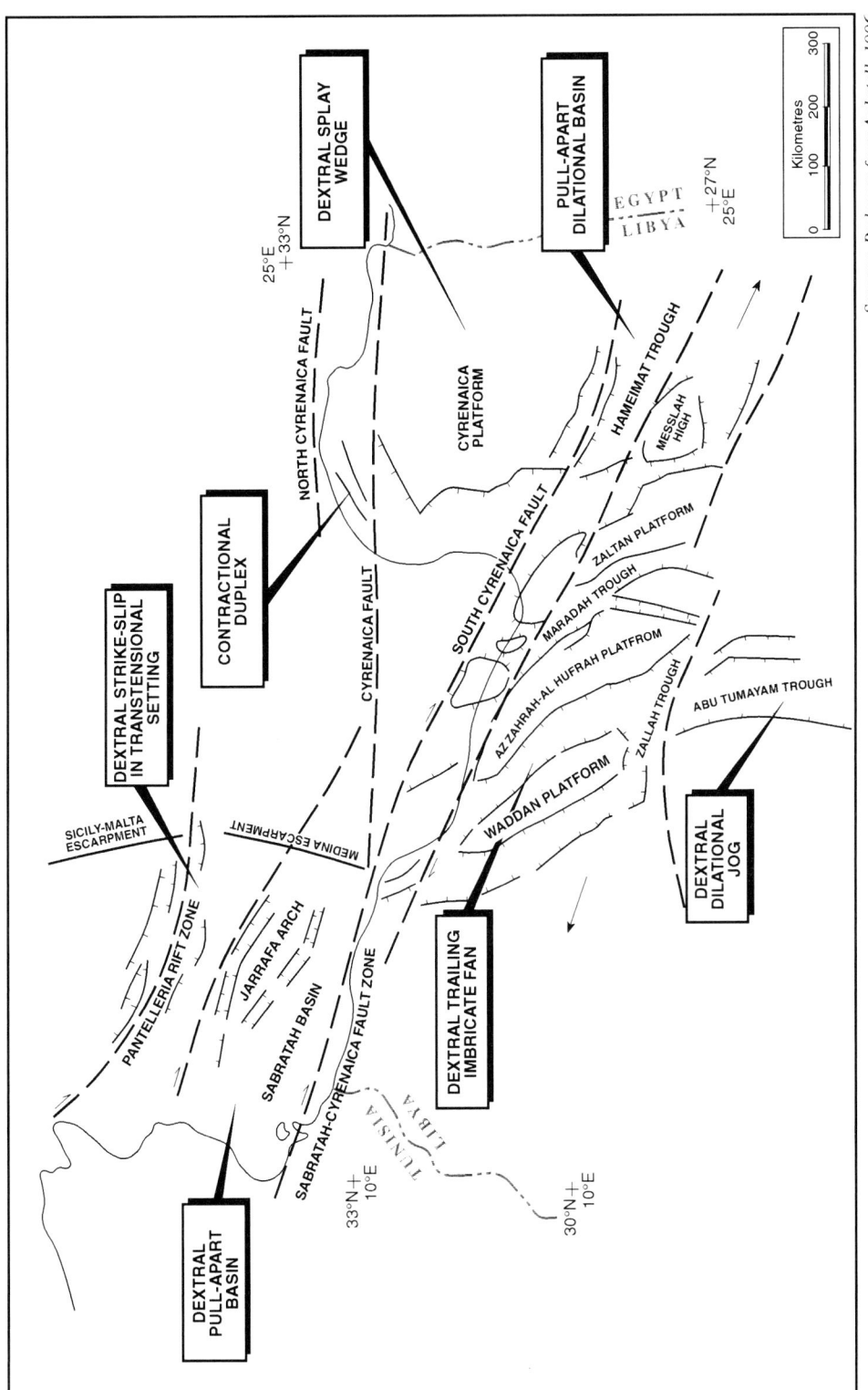

Figure 6.21   Northern Libya, Structural Synthesis

This map shows the structural interpretation developed by Anketell to explain the relationships between the Pelagian Block, the Sirt Basin, the Abu Tumayam Trough, Jabal al Akhdar Uplift, Cyrenaica Platform and Hameimat Trough. The relationships can be explained in terms of an overall dextral wrench model which developed progressively throughout the Mesozoic and culminated in the wrenching associated with the emplacement of the Calabrian Arc in the Mio-Pliocene. The model explains most of the observed features in the six main tectonic provinces of northern Libya.

the northeast. In offshore Cyrenaica, Oligocene sediments were deposited in pull-apart basins and subsidence continued in the Ajdabiya Trough.

Alpine tectonism culminated during the Miocene and dramatic tectonic consequences are visible in northern Libya. Reactivation of the Sabratah-Cyrenaica wrench zone led to foundering of the Pelagian Block (Figures 6.16, 6.18). Very thick Miocene sediments were deposited on the Pelagian Block, and over the Jifarah Basin. Rifting produced the horsts and grabens of the Pantelleria Rift Zone, and subsidence in the Sabratah and Misratah Basins. Subsidence also occurred in the Ajdabiya Trough and Binghazi Basin, and volcanic activity increased along the Al Haruj al Aswad-Gharyan axis.

The Messinian salinity crisis is well documented in the deep offshore area of the Sirt Embayment and indeed thick salt deposits can be traced on seismic data onto the thrust sheets of the Hellenic fore arc. Salt deposits have been found on the periphery of the basin at Sahabi in the Ajdabiya Trough, and in well B1-NC 35a on the Sabratah Basin, and deep canyons were cut both onshore and on the continental shelf. The emplacement of the Calabrian plate produced the Medina and Sirt wrenches on the Pelagian Block, numerous pull-apart grabens, and the Medina Escarpment. Subsidence continued in the Ajdabiya Trough, and its offshore extension, and volcanic activity produced subsea volcanoes which are preserved as seamounts. On the Cyrenaican Platform erosion unroofed Cretaceous rocks over a large area.

6.7   SUMMARY

It is evident from the above review and from the plate tectonic synthesis that seven main orogenic events have shaped the structure of Libya: the Pan-African, Caradocian, mid-Devonian, Hercynian, Aptian, Eocene and Miocene. These can all be related to the plate tectonic events outlined in chapter 2: the assembly of Pannotia, the separation of the Avalonian-Cadomian terranes from Gondwana, the initial collision of Laurasia and Gondwana, the culmination of the formation of Pangaea, the opening of the Tethys Ocean, the closing of the Tethys Ocean, and the Alpine orogeny, particularly the emplacement of the Calabrian and Hellenic arcs. Boote et al. assigned these events to four tectonic cycles, a lower Gondwanan (Cambrian to early Devonian), upper Gondwanan (mid-Devonian to Carboniferous), lower Tethyan (Permian to mid-Cretaceous) and upper Tethyan (late Cretaceous-recent). Craig, et al. preferred a six fold subdivision: Pan-African orogeny, Infracambrian extension, Palaeozoic extension and compression, Hercynian orogeny, Mesozoic extension and Alpine orogeny. The numerous other unconformities and hiatuses are mostly attributable to local tectonics or to eustatic sea level changes. Many authors have related tectonic events in Libya to well known events elsewhere such as the Caledonian, Taconian, and Laramian, but this is a questionable practice, and does not serve any useful purpose.[105]

Baird et al. published an excellent review of the tectonic history of the Sirt Basin in which they recognised four tectonic phases: phase 1, the entire pre-Cenomanian interval, which they regarded as essentially pre-rift. Phase 2, Cenomanian to Campanian, is distinguished by rapid subsidence within the Sirt Basin grabens, which they regard as the main rift phase. Phase 3, Maastrichtian to late Eocene, is characterised by 'macrobasinal subsidence' in broad sag basins. Phase 4, early Oligocene to recent, is dominated by major subsidence in the greater Ajdabiya Trough area, and regional tilting (Figure 6.10).[106]

Anketell developed a bold tectonic model for the whole of northern Libya in which he was able to relate one area to another. He believed that the tectonics of all northern Libya can be explained in terms of a shear model based on two principal shear zones, a northern Sabratah-Cyrenaica Fault extending from the southern margin of the Sabratah Basin, across the Sirt Embayment to the southern margin of the Cyrenaica Platform, and a southern fault extending from the Al Qarqaf Arch to the southern margin of the Sarir Trough (Figure 6.21). According to this model the Sirt Basin can be explained as a 'dextral trailing imbricate fan' within a dextral wrench zone. This bears a similar relationship to the Sabratah-Cyrenaica fault as the much smaller Wadi Ghan fault zone in the Jifarah Basin. The Abu Tumaym Trough which has a north-south orientation, was explained as a 'dextral dilational jog' in which east-west extension has taken place between sinistral wrench faults. The Cyrenaican Platform was interpreted as a 'dextral splay wedge' between two converging faults, and the Jabal Akhdar Uplift was considered as a 'contractional duplex'. Offshore the Pelagian Block contains Miocene pull-apart grabens caused by dextral shearing of the Calabrian arc and the consequent opening of the Medina and Sirt Wrenches. Although these rela-

tionships are complex they do provide a framework for interpreting the area as a whole, and for relating highly variable tectonic provinces to a common unified model.[107]

The above review demonstrates that whereas the structure of southern Libya exhibits strong Pan-African influences, that of central Libya is largely attributable to Hercynian tectonics. Northern Libya shows the increasing influence of Tethyan and Alpine tectonics, and in the Libyan offshore the Alpine phase can be differentiated into a western province (the Pelagian Block) controlled by the Calabrian Arc, and an eastern province, controlled by the Hellenic Arc north of Cyrenaica, with the Sirt Embayment representing a typical slope and rise margin between the two arcs. Curiously although the Hellenic Arc approaches closer to Cyrenaica than does the the Calabrian Arc to northwest Libya, its tectonic effects appear to be less dramatic. This may be due to the fact that the Calabrian Arc has been emplaced laterally from the west and has induced major wrenching, whereas the Hellenic Arc has been emplaced from the north and its main effect has been compressional.

Chapter 7

PETROLEUM GEOCHEMISTRY

7.1   Introduction

Libyan oil varies greatly in character, from sweet, low-sulphur light oils to waxy, heavy, biodegraded and gassy crudes. These variations are due to the nature of the source rock, and the subsequent maturation history. There is no question that very effective source rocks are present in Libya; over 100 billion barrels of oil in place have been entrapped which represents only a small proportion of the amount generated. But in order to sensibly direct future exploration efforts it is necessary to understand how the hydrocarbons originated, where they came from and when they were produced. The process can be divided into five phases: formation of an organic-rich source rock, maturation of the organic material into hydro-carbons, expulsion from the source rock, migration through carrier beds, and entrapment in a hydrocarbon trap. This is an inherently wasteful process. Generation, expulsion and migration are all very inefficient processes and estimates based on world-wide data suggest that only a small percentage of generated oil reaches an effective trap. Some of the hydrocarbons remain in the source bed, some is lost *en route*, and some migrates to the surface without being trapped. Even then hydrocarbon traps will eventually fail, due to faulting, tilting, uplift and unroofing, or by deep burial and thermal destruction of the trapped hydrocarbons. Oil accumulations can also be destroyed by biodegradation. Undoubtedly the main cause of oilfield destruction is post-entrapment tectonism.

Macgregor estimated that the average median oilfield age is only 35 Ma, which explains why only twenty of the world's 350 giant fields were charged during the Palaeozoic. In practical terms 90% of global oil resources were emplaced during the last 75 Ma. Most of the oilfields reservoired in Palaeozoic reservoirs in Libya and Algeria were generated during the Mesozoic or Cainozoic. Indeed it is doubtful whether any of the present-day oil accumulations in Libya were emplaced during the Palaeozoic. The main cause of destruction of the Palaeozoic oil pools was the Hercynian tectonism which affected the whole of Libya during the late Carboniferous. Macgregor presented a case study from the Illizi Basin of Algeria which demonstrates a long history of trap destruction, not only during the Hercynian orogeny, but also as a result of late Cretaceous uplift which led to hydrodynamic flushing in the shallow horizons, and to gas flushing from over-mature source rocks in the deeper part of the basin. He estimated that the gross accumulation efficiency in the Illizi Basin is only 1-2%.[1]

According to a comprehensive review by Klemme and Ulmishek more than 90% of the world's discovered reserves were sourced from six stratigraphic intervals, Silurian (9%), upper Devonian-Tournaisian (8%), Pennsylvanian-lower Permian (8%), upper Jurassic (25%), Turonian (29%) and Oligo-Miocene (12.5%). Libya is anomalous in that no more than 15% of its reserves were derived from these six intervals. Over 90% of Libyan oil was sourced from the Campanian Sirt Shale and the Silurian Tanzuft Shale.[2]

KEROGEN

Type I	Amorphous, algal sapropel (alginite) HI 950 - 600 OI <50, Convertibility 80%
Type II	Waxy, herbaceous sapropel (exinite) HI 600 - 400 OI <50, Convertibility 60%
Type III	Woody humic (vitrinite) HI <150 OI >50, Convertibility 40%
Type IV	Inert fusinite (inertinite) HI <50 OI >50, No hydrocarbon potential

TOTAL ORGANIC CARBON (TOC)
% Organic carbon

<0.5	Very poor
0.5 - 1.0	Poor
1.0 - 2.0	Fair
2.0 - 4.0	Good
>4	Very good

PYROLYSIS

S1   Free hydrocarbons volatilised at 250°C
S2   Hydrocarbons cracked from kerogen at 430-460°C
S3 CO_2 produced at 550°C
Tmax Temperature of optimum generation of S2 peak

Potential Yield (S1+S2)ppm   Production Index $\frac{S1}{S1+S2}$

0-200	Poor	Hydrocarbon Index $\frac{S2}{TOC}$ mg/g
2,000-6,000	Fair	
6,000-20,000	Good	Oxygen Index $\frac{S3}{TOC}$ mg/g
>20,000	Very Good	

Principal geochemical indicators table:

STAGE OF MATURITY	PRODUCT	VITRINITE REFLECTANCE	SPORE COLOUR AND SCI	Tmax °C I	Tmax °C II	Tmax °C III	THERMAL ALTERATION INDEX	TIME-TEMP. INDEX	LEVEL OF OIL MATURITY
IMMATURE	BIOGENIC GAS	-0.15	1				1 (UNALTERED)	0	0
		0.2	2 LIGHT YELLOW			420			1
		0.3	3	400	430	425			
		0.35	3.5			430	1+		4.5
EARLY (MARGINALLY) MATURE	WET GAS	0.4	4 YELLOW						
		0.45	4.5	410		435	2- (MODERATELY ALTERED)	3	7.8
MID MATURE	ONSET (OIL WINDOW)	0.5	5	430	440	440	2	15	9.5
	HEAVY OILS / PEAK / MEDIUM OILS	0.6	6 ORANGE	440		450	2+	25	
	LIGHT OILS	0.7	7		450	460	3- (MUCH ALTERED)	75	11.6
	BASE	0.8		450	470		3	160	13.5
LATE MATURE	CONDENSATE	1.0	8 BROWN		480		3+		
		1.2	8.5		500		4 SEVERELY ALTERED		
		1.4							
POST MATURE	WET GAS	1.45 / 1.6 / 1.8 / 2.0	10 BLACK				5 META-MORPHOSED	1500	20
	DRY GAS	3 → 4							

Figure 7.1   Principal Geochemical Indicators

The table shows the principal geochemical indicators used to determine kerogen type, source rock richness and source rock maturity. Published scales vary to some extent, and individual parameters such as Tmax vary within narrow limits. Vitrinite reflectance and spore colouration values are subjective and imprecise. In general Tmax derived from pyrolysis gives the most reliable indication of maturity.

Source: standard texts, various sources

In order to understand the evaluation of source rocks it is necessary to briefly review the main analytical techniques for assessing the type, quantity and maturity of organic matter. Organic matter in sediments breaks down under mild diagenesis to form kerogen, which varies in hydrocarbon potential depending on its origin. Chemical analysis has led to the recognition of four main types of kerogen which can be distinguished on the basis of their hydrogen and oxygen ratios (Figure 7.1). Type I kerogen is derived largely from algal material and has a high hydrogen index and low oxygen index. It has up to 80% convertibility to mobile hydrocarbons, and yields a light, high-quality oil. Type II kerogen is derived from herbaceous organic matter which accumulated in a marine environment and has a hydrogen index of over 400 and a low oxygen ratio. It has a convertibility to petroleum of up to 60%, and produces a medium gravity oil and gas. This is the most common type of kerogen found in petroleum source rocks. Type III, often referred to as woody kerogen, is derived from distant terrestrial material which has often been extensively oxidised. It has a hydrogen index of less than 150 and an oxygen index of over 50. It has a convertibility to hydrocarbons of less than 40% and yields mostly gas, plus varying amounts of waxy oil. Type IV kerogen is derived from reworked organic matter or from highly oxidised material, and has a hydrogen index of less than 50 and an oxygen index of over 50. It has little or no hydrocarbon potential.

The quantity of organic matter in a sediment is expressed in terms of total organic carbon (TOC) which is determined by heating a sample (from which all carbonate fragments have been removed) to 1000°C and measuring the amount of CO_2 evolved. Values below 1% are usually regarded as poor, values over 2% as good, and over 4% as very good (Figure 7.1). The remaining potential of a source rock can be assessed by pyrolysis, which also gives an indication of the level of maturity of a sample.

The assessment of source rock maturity is the third critical factor, and the most difficult to quantify. Various methods have been developed but most are imprecise or subjective. A rough assessment can be made by determining the colour of kerogen, spores and other palynomorphs such as dinoflagellate cysts. This ranges from yellow in immature source rocks to black in over mature rocks, and the results can be referred to a thermal alteration index (TAI). This method is useful in over-mature gas-prone basins but is insensitive in the immature/mature range. Alternatively a microscope photometer can be used to determine vitrinite reflectance of coal macerals. In general, values below 0.5% are immature, and values between 0.5% and 1.3% are mature to late mature for oil generation (Figure 7.1). However this method was developed initially for measuring the rank of coals, and vitrinite is often rare or absent in many potential source rocks, and can often be reworked, in which case it will yield a false value. Chemical methods can also be used to determine maturity, such as the use of gas chromatography on sediment extracts or oil samples to determine a carbon preference index. It has been found however that CPI is very much influenced by kerogen type and the method is at best imprecise. Probably the most useful technique, which provides information on both maturity and remaining potential, is pyrolysis, a method which simulates maturation by heating a sample until it yields its component fractions. An automated method developed by Institute Francais du Petrole is widely used and provides several key values: S1 is the amount of free hydrocarbons volatilised at a temperature of 250°C, S2 is the amount of hydrocarbon cracked from kerogen at a temperature depending on maturity and kerogen type, and S3 is the amount of carbon dioxide produced after heating the sample to 550°C. The temperature of the maximum generation of hydrocarbon at the S2 peak is Tmax, which is critical as a maturity indicator. In general Tmax values between 400 and 435°C indicate an immature source rock, 435 to 460°C is mature, and over 460°C is overmature. These figures vary slightly depending on the type of kerogen (Figure 7.1).

These determinations have led to the concept of an oil generation zone (the 'oil window') above which the source rock is insufficiently mature to produce oil, and below which the oil generating capacity has ceased. The depth to the oil window varies considerably in different basins due to local heat flow conditions and burial history. A source 'kitchen' is the area within which a source rock is capable of generating oil. A measure of the maturation history of an area can be gained by using the time-temperature index (TTI) method devised by Lopatin and subsequently modified by Waples. TTI is a formula for expressing the summed effects of temperature changes during the entire burial history cycle, and is particularly useful in cases where burial history has been interrupted by uplift or erosion or by a change in geothermal gradient. Assuming accurate input data, by plotting TTI against inferred burial history a visual indication can be gained of when a source rock reached maturity, how long it remained within the oil window and whether it eventually became overmature and produced only gas (see, for example, Figure 7.6). These methods have allowed the recognition of five or six significant source rocks in Libya, of which the most important are the Campanian, Turonian and 'Nubian' shales in the Sirt Basin, the Silurian and middle Devonian shales in the Murzuq and Ghadamis Basins, and the early Eocene marls in the Sabratah Basin. These source rocks account for most of the oil so far found in Libya.

Compared with publications on stratigraphy and structure, relatively few studies have been published on petroleum geochemistry in Libya. By contrast, some of the pioneering work on geochemical analytical techniques was conducted in Algeria, and several seminal publications resulted from this work. Nevertheless, sufficient data are available on Libyan petroleum geochemistry to make a worthwhile synthesis possible, and this chapter will attempt to bring these data together. The topic will be treated by area, starting with the Palaeozoic basins and ending with the Sirt Basin, Cyrenaica and the offshore areas.[3]

7.2   Al Kufrah Basin

To date no significant hydrocarbons have been found in the Al Kufrah Basin and this is probably due to lack of effective source rock. Evaluation of the petroleum potential of the Kufrah Basin began in the 1950's when BP conducted fieldwork and aerial reconnaissance. Oasis drilled two wells, A1-71 and D1-71, on the northern rim of the basin in 1961/2, and Agip drilled two wells (A1 and B1-NC 43) in the basin in 1978/80, the results of which were published by Grignani, et al. Since then reconnaissance surveys have been conducted by Robertson Research, Agoco and Lasmo.[4]

Speculative source rocks may be present in pull-apart grabens in the Infracambrian, as in western Algeria, Saudi Arabia and Oman, but these have yet to be proved in the Kufrah Basin. No trace of the Melaz Shuqran shale has been found in the Kufrah Basin and it is probably absent due to Caradocian erosion. Fair TOC values were reported by Lüning et al. from shale partings within the Cambro-Ordovician section in wells A1 and B1-NC 43, but these are too thin to be of any significance. Over much of north Africa Silurian deposition began with the accumulation of radioactive shales during the Rhuddanian in isolated basins on the irregular Ordovician surface. These 'hot shales' were deposited in anoxic conditions prior to the main Silurian marine transgression and form excellent source rocks (Figure 7.3). In the Kufrah Basin, however, the Tanzuft interval is represented at outcrop, and in the two deep wells, by shallow-marine siltstones and shelfal shales never exceeding 130m in thickness, and showing no evidence of the hot shales which are so conspicuous in the Murzuq Basin. No kerogen or TOC data is available for the Tanzuft Formation in the Al Kufrah Basin, but the description of the shallow-marine siltstones encountered in the A1 and B1-NC 43 wells is suggestive of low organic carbon levels and highly oxidised kerogen. It may be, of course, that hot shales are present in depressions which have not yet been drilled, but examination of the depth-to-basement map of Bellini et al., based on magnetic data, shows that well B1-NC 43 was drilled in almost the deepest part of the basin, in exactly the location where Rhuddanian shales might be expected to occur (Figure 7.2). Furthermore it is worth noting that Grignani, et al. dated the grey-black shales overlying the Mamuniyat Formation in well KW2 as late Ordovician in age (Figure 3.14). If this is correct, the Rhuddanian interval must have been penetrated in this well without any indication of the presence of hot shales. According to Lüning et al. temperature gradient data from well A1-NC 43, and work conducted by Agip, suggested that the depth to the top of the oil window in the Kufrah Basin is about 2000m. The basal Silurian reaches this depth in both the northern and southern depocentres of the basin (Figure 7.2), and if hot shales are present they are likely to have reached the oil window. Lüning et al. quoted data suggesting that the source rocks entered peak maturity during the late Mesozoic. At the present state of knowledge there is no evidence of the presence of hot shales in the Kufrah Basin, although the possibility cannot entirely be ruled out.[5]

7.3   Murzuq Basin

In contrast to the Al Kufrah Basin the Murzuq Basin contains one proven mature

source rock and at least two other potential source rocks, and reserves of about 1500 MMB of oil. Gas-oil ratios are low (10-15scf/bbl), and oil gravity within the Ordovician reservoir ranges from 36-45°API. Rhuddanian hot shales were deposited in depressions on the Ordovician surface (Figure 7.3) prior to the main Silurian marine transgression. The Tanzuft Shales reach a thickness of 350 to 475m on the western outcrops of the basin, but the hot shales are limited to a relatively thin interval at the base of the formation interval. The hot shales corresponds to a highly radioactive zone caused by high levels of uranium which is conspicuous on wireline logs. Lüning et al. published a spectral gamma-ray log from well C1-NC 174 which showed gamma-ray values up to 800 API units. It is interesting that in the underlying Ordovician thin high-radioactive zones are caused by high levels of thorium and potassium, rather than uranium. Lüning, et al. presented an isopach map which shows that the hot shales are absent over large areas in the northern part of the basin, and are confined to topographic lows which correspond closely to the areas of maximum total Tanzuft thickness. In the NC 174 area the basal hot shales range in thickness from zero to more than 15m, but the thickest shales are confined to a palaeovalley less than 20km wide. An isopach map covering a larger area was published by Echikh and Sola which shows hot shale thicknesses of up to 30m, and extending as far south as well H1-NC 58 (Figure 7.4). Seismic amplitude data can be tuned to give an indication of areas where hot shales are present, and Lüning, et al. published a hot shale distribution map of concession NC 174 based on seismic data, which corresponds closely to well data (Figure 7.5). One or two examples have been found of thin bands of hot shale higher in the section in other areas, but these appear to be very local occurrences.[6]

Source rock richness in the hot shales reaches 16.7% TOC in concessions NC 174 and

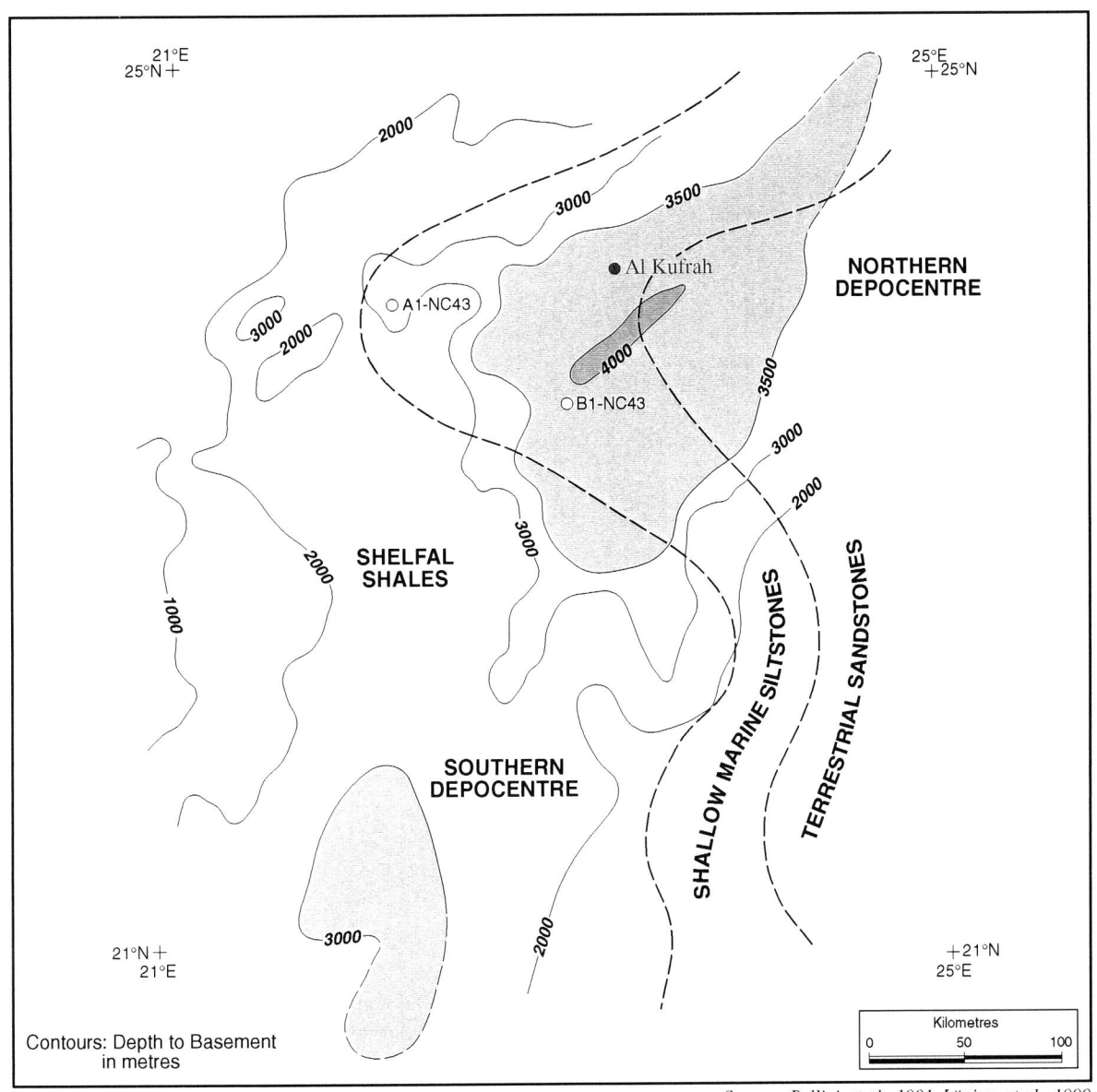

Source: Bellini, et al., 1991, Lüning, et al., 1999

Figure 7.2   Al Kufrah Basin

This map illustrates the source rock dilemma in Al Kufrah Basin. Rhuddanian black shales were not encountered in the two deep Agip wells, even though the B1 well was drilled close to the deepest part of the basin. Inferred Tanzuft lithology trends are shown based on the two wells and outcrop data. The contours show depth to basement based on magnetic data. Two depocentres are evident but with the apparent absence of effective source rocks are unlikely to have functioned as source kitchens.

NC 115, compared with values in the normal shales of about 0.5% to a maximum of 1.28%.

Meister, et al. reported TOC values up to 3.45% in concession NC 58. Pyrolysis data from wells

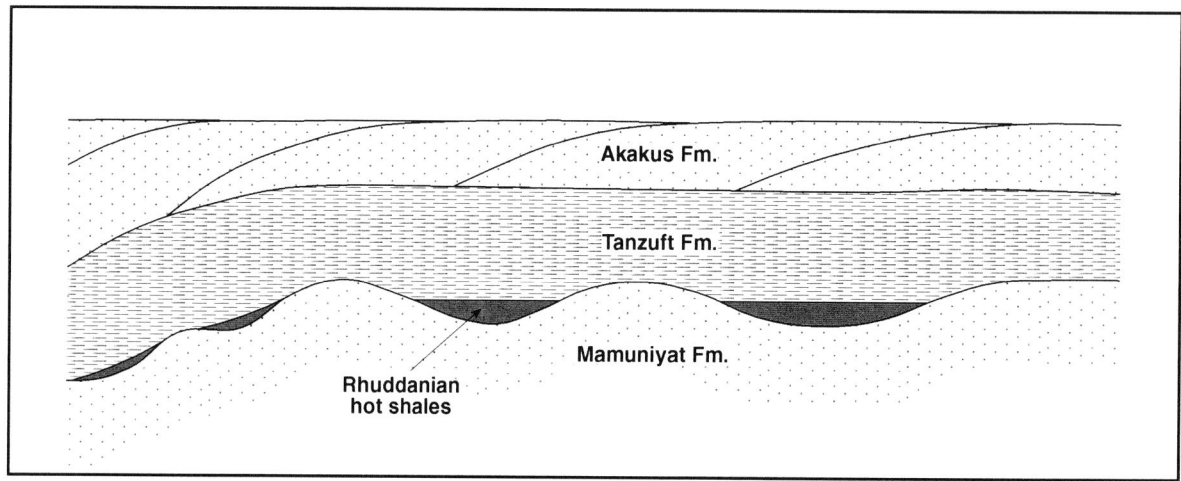

Source: Lüning, et al., 1999

Figure 7.3   Model for Hot Shale Distribution

The Silurian transgression introduced marine shales onto the irregular Ordovician post-glacial surface. The initial sediments deposited during the Rhuddanian stage were black anaerobic shales with a high uranium content, which form excellent source rocks. As the transgression spread the localised depressions were flooded by open marine shales with a much lower organic content. This explains the irregular distribution of the hot shales within the Murzuq Basin.

in which the Tanzuft shale is within the oil window give relatively low S2 values, but this is because most of the hydrocarbon has already been generated. Well E1-NC 58, in which the Tanzuft shale is still immature, gives values of 30,000-40,000 ppm which indicates excellent potential, and Davidson, et al. reported a value of 64,000ppm from concession NC 174.

Maturity data was published by Meister et al. Davidson et al., Aziz and Echikh and Sola. Vitrinite reflectance data suggests that the Tanzuft Formation in wells D1-NC 58 and H1-NC 58 in the depth range 2130m to 2330m is still at an early stage of maturity, whilst in C1-NC 58 at a depth of only 1170m the Tanzuft is at a late stage of maturity. This apparent contradiction illustrates the uncertainty of the vitrinite method. Data from the NC 101 area suggests that the Tanzuft source rock reached early maturity during the Carboniferous but may not have started generating hydrocarbons until the Mesozoic. In concession NC 115 the Tanzuft

source rock is currently at the late mature stage, although it may not have entered the oil window until the Cainozoic. Burial history modelling by Aziz however, based on the assumption of rapid burial during the Permian, suggested that oil generation in concession NC 115 began in the Carboniferous-Permian. Echikh and Sola believed that the high level of maturity in this area was due to a heat pulse caused by tectonism and volcanism during the Eocene. The present day geothermal gradient map of the Murzuq Basin shows a high heat flow trend extending southeastwards from the NC 115 area, which may indicate the area most affected by the Eocene heating event (Figure 7.4). Davidson, et al. published results of maturity modelling on well B1-NC 174 which suggest that the Tanzuft source rock reached mid-maturity (started to generate hydrocarbons) during the Cretaceous but was uplifted above the oil window during the Eocene (Figure 7.6). Aziz, on the other hand, in modelling well

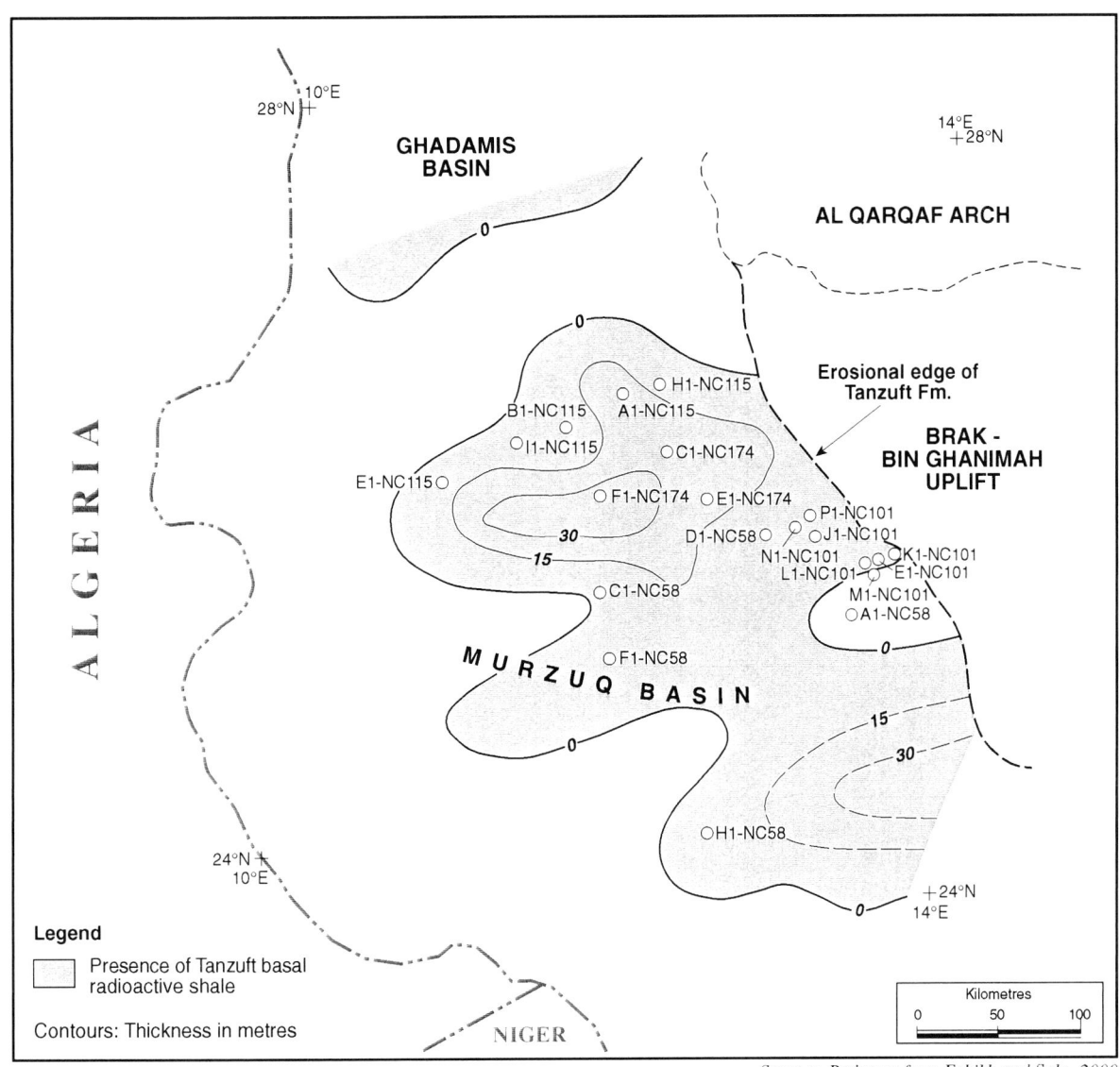

Source: Redrawn from Echikh and Sola, 2000

Figure 7.4   Murzuq Basin

This map shows the generalised distribution and thickness of the Rhuddanian hot shale in the Murzuq Basin based on well data. Maturity data shows that the northern depocentre is within the oil window and the area shown represents the approximate location of the source kitchen. The area east of well H1-NC 58 is speculative.

H1-NC 115 assumed a much greater amount of uplift during the Hercynian orogeny, and his model suggests that oil generation began in the Carboniferous to Permian.[7]

Although data is sparse it seems likely that most of the oil found in the Murzuq Basin originated from the Tanzuft hot shales and that a source kitchen is present in the north-central part of the basin. Oil to source correlations indicate that the Tanzuft hot shale was the source for all of the oil trapped in the NC 115 and NC 174 fields, although subtle differences

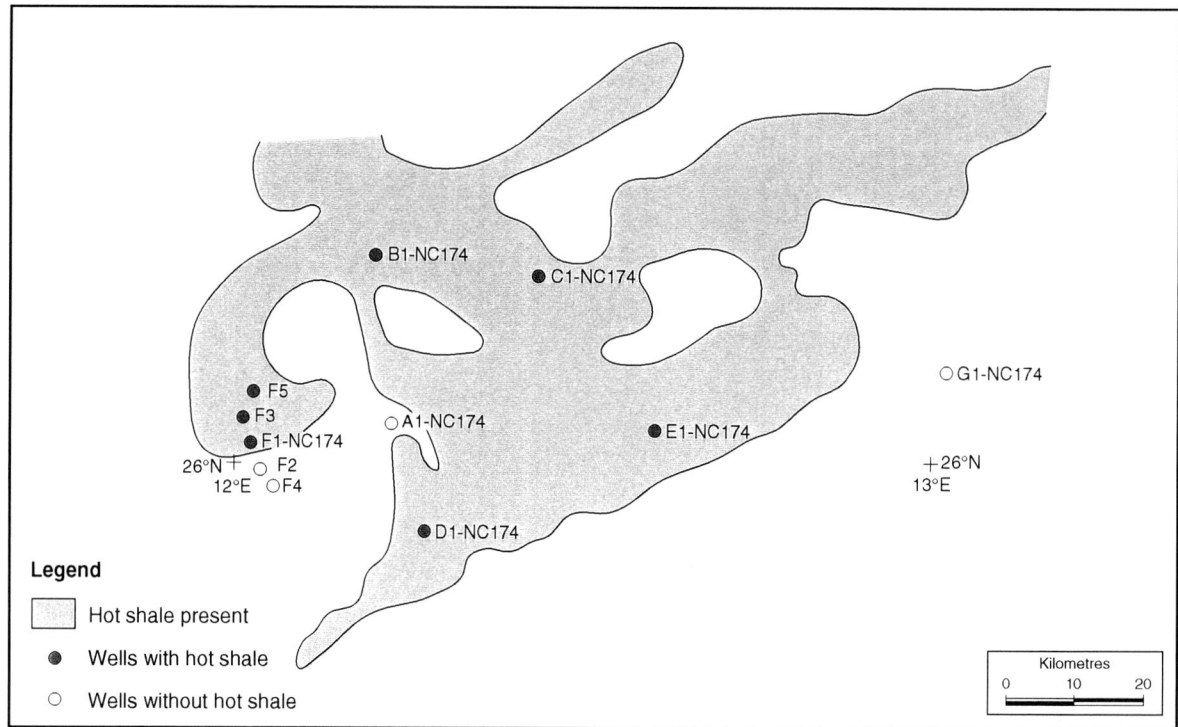

Source: Lüning, et al., 1999

Figure 7.5   Distribution of Tanzuft Hot Shales in Concession NC 174

The distribution of the Rhuddanian hot shale in concession NC 174 illustrates the irregular nature of the Ordovician surface on which it was deposited. It is restricted to the topographic valleys and depressions left after the Ordovician glaciation. It is overlain by non-radioactive shales of the Tanzuft Formation which are organically lean. The map is based on seismic amplitude data which is confirmed by evidence from the wells.

in biomarker profiles suggest that several sub-families may be present. Available data shows that the depth to the oil window is about 2150m and that the Tanzuft Formation entered the oil window no earlier than the late Jurassic and in many cases not before the Cretaceous. In some areas Eocene uplift elevated the source rock above the oil generating threshhold and generation ceased. In other areas the source rock is still generating hydrocarbons. Only in the deepest part of the basin (A1, B1-NC 58) did the Tanzuft shales reach overmaturity. All of the oilfields discovered to date are located in late Ordovician reservoirs near to the source kitchen. Migration took place by lateral diffusion from the source shales into adjacent Ordovician traps

and there is no evidence of long-range migration. The oils found in these fields are light oils with almost no gas/condensate, and with gravities ranging from 34 to 45°API.[8]

Other potential source rocks are present in the Murzuq Basin. Meister, et al. presented data showing fair to good TOC values in Devonian and Carboniferous shales, but in most areas these source rocks have not reached the oil generation zone. Vitrinite reflectance values implying mature Carboniferous source rocks in wells A1-NC 58 and A1-76 are probably spurious. Pyrolysis yields for these rocks range from poor to good, with values of 2500 to 3670ppm for the Devonian, and 3200 to 8580 ppm for the Carboniferous. Analyses conducted on crude oil samples from a

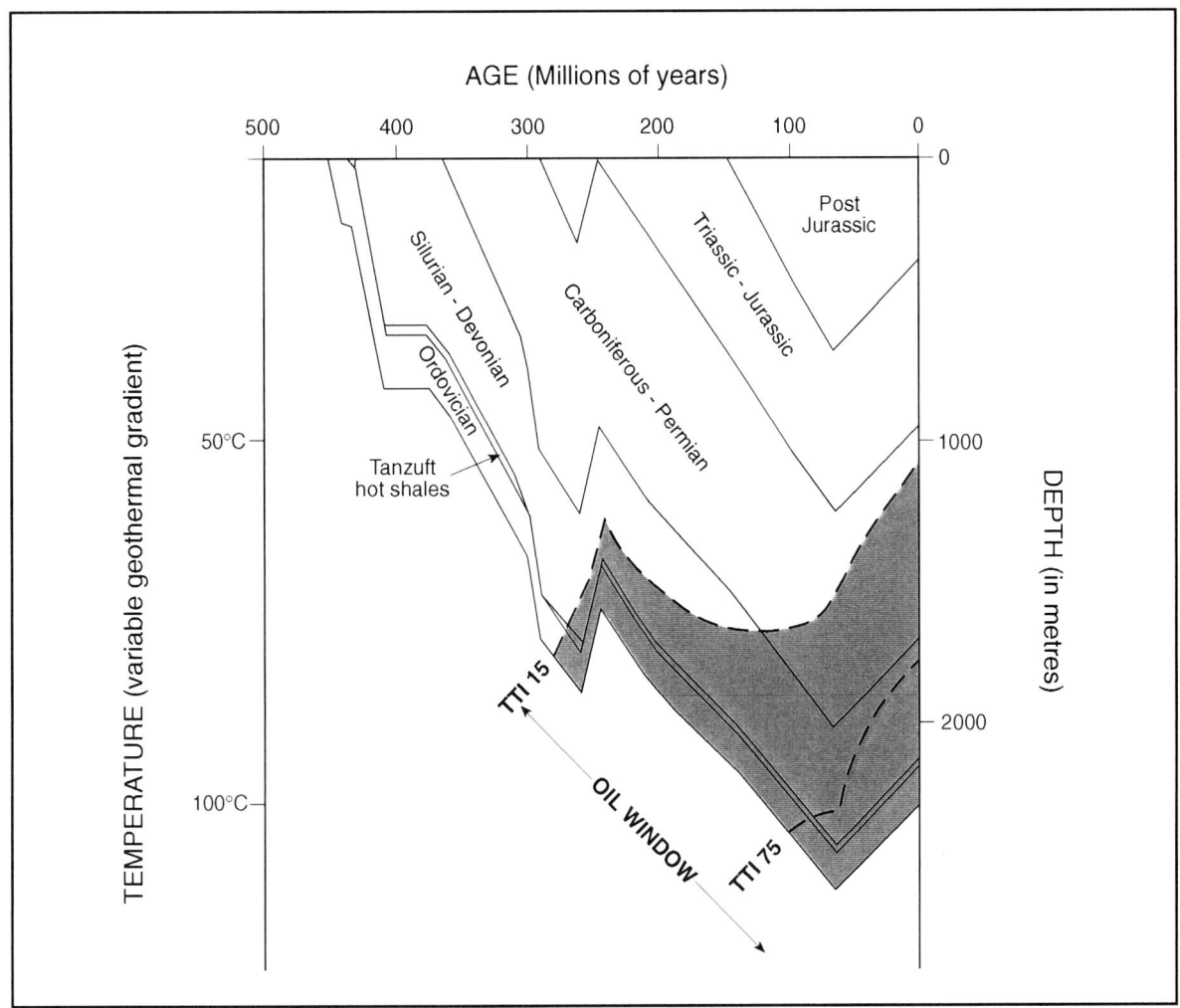

Source: Modified from Davidson, et al., 2000

Figure 7.6   Burial History/TTI Plot for Well B1-NC 174

Well B1-NC 174 is located in the northern Murzuq Basin, northeast of the Elephant field. It was drilled to a depth of 2293m. The modelling of this plot assumes uplift during the Hercynian orogeny and during the Cainozoic, and the geothemal gradient has been adjusted accordingly. The Silurian hot shales entered the oil window during the Permian, but did not reach peak maturity until the late Cretaceous and are still in the oil window at the present time.

Devonian reservoir in well A1-NC 58 yielded evidence that they originated from a Silurian source rock. The occurrence of gas and condensate at Atshan in Devonian and Carboniferous reservoirs, mentioned by Meister, et al. is not easy to explain, but Atshan is located on the Al Qarqaf nose and the hydrocarbons may have originated from the Ghadamis Basin rather than the Murzuq Basin.[9]

7.4   Ghadamis Basin

The Ghadamis Basin contains two mature source rocks which have generated hydrocarbons, and a minor potential source rock. By contast to the Murzuq Basin, the hydrocarbon accumulations are often gas-rich and many pools contain gas and condensate. The principal source rock, as in the Murzuq Basin, is the Silurian Tanzuft Formation. Hot shales are developed at the base of the formation, and in the Ghadamis Basin a second thin radioactive zone occurs higher in the section but with limited areal extent. This second hot shale has been dated by graptolites as Telychian in age in well ESR-1 in eastern Algeria. In Libya its distribution appears to be limited to the area of wells A1-NC 179, B1-NC 118, EE1-NC 7A, Z10-NC 5, E1-60, H1-60 and A1-NC 40B. The Telychian hot shale has also been found in wells in southern Tunisia. Detailed isopach maps, like those in concession NC 174 in the Murzuq Basin, have not yet been published for the Rhuddanian hot shales, but Echikh and Sola presented a regional map showing the distribution of the hot shales across the basin (Figure 7.7). This map shows a distinctly southwest-northeast alignment, with the hot shales concentrated in three main depocentres – the basin centre near Ghadamis, the Zamzam Depression, and a southern depocentre extending from Zarzaitine in Algeria to well M1-1 in Libya. The southern depocentre is separated from the main basin by a large subsurface ridge in the area between the Al Wafaa and Tahara fields where the basal hot shale is absent. The ridge presumably represents an area which was emergent during the deposition of the early Silurian hot shale. In the depocentres the thickness of the basal hot shale, according to Echikh and Sola, reaches 60m south of the Tahara Ridge, 60m in the Zamzam Depression, and 90m in the basin centre. These thicknesses are greatly in excess of those recorded in the Murzuq Basin, and may reflect a better development of the Rhuddanian hot shale in the Ghadamis Basin, or perhaps the use by Echikh and Sola of a different gamma- ray cut-

off value. Sikander reported a maximum net thickness of 45m of hot shale in the Ghadamis Basin. Daniels and Emme indicated that the richest interval is within the basal 35m, Lüning et al. reported a figure of 25m for wells in the NC 7 concession, and Bracaccia gave a figure of 14m for a well in concession NC 40. Kerogen from this interval is largely type II with some type I, and TOC values of up to 17% have been reported from the basal hot shales. Kerogen content towards the southern periphery of the basin appears to be more terrestrial in character. TOC values in the deepest part of the basin have been reduced by up to 50% by the thermal effects of over maturity. Extracts from the basal hot shales give light carbon isotope values of −28.8 to −30.8‰. Well A1-NC 118 contains oils derived from the mature Tanzuft source kitchen. Although the oils are genetically similar they show a range of gravity from 54 to 36°API corresponding to increasing maturity with depth.[10]

The Tanzuft source rock has been deeply buried in the central Ghadamis Basin beneath a thick Palaeozoic and Mesozoic cover, and maturity levels are generally high (Figure 7.8). According to Boote et al. the Tanzuft source rock is within the oil window over most of the basin (including the Zamzam Depression), and in the deepest part of the basin near Ghadamis is over-mature. This is supported by data from Algeria and Tunisia which shows an over-mature zone extending westwards from Ghadamis and flanked north and south by late and middle-mature zones extending as far north as El Borma and as far south as Edjeleh (Figure 7.8). Tmax values reported by Hangari et al. from wells in the northern part of the basin gave values of 432 to 439°C, which confirms peak maturity in this area. They also reported potential yield values of 6870 to 25000 ppm. Anomalous late mature values from around Alrar and Al Wafaa may be due to high heat flow generated by laccolithic intrusions which

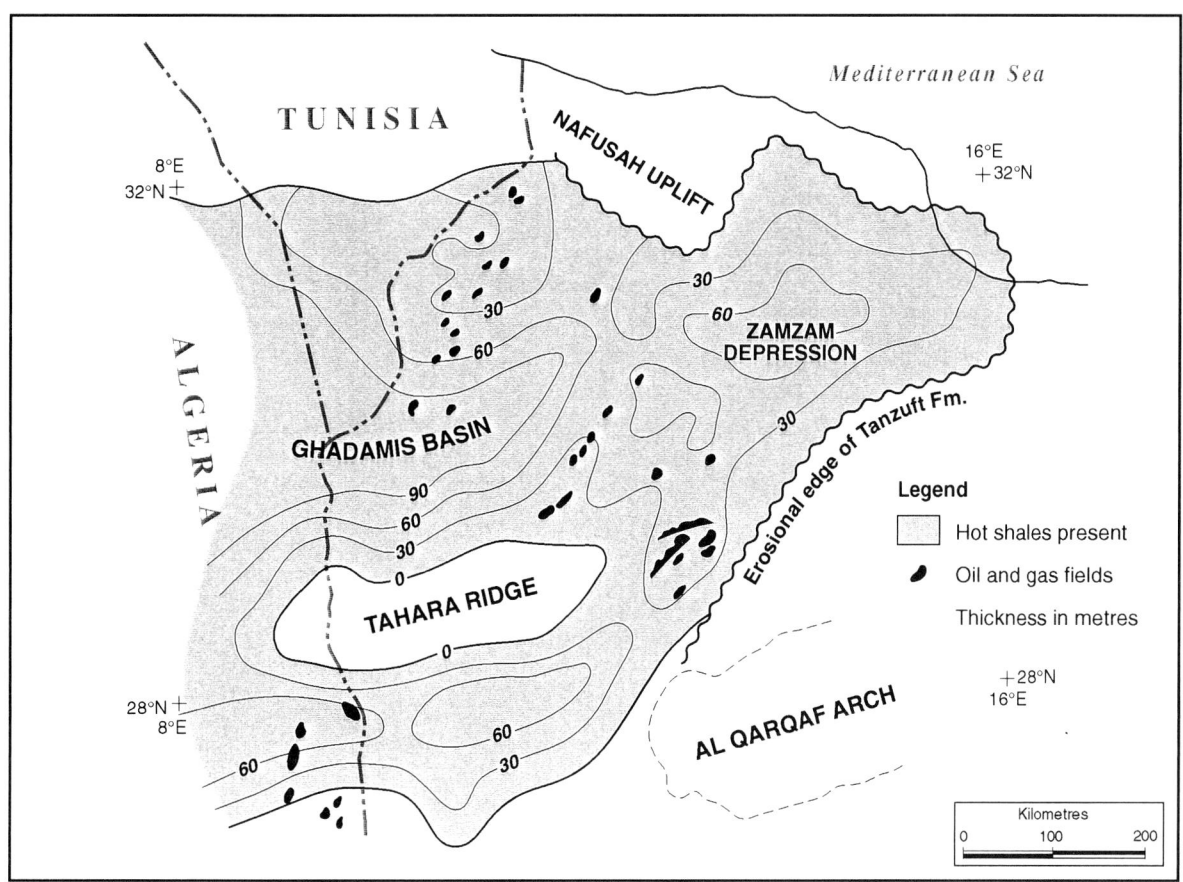

Figure 7.7   Ghadamis Basin, Distribution of Silurian Hot Shales

This map shows the distribution and thickness of the basal Silurian hot shales in the Ghadamis Basin. The thickness is greater than in the Murzuq Basin but also shows the effect of pre-Silurian topographic relief. Hot shales are absent on the Tahara Ridge, but thicken into the Ghadamis Basin and Zamzam Depression. The hot shales are over-mature in the deep Ghadamis Basin, but mature over much of the rest of the basin. The Tanzuft Formation is truncated to the east by the Hercynian unconformity.

have been reported from Stah and Mereksen in Algeria. According to basin modelling calculations, peak oil generation in the centre of the basin occurred between the early Carboniferous and late Jurassic, and towards the periphery of the basin at progressively later times. Oil generation during the Palaeozoic was affected, and in some cases halted by the Hercynian uplift. Peak generation on the basin flank, in the Al Kabir/concession 66 area, probably occurred during the mid-Cretaceous. Using data from

84 wells Daniels and Emme calculated that over 1400 billion barrels of oil have been expelled from Silurian source shales in the Algerian part of the Ghadamis and Illizi Basins. Lüning estimated that the Tanzuft hot shales account for 90% of all the hydrocarbons generated from Palaeozoic rocks in north Africa. A study on oil samples from the Tiji field area was conducted by Byramjee and Vasse. They concluded that the naphtheno-paraffinic oils reservoired in the Akakus reservoir on the Tiji field and other

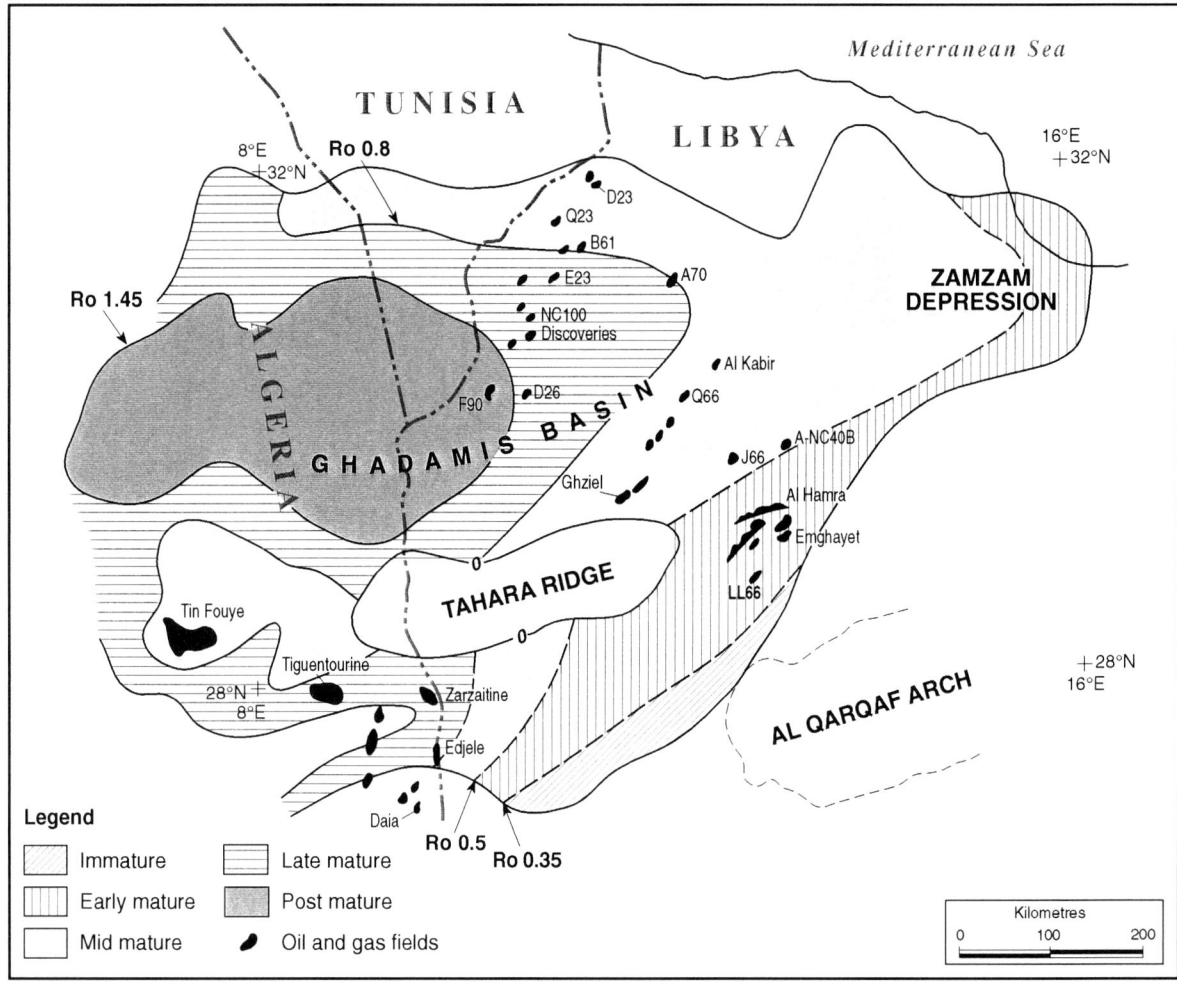

Sources: Daniels and Emme, 1995. Echikh, 1998. Boote, et al., 1998

Figure 7.8   Ghadamis Basin, Maturity Tanzuft Formation Hot Shale

Maturity levels for the basal Tanzuft hot shale have been established in Algeria and are projected into Libya using the vitrinite equivalent values of Boote, et al. The source rock is overmature in the centre of the basin, but eastwards in the shallower part of the basin it becomes progressively less mature.

small structural traps further south, were derived from the Tanzuft source rock by northward migration. The oils are chemically similar to those at Hassi Messaoud in Algeria, but the level of maturity is less at Tiji.[11]

A second source rock has been identified in the Middle and Upper Devonian, with the best development in shales and shaly carbonates of Frasnian age (Figure 7.9). The distribution of

this source rock is limited, due to Hercynian truncation to the north and pre-Visean truncation to the south. In Algeria the Frasnian hot shales reach 200m in thickness, but the average thickness is considerably less. In Libya the effective thickness is generally less than 15m. TOC values average 2 to 4%, but a value of 14% has been reported from Algeria. Eighty-five percent of the kerogen belongs to types I and II,

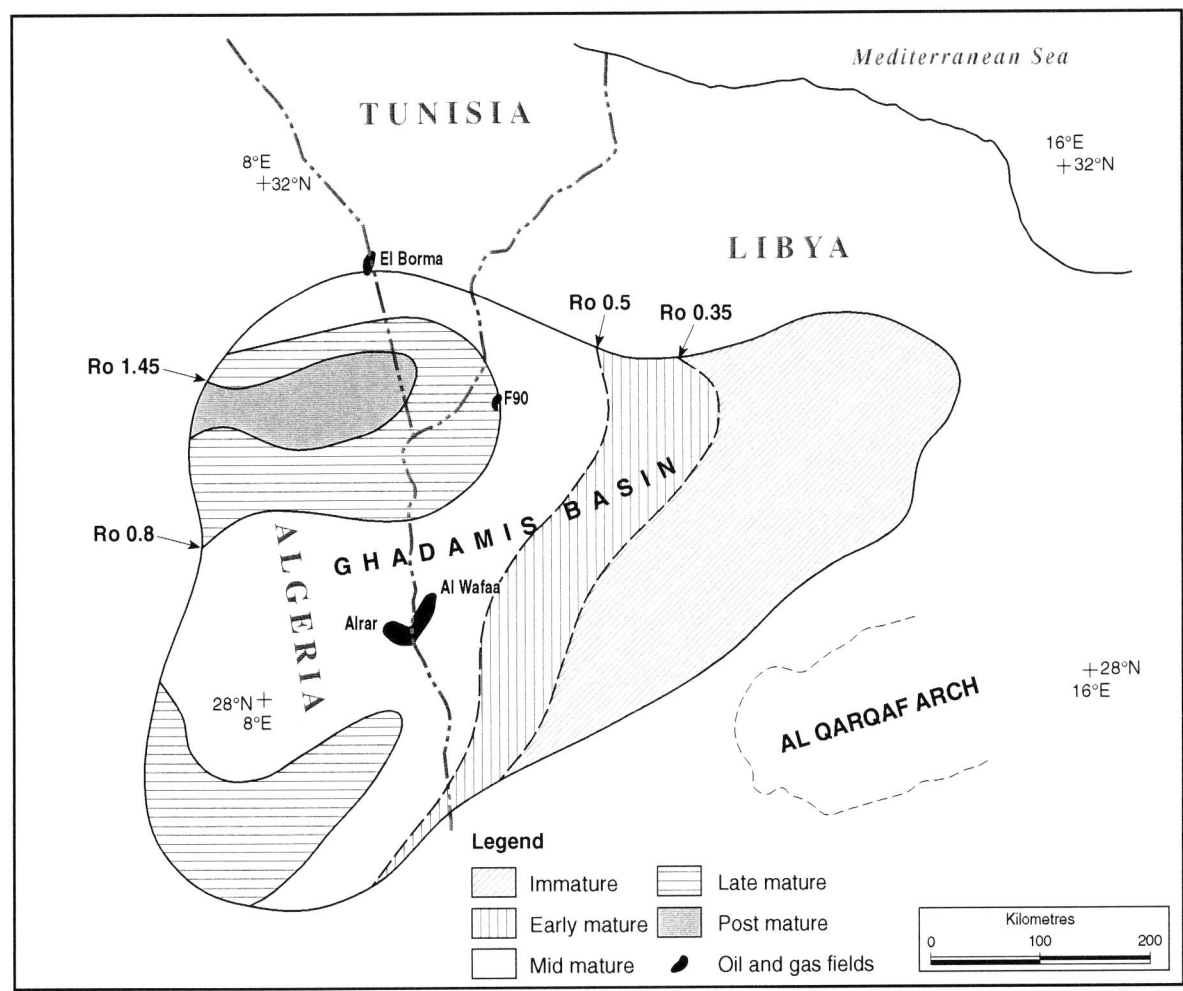

Sources: Daniels and Emme, 1995, Echikh, 1998, Belhaj, 1996

Figure 7.9   Ghadamis Basin, Frasnian Source Rock, Distribution and Maturity

The Frasnian source rock is less important than the Tanzuft source rock, but it nevertheless sourced several fields in Libya, Tunisia and Algeria, including Alrar, Al Wafaa, F-90 and El Borma. Maturity levels have been established in Algeria and are projected into Libya using the same depth to the oil window as established in Algeria.

except on the Tihemboka High where there is an increase in terrestrial type III kerogen. Potential yield figures of 3400 to 9550 ppm have been published by Hangari, et al. Maturity values, derived from pyrolysis data, spore colouration, and vitrinite reflectance, show that the areas around Al Wafaa and concession NC 100 are in the peak oil generation zone whilst in the Ghadamis area it is in the late oil generation zone (Figure 7.9). Maturity levels decrease towards the Al Qarqaf Arch and towards the northern truncation limit. The time of peak oil generation, based on modelling studies, was from early Cretaceous to Palaeocene with maximum generation during the mid-Cretaceous. In the Al Wafaa area peak generation may have been as late as Eocene. Daniels and Emme calculated that over 700 billion barrels of oil have been expelled from Devonian source rocks in the Algerian sector of the Ghadamis/Illizi Basins. Oil to source correlations suggest that Devonian shales sourced both the El Borma field on the Tunisian/Algerian border, and the Alrar/Al Wafaa fields on the Algerian-Libyan border (Figure 7.9). Byramjee and Vasse recognised two types of oil in the Palaeozoic of the northern Sahara. One of these types, characterised by very high paraffin and low aromatic content, probably correlates to the Frasnian source rock.[12]

Shales within the lower Carboniferous have some source rock potential, but are generally immature. TOC values of 1.75% were reported from well A1-NC 118, with Tmax of 432°C, indicative of very early maturity, although it could reach the oil window in the centre of the basin. The effective thickness is difficult to quantify but is likely to be small. Oil gravity in the Carboniferous reservoirs in the Atshan field exceed 47°API and the accumulation is largely gas-condensate. The shales of the Cambro-Ordovician are regarded as having little or no source potential since the kerogen is heavily oxidised.[13]

7.5 Sirt Basin

The Sirt Basin is the most prolific oil province in north Africa, with about 117 billion barrels of proven oil in place. The oils are generally sweet, with sulphur content between 0.15 and 0.66%, and with relatively little gas. Sixteen of the twenty-one major oilfields are undersaturated. Oil gravity is usually within the range 44 to 32° API. Oil pools in the lower Cretaceous are paraffinic with pour points around 38°C. Accumulations with a Palaeocene top seal have pour points up to 13°C, and those with Oligocene and Eocene top seals have a pour point less than 9°C. Oil accumulations have been found from depths of 700m to as deep as 4000m, and within a temperature range of 52°C to as

high as 143°C. The temperature of the Hutaybah gas pool reservoir is 150°C. Geothermal gradient in the Sirt Basin averages 25.5°C/km. Ibrahim studied geothermal gradients in the Maradah Trough and claimed a correlation between high geothermal gradients and hydrocarbon accumulations. This association was based on the belief that high geothermal gradients indicate subsurface fluid migration focussed into shallower reservoir closures. He claimed that 69% of the discoveries in the area studied were associated with anomalously high geothermal gradients.[15]

Most of the oil in the Sirt Basin was derived from the Campanian Sirt Shale, a rich and thick source rock with excellent characteristics. In West Libya the effective thickness of the Tanzuft hot shale averages between 25m and 45m. By contrast, in the Sirt Basin the effective thickness of the Sirt Shale source rock is frequently over 250m, and in places reaches 700m.[14] Baird et al. pointed out the interesting fact that the Sirt Shale does not fall into any of the six stratigraphic intervals which account for 90% of the world's oil, nor is a subsiding graben system a common habitat for source rock deposition. In these respects the Campanian source rocks of the Sirt Basin are unusual, if not unique. The Sirt Shale is a dark-brown to black laminated organic shale, rich in planktonic foraminifera. This implies that the upper water layers were not anoxic, but the subsiding troughs in which the shales accumulated may have been separated from the open sea by sills which led to water stratification and anoxic conditions on the stagnant floor of the troughs. The lack of bioturbation in the Sirt Shale supports the idea that the Campanian sea floor was an unfavourable environment for burrowing organisms. In the eastern Sirt Basin the lower part of the Sirt Shale passes into a dark-brown, argillaceous micritic limestone, the Taqrifat Limestone, which also provides evidence of an oxygen-deficient, low-energy environment, but with limited areal extent.

In the Sirt Shale the distribution of kerogen facies is closely related to the Campanian palaeogeography. Type IV kerogen, inert oxidised plant material, is found on the highest areas of the emergent platforms, particularly on the Az Zahrah-Al Hufrah, Al Jahamah, Zaltan and Waddan Platforms. This material had prolonged exposure to oxidising conditions. Kerogen type grades from type III around the trough margins, to type II in the centre of the troughs (Figure 7.10). Type I kerogen has not been found in the Sirt Shale.

A strong terrestrial content for the kerogen content of the Sirt Shale is also indicated by carbon isotope data, although the carbon isotopes become heavier towards the main depocentres. TOC values average between 2 and 5%, but occasionally exceed 10%. El Alami et al. produced a map showing the thickness of effective source rock within the Sirt Shale using a cut off of 1% TOC (Figure 7.11). This shows four main depocentres: the northern Zallah Trough with up to 300m of effective source rock, a small

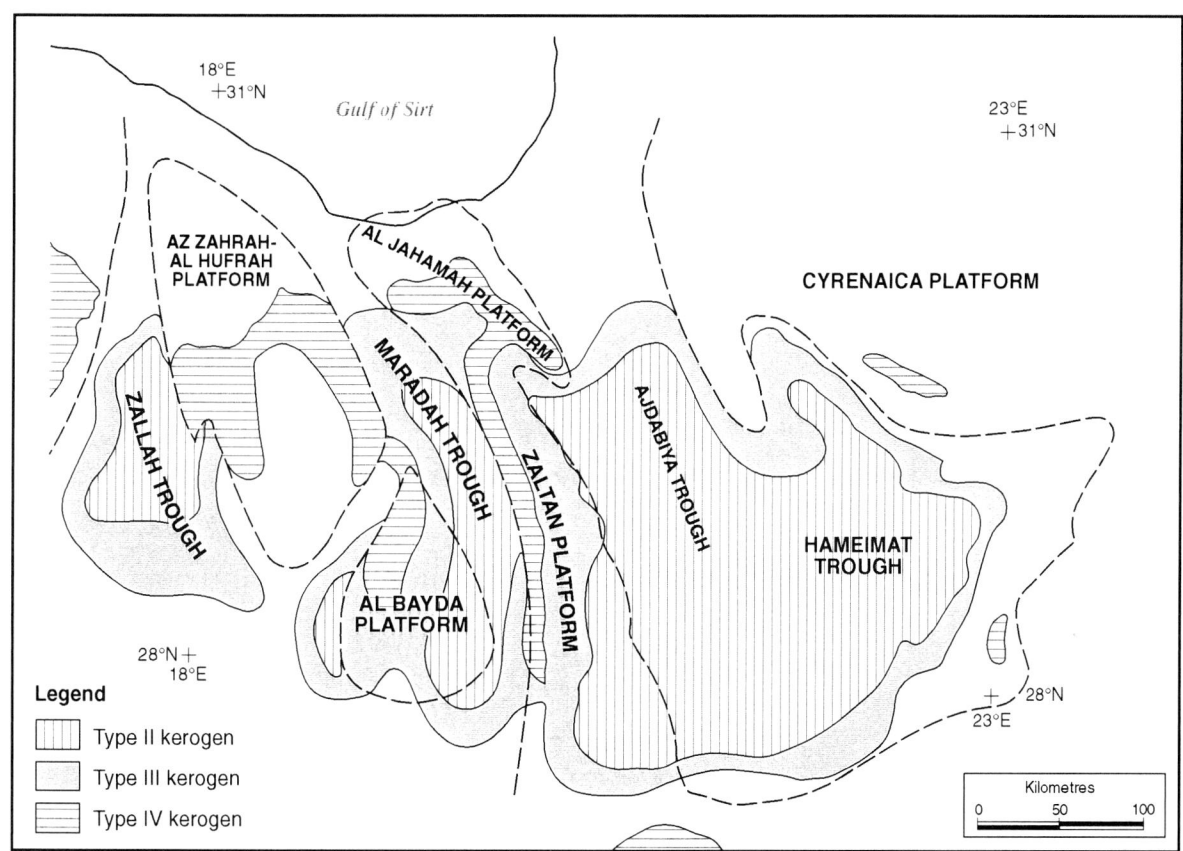

Source: Modified from El Alami, et al. ,1989

Figure 7.10   Sirt Basin, Campanian Source Rock, Kerogen Facies

Kerogen facies in the Sirt Formation was greatly influenced by Campanian palaeogeography. Subsidence within silled grabens allowed anoxic conditions to develop in which type II kerogen formed, whereas the shallower marine areas were open to extensive oxidation which produced type III kerogen. The upland areas were totally oxidised and produced type IV kerogen.

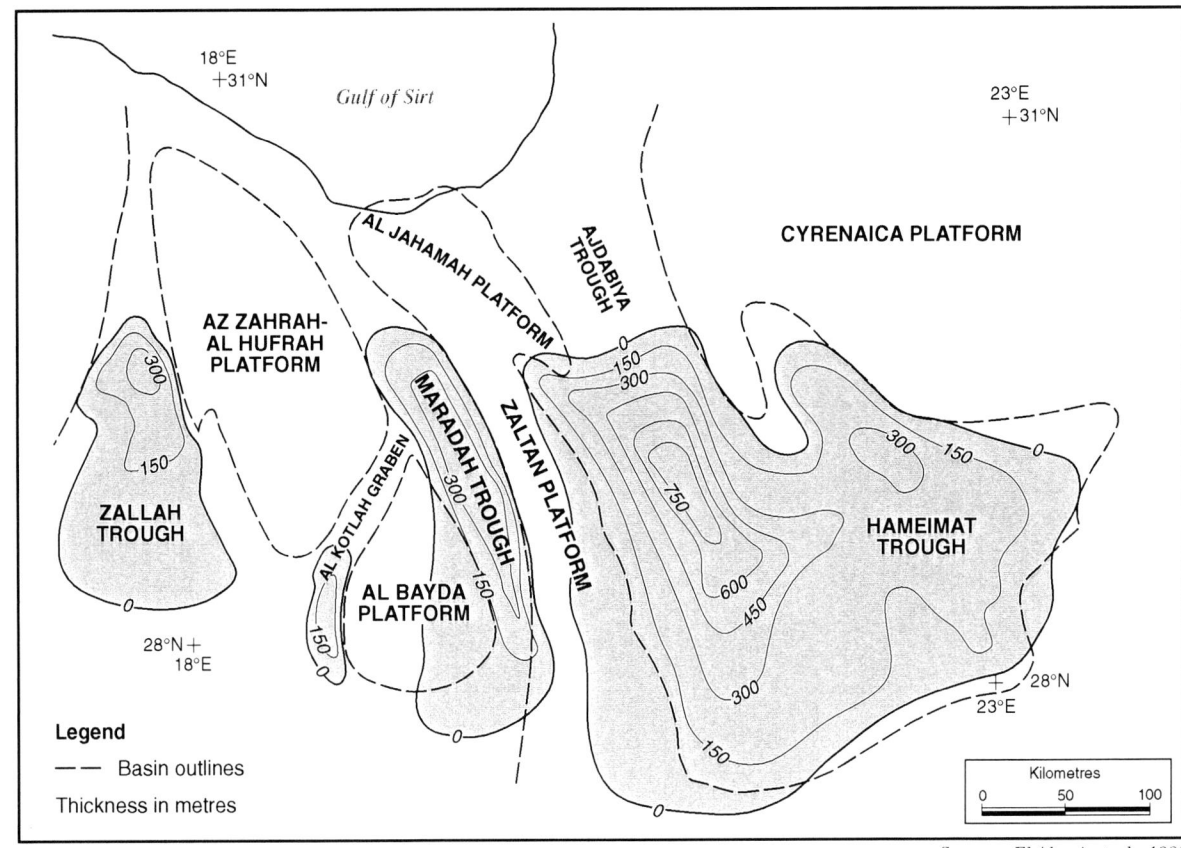

Figure 7.11   Sirt Basin, Effective Thickness of Sirt Shale Source Rock (>1% TOC)

This map shows the distribution and thickness of the Campanian Sirt Shale in the Sirt Basin containing >1% TOC. The main areas of rich source rock are in the northern Zallah Trough, Al Kotlah Graben, southern Maradah Trough and Ajdabiya Trough. The Sirt Shale is lean over the entire northern part of the basin, and in the southern Zallah Trough, and is thin or absent over the platform areas. The maximum development of effective source shales is 750m in the Ajdabiya Trough.

depocentre in the Al Kotlah Graben with 150m, the southern Maradah Trough with 350m, and the southern Ajdabiyah Trough with up to 750m of effective source rock. In the eastern Sirt embayment the Hameimat Trough shows an effective source rock thickness of 150 to 200m. It is worth noting that, according to El Alami et al., the entire northern part of the basin is lean due partly to unsuitable kerogen facies and partly to the high level of maturity. However, as mentioned below, this situation does not extend into the

offshore area, where upper Cretaceous shales have been shown to have fair source potential. El Alami et al. did not include the southern Zallah Trough and Abu Tumayam Trough in their study.[16]

Maturity determinations have been carried out using spore colouration and vitrinite reflectance data. This work allowed the identification of six source kitchens in which the Sirt Shale has reached, and in some cases passed, optimum maturity (Figure 7.12). Depth to the oil window in most of these areas ranges from

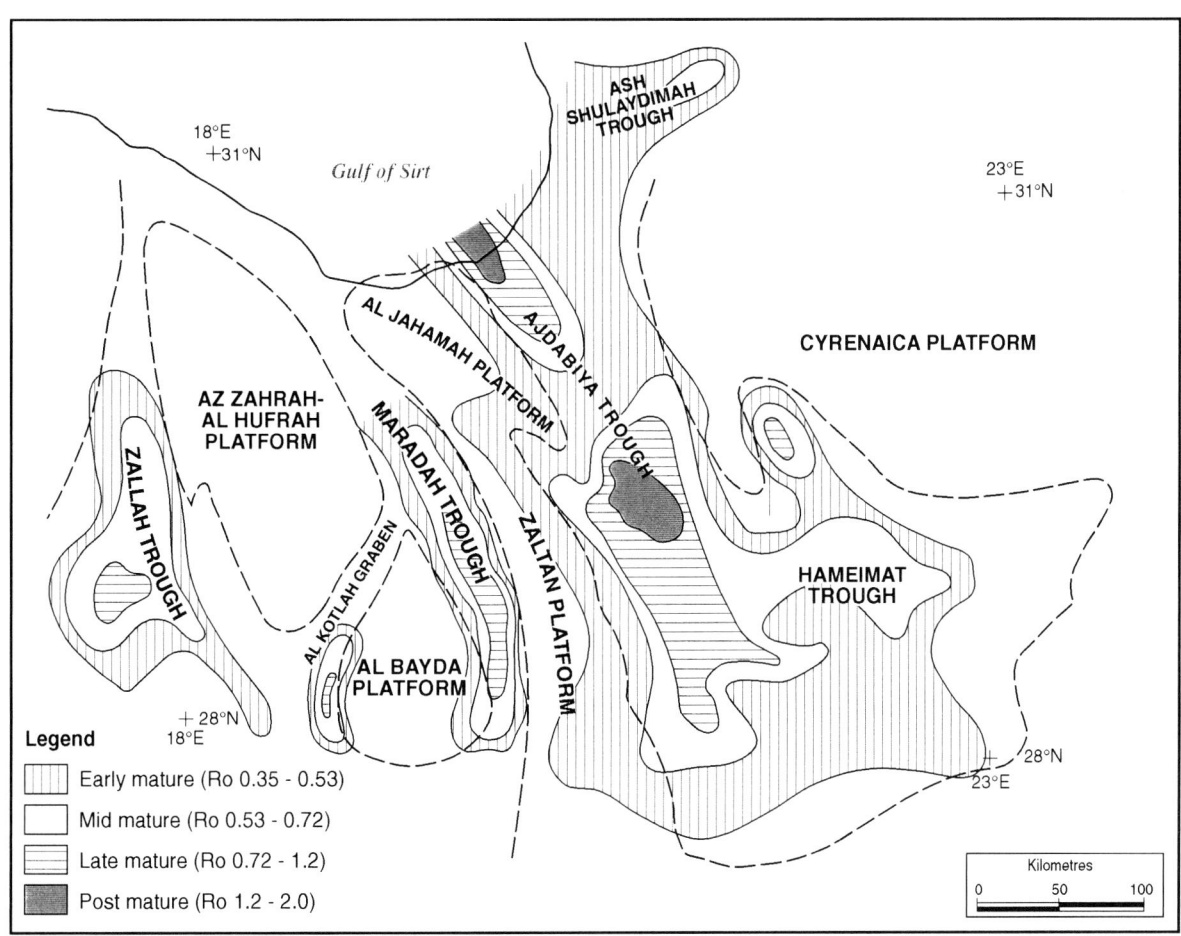

Source: El Alami, et al., 1989

Figure 7.12   Sirt Basin, Campanian Sirt Shale, Maturity and Source Kitchens

Four main source kitchens have been identified for the Campanian source rock in the Sirt Basin: the Zallah Trough, Al Kotlah Graben, Maradah Trough, and Ajdabiya/Hameimat Troughs. Large volumes of hydrocarbons have been generated from these source kitchens, with peak generation during the Oligo-Miocene. In most of these areas generation is still active, but in the deep Ajdabiya Trough the Sirt Shale is now in the post-mature zone.

3000 to 3500m. The six areas are the northern Zallah Trough, Al Kotlah Graben, southern Maradah Trough, southern Ajdabiya, northern Ajdabiya and Hameimat troughs. A late mature stage of gas/condensate generation has been reached in both of the Ajdabiya source kitchens. Combining data from these various sources it is evident that the Al Kotlah, southern Maradah and Ajdabiya/Hameimat Troughs all represent source kitchens from

which large amounts of oil have been generated. The Zallah Trough, on the other hand, has only a small area of thick effective source rock which for the most part is early mature, which suggests that hydrocarbon generation from this area has been modest. By contrast, the platforms areas and the northern part of the basin are either immature or lean. A detailed study on well A1-NC 157 south of the Harash field broadly confirmed these conclusions. TOC in the Sirt Shale reaches 3.63%, with potential yield values up to 2420 ppm. Kerogen is dominantly type II and the best quality source rocks were concentrated towards the top of the Sirt Formation. The top of the oil window was estimated at a depth of 2800m, with peak generation below 3250m.[17]

In addition to the Sirt Shale several other potential source rocks have been identified, particularly in the eastern Sirt Basin. Work on oil to source correlations using carbon isotope plots of aromatics against paraffin-naphthenes shows three distinct families with good correlation between source rock and oil samples. The main group comprises source rock extracts from the Sirt Shale and oils generated from this source, representing type II and III kerogen deposited in a marine reducing environment. A second group of isotopically heavier extracts comes from the Turonian and Lower Cretaceous source rocks of the Hameimat Basin. These are waxy crudes indicative of a lagoonal or lacustrine environment. The third group of isotopically lighter extracts correlates to Triassic shales encountered in wells in the Maragh Graben, and oils reservoired in the Amal, Jakharrah and As Sarah fields. The isotopic signature of the latter group suggests land derived material deposited in a highly reducing environment. Mass fragmentogram plots confirm the presence of three principal oil families with significant differences in triterpane and sterane ratios, and chromatograph plots of oils and rock extracts also fall into three groups.[18]

More recent studies by Burwood suggested that in addition to the three source intervals mentioned above, viable source shales are also present in several other formations. Burwood screened 50,000m of samples from 43 wells in the eastern Sirt Basin to determine source rock quality and chemical characteristics. There is no evidence of oil biodegradation in this area, and oil gravity varies within a narrow range,

typically 30 to 39° API. The most striking feature of the eastern Sirt embayment is the high waxiness of some of the oils, particularly at Jakarrah, Tuama and Abu Attiffel, plus some other pools such as Amal-Nafurah and Sarir C. Burwood identified seven potential source rock in the eastern Sirt embayment, as shown on Figure 7.15, and determined their defining characteristics. He found no indication of Tanzuft source potential in the wells examined, even in wells such as A1-46 near Antlat, where a very extensive Silurian shale section was penetrated. The shales are mature but in the absence of a hot shale interval potential yield is effectively zero. Triassic shales in the Maragh Graben, now known to be of Anisian age, contain a 30m organic-rich zone in which TOC values average 4.8%. The shales are immature in the area studied and because of their kinetic properties a high thermal regime would be required in order to generate hydrocarbons from these shales. The very low carbon-13 isotope values of these shales is diagnostic. The lower Cretaceous lacustrine shales within the Nubian sequence were generally found to have a significant sand content and to be rather lean in the wells studied. TOC values ranged from 1.9 to 2.5%. A very wide range of carbon isotope values has been found in the Nubian Shale, from depleted to highly enriched, implying that the kerogen content was derived from a very diverse palaeoflora. In the upper Cretaceous organic-rich shales occur at several levels from the Turonian to the early Maastrichtian. El Alami referred the pre-Taqrifat shales to the Etel Formation whereas Burwood referred them to the Rachmat Formation. Burwood showed that the Rachmat marine shales exhibit good source characteristics in certain wells such as MM1-59 and D1-NC 170 near the Messlah High. TOC values average 2.1%, and the effective thickness can reach 260m. The overlying Taqrifat Formation shows up to 65m of effective thickness in well MM1-59, with average TOC of 1.9%. The

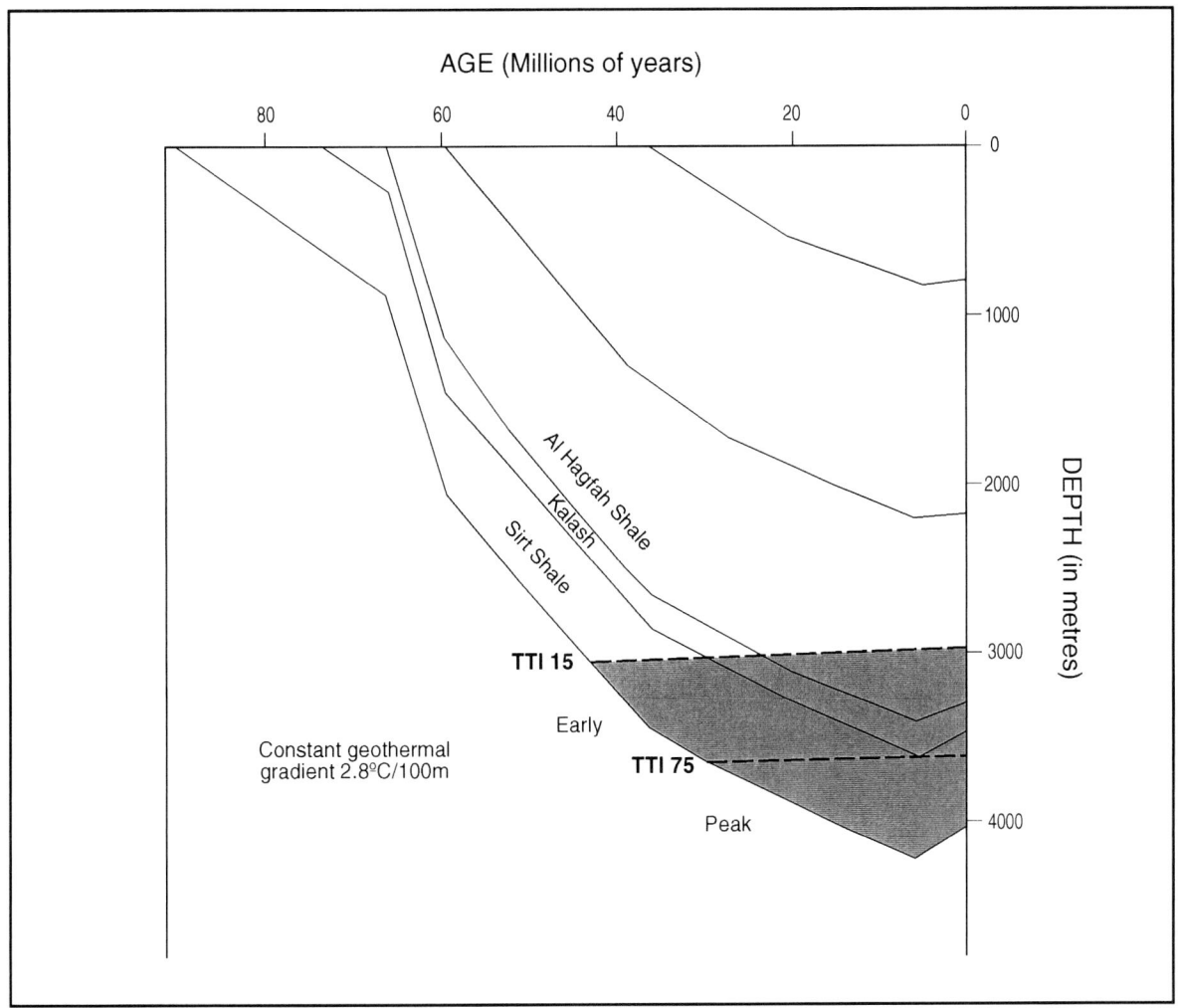

Figure 7.13   Burial History/TTI Plot, Maradah Trough, adjacent to Ar Raqubah Field

This is a pseudo-burial depth plot for the Maradah Trough close to the Ar Raqubah field. It assumes a constant geothermal gradient of 2.8°C/100m, and slight uplift during the last 4 million years. The geothermal gradient is based on well data from the Maradah Trough. The Sirt Shale entered the oil window during the mid-Eocene and reached peak generation at the beginning of the Oligocene. The Raqubah oil accumulation was sourced from the Maradah Trough by migration along the trough margin fault.

Campanian Sirt Shale has similar characteristics to the Rachmat and Taqrifat but, with an effective thickness of up to 760m and a very wide areal distribution, undoubtedly represents the principal source rock of the eastern Sirt Basin. Although the carbon isotope values of the Rachmat and Sirt Shale are similar, it is generally possible to identify extracts from the Sirt Shale by their slightly more depleted carbon-13 values. Within the Palaeocene succession Burwood found minor source potential in the Hagfah, Khayir and Harash shales. The Eocene Jalu Formation contains a highly oil-prone interval in the southern

Ajdabiya Trough, and the Antlat Formation shows a 20m section of oil-prone source rock in well B1-NC 129. These minor source rocks are immature in the eastern Sirt embayment but are likely to be mature in the Ajdabiya and Ash Shulaydimah troughs. On the basis of these studies Burwood recognised nine significant oil families, which contributed to the petroleum systems discussed in chapter 8.[19]

The Sirt Shale source rock is widespread over the basin, the lower Cretaceous source rock is best developed in the southeastern part of the basin, and the Triassic source rock is largely confined to the Maragh Graben. Modelling work by Ghori and Mohammed on wells U1-41 in the Ash Shulaydimah Trough, K1-97 in the Hameimat Trough and L4-51 in the Maragh Graben confirmed the source characteristics of the Triassic, Nubian and upper Cretaceous source rocks. The work of Burwood however suggests a more complex picture with Palaeocene and Eocene shales generating oil in the Ajdabiya Trough. He cited the cases of well K1-12 near Intisar which he determined was likely to have been sourced from the Palaeocene Harash Formation, and well B1-NC 152 near Antlat in which the oil is chemically correlatable to the Eocene Antlat Shale. The recognition of effective Palaeocene and Eocene source rocks in the Ajdabiya Trough opens up major new possibilities for oil generation deeper in the Ajdabiya Trough. This topic is discussed further in chapter 8.[20]

The timing of migration has been studied in several areas. In the Maradah Trough, adjacent to the Ar Raqubah field, modelling studies combining burial history and Lopatin's Time Temperature Index (TTI) have demonstrated that the Sirt Shale entered the early phase of oil generation during the mid-Eocene and reached peak generation during the Oligocene (Figure 7.13). In the southern Ajdabiya Trough similar studies show that peak oil generation did not occur until the mid-Miocene (Figure 7.14),

and in many areas active generation is still taking place today. In the eastern Sirt embayment Ghori and Mohammed determined the timing of onset of oil generation as Eocene in the deep troughs, to Oligocene in the shallower areas. A study of eleven wells in the Hameimat Trough conducted by Bender et al. showed that the depth to the top of the mature zone is 3500m, peak generation 4100m, and top of the over-mature zone 4570m. Oil generation began in the deepest part of the Hameimat Trough around Abu Attiffel during the Palaeocene, and this area is now over-mature. Wells on the Abu Attiffel field show a higher gas-oil ratio and higher oil gravity than at Messlah and As Sarir. The flanks of the Hameimat Trough entered the oil window during the Oligocene and are still generating hydrocarbons at the present day. The main hydrocarbon generating kitchen of the eastern Sirt embayment is located in the centre of the Hameimat Trough, with three very small isolated kitchens further south. The waxy crudes found in the Sarir C area may have been generated from these small kitchens.[21]

The waxy oils which characterise the Abu Attiffel field are believed to have originated from either the Nubian or Turonian-Santonian shales, but opinion is divided on which of these is the most effective source rock. Data published by El Alami, showed TOC values from the Nubian lacustrine shales (the Variegated Shale) averaging 0.94%, with an average potential yield of only 1710 ppm. The kerogen is mostly woody type III and is gas prone, and El Alami believed that the Nubian Shale is not an effective source rock in this area. El Alami's investigation of the Turonian-Santonian shales in the Abu Attiffel area showed them to be substantially thicker than the Nubian shales, lagoonal in character, and with TOC values averaging 2.85%. The kerogen is largely amorphous algal type II and the shales have a high production index. Analyses of the Taqrifat

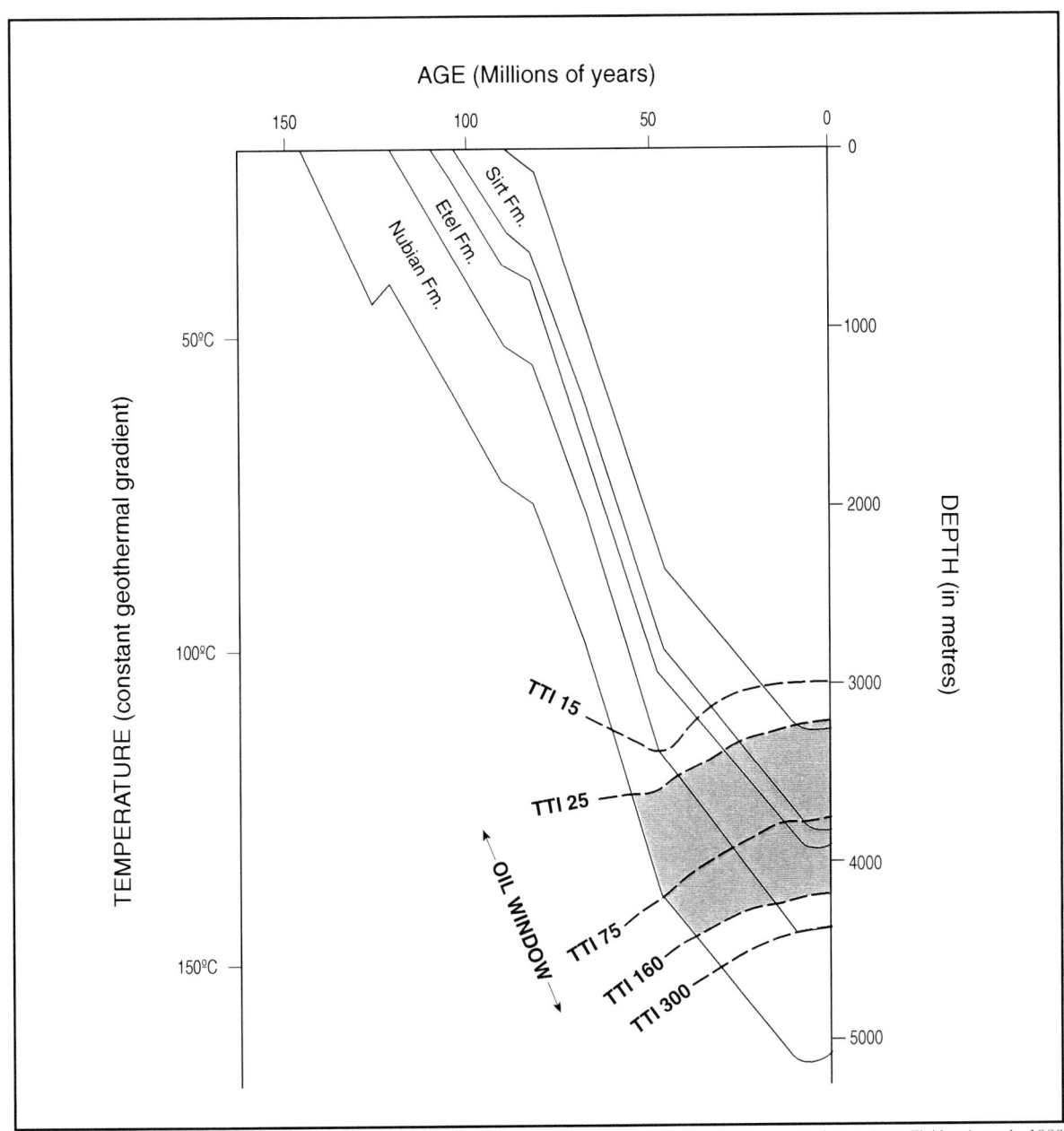

Figure 7.14   Burial History/TTI Plot for Well DD2a-59

Well DD2a-59 is located in the southern Ajdabiya Trough, SW of the Jalu field. It was drilled to a depth of 4476m. Modelling studies show very rapid subsidence during the Cainozoic, with the Sirt Shale entering the oil window in the Oligocene and reaching peak generation during the Miocene. The source rock is still in the oil generation window at the present time.

Limestone showed poor source potential with average TOC of 0.54% and potential yield of less than 1200ppm. According to El Alami, at Abu Attiffel the Nubian Shale is mid to post-mature with gas generating potential whereas the Turonian-Santonian Etel-Rachmat Shale is mid- to late mature with light oil generating potential. He concluded that the Turonian shale was the more effective of the two source rocks in the eastern Sirt Basin. Mention was made in chapter 4 of the Turonian anoxic event during which very high levels of atmospheric CO_2 produced a 'greenhouse effect'. El Alami envisaged a situation in which abundant algae and marine grass flourished in lagoonal conditions alternating with hyper-saline conditions which provided ideal conditions for the preservation of the organic material. Other authors, however, disagree with the view that the Nubian (Aptian) source rock is ineffective in the eastern Sirt embayment. In concession 82 in the Hameimat Trough Bu Argoub and Thusu recognised good quality Aptian source rocks, equivalent to the Variegated Shale, rich in terrestrial flora and microflora along with soft tissue which appear to be a good source of waxy oil. Burwood also favoured a Lower Cretaceous lacustrine source rock as the principal source for the Abu Attiffel and Nakhla fields.[22]

Studies on thermal maturity were conducted by Gumati and Schamel using data from 28 wells. They found that geothermal gradients were relatively low in the deep troughs of the Sirt Basin compared with the platforms. Vitrinite reflectance data from well AA1-6 in the Ajdabiya Trough shows that the section is immature down to the mid-Palaeocene, early mature in the early Palaeocene-Maastrichtian and late mature or post mature in the Campanian. This suggests that there is little potential for generating hydrocarbons from Cainozoic shales, even assuming adequate richness, anywhere except in the deep Ajdabiya Trough. In the western Sirt Basin, observed

levels of matuity from well QQQ1-11 showed that the Cainozoic is immature but the upper Cretaceous is highly mature. This suggests either that the geothermal gradient was higher during the late Cretaceous, or that mid-Cainozoic uplift was followed by the removal of at least 1000m of section on the western margin of the basin.[23]

Several other formations have been cited as potential source rocks in local areas in the Sirt Basin. The Silurian Tanzuft Shale, although discouraging in the east, may have potential on the western periphery of the basin where hot shales are more likely to be developed, both in the Zamzam Depression and possibly in the Al Fuqaha/Al Haruj al Aswad area. The Palaeocene has already been mentioned. Shales are present in the Hagfah, Harash and Khayir formations and Burwood's inferences in relation to well K1-12 near Intisar have been previously discussed. The Eocene Antlat Formation has fair oil potential in well B1-NC 129 and almost certainly becomes mature in the Ajdabiya Trough. Brady et al. believed that the Sheterat Shale of Thanetian age was the main source rock for the Intisar fields, but no supporting evidence was presented, and it seems most unlikely given the thinness of the formation and its probable lack of maturity in this area. Analyses performed on samples from well A1-NC 157 on the southern flank of the Ajdabiya Trough show that the Palaeocene Khayir and Khalifah formations and the Maastrichtian Kalash Formation all have very low TOC values and high oxygen indices, indicative of an unfavourable source rock environment. In the same well, analyses of the Rachmat and Lidam formations, which underlie the Sirt Shale, showed both formations to be lean and dominated by terrestrial type III kerogen. Parsons et al. presented TOC values for various potential source rocks in the central Sirt Basin. Average TOC values are 1.91% for the Sirt Shale, 0.95% for the Hagfah Shale,

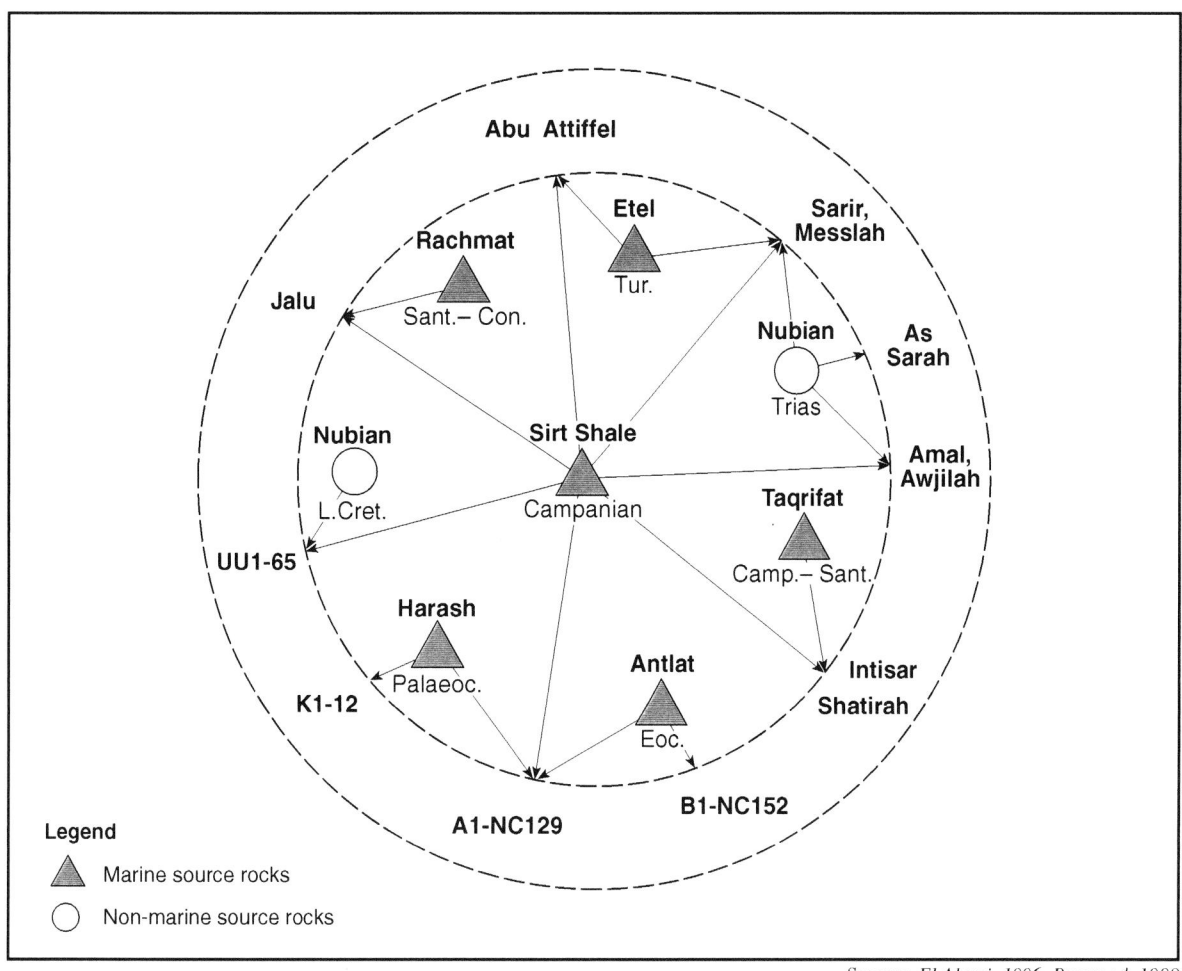

Source: El Alami, 1996, Burwood, 1998

Figure 7.15   Eastern Sirt Basin, Oil Families

In addition to the Sirt Shale the eastern Sirt Basin contains a number of other source rocks which have generated hydrocarbons in local areas. This diagram summarises the additional source rocks and illustrates a number of hybrid systems. The associations are based on carbon isotope studies, oil to source correlations and multivariate oil data analysis.

0.87% for the early Eocene shale and 0.99% for the Oligocene shale.

A note of caution must be sounded, however. The work of Burwood suggests that the source rock characteristics of the eastern Sirt Basin are complex. Mixing of oils from different source rocks is common and difficult to interpret, and shales which in most areas appear to have little hydrocarbon potential can locally develop favourable characteristics. The relationship between oil families and petroleum systems is discussed further in chapter 8.[24]

7.6   Cyrenaica

Information is scarce on the petroleum geochemistry of Cyrenaica. No major hydrocarbon discoveries have been made north of the Amal field, but there is no doubt that several potential source rocks are present in the area. Ghori recognised the presence of potential source rocks in the shales of the Devonian and early Carboniferous on the Cyrenaica Platform. These rocks contain thick sequences of organic-rich shales with TOC values between 1% and 2%. In most cases however, it appears that the kerogen is mostly type III, and gas prone. Jurassic rocks on the Cyrenaica Platform contain organic shales, but are immature. Further north, in the Jabal al Akhdar Trough and Marmarica Basin, marine Jurassic and early Cretaceous rocks are present, containing thin shales with moderate organic content. The facies is predominently gas prone, although there is evidence of oil-prone kerogen further north. In the Ajdabiya Trough organic-rich shales are present in the Upper Cretaceous, Palaeocene and lower Eocene. The Upper Cretaceous shales show TOC values between 1% and 3% and the kerogen is type II and type III (Figure 7.16). Burwood interpreted the Eocene shales of the Ajdabiya Trough as being the source of the oil found in the B1-NC 152 well near Suluq, and probably contributed to the oil found at Antlat (Figure 7.15).[25]

A maturity plot on well A1-NC 92 (Figure 7.16) from the Cyrenaica Platform shows that the depth to the oil window is around 2200m, peak generation at 3000m and base of the oil window at 4000m. The late Devonian-early Carboniferous source rock entered the oil generation zone during the late Cretaceous and peak generation during the Cainozoic. The Devonian shales are now in the late mature stage of generation. The gas shows encountered in several wells on the Cyrenaica Platform were probably generated from the this source.[26]

Well A1-36 on the Jabal al Akhdar Uplift was also modelled for maturity (Figure 7.17). This area underwent rapid subsidence in the late Jurassic and early Cretaceous, but was elevated during the Santonian inversion. Subsidence during the Cainozoic was slow and the base

Cretaceous in this well is currently at 3300m. The depth to the top of the oil generation window is about 2200m and to peak generation 3500m. TTI data from well A1-36 suggests that generation from the Jurassic shales could have begun before the Santonian inversion, but peak generation was not reached until the Palaeocene, with optimum generation during the late Cainozoic. The Cretaceous section is still in the early mature phase. Unfortunately the organic rich zones are thin. Further west, where the structural inversion has been much more pronounced, as in well A1-18, depth to the top of the oil window is only about 1220m, but even in this well the basal Cretaceous is barely within the oil window.[27]

In the Ash Shulaydimah Trough well U1-41 shows a depth to the top of the oil generation zone of 2500m, peak generation at 3500m and base of the oil window at 4150m. The top of the upper Cretaceous source rock is now at 4200m in the gas generating zone. Peak generation occurred during the Miocene. The lower Eocene Antlat shale is immature on the periphery of the Cyrenaica Platform, but in the Ajdabiya Trough near the Antlat field it is within the oil window, and as mentioned above, according to Burwood it has contributed to several small oil accumulations in the area. It appears likely that the oil found in the Eocene reservoirs in the Ash Shulaydimah Trough are hybrid oils from upper Cretaceous, Eocene and possibly Palaeocene source rocks.[28]

Ghori made no mention of the Silurian shales which are known to reach a thickness of 470m in well A1-46, and are presumed to underlie the entire Cyrenaica Platform. These have been described as dark grey pyritic shales, with some silty intervals, rich in graptolites, acritarchs and chitinozoa, deposited in a lagoonal setting. This description suggests some source potential, but possibly the Silurian shales are over-mature in much of the area. However at least one well with Silurian oil shows and source potential is

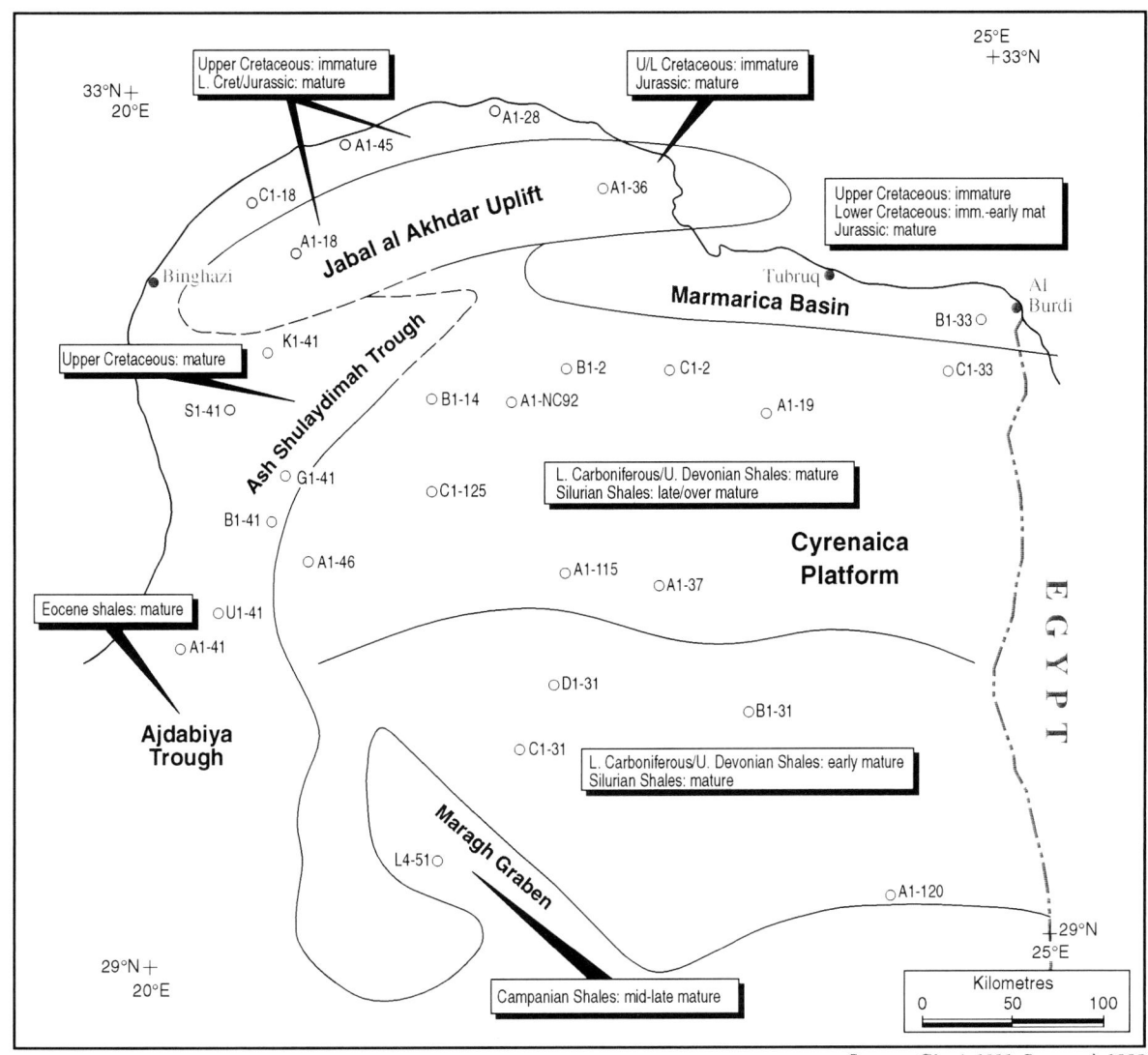

Figure 7.16   Cyrenaica, Source Rock Maturity

On the Cyrenaica Platform Lower Carboniferous/Upper Devonian shales have fair to good TOC values but are predominently gas prone. Silurian shales are also well developed but TOC values are questionable. In the Marmarica Basin and Jabal al Akhdar region thin gas prone shales are present in the Jurassic and Lower Cretaceous section, and in the Ash Shulaydimah Trough and Maragh Graben Campanian source shales are well developed. Generalised maturity data are indicated on the map. Burwood proposed a number of additional source rocks in the Ajdabiya Trough (Figure 7.15) which are relevant to the western and southern margins of Cyrenaica.

present on the western edge of the Cyrenaica Platform. Sola and Ozcicek published a map, based on proprietary data, showing prospective areas for a number of potential source rocks, and wells which have penetrated potential source rocks. They highlighted the area around well A1-NC 92 and another around well A1-115 as favourable for Palaeozoic source rocks, the

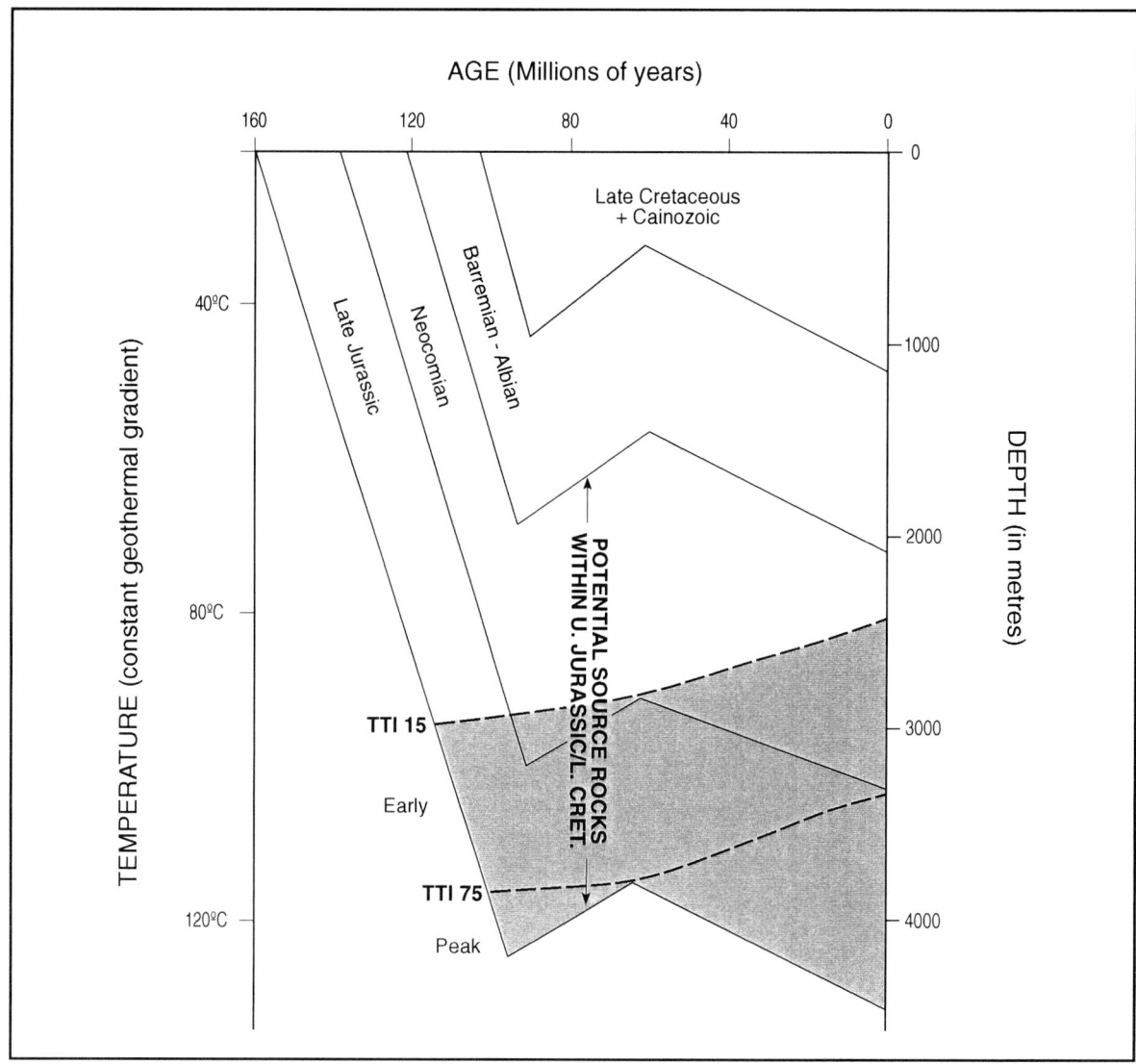

Source: Ghori, 1991

Figure 7.17   Burial History/TTI Plot for Well A1-36

Well A1-36 is located in the Jabal al Akhdar Trough which was inverted during the Santonian. The well was drilled to a depth of 4422m. The late Jurassic and early Cretaceous sections contain a few thin shales which contain gas prone kerogen. Modelling studies suggest that the Jurassic interval reached the oil window prior to the Santonian inversion, but peak generation was not reached until the Palaeocene, and then only for the Jurassic.

Marmarica Basin and the area centred on Binghazi as favourable for Mesozoic source rocks, and the area around Antlat as favourable for Eocene source rocks.[29]

Ghori concluded that the Devonian shales represented the best potential source rocks in Cyrenaica, with good Cretaceous and possibly younger potential in the Ash Shulaydimah Trough. As in the case of the eastern Sirt embayment, the source rock potential of

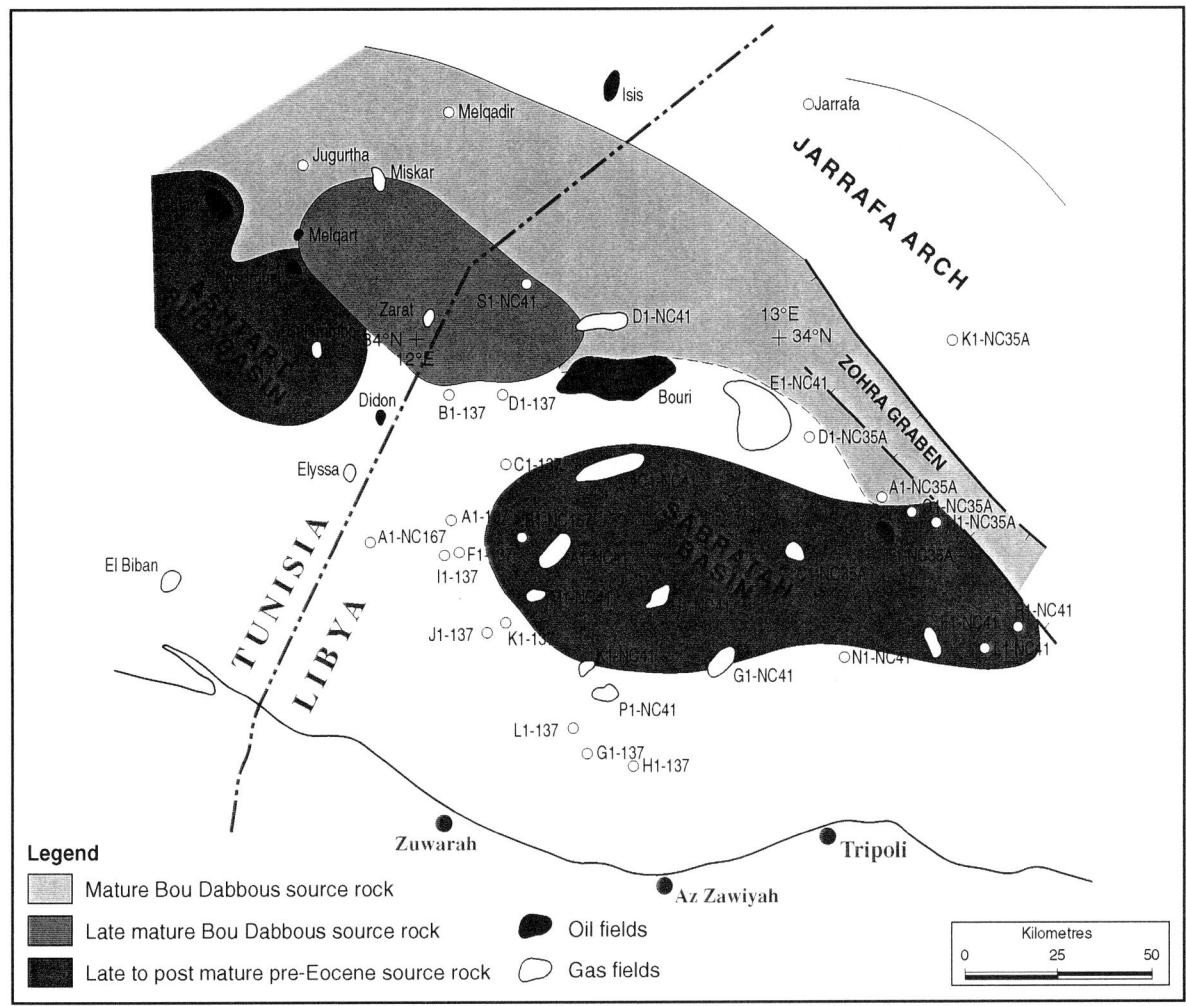

Source: Tunisia: Bishop, 1988, Racey et al., 2001, Bailey, et al., 1989
Libya: Anketell and Mriheel, 2000, Sbeta 1990, El Ghoul, 1991

Figure 7.18   Sabratah Basin,
Eocene and pre-Eocene Source Rocks, Maturity Map

This is a speculative map based on incomplete data which suggests that the principal oil source rock in the Sabratah Basin is the Ypresian Bou Dabbous Formation, the deep-water equivalent of the Jdeir Formation. The source rock appears to be late mature in the Miskar-Zarat area and mature to both east and west. The formation is confined to the area basinward of the Eocene carbonate ramp. The principal gas source rock is probably within the Cretaceous which is late mature to post-mature in the Sabratah and Ashtart depocentres.

Cyrenaica is probably much more complex than indicated above, and the work of Burwood et al. gives a tantalising glimpse of the complexities which may exist. It is to be hoped that oil companies will eventually permit more analytical data to be published.

7.7   Offshore

The Pelagian Block is a significant petroleum province with over twenty discoveries, most of which contain gas-condensate with or without an oil rim. Reservoirs range in age from Jurassic (Jerba) to

Miocene (Tazerka), but the principal reservoir is the lower Eocene nummulitic carbonate which contains 80% of the offshore hydrocarbons so far discovered. The lower Eocene reservoirs form part of the Farwah Group in Libya and the Metlaoui Group in Tunisia (Figures 5.2 and 5.7). Smaller discoveries have been made in Cretaceous carbonates (Isis, Miskar). The Eocene trend extends from Sidi el Itayem onshore Tunisia, through the Ashtart, Hasdrubal, Zarat, Salammbo, and Didon discoveries in the Tunisian offshore to the giant Bouri field, E-NC 41, C-NC 41, A-NC 41, and other smaller discoveries in Libyan waters (Figure 8.22). The group contains a range of facies from restricted evaporites through algal limestones, nummulitic limestones to deeper water globigerinid marls. In the Libyan area most of the hydrocarbon bearing structures are salt-related, whilst in Tunisia most of the structures are syn-sedimentary horsts or faulted anticlines on a seaward dipping ramp.[30]

Geochemical studies in Tunisia indicate that the principal source rock for the Eocene hydrocarbon accumulations is the Ypresian Bou Dabbous Formation, which is the deep-water lateral equivalent of the El Garia reservoir (Figure 7.18). In Libya the nummulitic reservoir is the Jdeir Formation and the deep water equivalent was formerly thought to be the Hallab Formation. However El Ghoul has shown that the Hallab Formation is largely Bartonian in age and cannot therefore be a time equivalent of the Bou Dabbous Formation. The equivalent of the Bou Dabbous Formation in Libyan territory is the Bilal Formation.[31]

The Bou Dabbous Formation contains amorphous Type II marine kerogen with TOC values between 0.5 to 2.5%. Biomarker evidence proves that condensate in Eocene reservoirs on the Hasdrubal, Ashtart and Melqart fields in Tunisia is derived from organic matter in the Bou Dabbous Formation (Figure 7.18). At Hasdrubal, which is probably typical of most of the Eocene fields including Bouri, analysis of oil samples shows a mature unbiodegraded oil derived from a marine source rock, and carbon isotope ratios show close similarities with other oils known to have been sourced from the Bou Dabbous Formation. Furthermore, molecular indicators suggest a migration distance of no more than 5km. Gas analyses from the same field provide evidence of slight cracking and generation in the peak to late maturity zones. Maturity modelling at Hasdrubal indicates the presence of adjacent source kitchens in which the Bou Dabbous source rock is in the peak to late-mature window (Figure 7.19). Oil generation began in the early Miocene with peak generation in the late Miocene. Oil and wet gas were sourced at the same time, but the more mobile gas reached the trap first. Studies on whole oils and extracts from the Jdeir Formation and Reineche Member from the Belina 1 well in concession NC 167 in the Libyan part of the basin showed the presence of wet gas in the Jdeir Formation which can be classified as a high thermogenic condensate-associated gas. The Reineche Member gas showed lower maturity and was classified as a migrated thermogenic petroleum-associated gas. Variations in maturity and composition at Belina strongly suggest different origins for the hydrocarbons and mixing within the reservoirs.[32]

A study of the Metlaoui Group in Tunisia, based on data from twenty offshore wells, revealed an unusually high heat flow (mean value 86mWm), much higher than in the Pantelleria and Linosa grabens further north. The high heat flow is attributable to recent wrenching and volcanism in the southern part of the Pelagian Block. It is also locally affected by the presence of salt walls and pillars. The depth to the oil window in this area varies from 1800m to 1900m for early mature to 2500m to 2600m for peak oil generation, and 3500m for the base of the oil

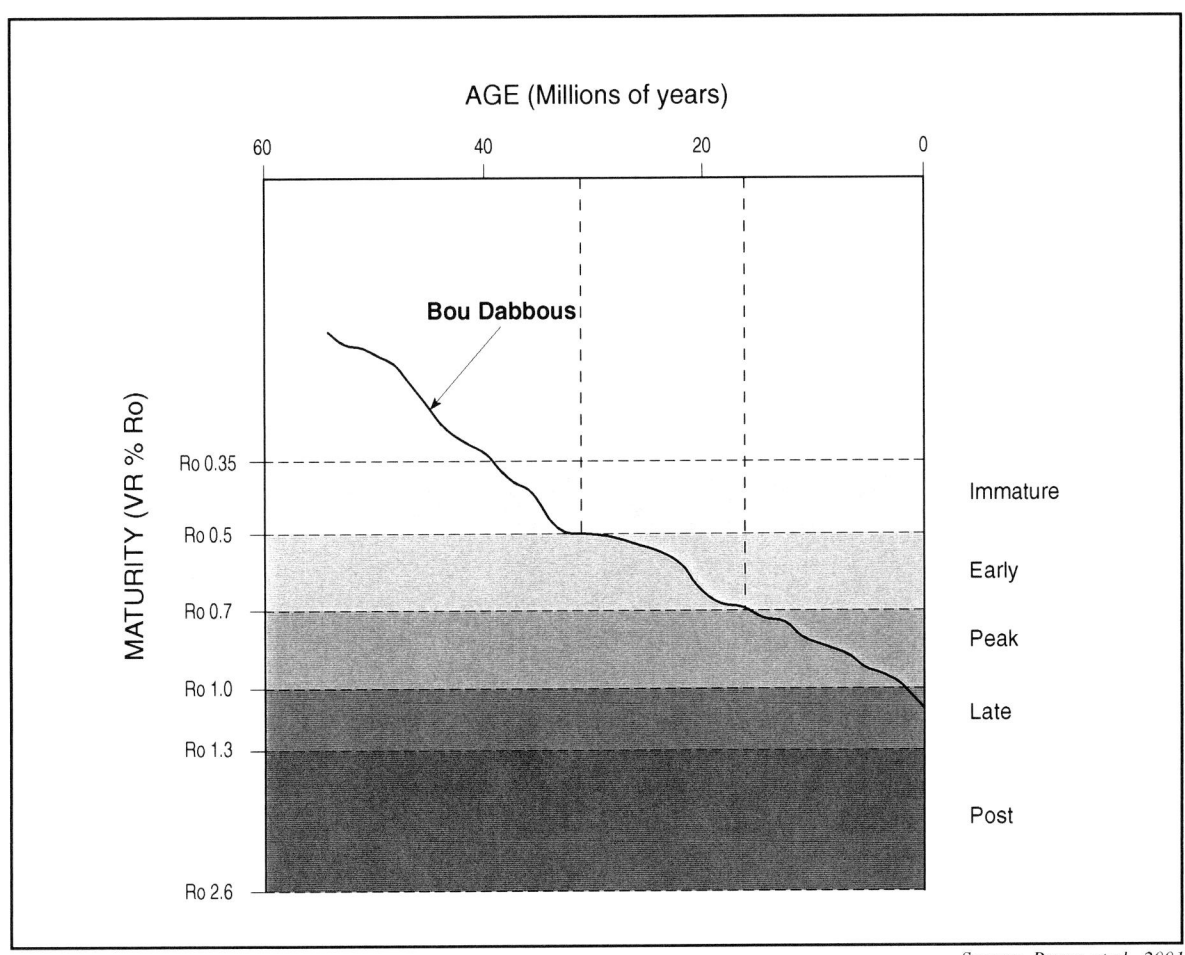

AGE (Millions of years)

MATURITY (VR % Ro)

Bou Dabbous

Immature

Early

Peak

Late

Post

Source: Racey, et al., 2001

Figure 7.19   Maturity Plot, Hasdrubal Field, Tunisia

The Hasdrubal field is located 80km northwest of Bouri, and is believed to be analogous to Bouri, both in terms of source rock richness and maturity. The source rock in each case is the Bou Dabbous Formation, a basinward marly equivalent of the El Haria (Jdeir) Formation of Ypresian age. The maturity plot shows the formation to have entered the oil window during the Oligocene and reached peak generation during the early Miocene.

zone. There is an increase in the level of maturity of the Metlaoui Group from the Reshef well in the north to the Salammbo field in the south. The Bou Dabbous source rock contains brown to black limestone horizons containing amorphous kerogen compatible with the high wax paraffinic crudes found at Ashtart, Melqart and Hasdrubal. Algal mats from sebkhas adjacent to restricted lagoons may also form a second potential source rock

within the Metlaoui Group which sourced the Didon, Hammon and Salammbo pools.[33] Data from the Libyan sector is scarcer but the situation at Bouri and the other Farwah discoveries is believed to be similar to Hasdrubal, with generation from the adjacent deeper-water marls (which interfinger between the nummulitic shoals), short-distance migration, and emplacement during the Miocene. Bernasconi et al. presented a series of maps of the Bouri field which showed the presence of open shallow-platform and deeper-platform facies immediately to the north of the field. Combining the data from Tunisia and from Bouri it seems most likely that the Bouri field was sourced from mature Bou Dabbous shales located to the north of the field (Figure 7.18). Anketell and Mriheel presented a burial history/maturity plot (Figure 7.20) for a well in the Sabratah Basin which is believed to be from the Bouri area. This shows that the Bou Dabbous source rock entered the oil window during the mid-Miocene but has not yet reached peak maturity. Possibly their data

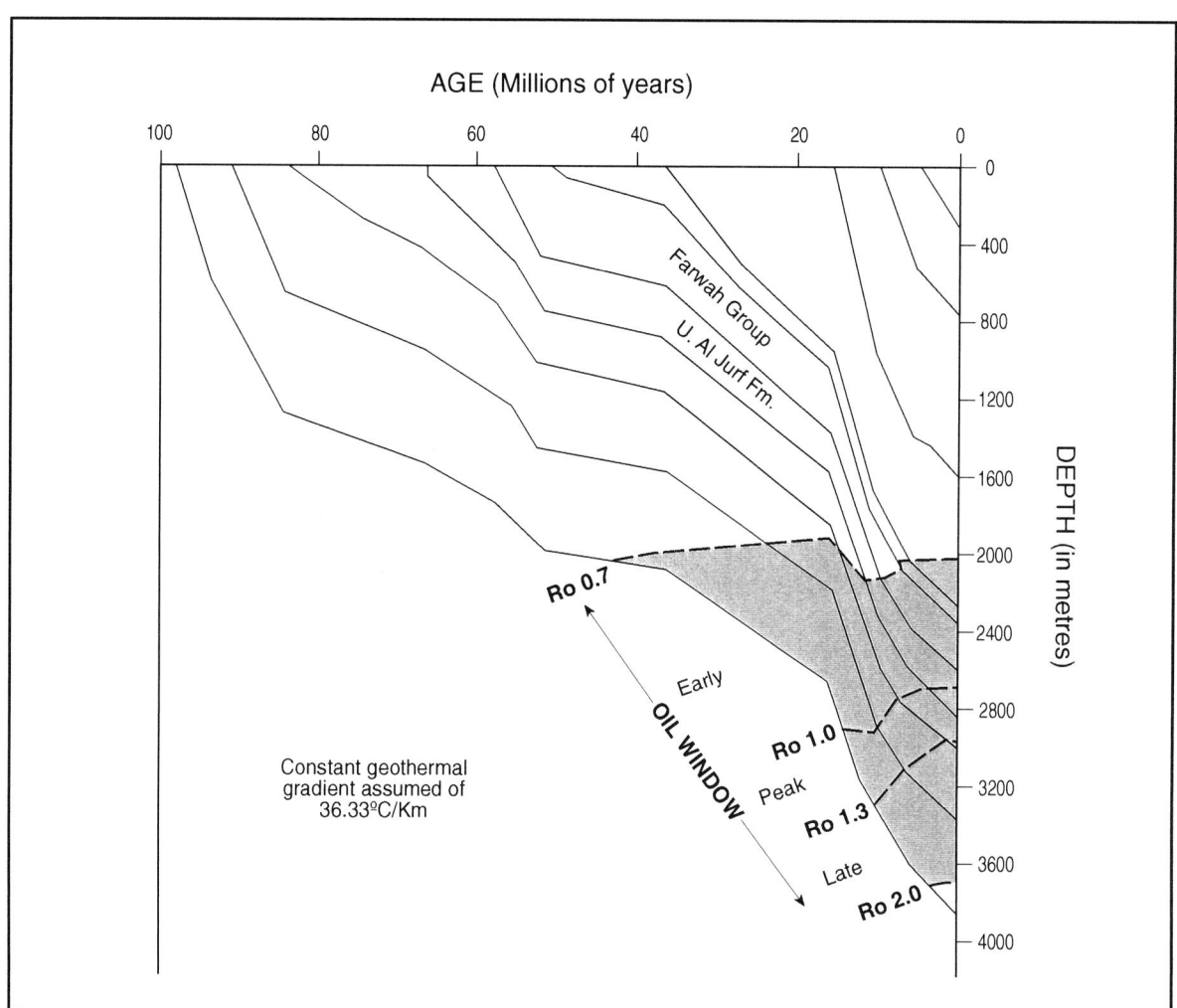

Source: Anketell and Mriheel, 2000

Figure 7.20   Burial History/Maturity Plot, Sabratah Basin

This plot shows the modelled burial history for the Libyan side of the Sabratah Basin. The well location has not been disclosed but is assumed to be in the Bouri area. The plot uses a constant geothermal gradient of 36.33°C/km, and shows the Eocene source rock to have entered the oil window during the late Miocene and to have barely reached the peak oil generation window. Compared with data from Tunisia and the character of the Bouri oil the assumed geothermal gradient may not adequately reflect the high heat flow in this area.

did not take sufficient account of the high heat flow in this area which was mentioned above.[34]

Eocene source rocks are not present in the Sabratah Basin south of the Eocene nummulitic trend, but evidence from Tunisia suggests that several potential source rocks are present within the upper Cretaceous (Figure 8.21). The Turonian Bahloul Formation is probably the richest hydrocarbon source rock within the upper Cretaceous. An impression of the location and depth of the Cretaceous source kitchens in the Sabratah Basin can be gained from the base Cainozoic structure map presented by El Ghoul and the structure map of the Bouri field presented by Bernasconi et al. El Ghoul's map shows a major depocentre in the Sabratah Basin, shallowing to the north towards Bouri. This depocentre is likely to indicate the location of the Cretaceous source kitchen (Figure 7.18). The major gas-condensate discoveries of D1-NC 41, E1-NC 41, C1-NC 41 and A1-NC 41 on the northern flank of the depocentre were probably sourced from this kitchen. In the centre of the basin, the base Cainozoic is at a depth of 5330m and this probably explains the presence of gas in the C1-NC 35a, F1-NC 41, G1-NC 41, H1-NC 41, M1-NC 41, K1-NC 41 and P1-NC 41 wells, which are all located in the deepest part of the basin. These conclusions are speculative, but appear to correspond with the available data.[35]

Little has been published on other potential source rocks in the Sabratah Basin, but a good deal of information is available from Tunisia. The Isis discovery in Tunisia, reservoired in Cenomanian rudist carbonates, was sourced from the Cenomanian Fahdene shales. Several other Cretaceous shales and marls have source potential, for example the bituminous marly carbonates of the Turonian Bahloul Formation, and the black shales of the Turonian-Campanian Kef and Aleg formations (Figure 8.21). According to Burwood, et al. however,

screening of these rocks has so far failed to reveal any significant Cretaceous source potential. The Maastrichtian-Palaeocene Al Jurf Formation (El Haria in Tunisia) was identified by Hammuda et al. as a possible source rock. The early Eocene bituminous phosphatic shales of the Chouabine Formation in Tunisia shows a strong radioactive response on logs, but this could be due to the presence of the phosphates. Post-Ypresian shales such as the Hallab and Ghalil formations in Libya and the Souar and Salammbo formations in Tunisia may also have some source rock potential, although these rocks are probably immature. There are still many unanswered questions on the source rocks of the northwest Libyan offshore and the release of further data on geochemistry of the offshore source rocks would be of great benefit to the exploration effort in this area.[36]

In the Sirt Embayment analysis of wells in former offshore concession 87 on the An Nuwfaliyah High revealed good source potential within the upper Cretaceous, and it has been suggested that the oil shows in wells A1-10, B1-10, A1-NC 157 and A1-106 on the Waddan Platform may have been derived from upper Cretaceous source rocks offshore (Figure 6.16). The Santonian-Campanian interval yielded TOC values of 1.1% to 1.8%, with predominantly algal kerogen, and potential yield figures of 3250 to 5190 ppm were obtained. The Turonian Etel Formation gave TOC values of 1.4% to 1.5% and potential yield of 3480 to 4310 ppm. Thin organic rich zones were also identified within the Cenomanian Lidam Formation with up to 2.2% TOC and 8050 ppm potential yield. Oil to source correlations show a good match between the oil recovered from well B1-10 with the Santonian-Campanian source rocks in wells A1-87 and C1-87. The upper Cretaceous source rocks are generally immature in the wells studied and the depth to the oil window is about 3000m and to peak generation about 3650m. Such conditions are likely to be present in the offshore extension of the Maradah Trough. These data indicate that the lean character of the Sirt Shale in the northern Sirt Basin changes to a richer kerogen facies in the offshore. This has a significant impact on the prospectivity of the coastal area around the Gulf of Sirt, which is discussed further in chapter 8.[37]

Further east, gas was discovered in Cretaceous reservoirs in wells B1-88 and D1-88 at a depth of about 3350m. These wells are located on the northern flank of the Al Jahamah Platform and were most probably sourced from the Sirt Shale in the Ajdabiya Trough which in this area is likely to contain dominently type III kerogen and which may be in the late mature window. In the offshore area northwest of Binghazi, well A1-NC 120 tested 5263 bopd of 36°API oil from marine uppermost lower Cretaceous and Turonian carbonates at a depth of 2436m. In the absence of any published information it is only possible to speculate on the source of this oil. The well is known to have penetrated a section of Hauterivian-Valanginian age (the Qahash Formation) which was described as varicoloured carbonaceous siltstone and sandy shale containing foraminifera and macrofossil fragments. The well bottomed in platform carbonates of late and middle Jurassic age. The source potential of these sequences has not been disclosed.[38]

Chapter 8

PETROLEUM SYSTEMS

8.1 Introduction

Having reviewed the stratigraphy, structural evolution and petroleum geochemistry of Libya, it is now possible to consider its petroleum systems. The concept of petroleum systems was formulated by Magoon and Dow in 1994. They sought to identify all of the controlling factors which determine the presence of oil and gas in a trap, including recognition of the source rock, the conditions of hydrocarbon generation, the hydrodynamics and timing of hydrocarbon migration, the means of entrapment and any subsequent events which may have affected the accumulation. Demaison and Huizinga expressed these factors in terms of three basic processes: charge, migration and entrapment.

As previously outlined, very large volumes of oil and gas have been discovered in Libya, and a brief overview of Libyan oil reserves (excluding gas) was given in chapter 1. Using available published sources, a realistic figure for total original oil in place appears to be about 135 billion barrels, and total original recoverable reserves about 40.9 billion barrels. This can be broken down by petroleum system (Figure 8.1), by region (Figure 8.2), or by age of reservoir (Figure 8.3). Although these figures are estimates, they are sufficiently reliable to demonstrate the overwhelming importance of the Sirt Basin, which contains 89% of the total oil reserves of Libya, and of the two principal reservoirs in the Sirt Basin, the Palaeocene carbonates and the Nubian sands. The Palaeocene carbonates, which include the Zaltan, Dayfah and Intisar fields, contain one-third of all the oil in Libya, and the Nubian Sands, which include the As Sarir and Messlah fields, hosts a further quarter. It is also worth repeating that although almost 500 exploration wells have tested some amount of hydrocarbon, 85% of the original recoverable reserves are contained in just 21 giant fields. Of the 1500 or so wildcat wells drilled in Libya about 400 have given some indication of hydrocarbons, and about 150 can be regarded as significant discoveries with development potential. This gives an overall success rate of one 1 in 3.75 for hydrocarbon shows, and 1 in 10 for significant discoveries. In areas such as the Murzuq Basin the success rate is considerably higher than the national average. The preeminence of the Sirt Basin is due to two main factors, the quality and thickness (up to 750m) of the effective Sirt Shale source rock, and the late timing of oil generation and migration. Nevertheless, as pointed out in the last chapter, there are considerable areas of the Sirt Basin where the oil generating potential is poor, due to leanness of the source rock or to unsuitable kerogen type.

The other basins are much less prolific. The Ghadamis Basin hosts many small accumulations in both structural and stratigraphic traps, but the thickness of the Silurian hot shales rarely exceeds 100m, and a proportion of the oil was generated before the Hercynian orogeny and has subsequently been lost as a result of trap destruction and leakage. Furthermore, a considerable proportion of Tanzuft source rock in the central Ghadamis Basin is now in the gas generating zone and its oil generating capacity is exhausted. The question arises why no fields

Basin	Petroleum System	Core Area	Approx. orig. reserves (MMB)
Murzuq	Tanzuft - Mamuniyat	Elephant	1000
Ghadamis	Tanzuft - Mamuniyat	Al Hamra	20
	Tanzuft - Akakus/Tadrart	Conc. 66	350
	M/U Devonian - Awaynat Wanin	Al Hamra	550
Western Sirt Basin	Sirt Shale - Palaeocene/Eocene	Zallah, Ghani	750
Central Sirt Basin	Sirt Shale - Palaeocene	Az Zahrah-Al Hufrah	3850
Western Ajdabiya Trough	Sirt Shale - Upper Cretaceous	Wahah	4050
	Sirt Shale - Palaeocene	Nasser, Dayfah	4500
	Sirt Shale - Eocene/Oligocene	Zaltan Platform	100
Eastern Ajdabiya Trough	Sirt Shale - Nubian	Kalanshiyu High	500
	Sirt Shale - Palaeocene	Intisar	3350
	Sirt Shale - Eocene	Antlat	150
Eastern Sirt Embayment	Sirt Shale - Nubian	Sarir	8500
	Sirt Shale - Eocene/Oligocene	Jalu	4500
	Triassic - Amal/Maragh Fm.	Jakharrah	450
	Hybrid	Amal, Nafurah	6000
Offshore	Eocene - Farwah	Bouri	1500
	Other petroleum systems		1770
			40,890

Sources: Refer to text. Reserve figures are modified from Boote, et al., 1998
Campbell, 1991, Thomas, 1995a, b, and Baird, et al., 1996

Figure 8.1   Libya, Principal Petroleum Systems

This chart shows the principal petroleum systems in Libya, as described in chapter 8. There are minor systems which are not shown. Many systems have charged more than one reservoir and several systems contain a mix of oils from different source rocks. The reserve figures are approximations based on published data.

comparable to Hassi Messaoud have been found in the Libyan sector of the Ghadamis Basin. Hassi Messaoud is the largest oilfield in north Africa. It is reservoired in Cambrian quartzitic sandstones equivalent in age to the Hasawnah Formation in Libya. The field owes its existence to a combination of factors. It is located on a prominent ridge on which the Cambrian reservoir immediately underlies the Hercynian unconformity. The reservoir has been heavily weathered which accounts for the fair reservoir properties developed in what is otherwise an impermeable quartzite. Nevertheless the recovery factor is only 21%. Original oil in place has been quoted as 47 billion barrels of which 10.2 billion barrels is recoverable. Oil was generated from basal Silurian shales, mostly to the west of the ridge, and in the absence of any significant traps migrated updip along the Hercynian unconformity and was trapped beneath Mesozoic evaporites. Unfortunately these conditions are unique. There is no analogy in the Libyan sector of the Ghadamis Basin to the Hassi Messaoud Arch and the associated trapping conditions, and no weathering enhancement of the Hasawnah Formation beneath the Hercynian unconformity has yet been documented in Libya.

The success rate in the Murzuq Basin has been high, and several major discoveries were made during the 1980s and 1990s. However the Silurian Tanzuft Formation is not present over considerable parts of the basin, the distribution of the hot shale source rock is discontinuous and patchy, and its thickness does not exceed 45m. Consequently the potential of the Murzuq Basin is limited in comparison to the Sirt Basin. The amount of oil generated from the Tanzuft hot shales in the Murzuq Basin is probably not more than 5% of that generated by the Sirt Shale in the Sirt Basin.

No hydrocarbons have yet been discovered in the Kufrah Basin and very few wells have yet been drilled. Indications from both well evidence and outcrop studies suggest that the Tanzuft Shale develops a more silty character in the Kufrah Basin and that the hot shale horizon is not present. Future drilling may prove the presence of hot shales in areas so far untested, but the deepest part of the basin appears to have been drilled with no indication of hot shales. Consequently, at the present time, there are no viable petroleum systems in the Kufrah Basin.

The first exploratory well in Libya was drilled on the Qasr al Ahrar anticline in the Jabal al Akhdar in 1956. Since then 77 exploratory wells have been drilled in Cyrenaica, and the results have been disappointing. In spite of several shows, the only significant success has been in the Ash Shulaydimah Trough where several discoveries have been made in Palaeocene and Eocene reservoirs at Antlat, and in wells A1, B1 and C1-NC 129. These oils, as discussed in chapter 7, are hybrid oils derived from Cretaceous, Palaeocene and Eocene source rocks. Excluding the Ash Shulaydimah Trough and the offshore area, no effective petroleum system has yet been established in Cyrenaica despite the presence of numerous shows. This is probably due to a combination of factors, including questionable source rocks and ineffective seals. On the Cyrenaica Platform shales within the Devonian represent late-mature potential gas-prone source rocks, and at least one well, on the western margin of the Cyrenaica Platform encountered oil shows within the Tanzuft Formation. Over a large part of the platform, however, the Tanzuft shale is deeply buried and probably post-mature. In the Jabal al Akhdar, oil source potential has been identified in Jurassic and basal Cretaceous shales particularly in the western coastal region around Binghazi which have been modelled as reaching peak generation during the late Cainozoic. Although only minor shows have been reported onshore, these shales may have been the source of the oil tested from the lower Cretaceous in offshore well A1-NC 120.

In the northwestern offshore the Bouri discovery is one of the largest in the Mediterranean. It is reservoired in lower Eocene nummulitic shoals which developed on a salt-supported ridge, and is sourced by Ypresian shales deposited in a more basinal setting. Numerous other discoveries have been made in the area surrounding Bouri, but these discoveries are characterised by high amounts of gas and most of the fields contain gas-condensate or gas with a thin oil rim. The geothermal gradient is high in the Sabratah Basin as a result of Miocene tectonism and volcanic activity, and consequently the depth to the oil window in the Bouri area is only about

Region	Original oil in place MMB	Original recoverable reserves MMB	Number of discoveries	% of total reserves
Sirt Basin	117,000	36,700	345	89.8
Murzuq Basin	5,200	1,600	20	3.9
Offshore	8,200	1,500	25	3.7
Ghadamis Basin	3,850	950	90	2.3
Cyrenaica	520	140	8	0.3
Total	134,770	40,890	488	100

Sources: Compiled from Campbell, 1991, 1997, Gurney, 1996, Thomas, !995a,1995b Boote, et al., 1998, Baird, et al., 1996, Oil & Gas Journal, World Oil US Energy Information Administration, Author's estimates

Figure 8.2   Libya, Total Oil Reserves by Region

These figures are inevitably imprecise since reserve estimation is a subjective process. However using the sources mentioned above, they are believed to represent a fair estimate of Libyan oil reserves (excluding gas). The table shows the total dominance of the Sirt Basin which is partly attributable to the late generation of oil in this area. Almost certainly large volumes of early generated oil from the Murzuq and Ghadamis Basins have been lost as a result of trap destruction. The figures for Cyrenaica include the Ash Shulaydimah discoveries, but not the Amal and Maragh Graben discoveries which are included with the Sirt Basin.

Age	Original oil in place MMB	Original recoverable reserves MMB	Number of discoveries	% of total reserves
Eocene + Oligocene	15,390	5,770	72	14.2
Palaeocene	41,370	13,295	95	32.5
Upper Cretaceous + contiguous Precambrian	30,485	8,164	106	20.0
Lower Cretaceous	28,720	9,229	88	22.5
Palaeozoics + Quartzites	18,805	4,432	127	10.8
Total	134,770	40,890	488	100

Sources: As figure 8.2

Figure 8.3   Libya, Total Oil Reserves, by Reservoir Age

Despite their apparent precision these figures are based on fragmentary data and should be regarded as generalised estimates rather than precise data. Nevertheless they show the dominance of the Palaeocene carbonates and the Lower Cretaceous Nubian Sands which between them account for 55% of all the oil reserves in Libya.

2000m. The gas, particularly in the fields south of Bouri, was probably derived from upper Cretaceous source rocks deep in the Sabratah Basin. Other fields, particularly in Tunisia, are reservoired in formations ranging from Cretaceous to Miocene in age, and sourced by a variety of local source rocks, but these accumulations are minor compared to the lower Eocene Farwah/Metlaoui system. The only other significant discovery in the Libyan offshore is the A1-NC 120 well off Binghazi, drilled by Agip in 1983. Oil is reservoired in early Cretacous marine carbonates and, as mentioned above, is presumed to have been sourced from Neocomian or Jurassic shales.

Seventeen major petroleum systems have been identified in Libya (Figure 8.1). There are many more minor systems, and instances in which several reservoirs have been charged from one source rock. There are other instances where mixing of oils from more than one source rock has taken place as indicated on Figures 7.15 and 8.18. The petroleum systems of Libya will be reviewed by area, starting with the Palaeozoic basins and ending with the offshore.[1]

8.2   Murzuq Basin

Aziz estimated that the Tanzuft hot shale in the Murzuq Basin had the potential to generate, migrate and entrap 40 billion barrels of oil. What the term 'and entrap' implies in this context is not clear. Discoveries to date total about 5200 MMB oil in place and 1600 MMB recoverable. Only one effective petroleum system has been found in the Murzuq Basin to date, but others may remain to be discovered (Figure 8.4). The basin is a cratonic sag basin which developed during the Palaeozoic, was severely affected by mid-Devonian tectonism and subsequently by the Hercynian orogeny during which the Al Qarqaf Arch was formed, and by further subsidence during the Mesozoic/Cainozoic. A conspicuous feature of

the Murzuq Basin is the shifting of depocentres through time, which has influenced both source rock maturity and reservoir distribution. The following review is based principally on the work of Echikh and Sola.[2]

8.2.1 Tanzuft-Mamuniyat Petroleum System

Reservoir

The Mamuniyat Formation in the Murzuq Basin consists of four main facies. In the centre of the basin a quarzititic facies is present. These rocks have poor reservoir quality with porosities of 2 to 7%. The grains show extensive quartz overgrowth as a result of deep-basin diagenesis during the Carboniferous and Mesozoic. In areas of significant tectonic activity such as the Tumarolin, Wadi Zalaylan and Bi'r Tazit wrench zones (Figure 8.4), the quality of the quartzitic reservoir is enhanced by intense fracturing. A quartzitic facies is also locally present at outcrop near Ghat. In wells F1, E1, G1 and H1-NC 58, in the southwestern part of the basin, which was not buried as deeply, a silty facies is developed in which porosities reach 15 to 20%. In the northern part of the basin in the area of the NC 115 wells a shallow marine sandstone facies is present with porosities between 8 and 12%, and reaching 25% in well A1-NC 115, but reservoir quality varies rapidly. The most striking facies, only locally developed in the B-NC 115 field, is known as the 'periglacial' horizon, and comprises coarse-grained to conglomeratic, poorly-cemented sandstones with excellent reservoir characteristics, and porosities up to 20% and permeabilities of 2000 to 3000mD. These facies collectively represent an association of shallow-marine clastics some distance seaward of a continental ice sheet. However, reservoir quality in the Mamuniyat Formation varies rapidly within a few hundred

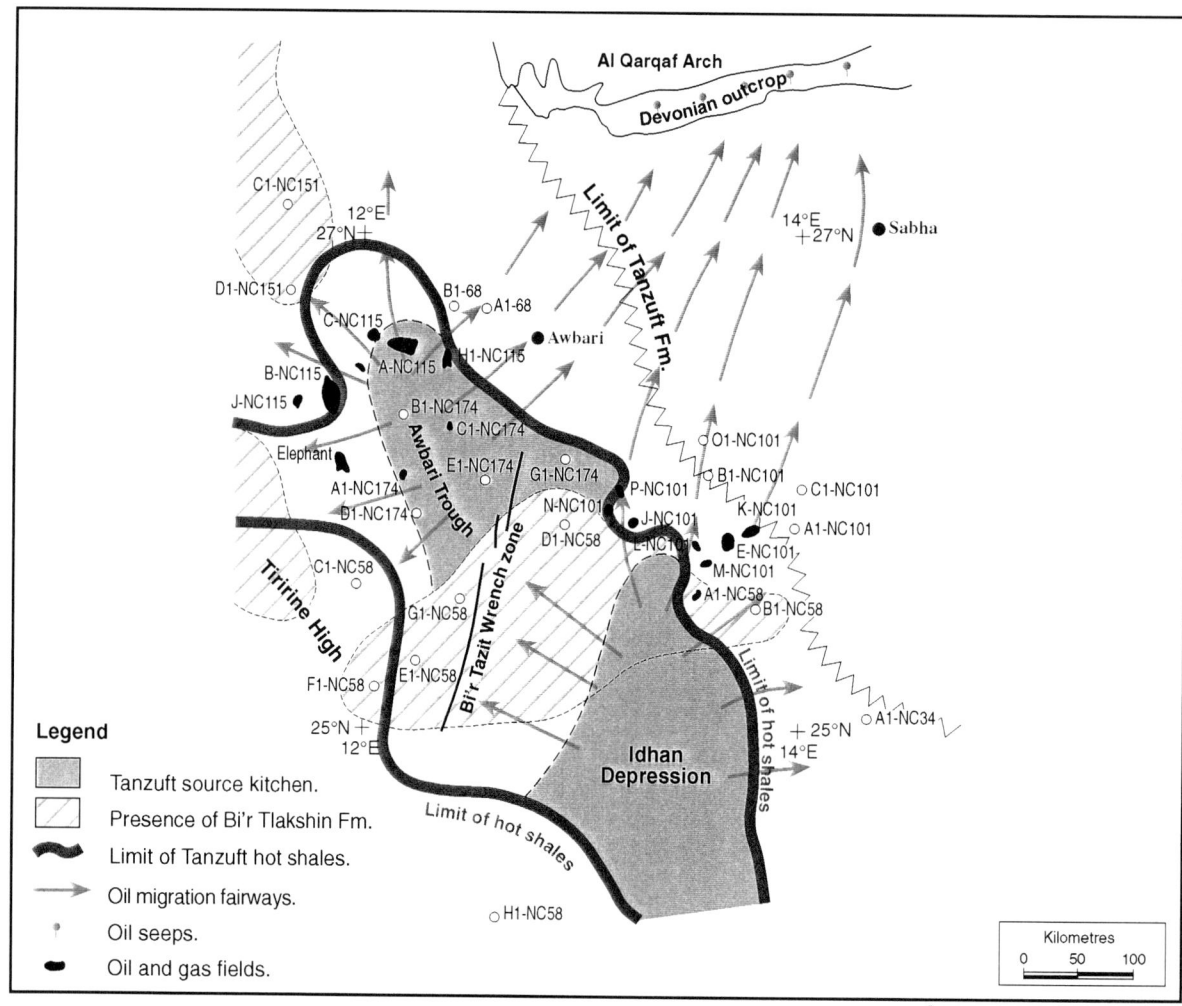

Figure 8.4   Murzuq Basin, Tanzuft-Mamuniyat Petroleum System

The Tanzuft-Mamuniyat petroleum system is the only system so far discovered in the Murzuq Basin. Good quality source shales are present in both the Idhan Depression and Awbari Trough, and both of these areas are mature source kitchens. Hydrocarbon generation began in the Mesozoic but peaked in the Cainozoic. Oil migrated into the subjacent Mamuniyat Formation sandstones and was trapped in updip structures, which are Palaeozoic and Mesozoic anticlines, some of which are faulted, and palaeotopographic features ('buried hills') on the irregular Ordovician surface which are draped by Tanzuft Shales. Oil has not been found where the Tanzuft seal is missing or where the Bi'r Tlakshin Formation intervenes between the source rock and the reservoir.

metres, as demonstrated in the F-NC 115, A-NC 115 and E-NC 101 fields. Meister, et al. suggested a gradual increase in porosity in the upper Mamuniyat from less than 5% in the basin centre to 10 or 15% towards the NC 115 area, and 25% at Ghat. This is probably too

much of a simplification. There is no doubt however that reservoir quality in the NC 101 fields on the northern margin of the Idhan Depression is worse than in the NC 115 and NC 174 areas in the Awbari Trough which are at shallower depths. Davidson et al. presented a diagentic sequence for the Mamuniyat sandstones in concession NC 174 which showed porosity reduction as a result of quartz overgrowths and the growth of fibrous illite, and porosity enhancement caused by dissolution of feldspars. They also found two distinct permeability groupings which they attributed to differences in the original clay content of the reservoir. Permeabilities in the clean sandstones are between 100 and 1000mD, but in the clay-rich sandstones average only 0.1 to 1mD. The Elephant field, discovered in 1997, proved 100m of oil pay with average porosity of 16%. Reserves of 561MMB have been published for this field and a peak production rate of 150,000 bopd is anticipated. The most northerly discovery in the Mamuniyat reservoir of the Murzuq Basin is the B1-NC 186 well drilled by Total in 2001 which tested 1300 barrels of oil per day with a gravity of 40°API.[3]

Source Rock

The geochemistry of the Tanzuft source rock was discussed in chapter 7. The Tanzuft Formation is limited in distribution due to mid-Devonian and post-Hercynian erosion, and the basal hot shale is restricted to topographic depressions on the eroded Ordovician surface. The hot shale is absent on the Tirirene High and over much of the western part of the basin. The maximum thickness of the hot shales is about 45m. Hot shales are also absent in areas where the sand content is high in the southern and eastern parts of the basin, for example in well H1-NC 58. TOC values of 3 to 8% are typical, but can reach 17%, and Figure 7.4 shows the generalised distribution of the hot shales in the Murzuq Basin. In general the thickness and quality of the hot shales deteriorates southwards. The depressions in which the hot shales accumulated and where anoxic conditions prevailed were formed by erosion of the Ordovician surface prior to the Silurian trans-gression. It is noteworthy that whilst amorphous kerogen has been found in wells D1 and A1-NC 58 in the Idhan Depression, much more terrestrial kerogen is present in wells E1, F1 and H1-NC 58 which are located further south. The vitrinite content of the sediments is very low. Despite some questionable maturity data, it is clear that the source shales have reached maturity in the Idhan Depression and this kitchen has sourced the NC 101 fields (Figure 8.4). Depth to the oil window is about 2100m. In the Awbari Trough the high maturity values at relatively shallow depth may be attributable to high heat flow as a result of Tertiary magmatic activity which is widespread in both western Libya and eastern Algeria. Present day geothermal gradient is highest along an axis passing through the NC 115 fields and close to the NC 101 fields. An oil generation plot for well B1-NC 174 (Figure 7.6) shows peak generation in the early Cainozoic but subsequent elevation above the oil window as a result of uplift during the late Cainozoic. Oil gravity on the Elephant field is 38°API. Modelling studies by Repsol, on the other hand, suggested that oil generation began in the Carboniferous/Permian, based on rapid subsidence during the Permian.[4]

Seals

Top seal for the Mamuniyat reservoir is provided by the Bi'r Tlakshin shales, Tanzuft shales or Devonian shales. The Bi'r Tlakshin has a patchy distribution, and its lithology is very variable, but it appears to represent an effective seal in most areas where it is developed. From a petroleum point of view the presence of the Bi'r Tlakshin Formation is a negative factor since it forms a barrier between the source rock and the Mamuniyat reservoir (Figures 3.13, 8.4 and 8.5). The Tanzuft Formation provides an effective seal in the western part of the basin but becomes more sandy to the south and east, reaching a sand

362

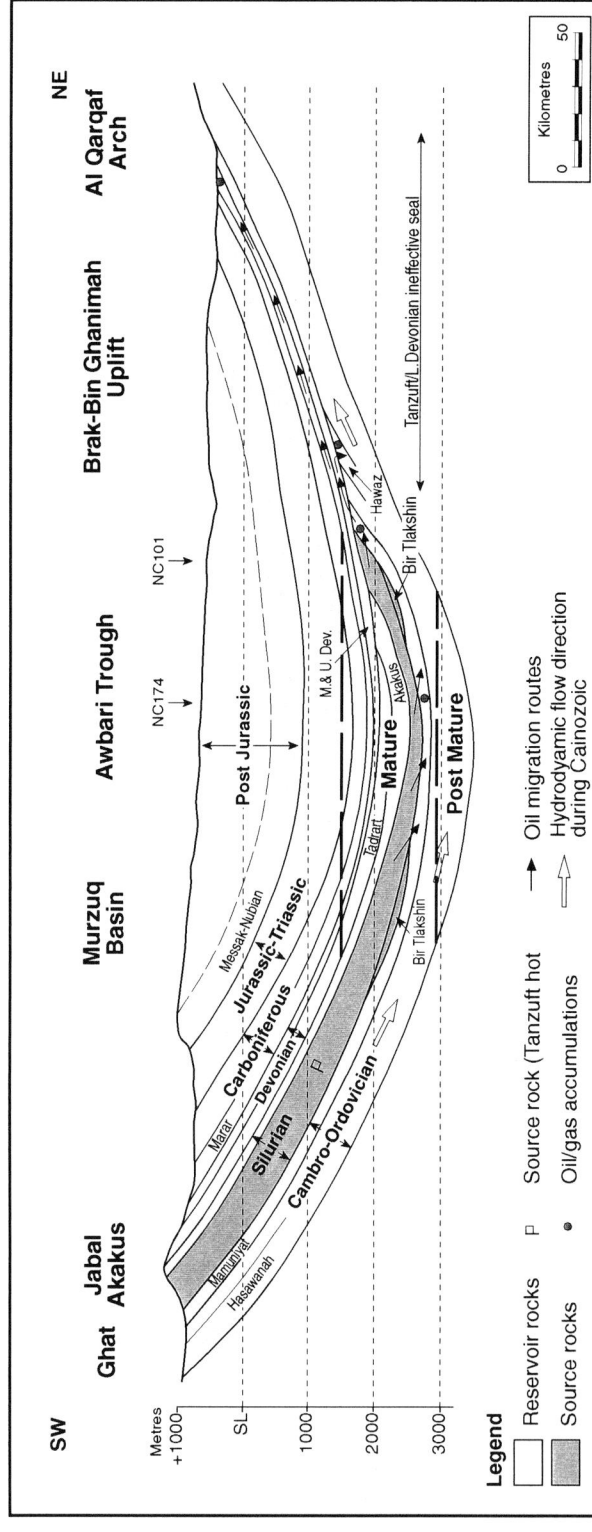

Figure 8.5 Murzuq Basin, Tanzuft-Mamuniyat Petroleum System

Oil was generated from the Tanzuft hot shale source kitchen and migrated into the underlying porous sandstones of the Mamuniyat Formation. The Mamuniyat does not appear to have been charged in areas where the Bi'r Tlakshin Formation is present. East of the NC 101 fields no effective seal is present and oil leaked into the Middle and Upper Devonian sandstones which contain oil seeps on the flank of the Al Qarqaf Arch. The direction of hydrodynamic flow during the Cainozoic was from southwest to northeast.

content of 20%. In these sandy areas the integrity of the Tanzuft Formation as a seal is questionable. In the eastern part of the basin, where the Tanzuft Formation is absent, the Mamuniyat reservoir is overlain by Devonian sands and shales which generally do not provide a good seal. Seeps in Devonian outcrops on the southern flank of the Al Qarqaf Arch suggest that leakage has taken place in this area (Figure 8.5). The NC 101 fields are located very close to the pinchout limit of the Tanzuft Shales and seal in this area is a problem. Faults are frequently associated with leakage, particularly on the Tiririne High which is heavily faulted by Mesozoic faults, and on the Brak-Bin Ghanimah Uplift where the Devonian is faulted. Fault leakage may explain why the D1-NC 174 well was dry. However in areas where reverse faulting is evident, as in the H-NC 115 and Elephant structures, the oil pool is trapped against the fault which is clearly sealing. The major wrench faults also seem to act as barriers to migration and the Bi'r Tazit fault appears to have prevented oil migration from the Idhan Depression onto the Tiririne High (Figure 8.4).[5]

Trap

The main trapping mechanism in the Murzuq Basin is structural, but palaeogeomorphic and stratigraphic traps are also present, and permeability barriers may also form traps in a few cases. The Elephant field (F-NC 174) is a large block faulted anticline with rollover into a steep reverse fault. The NC 101 and NC 58 structures are predominantly small pre-Hercynian anticlines with low relief. Some traps are faulted as in well J1-NC 115. In this case, the fault intersect appears to represent the spill-point and the fault plane is not an effective seal. Reverse faults form the trap in the H-NC 115, C-NC 115 fields, and on the Elephant fields, and in these cases the faults are sealing. Palaeotopographic features of possible glacial origin, draped and

sealed by Tanzuft Shales form excellent traps as in the B-NC 115, E-NC 101 and K-NC 101 structures. Traps formed by permeability barriers are known in the Mamuniyat Formation in Algeria and the rapid variation in reservoir quality as proved on the J-NC 101 structure suggests that similar traps may be present in Libya. In the Devonian, the complex nature of the mid-Devonian unconformity presents stratigraphic trap possibilities (Figure 8.5).[6]

Summary

Tanzuft hot shales with high TOC values are well developed in the Idhan Depression and Awbari Trough. Both these areas are effective source kitchens in which the source rocks are early mature to late mature. Major subsidence during the Carboniferous was followed by Hercynian uplift and by continued subsidence during the Mesozoic. Generation began in the Mesozoic but in the case of the Awbari Trough did not reach peak generation until the Cainozoic. Oil was expelled and migrated into the subjacent Mamuniyat sandstone reservoir. The incised valleys filled with periglacial clastics may have provided additional migration pathways. Hydrocarbons were trapped in a variety of structures including Palaeozoic and Mesozoic anticlines, some affected by reverse faulting, and remnant palaeotopographic structures of possible glacial origin draped by Silurian shales. Subsequent diagenesis may also have produced local traps. Where the B'ir Tlakshin Formation is present, communication between the source rock and the reservoir was effectively prevented and no oil has been found in these areas. Where the Tanzuft Formation is thin or absent, oil has leaked into Devonian sandstones and most of this oil has been lost, as proved by Devonian oil seeps on the flank of the Al Qarqaf Arch. No oil has been found in the southern part of the basin where the Tanzuft Formation is sandy and the hot shales are not developed. The Bi'r Tazit wrench fault appears to have prevented migration from the Idhan Depression onto the adjacent Tiririne High (Figure 8.4).

Nine commercial and six sub-commercial oil discoveries have been made to date, including the giant Elephant field in concession NC 174, three large fields in concession NC 115 (A, B and H), plus smaller discoveries in concession NC 101. Oil gravity is generally in the range of 34 to 47°API. Meister, et al. claimed to recognise a trend from medium gravity oils deeper in the basin to lighter gravity oils and condensate on the basin margin. Gas-oil ratios within the basin are generally very low. The NC 101 fields are characterised by fair reservoir quality with net pay thicknesses of 15 to 45m, and well productivity of 400 to 500 BOPD. Only the E1-NC 101 field has rather better characteristics with a potential flow rate of 1260 BOPD. By contrast the NC 115 fields have net pay thicknesses of 30 to 40m with wells which can flow at rates up to 3150 BOPD. The Elephant field tested oil at a rate of 8400 BOPD from a net pay section of 100m. The fields discovered to date lie in two groups. The NC 115 and NC 174 fields are located around the northern margin of the Awbari Trough at a depth of about 1500 to 2000m and the NC 101 fields are located around the northern margin of the Idhan Depression at a depth of about 2500 to 3000m. These areas share three key characteristics: hot shales are present in mature kitchens close by, the Bi'r Tlakshin Formation is absent allowing ease of communication between the source rock and the reservoir, and presence of the Tanzuft shales as a top seal. There is no evidence of long range migration within the Mamuniyat reservoir as proposed by Meister, et al. The presence of fresh water in the Mamuniyat reservoir is indicative of flushing, and although the system is probably inactive at present, it might have resulted in the removal of gas and light hydrocarbons in the past.[7]

8.2.2 Other Reservoirs and Source Rocks

The Hasawnah Formation has modest reservoir quality due to the presence of diagenetic clays and quartz overgrowths. On the A-NC 115 field, porosity ranges from 11 to 13% with 2 to 750mD permeability, but these characteristics are exceptional. The Hawaz Formation represents a secondary target in the Murzuq Basin and thin oil zones have been found in one or two wells, particularly in situations where the Mamuniyat Formation is absent and the Hawaz is in contact with the Tanzuft source rock. Porosity in the lower part of the formation reaches 16% with permeability as high as 900 mD. Repsol discovered oil in the Hawaz Formation in well A1-NC 186, north of the NC 115 fields, with a flow rate of 2500 BOPD, but generally reservoir quality is inferior to the Mamuniyat. The Melaz Shuqran Formation may have source potential, but details of the thickness, quality and kerogen facies have not been published. Radioactive shales have also been found within the Mamuniyat sequence, but the characteristics of these shales have not been evaluated.

The Devonian sandstones generally have good reservoir characteristics with porosities typically between 12 and 23%. Oil has been tested from Devonian sandstones in wells A1-NC 58, P1-NC 101 and C1-NC 115, and shows have been encountered in many other wells, but flow rates rarely exceed 300 BOPD. This oil was derived from the Tanzuft Shales by leakage into the Devonian (Figure 8.5). TOC values of 2 to 4% have been found in Devonian shales in wells A1-NC 58 and A1-NC 34, but these shales are immature. Similarly in concession NC 115, Carboniferous shales have been reported as having high TOC values. At Atshan, on the saddle between the Murzuq and Ghadamis Basins, gas, light oil and condensate have been found in Carboniferous sandstones with net pay thicknesses from 1.5 to 15m, and porosities in the range 15-25%. The provenance of these hydrocarbons has not yet been established, although it seems most likely that they originated by long range migration from the Ghadamis Basin.[8]

8.3   Ghadamis Basin

The Ghadamis Basin extends into Algeria, Tunisia and Libya and includes the Illizi Basin in Algeria. Daniels and Emme calculated that 2.1 trillion barrels of oil-equivalent have been generated from the Algerian sector of the Ghadamis Basin (1.4 trillion Tanzuft and 0.7 trillion Devonian). A pro-rated figure for Libya would be in the order of 1.22 trillion barrels. The Ghadamis Basin is unusual in containing relatively modest amounts of hydrocarbons in relation to the amount of hydrocarbons generated. This is largely due to the timing of hydrocarbon generation, and to extensive flushing during the late Cainozoic. Generation began earlier in the Ghadamis Basin than in the Murzuq Basin due to deeper burial during the Palaeozoic and Mesozoic. It is unlikely that any of the hydrocarbons generated during the period before the Hercynian orogeny have survived, and even that generated during the Jurassic and early Cretaceous has probably suffered from extensive leakage. By the end of the Mesozoic the Tanzuft source rock was already late mature in the centre of the basin, and it is likely that most of the hydrocarbons that have survived were generated from the peripheral mature zone during the Cainozoic.[9]

Approximately 9.6 billion barrels of oil in place have been discovered in the basin, (less than 1% of the estimated amount generated) of which 5.7 billion barrels are in Algeria, 3.3 billion barrels in Libya and 0.6 billion barrels in Tunisia (Figure 8.6). Three principal petroleum systems have been identified, the Tanzuft-Mamuniyat, Tanzuft-Akakus-Tadrart,

and the Middle/Upper Devonian-Awaynat Wanin. There are also a number of sub-systems, such as the long range migration of oil into the Triassic reservoir in the Ghadamis area (Figure 8.7). The following discussion is concerned principally with the Libyan part of the basin.

No significant oil shows have yet been found in the Jifarah Basin to the north of the Nafusah-Dahar Uplift, although minor shows have been reported in the Bi'r al Jaja, Al Guidr and Kurrush Formations. Nevertheless the presence of Silurian and Middle Triassic black shales, in association with Triassic clastics suggests that this area is not totally devoid of potential.[10]

8.3.1   Tanzuft-Mamuniyat Petroleum System

In the Murzuq Basin the Tanzuft-Mamuniyat system is the only significant petroleum system, and it is important in the Illizi Basin of Algeria where over 1.2 billion barrels of oil in place is reservoired within the Mamuniyat reservoir, most notably on the Tin Fouyé-Tabankort field. The system generally has a high gas content. In the Libyan sector of the Ghadamis Basin however, relatively few wells have reached the Mamuniyat Formation, and only about 100 million barrels of oil-in-place have so far been found in the Mamuniyat reservoir, in three small pools on the Al Hamra High. In this case, oil generated from the Tanzuft hot shales has migrated laterally and updip into the subjacent Mamuniyat reservoir. The deltaic and periglacial sands which characterise the Mamuniyat Formation in the Murzuq Basin and Al Qarqaf Arch areas pass northwards into an open marine facies with carbonates and an increase in shale content, and in the northern part of the basin the sand is totally missing, due either to shale-out or to pre-Silurian erosion. The Ordovician play is therefore largely confined to the southern part of the basin.

Period	Original oil in place MMB	Original recoverable reserves MMB	Number of discoveries	% of total reserves
Triassic	68	24	4	2
Carboniferous			1	
Upper Devonian	602	180	12	20
Lower Devonian	1,172	346	30	36
Silurian	1,405	378	38	40
Ordovician	100	22	5	2
Total	3,347	950	90	100

Sources: Echikh, 1998, Boote, et al., 1998

Figure 8.6   Ghadamis Basin, Distribution of Reserves by Reservoir Age

The oil in place figures have been calculated from data provided by Echikh, and the original recoverable reserves have been adjusted to the estimate of Boote, et al., 1998. The number of discoveries are wildcat wells in which some oil was tested. The figures are somewhat less than those estimated in Figure 8.1. Echikh provided no figure for the Carboniferous, although the Atshan field is known to contain oil and gas in Carboniferous reservoirs.

Reservoir quality in the Mamuniyat sands decreases with depth, but is enhanced in certain areas by faulting and fracturing.[11]

8.3.2   Tanzuft-Akakus-Tadrart Petroleum System

Reservoirs

The Akakus Formation is well developed in the Ghadamis Basin, where it has been divided into lower, middle and upper units. It has been affected by widespread post-Silurian uplift and erosion around the basin margin, leaving progressively older Akakus sands subcropping the Devonian from the basin centre towards the margin. The Lower Akakus is best developed on the southern flank of the basin, north of the Al Qarqaf Arch where sand content is in the 60 to 80% range (Figure 8.7). Sand content decreases steadily to the north and in the Tiji area it is

around 20%. Porosities average 20 to 25% in the south declining to 12 to 15 % in the north. Permeability values are typically around 100mD. The Lower Akakus is the main producing unit within the Akakus Formation in Libya. In general terms the F6 reservoir (Figure 3.15), which is the single most important producing reservoir in the Algerian part of the basin, is equivalent to the combined Akakus-Tadrart sequence in Libya. Hydrodynamic activity has exerted a major influence on hydrocarbon distribution in the F6 reservoir in Algeria and water flushing is a significant problem. In the Tin Fouyé-Tabancort area some traps are attributed solely to hydrodynamic causes. The F5 reservoir in Algeria (Figure 3.15) is equivalent to the Wan Kasa Formation in Libya.[12]

Two reservoir units are present within the Lower Devonian, the Tadrart and Wan Kasa Formations, although the Wan Kasa is of only minor importance. The Tadrart has a wide distri-

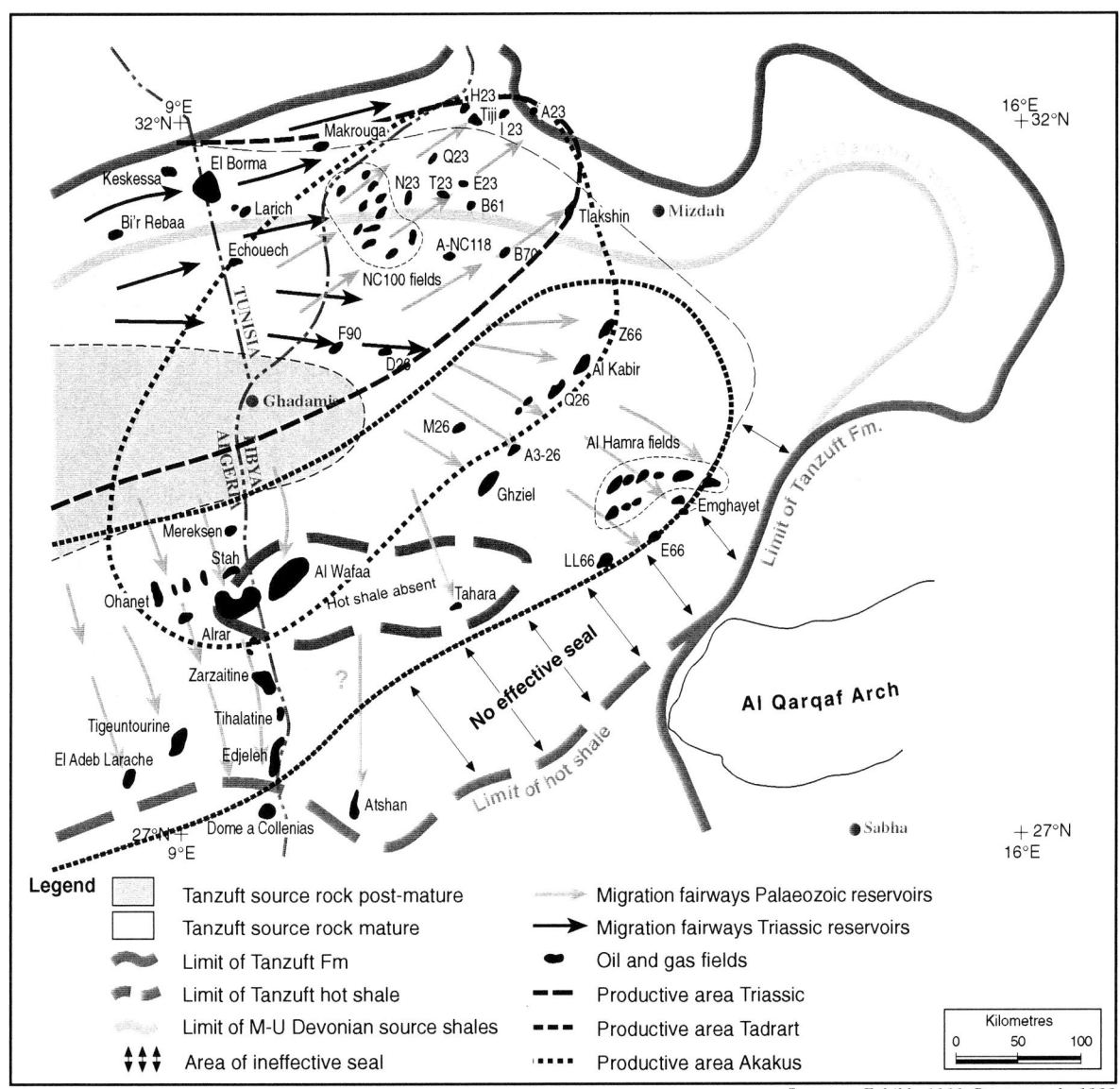

Figure 8.7   Ghadamis Basin, Petroleum Systems Summary Map

The Tanzuft source rock is post-mature in the centre of the basin, and most of the discovered hydrocarbons were generated from mature Tanzuft hot shales on the basin flanks during the Cainozoic. The Tanzuft-Mamuniyat petroleum system is under-explored in the Libyan sector of the Ghadamis Basin and has been penetrated by relatively few wells. It is overshadowed by the Tanzuft-Akakus-Tadrart system which contains 60% of the all the oil found in the basin. There are two main oil fairways: northeastwards towards the NC 100 and Tiji group of fields, and southeastwards towards the Al Kabir and Al Hamra group of fields. The Frasnian source rock charged the Alrar and Al Wafaa fields. Most of the oil found in Triassic reservoirs in the El Borma-Echouech-F 90 group of fields was derived by long range migration from the Devonian source rock with some contribution from the Tanzuft. Seals within the Akakus and Lower Devonian become ineffective towards the south and oil has leaked into Upper Devonian and Carboniferous reservoirs in these areas.

bution over the basin and reservoir characteristics are related to depth of burial, porosity varying from 15 to 20% in the south, 8 to 10% in the centre of the basin and 11 to 14% on the northern limb (Figure 8.7). Three diagenetic processes have been identified as affecting the reservoir quality of the Tadrart Formation. Quartz overgrowths occur in Tadrart sandstones with low chlorite content. Where chlorite is abundant however, secondary silicification is inhibited. The Tadrart sandstones are frequently cemented by carbonate, and porosity can be enhanced where the carbonate cement has been subsequently leached out. The presence of illite generally leads to reduction of permeability by the blocking of pore throats. Shah et al. attempted to explain porosity variations within the Silurian-Devonian reservoirs in terms of synsedimentary processes which winnowed out clay minerals from the more exposed parts of the reservoir sands prior to consolidation. The Wan Kasa Formation, which is equivalent to the F5 reservoir in Algeria, is more variable in character and shows significant facies changes. It was also more affected by the mid-Devonian tectonism and subsequent erosion. The best reservoir characteristics are found on the southern flank of the basin where the lower Wan Kasa shows a good development of nearshore and deltaic sands. Further north it passes into a more calcareous and shaly facies. In Algeria the F5 reservoir is productive on the El Abed Larache, Tiguentourine, Ouan Taradet and Tan Emellel fields (Figure 8.7). The upper Wan Kasa has little or no reservoir potential.[13]

Source Rock and Source Kitchens

The geochemistry of the Tanzuft source rock was discussed in chapter 7. The hot shales are generally thicker than in the Murzuq Basin and, due to deeper burial, exhibit a higher degree of maturity. A map showing the thickness of the Tanzuft hot shales is shown as Figure 7.7. The

basic characteristics are as follows. The hot shales average 35m to 45m in thickness. They are best developed in the Ghadamis and Zamzam areas and in a southern trough around Zarzaitine. Kerogen is principally type II and TOC values average 4 to 8% but exceptionally reach 17%. In the basin centre at Ghadamis the Tanzuft hot shales are post-mature, but the degree of maturity decreases towards the basin margin and on the Al Hamra ridge it is early mature. Generation began in the basin centre during the Carboniferous but was probably halted by the Hercynian uplift and did not resume again until the late Jurassic. Peak generation occurred during the Cretaceous. The oils produced from the Tanzuft hot shales are naphtheno-paraffinic and have a relatively high gas-oil ratio as a result of the high level of maturity. The Ghadamis Basin and its extension into the Zamzam Depression represents a large source kitchen, although the central part is post-mature and no longer has any oil generating capacity (Figure 8.7).[14]

Seals

Top seal for the Lower Akakus sandstone reservoir is provided by shales within the Akakus which are persistent over a large area in the north-central part of the basin. The Akakus is productive from the Al Kabir trend in the south to Tiji in the north (Figure 8.7). Towards the basin margin however the sealing shales are progressively removed by truncation at the post-Silurian unconformity, and in this area oil has leaked into the overlying Devonian (Figure 8.8). This is particularly the case in Algeria. The Tadrart reservoir is sealed by the shales of the Wan Kasa Formation and the overlying Emghayet Shale over a wide area from the Al Kabir and Al Hamra fields westwards into Algeria. Further south these formations develop a more sandy facies and become ineffective as seals (Figure 8.7). The Hercynian unconformity

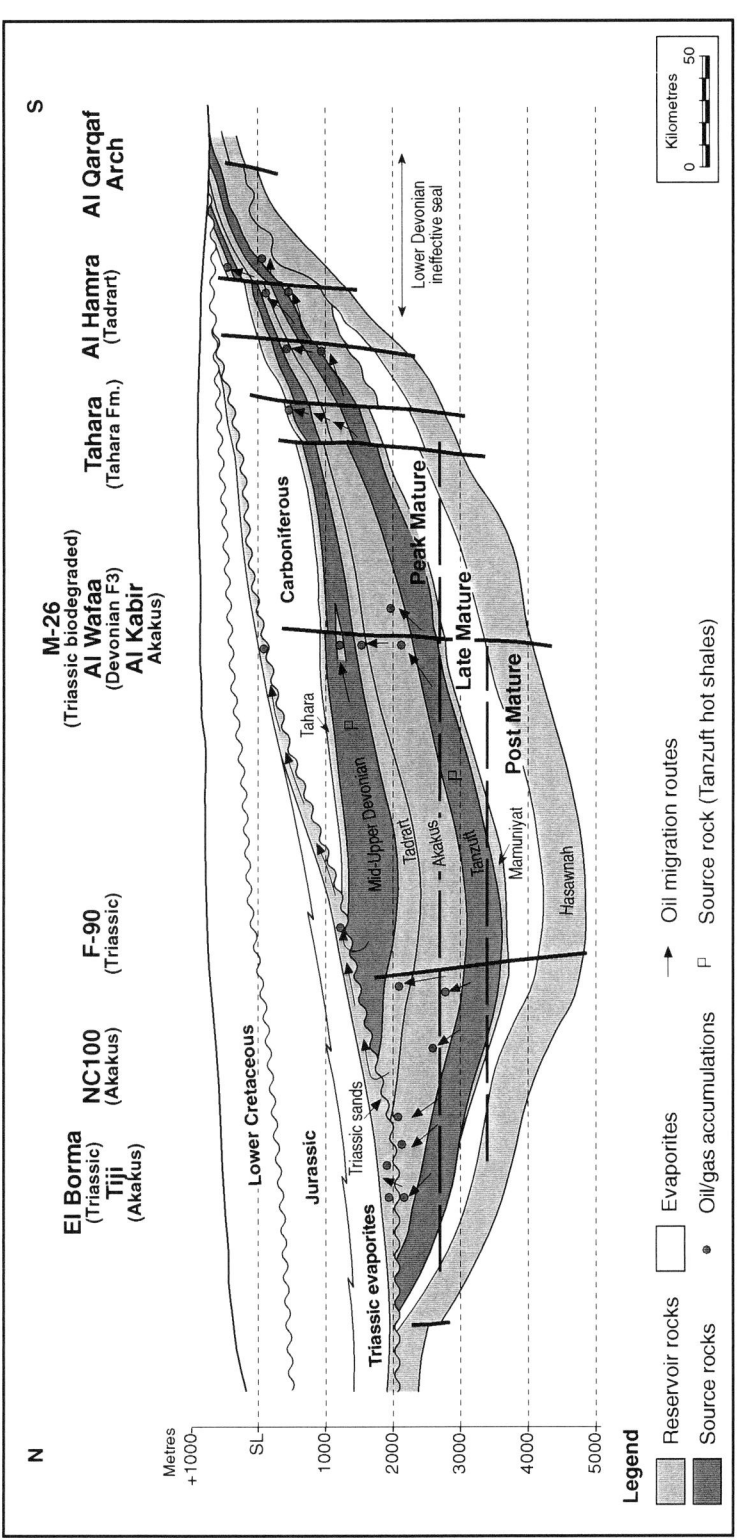

Figure 8.8   Ghadamis Basin, Petroleum Systems Summary

Source: Echikh, 1998

Source rocks are present in the Tanzuft Formation and in the Middle-Upper Devonian (primarily Frasnian). In the centre of the basin the Tanzuft Formation is post-mature. Oil and gas were generated from the basin flanks during the Mesozoic and Cainozoic and migrated into the overlying Akakus Formation, and via faults into the Tadrart Formation. Where the Lower Devonian seal is ineffective hydrocarbons leaked into the Tahara Formation and into the Carboniferous. Some oil was trapped beneath the Hercynian unconformity, but leakage into Triassic sandstones also occurred. The Devonian source rock charged the Alrar and Al Wafaa fields, and leaked into Triassic sands at the Hercynian unconformity, where it was trapped beneath an evaporite seal.

locally forms a seal where overlain by Triassic shales and fields such as Tiji are stratigraphically trapped beneath the unconformity. However, where Triassic sands overlie the unconformity major leakage has occurred, and the sands form excellent carrier beds and reservoirs. It is notable that Akakus and Tadrart oil pools are located at some distance from the unconformity (Figure 8.8). The basin is extensively faulted as a result of Devonian, Hercynian and Eocene tectonism and leakage along faults has undoubtedly occurred, particularly in heavily faulted areas. However there are numerous examples where both normal and reverse faults are known to be sealing and form an essential part of the trap. The integrity of faults as seals depends largely on the intensity of faulting and on the cross-fault juxtaposition of permeable and impermeable beds.[15]

Traps

The main trap type in the Ghadamis Basin is structural, although stratigraphic traps are also present, particularly beneath the Hercynian unconformity. Echikh recognised four types of structural traps. The L-NC 100 field in Libya and the Stah field in Algeria represent the most basic type of broad, low-relief, unfaulted anticlines within the Silurian and Devonian. Traps formed by normal faults generated during the Hercynian orogeny and which may represent reactivated basement faults are found on the J-NC 100 and T-NC 7 fields in Libya, and at Zarzaitine and Tin Fouyé in Algeria. In these cases the faults are sealing and the traps are fault-dependent. Reverse faulted traps have been described by Hammuda from the Al Hamra fields in Libya which were formed by compressional events during the Eocene, and again the faults are sealing. Hammuda also presented a Top Tadrart structure map of the southern Ghadamis Basin. The most typical stratigraphic trap is in porous sandstones trapped by shales beneath the Hercynian unconformity. The Tiji field on the northern limb of the basin is a good example in which the Akakus sandstone subcrops Permo-Triassic shales. Updip shale-out contributes to the trap on the Al Kabir field, and traps caused by facies changes within the Akakus reservoir occur on the G-NC 100 and Z-NC 100 fields.[16]

Summary

The Tanzuft-Akakus-Tadrart petroleum system is the most important system within the Ghadamis Basin containing 59% of all the oil found in the basin across the three countries. Oil was generated from the Tanzuft hot shale source rock beginning in the Carboniferous and continuing into the Cainozoic. Much of the early-generated hydrocarbons have been lost by leakage and trap destruction. The source rock was deeply buried during the Carboniferous and Mesozoic and is now post-mature in the basin centre. The source rocks buried deep in the basin have probably been in the gas window during much of the Cainozoic, and most of the oil has probably been generated from source rocks on the flanks of the basin. The depth distribution of fields in the basin shows the majority of fields to be in the range between 2000 to 2500m.[17]

Migration took place into a variety of structural and stratigraphic traps in reservoirs of late Silurian and early Devonian age along conduits within a generally deltaic sequence of rocks. The major seals are provided by shales of Emsian and Eifelian age, but in the south where these seals become ineffective, oil has leaked into Middle and Upper Devonian reservoirs. In many areas faults provide an additional migration path. On the northern rim of the basin, the Silurian-Devonian sequence is truncated by the Hercynian unconformity and in this region oil is stratigraphically trapped beneath the unconformity by overlying Permo-Triassic shales. During the late Cainozoic water flushing took place in the south of the basin by water entering the basin from the Al Qarqaf Arch, and the more permeable formations, particularly in the Lower Devonian have been partially flushed. Flushing has also been noticed in Triassic reservoirs in the F-90 field area and on the Al Kabir trend.[18]

Three major groups of fields are present in the Libyan part of the basin (Figure 8.7). The northern group including the Tiji, Tlakshin, NC 2 and NC 100 fields is reservoired in Akakus and Tadrart sands, and is trapped in small structural and stratigraphic traps, by migration from the southwest. The size of these fields may have been considerably reduced by flushing. This group of fields contains reserves of about 65 MMB oil equivalent. A second group of fields is aligned along a southwest-northeast trend extending from Ghziel to Al Kabir. These fields are reservoired in Akakus and Tadrart sands with some leakage into Upper Devonian and Carboniferous reservoirs, and were charged from the west. Hydrodynamic flushing has also affected this area. These fields contain reserves of about 250 MMB oil equivalent. The third group, the Al Hamra fields, are mostly reservoired in Devonian sands, since the Wan Kasa and Emghayet shales are not effective seals in this area, and were charged by relatively long distance migration from the northwest. It is likely that much oil has been lost in this area due to ineffective seals. Reserves in the Al Hamra fields are about 635MMB oil equivalent.[19]

8.3.3 Middle/Upper Devonian-Awaynat Wanin Petroleum System

A second petroleum system in the Ghadamis Basin occurs within the Middle and Upper Devonian (Figure 8.7). Several shales have source potential in this interval including the Emghayet shale (Eifelian) with TOC of about 1% and the Frasnian radioactive calcareous shales, which are known as the Cues Limestone in Libya, with TOC values up to 10%. Oil derived from the middle and upper Devonian source rocks is reservoired for the most part in sandstones of the Awaynat Wanin Formation, but has also leaked into younger formations (Figure 8.8). The geochemistry of the middle and upper Devonian source rock was discussed in chapter 7. In summary the Devonian source rock is less extensive than the Tanzuft and is less mature. TOC values average 2 to 4% and kerogen is type I and II. Maturity ranges from late mature in the basin centre to immature on the margin. Peak generation occurred during the Eocene. The source rock is high in paraffins and low in aromatics.[20]

Reservoir rocks within the Awaynat Wanin Formation are more variable than in the Lower Devonian. The individual sands are generally thin, locally restricted and reservoir quality is usually mediocre, except close to the source of sand input. The lowest sand, of Eifelian age, is best developed in Algeria where it is known as the F4 reservoir, and is productive at Zarzaitine, Edjeleh and Tiguentourine (Figure 8.7). This unit is not well developed in Libya. A second reservoir horizon in the Eifelian-Givetian sequence, known as the F3 reservoir in Algeria, extends into Libyan territory and contains large gas-condensate accumulations in stratigraphic traps at Alrar in Algeria and Al Wafaa in Libya. The sand pinches out updip and is sealed by the overlying Frasnian shales around a broad embayment. The fields have a large gas cap and a thin oil leg. The Tahara sand of Famennian age is equivalent to the F2 reservoir in Algeria. It has a widespread distribution over much of the basin and is gas bearing at Zarzaitine and Edjeleh, and oil bearing in the Tahara field in Libya. At Alrar the Frasnian source rock occurs 10 or 15m above the top of the F3 sand but possibly downdip the two units are in direct contact. The Tahara reservoir was charged by migration through the Awaynat Wanin sands and the oil was trapped by shales at the Devonian-Carboniferous boundary.[21]

8.3.4 Other Reservoirs and Source Rocks

Thin reservoirs have been found in Carboniferous sandstones, the M reservoirs in Libya and the D and B reservoirs in Algeria (Figure 3.16), but these units are shallow and generally insignificant. The first discovery ever made in Libya, on the Atshan structure, on the saddle between the Ghadamis and Murzuq Basin, contains gas and condensate in several thin Carboniferous sands. These sands were probably charged from Tanzuft source rocks in the Ghadamis Basin by migration through the Akakus and Devonian sandstones in the area where these units contain no effective seals. The Carboniferous also contains shales with TOC values of around 1.75%. These shales are mostly immature, but could possible reach maturity in the basin centre. Shales within the Tahara Formation contain herbaceous kerogen and have oil generating potential, but the shales are thin and immature over most of the region.[22]

Triassic sandstones form major reservoirs in Algeria and Tunisia where they are sealed by overlying evaporites. In many cases the fields developed over palaeotopographic highs on the eroded Palaeozoic surface. Several large fields are on production from Triassic reservoirs including El Borma, Bi'r Berkine, Rhourde Nouss and Nezla. Southern Tunisia is relatively under-explored compared with Algeria, and considerable potential remains in this area. Reservoir quality and lateral continuity is generally good since the sands represent the meander belt of a fluvial system. The best reservoirs are in point bars and braided fluvial sands, and quality deteriorates in the flood plain facies. This reservoir extends into Libyan territory to the north of Ghadamis and a number of small discoveries have been made including the F-90, D-26, A-23 and the biodegraded M-26 pool (Figure 8.7). Both the reservoir and the evaporite seal pinch out rapidly to the east. The charging of the Triassic oil pools is not a simple matter. It is likely that oil migrated from truncated Silurian and Devonian source rocks updip along the Hercynian unconformity and then into the overlying Triassic sandstones. Both source rocks however have been in the late or post-mature window in the centre of the basin since the Cretaceous, so it is likely that the oils reservoired in the Triassic were derived from the basin margin where the source rocks are still at peak maturity (Figure 8.8). Geochemical evidence suggests that the El Borma field was sourced from the Frasnian with perhaps a minor contribution from the Tanzuft source rock. Yet fields such as Keskessa (Algeria), Makhrouga (Tunisia) and A-23 (Libya) are located much closer to the Tanzuft subcrop. On the other hand a Devonian source would be more logical for the updip fields such as Rhourde Fekroun (Algeria), Echouech (Tunisia) and F-90 (Libya) since this would involve a shorter migration route. Further analytical work is required to solve this problem. There are no Triassic fields downdip of the Tanzuft subcrop.[23]

8.4 Sirt Basin

The Sirt Basin contains 89% of all the hydrocarbons discovered in Libya. This is primarily due to three factors, the Mesozoic-Cainozoic age of the basin, the presence of a rich and prolific source rock in the Upper Cretaceous Sirt Shale, and the late age of oil generation and migration (mostly Cainozoic, and much of it late Cainozoic). In a review published in 1980 Parsons et al. categorised the hydrocarbon discoveries at that time according to reservoir age, trap type, reservoir type, age of top seal, reservoir depth and reservoir temperature (Figure 8.9). Whilst many discoveries have been

SIRT BASIN, CLASSIFICATION OF HYDROCARBON DISCOVERIES, 1980.					
By age of top seal	By reservoir age	By trap type	By reservoir type	By reservoir depth	By reservoir temperature ºC
Oligocene 8.4%	Oligocene-Eocene 8.4%	Structural 83.7%	Carbonate 42.2%	0 - 600m -	10 - 38ºC 14.3%
U. Palaeocene 20.7%	Palaeocene 33.8%	Stratigraphic 16.3%	Clastic 57.8%	600 - 1200m 14%	38 - 66ºC 23%
L. Palaeocene 27.1%	U.Cretaceous/Palaeozoic 28.3%			1200 - 1800m 25%	66 - 93ºC 53%
Cretaceous 43.8%				1800 - 2400m 6%	93 - 121ºC 8%
	Cretaceous sandstones 25.5%			2400 - 3200m 49%	121 - 149ºC 2%
				Below 3200m 6%	

Source: Parsons, et al., 1980

Figure 8.9   Sirt Basin, Classification of Hydrocarbon Discoveries

In 1980 Parsons, et al., classified the hydrocarbon discoveries of the Sirt Basin according to age of top seal, age of reservoir, trap type, reservoir type, depth and reservoir temperature. Although numerous discoveries have been made since then the percentages are still substantially correct.

made since then, the data is still generally valid. They showed that the Lower Cretaceous (Nubian) Sandstone is the most prolific reservoir, followed by Palaeocene carbonates and then Upper Cretaceous clastics. More than 80% of the oil is found in structural traps, and almost 50% of the fields are in the depth range 2400 to 3200m and a temperature range of 66 to 93°C. A more recent review by Baird et al. classified the known reserves by reservoir and largely confirmed the earlier findings, but provided much more detailed information on individual reservoirs (Figure 8.10). The Sirt Basin can be characterised by four main features: excellent reservoirs in both clastics and carbonates, major shale and evaporite seals throughout the succession, an abundance of structures mostly related to the tensional regime which dominated the history of the basin, and a thick and mature oil prone source rock.[24]

No estimates have been published of the amount of oil generated in the Sirt Basin, but even assuming a 2% trapping efficiency (which is high) it must amount to at least 5.8 trillion barrels. This would imply a generating capacity of about 0.25 barrels per m³ of source rock with more than 1% TOC. To date approximately 117 billion barrels of oil in place have been discovered in the Sirt Basin of which 93 billion barrels are contained in nineteen giant fields. Although the Sirt Shale is by far the most important source rock in the basin, several other sequences also with source potential have been recognised, including the Triassic, Nubian and Turonian shales, as discussed in chapter 7. The Sirt Basin petroleum systems will be discussed in five areas: the western area centred on the Zallah Trough, the Maradah Trough, western Ajdabiyah Trough, eastern Ajdabiya Trough and eastern Sirt embayment.[25]

Numerous authors have documented the progressive subsidence of the Sirt Basin during the Cainozoic, and Roohi showed that the major period of subsidence and eastward tilting occurred after mid-Eocene times. Geochemical modelling in the Sirt Basin shows peak generation starting during the late Eocene-Oligocene with migration during the late

Age	Reservoirs	Reserves MMB	%
Oligocene, Rupelian	Arida	1,468	4
Eocene, Lutetian	Jalu	1,834	5
Eocene, Ypresian	Al Jir, Facha	734	2
Palaeocene, Thanetian	Zaltan, U. Sabil, Harash	6,239	17
Palaeocene, L. Thanetian	Az Zahrah	1,101	3
Palaeocene, Montian	Al Bayda	734	2
Palaeocene, Danian	Dayfah, U. Satal	5,140	14
U. Cret., Maastrichtian	Wahah, Kalash	4,038	11
U. Cret., Camp.-Con. + Precambrian	Taqrifat + Basement	2,203	6
U. Cret. Cenomanian	Bahi, Maragh	1,834	5
Lower Cretaceous	Upper Nubian	2,203	6
Lower Cretaceous	Lower and Middle Nubian	6,973	19
Mesozoic/Palaeozoic	Quartzites	1,834	5
	Minor reservoirs	365	1
Total		36,700	100

Source: Baird, et al., 1996

Figure 8.10   Sirt Basin, Distribution of Reserves, by Reservoir

This chart shows the distribution of oil reserves by reservoir. It assumes total original reserves of 36.7 billion barrels for the Sirt Basin, excluding gas (see text). The percentage breakdown by reservoir is from Baird, et al., 1996.

Oligocene and Miocene. It is evident that due to the eastward tilting of the basin oil migration from the source kitchens to the traps in the western Sirt Basin has been principally from northeast to southwest. Hydrocarbons frequently migrated onto the adjacent platforms. Indeed much more oil has been found on the platforms than in the troughs. Migration onto the platforms was achieved by flowage up fault conduits on the eastern boundary fault of the platforms, and into porous carrier beds. Westward migration then occurred, until a trapping configuration was encountered, and

frequently hydrocarbons migrated as far as the western margin of the platform. The north-central area contains several gas fields which were sourced from gas-prone source kitchens in the Ajdabiya and Maradah troughs.[26]

Western Sirt Basin

The western Sirt Basin comprises the tectonic elements of the Dur al Abd Trough, Zallah Trough, Al Kotlah Graben and Abu Tumayam Trough (Figure 8.11). The Hun Graben is not included since the Sirt Shale is not

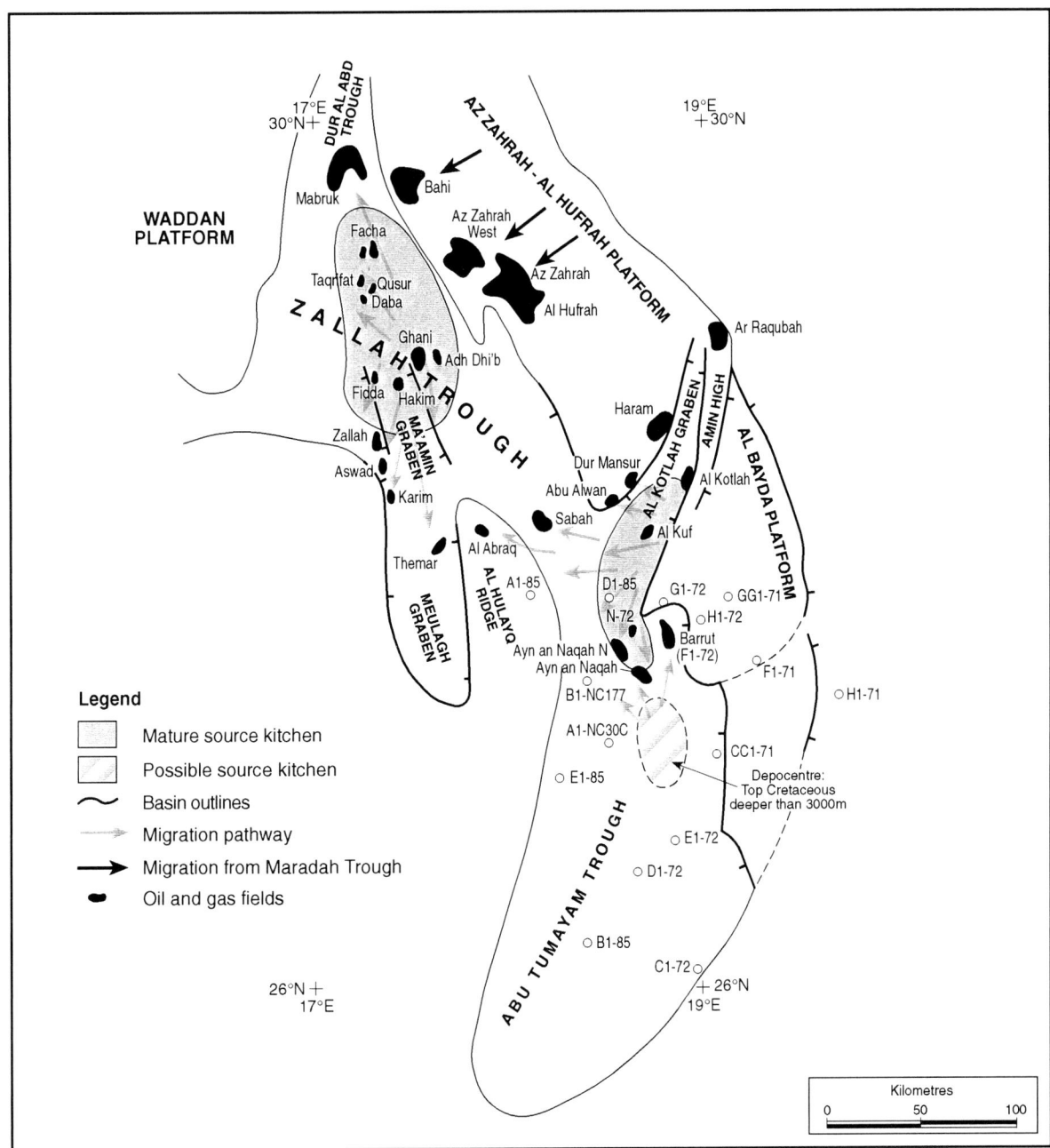

Sources: El Alami, et al., 1989, Johnson and Nicoud, 1996, Schroter, 1996, Knytl, et al., 1996

Figure 8.11   Zallah Trough and Abu Tumayam Trough Petroleum Systems

Two Sirt Shale source kitchens are present in the the Zallah Trough and its southern extension the Abu Tumayam Trough. Hydrocarbons migrated along fault conduits and into two principal reservoirs, the Palaeocene Farrud Member and the early Eocene Facha dolomite. The former is sealed by the Khalifah shales and the latter by the Hun evaporites. Oil from the northern kitchen sourced the Mabruk, Facha, Ghani and Zallah fields, and the southern kitchen sourced the Haram, Al Kotlah, Al Kuf and Ayn an Naqah fields, and probably also the Sabah field.

an effective source rock in this area. The possibility of Silurian source rocks in the western Sirt Basin is discussed below. To date no major oil shows have been found in the Sirt Basin west of the Zallah Trough. A significant source kitchen is present in the northern Zallah Trough and Dur al Abd Trough extending from the Mabruk field to south of the Zallah field. It should be noted however that the level of maturity in the northern area where the source shales are richest is early mature, and peak maturity is only found in the Zallah area, where the source shales are relatively thin. This suggests that the Zallah Trough may have only generated a modest amount of hydrocarbons compared with the deeper basins to the east, and it is worth pointing out that very little oil has been found on the Waddan Platform to the west. The only exceptions are the oil shows encountered in the A1-10 and B1-10 wells on the Waddan Platform which are reservoired in carbonates of the Kalash Formation. The Kalash in this area shows porosities between 22% to 33%, but permeability is only 0.5 to 0.9mD. Analysis of wells in the neighbouring offshore area suggests that this oil originated from Cretaceous source rocks to the north rather than from the southeast. Indeed a good correlation was obtained between the oil recovered on test from well B1-10 with extracts from Turonian shales in offshore well C1-87. Five groups of oilfields are present in the western troughs, the isolated Mabruk field in the north, the Facha, Taqrifat, Qusur group, the Ghani, Fidda, Adh Dhi'b, Hakim group, and the Zallah, Aswad, Safsaf group. The Sabah field is located in the 'Ayn an Naga sub-basin in the extreme south. The principal petroleum system in this area is the Sirt Shale-Palaeocene/Eocene system.[27]

8.4.1 The Zallah Trough Sirte Shale-Palaeocene/Eocene Petroleum System

Reservoirs

The Lower Palaeocene Farrud Member is well developed on the Ghani group of fields where it is the principal reservoir. Porosity is high due to dolomitization and leaching, particularly on the ancient shoal areas, where porosity reaches 24%. These areas have production potential of up to 5000 BOPD per well compared with less than 3000 BOPD per well in the inter-shoal areas. Total production from the Ghani field in 1991 was about 100,000 BOPD from three pay-zones. The Sabah field produces from the same formation. In the Upper Palaeocene the Az Zahrah Formation (Figure 5.2) has fair reservoir quality and is hydrocarbon bearing at Mabruk and on the Facha-Taqrifat group of fields. The Mabruk field has recently been developed by Total. The Zaltan carbonate however is not well developed in the Zallah Trough, being represented by moderately deep-marine argillaceous and micritic limestones.[28]

In the Eocene the Facha Dolomite Member has excellent reservoir quality with a thickness of 100 to 150m and is oil bearing on the Zallah group of fields which have been on production since 1979. It represents a restricted environment of deposition transitional between the open marine Mesdar Limestone and the evaporites of the Hun Member. As in the Farrud Member, porosity has been enhanced by the dolomitization process.[29]

Source Rocks

The petroleum geochemistry of the Zallah Trough was discussed in chapter 7. A source kitchen is present in the northern Zallah Trough and in the Facha area the thickness of effective source rocks reaches 330m, but it is early mature and probably has not generated large amounts of hydrocarbons (Figure 8.11). In the Zallah area the thickness of effective source rocks is around 100m and in this area it is at peak maturity. The oil reservoired in the Zallah and Ghani groups of fields was generated from this kitchen. El Alami et al. suggested that the Az Zahrah-Al Hufrah field was sourced from the Zallah Trough kitchen, by migration up the platform bounding fault, but given the Neogene subsidence history of the area and the location of the Az Zahrah-Al Hufrah field adjacent to the early mature source kitchen, it is difficult to see how these large fields could have been charged from the Zallah Trough. The generating capacity of the immature source kitchen in this area was probably insufficient to fill the giant Al Zahrah-Al Hufrah complex. It is more likely, as suggested by Roohi, that they were charged by migration from the Maradah Trough.[30]

Seals

Two major seals are present in the Palaeocene-Eocene sequences in the Zallah Trough. In the Palaeocene the Rabia and Khalifah shales form an extensive regional seal for the Farrud and Az Zahrah carbonates, and in the Eocene the thick evaporites of the Hun Evaporite Member provide top seal for the Facha Dolomite. The evaporites are over 600m thick in the Zallah Trough. The area is heavily faulted, particularly in the south in the Meulagh Graben and on the Al Hulayq Spur. Examples of sealing faults can be found, as for example when the Facha reservoir is displaced against the Hun evaporites. The Ghani field is dependent on fault seal for closure on its western flank, but other faults appear to have leaked, and probably acted as conduits for hydrocarbon migration.[31]

Traps

The majority of the hydrocarbon traps in the Zallah Trough are structural traps formed during Eocene tectonism. Whilst the eastern part of the basin is relatively undisturbed the western flank is heavily faulted and deformed, and most of the traps are located in this area. The northern group of fields – Mabruk, Facha-Taqrifat – are located updip of the main depocentre of the basin, and are basically faulted anticlines with a northwesterly trend. The Ghani group of fields are located deeper in the trough in the Ar Ramlah syncline, and are mostly fault closures on the western margin of tilted fault blocks. The Hakim and Fidda fields are located in the Ma'amir Graben which represents the western and deepest part of the Zallah Trough (Figure 8.11). The Zallah group of fields are larger anticlinal structures, partially fault dependent on a terrace to the west of the Ma'amir Graben. The structure becomes extremely complex in the Meulagh Graben and compressional flower structures are characteristic of the Abraq-Themar area. The graben passes beneath the volcanic cover of Al Haruj al Aswad. The Sabah field is a faulted anticline, at shallower depth, in the narrow sub-basin linking the Zallah Trough to the Abu Tumayam Trough.[32]

Summary

The Zallah Trough contains a significant source kitchen for the Sirt Shale which generated hydrocarbons from the late Eocene to the present time. Migration took place from the centre of the basin and charged the Zallah group of fields to the west, and the Ghani group of fields to the north. Some oil migrated into the Meulagh Graben where sub-commercial reserves have been found at Themar. In the northern part of the basin hydrocarbons were generated from less mature source rocks which migrated northwestwards and charged the Facha group of fields and Mabruk. Some oil may have leaked up the western boundary fault of the Az Zahrah-Al Hufrah Platform but the major fields on the platform were probably sourced from the east.

8.4.2 Western Sirt Basin, Other Reservoirs and Source Rocks

No significant hydrocarbons have been found in Cretaceous reservoirs in the Zallah Trough. This is probably due to the absence of good reservoirs above the Sirt Shale. The Kalash Formation constitutes a viable reservoir only on the adjacent platforms where nearshore facies are developed. In the Zallah Trough it is a tight chalky limestone with very poor reservoir characteristics. As far as the deeper reservoirs are concerned, the problem seems to be the lack of an effective conduit between the source rock and reservoirs beneath the Sirt Shale.

The Zallah Trough connects southwards with Al Kotlah Graben and the Abu Tumayam Trough. A source kitchen is present in the Al Kotlah Graben which contains rich Sirt Shale source rocks which have reached peak maturity in the area between the Kuf and Ayn an Naqah fields (Figure 8.11). This source kitchen has charged the fields of Al Kuf, Al Kotlah, Haram, and Dur-Mansur in the Al Kotlah Graben and on the southern tip of the Az Zahrah-Al Hufrah Platform. The Abu Alwan field however shows some differences. Oil samples from this field show a different chemical signature and carbon isotope values compared with samples from Warid, Haram, Al Kotlah and Al Kuf. These differences may be attributable to kerogen variations in the area which sourced the Abu Alwan field. These fields are reservoired in a variety of different formations including quartzites

(Haram), basal marine Cretaceous sands (Abu Alwan, Dur-Mansur) and Samah dolomites (Al Kuf, Al Kotlah). This kitchen probably also sourced the Samah field and possibly the Al Abraq oil pool on the Al Hulayq Ridge, although this would imply migration over a distance of almost 100km. The giant Ar Raqubah field however was most likely sourced from the Maradah Trough.[33]

Further south, in the northern part of the Abu Tumayam Trough Mobil reported good shows (mostly gas) from well F1-72 in 1966, and when Veba resumed exploration in this area in 1993 several additional discoveries were made, culminating in the discovery of the Ayn an Naqah pool in concession 72 and further discoveries in concession NC 177. The B1-NC 177 discovery tested oil at a rate of 728 BOPD from Facha dolomites with a potential flow rate of 1800 to 2500 BOPD with an oil gravity of 42° API. A development plan has been submitted for this field. The structures in this area are mostly faulted anticlines with a northwest-southeast trend. These fields were also sourced from the Al Kotlah source kitchen, with the possible exception of the Barrut field.[34]

The Abu Tumayam Trough is very under-explored and the question arises whether hydrocarbon potential continues to the south. Unfortunately this area was not sampled by El Alami et al. in their geochemical evaluation of the Sirt Basin, and no geochemical data has been published from this area. According to Wennekers et al. the depocentre of the trough is located south of the 'Ayn an Naqah discoveries which suggests that the Sirt Shales should be mature in this area (Figure 8.11). As far as the reservoirs and seals are concerned, Abugares published a cross-section showing that the Facha Dolomite extends to the south of well D1-72 (Figure 8.11), but pinches out before reaching well H1-85. The Farrud Member was shown by Bezan, et al. as extending over the entire Abu Tumayam Trough. There could therefore be further potential in the central part of the Abu Tumayam Basin.[35]

Central Sirt Basin

The Central Sirt Basin, for the purposes of this section, comprises the source kitchen of the Maradah Trough and the oil fields sourced from it. Unlike the Zallah Trough most of the oilfields charged from the Maradah Trough are located on the adjacent platforms, particularly the Al Zahrah-Al Hufrah and the Al Bayda Platforms. Only one major discovery has been made in the trough, the Wadi field, which is a large horst located in the eastern part of the trough. Roohi estimated that 12 billion barrels of oil in place and 3 trillion cubic feet of gas in place have been discovered in areas sourced from the Maradah Trough, which suggests that at least 600 billion barrels of oil-equivalent have been generated from the trough. The fields on the Az Zahrah-Al Hufrah Platform were charged by migration up the western boundary fault of the Maradah Trough and into carrier beds of mainly Palaeocene age (Figure 8.13). As discussed previously, the platform had already been tilted towards the northeast by the time of peak oil generation during the Oligo-Miocene, directing oil migration towards the southwest. This involves relatively long migration routes to the Bahi, Az Zahrah and Al Hufrah oilfields. It has long been assumed that oil also migrated up the eastern boundary fault of the trough and sourced the Lahib, Zaltan, Jabal, Wahah and Dayfah fields, and this view was supported by El Alami et al. This however remains to be proved. Given the structural configuration at the time of oil migration it is easier to envisage that these fields were charged by long range migration from the Ajdabiya Trough, and this is supported to some extent by the progressive rise in oil-water contacts in fields on the Zaltan Platform from northeast to southwest (Figure 8.16). However, the major fields on the Zaltan Platform are located very close to the eastern boundary fault of the Maradah Trough and it is likely that some oil migrated from the Maradah Trough up the fault plane and through fractured quartzites. These fields are discussed in section 8.4.5. Unfortunately the source facies of the Sirt Shale is very similar in the Maradah Trough to that of the Ajdabiya Trough and it has not yet been possible to chemically separate the oils generated from these two kitchens. [36]

As discussed in chapter 7 the Sirt Shale becomes progressively leaner in the northern Maradah Trough and its generating capacity is correspondingly reduced. The Wadi field is reservoired in quartzites which are partly Nubian equivalent and partly Ordovician. The Ar Raqubah and Bazzuzi fields are reservoired in Upper Cretaceous Wahah Sands, but almost all of the other accumulations sourced from the Maradah Trough are trapped in Palaeocene carbonates. The principal petroleum system in the central Sirt Basin is therefore the Sirt Shale-Palaeocene system.

8.4.3 Maradah Trough, Sirt Shale-Palaeocene Petroleum System

Reservoirs

The palaeogeography of the Az Zahrah-Al Hufrah Platform was discussed in chapters 5 and 7 and is shown on Figure 8.12. Essentially the pre-Upper Cretaceous surface was onlapped by Upper Cretaceous formations from both north and south with the crestal area between Az Zahrah and the Manzilah Ridge remaining emergent until Maastrichtian/Danian times. Palaeocene deposition was influenced by the same palaeogeography and shales were deposited to north and south, whilst carbonates are confined largely to the structural highs. This differentiation continued almost until the close of the Palaeocene. The passage from limestone

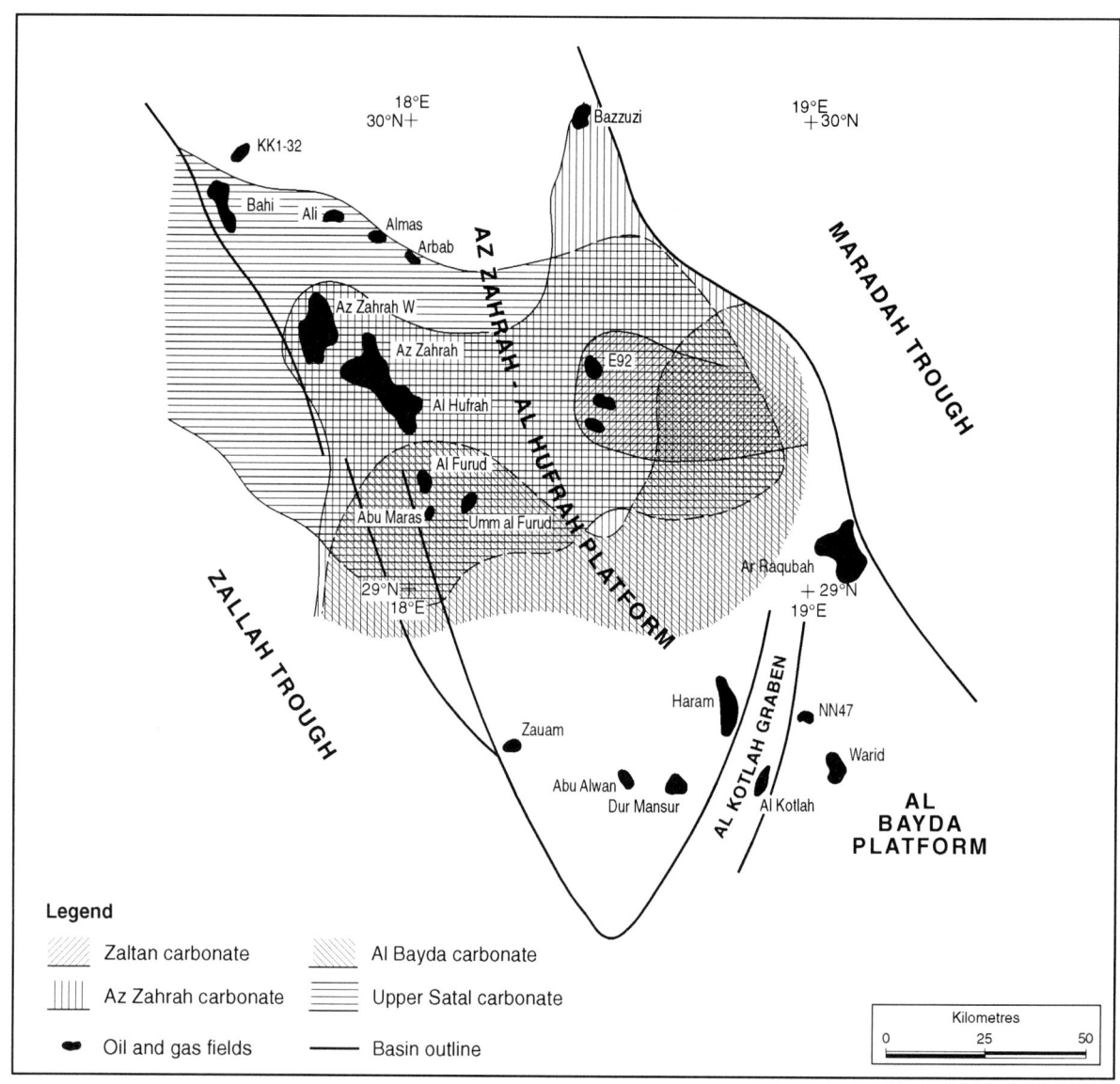

Figure 8.12   Sirt Basin, Az Zahrah-Al Hufrah Platform, Limit of Palaeocene Carbonate Reservoirs

This map shows the effective limit of good quality reservoir facies within the four Palaeocene carbonates on the Az Zahrah-Al Hufrah Platform. Both the Az Zahrah West and E-92 pools are stratigraphically trapped at the updip margin of the Az Zahrah and Zaltan Limestones. All four limestone reservoirs pass into argillaceous and micritic facies beyond the limits shown.

to shale frequently occurs within a distance of a few kilometres and appears to represent a ramp configuration rather than an abrupt platform margin. Four carbonate reservoirs are present,

separated by shales: the Upper Satal, Al Bayda, Az Zahrah and Zaltan Limestones.[37]

The Upper Satal Member, of Danian age, has excellent reservoir characteristics. It has a thickness of 130m with porosity values in the order of 20-25%. It extends westwards from the Manzilah Ridge to west of the Mabruk field, and contains oil in the Bahi, Az Zahrah, Al Hufrah, Ali, Arbab and Almas fields (Figure 8.12). The Al Bayda Formation is well developed on the Manzilah Ridge where it forms part of a continuous carbonate sequence extending up into the Thanetian. Towards the Al Hufrah field the upper member shales-out as the Rabia Shale Member and the lower (Thalith) member deteriorates as an effective reservoir. Oil shows are present in the Thalith Member on several fields, but it is the least important of the four reservoirs. The Az Zahrah (Dahra) Limestone, known further west as the Mabruk Limestone, has excellent reservoir characteristics with a thickness of 90m and porosity up to 25%. It is the major producing horizon on the Az Zahrah, Al Hufrah, Al Furud, Abu Maras, and Umm Furud fields Figure 8.12). Reservoir quality deteriorates rapidly to the west and north of these fields, introducing a stratigraphic element to the trap at Az Zahrah and a deterioration of reservoir quality at Mabruk. The Zaltan Limestone is a poor reservoir over much of the Az Zahrah-Al Hufrah Platform, being represented by argillaceous and micritic limestones, but on the eastern part of the Al Zahrah-Al Hufrah High and on the southern Manzilah Ridge a higher energy facies is developed. The thickness of the Zaltan Limestone in this area is 60 to 100m and porosity averages 20 to 25%. This limited area represents an isolated shoal which shales out rapidly to the west. The formation is hydrocarbon bearing only in the E-92, A1-NC 18 and A1-NC 99 area (Figure 8.12).[38]

The Al Bayda Platform had a similar history. The pre-Upper Cretaceous surface was progressively onlapped by Upper Cretaceous sediments with the Amin High remaining emergent until the Danian (Figure 8.14). Palaeocene deposition reflects the underlying structure. The same four carbonate reservoirs are present as on the Az Zahrah-Al Hufrah Platform but in this area the Al Bayda Formation carbonates form the principal reservoirs. The Upper Satal Member is best developed on the western part of the platform and passes abruptly into Al Hagfah shales to the east. It is well developed only on the Al Bayda and Warid fields where it has a thickness of 130m. The Al Bayda Formation attains its optimum development in this area with two distinct carbonate reservoirs, the Lower Al Bayda (also known locally as the Meem, Zaqqut, or M-59 pay zone) and the Upper Al Bayda (also known as the Al Awra or Ora Member). Both represent shoal or shallow-marine limestones with fair reservoir characteristics. The lower member has a thickness of 75m over the western area with porosity in the order of 16-22%, but it thins towards the east where it is only 60m thick on the Eteila field (Figure 8.14). This unit is oil bearing on the Al Bayda, Warid, Tibisti (Ora) and Zaqqut fields. The upper member is best developed on the Al Bayda and Tibisti (Ora) fields and has a constant thickness of about 65m over much of the platform. It is productive at Al Bayda, Warid, and several smaller fields. The Al Bayda Formation is also considered prospective on the horst south of the Al Bayda Platform in the area of wells E1-71, F1-71 and M1-71.The Az Zahrah Formation is well developed only on the western part of the Al Bayda Platform. It reaches a thickness of 120m in the area of the Khalifah and Al Bayda fields, but shales out before reaching the Samah and Tibisti (Ora) fields. It is oil bearing in only two fields on the western margin of the platform. The Zaltan Limestone forms a blanket over the entire area with a fairly constant thickness of about 130m over both the platform and the adjacent troughs. Reservoir quality is fair to good. The Zaltan Limestone contains oil at Al Bayda, Sarab and Labiba.[39]

Source Rocks

The Sirt Shale source rock in the Maradah Trough was discussed in chapter 7. The effective thickness (with TOC greater than 1%) reaches 330m, and the shales have reached peak maturity over a large part of the trough (Figure 8.14). However it should be noted that the type II kerogen is only present in the southern part of the trough and north of the Bazzuzi field it is dominantly type III. There is some gas generating capacity further north, but little in the way of oil potential. Gardner, et al. speculated about the source potential of the area north of the Az Zahrah High which they called the Harawah Trough, but given the type of kerogen found in this area, it does not appear promising. Ibrahim indicated a number of geothermal gradient anomalies in the Maradah Trough and there is some evidence of elevated geothermal gradients in the trough west of the Attahadi field.

Some of the fields, particularly at the end of the migration routes, are oversaturated and have a considerable gas cap. The Sirt Shales have sourced most of the oil found on the Az Zahrah-Al Hufrah and Al Bayda Platforms, and may have contributed to some of the large fields on the western margin of the Zaltan Platform, although this has yet to be proved. The western boundary faults of the Maradah Trough acted as conduits for the migration of oil which entered carrier beds on the platforms and migrated westwards (Figure 8.14). The migration routes are often complex, particularly on the Az Zahrah-Al Hufrah Platform which is bounded on the east by the Manzilah Ridge (Figure 8.13). Roohi has shown that although the ridge was structurally high until middle Eocene times it subsided rapidly thereafter. The ridge acted as a barrier over which hydrocarbons had to pass before reaching the platform, and due to the late onlap of the ridge during the Danian, and its post Eocene subsidence, entry onto the platform was significantly restricted. This explains why oil has not been found in Cretaceous reservoirs in the area west of the ridge (Figure 8.13). Beyond the limits of the ridge however, at Ar Raqubah and Bazzuzi, hydrocarbons are present in Cretaceous reservoirs (Figure 8.14).[40]

Seals

Seals on the two platforms are provided by shales within the Palaeocene, most notably the Khalifah Shale which is present over the entire area. As discussed above, Palaeocene sedimentology was greatly influenced by water depth, and on the structural highs such as the Manzilah Ridge the carbonate sequence is continuous from the Upper Satal to the Az Zahrah Formation, with the combined reservoirs being sealed by the Khalifah Shale. In the Az Zahrah-Al Hufrah area however, the Rabia Shale Member provides a seal for the Satal and Lower Al Bayda units and only the Az Zahrah Formation is sealed by the Khalifah (Figure 8.13). In the Ar Raqubah and Bazzuzi area the Zaltan Limestone is sealed by shales beneath the Harash Formation, but on the structural highs only a very thin argillaceous limestone separates the Zaltan from the Harash, and this may explain why no significant oil has been found in the Zaltan Formation in this area. One or two examples of sealing faults can be demonstrated. A major fault extends from west of the Az Zahrah field to west of the Zauam field. No oil has been found on the platform west of this fault which suggests that it acted as a barrier to hydrocarbon migration (Figure 8.14).

On the Al Bayda Platform the Upper Satal Member on the Al Bayda field is sealed by the Thalith Member, and the Al Bayda and Az Zahrah Formations are sealed by the Khalifah Shales. The Zaltan Limestone is capped by the Harash Formation which is not an effective seal.[41]

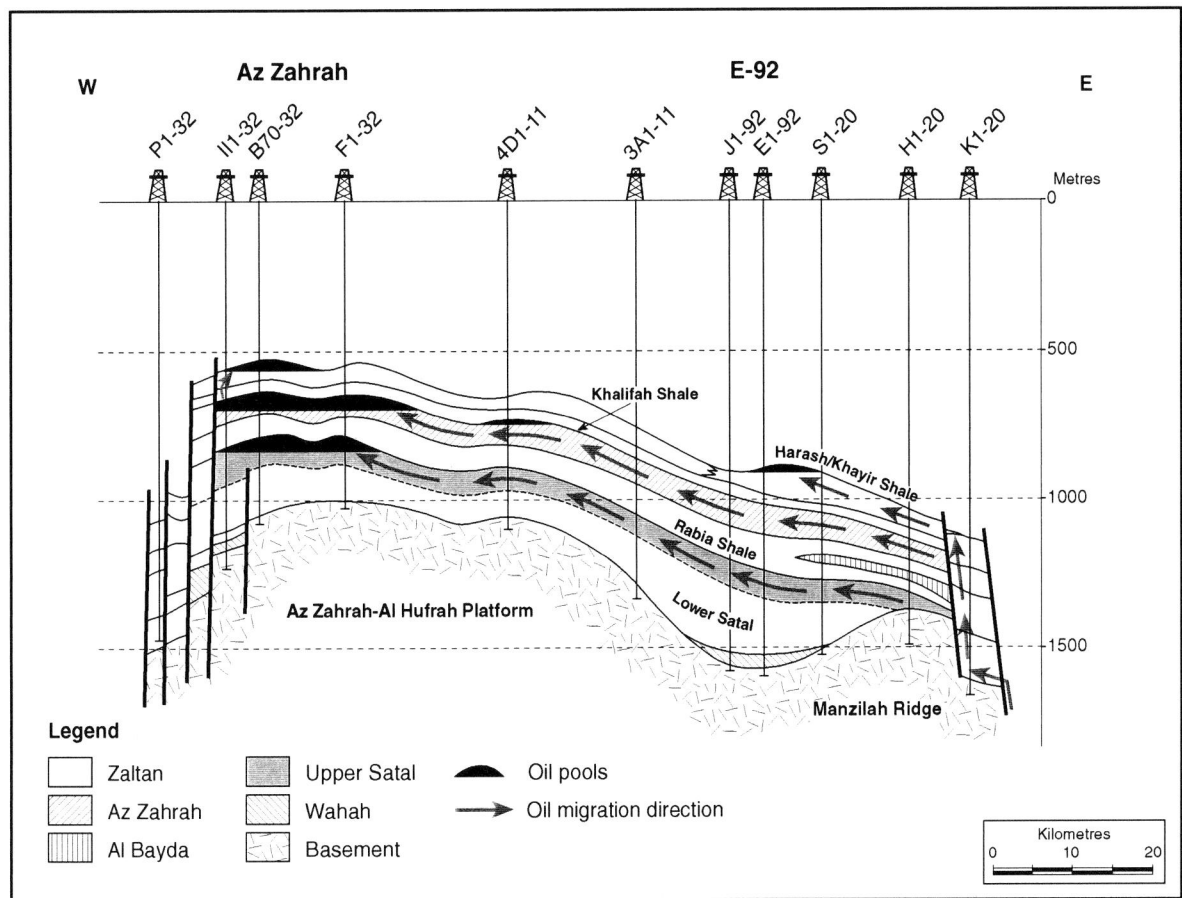

Source: Roohi, 1996b

Figure 8.13   Az Zahrah-Al Hufrah Platform, Hydrocarbon Migration

Oil generated in the Maradah Trough migrated via faults and into Palaeocene carbonates on the Az Zahrah-Al Hufrah Platform. Some was trapped in stratigraphic traps as at E-92, but most migrated to the crest of the platform where it was trapped in low relief closures. The Rabia, Khalifah and Harash-Khayir Shales form local seals. The section shows how the Manzilah Ridge prevented oil from entering Upper Cretaceous reservoirs in this area.

Traps

Most of the traps on the Az Zahrah-Al Hufrah and Al Bayda Platforms are structural, resulting from mid-Eocene tectonic activity. Most are low relief anticlines with a northwestern orientation. On the Al Bayda Platform, by contrast, the traps have a northeastward trend similar to that of the adjacent Al Kotlah Graben and Abu Tumayam Trough to the south and west. Most of the structures appear to be filled to spill point. The complex geometry of the Palaeocene carbonates provides scope for stratigraphic traps. The Az Zahrah field is partly attributable to stratigraphic trapping caused by the rapid shale out of the Az Zahrah Formation to the west, and the E-92, A-NC 18, A-NC 99 complex are trapped by a similar shale out in the Zaltan Limestone (Figure 8.13). The Bahi field may be partly fault dependent on its northern margin.[42]

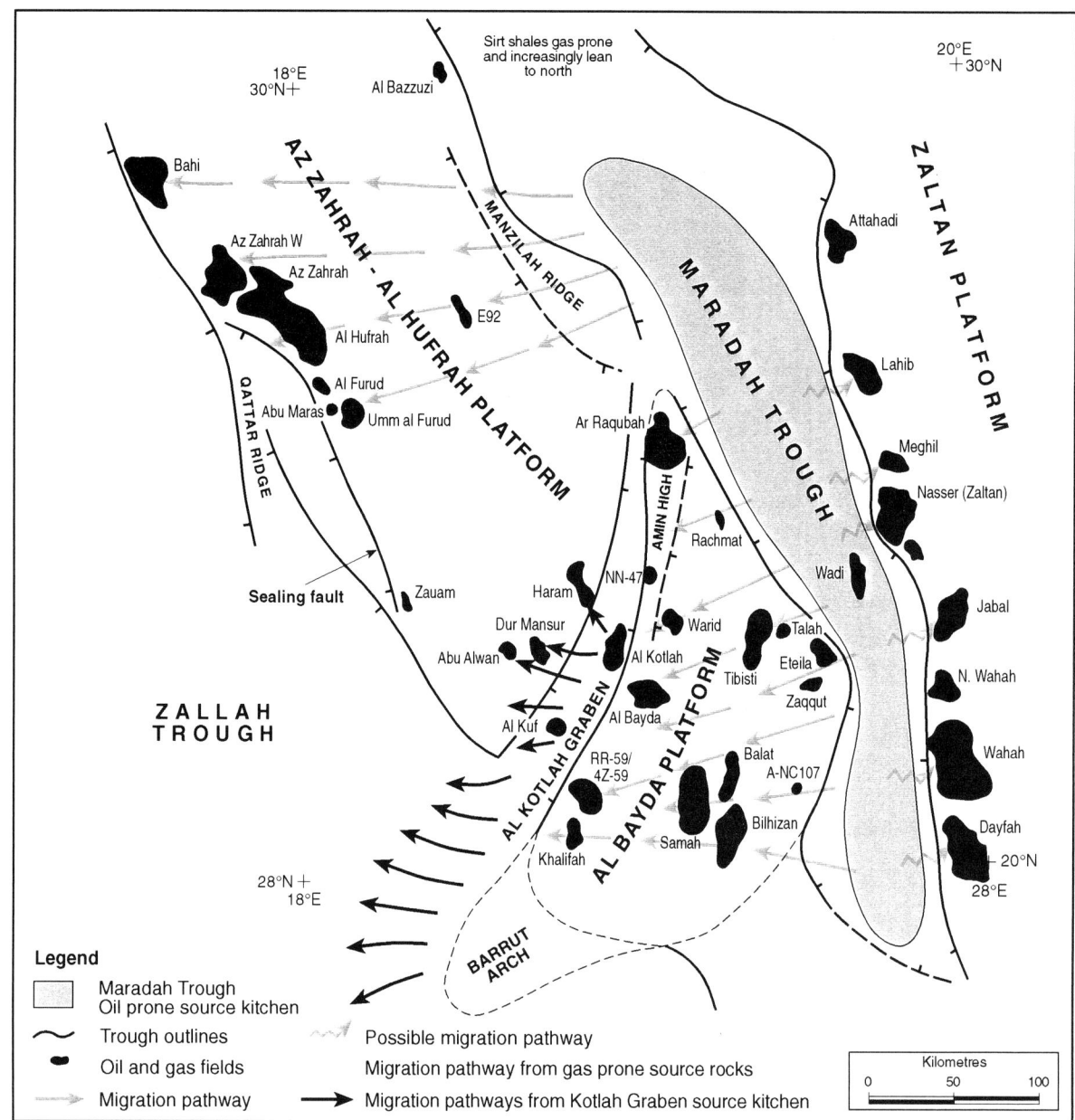

Figure 8.14   Maradah Trough Petroleum Systems

A major Sirt Shale source kitchen is present in the Maradah Trough which has generated huge amounts of hydro-carbons, much of which has migrated onto the Az Zahrah-Al Hufrah and Al Bayda Platforms. Only one major oil pool has been found in the trough itself. Migration onto the Az Zahrah-Al Hufrah Platform was controlled by the Manzilah Ridge which effectively prevented oil from entering Cretaceous reservoirs to the west. Further north the source rock becomes gas prone and lean. Hydrocarbons generated from the southern Maradah Trough migrated onto the Al Bayda Platform, and into the giant Raqubah field. The question of how much oil migrated from the Maradah Trough onto the Zaltan Platform is still unresolved.

Summary

The Maradah Trough represents a major source kitchen for the Sirt Shale and large volumes of hydrocarbons have been generated from it, although the source rock becomes lean further north. Much of the oil and gas migrated onto the adjacent platforms, particularly to the west, along fault conduits and into carrier beds which rise to the southwest. Four carbonate reservoirs are present in the Palaeocene on the platforms, of which the Al Bayda and Az Zahrah Formations are the most important. Migration paths are often complex, especially on the Az Zahrah-Al Hufrah Platform where they are complicated by the presence of the Manzilah Ridge, which restricted migration into the various reservoirs. Oil moved relatively freely through the carbonate carrier beds, filling the structures to spill point *en route*, and some was trapped in stratigraphic traps caused by shale out of the carbonate reservoir. Much of the oil reached the western margin of the platform, a distance of about 100km from the source kitchen.

The situation on the Al Bayda Platform is different in several respects from the Az Zahrah-Al Hufrah Platform. The structural grain is northeast-southwest rather than northwest-southeast, the reservoirs tend to decrease in quality towards the east, and the structural equivalent of the Manzilah Ridge (the Amin High) is located along the northwestern margin of the platform. Nevertheless there are similarities. Hydrocarbons generated in the Maradah Trough migrated up the western boundary fault and into a variety of carrier beds, the most important of which are the Al Bayda Formation carbonates. Oil accumulations are much more evenly spread over this platform than on the Az Zahrah-Al Hufrah Platform. This is due to several factors, the pre-Upper Cretaceous palaeogeography which controlled the Upper Cretaceous transgression and to a lesser extent the facies of the overlying Palaeocene deposits, plus the mid-Eocene tectonic events which produced the majority of the traps.

8.4.4 Central Sirt Basin, Other Reservoirs and Source Rocks

Two of the largest fields sourced from the Maradah Trough are not reservoired in Palaeocene reservoirs. Ar Raqubah field is reservoired in Upper Cretaceous Wahah calcarenites, and the Wadi field, within the Maradah Trough, is resevoired in fractured quartzites of Lower Cretaceous and Ordovician age.

Ar Raqubah field is a giant field containing 1875 MMB original oil in place and 750 MMB recoverable reserves in Wahah nearshore calcarenites. In several respects the field is unique. It is located on an anticlinal structure immediately east of the Kotlah Graben and adjacent to the Maradah Trough boundary fault. The structure was emergent during the Upper Cretaceous but was onlapped from the south by bioclastic limestones and calcarenites during the Maastrichtian. The Wahah reservoir, however, pinches out on the southern flank of the structure, and is absent on the northern limb. The structure subsided during the early Palaeocene and was covered by Al Hagfah Shales which form a seal for the oil accumulation. The structure was folded and faulted during the mid-Eocene and charged with hydrocarbons during the Oligo-Miocene. The field contains a 207m hydrocarbon column extending from the Wahah into the underlying quartzites. There is a 148m oil column with an overlying gas cap, and oil gravity is 42°API. The original gas-oil ratio was 935 scf/bbl. Reservoir quality is very variable. The Wahah reservoir becomes progressively more shaly downdip, and on the field itself low permeability zones divide the field into a number of vertical and horizontal segments. The low permeability zones are due to a combination of depositional facies, subsequent diagenesis and faulting. Closure to the northwest is provided by a major sealing fault. Porosity in the Wahah reservoir averages 17% with permeability around 200 mD. Porosity in the weathered and fractured quartzites is about 2%. In addition, gas is preserved in the Lower Al Bayda carbonate on the Raqubah field. The field was charged from the Maradah Trough by oil migrating up the western boundary fault. It entered the structure along the unconformity at the base of the Hagfah Shales and through fractures within the quartzites. A gravity survey over the Raqubah field confirmed the presence of a basement high, and the gentle southerly dip. It also showed the thickness of basin-fill to reach 5000m in the adjacent troughs and showed the continuation of the Kotlah Graben to the west of the Raqubah field. As mentioned above, the Manzilah Ridge acted as a barrier for the migration of oil into Cretaceous sediments further north, but beyond the northern end of the ridge another pool, analogous to Raqubah, but much smaller, is preserved at Bazzuzi. This field is gas prone due to the kerogen facies and late maturity of the source rock in the northern Maradah Trough.[43]

The Wadi field is a major tilted horst block located within the Maradah Trough (Figure 8.14). It contains oil in fractured quartzitic sandstones which have been dated as Lower Cretaceous in the upper part and Ordovician in the lower part. The quartzite sequence contains a shale member in the middle which El Hawat et al. compared to the middle Nubian shale in the eastern Sirt Basin. The quartzites are covered by Lidam Formation dolomites and shales, and then by Senonian shales and evaporites which provide the seal. The structure was charged by lateral migration from the Sirt Shale source rocks, which are downthrown on both sides of the horst. So far, the Wadi field is unique as the only oil bearing structure within the Maradah Trough, although

gravity data presented by El Batroukh and Zentani suggests the presence of similar structures further north.[44]

On the Al Bayda Platform there are a greater number of oil pools reservoired in Upper Cretaceous reservoirs than on the Az Zahrah-Al Hufrah Platform. In the south, at Samah, Bilhizan and Al Balat oil is present in basal Upper Cretaceous sands, and at Khalifah and a number of small fields further west oil is found in the Wahah Formation. This reflects the early transgression of marine Upper Cretaceous rocks in this area, and the unrestricted access to these rocks for oil migrating from the southern Maradah Trough (Figure 8.14).[45]

On the eastern side of the Maradah Trough Hammuda described sandbars and sandspits of lower Campanian age on the platform margin of the Zaltan Platform south of Dayfah, and Jones and El Ghoul referred to sand wedges of Maastrichtian age on the downthrown side of the platform margin fault in the Nasser area. Minor amounts of oil were found in a sandbar deposit in the G1-71 well. The sand wedges on the downthrown side of the Zaltan Platform bounding fault can be regarded as Wahah equivalents in a unique setting. In the YYY1-6 well the calcarenites reach a thickness of 740m. As well as providing potential reservoir rocks this sequence may also provide the conduit by which some oil migrated from the Maradah Trough onto the Zaltan Platform. The provenance of the oil reservoired at Zaltan, Wahah and Dayfah is still unresolved. For the purposes of this review these fields will be considered in the following section.[46]

The possibility of other source rocks in the Maradah Trough has frequently been addressed. Turonian shales of the Etel Formation, Danian shales of the Al Hagfah Formation and even Silurian Tanzuft shales have been suggested as potential source rocks. However there is no evidence of the presence of Silurian source rocks in the trough, and as Parsons et al. pointed out, the Hagfah Shale is lean and probably immature over much of the trough. The Turonian shale may have minor potential but is totally overshadowed by the Sirt Shale.[47]

Ajdabiya Trough

Due to a fortuitous combination of circumstances the Ajdabiya Trough is the most prolific part of the Sirt Basin and is probably responsible for 60% of all the oil found in the Sirt Basin. The trough is the largest, deepest and most important source kitchen in Libya. The evolution of the trough was discussed in chapter 6. Subsidence has been rapid and continuous since the mid-Cretaceous and has continued to the present day. The thickness of the Oligo-Miocene section in the centre of the trough is over 3000m. The unique character of the Ajdabiya Trough is due to four main factors. A major source kitchen is present in the Ajdabiya Trough and enormous amounts of oil and gas have been generated from it. Some has been trapped in structures within the trough, but a large proportion has migrated westwards onto the Zaltan Platform (Figure 8.15). In addition, since the trough occupies the axis of the basin, large volumes of oil have also migrated eastwards towards the Cyrenaica Platform and eastern Sirt embayment. Secondly, because of the greater thickness of Neogene overburden compared with the rest of the basin, hydrocarbons are reservoired not only in Cretaceous and Palaeocene reservoirs but also in Eocene and Oligocene reservoirs. Major regional seals are developed in the early Palaeocene, late Palaeocene, late Eocene and Miocene, and major structural and stratigraphic traps were already in place by the time of peak oil generation and expulsion. For the sake of convenience the petroleum systems of the Ajdabiya Trough will be considered by area. Migration to the west produced three petroleum systems: the Sirt Shale-Upper Cretaceous, Sirt Shale-Palaeocene and Sirt Shale-Eocene/Oligocene systems. Migration to the east produced three main systems: the Sirt Shale-Nubian system in the south, Sirt Shale-Palaeocene system in the centre, and Sirt Shale-Eocene system in the north.

Western Ajdabiya Trough

For the purposes of this review the Western Ajdabiya Trough comprises the area west of Intisar, and the Zaltan Platform which received much of its hydrocarbon charge from the Ajdabiyah Trough (Figure 8.15).

8.4.5 Sirt Shale-Upper Cretaceous Petroleum System

Reservoir

The principal Upper Cretaceous reservoir on the Zaltan Platform is the Wahah Limestone of Maastrichtian age which is productive on the Wahah, Jabal, SE Nasser, Lahib, As Surah and Hutaybah fields. In all of these fields the Wahah Formation represents a fringing apron of calcarenites and nearshore limestones which developed around residual islands during the latter stages of the Upper Cretaceous marine transgression. El Ghoul provided a reconstruction of the palaeogeography in his study of the SE Nasser field. In all cases the Wahah passes downdip into more argillaceous facies representing deeper-water deposition, and in many cases the crests of the structures are bald. On the SE Nasser field the thickness of the Wahah Limestone ranges from zero to 107m within a distance of 2km. The relationship between the Wahah and Kalash formations has been a subject of much discussion. Jones demonstrated the presence of an unconformity between the two on the central part of the Zaltan Platform. At Hutaybah however, the situation is less clear and the two formations appear to interfinger. Porosity averages 17% and permeability 200 mD. The oil pools frequently extend downwards into

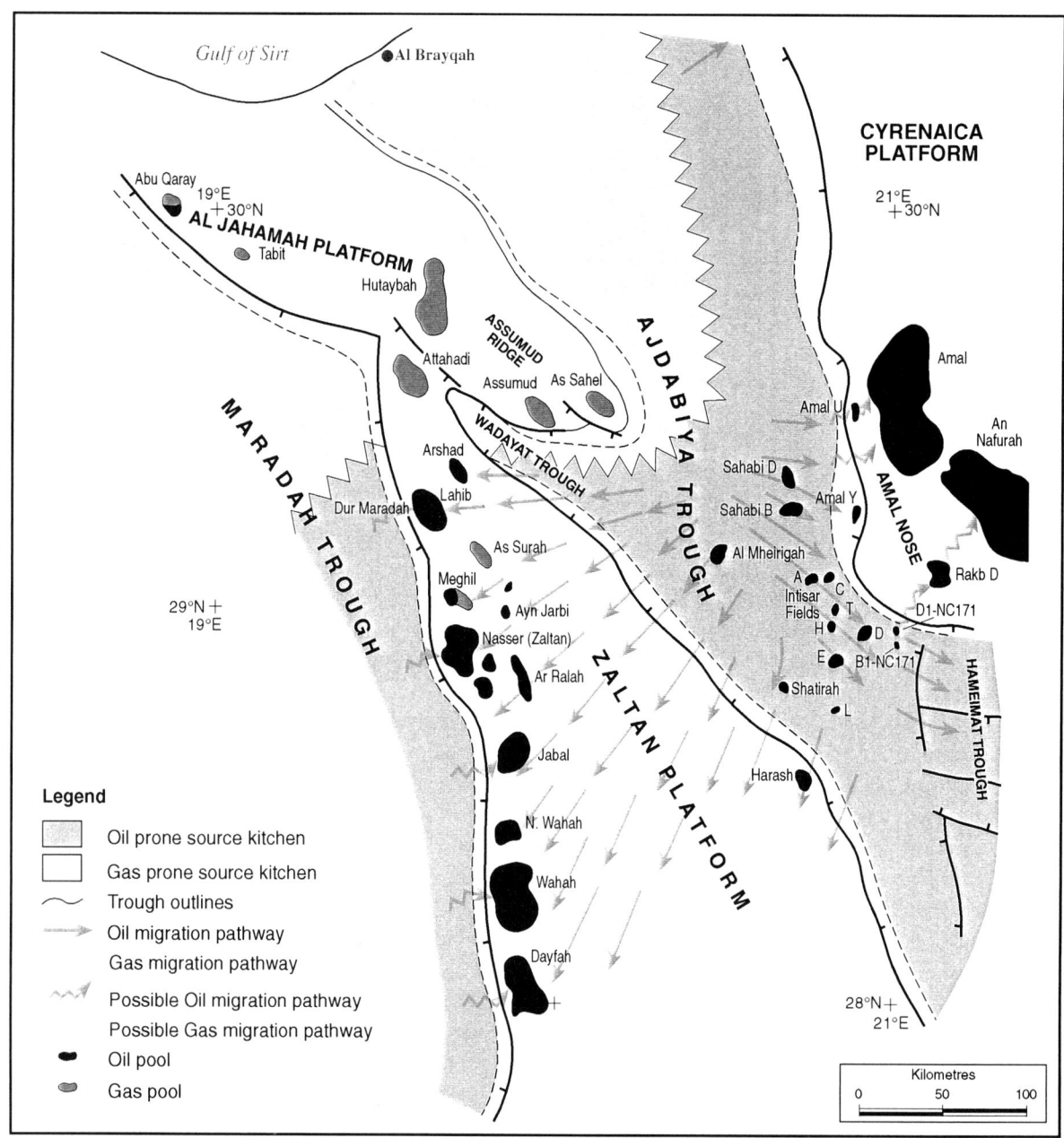

Sources: El Alami, et al., 1989, Roohi, 1996a

Figure 8.15   Western Ajdabiya Trough Petroleum Systems

The Sirt Shale source kitchen in the Ajdabiya Trough is very extensive and has probably sourced most of the fields on the Zaltan Platform. The gas fields in the north were sourced from the gas prone kitchen in the northern Ajdabiya Trough. The Wahah Limestone, sealed by the Kalash Limestone and Al Hagfah Shale, is the main carrier bed and is also a major reservoir. Locally, at Nasser and Dayfah where the Lower Palaeocene retained some permeability, oil was able to migrate into the overlying Palaeocene carbonates. On the Jahamah Platform and Assumud Ridge the lack of seals allowed hydrocarbons to migrate into Eocene and Oligocene reservoirs.

fractured quartzites of uncertain age. Porosity in the quartzites is usually less than 2%, and permeability is dependent on the amount of fracturing. SE Nasser has an oil column of 58m, but the reservoir thickness varies very rapidly.[48]

Source Rocks

The source rocks of the Ajdabiya Trough were discussed in chapter 7. In essence a Sirt Shale source kitchen is present in the trough in which the effective thickness of organic shales (TOC over 1%) reaches 760m in the Intisar area. Depth to peak oil generation is about 3800m and to the top of the gas window at about 4200m. The mature source area extends over a length of 190km and covers an area of about 7000km^2 (Figure 8.15). Little is known about the Sirt Shale in the northern Ajdabiya Trough due to its depth, but the indications are that it becomes lean and passes into a type III kerogen facies. Major generation began in the Eocene and is still continuing today. Hydrocarbons entered into the Wahah reservoir, which overlies and sometimes interfingers with the Sirt Shale source rock, by lateral and updip migration, assisted by faults. El Alami suggested that the gas in the Hutaybah, Meghil and As Surah fields migrated from the northern gas-prone source kitchen via the Wadayat Trough.[49]

Seals

Top seal for the Wahah reservoir on the Zaltan Platform is provided by the tight micritic limestones of the Kalash Formation and the shales of the Danian Al Hagfah Formation (Figure 8.16). These units provide a thick regional seal which extends over most of the platform. There are however local exceptions. At both the Nasser and Dayfah fields topographic relief during the Maastrichtian and Danian was such that porous carbonates were deposited during this period, which allowed hydrocarbons to migrate into overlying Palaeocene carbonate reservoirs. The large faults which mark the western boundary of the Zaltan Platform are sealing at Wahah level and most of the Wahah fields are partially fault dependent, although it is likely that some oil migrated up these faults and through fractures in the underlying quartzites into the fields on the western margin of the platform.[50]

Traps

The Wahah fields are unusual in that they invariably reflect pre-Upper Cretaceous highs with a fringing apron of porous calcarenites, sealed by early Palaeocene shales. The core of the structure is usually formed of quartzites, and the crest is frequently bald. Vertical closure is typically 100 to 150m, and the spill point is usually to the east or northeast. The structures are deceptive since not all of the reservoir sequence within closure is Wahah Formation. Indeed, because the Wahah Limestone frequently pinches out on the flank of the structure, it is not easy to define the pinch-out edge of the Wahah Formation on seismic data, a problem which was highlighted by El Ghoul.[51]

Summary

Large amounts of oil and gas are reservoired within the Wahah Formation on the Zaltan Platform. The Wahah field plus satellites contains original oil in place of over 3 billion barrels, and collectively the Wahah fields on the Zaltan Platform contain an estimated 5.8 billion barrels of oil. In addition about 5TCF of original gas in place is reservoired on the Hutaybah, Meghil and As Surah fields. In the absence of any chemical proof it is generally assumed that the bulk of these hydrocarbons were sourced from the Ajdabiya Trough and migrated along relatively long migration routes onto the Zaltan Platform. This is supported to some extent by the progressive

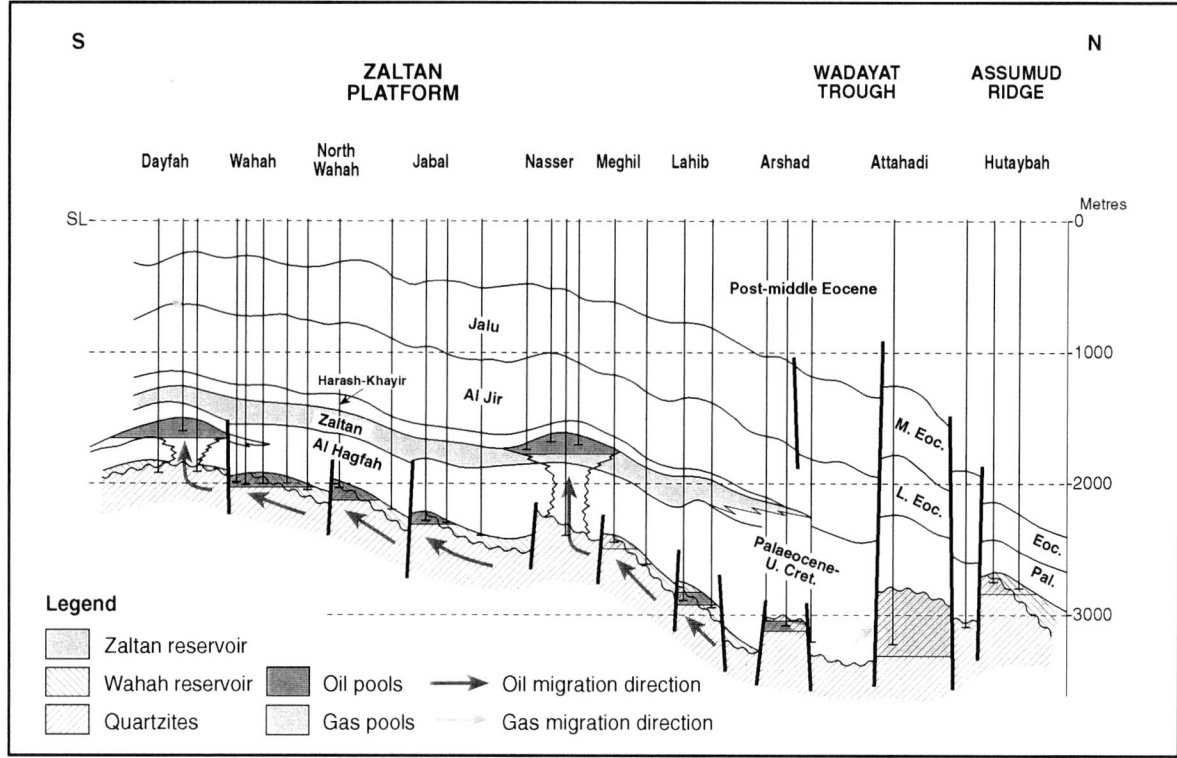

Source: Modified from Roohi, 1996a

Figure 8.16   Zaltan Platform, Hydrocarbon Accumulations

This cross-section from Hutaybah to Dayfah shows the progressive rise in oil-water contacts from north to south, although hydrocarbons actually migrated from northeast to southwest. Hydrocarbons migrated through Upper Cretaceous and fractured basement carrier beds, and most of the oil pools are sealed beneath Palaeocene shales and reservoired in the Wahah. At Nasser and Dayfah however oil was able to migrate through Hagfah shallow-water carbonates into major Palaeocene carbonate shoals. Both the Zaltan and the Wahah reservoirs are absent in the Wadayat Trough. In this area gas is reservoired in fractured quartzites.

stepping-up of oil-water contacts on the Zaltan Platform from northeast to southwest, although this could be enhanced by post-migration tilting of the Zaltan Platform to the northeast (Figure 8.16). There was probably also an as-yet-unquantified contribution from the Maradah Trough source kitchen. Generation probably began during the Eocene but peaked during the Oligocene-Miocene. Most of the structures on which Wahah Formation is present are pre-Upper Cretaceous structural highs, and the Wahah Formation represents a nearshore apron formed during the Maastichtian marine transgression.

8.4.6 Sirt Shale-Palaeocene Petroleum System

Reservoirs

The palaeogeography of the Palaeocene carbonate shelves was discussed in chapter 5. Unlike the Az Zahrah-Al Hufrah Platform the Palaeocene carbonate reservoirs associated with the Ajdabiya Trough are local developments of reefal and high energy carbonates, with spectacular reservoir properties. Two areas in particular represent giant oil accumulations which between them contain around 14.1 billion barrels of original oil in place, 30% of all the oil found in the Sirt Basin. These are the Nasser (Zaltan), and Dayfah (Defa) fields (Figure 8.15). Data on the Nasser field has been published by Fraser and Bebout and Pendexter. It is located close to the northern limit of the Zaltan carbonate shelf. The field developed on a pre-Palaeocene topographic high which is bald of Wahah sediments. Porous lime muds accumulated on the high which pass laterally into non-porous Al Hagfah Shales. The structure gradually widened through time and took on the aspect of an atoll with a central lagoon. Growth of the structure culminated in a thick and widespread high energy carbonate shoal (the Zaltan Formation) which forms the main reservoir of the field. The progressive expansion of the Nasser carbonate complex produced a characteristic mushroom shape in cross-section, with the central core representing a conduit by which oil migrated into the Zaltan carbonate reservoir (Figure 8.16). The oil-water contact is at about 1600m subsea, within the Zaltan Formation, and the gross oil column is 90m. The Zaltan reservoir extends as far as the Meghil field. Porosity is highest in the cor-algal micrite and *Discocyclina* calcarenite facies and has subsequently been enhanced by ground water leaching to over 40% in some places. There is a strong water drive and injection to maintain reservoir pressure has not been required. Thomas quoted original oil in place figures of 6300 MMB and recoverable reserves of 2200 MMB.[52]

The Dayfah field bears some similarities to the Nasser field, but the main reservoir is within the Danian and early Montian which is equivalent to the Al Hagfah Shale and Lower Al Bayda Member. The Dayfah carbonate is 170m thick, of which about 90m is above the oil-water contact, but Roohi suggested that the oil-water contact is tilted towards the northeast. The reservoir comprises a carbonate shoal in which corals and algae form the framebuilders, with abundant molluscs and foraminifera in a bioclastic calcarenite matrix. Porosity is in the range 25 to 30% and permeability is commonly over 200 mD. Halbouty quoted original reserves of 2000 MMB which implies original oil in place of over 5000 MMB. Thomas gave somewhat smaller figures.[53]

The only other Palaeocene oil field on the Zaltan Platform sourced from the Ajdabiya Trough is the small Harash field on the north-eastern margin of the platform.

Source Rocks

The source rock for the Palaeocene reservoirs on the Zaltan Platform is the Sirt Shale, but in this case the Maradah Trough may be the principal provenance of the oil. The Zaltan and Dayfah fields were charged by migration through Upper Cretaceous aquifers with vertical migration into porous Palaeocene carbonates on the two fields. The lagoonal facies at Nasser appears to have acted as a conduit connecting the upper Cretaceous aquifer to the Upper Palaeocene reservoir.[54]

Seals

On the Zaltan Platform the caprock for the Zaltan Limestone is formed by a blanket of argillaceous micrites at the top of the Zaltan Formation. Further oil reserves are contained within thin limestones in the Harash Formation which in turn are sealed by shales of the Harash and Khayir Formations. As mentioned above the lower part of the Zaltan complex is encased in Al Hagfah Shale up to the level of the Khalifah Formation. On the Dayfah field the top seal is provided by the argillaceous carbonates of the Upper Al Bayda Member and the overlying Khalifah Shale, whilst the Al Hagfah Shale provides lateral seal.[55]

The question arises why there are not more hydrocarbon accumulations in Palaeocene carbonates on the Zaltan Platform. The source kitchens are large and prolific and the Palaeocene reservoirs at Nasser and Dayfah are of excellent quality. Why then is the situation so different to that of the Az Zahrah-Al Hufrah and Al Bayda Platforms? The explanation is probably due to a combination of factors. It appears that hydrocarbons migrated from the kitchen to the platform through Upper Cretaceous carrier beds, particularly the Wahah and Kalash Formations. The carrier beds are covered by the Al Hagfah Shale which in this area acts as a regional seal (Figure 8.15). The Nasser and Dayfah fields appear to be unique in providing a conduit through the Hagfah Shales into the Palaeocene carbonates. The Upper Satal, Al Bayda and Az Zahrah carbonates, which are so well developed on the Az Zahrah-Al Hufrah and Al Bayda Platforms, are effectively absent on the Zaltan Platform. Apart from the localised Dayfah Limestone only the Zaltan Limestone is well developed and even that is restricted to the southern part of the platform.

Traps

Traps in the Palaeocene on the Zaltan Platform and in the Ajdabiya Trough are different in character to the structural traps on the Az Zahrah-Al Hufrah and Al Bayda Platforms. The major Palaeocene accumulations in this area are in stratigraphic traps associated with reefs or shoals rather than in structural closures or platform-margin stratigraphic traps which characterise the Az Zahrah-Al Hufrah Platform. The Nasser structure originated as a shoal, evolved into an atoll and then spread laterally as a high-energy carbonate bank. Dayfah also developed as a carbonate bank but at a lower stratigraphic level. These are unusual trapping mechanisms which developed under particular environmental conditions. Much effort has been expended in searching for other areas where such conditions might exist, but so far without success. Structural closures within the Palaeocene are not uncommon, for example on the Al Jahamah and Sabil Platforms, but have yet to prove effective as hydrocarbon traps, either due to problems of charging, or to lack of effective seal.[56]

Summary

The Palaeocene on the Zaltan Platform contains some of the largest oil pools in the Sirt Basin, but surprisingly in rare and unusual settings. Neither of the two main accumulations is preserved in structural or stratigraphic traps as on the Az Zahrah-Al Hufrah and Al Bayda Platforms, but in structures which developed under unique and fortuitous depositional conditions. Indeed Palaeocene carbonates are generally poorly developed on the Zaltan Platform. Only one major carbonate is present, and that is confined to the southern part of the Zaltan Platform. Nevertheless the reefs, shoals and carbonate banks developed at Nasser and Dayfah represent magnificent oilfields with enormous reserves.[57]

8.4.7 Sirt Shale-Eocene/Oligocene Petroleum System

This system is best developed in the Hameimat Trough, which is covered in section 8.4.13, but two interesting examples are also present in the Ajdabiya Trough which merit a brief mention.

The principal reservoir is the Jalu (Gialo) Formation of Lutetian age. West of the Ajdabiya Trough a shoreline trend extends from As Sahel northwestwards to Abu Quray on the Jahamah Platform and similar conditions extend southwards on the Zaltan Platform (Figure 5.8).

The Jalu Formation is a massive nummultic limestone up to 480m thick capped by shales of Bartonian/Priabonian age. On the Assumood Ridge the formation contains gas in structural closures at As Sahel and Assumud, and further west small amounts of gas at Tabit and oil at Abu Quray (Figure 8.15). Abu Quray also contains gas in Arida Sands of Oligocene age. This is the only known example of hydrocarbons being present in an Oligocene reservoir outside the eastern Sirt embayment. The accumulation however is very small. Oil has also been found in poor quality Jalu carbonates on the Zaltan Platform at Dur Maradah, Nasser, and BBBB1-6. It is assumed that the gas was generated from the gas prone Sirt Shale source rock of the northern Ajdabiya Trough and migrated onto the Assumood Ridge from the northeast through late Cretaceous, Palaeocene and early Eocene carbonates before being trapped beneath the Awjilah Shale which is the principal regional seal in the area. The oil on the Zaltan Platform probably originated from the oil prone source facies further south in the Ajdabiya Trough (Figure 8.15). The As Sahel and Assumood gas pools were put on production in 1991 and 1993.[58]

8.4.8 Other Reservoirs

The Wadayat Trough (Figure 8.15) was mentioned in chapter 6 as an embayment between the Assumud Ridge and the Zaltan Platform. It seems to separate the predominantly oil-prone reservoirs to the south from the gas-prone reservoirs to the north. It also represents a structural saddle between the Zaltan Platform north of the Lahib field and the Assumud Ridge at Hutaybah (Figure 8.16). This area was a structural low during the late Cretaceous and early Cainozoic, and neither Wahah Formation nor Palaeocene carbonates are present in reservoir facies in this area. A large horst is present within the low in which fractured quartzites of Cambrian

age form the core of the horst. Appraisal by Sirte Oil Company showed the horst to contain a giant gas field which was appropriately named Attahadi (challenge), with original gas in place of 9 to 10 TCF. The field has been developed and production was scheduled to begin in 2000/2001 at a rate of 250 to 350 MMCFGPD. The gas is believed to have been sourced from the gas-prone Sirt Shale source rock in the northern Ajdabiya Trough and to have migrated along the Wadayat Trough into the horst (Figure 8.15). Seal is provided by shales of Cenomanian age. Alternatively, but less likely, it is possible that the gas was generated in the northern Maradah Trough and migrated eastwards. The Attahaddi field is one of the few fields in Libya to be reservoired in proven Cambrian quartzites. Further south the small Arshad oilfield has produced small amounts of oil from the quartzites, but whether these quartzites are also Cambrian in age has not yet been determined.[59]

Eastern Ajdabiya Trough

For the purposes of this review the eastern Ajdabiya Trough comprises the Nubian Kalanshiyu province in the south, the Palaeocene Intisar reef province in the centre, and the Eocene Ash Shulaydimah Trough in the north. The Jalu, Amal, Nafurah, Messlah, Sarir and Hameimat areas are considered in the section on the eastern Sirt embayment.

8.4.9 Sirt Shale-Nubian
Petroleum System

Reservoir

The Nubian (Sarir) Sand is without doubt the best clastic reservoir in Libya. The formation passes from a quartzitic facies in the central Ajdabiya Trough to a highly porous fluvial sandstone in the southern Ajdabiya Trough. The preservation of the Nubian Sands in the southeastern Sirt Basin was controlled by the regional palaeogeography. Ibrahim and Ambrose both published isopach maps of the Nubian sequence which show the presence of large palaeotopographic highs, such as the Messlah, Jalu, Harash and T1-59 highs, which were never covered by Nubian Sandstone, whereas thick Nubian Sands are developed in the intervening troughs (Figure 8.17). However Albian tectonic events resulted in transpressional pop-up of the Kalanshiyu High, and subsequent erosion led to the removal of the Upper Sarir Sandstone from the crest of the high as seen on the Kalanshiyu (EE-59) field.[60]

According to Rossi et al. the sandstone represents an association of alluvial fans, braided streams, meandering streams and lacustrine deposits. The correlation problems associated with the Sarir Sandstone were mentioned in chapter 4. Ibrahim and El Hawat favoured a three-fold subdivision into Upper Sarir Sandstone, Middle Shale and Lower Sarir Sandstone. Ambrose however suggested that this scheme was flawed and has resulted in major correlation errors. He recognised two shale horizons, and proposed a five-fold subdivision of Upper Sandstone, Variegated Shale, Middle Sandstone, Red Shale, and Lower Sandstone. This five-fold sequence is shown in well 5I3-59. This revision has major implications for the distribution of the two principal sandstones.[61]

The characteristics of these units were described in chapter 4. Both the Upper and Middle Sarir Sandstones are important reservoirs in the Kalanshiyu area. Porosity is around 20% with permeabilities occasionally reaching 200 mD. The Lower Sarir Sandstone in this area is thin and reservoir quality is erratic. On the western margin of the Messlah High volcanic flows are present within the Nubian sequence and these have contributed large amounts of clay minerals to the Nubian Sands which have reduced reservoir quality on the O-80 and T-80 fields.[62]

Source

The Sirt Shale is the principal source rock in the southern Ajdabiya Trough. TOC content in this area ranges from 2 to 5% and kerogen is principally type II. It is mature in the southern Ajdabiya Trough and perhaps even late mature in the Kalanshiyu Trough (Figure 8.17). However in the shallow Rachmat Trough it is immature. The southern Ajdabiya source kitchen almost certainly sourced the Chadar (A-126) field, the three fields on the Kalanshiyu High, Kalanshiyu (EE-59), 5I-59 and 4T-59, and the two fields on the western margin of the Messlah High, 3U-59/O-80 and T-80, although the small source kitchen in the Kalanshiyu Trough may also have contributed to the fields on the Kalanshiyu High. These fields are gas-rich and it has been suggested that late-stage gas progressively displaced oil from these fields from west to east. Migration from the Campanian source rock into the stratigraphically lower Nubian Sands was facilitated by fault conduits and by relay ramps which are well developed around the main palaeotopographic and structural highs. Ambrose made a case for the intra-Sarir Variegated shale as a secondary source rock in the Kalanshiyu area. These shales yield a waxy extract, and may be linked to the high-wax oils found in the O-80 field. Ambrose also speculated that Triassic shales could be mature in the Rachmat Trough.[63]

Seals

The Nubian (Sarir) Sandstone in the southeastern Ajdabiya Trough is sealed by Upper Cretaceous shales of the Etel, Rachmat and Sirt Formations. In the Kalanshiyu Trough thick evaporites are developed in the Cenomanian, but are absent elsewhere. The area is heavily faulted and the faults provide effective seals where shales are juxtaposed against the Nubian Sands.[64]

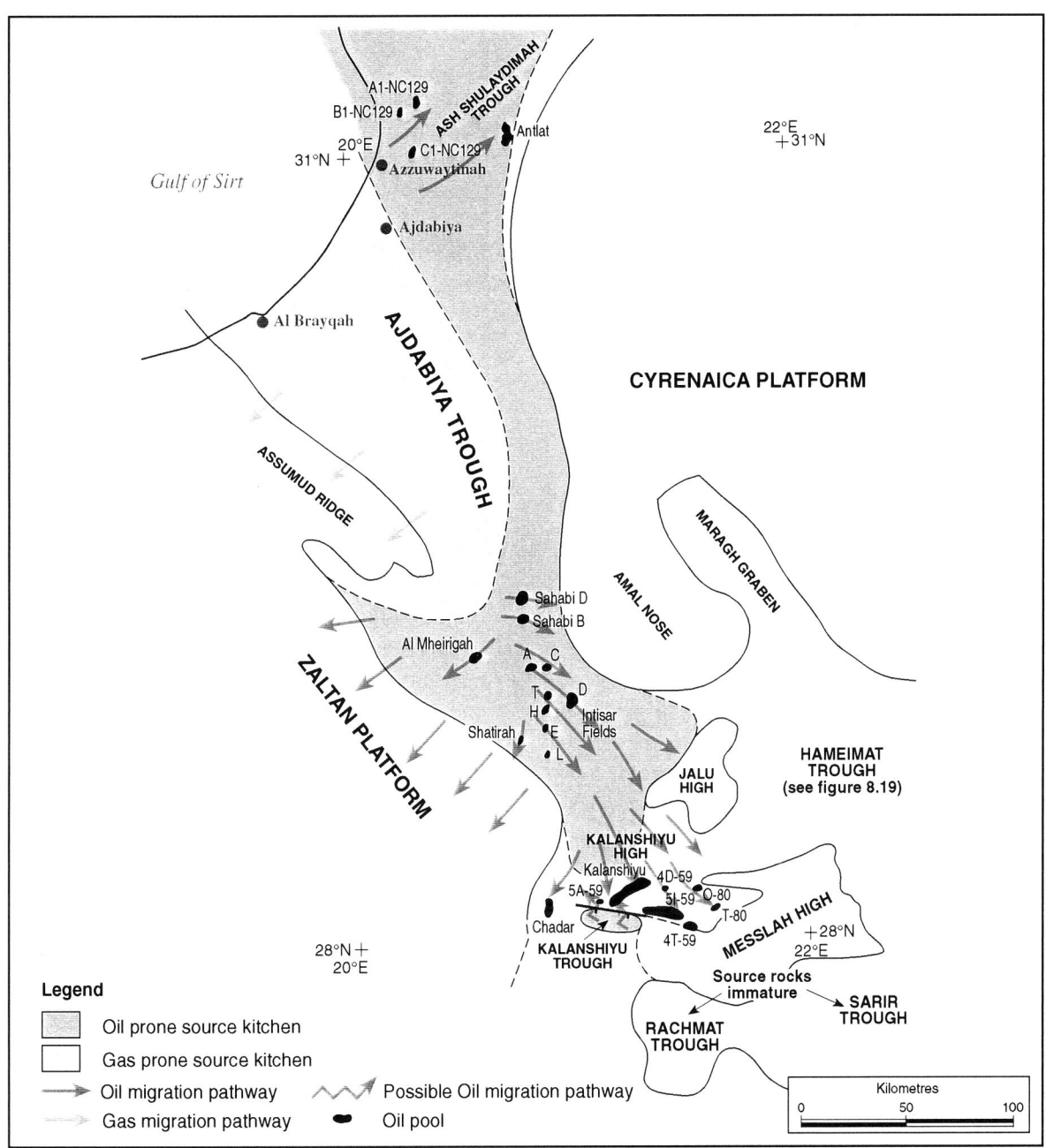

Sources: El Alami, et al., 1989, Ibrahim, 1991, Ambrose, 2000

Figure 8.17   Eastern Ajdabiya Trough Petroleum Systems

The Sirt Shale source kitchen extends for 300km from the Ash Shulaydimah Trough to the Kalanshiyu Trough. The kitchen has sourced the Nubian fields in the Kalanshiyu area, the Palaeocene reefs in the Intisar area and the Eocene carbonates in the Antlat area, although Palaeocene and Eocene source rocks may also have contributed. The Kalanshiyu fields were mostly charged by migration along faults, whilst the Intisar and Antelat fields were charged by migration through porous carbonates. In each case seal is provided by regionally extensive shales.

Traps

The traps which affect the Nubian Sandstone in the southeastern Ajdabiya Trough are mostly structural and are related to Albian compressional tectonics. The Kalanshiyu High is a horst which was inverted by transpressional tectonics during the Albian. The east-west faults which bound the horst continued to be active until the early Campanian. The Kalanshiyu, 5I-59 and 4T-59 fields are all culminations on the Kalanshiyu High. A similar tectonic regime characterises the western margin of the Messlah High, and the O-80 field is also an inverted horst formed by the reactivation of transpressional faults (Figure 8.17). The T-80 pool, on the other hand, is a simple anticlinal closure located within a re-entrant on the western margin of the Messlah High.[65]

Summary

The Sirt-Shale-Nubian petroleum system in the southeastern Ajdabiya Trough contains over 500 MMB of original oil in place in six main fields which were sourced by the Sirt Shale from the southern Ajdabiya Trough. These fields have not yet been put on production although development is planned. The accumulations vary from light, low-sulphur crudes to gassy and waxy crudes which suggests a mixed origin for these oils, perhaps involving a contribution from the waxy intra-Nubian source rock, and a late charge of gas from source rocks in the Kalanshiyu Trough.

8.4.10 Sirt Shale-Palaeocene Petroleum System

Reservoir

The Intisar, Sahabi and Al Mheirigah reefs are located within the Ajdabiya Trough in the shale embayment between the Upper Sabil and Zaltan carbonate shelves (Figures 5.5 and 8.17). The Intisar reefs were described by Terry and Williams and subsequently by Brady et al. In the axis of the Ajdabiya Trough rapid subsidence during the late Palaeocene led to a situation in which carbonate deposition was replaced by shale deposition. However, in a few isolated areas reef growth was able to keep pace with subsidence, and pinnacle reefs grew to a height of 385m. Individual reefs reach 5km in diameter. The age of the reefs is Thanetian, which makes them time-equivalent to the Zaltan Formation on the Zaltan Platform. The Intisar reefs are composed of coral and algal framebuilders with a bioclastic and micritic matrix. Porosity averages 22% and permeability averages 87 mD, but can reach as much as 500 mD. The gross oil column in the 'D' reef is 291m thick. Oil gravity is 40°API and the original gas-oil ratio was 509 scf/bbl. The presence of ten reefs has been established of which the Intisar 'A' and 'D' structures represent giant fields. Total oil in place for the whole complex is over 4 billion barrels. In addition to the reefs at Intisar, much smaller reefs have been found at Al Mheirigah, As Sahabi and Shatirah. The discovery well of the Intisar 'D' field tested oil at a rate of 75,000 bopd, which is impressive by any standards, and Brady, et al. claimed a recovery factor of 75% for the field. This perhaps explains why Occidental shipped out the first cargo of crude from Intisar in less than a year from the date of the initial discovery.[66]

No significant oil has been found in the Palaeocene on the Sabil shelf, largely due to the lack of seals.

Source Rock

At Intisar Brady et al. believed the reefs to be sourced from the late Palaeocene Sheterat Shale but, as discussed in chapter 7, this is most unlikely since the Sheterat has neither the capacity nor the organic richness to generate the large amounts of oil found in the Intisar reefs. Similarly the Al Khayir Shale is not a viable candidate since it has been shown to be lean. Geochemical evidence suggests that the Intisar oil was derived from Upper Cretaceous shales, which are the only source rocks with adequate potential in this area. Whether the oil was derived from the Sirt Shale or the subjacent Taqrifat/Rachmat sequence, as favoured by Burwood et al., remains to be resolved. Vertical migration probably took place through the Kalash and Lower Sabil Limestones and through the Sheterat Shale which, as suggested by Barr and Weegar, is probably not an effective seal in this area. Burwood et al. also examined oil extracts from well K1-12, drilled by Mobil close to the Intisar 'A' field. They concluded that the most likely source for these extracts was the marine Palaeocene Harash shale, despite its apparent lean character in most of the areas it has been analysed (Figure 8.18).[67]

Seal

The Intisar and adjacent structures are pinnacle reefs. Top seal and lateral seal is provided by thick shales overlying the Upper Sabil carbonate shelf on which the reefs grew. These shales equate to the Harash and Khayir Formations.[68]

Trap

The reefs at Intisar, As Sahabi, Al Mheirigah and Shatirah are stratigraphic traps. In each case reefal framebuilders became established on small swells on the sea floor and reefal growth was able to keep pace with subsidence. They are surrounded by Harash and Khayir Shales which form both the seal and trap. Oil was generated from the upper Cretaceous shales during the Oligocene/Miocene, migrated through the Kalash and Lower Sabil carbonates and into the overlying reefs. Most of the structures appear to have been filled to spill-point, although some of the reefs have developed rim-synclines which

may have affected the migration pathways. At least one reef in the complex is barren of oil.[69]

Summary

Oil bearing pinnacle reefs of Palaeocene age are present in the re-entrant between the Sabil and Zaltan Platforms. This group of fields is unique in Libya. This is the only area so far discovered in which pinnacle reefs apparently flourished in the rapidly subsiding south Ajdabiya Trough. About a dozen reefs have been discovered of which two represent giant fields. Collectively they contain over 4 billion barrels of original oil in place. They were sourced from upper Cretaceous shales and are trapped and sealed beneath the shales of the Harash and Khayir Formations.

8.4.11 Sirt Shale-Eocene Petroleum System

On the eastern side of the Ajdabiya Trough oil has been discovered in Eocene Jalu carbonates (with local formation names) at Antlat, A-NC 129, B-NC 129 and C-NC 129 (Figure 8.17). These reservoirs form part of the great development of Eocene carbonates in the Ash Shulaydimah Trough between the Cyrenaica Platform and the Jabal al Akhdar Uplift which was described in chapter 5. In this case the oil is trapped in small faulted structural closures formed during the late Cainozoic tectonic cycle and sealed by late Eocene shales. Reservoir quality is generally fair to poor and these discoveries have not yet been developed. Burial-depth and TTI data prepared by Ghori showed that the Sirt Shale is the most likely source rock in this area, and that it is within the peak oil generation window in the Antlat area. However, Burwood et al. suggested that the oil at Antlat represents a hybrid system with a contributions from

Petroleum System		Oil Family	Source type	Representative oil pools
Source	Reservoir			
Sirt Shale + ? Turonian	Nubian	1	Marine Marine	Abu Attiffel, Ar Raml
Sirt Shale + ? Turonian + ? Triassic	Nubian	2	Marine Marine Lacustrine	Sarir, Messlah
Sirt Shale + ? Rachmat	Eocene	3	Marine Marine	Jalu
Triassic + Sirt Shale	Maragh/Amal	4	Lacustrine Marine	Amal, Nafurah, Awjilah
Triassic	Maragh/Amal	5	Lacustrine	Jakharrah, As Sarah, Libba
Sirt Shale + ? Taqrifat	Upper Sabil	6	Marine Marine	Intisar, Shatirah
Sirt Shale + ? Harash + ? Eocene	Upper Eocene	7	Marine Marine Marine	A1-NC129, ? Antlat
Sirt Shale + Nubian	Nubian	8	Marine Lacustrine	UU65, VV65, JJ65
Harash (Palaeocene)	Upper Sabil	9	Marine	K1-12 (near Intisar)
Eocene	Upper Eocene	10	Marine	B1-NC152

Sources: Burwood, et al., 2000, El Alami, 1996, Ambrose, 2000, Gras and Thusu 1998

Figure 8.18   Eastern Sirt Embayment, Petroleum Systems

This chart is a listing of potential petroleum systems in the eastern Sirt Embayment. The Sirt Shale is the dominant source rock, but many of the reservoired oils appear to be hybrid systems with contributions from Turonian, Nubian (Variegated Shale) or Triassic source rocks. Burwood et al., proposed that Palaeocene and Eocene shales are also significant source rocks in the Ajdabiya Trough. Oil families 6 to 10 are from Burwood et al.

Palaeocene and Eocene source rocks which although not mature at Antlat, may reach maturity further west in the Ajdabiya Trough (Figure 8.18). The fields were most probably charged by migration through late Cretaceous and Palaeocene carbonates and along the many late Eocene and Oligocene faults which mark the eastern margin of the Ajdabiya Trough in this area. Burwood et al. also suggested that the small pool discovered by Agoco in 1991 in well B1-NC 152 may have been completely sourced from Eocene shales. If so, this would represent the first example of a viable Eocene source rock.

It is becoming evident that the source rock problems in the eastern Sirt Basin are more complex than was originally thought. Recent work suggests that in the Ajdabiya Trough Palaeocene and Eocene shales may be sufficiently mature to have generated hydrocarbons, and this raises the possibility that deeper in the Ajdabiya Trough these shales may be significant source rocks (Figure 8.18). More work is needed to determine the regional characteristics of these shales.[70]

Eastern Sirt Embayment

The eastern Sirt Embayment is the most distinctive province within the Sirt Basin in terms of structural evolution, reservoirs and source rocks. It represents the eastern arm of the Sirt Basin, and has a predominantly east-west structural grain reflecting the underlying Palaeozoic structure. There is clear evidence of rifting during the Triassic, and the major horsts at Messlah, Jalu and South Sarir were formed as a result of pull-apart tectonics during the Neocomian. Deposition of the Nubian (Sarir) Sandstone was determined by the resultant palaeogeography. The Nubian Sandstone is the principal reservoir in the eastern Sirt Embayment, and the Sarir, Messlah and Abu Attiffel giant fields are all reservoired in Nubian Sandstone. About forty Nubian discoveries have been made with collective original reserves of about 9 billion barrels. Further north on the Amal Nose giant hydrocarbon accumulations are present on structural highs reservoired mostly in quartzitic sandstones and basal Cretaceous clastics, which have been sourced by hybrid petroleum systems. Triassic source rocks in the Maragh Graben have sourced fields within the graben and have contributed to the fields on the Amal Nose. Most of the oil in the eastern Sirt Embayment has been sourced from the Sirt Shale, but shales within the Turonian, Lower Cretaceous and Triassic, also have source

potential and have made significant contibutions in some areas. For the sake of convenience some authors subdivide the pre-Upper Cretaceous sequence of the eastern Sirt embayment into several informal divisions, PUC A generally refers to the lower Cretaceous and Triassic sequences and PUC B to the Palaeozoic, although other authors prefer to subdivide the pre-Mesozoic into PUC B (late Permian) PUC C (Cambro-Ordovician age) and PUC D (unknown age).

The quantification of petroleum systems in the eastern Sirt Basin is complicated by the numerous hybrid systems caused by mixing of oils from the four principal sources. Ambrose recognised three petroleum systems within the Nubian province whereas Burwood et al. listed ten petroleum systems for the wider area of the eastern Sirt Basin, including potential Palaeocene and Eocene source rocks (Figure 8.18). In the following review three main systems are analysed, plus a number of hybrid systems, but it will become evident that mixing and leakage into other reservoirs has considerably blurred this simple categorisation.[71]

8.4.12 Upper Cretaceous-Nubian Petroleum System

Reservoir

The Hameimat Trough formed a major sedimentary depocentre during the deposition of the Nubian Sandstone, with braided fluvial systems entering the basin from all directions. The principal palaeovalleys, however entered from the west through the Sarir Trough and north of the Messlah High. Alluvial fans and fan deltas have been found on the flanks of the major horsts, particularly on the eastern flank of the Messlah High and the northern flank of the South Sarir High (Figure 8.19). According to Ambrose the Nubian Sandstone reaches a

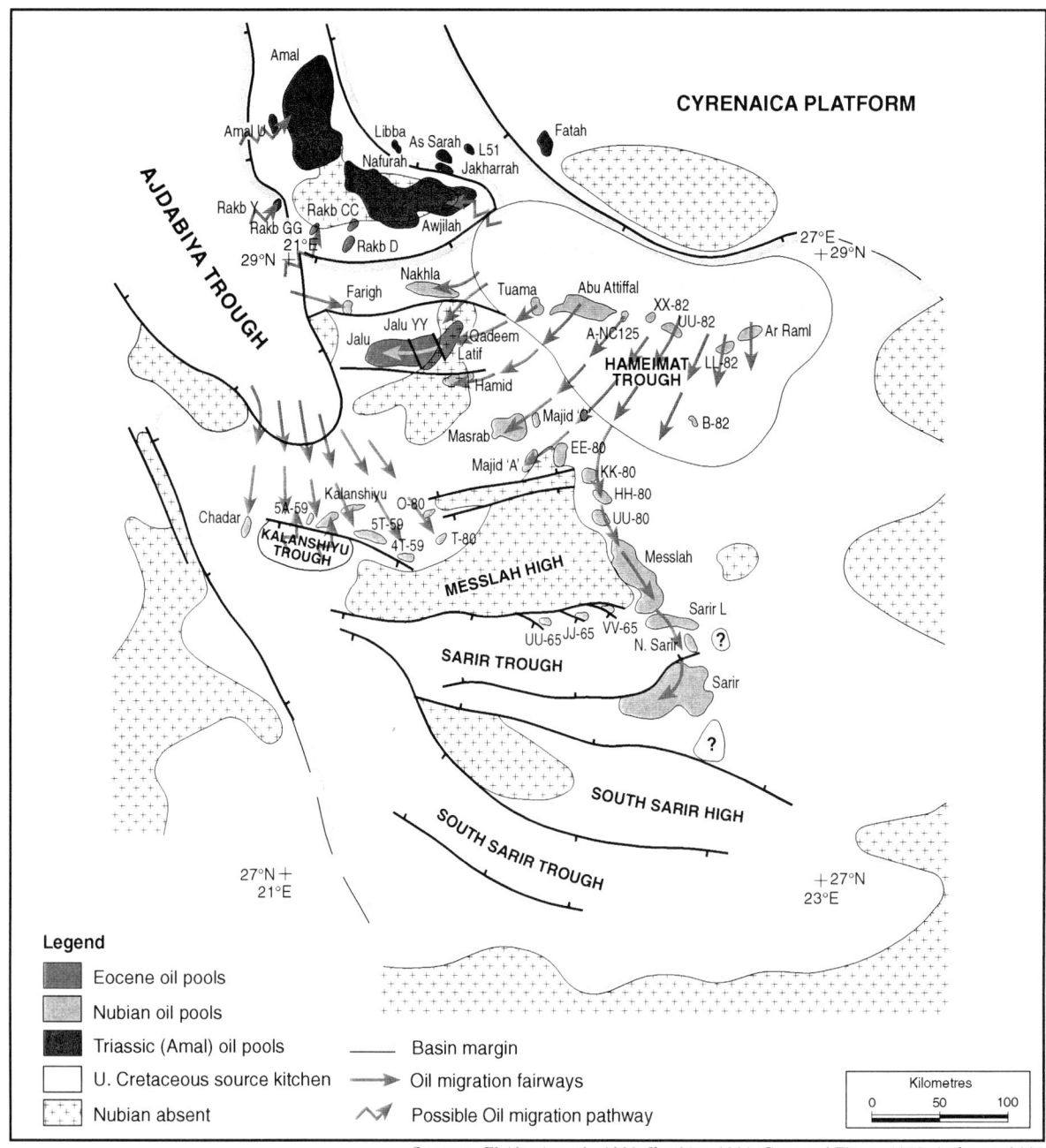

Sources: El Alami, et al., 1989, Ibrahim, 1991, Gras and Thusu, 1998, Ambrose, 2000

Figure 8.19 Eastern Sirt Embayment, Sirt Shale Petroleum System

The Sirt Shale source kitchen in the Hameimat Trough has sourced most of the Nubian fields to the south and west, along relatively long migration fairways. On the faulted saddle between the Ajdabiya Trough and Hameimat Trough oil migrated up faults into Eocene carbonate reservoirs on the Jalu trend. The Amal and Nafurah/Awjilah fields also have received some oil from Sirt Shale source kitchens. Other small source kitchens may be present in the Kalanshiyu Trough and to the east of Sarir, but the maturity of these rocks is has not yet been proved.

thickness of 2400m in the Hameimat depocentre and 1500m in the Sarir Trough. Along the main palaeovalleys it averages about 300m in thickness. It is absent over the major horsts. Mansour and Magairhy illustrated the irregular nature of the topography in the Jalu area where the thickness of the Nubian varies from zero to 790m within a distance of 6.5km. In concession 82, in the centre of the Hameimat Trough, the top of the Nubian Sandstone is at a depth of more than than 4500m. This has led to porosity reduction and overpressuring within the overlying evaporites. Ambrose subdivided the Nubian (Sarir) Sandstone into three sandstone members separated by two shale horizons. The complete sequence is present in the Hameimat Trough, but the upper units are progressively truncated on the flanks of the horsts by pre-Cenomanian erosion, and at both Messlah and Sarir the Upper Sandstone and Variegated Shale are missing. The Middle Sarir Sandstone has the best reservoir characteristics. Descriptions have been published of the three principal fields, Sarir, Messlah and Abu Attiffel (Figure 8.19).[72]

At Sarir the Middle Sarir Sandstone forms the principal reservoir. It has been divided into five reservoir members of which member 4 (a friable sheet sand about 30m thick) is the most permeable. Intergranular porosity ranges from 15 to 20% with an average of 18 to 19%. Average permeability is 200 to 300 mD. The total thickness of the Middle Sarir Sandstone is 300m, but the oil column is only 90m thick. The Lower Sarir Sandstone is barely present above the oil-water contact and plays no part in the oil reservoir. The Upper Sarir Sandstone and Variegated Shale are absent due to pre-Cenomanian erosion. The field covers an area of 384km[2] and contains original oil in place of 12 billion barrels and original recoverable reserves of 6.5 billion barrels.[73]

The Messlah oil pool is reservoired in a stratigraphic trap on the eastern flank of the Messlah High. As at Sarir, the Upper Sarir Sandstone and Variegated Shale are absent due to pre-Cenomanian erosion and the oil is trapped in the Middle Sarir Sandstone. The reservoir thins from 170m in well HH30-65 to zero at the pinch out edge. The original oil-water contact was at 2549m subsea, and the oil column in the downflank wells reaches 122m, with a maximum net pay of 60m adjacent to the southern boundary fault. Porosity averages 17% and permeability 500 mD. The Lower Sarir Sandstone is productive in some of the updip wells, but is much less important than the Middle Sarir main pay sand. The Lower Sand is not present in the northern part of the field. The field covers an area of 200km[2]. Original oil in place is around 3 billion barrels with 1.5 billion barrels of recoverable reserves.[74]

The Abu Attiffel field is located deep in the centre of the Hameimat Trough (Figure 8.19). The crest of the field is at 3886m subsea and the oil-water contact at 4336m subsea. The Nubian Sandstone in this area represents a mix of sedimentary environments reflecting its location in the centre of a large intracratonic basin. Both braided and higher-energy meandering systems are present, with numerous shale intervals representing temporary lakes. Sand bars, channel fill, levees and floodplain deposits have been recognised. Production is from the Upper Sarir Sandstone and the Variegated Shale forms the effective base of the oil pool. The Abu Attiffel reservoir has been divided into six zones which broadly represent sedimentological cycles. Reservoir characteristics are very variable. Porosity values range between 8 and 16%, and permeability ranges from zero to over 1000mD. Due to limited water drive, water injection has been required to maintain reservoir pressure. Reserves at Abu Attiffel have been estimated at 1.2 billion barrels original recoverable, suggesting an original oil in place figure of about 3.6 billion barrels.[75]

The Upper Sarir Sandstone is also oil bearing in the UU-65 pool on the southern margin of the Messlah High (Figure 8.19), and in this field the upper beds of the Middle Sarir Sandstone are plugged by a bituminous tar mat. Burwood et al. suggested that UU-65, and neighbouring pools, were sourced from a nearby local lacustrine source kitchen.[76]

Source

The Campanian Sirt Shale reaches an effective thickness of 300m (greater than 1% TOC) in the Hameimat Trough and has reached peak maturity in the basin centre. The depth to the oil generation window is about 3350m. Towards the south, however, both the effective thickness and the level of maturity decrease. At Masrab the thickness is 170m, and the source rock is early mature. At Sarir there is no effective source rock and the Sirt Shale is immature. It is evident that oil has migrated southwards from the basin centre within the Nubian aquifer and has been progressively trapped in a series of traps along the eastern margin of the Messlah High within a fairway ending at Sarir (Figure 8.19). A second fairway charged the Masrab field and the A-105 and Z-80 fields, and a third fairway to the west charged the Nakhla field. Burial history studies show that the Sirt Shale entered the oil generation window during the mid-Eocene and reached peak generation in the early Miocene. Bu Argoub studied the Upper Cretaceous shales at Sarir and found marine amorphous kerogen within the Turonian shales with oil generating potential. In this area however the Turonian shales are immature. A study by Bender, et al. on thirteen wells in the eastern Sirt Embayment demonstrated the limit of the Hameimat Trough source kitchen for the Sirt Shale but showed an extension along the fairway towards Sarir. They placed the top of the oil zone at 3500m, peak generation at 4100m and post mature zone at

4570m. This suggests that the centre of the Hameimat Trough at the present time is post-mature and gas prone.

The oil at Sarir has a gravity of 36.8°API, an initial GOR of 143 scf/bbl and a sulphur content of 0.25%. Pour point is high, ranging from 13 to 24°C, and wax content is 19%. The values at Messlah are oil gravity of 39.6° API, an original gas-oil ratio of 380 scf/bbl, pour point of 9°C and wax content of 8%. Both Sarir and Messlah show evidence of a tar mat at the oil-water interface. Oil to source correlations based on carbon isotopes show that the Messlah, Sarir L and Sarir C fields were overwhelmingly sourced from the Sirt Shale. In order to explain the high wax content it has been suggested that both Sarir and Messlah contain a contribution from a second, presumably lacustrine source rock, but the precise provenance of this waxy crude has not yet been established. The Sarir field has a tar mat at the base of the oil column which may be due to a miscibility reaction between the two oils.

The oil at Abu Attiffel is a highly paraffinic undersaturated oil with a gravity of 42° API, pour point of 36°C and original gas-oil ratio of 1284 scf/bbl. A study of the geochemistry of the Abu Attiffel area by El Alami showed that the most likely source of the waxy crude is the Turonian Etel Shale which is much richer than the Nubian Variegated Shale. The source for the small fields on the southern margin of the Messlah High is unclear. Ambrose proposed a local Triassic source kitchen in the Sarir Trough, but Burwood et al. favoured the intra-Nubian shale as the source for the waxy crude at Sarir and the UU-65 group of fields. The question remains whether the intra-Nubian shale is sufficiently rich or sufficiently mature in this area to generate significant hydrocarbons.[77]

Seal

The Nubian fields, such as Abu Attiffel, in the Hameimat Trough are sealed by Upper Cretaceous shales and evaporites belonging to the Etel Formation. The evaporites reach a thickness of 365m in the Hameimat Trough. The seal between the Nubian Sandstone and the Lidam Formation is often thin or absent and oil has leaked into the Lidam dolomites, as at Masrab. The Nubian Sandstone is absent on the Hamid field and oil has migrated along the unconformity and into the Lidam reservoir. Oil has also leaked into the Lidam Formation at C-NC 59 and on the 4U-59 structure. Further south, in the Sarir area, shales of the Rakb Formation form the top seal, and lateral closure to the northwest is formed by fault seal in which the Sarir Sandstone is juxtaposed against Rakb Shale. The Rakb Shale also forms the top seal for the Messlah field and the small oil pools fringing the Messlah High.

A study of the Upper Cretaceous sequence overlying the Nubian sandstone at Sarir was conducted by Bu Argoub. He determined that the earliest overlying sediments are the Transgressive Sandstone Unit of Cenomanian age, followed by shales of Turonian to Campanian age.[78]

Trap

Gras and Thusu recognised six main trapping mechanisms in the eastern Sirt Basin which they divided into basinal and marginal types. The Sarir C field has four-way dip closure over an intrabasinal high, although closure to the north and southwest is fault assisted. At Abu Attiffel oil is trapped in a major fault-bounded horst over an intrabasinal high. The structure is tilted towards the north and major faults form the northern and southern boundaries of the field. Smaller oil accumulations such as Ar Raml, UU-82 and XX-82 are similar tilted fault blocks in a basinal setting. Masrab is an intra-basin high, without major faulting, located within the palaeovalley extending along the northern margin of the Messlah High. The Messlah field represents a fault-assisted stratigraphic trap. The

Nubian reservoir pinches out on the eastern flank of the Messlah High, and is trapped between the overlying Upper Cretaceous shales and the underlying crystalline basement with fault closure along the southern margin. This type of trap is difficult to delimit due to the inability of seismic techniques to adequately define the pinch-out edge. Around the margin of the Messlah High other small pools are trapped on faulted ramps against the high. These include the EE-80, Majid and Z-80 pools on the northern flank, the KK-80, HH-80 and UU-80 pools on the eastern margin and the VV-65, JJ-65 and UU-65 pools on the southern margin. On the Majid 'A' field, oil is trapped in a thin wedge of Nubian Sandstone which pinches out against the northern spur of the Messlah High, with closure to the south provided by a fault (Figure 8.19). The gross oil column reaches 87m with average porosity of 17%. A tar mat is present at the base of the oil column. Original reserves were 19 million barrels but the field is now effectively depleted.[79]

Summary

The Nubian play in the eastern Sirt Basin is one of the major petroleum provinces of Libya. It contains around 9 billion barrels of original reserves in over 40 accumulations, of which three – Sarir C, Messlah and Abu Attiffel – are giant fields. The reservoir is a complex of fluvial sands deposited in braided streams draining into a major interior basin. The oil was mostly sourced by upper Cretaceous shales in the Hameimat Trough and migrated along a series of fairways through the Nubian Sandstone, and was trapped in structural and stratigraphic traps over a distance of about 100km. Several of the oil pools have a high wax content and it is likely that other source rocks have contributed to these pools. The Turonian and Triassic shales are the most likely sources of waxy crude. The intra-Nubian shales appear to be too lean to have generated significant volumes of oil, although in local areas they may be richer.

8.4.13 Upper CretaceousEocene/ Oligocene Petroleum System

A structural saddle separates the southern Ajdabiya Trough from the Hameimat Trough and the Jalu Horst is located within this saddle (Figure 8.20). Part of the horst was high and remained uncovered during the deposition of the Nubian Sandstone. The area was affected by mid-Cretaceous uplift and erosion, and during the Eocene and Oligocene an east-west shoreline trend became established through this area producing excellent reservoirs in littoral carbonates (Jalu Formation) and sandstones (Arida or Chadra Formations). These reservoirs have excellent porosity and permeability characteristics. Isopach maps of these formations in the Jalu area have been published by Mansour and Magairhy. The present-day horst was formed by late Eocene tectonism. These reservoirs contain large oil reserves. Thomas and Campbell estimated original reserves for the Jalu field of 4 billion barrels of oil. Original oil in place is probably around 9 billion barrels, although Thomas quoted a figure of 14 billion barrels which appears excessive. Several other fields have been discovered on the same trend, YY-59, Latif and Qadeem to the east and 4M-59 to the west. A review of the Latif/Qadeem area was published by Mansour and Magairhy. They showed that in addition to the Eocene and Oligocene reservoirs these fields also contain oil in deeper pools, including the Lidam and Taqrifat carbonates. The upper Cretaceous shales are the only source rocks with sufficient richness, thickness and maturity to have generated this oil, although Burwood et al. suggested that the Jalu oil was probably derived from the Rachmat Formation rather than the Sirt Shale (Figure 8.18). The oil is assumed to have originated from the Hameimat Trough and to

have migrated into the Eocene and Oligocene reservoirs during the Miocene via faults bounding the horst (Figure 8.20). It has been speculated that the Palaeocene Hagfah Shales or shales within the Eocene could represent potential source rocks, but the Palaeocene is generally lean, and the Eocene is immature in this area. The oil pool is trapped in several separate reservoir zones within the Jalu carbonate and Arida Sandstone and is sealed by tight micritic carbonates in the Jalu Formation, shales of the Awjilah Formation, and early Miocene shales overlying the Oligocene sandstones. On the structural crests, the Awjilah seal is frequently ineffective as at Jalu, and oil has leaked into the overlying Oligocene reservoirs. In the Maragh Graben oil shows have been encountered in Oligocene sandstones in wells D1-96 and F1-96. Despite porosities of 20% and horizontal permeablity of 1000mD reservoir performance is poor. This has been attributed to poor vertical permeability (average 100mD) caused by intense flaser bedding within the reservoir. Vertical seal is provided in this area by the Arida Shale Member immediately overlying the Chadra reservoir unit.[80]

8.4.14 Maragh Graben, Triassic-Amal Petroleum System

Triassic source rocks in the eastern Sirt Basin, and particularly in the Maragh Graben, were first recognised in 1982 when Triassic fossils were found in lacustrine shales in well L4-51. These Middle Triassic shales contain type II and type III kerogen and occur within the Amal Formation, long thought to be Cambro-Ordovician in age. The Triassic shales produce a waxy crude, but with a carbon isotope signature different from that of the Turonian and Nubian Variegated Shale. TOC content is fair to good. In the Hameimat Trough these shales are probably late or post-mature, but in the Maragh Graben thermal modelling on well L4-51 by

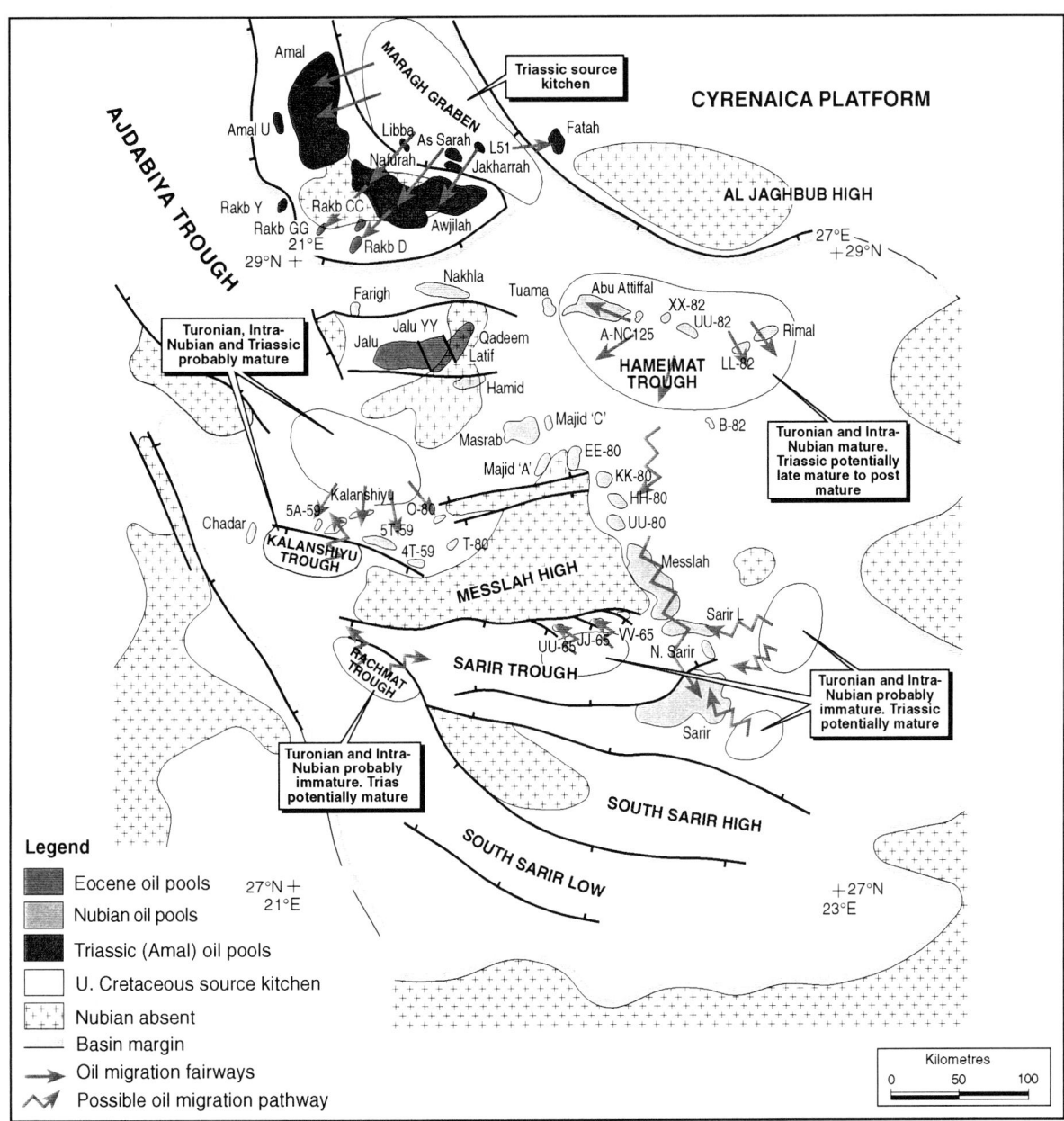

Sources: El Alami, et al., 1989, El Alami, 1996, Gras and Thusu, 1998, Ambrose, 2000

Figure 8.20   Eastern Sirt Embayment, Pre-Campanian Source Rocks

Beneath the Sirt Shale potential source rocks have been identified in the Turonian, intra-Nubian (Variegated Shale) and Triassic. El Alami dismissed the intra-Nubian shales as being lean and insignificant. The Turonian and Triassic shales both produce waxy oils, and oils of this type have been found at Amal/Nafurah-Awjilah, Abu Attiffel, the fields on the Kalanshiyu High, Messlah and Sarir. Many of these oils appear to be mixtures of non-waxy oils from the Sirt Shale and waxy oils from the Turonian or Triassic. The Amal and Maragh Graben fields were probably charged from the Triassic shales in the Maragh Graben. The waxy oils at Messlah and Sarir may have been derived from Turonian shales in the Hameimat Trough or from small local source kitchens.

Ghori and Mohammed showed the Triassic shales to be at peak to late maturity. Burwood et al. identified the Triassic source rock as the principal source rock for the As Sarah field and presumably also for the Libba and Jakarrah fields (Figure 8.20). The As Sarah field is located on a fault block tilted towards the north, and the sedimentology of the field was described by Kuehn who identified a braided to meandering fluvial system with overbank deposits and numerous palaeosols. The reservoir is sealed by black shales of Upper Triassic age. The reservoir characteristics of the As Sarah Triassic reservoir were examined by Ball, et al. They demonstrated that the reservoir shows widely varying poroperm characteristics. Porosity averages around 15% but permeability values fluctuate between 50 and 800mD, which is characteristic of this type of deposit. A tar-mat is present at the oil-water interface. Examination of the reservoir with a Formation Microscanner log shows numerous fractures, the majority of which are healed and sealing. Migration is assumed to have taken place from the Maragh Graben source kitchen into the adjacent Amal Formation starting in the late Eocene.[81]

8.4.15 Hybrid Petroleum Systems

It is generally agreed that three effective source rocks are present in the eastern Sirt Embayment: the upper Cretaceous marine shales of Campanian and Turonian age, and Triassic lacustrine shales. The petroleum systems involving the Sirt Shale and the Triassic source rocks have been discussed above. It now remains to discuss the hybrid systems which have resulted from the mixing of oils from the three source rocks. Burwood recognised ten oil families, but some of these are of minor importance and others are equivocal. This section will concentrate on the commercially most important systems.[82]

The giant Amal field is located on the Amal Nose, a basement ridge extending between the Ajdabiya Trough and the Maragh Graben (Figure 8.20). The characteristics of the field were described by Roberts. It is now known that the main clastic reservoir on the field, the Amal Formation, is at least partly Triassic in age. The field covers 400km^2 and plunges towards the north. Southwards it wedges-out onto a basement ridge. The oil pool extends into the flanking Maragh Formation of Upper Cretaceous age which onlaps the Amal Formation on the northern part of the field. The gross oil column reaches 180m with an oil-water contact at 3125m subsea. The Amal Formation ranges from feldspathic sandstone to quartzite and reservoir quality is poor to fair. The Maragh Formation has better characteristics, although it contains considerable amounts of interstitial clay and numerous shaly intervals. The Amal crude is an undersaturated paraffinic oil with a gravity of 35°API and a pour point of 18°C. Burwood suggested that the oil reservoired at Amal is the result of mixed system charging and that it shows elements of both the Sirt Shale and Triassic source rocks. The Sirt Shale charge was most likely derived from the Ajdabiya Trough source kitchen to the west, with a significant contribution from the Triassic shales in the Maragh Trough. Top seal is provided by the shales of the Rakb Formation. Original recoverable reserves were estimated by Campbell and Thomas as 4.25 billion barrels, suggesting an original oil in place figure of around 17 billion barrels.[83]

The Nafurah-Awjilah field is located on the Amal nose to the southeast of the Amal field. The accumulation forms a single field, but spans the boundary between concession 51 and concession 102, hence the two names. The northern part (Nafurah) was described by Belazi and the southern part (Awjilah) by Williams. The field is located on the crest of the Amal basement nose, and the D2-102 well proved the crest to be bald, with Campanian shales resting directly on granitic basement. The structure has four-way dip closure with the spill-point towards the northeast, where the western margin fault of the Maragh Graben terminates the field (Figure 8.20). On the flank of the structure three major reservoirs are present, the Amal quartzitic sandstone, the Maragh sandstone and the Taqrifat carbonate, and oil is also present in fractured granitic basement. Other oil pools are present in the overlying Palaeocene and Eocene carbonates. In all, oil has been found in eighteen different stratigraphic intervals, although 90% of the reserves are contained within the contiguous Taqrifat/ Maragh/Amal and basement reservoirs. The higher oil pools represent leakage along peripheral faults. As with the Amal field, the Nafurah-Awjilah oils show evidence of mixing from the Sirt Shale source kitchen to the west and from the Triassic source kitchen to the east. According to burial history/TTI plots, generation began in the late Eocene and has continued to the present time. The oil trapped in the Palaeocene and Eocene reservoirs most likely represents leakage via faults, particularly the fault marking the southwestern margin of the Maragh Graben, which is known to extend into the Upper Eocene. Seals are provided by the Rakb Shale (Upper Cretaceous), Khayir Shale (late Palaeocene), Awjilah Shale (late Eocene) and early Miocene shales. Estimates of original reserves vary from 1.5 billion barrels (Thomas), to 2.2 billion barrels (Burwood et al.). Likely original oil in place is of the order of 9 billion barrels.[84]

8.5   Offshore

Pratsch recognised ten petroleum systems in Tunisia of which four are relevant to the offshore, and two have major significance (Figure 8.21). In the northwest Libyan offshore, however, only one viable petroleum system has been established. The Eocene-Farwah petroleum system nevertheless includes the Bouri field, the largest offshore oilfield in the Mediterranean, and several smaller accumulations have been found on the same trend. The structure of the Pelagian Block and the Sabratah Basin was described in chapter 6 and the geochemistry of the area in chapter 7. In essence, the Pelagian Block represents a fragment of the north African plate which was severely affected by mid-Cainozoic Mediterranean tectonism and was characterised by very rapid subsidence during the Oligo-Miocene. The principal reservoirs in the Libyan sector occur within the Eocene, and the most likely source rock for these accumulations is the lower Eocene Bou Dabbous shale, but the gas and condensate in the centre of the Sabratah Basin is probably derived from deeper source rocks.[85]

8.5.1 Eocene-Farwah Petroleum System

Reservoir

The Farwah Group of Ypresian age includes a nummulitic limestone, the Jdeir Formation, which forms an excellent reservoir, and accounts for 85% of the productive reservoir at Bouri. The remaining 15% is contained within the underlying Jirani Dolomite. The reservoir represents a nummulitic bank which accumulated in a ramp setting on a local seafloor swell caused by underlying salt tectonics. The formation contains a variety of facies ranging from tidal flat and restricted shallow platform to open shallow platform with nummulitic banks and deeper platform to the north. The high-energy ramp facies found at Bouri extends westwards towards the Ashtart field in Tunisia and eastwards to well E1-NC 35a. Northwards it passes into fore-bank facies and southwards into low-energy, back-bank facies (Figure 8.22). The quality of the Jdeir reservoir has been modified by diagenesis. Calcite cementation occurred in the early stages of burial, but subsequent fresh water leaching

Age	Source Rocks	Seals	Lithology	Reservoirs	Oil and gas pools
Miocene		Oued Bel Khedim (Marsa Zouaghah)		Birsa (Al Mayah)	Tazerka, Oudna, Dougga Maamoura, Yasmine
Oligocene		Salammbo (Ra's abd Jalil)		Ketatna (Dirbal)	Halk el Menzil
Eocene	Bou Dabbous	Souar (Ghalil) / Chouabine (Bilal)		Reineche (Dahman) / Metlaoui (Farwah) El Garia (Jdeir) (Jirani)	Bouri, Ashtart, Sidi El Itayem, Miskar, Didon, Hasdrubal
Palaeocene	El Haria (U. Al Jurf)				
Maastrichtian		(L. Al Jurf)		Abiod (Abu Isa)	Tazerka, Maamoura Miskar
Campanian	Aleg (Jamil)				
Santonian					
Coniacian	Bahloul	Kef		Douleb	Miskar, Elyssa, Jugurtha
Turonian				Bireno	Rhemoura, Elyssa, Mahares Miskar, Gremda, El Ain
Cenomanian	Fahdene (Al Algah)			Zebbag	Elyssa, Ezzaouia, El Biban Isis, Didon
Albian				Serdj	
Aptian					Douleb, Tamesmida, Semmama
Barremian		Sidi Khalif		Lower Cretaceous carbonates	
Haut.-Valanginian					Robbana, Cap Bon
Berriasian					
Jurassic		Evaporites		Nara	El Biban

Sources: Pratsch, 1990, Hughes and Reed, 1995, Yukler, et al., 1995

Figure 8.21   Northwestern Offshore Petroleum Systems

This chart illustrates the many potential source rocks and reservoirs on the Pelagian Block. Since it is based largely on Tunisian data the nomenclature shows Tunisian stratigraphic names with the Libyan equivalents in brackets. In practical terms the two principal source rocks are the Bou Dabbous and the Bahloul shales, and the principal reservoirs are the Eocene and Cretaceous carbonates. In Tunisia hydrocarbons have been found at ten different stratigraphic levels, but in Libyan territory so far only the Farwah reservoir is productive. The Farwah contains the giant Bouri field sourced from Eocene shales and several large gas-condensate fields which were probably sourced from the Cretaceous.

has enhanced reservoir quality. The final burial stage led to compaction, but some late stage dissolution occurred during the migration of the hydrocarbons. The resultant porosity ranges from 10 to 25 % with an average of 16%, and permeability values up to 2000 mD have been reported. The Jirani dolomite contains good vuggy porosity with typical values of 16 to 22%, but permeability is very variable. Published estimates of original reserves at Bouri

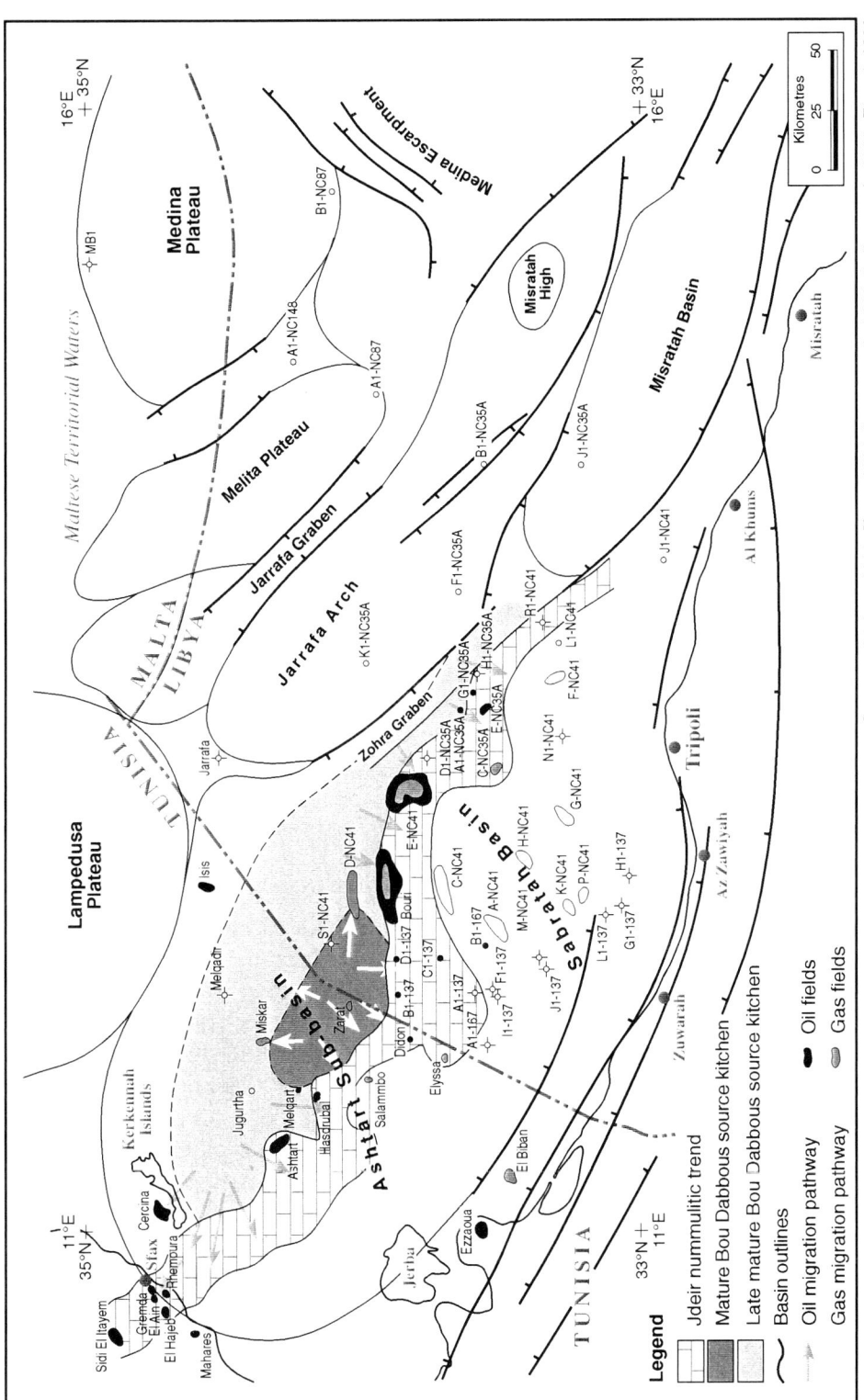

Sources: *Bishop, 1988, Sbeta, 1990, Racey, et al., 2001, Bailey, et al., 1989, Finetti, 1982*

Figure 8.22   Offshore, Eocene-Farwah Petroleum System

The Farwah (Metlaoui) Group contains the Jdeir (El Garia) reservoir of early Eocene age. The Jdeir Formation forms part of the ramp margin of a shallow shelf and is characterised by the presence of nummulitic shoals which represent storm deposits on the seaward ramp margin. The trend extends from Sfax on the Tunisian coast to the southern flank of the Misratah Basin. The trend contains the giant Bouri field plus several smaller discoveries. The principal source rock is the Bou Dabbous shale of Eocene age which is present seaward of the ramp. The source rock is in the peak oil generation zone north of Bouri and Ashtart, but probably within the late oil generation window at Zarat. Generation and migration occurred during the Oligocene and Miocene.

vary widely from 670 million barrels to 5 billion barrels. A figure towards the lower end of this range is most likely. Recent improvements in measuring gas while drilling have been used at Bouri to assess cap rock efficiency (sudden increase in gas as the reservoir is penetrated), porosity characteristics (high gas values correlate with good porosity), depth of the gas-oil contact, and the differentiation in the reservoir of gas-condensate from oil (significant changes in gas ratios).[86]

In Libyan waters Sbeta mapped the distribution of the Jdeir Formation, and showed the fore-bank facies to extend from D1-NC 41 eastwards to A1-NC 35a and beyond. The high-energy nummulitic bank facies extends from B1-137 through the Bouri field, D1, C1 and E1-NC 35a and north of well R1-NC 41. The low-energy back-bank facies occupies the area to the south and includes wells I1, F1, K1, E1, J1 and L1-137, and wells A1, C1, H1 and G1-NC 41. Further east, the N1, F1 and L1-NC 41 wells show a local development of bryozoan limestones (the Tajoura Formation) within the back-bank facies zone. A shoreline facies of the Jdeir Formation (Taljah Formation) is present in the nearshore wells G1 and H1-137 (Figure 8.22).[87]

In Tunisia, the equivalent of the Farwah Group is the Metlaoui Group which contains hydrocarbons in several fields such as Ashtart, Zarat, Didon, Hasdrubal and as far west as the onshore Sidi El Itayem field. The group has been extensively studied. Ashtart, which went on stream in 1974, has original oil-in-place of 720 million barrels and 245 million barrels of original recoverable reserves. Loucks et al. developed a facies model of the Metlaoui carbonate ramp in Tunisia which shows a transition from tidal flat (Fais Formation), through shallow restricted lagoon (Ain Merhotta Formation), high-energy shoal (up-dip El Garia Formation), low-relief nummulitic bank (down-dip El Garia Formation), open-marine deep shelf (Ousselat Member) to basinal (Bou Dabbous Formation). This corresponds closely to the situation found in the Libyan offshore. These authors showed that in Tunisia porosity within the El Garia Formation averages 15.4% and permeability 5.6 mD, although permeabilities as high as 3400 mD have been recorded. The best reservoir quality is found in the nummulitic gravels, although reservoir properties vary rapidly. This has been partly attributed to the effects of storm processes during deposition. High permeability zones near the base of the formation probably provide the conduit by which the hydrocarbons entered the formation. Studies in Tunisia have suggested that the nummulitic bank is not continuous. Analysis of drilling results in the Hasdrubal area shows several wells on the 'nummulite trend' as being dry or encountering poor quality reservoir. A model, developed by Bailey, et al, and refined by Racey, et al. suggested that at Hasdrubal the nummulitic mounds show evidence of redeposition in a mid-ramp setting under the influence of storms.[88]

Source Rock

According to Ricchiuto and Pajola the principal source rock for the oil accumulations in concession NC 41 is the early Eocene anoxic shale sequence known in Tunisia as the Bou Dabbous Formation which represents the deep-water equivalent of the Jdeir Formation (Figure 8.22). This formation contains amorphous marine type I and type II kerogen and oil to source correlations based on carbon isotopes indicate a close match. Biomarker and kerogen typing evidence proves the Bou Dabbous to be the source of both the oil and gas reservoired in the Hasdrubal field, which is located between Bouri and Ashtart. TOC values up to 2.5 % have been recorded from Tunisia. The typical Bou Dabbous oil found at Ashtart, Melqart and Hasdrubal is a high wax paraffinic oil derived from amorphous marine kerogen. Depth to the oil window is about 1800m, to peak generation 2500m and base of the oil window 3500m. The Bou Dabbous shales are within the oil generation zone around the Bouri field, and burial history modelling suggests that generation occurred in the late Miocene. The Bouri field is located at a depth of 2500m and the Ashtart field at a depth of about 3000m. Only in the deepest part of the basin does the Bou Dabbous reach the gas generation zone. The extensive gas deposits in concession NC 41, estimated by some authors at over 20 TCF, were probably generated from a deeper source rock, most likely the Turonian Bahloul Formation. According to Baric and Cota, oil and gas samples from the Eocene in well B1-NC 167 show evidence of mixing and varying levels of maturity, but carbon isotope analysis on oil and gas samples from the Hasdrubal field suggests a common origin from the late oil generation zone of the Bou Dabbous source rock. Anketell and Mriheel took the view that the extensive gas condensate deposits were produced by secondary thermal cracking of early generated hydrocarbons from Eocene source rocks in the centre of the basin. The distribution of the Bou Dabbous source rock in Libya has not been adequately documented. The situation is complicated by the stratigraphic revision of El Ghoul who demonstrated that the Hallab Formation is not the Libyan equivalent of the Bou Dabbous, and the maps of Sbeta give only a general indication that the deep water equivalent of the Jdeir Formation extends along the margin of the carbonate ramp and into the Zohra Graben (Figure 8.22). The northward and eastward limits of the Bou Dabbous as an effective source rock remain undefined.[89]

Seal

Top seal for the Jdeir nummulitic formation at Bouri is provided by shales of the Ghalil Formation which are equivalent to the deep-water Souar Formation of Tunisia. The Ghalil Formation passes shorewards into the Harshah Formation and the facies distribution of this unit has been mapped by Sbeta who demonstrated a gradual passage from outer shelf shales and mudstones to intra-tidal shaly dolomites, anhydrites and mudstones. On the Hasdrubal field in Tunisia a compact micrite at the base of the Cherahil Formation (equivalent to the Harshah Formation in Libya) provides the top seal. In areas where the Cherahil Formation is fractured hydrocarbons have migrated upwards and are ultimately sealed beneath the Oligocene Salammbo Formation (Ra's Abd Jalil Formation in Libya). The nummulitic mounds are draped over existing topography and the drape provides a measure of lateral seal as can be demonstrated at both Bouri and Hasdrubal. The bounding faults on the Hasdrubal horst are clearly sealing.[90]

Trap

The trap at Bouri and most of the hydrocarbon-bearing structures in the northwestern offshore is formed by simple four-way dip closure caused by drape over an existing topographic high. The topographic high in most cases is caused by salt diapirs, salt swells or salt walls in the underlying section. This is particularly the case in concessions 137 and NC 41 where ten hydrocarbon accumulations have been found in this type of trap. Relief on the top reservoir horizon at Bouri amounts to 230m with the spill point at 2592m subsea. In Tunisia the Hasdrubal field is a horst block, the faults of which were initiated during the late Cretaceous and attained their present configuration during the early Oligocene. A stratigraphic pinch-out forms the northern limit of the field.[91]

Summary

The Eocene-Farwah petroleum system contains the giant Bouri field plus several smaller fields in Libya and Tunisia. The Farwah sequence includes an Eocene nummultic reservoir, the Jdeir Formation, which represents a nummulitic bank deposited in a ramp setting and in some cases reworked and redeposited by storm action. The trend is discontinuous and several dry holes have been drilled on the trend in areas where the nummulitic bank is not well developed. The banks are preferentially located on topographic highs which in many cases are caused by underlying salt tectonics. The principal source rock is the Bou Dabbous Formation, a deep-water equivalent of the Jdeir Formation, although some of the gas condensate found in the Farwah Group within the Sabratah Basin may have been generated from Cretaceous source rocks, particularly the Turonian Bahloul Formation (Lower Makhbaz Formation in Libya). Seal is provided by shales of the Ghalil and Harshah Formations. The principal traps are anticlines formed by drape over nummulitic mounds and deeper salt structures. Horsts and tilted fault blocks are also found. Oil generation was very late, and most of the migration appears to have occurred within the last 10 million years.

Three aspects of the Eocene-Farwah petroleum system remain unresolved. The eastern limit of the nummultic trend is undefined. As described in chapter 5, Eocene nummulitic limestones are found on the Jahamah Platform and it is intriguing to speculate that the trend may continue between these two areas. The distribution and maturity characteristics of both the Bou Dabbous and Bahloul source rocks are ill defined in Libya, and until more information becomes available the remaining potential of this area remains unclear. Thirdly, the origin of the abundant gas-condensate in the basin is still problematic.

South of the main Jdeir trend it is likely that late and post-mature Cretaceous source rocks generated most of the gas-condensate trapped in the southern part of the basin.[92]

8.5.2 Other Reservoirs and Source Rocks

Significant discoveries have been made in Cretaceous carbonates in the Tunisian offshore (Figure 8.21). After the Eocene petroleum system the second most important system in Tunisia is the Bahloul-Zebbag system where basal Turonian black shales of the Bahloul Formation have sourced oolitic reservoirs at the top of the Cenomanian-Turonian Zebbag Formation. The facies distribution of the Zebbag Formation is complex, but there appears to be a general trend from nearshore in the west through transitional to deeper outer platform near the Libyan boundary. The reservoir is missing in some areas, such as on part of Jerba Island, due to salt tectonics. The reservoir facies contains foraminiferal, rudistid and oolitic limestones, calcarenites and dolomites deposited in a shallow subtidal environment. Porosity in the foraminiferal and rudistid limestones averages 12%, but can exceptionally reach 40%. The early Turonian Bahloul Formation is a dark-grey laminated, globigerinid marl which is a proven source rock in Tunisia, although Burwood et al. threw some doubts on its effectiveness. In Italy equivalent rocks have TOC values as high as 22%. The Isis gas field in Tunisia is reservoired in Zebbag carbonates and hydrocarbons have also been found in this reservoir at Elyssa, El Biban, Didon and Ezzaouia (Figure 8.23).[93]

Hydrocarbons have been found in several other Upper Cretaceous carbonate reservoirs in Tunisia (Figure 8.21). The Turonian Bireno Formation contains oil and gas at Elyssa, Miskar, and several minor fields, The Santonian Douleb Formation hosts hydrocarbons at

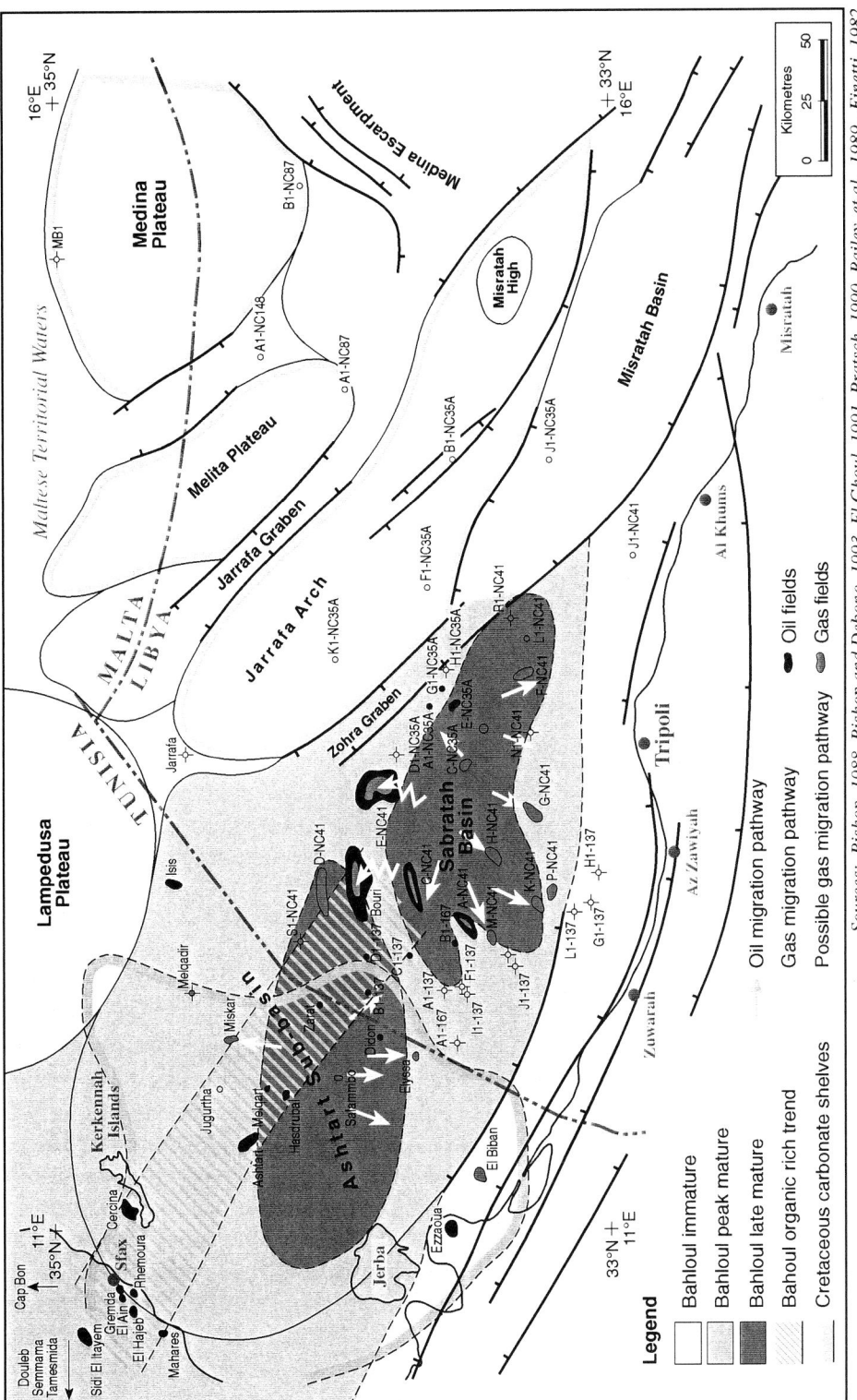

Figure 8.23   Offshore, Pre-Eocene Petroleum Systems

Sources: Bishop, 1988, Bishop and Debono, 1993, El Ghoul, 1991, Prutsch, 1990, Bailey, et al., 1989., Finetti, 1982

The Turonian Bahloul Formation is the principal Cretaceous source rock. The shales are late-mature in the Ashtart sub-basin and a similar late-mature kitchen is probably present in the Sabratah Basin south of Bouri. These kitchens have probably generated the large volumes of gas and condensate found in reservoirs ranging from Upper Cretaceous to Eocene in age. The Bahloul is particularly rich in organic matter in a belt extending from Sfax to Bouri. Around the basin margin the peak-mature shales have probably sourced the oils found at Isis, Rhemoura, Grenda, and El Ain. Other organic-rich shales may also have generated hydrocarbons but their potential has not yet been quantified. Nothing has yet been published on pre-Eocene source rocks in the Libyan offshore.

Jugurtha, Miskar and Elyssa, and the Maastrichtian Abiod Formation contains gas at Miskar, Tazerka and several smaller fields. The Bahloul shales are the likely source for all these accumulations. In Libya the Upper Cretaceous carbonate platforms extend as far east as the Medina Escarpment, and form the principal target on the Jarrafa Arch and Melita Plateau (Figure 8.23). Drilling results to date have been disappointing. Seven wildcat wells have been drilled in this area without success. Projecting the maps produced by Bishop from the Tunisian offshore it seems likely that the Cretaceous carbonates in Libya belong to a predominantly outer shelf setting where reservoir quality is likely to be mediocre. The presence of effective source rocks may also be a problem in this area. Similar results have been obtained in Maltese waters where four dry holes have been drilled. Nevertheless Bishop and Debono suggested that shelf edge calcarenites and rudistid reefs could be present in these areas, and the carbonate platforms cannot yet be entirely written off.[94]

Platform carbonates also dominate the early Cretaceous sequence on the Pelagian Shelf, and a few small discoveries have been made in pre-Cenomanian reservoirs. In Tunisia Aptian carbonates of the Serj Formation contain oil at Douleb, Semmama and Tamesmida, and Hauterivian-Valanginian carbonates contain hydrocarbons at Robbana and Cap Bon (Figure 8.23). The El Biban field southeast of Jerba contains oil, gas and condensate in the Middle and Upper Jurassic Nara Formation, and the large fields off southeast Sicily produce from Jurassic and Triassic carbonates. Seals for the Cretaceous hydrocarbon accumulations in Tunisia are provided by the Aleg (Jamil Formation in Libya) and El Haria (Al Jurf) shales. Within the Libyan sector of the Pelagian Shelf the early Cretaceous section has been penetrated as deep as the Neocomian but so far without success. The Neocomian-Barremian Turghat Formation represents an inner shallow shelf environment of deposition with considerably better reservoir characteristics than the Upper Cretaceous carbonates, but no data has yet been published on the eastward limit of the Turghat Formation.[95]

As far as source rocks are concerned, the most important source rock in the Cretaceous is the Lower Turonian Bahloul Formation. The Bahloul black limestone contains an average 3.9% TOC rising to a maximum of 8%. A zone of organically rich Bahloul source rock extends into Libyan territory in the Bouri area extending along a northwest-southeast trend (Figure 8.23). The eastern extension of the source facies in Libya is as yet undefined. The depth to the top of the oil window for the Bahloul Formation is around 2500m and to the top of the gas zone about 4000m. Over most of the Ashtart sub-basin and in the Sabratah Basin the Bahloul Formaton is within the gas generation zone and is the likely source of the Miskar gas field (Turonian reservoir) and the Eocene gas-condensate discoveries in concessions 137 and NC 41 (Figure 8.23). The majority of the gas-condensate accumulations are located over salt swells and salt walls which have generated fractures and faults in the Upper Cretaceous and Palaeocene section thereby providing migration routes for hydrocarbons to pass from the Turonian source rock to the younger reservoirs.[96]

Analyses of many samples from different shales between the middle Cretaceous and Eocene from the Tunisian offshore show similar geochemical characteristics and are difficult to separate chemically. In addition to the Bou Dabbous and Bahloul source rocks already discussed, many other shales have been mentioned in published accounts as proven or potential source rocks (Figure 8.21). These include the Campanian Aleg shales (Jamil Formation in Libya), the Palaeocene El Haria shales (Al Jurf Formation in Libya), and the Albian-Cenomanian Fahdene shales (Al Algah

Formation in Libya). Other possible source rocks may be present in the early Cretaceous and Jurassic. No systematic analyses have been published of these potential source rocks but they are believed to be of only local interest.[97]

Burwood et al. suggested some novel correlations between source rocks and reservoirs in the Tunisian offshore. They claimed that 'Miskar-style' gas/condensate pools were derived from gas prone Bou Dabbous source rocks, that the Elyssa accumulation was derived from the Palaeocene El Haria shales, and the Isis pool from Middle Jurassic Nara source rocks. These abnormal associations remain to be demonstrated.[98]

8.5.3 Sirt Embayment and Eastern Offshore

East of the Pelagian Shelf the geology of the continental margin changes dramatically. The shallow shelf is replaced in the Gulf of Sirt by a typical continental margin of slope and rise. In effect, the offshore Gulf of Sirt is divided into two provinces by the Sabratah-Cyrenaica Fault Zone (Figure 6.16). The area south of the fault is a continuation of the Sirt Basin into the offshore. The area north of the fault contains a wedge of Cainozoic and Upper Cretaceous rocks overlying faulted and truncated pre-Upper Cretaceous rocks which include Jurassic, Triassic and Permian marine sequences. A few wells have been drilled in the area south of the fault, but the area to the north has been penetrated by only one well, A1-NC 42, drilled by Agip in 1985 (see Figure 6.16 for location). The well was drilled in a water depth of about 900m to a depth of 3746m. A stratigraphic column of the well was published by Belhaj which shows a thick Cainozoic section overlying a thin Cretaceous sequence interrupted by several unconformities, and a non-marine Jurassic section overlying undifferentiated Palaeozoics. This is similar to the

sequence revealed by seismic data shown on Figure 6.20. Although minor discoveries have been made in the nearshore area, mostly gas in Cretaceous reservoirs, no viable petroleum system has yet been established in the area. Geochemical screening of wells C1-87, A1-87 and several neighbouring wells has shown good source rock characteristics in the Lidam, Etel, Rachmat and Sirt Formations. Far from being lean as originally believed, these rocks reveal TOC values up to 2.19%, and potential yield values reaching 8050 ppm. However, these rocks are immature to marginally mature and would require burial to 3500m to reach peak generation. These conditions are not found south of the Sabratah-Cyrenaica Fault, but they are certainly present to the north, which significantly increases the attractiveness of the Upper Sirt Slope north of the An Nuwfaliyah High. In addition, Palaeocene and Eocene shales should be within the oil generation window in the offshore extension of the Ajdabiya Trough and, depending on their organic content and kerogen facies, could represent potential source rocks not only in this area but also conceivably in the Misratah Basin where the Eocene reaches a depth of 3800m. Only one well, J1-NC 35a, has been drilled in the basin, which extends over an area of 4000km^2 (Figure 8.23).[99]

Further east the A1-NC 120 discovery off Binghazi in 1984 raised hopes of a significant new oil province, but the follow-up was was dry, and a further well, A1-NC 173, drilled by Lasmo in 1996, was also dry (Figure 6.16). The discovery is located on the edge of the Cyrenaica Platform where it extends into the offshore as the Cyrenaica Ridge. The well tested 5,263 bopd of 36°API oil from a multiple pay section of Lower Cretaceous age at a depth of 2440m, with possible reserves of 500 million barrels. The source of the oil is likely to be the Jurassic or basal Cretaceous shales mentioned by Ghori as being oil prone and mature in the western coastal region of Cyrenaica. No data on source rock potential has been released from the A1-NC 120 well.[100]

The northern coast of Cyrenaica is marked by a very narrow continental shelf and steep slope. Only the B1-NC 120 and A1-NC 128 wells have been drilled in the entire northern Cyrenaica offshore area, and little is known about the geology of this region. It is possible that potential Jurassic and early Cretaceous marine source rocks are present in the Marmarica Basin, and similar rocks may be mature in the deeper offshore area. Further exploration of this area is required before it can be written off as non-prospective.

Chapter 9

POSTSCRIPT:

WHERE ARE THE REMAINING UNDISCOVERED RESERVES?

9.1 Introduction

Chapter 8 reviewed seventeen established petroleum systems in Libya which account for most of the 134 billion barrels of original oil in place so far discovered in Libya. The question now arises how much oil remains to be discovered and where is it? Campbell published an estimate in 1997 giving the following statistics for Libya. Libya was ranked 11th in the world in terms of ultimate oil endowment with cumulative production to the end of 1996 of 20.38 billion barrels, estimated remaining reserves of 22.78 billion barrels, and 2.84 billion barrels yet to find, giving an ultimate endowment of 46 billion barrels. This equates to an original oil in place figure of about 144 billion barrels. The Libyan government would regard this figure as a considerable under-estimate. They published a remaining reserves figure of 29.5 billion barrels, but have not released their estimate of yet-to-find oil. The above figures are for oil alone and do not include gas equivalent. According to Campbell, the mid-point of depletion will be reached in 2001. The prediction of yet-to-find oil is based on mathematical models utilising worldwide experience of depletion patterns versus new discoveries, in basins of varying exploration maturity. These figures include oil recoverable by normal waterflood, pressure maintenance and artificial lift by conventional pumps. Of course, improved production technology may improve recovery from some fields, which is the basis for the government's higher reserve estimate, but as mentioned in chapter 1, it is unrealitic to assume

that enhanced oil recovery techniques can be applied to all fields. Given that a finite amount of oil remains to be found, the equation which governs the remaining productive life of an area is the balance between production and the addition of new reserves, which in turn is related to the intensity of exploration drilling. As mentioned in chapter 1, exploration drilling in Libya has been at a relatively modest level during the last twenty-five years which in practice has led to conservation of Libya's oil reserves.[1]

9.2 Yet-to-find Oil

Assuming the validity of Campbell's figures, and making allowances for discoveries since 1997 a further 7.5 billion barrels oil in place or 2.5 billion barrels recoverable is yet-to-be-found. Inevitably much of this oil will be found in established areas and in established petroleum systems. Nevertheless explorationists are ever seeking to open new frontiers, and it is likely that a proportion of the yet-to-find oil will be found in new areas. The discoveries in the Murzuq Basin between 1982 and 1985 opened up an area which was previously considered marginal, and the Bouri discovery in 1976 opened up the northwestern offshore. There are several under-explored areas which have the potential for contributing to the yet-to-find oil in Libya. This short postscript will briefly review some of these areas. The results are summarised on Figure 9.1 and the accompanying table.

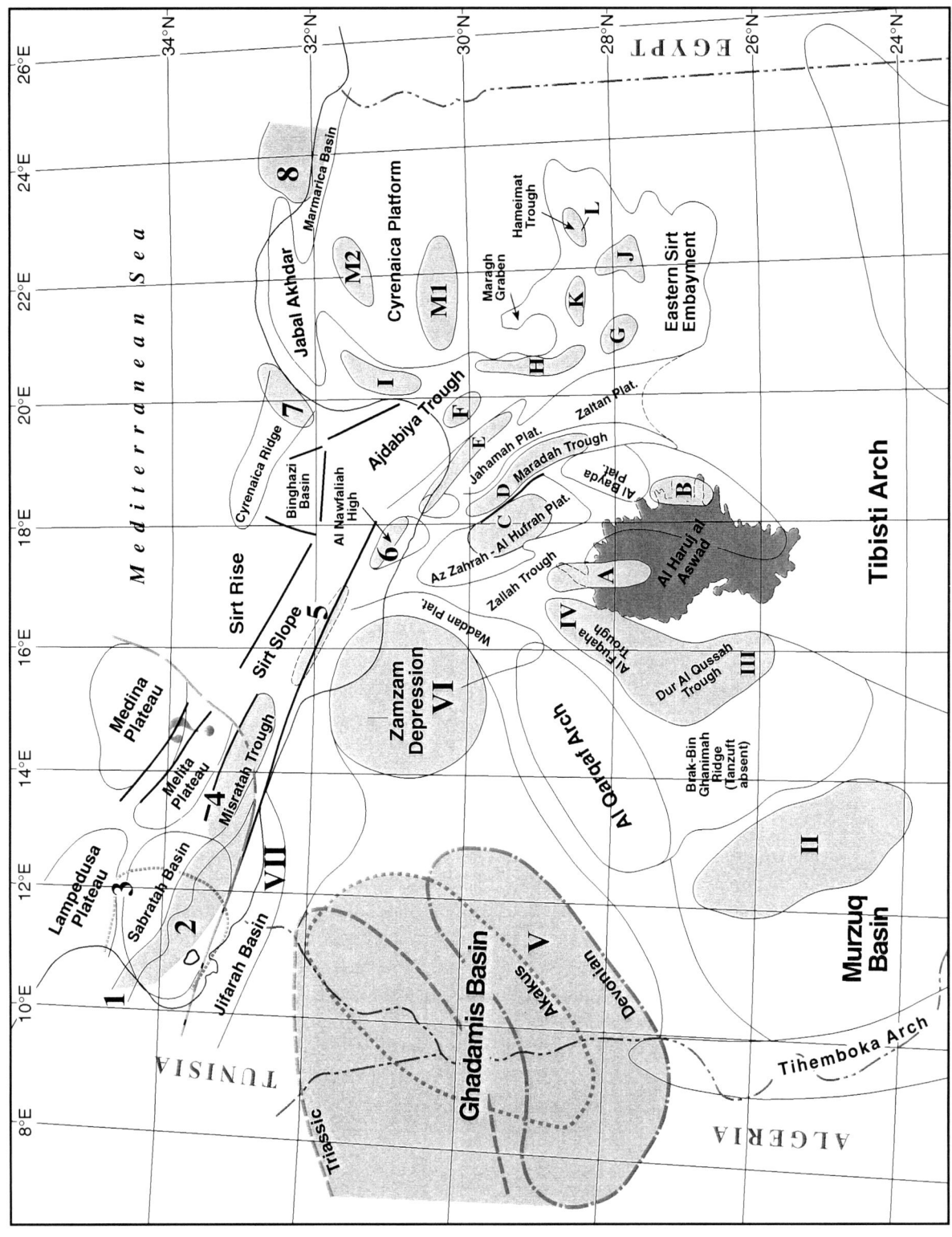

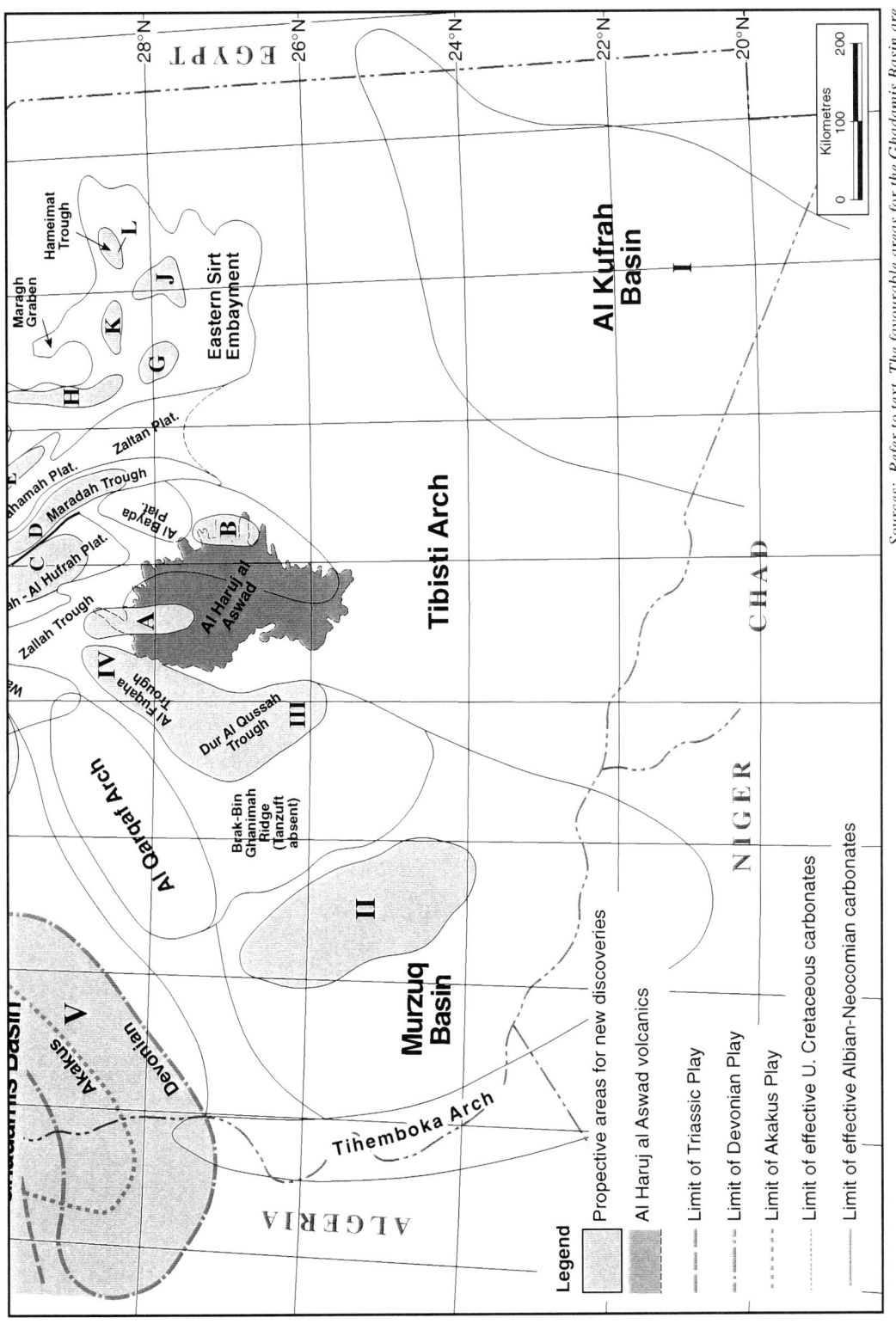

Sources: Refer to text. The favourable areas for the Ghadamis Basin are from Echikh, 1998, and for Cyrenaica from Sola and Ozcicek, 1990

Figure 9.1   Libya, Prospective Areas for New Discoveries

This map illustrates the text of chapter 9. It is an attempt to show under-explored areas which have some potential for new discoveries. Roman numerals refer to Palaeozoic basins, letters to the Sirt Basin and Arabic numerals for the offshore. For explanation see accompanying panel.

Explanation of Figure 9.1

Palaeozoic Basins

I   Al Kufrah Basin. Abundant reservoirs, traps and seals, but viable source rocks not yet encountered.

II   Murzuq Basin. Margins of Awbari Trough and Idhan Depressions prospective for Ordovician reservoirs.

III   Dur al Qussah Depression. Prospective for Palaeozoic reservoirs sourced by Tanzuft Shale.

IV   Al Fuqaha Depression. Prospective for Palaeozoic reservoirs sourced by Tanzuft Shale.

V   Ghadamis Basin. Favourable areas for the Akakus, Devonian and Triassic reservoirs are shown. Ordovician reservoirs are also prospective beneath the Tanzuft Shale.

VI   Zamzam Depression. Prospective for Palaeozoic reservoirs sourced by Tanzuft Shale, and for Cretaceous reservoirs on the Waddan Platform sourced from the Sirte Shale to the north.

VII   Jifarah Basin. Prospective for Triassic and Akakus reservoirs sourced by Tanzuft Shale.

Sirt Basin

A   Meulagh Graben. Extension beneath Al Haruj al Aswad volcanics. Prospective for Palaeocene and Eocene carbonates.

B   Abu Tumayam Trough. Depocentre south of Ayn an Naqah discovery. Eocene potential.

C   Az Zahrah-Al Hufrah Platform. Stratigraphic traps in Palaeocene carbonates.

D   Maradah Trough. Prospective for large horsts with fractured quartzite reservoirs. Deep and gas prone. Additional possibility of Upper Cretaceous sand-wedges on platform margins.

E   Jahamah Platform. Prospective for gas in Eocene, Cretaceous and quartzite reservoirs. Sourced from gas-prone kitchen to north.

F   Al Brayqah Horst and similar intra-basin structures. Prospective for Eocene and Palaeocene carbonates sourced from possible Eocene source rocks.

G   Kalanshiyu High and southern Ajdabiya Trough. Prospective for Nubian reservoirs including stratigraphic pinch-out traps.

H   Intisar reef trend. Potential for further pinnacle reef discoveries.

I   Ash Shulaydimah Trough. Prospective for Eocene carbonate reservoirs

J   Messlah area. Prospective for stratigraphic traps in Upper Sarir Sandstone.

K   Jalu-Hamid trend. Prospective for further discoveries in various reservoirs from Cenomanian to Oligocene in age.

L   Hameimat Trough. Prospective for further deep Nubian discoveries.

M1, M2   Cyrenaica Platform. Prospective for Palaeozoic gas discoveries.

Offshore

1   Sabratah Basin. Eocene Farwah nummulitic ramp trend.

2   Sabratah Basin. Gas-condensate potential sourced from Cretaceous shales.

3   Pelagian Block. Prospective for Cretaceous carbonates. Reservoir quality and source rock potential uncertain.

4   Misratah Basin. Speculative potential for Cretaceous carbonates.

5   Sirt Slope. Full marine Mesozoic sequence. Speculative potential in various reservoirs sourced by Cretaceous and Jurassic marine shales.

6   An Nuwfaliah High. Prospective for gas in Eocene, Cretaceous and quartzite reservoirs. Sourced from gas-prone kitchen to north.

7   Binghazi offshore. Prospective for Lower Cretaceous carbonate reservoirs sourced by Jurassic or early Cretaceous source rocks.

8   Marmarica Basin. Prospective for Lower Cretaceous carbonate reservoirs sourced by Jurassic or early Cretaceous source rocks.

9.3 Al Kufrah Basin

To date no oil or significant oil shows have been found in the Kufrah Basin (Figure 7.2 and Area I on Figure 9.1). It will be clear from previous chapters that there is no lack of reservoirs or structures within the basin. The principal problem is that of source rock. Lüning et al. presented a detailed review of the Tanzuft Formation in the Al Kufrah Basin based on both field work and limited well data. They concluded that in general the facies of the Tanzuft Formation was not conducive to the development of a hot shale facies, and sedimentological data supports this conclusion. However they did not rule out the possibility of hot shales being present in local depocentres, in the same way that hot shales are preserved in local depressions in the Murzuq Basin. The question then arises that if hot shales were encountered would they be mature. Lüning, and his associates reviewed the burial history of the basin and concluded that the Tanzuft shales probably reach maturity in the basin centre. Using the analogy of Oman they also suggested that other potential source rocks could be present in Infra-Cambrian rocks preserved within basement grabens. It is nevertheless a long shot, and the probability is that effective source rocks are not present in the Kufrah Basin.[2]

9.4 Murzuq Basin

Echikh and Sola presented an excellent analysis of the factors which control petroleum systems in the Murzuq Basin. The key elements are the presence of hot shales, which are confined to the north-central part of the basin, the extent of the source kitchen, the absence of Bi'r Tlakshin shales which form a barrier between the source rock and the Mamuniyat reservoir, and the presence of good quality reservoirs. To date two oil migration fairways

have been discovered which produced the oil accumulations on the northern margin of the Awbari Trough and Idhan Depression (Figure 8.4). There may be others: south of the Elephant field, southeast of the H-NC 115 field, and east and west of the Idhan Depression (Area II on Figure 9.1). It is also worth noting that where the Mamuniyat and Melaz Shuqran Formations are absent, the Hawaz Formation forms a good alternative reservoir.

Further east the Dur al Qussah Trough is essentially undrilled (Area III on Figure 9.1). The thick Lower Palaeozoic sequences were described long ago by Klitzsch and more recently detailed biostratigraphy has shown the presence of a Tanzuft Shale re-entrant extending northwards to Al Fuqaha and as far as wells C1-44 and A1-11 (Area IV on Figure 9.1). Details are lacking on the richness and maturity of the Tanzuft shales in this area, but if favourable the Dur al Qussah Trough could have considerable potential.[3]

9.5 Ghadamis Basin

The Ghadamis Basin contains two principal petroleum systems in the Libyan sector of the basin which have been well documented by Echikh. The Tanzuft source rock charged Akakus and Devonian reservoirs in the north around Tiji and on the Al Kabir trend, and in the east in the Al Hamra fields (Figure 8.7). The Devonian source rocks charged the Alrar-Al Wafaa fields and probably the Triassic fields in the Ghadamis area. No doubt further discoveries will be made in these, and intervening areas. The most striking feature however, is the virtual absence of significant oil discoveries in Ordovician reservoirs. Ordovician reservoirs form the main producing horizons in the Murzuq Basin and in the Algerian sector of the Ghadamis Basin, and it is remarkable that no major Ordovician discoveries have been made in Libya. It is clear that exploration drilling to date has concentrated on the shallower objectives and relatively few wells have evaluated the Ordovician potential, particularly in the basin centre. This remains a major under-explored objective (Area V on Figure 9.1). Similarly, although a few Trissic discoveries have been made northeast of Ghadamis, the Triassic potential has not been given the attention it deserves. Recent discoveries in Algeria should encourage further evaluation of this play. The discovery of the Al Wafaa field in 1990 in a large stratigraphic trap in the Devonian F3 sand suggests that similar traps could be present around the pinch-out magin of the sand and perhaps in other Devonian sands which are known to pinch out further south.[4]

The geology of the Zamzam Depression has been mentioned many times in earlier chapters. Based on detailed biostratigraphic work Belhaj was able to demonstrate a full sequence of Palaeozoic rocks in what amounts to a separate sub-basin of the Ghadamis Basin, extending beneath the Hun Graben and Waddan Platform (Area VI on Figure 9.1). Tanzuft shales are present with potential reservoirs in both Ordovician and Akakus-Devonian sequences. This area has been strangely neglected since the earliest days of exploration drilling in the 1960's. It may be that the Tanzuft shales are ineffective source rocks in this area due either to lack of hot shales or inadequate maturity, but oil shows in several old wells suggest otherwise and this area merits further investigation.[5]

No major discoveries have yet been made in the Jifarah Basin to the north of the Jabal Nafusah Uplift despite hydrocarbon shows in Permian and Triassic reservoirs. The basin contains good reservoirs, particularly the Triassic Kurrush Formation and the Silurian Akakus Formation, and the Tanzuft hot shale source rock is present in the area, plus a black shale interval within the Triassic. Fault and unconformity traps are present throughout the basin, and the Al Aziziyah fault system is par-ticularly attractive since it is likely that Triassic reservoir rocks are juxtaposed against Tanzuft source rocks. This area cannot yet be written off.[6]

9.6 Western Sirt Basin

Evidence presented by El Alami et al. suggests that no effective Sirt Shale source rock is present west of the Zallah Trough, and drilling results from the Hun Graben would appear to support this view. It is curious that whereas the source kitchens deeper in the Sirt Basin have conspicuously charged traps on the adjacent platforms there appears to have been virtually no migration of oil from the Zallah Trough onto the Waddan Platform. This is most likely due to the relatively small volume of the Zallah Trough source kitchen. Perhaps insufficient hydrocarbons were generated to charge any but the structures within the trough, with perhaps some migration into the fields on the western edge of the Az Zahrah-Al Hufrah Platform, although Roohi thought this unlikely. The Zallah Trough is one of the most complex areas of the Sirt Basin, containing a number of elements which differ markedly in structural style. The main petroleum system in the Zallah Trough is the Sirt Shale source rock and reservoirs in the Palaeocene and Eocene. The main source kitchen charged fields to the north along a fairway from Daba to Mabruk (Figure 8.11). Fairways to the south sourced fields in the main trough axis, in the Ma'amin Graben and on the western flank of the trough. The Themar and Al Abraq discoveries were made beneath the Al Haruj al Aswad lavas, and Knytl et al. and Schröter both suggested that the Meulagh Graben continues to the south beneath the volcanics (Area A on Figure 9.1). Because of the difficulties of obtaining good quality seismic data below the volcanics this area is very poorly known. Only a handful of wells has been drilled through the volcanic sequence and they may not have been optimally located. This area has significant untested potential. In addition Klitzsch suggested that a remnant of the Kalanshiyu Trough, containing Lower Palaeozoic sequences similar to those in the Dur al Qussah Trough may also have been preserved beneath Al Haruj al Aswad. This concept does not appear to have been tested.[7]

Further south the Kotlah source kitchen sourced accumulations in the Kotlah Graben and on the adjacent platforms, and the kitchen probably extends southwards into the Abu Tumayam Trough (Figure 8.11). This is an area which has seen significant exploration efforts in recent years resulting in the Barrut and Ayn an Naqah discoveries. Whether these fields were charged from the Al Kotlah source kitchen or a kitchen located further south is not clear, but structurally the depocentre of the Abu Tumayam Trough is located south of Ayn an Naqah, and this area has been very lightly explored (Area B on Figure 9.1). No data on source rocks has been published from the Abu Tumayam Trough, but it has many similarities with the Zallah Trough, and in the absence of evidence to the contrary there is every reason to anticipate a viable source kitchen in the Abu Tumayam Basin. Well evidence shows good quality reservoirs sealed beneath thick evaporites and this area may have considerable potential.[8]

9.7   Maradah Trough

A major Sirt Shale source kitchen is located in the Maradah Trough which has generated large volumes of hydrocarbons which have sourced primarily Palaeocene reservoirs on the Az Zahrah-Al Hufrah and Al Bayda Platforms, and probably contributed to some of the crestal fields on the Zaltan Platform including Nasser and Dayfah. El Alami et al. showed that the Sirt Shale passes into a gas-prone kerogen facies to the north, and this is borne out by the gas accumulation at Bazzuzi and by gas shows in several

wells in the trough. The problems of access for hydrocarbons to Cretaceous reservoirs on the Az Zahrah-Al Hufrah Platform due the the presence of the Manzilah Ridge were discussed in chapter 8. Only to the north and south of the ridge have hydrocarbons charged the Wahah reservoir. The presence of several stratigraphic and partial stratigraphic traps in Palaeocene carbonates on the platform suggests that other similar traps may remain to be discovered or existing traps may prove to be larger than currently mapped (Area C on Figure 9.1). Sedimentological evidence shows that the Palaeocene carbonate shelves in the western Sirt Basin have ramp-type margins, and the possiblility of finding high-energy barrier-reef type margins is remote.[9]

Hallett and El Ghoul reviewed the deep potential of the Maradah Trough and pointed out the possibility of analogues of the Wadi field being present further north. Large structures have been mapped in this area using gravity data, and none of these structures has yet been drilled (Area D on Figure 9.1). El Hawat et al. showed that part of the reservoir at Wadi is Lower Cretaceous in age, and it is likely that these Nubian-equivalent fractured quartzites are present throughout the trough. Ibrahim studied geothermal gradients in the trough and, as described in chapter 7, developed a predictive method for determining prospective areas. Sand wedges on the downthrown side of the platform margin faults were described by El Ghoul, Jones and Hammuda, and although oil has yet to be found in these wedges, they represent excellent prospective reservoirs. The deep potential of the trough is virtually unexplored particularly in the area north of Ar Raqubah. It is likely to be gas prone and very deep but some of the structures, as shown by the gravity data, are very large.[10]

9.8   Western Ajdabiya Trough

The Ajdabiya Trough has probably generated more hydrocarbons than any other comparable

424

size area in Libya. Seven petroleum systems
were described in chapter 8 and inevitably
further discoveries will be made in this area.
Because of the depth of the Ajdabiya Trough
little is known about the hydrocarbon potential
of the northern part of the basin. El Alami et al.
suggested that the Sirt Shale source rock is lean
and probably gas prone in this area, and the gas
accumulations at Attahadi, Hutaybah, As Sahel
and Assumud support this view. A similar
situation is present in the near offshore area
where gas is present in several wells on the An
Nufaliyah High. There may be further scope for
similar gas accumulations along the seaward
margin of the Al Jahamah, Shammar and An
Nuwfaliyah Platforms (Area 6 on Figure 9.1).
An Eocene and Oligocene shoreline trend passes
through this area which may possibly join the
Farwah trend in the northwestern offshore
(Area E on Figure 9.1). This is a prolific trend
which contains gas at As Sahel and Assumood
and several smaller discoveries, and east of the
Ajdabiya Trough a similar trend is present in the
Jalu area. Further significant discoveries on this
trend cannot be discounted.

In the Ajdabiya Trough Wennekers et al.
pointed out the presence of a large intra-basinal
high which they called the Al Brayqah Horst
(Area F on Figure 9.1). This feature could
contain reservoirs and traps within Palaeocene
and Eocene reservoirs, and Cretacous reservoirs
may be present at depth. Burwood, et al.
suggested that in this area Palaeocene and
Eocene shales are probably mature and oil
prone, and the area has considerable
unevaluated potential.[11]

9.9 Eastern Ajdabiya Trough

The eastern Ajdabiya Trough contains three
main petroleum systems primarily sourced from
the upper Cretaceous, but with significant
additions from other source rocks, and
reservoired in Nubian Sands in the south,

Palaeocene reefs in the centre, and in Eocene
carbonates in the north.

The Nubian discoveries on the Kalanshiyu
High surprisingly have not yet been developed
despite their considerable reserves. These are
gas-rich fields sourced from the southern
Ajdabiya Trough with a possible contribution
from older source rocks in the Kalanshiyu
Trough. This is a prolific oil province and the
likilihood of further discoveries is high (Area G
on Figure 9.1). Several pools around the
Messlah High are trapped in faulted ramp
structures with an up-dip pinchout. Additional
pools of this type are likely around the margins
of basement highs.[12]

The prolific Intisar reef plays appear to be
limited to a small area in the eastern Ajdabiya
Trough, determined by the palaeogeography of
the adjacent platforms and the southern
Ajdabiya Trough. Much effort has been
expended in the search for similar reefs in other
areas. In principle there is no reason why other
reef areas should not be present seaward of the
margin of the Palaeocene carbonate platforms,
but the conditions for reef growth are very
specific and fortuitous, and to date no other
areas have been found (Area H on Figure 9.1).
Evidence from the Sabil Shelf suggests that
during the Palaeocene its southern margin may
have been formed by a barrier-reef which
implies the presence of high-energy porous
carbonates.[13]

On the eastern magin of the Ajdabiyah
Trough, north of the reef area, Eocene
carbonates are well developed, particularly in
the Ash Shulaydimah Trough and several small
discoveries have been made in this area (Area I
on Figure 9.1). The Ajdabiya Trough is one of
the few areas in which Eocene source rocks may
be mature. Ghori claimed that the Sirt Shale was
the only source rock which was sufficiently rich
and mature to charge the Eocene reservoirs, but
Burwood disagreed and on the basis of carbon
isotope correlations suggested an Eocene source

for some of the oil shows in the Antlat area. If this is true it considerably enhances the attractiveness, not only of this area, but also of structures within the Ajdabiya Trough such as the Al Brayqah Horst.[14]

9.10   Eastern Sirt Embayment

The eastern Sirt Embayment contains the most complex petroleum systems in Libya. This is due to the presence of several potential source rocks, and evidence of considerable mixing of oils from different sources. The embayment contains the giant fields of Jalu, Sarir, Messlah and Abu Attiffel, and the adjacent highs host the Amal and An Nafurah-Awjilah fields. The density of exploration drilling is higher in this area than anywhere else in Libya, and the eastern Sirt Embayment can be regarded as a mature exploration area.

Two significant advances have been made in recent years. In the Maragh Graben the Amal Formation, which was previously regarded as Cambro-Ordovician in age, has been dated by palynology as Triassic, at least in the upper part. This has radically changed the concept of the petroleum system in the Maragh Graben. The Jakarrah, As Sarah and Libba fields are now regarded as being both sourced and reservoired in Triassic rocks, and the Amal and Nafurah-Awjilah fields contain a hybrid oil derived from both Sirt Shale and Triassic source rocks. The second significant advance was the recognition by Ambrose of a miscorrelation of the Nubian Variegated Shale by some previous authors. He maintained that the Upper Sarir Sand is truncated by the Cenomanian unconformity before reaching the eastern margin of the Messlah High, providing a pinch-out play which had not previously been recognised (Area J on Figure 9.1). It seems probable that most of the future discoveries will be in subtle traps of this type. Mansour and Magairhy produced a detailed analysis of the Upper Cretaceous and Cainozoic potential of the Hameimat Trough and recognised some remaining potential in post-Nubian reservoirs, particularly around the Jalu-Hamid trend (Area K on Figure 9.1). At least five horizons have reservoir potential in this area, ranging in age from Cenomanian to Oligocene. Exploration of the deep Nubian play in the centre of the Hameimat Trough is continuing and further discoveries in this area are likely (Area L on Figure 9.1). The Nubian is buried to a depth of 4000m beneath a thick evaporite cover.[15]

9.11   Cyrenaica

With the exception of the Ash Shulaydimah Trough and the offshore area no effective petroleum system has yet been established in Cyrenaica. This is probably attributable to questionable source rocks and ineffective seals. Ghori described late Devonian gas-prone source rocks on the Cyrenaica Platform, but these source rocks are now post-mature. Ghori did not discuss the potential of the Tanzuft Shale which is present in several wells such as A1-46, D1-31 and B1-31, and yielded oil shows in some of these wells, but these rocks are also likely to be post-mature further to the northeast and possibly exhausted. In the Jabal al Akhdar Jurassic and early Cretaceous oil-prone shales have reached peak maturity in the western coastal area. No significant shows have been found onshore, but these rocks probably sourced the offshore A1-NC 120 oil pool. Sola and Ozcicek produced a hydrocarbon potential map for Cyrenaica which showed the most favourable area for Cainozoic (primarily Eocene) hydrocarbon accumulations to be the Ash Shulaydimah Trough, for Mesozoic accumulations (Cretaceous and Jurassic) the Binghazi area and Marmarica Basin, and for Palaeozoic (gas) accumulations the Cyrenaica Platform (Areas M1 and M2 on Figure 9.1).[16]

9.12 Offshore

To date only the Eocene-Farwah petroleum system has proved viable in the Libyan northwestern offshore. The Farwah nummulitic trend passes across the shelf from Sfax on the coast of Tunisia almost to Al Khums in Libya (Area 1 on Figure 9.1). The nummulitic bank is backed by restricted low-energy shelf deposits, and passes seawards into outer shelf shales which represent the principal source rock for the Farwah trend. Shoreward of the nummulitic trend several large gas-condensate discoveries have been made which, on the basis of available evidence, were most likely sourced from deeply-buried upper Cretaceous source rocks (Area 2 on Figure 9.1). Many of the discoveries in the Sabratah Basin are in structures generated by salt tectonics.

It is necessary to investigate why none of the numerous pre-Eocene systems found in Tunisia have proved productive in Libya. In general, the sediments on the Pelagian Shelf pass from littoral and nearshore facies near the present coastline through shallow platform to outer platform deposits towards the outer margin of the Pelagian Shelf. Bishop showed the limit of good quality Turonian Zebbag reservoir close to the Miskar field (Figure 8.23). East of this line the formation passes into outer platform carbonates, and this may represent a model for other upper Cretaceous carbonates.[17]

Detailed facies analyses of the Cretaceous carbonates in Libyan territory have not yet been published, but the descriptions of Hammuda, et al. of the Upper Cretaceous formations are not encouraging. The Abu Isa Formation of Maastrictian-Campanian age is a dense micritic limestone with abundant planktonic foraminifera. The Makhbaz Formation of Coniacian to late Cenomanian age is a micritic and shaly limestone, and the Alalgah Formation of Cenomanian age is composed of restricted shelf dolomites and anhydrites. This probably explains why the equivalents of the Douleb,

Bireno and Zebbag carbonates have not yielded any success in Libya.[18]

The situation is more promising in the early Cretaceous where the Turghat Formation represents an inner shallow-platform carbonate, with no planktonic foraminifera, and fair reservoir characteristics. The Turghat Formation probably spans most of the early Cretaceous. Early Cretaceous carbonates seem to offer the best possibilty of success on the carbonate platforms in the Libyan offshore, but relatively few wells have been drilled to test this objective (Area 3 on Figure 9.1). A related question is whether effective source rocks are present in the outer shelf area. Bishop showed a zone of organic rich sediments within the early Turonian Bahoul Formation extending into Libyan territory near Bouri, but information is lacking on its further eastward extension. The equivalent Makhbaz Formation is not encouraging, but the dark-brown marls and shales of the Santonian Jamil Formation may represent a possible alternative. No wells have penetrated the pre-Cretaceous section in the outer shelf area and nothing is known of reservoir or source rock potential in this area.[19]

On the southern margin of the Sabratah Basin the L1-137 well penetrated to the late Triassic and found several good quality clastic reservoirs in the pre-Cenomanian sequence, and south of the Sabratah-Cyrenaica fault zone wells in the Jifarrah Basin, such as B1-23 and J1-23, penetrated into the Permian, but north of well L1-137 these formations are very deep.[20]

The Misratah Basin is virtually undrilled. Only one well, J1-NC 35a, was drilled in the graben near the northern flank of the basin. According to Del Ben and Finetti the Mesozoic and Cainozoic sequences thicken dramatically to the north of well J1-NC 41 and the basin contains 7000m of post-Palaeozoic sediments (Figure 6.20). The nature of the basin fill can only be inferred, but could contain both source rocks and reservoirs particularly within the

Cretaceous. The Eocene nummulitic trend may continue along the platform edge to the south of the basin (Area 4 on Figure 9.1).[21]

In the Gulf of Sirt small gas pools have been found on the offshore northern flank of An Nufaliyah High, probably sourced from upper Cretaceous shales to the north (Area 6 on Figure 9.1). Geochemical screening of these wells showed fair organic content in the upper Cretaceous shales and these should be mature in the offshore area north of the Waddan Platform (Area 5 on Figure 9.1). The area north of the Sabratah-Cyrenaica fault zone has only been penetrated by one well, A1-NC 42, and details of potential reservoirs or source rocks have not been released. Marine Lower Cretaceous, Jurassic and Triassic rocks are present north of the fault, according to seismic interpretation, but nothing is known of their hydrocarbon potential. Water depth quickly reaches 1000m in this area.[22]

The A1-NC 120 well offshore Binghazi tested oil from Lower Cretaceous carbonates which were probably sourced by Jurassic or early Cretaceous marine shales. No information is available on reservoir quality but the well demonstrated the presence of viable source rocks offshore Cyrenaica. Only two other wells have been drilled in this area and the possibility of further discoveries cannot be ruled out (Area 7 on Figure 9.1). No wells have been drilled in the offshore north of Cyrenaica, but east of Darnah the Marmarica Basin shows some similarities to the Binghazi offshore. Thick early Cretaceous and Jurassic carbonates were drilled in well B1-33, but if shales are present deeper in the basin the area offshore Tubruq could merit further investigation (Area 8 on Figure 9.1).[23]

NOTES

Chapter 1

1 First geological map of Libya: Vinassa de Regny, 1912.
2 Bibliography of the Geology of Libya: Ettalhi et al. 1978, p. 38-44
3 Waddams, 1980, p. 27.
4 First indication of oil: Crema, 1926.
5 Desio on Sirt Basin prospectivity: Wright, 1981, p. 220.
6 Wright, 1981, p. 48-74.
7 Waddams, 1980, p. 57.
8 1955 Petroleum Law: Waddams, 1980, p. 57-70, and Gurney, 1996, p. 33-41.
9 Annual summaries of petroleum developments in Libya were published in successive volumes of the Bulletin of the American Association of Petroleum Geologists up to 1988. References are given by year, volume and page. For first licence awards see 1956, vol. 40, p. 1611
10 BAAPG, annual reviews, 1957-1961.
11 BAAPG, annual reviews, 1957-1961.
12 First well under 1955 Petroleum Law, BAAPG, 1957, vol. 41, p. 1568, 1571.
13 BAAPG, 1959, vol. 43, p. 1653-1657.
14 BAAPG, 1959, vol. 43, p. 1653-1657.
15 BAAPG, 1959, vol. 43, p. 1653-1657, and BAAPG, 1960, vol. 44, p. 1122-1126.
16 Zaltan flowrate: BAAPG, 1960, vol. 44, p. 1122, field description: Fraser, 1967, p. 259-264.
17 BAAPG, vols. 43 and 44, op. cit.
18 BAAPG, 1961, vol. 45, p. 1163-1168, and 1962, vol. 46, p. 1215-1221.
19 Sarir discovery: BAAPG, 1962, vol. 46, p. 1218, field description: Gillespie and Sanford, 1967, p. 181-193.
20 BAAPG, 1962, vol. 46, p. 1215-1221, and 1963, vol. 47, p. 1368-1377.
21 BAAPG, vols. 46 and 47, op.cit.
22 Waddams, 1980, p. 78-81.
23 Waddams, 1980, p. 101-114.
24 Waddams, 1980, p. 117-124.
25 Waddams, 1980, p. 137-151. Waddams was employed by the Ministry of Petroleum Affairs, and was involved in the negotiations which led to the 1965 Petroleum Law.
26 BAAPG, vols. 46 and 47, op. cit. 1964, vol. 48, p. 1645-1653. 1965, vol. 49, p. 1244-1251. 1966, vol. 50, p. 1691-1701. 1967, vol. 51, p. 1571-1579.
27 BAAPG, vols 46-51, op. cit.
28 Waddams, 1980, p. 198.
29 BAAPG, 1969, vol. 53, p. 1714
30 BAAPG, volumes for relevant years.
31 Waddams, 1980, p. 197.
32 Waddams, 1980, p. 199-200.
33 Waddams, 1980, p. 165-169.
34 Wright, 1981, p.232.
35 Waddams, 1980, p. 176-180.
36 Occidental discoveries: Hammer, 1987, p. 337-344.
37 Hammer, 1987, p. 340.
38 BAAPG, 1967, vol. 51, p. 1571-1579. 1968, vol. 52, p. 1495-1504. 1969, vol. 53, p. 1709-1717. 1970, vol. 54, p. 1466-1473.
39 Waddams, 1980, p. 206.
40 Wright, 1981, p. 141-142.
41 Hammer, 1987, p. 344-349.
42 Waddams, 1980, p. 232.
43 Waddams, 1980, p. 231-236.
44 Libyan Producers' Agreement: Waddams, 1980, p. 240.
45 BP Nationalisation: Waddams, 1980, p. 251-253.
46 Waddams, 1980, p. 254-255.
47 Wright, 1981, p. 236.
48 Waddams, 1980, p. 256.
49 BAAPG, 1975, vol. 59, p. 1873-1874, and 1976, vol. 60, p. 1769-1770
50 BAAPG, 1971, vol. 55, p. 1542.
51 Boundary disputes. Wright, 1981, p. 250.
52 Waddams, 1980, p. 274.
53 Wright, 1981, p. 227.
54 Production Sharing Agreements: Waddams, 1980, p. 260-263. New concessions: BAAPG, 1975, vol. 59, p. 1873-1874.
55 BAAPG, 1977, vol. 61, p.1701 and 1978, vol. 62, p. 1798-1799.
56 Oil price 1975-1979: Gurney,. 1996, p. 133.
57 Al Brayqah refinery: Waddams, 1980, p. 201-202. Az Zawia refinery: Gurney 1996, p. 151-152. Ra's Lanuf refinery: Gurney, 1996, p. 153-154.
58 Waddams, 1980, p. 281-282.
59 Gurney, 1996, p. 203-206.
60 Waddams, 1980, p.280.
61 BAAPG, 1976, vol. 60, to 1980, vol. 64.

62   BAAPG, 1978, vol. 62, p. 1798.

63   Thomas, 1995, p. 100.

64   BAAPG *Annual Petroleum Developments in North Africa* volumes 60 to 64.

65   Production: Gurney, 1996, p. 92. Fields on stream: BAAPG, vols. 60 to 64.

66   Background: Yergin, 1991, p. 678-680. Political situation: US Energy Information Administration, Jan. 2000 World Oil Market and Oil Price Chronologies 1970-1999. EIA Web site.

67   Yergin, 1991, p. 710-714.

68   Gurney, 1996, p. 226.

69   Gurney, 1996, p.71.

70   The figure of 2.7 billion barrels is derived from the annual production figures quoted in the BAAPG annual volumes up to 1982, with some interpolated values.

71   Yergin, 1991, p. 746-747.

72   Yergin, 1991, p. 748-752.

73   Yergin, 1991, p. 750-751.

74   Gurney, 1996, p. 69-70.

75   Gurney, 1996, p. 69-70. BAAPG, 1982, vol. 66, p. 2164-2165.

76   Gurney, 1996, p. 86-87. Thomas, 1995, p.100.

77   BAAPG, 1980, vol. 64, to 1986, vol. 70.

78   US Energy Information Administration, Jan. 2000, op. cit. p. 14-15.

79   The Great Man-Made River project was initiated in the 1970's following the discovery of large fresh-water aquifers in the Kufrah Basin. It was designed to bring fresh-water to the populated coastal strip via a 4m diameter concrete and steel pipeline. The project is planned to supply 2 million m^3 of water per day from 270 wells in the As Sarir-Tazirbu area, and 1.1 million m^3 per day from 308 wells in the Jabal Hasawnah area. The first two phases have been completed and the third phase is scheduled for completion in 2007. Binghazi and Tripoli are already receiving water from the GMR and the large surplus is scheduled for agricultural use.

80   Gurney, 1996, p. 72-80, 181-182.

81   Sanctions: Gurney, 1996, p. 226-230. Revenue losses: US Energy Information Administration, July 1999, Country Review: Libya, EIA Website.

82   US Energy Information Administration, July 1999, Libya.

83   Petroleum Exploration Society of Great Britain Newsletter, Elephant field: May 1999, p. 51, Ayn an Naqah field: Aug-Sept 1999, p. 71

84   US Energy Information Administration, July, 1999, Libya.

85   US EIA Jan. 2000, World Oil Market, p. 24-30.

86   US EIA July 1999, Libya.

87   NOC Press Release, December 1999.

88   US EIA, July 1999, Libya.

89   Thomas, 1995, p.103.

90   Libyan oil reserves: Campbell, 1991, p. 260-261. Total figures updated: Campbell, 1997.

91   These data are compiled from the annual field production figures given in the BAAPG annual volumes, plus the author's estimates.

92   Gurney, 1996, p. 98.

93   Yet-to-be-discovered oil: Campbell, 1997.

94   Waddams, 1980, p. 199-200 and 288-292.

95   Gurney, 1996, p. 178-186.

96   Gurney, 1996, p. 179.

97   Petroleum Exploration Society of Great Britain, Newsletter, October 1999, p. 59.

Chapter 2

1   De Wit, et al. 1988. Unrug, 1997, p. 1-6.

2   Rodinia: The name is derived from the Russian 'to beget'. Map: Unrug, et al. 1996. Description: Unrug, 1997, p. 1-6.
The name Rodinia was introduced by McMenamin M.A.S. and McMenamin D.L.S., 1990 in The Emergence of Animals: the Cambrian Breakthrough. Columbia University Press, New York, 1990. The term was popularised by P. Hoffman, 1991, Did the breakup of Laurentia turn Gondwana inside-out? Science, vol. 252, p. 1409. A summary review was given in *The Garden of Ediacara* by M.A.S. McMenamin, Columbia University Press, 1998.

3   Vail, 1991, p. 2261-2264.

4   Vail, 1991, p. 2263. Grubic, 1984, p. 13-14. Geological Map of Libya, 1985, scale 1:1,000,000. Bin Ghanimah batholith: Ghuma and Rogers, 1978, p. 1351-1358. Fullagar, 1980, p. 1051-1058. Ghuma and Rogers, 1980, p. 1059-1064.

5   Unrug, et al. 1996

6   Vail, 1991, p. 2261-2263.

7   Unrug, et al. 1996. Pannotia: Powell, C. McA., 1995, *Are Neoproterozoic glacial deposits preserved on the margin of Laurentia related to fragmentation of two supercontinents?* Geology, vol. 23. p. 1053-1054.

8   Unrug, et al. 1996.

9   Unrug, 1997, p. 1-6.

10   Vail, 1991, p. 2262. Schurmann, 1974.

11   Unrug, 1997, p. 1-6. De Wit, et al. 1988.

12   De Wit, et al. 1988. Vail, 1991, p. 2261-2263. Sutcliffe, et al. 2000, p. 2.

13   Detachment of the Avalonian, Cadomian and Cimmerian terranes: Unrug, 1997, p. 4, and Unrug, et al. 1999, p. 133-157. Iapetus Ocean: Harland and Gayer, 1972. *The Arctic Caledonides and earlier oceans.* Geological Magazine, vol. 109, p. 289-314.

14   Klitzsch, 1971, p. 253-262. Boote, et al. 1998, p. 9-15. Separation of Avalonian and Cadomian terranes: Unrug, et al. 1999, p. 133-157.

15   Beuf, 1969, p. 103-116. Beuf, et al. 1971. Brenchley, et al. 1994, p. 295-299. Semtner and Klitzsch, 1994, p. 743-751. Sutcliffe, et al. 2000. p. 3.

16 Luning, et al. 2000a, p. 121-200.

17 Boote, et al. 1998, p. 16.

18 Boote, et al. 1998, p. 16-17.

19 Klitzsch, 1971, p. 253-262. Boote, et al. 1998, p. 17.

20 De Wit, et al. 1988. Boote, et al. 1998, p. 17.

21 Break up of Pangaea and the establishment of Tethys: Ricou, 1994, p. 169-218, and Dercourt, et al, 1986, p. 241-315. Transit plate: Ricou, 1994, p. 195. Early oceanic crust in the eastern Mediterranean: Ricou, 1994, p. 195.

22 Wilson and Guiraud, 1998, p. 238.

23 Mamgain, 1980. p. 94-96. Kilian, 1931, p. 109-111.

24 Megashear: Ricou, 1994, p. 196.

25 Ricou, 1994, p. 195. Morgan, et al. 1998, p. 408.

26 Wilson and Guiraud, 1998, p. 241. Banerjee, 1980, p. 173-174. Thusu 1996, p. 455-474.

27 Break up of Gondwana: Ricou, 1994. Establishment of the Tethyan seaway: Dercourt, el al. 1986. Hot spots: Fouad, 1991, p. 2452-2459. Relative motions: Livermore and Smith, 1985, p. 93.

28 Dercourt, et al. 1986.

29 Dercourt, et al. 1986, plates 2 and 3. Morgan, et al. 1998, p. 408.

30 Tunisia: Morgan, et al. 1998, p. 405-422. Libya: Banerjee, 1980, p. 174. Algeria: Boote, et al. 1998, p. 21.

31 Dercourt, et al. 1986, plate 4. Dewey, et al. 1989, p. 265-283. Livermore and Smith, 1985, p. 83.

32 Dercourt, et al. 1986, plate 4. Libya rift phase: Gras and Thusu, 1998, p.322. Collapse of the Sirt Arch: El Makhrouf, 1996, p. 107-121. Gulf of Sirt: Finetti and Del Ben, 2000, p. 42. Hot spot: Van Houten, 1983, p. 115-118. Triple junction: Burke, 1977, p. 371-396. Harding, 1984, p. 333-362. Anketell, 1996, p. 83-84.

33 Wilson and Guiraud, 1998, p. 245.

34 Ricou, 1994, figs. 18-20.

35 Dercourt, et al. 1986, plate 5.

36 Guiraud, 1998, p. 222. Boote, et al. 1998, p. 21.

37 Dercourt, et al. 1986, plate 6.

38 Dercourt, et al. 1986, plate 7. Two plates: Burke and Dewey, 1974, p. 313-316. Anketell and Kumati, 1991, p. 2353-2370. Stress fields: Schäfer, et al., 1980, p. 907-922. Three plates: El Makhrouf, 1996, p. 116.

39 Eocene volcanics: Woller and Fediuk, 1980, p. 1081-1096. Jabal Awaynat ring-dykes: Andre, et al. 1991, p. 2511-2528.

40 Dercourt, et al, 1986, plate 8.

41 Dercourt, et al. 1986, plate 9. Jongsma, 1991, p. 2331-2352. Anketell, 1996, p. 83.

42 Dercourt, et al. 1986, plate 10. Woodside, 1991, p. 2319-2329.

43 Volcanics: Wilson and Guiraud, 1998, p. 249. Stress field: Schäfer, et al., 1980, p. 907-922. Messinian crisis: Barr and Walker, 1973, p. 1244-1251.

44 Schäfer, et al. 1980, p. 907-922. Westaway, 1996, p. 89-100.

Chapter 3

1 Early exploration: Beyrich, 1852, p. 143-161. Fezzan geological map: Rolland, 1880, p. 508-551. Krumbeck, 1906, p. 51-136. Cyrenaica: Gregory, 1911, p. 572-615. Prof. Gregory was a member of the Jewish Territorial Commission sent out in 1909 to examine the territory proposed for a Jewish settlement in Cyrenaica.

2 A comprehensive bibliography of Italian publications is given in Ettalhi, et al. 1978.

3 Sahabi palaeontology: Boaz, et al. 1982, p. 3-142, and Boaz, et al. 1987, p.4-395.

4 Ettalhi, et al., 1978, p. 38-44.

5 Ettalhi, et al., 1978, under the various authors.

6 Ettalhi, et al., 1978, p. 55.

7 Ettalhi, et al., 1978, under the various authors

8 Burollet, 1960, p. 1-62.

9 See discussion in the introduction to Barr and Weegar, 1972, p. 3-4.

10 Massa and Collomb, 1960, p. 65-73. Jacque, 1962, p. 1-44. Freulon, 1964, p. 1-198, de Lestang, 1965, p. 479-488.

11 First Saharan Symposium: Klitzsch, 1963, p. 97-113. Burollet, 1963, p. 219-227. Jordi and Lonfat, 1963, p. 114-122. Mennig, 1963, p. 186-201. Vittimberga and Cardello, 1963, p. 228-239. Publications: Maradah: Selley, 1966, 30p. Gemini: Pesce, 1968, 81p. Microfacies: Lehmann, et al. 1968, 80p. Jabal Nafusah: Hammuda, 1969, 74p.

12 Zaltan: Fraser, 1967, p. 259-264. As Sarir: Gillespie and Sanford, 1967, p. 181-193. Intisar A: Terry and Williams, 1969, p. 31-48. Amal: Roberts, 1970. p. 438-448. Awjilah: Williams, 1972. p. 623-632.

13 Barr and Weegar, 1972, p.11-179.

14 Structural history: Klitzsch, 1971, p. 253-262. Jabal Nafusah: Assereto and Benelli, 1971, p. 37-86, Hammuda, 1971, p. 87-98. Wadi ash Shati: Goudarzi, 1971, p. 489-500.

15 Industrial Research Centre, Tripoli. Geological Map of Libya, scale 1:250,000, 61 sheets, 1974 to 1993, each sheet with explanatory booklet.

16 IRC Bulletins - Bulletin 11: Ettalhi, et al. 1978. Bulletin 12: Megerisi and Mamgain, 1980. Bulletin 13: Banerjee, 1980. Bulletin 14: Mamgain, 1980.

17 Southern Basins: Bellini and Massa, 1980, p. 3-56. Cyrenaica biostratigraphy: Barr and Berggren, 1980, p. 163-192. Messinian event: van Hinte, et al. 1980, p. 205-244. Eastern Sirt embayment: Benfield and Wright, 1980, p. 463-500.

18 National Atlas of the Socialist People's Libyan Arab Jamahiriya, 1978, M.M. Unis, Director General.

19 Offshore Nomenclature: Hammuda, et al. 1985, 165p. Tunisia: Fournie, 1978, p. 97-148.

20 OAPEC Nomenclature: Lababidi and Hamdan, 1985, 171p.

21 Kufra: Grignani, et al. 1991, p. 1159-1228, Bellini, 1991, p. 2155-2184. Murzuq: Pierobon, 1991, p. 1766-

1784. Cyrenaica: Duronio, et al. 1991, p. 1589-1620. Hamada al Hamra: Nairn and Salaj, 1991, p. 1621-1636. Precambrian: Vail, 1991, p. 2159-2268.

22 Eastern Ghadamis Basin: Belhaj, 1996, p. 57-96. Palaeocene: Bezan, 1996, p. 97-118. Amal Formation: Thusu, 1996, p. 455-474. Syntheses: Wennekers, et al. 1996, p. 3-36. Baird, et al. 1996, p. 3-56. Anketell, 1996, p. 57-87.

Sabha Conference: Hot shales: Lüning, et al. 2000, p. 151-174. Mamuniyat: Davidson, et al. 2000, p. 295-320. Aziz, 2000, p. 349-368. Blanpied, et al. 2000, p. 485-508. Murzuq stratigraphy: Echikh and Sola, 2000, p. 175-222.

Symposium on the Hydrocarbon Geology of North Africa. London, November, 1995. Proceedings published as Geological Society Special Publication No 132, edited by D.S. Macgregor, R.T. J Moody and D.D. Clark-Lowes, 1998, 442pp.

Conference on Petroleum Systems and evolving technologies in African E&P, London, 16-18 May, 2000. Book of abstracts issued.

Second Symposium on the Sedimentary Basins of Libya: The Geology of Northwest Libya, Tripoli, 6-8 November 2000. Book of abstracts issued.

23 Trace fossils: Seilacher, 1991, p. 1565-1581. Cyrenaica: *Palynostratigraphy of northeast Libya*, 1985, edited by B. Thusu and B. Owens. J. Micropalaeontology, vol. 4. *Subsurface palynostratigraphy of northeast Libya*, 1988, edited by A. El-Arnauti, B. Owens and B. Thusu. Garyounis Univ. Pubs. Benghazi. Eastern Ghadamis/western Sirt Basins: Belhaj, 1996, p. 57-96.

24 Sequence stratigraphy: a good example of the application of sequence stratigraphy to the Palaeozoic of Libya is provided by Sutcliffe, et al. 2000, p. 14-69. Radiometric dating: see for example Cahen, et al. 1984, p. 255-269.

25 Vail, 1991, p. 2260-2268. Cahen, et al. 1984.

26 Goodell, 1991, p. 2630. Aziz, 2000, p. 352.

27 Vail, 1991, p. 2261-2262. Outcrops: Radulovic, 1984b, Grubic, 1984a, Grubic, 1984b.

28 Wacrenier, 1959, p. 281-288. Klitzsch, 1966, p. 1-18. Ghuma and Rogers, 1976, p. 904. Assaf, 1991, p. 2287-2293. El Makhrouf and Fullagar, 2000, p. 379-395.

29 Ghuma and Rogers, 1976, p. 904. Vail, 1991, p. 2263. Jurak, 1978, p. 17-18. Oun, et al. 2000, p. 71. Geological Map of Libya, 1985, scale 1,000,000.

30 Al Awaynat: Menchikoff, 1927, p. 337-354. Sandford, 1935, p. 323-381. Mahrholz, 1968. Klerkx, 1980, p. 901-906. Shurmann, 1974, p. 325-328, Cahen, et al. 1984, p. 255-269.

31 Baudet, 1988, p. 17-26. El-Arnauti and Shelmani, 1988, p. 7.

32 Schurmann, 1974, p. 325-328. Cahen, et al. 1984, p. 266. Vail, 1991, p. 2265. Shackleton and Grant, 1974, p. 3-6.

33 Williams, 1971, p. 504.

34 Beuf, et al. 1971. Burollet, 1960, p. 24-25. Bellini and Massa, 1980, p. 3-6.

35 Boote, et al. 1998, p. 7-68.

36 Early Palaeozoic sand wedge: Burke and Kraus, 1998, p. 42. The detachment of the Avalonian and Cadomian terranes: Unrug, 1991, p. 133-157. Ordovician glaciation: Beuf, et al. 1971, especially part 2, chapter 4.

37 Olochi Sandstone: Dalloni, 1934, revised by Lelubre, 1946b, p. 429-431. Mourizidie Formation: Jacqué, 1962, p. 1-43. Sinha and Ben Rahuma, 2000, p. 89. Bi'r Bayai Formation: unpublished name of Selley, reported by El Arnauti and Shelmani, 1988, p. 7. D1-NC 115: Aziz, 2000, p. 352. Cyrenaica: El Arnauti and Shelmani, 1988, p. 7. Radiometric dating: Legrand, 1985, Selley, 1997a, p. 4.

38 Boote, et al. 1998, fig.1.

39 Klitzsch, 1971, p. 253-262.

40 Burollet, 1960, Massa and Collomb, 1960, p. 65-73.

41 Jurak, 1978, p. 19-22.

42 Cepek, 1980, p. 375-382.

43 Outcrops: Gundobin, 1985, Woller, 1978, Parizek, et al. 1984, Seidl and Rohlich, 1984, Woller, 1984. Palaeohydraulics: Mayouf and Turner, 2000, p. 65.

44 Outcrops: Protic, 1984b, Radulovic, 1984a, Radulovic, 1984b, Grubic, 1984a, Galecic, 1984, Grubic, 1984b. See also Grubic, 1991, p. 1529-1564. Clark-Lowes and Ward, 1991, p. 2099-2153.

45 Pierobon, 1991, p. 1769. The sedimentological work was done by J.C. de Castro of Braspetro.

46 Klitzsch, 1963, p. 97-118. Klitzsch, 1971, p. 256-257.

47 Burollet, 1963, p. 1537-1545. Local nomenclature: Vittimberga and Cardello, 1963, p. 228-240. A1-NC 43 and B1-NC 43: Grignani, et al. 1991, p. 1159-1227. Turner, 1980, p. 351-374.

48 Deunff and Massa, 1975, p.21-24. Belhaj, 1996, p. 65. Echikh, 1998, figure 6.

49 Barr and Weegar, 1972, p. 14-18 (Amal Formation), 97-105 (Hofra Formation).

50 Bonnefous, 1972, p. 225-261.

51 Robertson study: Wennekers, et al. 1996, p. 17. Wadi field: El Hawat, et al., 1996, p. 3-31. Attahadi: Rasul, 2000, p. 64. PUC B, Jakharrah: Sinha and Eland, 1996, p. 49-60.

52 El Arnauti and Shelmani, 1988, p. 7-9.

53 Boote, et al., 1998, fig. 1.

54 Havlicek and Massa, 1973, p. 267-290.

55 Gundobin, 1985.

56 Parizek, et al. 1984. Seilacher, 1991, p. 1565-1581. Seilacher, 2000, p. 237-258.

57 West Murzuk outcrops: Protic, 1984b, Radulovic, 1984a, Radulovic, 1984b. Braspetro wells: Pierobon, 1991, p. 1769.

58 Deunff and Massa, 1975.

59 Banerjee, 1980, p. 229. Bellini and Massa, 1978, 2nd Symposium on the Geology of Libya, abstracts, p. 15-

17. Echikh, 1998, figure 6.

60 Belhaj, 1996, p. 65-66

61 El Arnauti and Shelmani, 1988, p. 6-9.

62 Lababidi and Hamdan, 1985, fig. 4.

63 Massa and Collomb, 1960, p. 68-69. Collomb, 1962, p. 7-35.

64 Vos, 1981, p. 153-170.

65 West Murzuk outcrops: Gundobin, 1985, Parizek, et al., 1984, Protic, 1984b, Radulovic, 1984a, Radulovic, 1984b. Braspetro wells: Pierobon, 1991, p. 1769. NC 115: Echikh and Sola, 2000, fig. 4.

66 Bellini, et al. 1991, p. 2155-2184.

67 Deunff and Massa, 1975, p.25

68 Belhaj, 1996, p. 65.

69 Lababidi and Hamdan, 1985, fig. 4.

70 Unrug, 1999, op. cit. p. 133-157.

71 Massa and Collomb, 1960, p. 69.

72 Collomb, 1962. Havlicek and Massa, 1973.

73 Collomb, 1962, p. 7-35. Blanpied, et al. 1998, p. 43. McDougall and Martin, 2000, p. 223-236. El Hawat and Bezan, 1998.

74 Outcrops: Protic, 1984b, p. 23. Murzuq Basin wells: Pierobon, 1991, p. 1771. Echikh and Sola, 2000, p 179.

75 Banerjee, 1980, p. 170. Bellini, et al. 1991, p. 2160-2161.

76 Deunff and Massa, 1975, p. 23.

77 Belhaj, 1996, p. 65-65. Land plants: Gray, et al., 1982, p. 197-201.

78 Lababidi and Hamdan, 1985, fig. 5. Echikh, 1998, figure 6.

79 Hill and Molyneux, 1988, p. 27-44. Paris, 1988a, p. 61-73.

80 Outcrops: Radulovic, 1984b. p. 28-30. Abugares and Ramaekers, 1993, p. 67. Sutcliffe, et al. 2000, p. 26-50.

81 Outcrops: Radulovic, 1984a, p. 27-29. Grubic, et al., 1991, p. 1538. El Hawat and Bezan, 1998.

82 Abugares and Ramaekers, 1993, p. 42-44.

83 Massa and Collomb, 1960, p. 69. Parizek, et al., 1984, p.30.

84 Collomb, 1962, Havlicek and Massa, 1973.

85 Boote et al, 1998, p. 16. Blanpied, et al. 1998, p. 43. Rubino, et al. 2000, p. 80. Deynoux, et al. 2000, p. 27. Smart, 2000, p. 397-416. Whittington and Walker, 1998, p. 42. Gravity slide: Glover, et al. 1998b, p. 21. Fitches, 1998, p. 45-46. El Hawat and Bezan, 1998. El Hawat, et al. 2000, p. 36.

86 Outcrops: Gundobin, 1985, Parizek, et al., 1984, Seidl and Rohlich, 1984, Protic, 1984b, Radulovic, 1984a, Radulovic, 1984b. Sequences: McDougall and Martin, 2000, p. 223-236. Wells: Pierobon, 1991, p. 1771-1772. NC 115: Aziz, 2000, p. 358. Lithofacies: Echikh and Sola, 2000, p. 181. Provenance: Fello and Turner, 2001.

87 Bellini and Massa, 1980, p. 24. A1-77, etc: Echikh and Sola, 2000, p. 181.

88 Bellini and Massa, 1980, fig.2. Bellini, 1991, p. 2161-2171. Grignani, et al., 1991, p. 1163. Turner, 1980, fig. 2.

89 Deunff and Massa, 1975, p. 21-24. Bergstrom and Massa, 1991, p. 1323-1342. Bryozoan mounds: Massa and Bourrouilh, 2000, p. 63. Echikh, 1998, p. 116.

90 Belhaj, 1996, p. 66-67. Bracaccia, et al., 1991, p. 1735.

91 El Arnauti and Shelmani, 1988, p. 9. Hill and Molyneux, 1988, p. 27-44. Molyneux, 1988, p. 45-59.

92 Lababidi and Hamdan, 1984, fig. 4.

93 Echikh, 1998, p. 114-115.

94 Gonwanan glaciations: Hambrey and Harland, 1981, p. 36-191. Migration of glacial centres: Caputo and Crowell, 1985, p. 1020-1036. Murzuq Basin: Boote, et al., 1998, p. 52-53. Abugares and Ramaekers, 1993, p. 62-67. Echikh, 1998, fig. 6. Sutcliffe, et al. 2000, p. 28-51. Sutcliffe, et al. 2001. Smart, 2000, p. 397-416, Aziz, 2000, p. 349-368. Glover, et al. 2000, p. 417-430. Blanpied, et al. 2000, p. 485-508.

95 Recent work: Semtner and Klitzsch, 1994, p. 743-751, Brenchley, et al. 1994, p. 295-298. Tiguentourine: Hirst, et al. 2001.

96 El Arnauti and Shelmani, 1988. p. 10.

97 Iyadar: Massa and Jaeger, 1971, p. 313-321. Deunff and Massa, 1975. Bi'r Tlakshin: Echikh and Sola, 2000, p. 182-185. Blanpied, et al. 2000, p. 485-508. Fekirine and Abdallah, 1998, p. 87-108.

98 Desio, 1936a, p.116-120.

99 Klitzsch, 1965, p. 605-607. Klitzsch, 1969, p. 83-90.

100 Outcrops: Protic, 1984b, Radulovic, 1984a, Radulovic, 1984b, Galecic, 1984, Grubic, 1984b. Tempestites: de Castro, et al., 1991, p. 1757-1765. Radioactive minerals: Almehdi, et al., 1991, p. 2645-2658. E. Murzuk: Klitzsch, 1968, p. 497. Bellini and Massa, 1980, p. 25.

101 Bellini and Massa, 1980, p. 13-16.

102 Abugares and Ramaekers, 1993, p. 47-48. Lüning, et al. 2000, p. 131, 169-170. Oxygen-poor Rhuddanian: Lüning, 2000, p. 54.

103 Alum deposits at Ghat: Lefranc, 1991, p. 2605-2617. Abugares and Ramaekers, 1993, p. 48-49. For a recent survey of the mystique of alum see *The Floating Egg*, by Roger Osborne, 1998, Jonathan Cape, London, 372p.

104 Pierobon, 1991, p. 1773-1775. Mamgain, 1980, p. 34. Wennekers et. al. 1996, p. 10. Abugares and Ramaekers, 1993, p. 49. Bellini and Massa, 1980, p. 15. Lüning, et al. 2000, p. 164-168. Echikh and Sola, 2000, p. 186-187.

105 Bellini and Massa, 1980, p. 25

106 Bellini and Massa, 1980, p. 26. Fürst and Klitzsch, 1963, figure 4. Klitzsch, 1971, figure 2. Wennekers, et al. 1996, figure 4.

107 Grignani, et al., 1991, p. 1159-1227. Turner, 1980, fig. 2. Lüning, et al. 2000, p. 171.

108 Woller, 1984, p. 21, and Gundobin, 1985, Parizek, et al., 1984.

109 Massa and Jaeger, 1971, p. 313-321. Jaeger, et al. 1975, p. 68-77. F1-66: Lüning, et al. 2000, p. 171.

110 Belhaj, 1996, p. 69-70. Bracaccia, et al., 1991. p. 1736-

1737. Thusu, 1996, p. 463-465.

[111] Hill and Molyneux, 1988, p. 27-44. Paris, 1988a, p. 45-60. Richardson, 1988, p. 89-110. El Arnauti and Shelmani, 1988, fig. 5.

[112] Hoffmeister, 1959, p. 331-334. Richardson and Ioannides, 1973, p. 257-307.

[113] Tanzuft organic shales: Luning, et al. 2000, p. 164-169.

[114] Desio, 1936b, p. 319-356. Klitzsch, 1965, p. 605-607. Klitzsch, 1969, p. 83-90.

[115] Abugares and Ramaekers, 1993, p. 49. Massa and Jaeger, 1971, fig. 8. Seilacher, 1991, p. 1565-1581. Seilacher, 1996, p. 523-530.

[116] Outcrops: Protic, 1984b, Radulovic, 1984a, Radulovic, 1984b, Galecic, 1984, Grubic, 1984b. Matgrounds: Pfluger, 1998. Horizon d'Iknouen: Freulon, 1964, p. 102, 107.

[117] Pierobon, 1991, p. 1775. Echikh and Sola, 2000, p. 189, 202.

[118] Klitzsch, et al., 1973, p. 2465-2467. Mamgain, 1980, p. 38-39. Belhaj, 1996, p. 69-72.

[119] Pallas, 1980, cross sections 1 and 2. Vascular plants, Daber, 1971, p. 35-40. Lejal-Nicol, 1991, p. 1343-1349. C1-44: Thusu, 1996, p. 465-466.

[120] Grignani, et al. 1991, p. 1163-1164. Turner, 1980, fig. 2.

[121] Massa and Jaeger, 1971, p. 313-321. Mamgain, 1980, p. 38-41. Echikh, 1998, fig. 3. Weyant and Massa, 1991, p. 1297-1322. Bani Walid water well: Tekbali and Wood, 1991, p. 1243-1273. L. Akakus petrology: Elfigih, 2000, p. 33. El Mehdawi, 2000, p. 40. Sghair and Anketell, 2001.

[122] Belhaj, 1996. p. 68-73. Bracaccia, et al. 1991, p. 1736-1741.

[123] El Arnauti and Shelmani, 1988, p. 10. Paris, 1988, p.73-88.

[124] Echikh, 1998, p. 115-116. Alem, et al., 1998, p. 175-186.

[125] Pallas, 1980, fig. 1. Klitzsch, 1971, fig. 2. Weyant and Massa, 1991, p.1299. Mergl and Massa, 2000, p. 41-88. Rubino and Blanpied, 2000, p. 321-348.

[126] Burollet, 1960, p. 16. Klitzsch, 1965, p. 605-607. Klitzsch, 1969, p. 87.

[127] Outcrops: Protic, 1984b, Radulovic, 1984a, Jakovljevic, 1984, Radulovic, 1984b, Galecic, 1984, Grubic, 1984b. Clark-Lowes and Ward, 1991, p. 2099-2153. Age: Belhaj, 2000, p. 127, 131. Clark-Lowes, 1978, p. 21-23.

[128] Pierobon, 1991, p. 1775. Sutcliffe, et al. 2000, p. 56.

[129] Inversion: Pierobon, 1991, p. 1772. Meister, et al. 1991, p. 2727. Sutcliffe, et al. 2000, p. 60. Klitzsch, 1966, p. 24. Mamgain, 1980, p. 94.

[130] Outcrops: Gundobin. 1985, Parizek, et al., 1984, Seidl and Rohlich, 1984.

[131] El Arnauti and Shelmani, 1988, p. 11. Grignani, et al. 1991, p. 1164-1165. Turner, 1980, fig. 2.

[132] Echikh, 1998, p. 116-117. Massa and Moreau-Benoit, 1976, p. 287-332. Weyant and Massa, 1991, p. 1300.

Belhaj, 2000, p. 125-133. Hammuda, 1980, p. 505.

[133] Belhaj, 1996, p. 73-76.

[134] El Arnauti and Shelmani, 1988, p. 10.

[135] Aliev, et al., 1971, 192. Echikh, 1998, p. 116-117. Alem, et al. 1998, p. 175-186.

[136] Borghi and Chiesa, 1940, p. 122-137. Klitzsch, 1965, p. 605-607. Klitzsch, 1969, p. 87 and fig. 2. Grubic, et al. 1991, p. 1546. Mergl and Massa, 1992.

[137] Outcrops: Protic, 1984b, Jakovljevic, 1984, Galecic, 1984, Grubic, 1984b. Clark-Lowes, 1978, p. 21-23.

[138] Pierobon, 1991, p. 1772-1777. Echikh and Sola, 2000 p. 189, 191. Klitzsch, 1966, p. 24. Meister, et al. 1991, p. 2727. Sutcliffe, et al. 2000, p. 60.

[139] Grignani, et al. 1991, p. 1164-1165. Massa and Bellini, 1980, p. 44.

[140] Massa and Moreau-Benoit, 1976, p. 287-332. Belhaj, 1996, p. 75-79.

[141] Streel, et al. 1988, p. 111-128. El Arnauti and Shelmani, 1988, p. 10. Paris, 1988, p. 73-88.

[142] Echikh, 1998, p. 116-117. Aliev, et al., 1971. p. 192.

[143] Borghi, 1939, p. 115-184. Lelubre, 1946a, p. 337-357. Massa and Collomb, 1960, p. 65-73. Echikh and Sola, 2000, p. 189, 191.

[144] Outcrop: Gundobin, 1985, p. 58-62. Vos, 1981, p. 67-88.

[145] Collomb, 1962, p. 7-35. Massa, Termier and Termier, 1974, p. 139-206. Massa and Moreau-Benoit, 1976, p. 287-332. Weyant and Massa, 1991, p. 1301-1303. Gundobin, 1985, p. 58-62. Parizek, et al., 1984, Seidl and Rohlich, 1984, Woller, 1984.

[146] Protic, 1984, p. 44-46. Furst and Klitzsch, 1963, p. 158-170. Bellini and Massa, 1980, fig. 18. Outcrops: Jakovljevic, 1984, Galecic, 1984, Grubic, 1984b.

[147] Goudarzi, 1971, p. 489-500. Turk, et al. 1980. p. 1019-1043. The French Study Group report was prepared for the Industrial Research Centre, Tripoli, in 1976, but never published. The title of the report is *'Studies for the development of the Wadi Shatti iron ore deposits, L.A.R.'* The new stratigraphic scheme was adopted on three sheets of the Geological map of Libya: Idri (Parizek et al., 1984), Sabha (Seidl and Rohlich, 1984), and Al Fuqaha (Woller, 1984). Facies associations: Sutcliffe, et al. 2000, p. 56. Blanpied and Rubino, 1998, p. 9.

[148] Parizek et al. 1984, p. 38-39. Mergl and Massa, 2000, p. 70. Rubino and Blanpied, 2000, p. 327.

[149] Parizek et al. 1984, p. 42-45. Mergl and Massa, 2000, p. 75. Rubino and Blanpied, 2000, p. 327.

[150] Seidl and Rohlich, 1984, p. 30-34. Rubino and Blanpied, 2000, p. 327.

[151] Seidl and Rohlich, 1984, p. 34-38. Rubino and Blanpied, 2000, p. 327.

[152] Seidl and Rohlich, 1984, p. 38-42. 'Paul's Garden': Hlustik, 1991, p. 1275-1284. Mergl and Massa, 2000, p. 76.

[153] Seidl and Rohlich, 1984, p. 44-48. Vavrdova, 1991, p. 1285-1296. Mergl and Massa, 2000, p. 76.

[154] Sedimentology, NC 58: Pierobon, 1991, p. 1778-1779. Awaynat Uplift: Sutcliffe, 2000, p. 63. Supercrop map: Echikh and Sola, 2000, p. 191.

[155] Furst and Klitzsch, 1963, p. 164, and figs. 3-4.

[156] Vittimberga and Cardello, 1963, p. 228-239. Klitzsch, 1966, p. 1-17. Turner, 1980, fig. 2.

[157] Grignani, 1991, p.1165

[158] Bellini and Massa, 1980, p. 17. Emgayat Shale: Shah, et al., 1993, p. 345-352. Weyant and Massa, 1991, p. 1301. Cues Limestone: Said, 1974. Anoxic conditions: Lüning, 2000, p. 54-55. Alrar field: Chaouchi, et al., 1998. Age dating: Elzaroug and Lashhab, 1998, p. 38-39. Isopachs: El-Rweimi, 1991, p. 2185-2193.

[159] Belhaj, 1996, p. 79-84. El Rweimi, 1991, p. 2188.

[160] Echikh, 1998, p. 117-118.

[161] Streel, et al. 1988, p. 111-128. El Arnauti and Shelmani, 1988, p. 10-11, fig. 6.

[162] Aliev, et al. 1971, p. 192-193. Echikh, 1998, p. 117. There is some disagreement about the age of the F4 and F3 reservoirs. The ages given in the text are from Echikh. However Aliev regarded both the F5 and F4 as Emsian in age, and Chaouchi gave an Emsian age for the F4 and an Eifelian-Givetian age for the F3. Chaouchi, et al. 1998, fig. 4. Elzaroug and Lashhab, 1998, p. 38.

[163] Burollet and Manderscheid, 1967, p. 285-302. Massa, Termier and Termier, 1974, p. 139-206. Seidl and Rohlich, 1984, tab. 1. Sedimentological features: de Castro, 1991, p. 1761. Sutcliffe, et al. 2000, p. 56, 63. Mergl and Massa, 2000, p. 41-88. Hassi, 1998. Isopach: El Rweimi, 1991, p. 2189. A1-NC 40B: El Harbi and Abuhamida, 2000, p. 35. El Mehdawi, 2000, p. 273-294. Strunian: Harland, et al., 1990, fig. 3.6b.

[164] Lelubre, 1948, p. 79-81. Gundobin, 1985, p. 63-69. Whitbread and Kelling, 1982, p. 1091-1107. Vachard and Massa, 1984, p. 167. Belhaj, 2000, p. 134, 140. El Harbi, 2000, p. 35. Mergl and Massa, 2000, p. 53-54.

[165] Outcrops: Berendeyev, 1985c, Roncevic, 1984, Komarnicki, 1984. Freulon, 1953, p. 233-234.

[166] Outcrops: Protic, 1984b, Radulovic, 1984a, Jakovljevic, 1984, Galecic, 1984, Grubic, 1984b.

[167] Pierobon, 1991, p. 1778-1779. Mamgain, 1980, p. 76. Aziz, 2000, p. 364.

[168] Outcrops: Parizek, et al., 1984, Seidl and Rohlich, 1984, Woller, 1984.

[169] Klitzsch, 1963, p. 97-118. Mamgain, 1980, p. 73-74.

[170] Vittimberga and Cardello, 1963, p. 228-240. Grignani et al. 1991, p. 1165-1166. Turner, 1980, fig. 2.

[171] Massa, et al. 1980, p. 429-442. Mamgain, 1980, p. 72. Swire and Gashgesh, 2000, p. 96. Belhaj, 2000, p. 137.

[172] Belhaj, 1996, p. 83-89, AAPG World Petroleum Developments, 1959.

[173] Loboziak and Clayton, 1988, p. 129-150. El Arnauti and Shelmani, 1988, p. 12-13, fig. 7.

[174] Aliev, et al. 1971, p. 56

[175] Lelubre, 1952, p. 109-148. Collomb, 1962, p.1-35.

[175 cont.] Berendeyev, 1985, p. 31-34. Vachard and Massa, 1984, p. 167-169. Mergl and Massa, 2000, p. 41-88.

[176] Outcrops: Gundobin, 1985, Protic, 1984a, Roncevic, 1984, Parizek, et al., 1984, Seidl and Rohlich, 1984, Woller, 1984.

[177] Outcrops: Protic, 1984b, Komarnicki, 1984, Jakovljevic, 1984, Galecic, 1984, Grubic, 1984b.

[178] Pierobon, 1991, p. 1778-1779. Mamgain, 1980, p. 79.

[179] Klitzsch, 1963, p. 25-26. Mamgain, 1980, p. 80-81.

[180] Bellini, 1991, figs. 2-4. Grignani, 1991, fig. 5.

[181] Bellini and Massa, 1980, p. 19-22. Belhaj, 1996, p. 86. Wennekers, et al, 1996, p. 16. Massa, Termier and Termier, 1974. Swire and Gashgesh, 2000, p. 96.

[182] Loboziak and Clayton, 1988, p. 129-150. El Arnauti and Shelmani, 1988, p.13, fig. 7.

[183] Aliev, et al. 1971,p. 63 and p. 195.

[184] Lelubre, 1952, p. 109-148. Collomb, 1962, p. 1-35. Berendeyev, 1985, p. 35-37. Vachard and Massa, 1984, p. 169. Mergl and Massa, 2000, p. 54-57.

[185] Outcrops: Gundobin, 1985, Protic, 1984a, Roncevic, 1984. Analcite, see Almehdi et al., 1991, p. 2654.

[186] Outcrops: Protic, 1984b, Komarnicki, 1984, Jakovljevic, 1984, Galecic, 1984, Grubic, 1984b.

[187] Pierobon, 1991, p. 1779.

[188] Klitzsch, 1966, p. 26. Bellini and Massa, 1980, p. 27. Grignani et al. 1991, p. 1176.

[189] Bellini and Massa, 1980, p. 19-21. Belhaj, 1996, p. 86. Massa, Termier and Termier, 1974, p. 161. Swire and Gashgesh, 2000, p. 96. Loboziak and Clayton, 1988, p. 129-150. El Arnouti and Shelmani, 1988, p. 13, fig. 7.

[190] Aliev, et al. 1971, p. 63 and p. 196.

[191] Kilian, 1931, p. 109-111. De Lapparent and Lelubre, 1948, p. 1237-1240.

[192] Mamgain, 1980, p. 86. Protic, 1984a, p. 26. Abugares and Ramaekers, 1993, p. 56.

[193] Outcrops: Berendeyev, 1985b, Berendeyev, 1985a, Roncevic, 1984, Komarnicki, 1984, Jakovljevic, 1984, Galecic, 1984, Grubic, 1984b. Banerjee, 1980, p. 81. Abugares and Ramaekers, 1993, p. 56. Pierobon, 1991, p. 1779-1780. Bellini and Massa, 1980, p. 22. Klitzsch, 1966, p. 26.

[194] Grignani, et al. 1991, p. 1166-1167. Loboziak and Clayton, 1988, p. 129-150. El Arnouti and Shelmani, 1988, p. 13, fig. 7.

[195] Bellini and Massa, 1980, p. 22.

[196] Aliev, et al. 1971, p. 65.

[197] Jabal Tebaga: Baird, 1967, p. 85-107. Kirchaou Permit: Benzarti, 1993, p. 297-313. Southern Tunisia: Kilani-Mazraoudi, et al. 1990, p. 273-291. Cyrenaica: Brugman, et al. 1988, p. 151-156. El Arnauti and Shelmani, 1988, p. 14, fig. 8. Offshore: Del Ben and Finetti, 1991, figures 2 and 3.

[198] Mennig, et al., 1963, table 1. Hammuda, et al. 1985, p. 47. Swire and Gashgesh, 2000, p. 96. Brugman and Visscher, 1988, p. 157-170. Brugman, et al. 1988, p. 151-156. El Arnauti and Shelmani, 1988, p. 14,

fig. 8. Tehenu Basin: Keeley and Massoud, 1998, p. 269.

[199] Mennig, et al., 1963, table 1. Kar, et al., 1972, p. 389-453. Hammuda et al. 1985. p. 43-48. Adloff, et al. 1986, p. 27-63. Swire and Gashgesh, 2000, p. 96.

[200] Wennekers, et al. 1996, p. 16, fig. 6. Sikander, 2000, p. 87. Meister, et al., 1991, p. 2725-2741. Pierobon, 1991, p. 1781.

[201] Grignani et al. 1991, p. 1166-1167.

[202] El Arnauti and Shelmani, 1988, p. 14.

[203] Bellini and Massa, 1980, p. 22. Aliev et al, 1971, p. 66.

Chapter 4

[1] Jabal Nafusah: Rubino, et al. 2000, p. 79. Triassic offshore: Del Ben and Finetti, 1991, figures 2 and 3. Cyrenaica: Shelmani, et al. 1992.

[2] Ouled Chebbi Formation: Mennig, et al., 1963, p. 188. Al Guidr Formation: Hammuda, et al. 1985, p. 14-18. Tunisia: Rubino, et al. 2000, p. 79.

[3] Belhaj, 1996, tab. 3. Tunisia: Brichant, 1952, p. 1456-1458.Aliev, et al. 1971, p. 69-70. Echikh, 1998, p. 118-119.

[4] Kurrush Formation: Desio, et al. 1960b, p. 273-323. Ra's Hamia Formation: Burollet, 1960, p. 44. IRC mapping: El Hinnawy and Cheshitev, 1975, p. 13-14. Rubino, et al. 2000, p. 79.

[5] Kurrush hypostratotype: Hammuda et al. 1985, p. 98-102. Sedimentology: Mennig, et al. 1963, p. 200. Ghadamis Basin: Sinha, 1980, figures 1-6. Eastern limit: Belhaj, 1996, fig.25. Kirchaou Sandstone: Busson, 1967b Turner, et al. 2001. Rheouis Formation: Burollet, 1956.

[6] Aliev et al. 1971, p. 69-70. Turner, et al. 2001. Sabaou and Turner, 2001. Echikh, 1998, p. 118-119. Turner, et al. 2001. Cyrenaica: El Arnauti and Shelmani, 1988, fig. 8. Brugman and Visscher, 1988, p. 158. Shelmani, et al. 1992, p. 233-240.

[7] Al Aziziyah Formation: Parona, 1914, p. 1-26. Type section: Christie, 1955. IRC revision: El Hinnawy and Cheshitev, 1975, p. 15-17. Sedimentology Assereto and Benelli, 1971, p. 37-86. Subdivision: Fatmi et al 1980, p. 62-63. Rubino, et al. 2000, p. 79.

[8] Hammuda, et al, 1985, p. 10-13. Banerjee, 1980, p. 14. Tunisia: Busson, 1967b, Burollet, 1956.

[9] Ghadamis Basin: Sinha, 1980, figures 1-6. Belhaj, 1996, p. 88. Cyrenaica: Brugman and Visscher, 1988, p. 160-163. Shelmani, et al. 1992, p. 233-240.

[10] Aliev, et al. 1971, p. 69-70. Echikh, 1998, p. 118-119.

[11] Abu Shaybah Formation: Christie, 1955. Revisions: Desio et al. 1960b, p. 273-323, Burollet, 1960, p. 44. IRC mapping: El Hinnawy and Cheshitev, 1975, p. 15-16. Smetana, 1975, p. 13-14. Age: El Hinnawy and Cheshitev, 1975, p. 10. Fatmi et al, 1980, tab. 1. Hammuda, et al. 1985, p. 4-6. Sedimentology:

Assereto and Benelli, 1971, p. 37-86.

[12] Hammuda et al. 1985, p. 4-6.

[13] Ghadamis Basin: Sinha, 1980, figure 5. SE of Tripoli: Belhaj, 1996, p. 85. Tunisia: Hammuda et al, 1985, p. 5. Burollet, 1978, p. 335, 344. Rubino, et al. 2000, p. 79.

[14] Aliev, et al, 1971, p. 69-70. Echikh, 1998, p. 118-119.

[15] As Sarah/Tuama fields: Kuehn, 1996, p. 251-261. Cyrenaica: Brugman and Visscher, 1988, p. 161-163. Shelmani, et al. 1992, p. 233-240.

[16] Continental Post Tassilien: Kilian, 1931, p. 109-111. Subdivision into Tiguentourine, Zarzaitine and Taouratine Formations: de Lapparent and Lelubre, 1948, p. 1104-1106.

[17] Zarzaitine Formation: de Lapparent and Lelubre, 1948, p. 1104-1106. Aliev, et al 1971, p. 72-73. Turner, et al. 2001. Subdivision: Lefranc, 1958, p. 1360-1363. Dinosaurs: de Lapparent, 1958, p. 1237-1240. Libyan outcrops: Protic, 1984, p. 29-32, Roncevic, 1984, p. 39-42, Berendeyev, 1985b, p. 24-26, Berendeyev, 1985a, p. 38-40.

[18] Murzuq Basin: Komarnicki, 1984, p. 34-36. Jakovljevic, 1984, p. 59-62. Galecic, 1984, p. 43-45. Summaries: Grubic et al, 1991, p. 1555-1557, Jakovljevic, 1991, p. 1584-1585.

[19] Pierobon, 1991, p.1780-1781. IRC Geological Map of Libya, 1:1,000,000, 1985. Aziz, 2000, p. 366.

[20] Grignani, et al, 1991, p. 1167, 1177, 1178.

[21] Brugman and Visscher, 1988, p. 158. Sinha and Eland, 1996, p. 49-60. Thusu, 1996, p. 455-474. Gras and Thusu, 1998, p. 322.

[22] Brugman and Visscher, 1988, p. 159. Thusu, 1996, p. 455-474. Gras and Thusu, 1998, p. 322. Gardiner, 1988, p. 259-266.

[23] Wilson and Guiraud, 1998, p. 238. Almond, 1991, p. 2495-2509. Cahen, et al. 1984, chap. 14. Massa and Delort, 1984, p. 1087-1096.

[24] Offshore: Sbeta, 1991, figure 3. Del Ben and Finetti, 1991, figures 2 and 3.

[25] Abu Ghaylan Formation: Christie, 1955, 1-60. Desio, et al. 1963, p. 1-126. Burollet, 1963, p. 1-19. Magnier, 1963, p. 89-94. Syndepositional slumping: Gray, 1971, p. 310.

[26] Christie, 1955. Assereto and Benelli, 1971, p. 62-66. El Hinnawy and Cheshitev, 1975, p. 18-21. Ra's Tahunah: Fatmi and Sbeta, 1991, p. 2230.

[27] Hammuda et al. 1985, p. 1-2.

[28] Sinha, 1980, figure 5. Belhaj, 1996, p. 91. Cyrenaica: Shelmani, et al. 1992, p. 233-240. Brugman and Visscher, 1988, p. 161.

[29] De Lapparent, 1952, p. 129-134. Christie, 1955.

[30] Bi'r al Ghanam Fm: De Lapparent, 1972. Bi'r al Ghanam Gypsum: El Hinnawy and Cheshitev, 1975, p. 22-23. Bu en Niran Member: Burollet, 1963b. Abreghs Gypsum: Burollet, 1963b. Banerjee, 1980, p. 61-63. Ra's Ajdir: Smetana, 1975, p. 14-16. Petrography: El Haddad, et al. 2000, p. 156. Reserves:

Gualtieri, 1959.

[31] Hammuda et al. 1985, p. 39-42. Banerjee, 1980, p. 61-64. Sinha, 1980, figures 2-5. Aliev et al. 1971, p. 72. Thusu, et al. 1988, p. 194. Shelmani, et al. 1992, p. 233-240.

[32] Desio, et al, 1960a, p. 65-114. Desio and Ronchetti, 1960, p. 173-196. El Hinnawy and Cheshitev, 1985, p. 23-24. Smetana, 1975, p. 21-22. Banerjee, 1980, p. 248-250.

[33] Pre-1975 concept: Hammuda, 1971, p. 87-98. 1975 revision: El Hinnawy and Cheshitev, 1975, p. 24-31. Post-1975 view: El Zouki, 1980, p. 393-418. Fatmi and Sbeta, 1991 p. 2227-2234.

[34] El Hinnawy and Cheshitev, 1975, p. 24-31. Smetana, 1975, p. 23-24. Antonovic, 1977, p. 14-16. Novovic, 1977b, p. 17-21. Palynology, Yifran: Al Jumaily, 2000, p. 8. El Zouki, 1980a, p. 397-399. El Zouki, 1980b, p. 419-426. Sinha, 1980, figures 2-5. Belhaj, 1996, p. 85.

[35] Hammuda, et al. 1985, p. 96.

[36] Thusu, et al, 1988, p. 183.

[37] Burollet, 1963b, El Hinnawy and Cheshitev, 1975, p. 29. Smetana, 1975, p. 26-27. Novovic, 1977b, p.18-20. Antonovic, 1977, p.14-17. Belhaj, 1996, p. 92. Sinha, 1980, figures 2-6. El Zouki, 1980, p. 393-418.

[38] Hammuda, et al. 1985, p. 96.

[39] Thusu, et al. 1988, p. 184.

[40] Duronio, et al. 1980, p. 1598-1616.

[41] Burollet, 1963b, Hammuda, 1971, p. 92. El Hinnawy and Cheshitev, 1975, p. 29-31. Antonovic, 1977, p. 16-17. Novovic, 1977b, p. 20-22. El Zouki, 1980a, p. 397-399. El Zouki, 1980b, p. 419-426. Belhaj, 1996, p. 85.

[42] Hammuda, et al. 1985, p. 96.

[43] Thusu, et al. 1988, p. 173, 184.

[44] Duronio, et al. 1980, p. 1598-1606.

[45] Duronio, et al. 1980, p. 1606-1616. A1-36: personal communication from B. Thusu.

[46] Banerjee, 1980, p. 83.

[47] de Lapparent and Lelubre, 1948, p. 1106-1108. Aliev, et al. 1971, p. 71-73.

[48] South Ghadamis outcrops: Berendeyev, 1984b, p. 27-30, Berendeyev, 1984a, p. 41-44. Similarity with Kiklah Formation: Batton, 1965, p. 7-97.

[49] Murzuq outcrops: Komarnicki, 1984, p. 41-43. Jakovljevic, 1984, p. 65-68. Galecic, 1984, p. 45-48. Summaries: Grubic, et al. 1991, p. 1557. Jakovljevic et al, 1991, p.1585.

[50] NC 115: Aziz, 2000, p. 366. Braspetro wells: Pierobon, 1991, p. 1780-1781. East flank: Al Maghrabi and Cheshitev, 1977. Sabha/Awbari: Jakovljevic, 1991, p. 1585. Sabha outcrops: Geological Map of Libya, 1985, scale 1:1,000,000. A1-73: Banerjee, 1980, p. 83. Central Murzuk: Wennekers, et al. 1996, p. 19. Kufrah: Grignani, et al, 1991, p. 1167.

[51] Wennekers, et al. 1996, p. 20.

[52] Thusu, et al. 1988, p. 197.

[53] Continental Intercalaire: Kilian, 1931. Djoua Group: Lefranc, 1958. Messak Formation: Klitzsch, 1963, p. 26-27. Nubian Sandstone: first reference Russeger, 1837, p. 665-669; first use in Libya: Desio, 1935.

[54] El Hinnawy and Cheshitev, 1975, p. 24-31. El Zouki, 1980, p. 394-406. Fatmi and Sbeta, 1991, p. 2227-2234.

[55] Christie, 1955, p. 17-18. Hammuda, 1971, p. 87-96. El Hinnawy and Cheshitev, 1975, p. 26-29. Novovic, 1977b, p. 20-22. Antonovic, 1977, p. 16-17. El Zouki, 1980, p. 406-412. Fatmi and Sbeta, 1991, p. 2227-2234. Tekbali, 2000, p. 70.

[56] El Zouki, 1980a, p. 406-412. El Zouki, 1980b, p. 419-426. Sinha, 1980, figure 5. Burollet, 1960.

[57] Wennekers, et al. 1996, p. 24. Sinha, 1980, figures 2-6. Belhaj, 1996, p. 85. Banerjee, 1980, p. 155-156.

[58] Hammuda et al. 1985, p. 95-97.

[59] Fournie, 1978, p. 102.

[60] Hammuda, et al. 1985, p. 154-162.

[61] Hammuda, et al. 1885, p. 109-112.

[62] Klitzsch, 1963, p. 26-27. Sabha spore assemblage: Tekbali, 1994, p. 297-311. Jakovljevic, et al. 1991, p. 1586.

[63] West flank: Komarnicki 1984, p. 45-48. Jakovljevic, 1984, p. 69-72. Galecic, 1984, p. 48-50. Grubic 1984b, p. 40-43. North flank: Stefek and Rohlich, 1984, p. 21-28. Mrazek, 1984, p. 15-20. Korab, 1984, p. 16-18. Previous authors: Klitzsch and Baird, 1969, p. 67-80. Klitzsch 1972, p. 483-494. Foggaras: Klitzsch and Baird, 1969, p. 73.

[64] Goudarzi, 1970, plate 5. Banerjee, 1980, p. 176-178. Geological map of Libya, 1985, 1:1,000,000. Braspetro: Pierobon, 1991, p. 1781.

[65] Lorenz, 1980, p. 383-392. Lorenz, 1987, p. 254-264. de Castro, see Pierobon, 1991, p. 1781. Lateritic weathering: Kallenbach, 1972, p. 302-322.

[66] Desio 1935. Russeger, 1837, p. 665-669. SW Egypt: Klitzsch, et al. 1979, p. 967-974. Barr and Weegar, 1972, p. 132-133. Sarir Sandstone: Gras and Thusu, 1998, p. 322. Ambrose, 2000, p. 165-191.

[67] Nalut: Novovic, 1977b. Offshore: Hammuda, et al., 1985, p. 93-97. Zamzam: Belhaj, 1996, p. 93-94. South flank Ghadamis Basin: Geol. Map of Libya, 1985, 1: 1,000,000. Djoua Group: Banerjee, 1980, p. 101-102.

[68] Grignani, et al. 1991, p. 1167. Sinha and Pandey, 1980, p. 631. Burollet, 1963, p. 219-227.

[69] Barr and Weegar, 1972, p. 132-133. Banerjee, 1980, p. 193. Charophytes and ostracods: Viterbo, 1968, p. 393-402.

[70] Ibrahim, 1991, p. 2766. Rossi, et al. 1991, p. 2211-2225. Ambrose, 2000, p. 175. Thusu, 1996, p. 459-462.

[71] El Hawat, 1996, p. 12-22. Abdulghader, 1996, p. 234-239.

[72] Hea, 1971, p. 107-126. Bonnefous, 1972, p. 225-261. Wennekers, et al. 1996, p. 17. El Hawat, et al. 1996, p. 8-12.

[73] Wennekers, et al. 1996, p. 24-25.

[74] Thusu, et al. 1988, p. 196-197.

[75] Jabal al Akhdar Trough: El Hawat and Shelmani, 1993, p. 11-13. Stratigraphic nomenclature and offshore wells: Duronio, et al. 1991, p. 1598-1616. Facies B1-18: Pallas, 1980, fig. 4.

[76] Duronio, et al. 1991, p. 1595.

[77] Onshore wells: Uwins and Batten, 1988, p. 219-224.

[78] Possible continental late Cretacous rocks in the Murzuq Basin: Wennekers, et al. 1996, p. 27.

[79] Christie, 1955, p. 19-20. El Hinnawy and Cheshitev, 1975, p. 31-35. El Bakai, 2000, p. 30.

[80] Original definition: Christie, 1955, p. 19-21. New type sections: El Hinnawy and Cheshitev, 1975, p. 30-35.

[81] Nalut: Novovic, 1977, p. 24-26. Mizdah: Antonovic, 1977, p. 16-19. Al Khums: Mann, 1975a, p. 19-26. Megerisi and Mamgain, 1980a, p. 4-6.

[82] Wennekers, et al. 1996, p. 26-28. Sinha, 1980, figures 2-6. Geological Map of Libya, 1:1,000,000, 1985. Belhaj, 1996, p. 85, 93, 94.

[83] Zaccagna, 1919, p. 1-70. Gharyan: El Hinnawy and Cheshitev, 1975, p. 36. Tunisian border: Novovic, 1977a, p. 11-12. Nalut: Novovic, 1977b, p. 25-26. Mizdah: Antonovic, 1977, p. 18-20. Megerisi and Mamgain, 1980a, p. 6-7. Wennekers, et al. 1996, p. 26. Sinha, 1980, figures 2-6. Dolomitization: Chaabani, et al. 2000, p. 25.

[84] Christie, 1955, p. 22-24. Type section: El Hinnawy and Cheshitev, 1975, p. 37-39. Banerjee, 1980, p. 214-216. Megerisi and Mamgain, 1980a, p. 7-8. Novovic, 1977a, p. 12-13. Novovic, 1977b, p. 28-29. Mizdah: Antonovic, 1977, p. 21-24. Bani Walid: Zivanovic, 1977, p. 16-18.

[85] Ghadamis: Rohlich, 1979, p. 20-22. Bi'r Amasin: Alexeyev, 1984a, p. 21-24. Wadi Tanarut: Alexeyev, 1984c, p. 20-22. Subsurface: Sinha, 1980, figures 2-6. Belhaj, 1996, p. 85, 93-94. Rohlich and Youshah, 1991, p. 2371-2380.

[86] Wadi ain Armas: Berendeyev, 1985b, p. 31-33. Hamadat Tanghirt: Berendeyev, 1985a, p. 45-48.

[87] Burollet, 1960, p. 21-21. Jordi and Lonfat, 1963, p. 116. Antonovic, 1977, p. 24-26.

[88] Jordi and Lonfat, 1963, p. 116. Banerjee, 1980, p. 178-184. Megerisi and Mamgain, 1980a, p. 9-12. Type section: Antonovic, 1977, p. 25-27. *Inoceramus assemblage:* Troger and Rohlich, 1991, p. 1357-1381. Troger, 2000, p. 98.

[89] Novovic, 1977a, p. 14-15. Novovic, 1977b, p. 31-32. Zivanovic, 1977, p. 20-22. Ghadamis and Darj: Rohlich, 1979, p. 24-27. Chaloupsky, 1979, p. 26-27. Alexeyev, 1985a, p. 25-28. Alexeyev, 1985c, p. 21-24.

[90] Alexeyev, 1985a, p. 30-31. Rohlich, 1979, p. 25-27. Novovic, 1977a, p. 15. Novovic, 1977b, p. 31-33. Antonovic, 1977, p. 27-28. Zivanovic, 1977, p. 23-26. Minor outcrops: Salaj, 1979, p. 19-20. Cepek, 1979, p. 23-26. Chaloupsky, 1979, p. 29-32. Alexeyev, 1985c, p. 25-28. Southern rim: Berendeyev, 1985b, p. 34-36. Berendeyev, 1985a, p. 49-52. Gundobin, 1985, p. 77-80. Wells: Sinha, 1980, p. 648-651.

[91] Belhaj, 1996, p. 85, 93-94.

[92] Nairn and Salaj, 1991, p. 1621-1623.

[93] Bin 'Affin Member: Furst, 1965, p. 1067-1073. Tmassah: Korab, 1984, p. 21-24. Al Fuqaha: Woller, 1984, p. 34-38. Al Washkah: Woller, 1978, p. 17-24. El Haddad, et al. 2000, p. 369-378.

[94] Bi'r al Ghurab Mbr: Nairn and Salaj, 1991, p. 1629-1630.

[95] Nairn and Salaj, 1991, p. 1630-1632.

[96] Jordi and Lonfat, 1963, p. 118. Nairn and Salaj, 1991, p. 1632-1635. Age L. Tar Member: Eliagoubi and Powell, 1980, p. 137-153.

[97] Jordi and Lonfat, 1963, p. 118.

[98] Woller, 1978, p. 17-20. Gundobin, 1985, p. 81-84.

[99] Western outcrop: Berendeyev, 1985b, p. 35-37. Alexeyev, 1985a, p. 34-38. Chaloupsky, 1979, p. 34-39. *Inoceramus:* Troger and Rohlich, 1991, p. 1358-1381. Troger, 2000, p. 98.

[100] Northern outcrops: Novovic, 1977b, p. 36-38. Antonovic, 1977, p. 29-32. Salaj, 1979, p. 21-30. Cepek, 1979, p. 27-33. Zivanovic, 1977, p. 26-29. Subsurface: Sinha, 1980, p. 648-651. Belhaj, 1996, p. 85, 94. Karstification: Jerzykiewicz, et al. 2000, p. 53.

[101] Hammuda, et al. 1985, p. 7-9. Tunisia: Fournie, 1978, p. 130, 139.

[102] Hammuda, et al. 1985, p. 103-105. Tunisia: Fournie, 1978, p. 130.

[103] Hammuda, et al. 1985, p. 84-85. Tunisia: Fournie, 1978, p. 137.

[104] Hammuda, et al. 1985, p. 54-55. Tunisia: Fournie, 1978, p. 129, 131, 133.

[105] Hammuda, et al. 1985, p. 20-24. Tunisia: Fournie, 1978, p. 112.

[106] Barr and Weegar, 1972, p. 34-36, Wennekers, et al. 1996, p. 28. Bu-Argoub, 1996, p. 428.

[107] Sghair, 1996, p. 65-81.

[108] Barr and Weegar, 1972, p. 124-127. A1-96: Thusu, 1996, p. 458.

[109] Sghair and El-Alami, 1996, p. 263-274.

[110] Heselden, et al. 1996, p. 197-210

[111] Barr and Weegar, 1972, p. 120-122. Banerjee, 1980, p. 162, Wennekers, et al. 1996, p. 28. Baird, et al. 1996, figure 4.

[112] El-Bakai, 1996, p. 83-97.

[113] Heselden, et al. 1996, p. 208-209. Mansour and Magairhy, 1996, p. 504.

[114] Barr and Weegar, 1972, p. 21-22. Banerjee, 1980, p. 41. Wennekers, et al. 1996, p. 28.

[115] Barr and Weegar, 1972, p. 58-60. Banerjee, 1980, p. 114. 5P1-59: Baird, et al. 1996, p. 18, 21. Evaporite basins: Wennekers, et al. 1996, p. 26, 28. Hameimat Trough: Mansour and Magairhy, 1996, p. 503. Barbieri, 1996, p. 192. Abdelhameed, 2000, p. 283. Palaeogeography: El Alami, et al. 1989, figure 3. As

438

Sarir: Bu-Argoub, 1996, p. 428. Greenhouse effect: Kuhnt, et al. 2000, p. 49. Lüning, 2000, p. 56-57. Lüning, 2001.

[116] Barr and Weegar, 1972, p. 134-136. Williams, 1972, p. 627. Banerjee, 1980, p. 222-223. Wennekers, et al. 1996, p. 30. Bu-Argoub, 1996, p. 428. Abdelhameed, 2000, p. 283.

[117] Barr and Weegar, 1972, p. 161-163. Banerjee, 1980, p. 246-247. Williams, 1968, p. 197-205. Williams, 1971, p. 506. Mansour and Magairhy, 1996, p. 515. El Alami, et al. 1989, p. 22. Barbieri, 1996, p. 189.

[118] Barr and Weegar, 1972, p. 156-158. Banerjee, p. 243-244. Wennekers, et al. 1996, p. 30. Baird, et al. 1996, p. 18-19. Bu-Argoub, 1996, p. 426. Barbieri, 1996, p. 191. Hammuda, 1980, p. 509-520.

[119] Barr and Weegar, 1972, p. 111-113. Banerjee, 1980, p. 149-150. Age: Eliagoubi and Powell, 1980, p. 137-153. As Sarir: Bu-Argoub, 1996, p. 428. Baair, et al. 2000, p. 12. Jabal az Zalmah: Lüning, et al. 2000, p. 721-731. Danian age: Jones, 1996, p. 179. A1-NC 29A: Tmalla, 1992, p. 542-552.

[120] Barr and Weegar, 1972, p. 148-149. Banerjee, 1980, p. 227-228.

[121] Barr and Weegar, 1972, p. 150-153. Banerjee, 1980, p. 230-231. Roohi, 1996b, p.442-444.

[122] Barr and Weegar, 1972, p. 165-167. Banerjee, 1980, p. 270. Eliagoubi and Powell, 1980, p. 137-153.

[123] Broughton, 1996, p. 391-418. El Ghoul, 1996, p. 137-154. Jones, 1996, p. 169-184.

[124] Kleinsmeide and van der Bergh, 1968, p. 115-124. Pietersz, 1968, p. 125-130. Barr, 1968, p. 131-148. Change of name: Megerisi and Mamgain, 1980a, p. 48. Western outcrop: Klen, 1974, p. 14-17. Eastern outcrop: Rohlich, 1974, p. 15-17. Recent review: Al Hawat and Shelmani, 1993, p. 38. A1-NC 120: Duronio, et al. 1991, p. 1603. A1-28 etc.: Uwins and Batten, 1988, p. 224-228.

[125] Kleinsmeide and van den Berg, 1968, p. 115-117. Barr and Weegar, 1972, p. 107-109. Type section: Klen, 1974, p. 15-17. Al Bayda: Rohlich, 1974, p. 15-17. Megerisi and Mamgain, 1980a, p. 49-50. Banerjee, 1980, p. 15-16. Rohlich, et al. 1996, p. 265-285. Duronio, et al. 1991, p. 1594, 1598. El Hawat and Shelmani, 1993, p. 14.

[126] Rohlich, 1974, p. 19-22.

[127] Al Bayda: Rohlich, 1974, p. 19-22. Erroneous mapping Zawayat Masus: Mazhar and Issawi, 1977, Bi'r Hakim: Swedan and Issawi, 1977. Revision: Megerisi and Mamgain, 1980c, p. 73-89. New map: IRC (1985) Geological Map of Libya, scale 1:1,000,000. Al Burdi: Schiettecatte, 1972, p. 59-64. A1-NC 120: Duronio, et al. p. 1598, 1601.

[128] Klen, 1974, p. 20-22. Banerjee, 1980, p. 263, Megerisi and Mamgain, 1980a, p. 53-54

[129] Pietersz, 1968, p. 125-130. Kleinsmiede and van den Berg, 1968, p. 115-124. Klen, 1974, p. 19. Rohlich, 1974, p. 24-25. Megerisi and Mamgain, 1980a, p. 52-53. Banerjee, 1980, p. 269-270. Duronio, et al. 1991, p. 1598, 1607.

[130] Barr and Hammuda, 1971, p. 27-38. Rohlich, 1978, p. 401-412. Barr, 1968, p. 137-140. El Hawat and Shelmani, 1993, p. 13-14, 48. Barr, 1972, p. 1-46. Banerjee, 1980, p. 24-25. El Mehaghag and Muftah, 1996, p. 501-520. A1-NC 120, A1-NC 128: Duronio, et al. 1991, p. 1594, 1598, 1607. A1-45: Uwins and Batten, 1988, p. 224-228. El Mehdawi, 1991, p. 1351-1355.

[131] Barr and Hammuda, 1971, p. 27-40. Rohlich, 1974, p. 28-29. Megerisi and Mamgain, 1980a, p. 54-55. El Mehaghag and Muftah, 1996, p. 501-520. El Mehaghag, 2000, p. 254. Hay, 1968, p. 149-158. Duronio, et al. 1991, p. 1598-1600.

Chapter 5

[1] Trans-Sahara seaway: Desio, 1971, p. 11-16. Reyment and Reyment, 1980, p. 245-254. El Sogher, 1996, p. 379-380.

[2] Zimam Formation: Jordi and Lonfat, 1963, p. 116-122. Revision: Nairn and Salaj, 1991, p. 1621-1636. Banerjee, 1980, p. 278-280. Megerisi and Mamgain, 1980a, p. 17-20. Danian foraminifera: Woller, 1978, p. 25-28.

[3] Geological Map of Libya, 1985, scale 1:1,000,000. Bani Walid: Zivanovic, 1977, p. 29-34. Shawa: Chaloupsky, 1979, p. 45-48. Al Qaryat al Gharbiyah: Salaj, 1979, p. 30-36. Al Qaryat ash Sharqiyah: Cepek, 1979, p. 36-40. Wadi Tanarut: Alexeyev, 1985c, p. 32-35. Habayt al Ghardaqah: Alekseyev, 1985b, p. 19-24. Qasr ash Shwayrif: Gundobin and Yevdokimov, 1985, p. 25-34. Qararat al Marar: Gundobin, 1985, p. 85-90. Jabal Hasawnah: Jurak, 1978, p. 29-33.

[4] Al Fuqaha: Woller, 1984, p. 38-44. Tmassah: Korab, 1984, p. 25-30. Gohrbandt, 1966, p. 38.

[5] Shurfah Formation,Bu Ras Member and Operculina Limestone: Jordi and Lonfat, 1963, p. 114-122. Banerjee, 1980, p. 235-237. Megerisi and Mamgain, 1980a, p. 21-26. Qaltah Member and Operculinoides Limestone: Burollet, 1960, p. 27. Ammur Member: Shakoor and Shagroni, 1984, p. 45-47. Foraminiferal fauna: Haynes, 1962, p. 90-97.

[6] Shurfah Formation outcrops: Bani Walid: Zivanovic, 1977, p. 35-38. Al Qaryat ash Sharqiyah: Cepek, 1979, p. 43-54. Habayt al Ghardaqah: Alexeyev, 1985b, p. 26-34. Qasr ash Swayrif: Gundobin and Yevdokimov, 1985, p. 35-44. Hun: Shakoor and Shagroni, 1984, p. 37-47. Al Washkah: Woller, 1978, p. 33-40. Al Fuqaha: Woller, 1984, p. 43-57. Tmassah: Korab, 1984, p. 31-40. Sarir Tibisti: Gohrbandt, 1966, p. 38. Jabal Nuqay: Furst, 1965, p. 1060-1088. Furst, 1968, p. 296-313.

[7] Al Jurf Formation: Hammuda, et al., 1985, p. 20-24.

Abutarruma, 2000, p. 19. Tunisia: Fournie, 1978, p. 112-113, 120-123.

8   Ehduz Formation: Hammuda, et al. 1985, p. 68-69. Tunisia: Fournie, 1978, p. 112-113.

9   Barr and Weegar, 1972, p. 89-92. Banerjee, 1980, p. 131-132. Sinha and Mriheel, 1996, p. 156-161. Bezan, 1996, p. 101-102.

10 Barr and Weegar, 1972, p. 51-53. Banerjee, 1980, p. 90-91. Bezan, et al. 1996, p. 143.

11 Barr and Weegar, 1972, p. 150-153. Banerjee, 1980, p. 230-231. Extent: Bezan, 1996, p. 101. Mouzughi, 1991, p. 1841-1854. Roohi, 1996b, p. 444.

12 Barr and Weegar, 1972, p. 140-143. Banerjee, 1980, p. 225-226. Bezan, 1996, p. 100-101.

13 Barr and Weegar, 1972, p. 37-39. Banerjee, 1980, p. 51-52. Montian Stage: Harland, et al. 1990, p. 61. Manzilah Ridge: Roohi, 1996b, p. 442-445.

14 Barr and Weegar, 1972, p. 42-44. Bezan, et al. 1996, p. 139, 142. Bezan, 1996, p. 101-105.

15 Rabia Member: Barr and Weegar, 1972, p. 40-42. Banerjee, 1980, p. 52.

16 Sabah area: Garea, 1996, p. 115-134. Garea and Braithwaite, 1996, p. 289-304. Farrud Member: Bezan, 1996, p. 103-106. Bezan, et al. 1996, p. 143-146. Ghani: Abushagur, 1991, p. 1827-1840. Abushagur and LeMone, 1991, p. 1871-1881. Al Bayda Platform: Mriheel, 1994, p. 65-66. Sinha and Mriheel, 1996, p. 158-166. Abu Tumayam Basin: Johnson and Nicoud, 1996, p. 211-221.

17 Bezan, 1996, p. 105. Bezan, et al. 1996, p. 148-150. Sinha and Mriheel, 1996, p. 161-169.

18 Barr and Weegar, 1972, p. 114-116. Banerjee, 1980, p. 151-152

19 Bezan, 1996, p. 108, 113. Roohi, 1996b, p. 442-445.

20 Barr and Weegar, 1972, p. 48-50. Banerjee, 87-88. Bezan, 1996, p. 105-113. Roohi, 1996b, p. 442-445.

21 Barr and Weegar, 1972, p. 154-155. Banerjee, 1980, p. 234. Bezan, 1996, p. 113.

22 Barr and Weegar, 1972, p. 106-107, 168-170. Banerjee, 1980, 143-144.

23 Bebout and Pendexter, 1975, p. 665-693.

24 Bezan, 1996, p. 109-115. Sinha and Mriheel, 1996, p. 173. Roohi, 1996b, p. 442-445.

25 Barr and Weegar, 1972, p. 144-147. Banerjee, 1980, p. 226-227. Bezan, 1996, p. 109.

26 Terry and Williams, 1969, p. 31-47. Brady, et al. 1980, p. 543-564. Gumati, 1992, p. 305-318.

27 Barr and Weegar, 1972, p. 93-94. Banerjee, 1980, p. 144-145.

28 Bezan, 1996, p. 112-115. Sinha and Mriheel, 1996, p. 176.

29 Spring and Hansen, 1998, p. 335-353. Haq, et al. 1987, p. 1156-1166.

30 Burollet, 1960, p. 26-28. Barr and Weegar, 1972, p. 117-119. Banerjee, 1980, p. 59-60.

31 Barr and Weegar, 1972, p. 117-119. Bezan, 1996, p. 111, 115-117. Sinha and Mriheel, 1996, p. 178.

32 Barr, 1968, p. 136-137. Rohlich, 1974, p. 29-32. Megerisi and Mamgain, 1980a, p. 55-56. Banerjee, 1980, p. 36-37. Barr and Berggren, 1980, p. 163-173. Thanetian age: Eliagoubi, 1980, p. 155-162. Nannofossils: El Mehaghag and Badi, 2000, p. 253. Duronio, et al. 1991, p. 1598-1599. Dolomites: Elwerfalli and Stow, 2000, p. 158.

33 Megerisi and Mamgain, 1980a, p. 70-76. Geological Map of Libya, 1985, scale 1:1,000,000. Ring intrusion, Jabal Awaynat: Flinn, et al. 1991, p. 2539-2557. Gharyan lavas: Busrewil and Wadsworth, 1980, p. 1095-1105.

34 Desio, 1935, p. 187, Furst, 1968, p. 296-313. Savage, 1969, p. 167-171. Wight, 1980, p. 309-325. Geological Map of Libya, 1985, scale 1:1,000,000.

35 Burollet, 1960, p. 26-28. Jordi and Lonfat, 1963, p. 114-121. Barr and Weegar, 1972, p. 117-119. Megerisi and Mamgain, 1980a, p. 27-32.

36 *Flosculina Limestone:* Burollet, 1960, p. 27. Furst, 1965, p. 1060-1088. Woller, 1978, p. 56-57.

37 Burollet, 1960, p. 27. Megerisi and Mamgain, 1980a, p. 30-31 Shakoor and Shagroni, 1984, p. 56-58. Dolomitization: Lashhab and West, 1996, p. 31-43.

38 Bani Walid: Zivanovic, 1977, p. 38-40. Al Qaddahiyah: Mijalkovic, 1977, p. 16-19. Al Qaryat al Sharqiyah: Cepek, 1979, p. 57-66. Abu Njaym: Said, 1981, p. 33-40. Hun: Shakoor and Shagroni, 1984, p. 49-58. Al Washkah: Woller, 1978, p. 41-52. Zallah: Vesely, 1985, p. 19-22. Al Fuqaha: Woller, 1984, p. 56-61. Al Haruj al Aswad: Busrewil and Suwesi, 1993, p. 13-14. Tmassah: Korab, 1984, p. 40-42. Waw al Kabir area: Wight, 1980, fig. 4. Fürst, 1968, p. 296-313. Geological Map of Libya, 1:1,000,000, 1985.

39 Al Jir Formation: Burollet, 1960, p. 27. Barr and Weegar, 1972, p. 75-85. Megerisi and Mamgain, 1980, p. 32-35. Mijalkovic, 1977, p. 18-20. Shakoor and Shagroni, 1984, p. 62.

40 Bin Isa Member: Burollet, 1960, p. 27. Megerisi and Mamgain, 1980a. Said, 1981, p. 43-44. *Orbitolites Member:* Burollet, 1960, p. 39. Bi'r Zaydan Member: Desio, 1943, p. 75. Megerisi and Mamgain. 1980. P. 33-34. Said, 1981, p. 45-46.

41 Al Qaddahiyah: Mijalkovic, 1977, p. 18-20. Abu Njaym: Said, 1981, p. 43-46. Hun: Shakoor and Shagroni, 1984, p. 60-63. Wadi bu ash Shaykh: Jurak, 1985, p. 18-24. Zallah: Vesely, 1985, p. 24-28. Abu Na'im: Zikmund, 1985, p. 19-22. Ash Sharit: Fürst, 1968, p. 296-313.

42 Wadi Thamat Formation: Desio, 1943, p. 78. Burollet, 1960, p. 51-52. Goudarzi, 1970, p. 38.

43 Al Gata Member: Goudarzi, 1970, p. 38. Megerisi and Mamgain, 1980a, p. 36.

44 Thmed al Qusur Member: Burollet, 1960, p. 51-52. Megerisi and Mamgain, 1980a, p. 36-37.

45 Qararat al Jifah Member: Goudarzi, 1970, p. 38. Megerisi and Mamgain, 1980a, p. 37-38.

46 Abu Njaym: Said, 1981, p. 47-50. An Nuwfaliyah:

Srivastava, Shagroni and Njoma, 1980, p. 25-34. Wadi bu ash Shaykh: Jurak, 1985, p. 24-44. Maradah: Mastera, 1985, p. 23-30. Zallah: Vesely, 1985, p. 29-40. Abu Na'im: Zikmund, 1985, p. 23-36. Al Hufrah area: Antetell and Kumati, 1991, p. 1883-1906.

47 Farwah Group: Hammuda, et al. 1985, p. 70-74. Sbeta, 1990, p. 42-56. Sbeta, 1991, p. 1929-1966. El Ghoul, 1991, p. 1637-1655.

48 Bilal Formation: Hammuda, et al. 1985, p. 36-38. Revision: El Ghoul, 1991, p. 1645-1647 and fig. 3. Sbeta, 1990, p. 47. Fournie, 1978, p. 115-118.

49 Jdeir Formation: Hammuda, et al. 1985, p. 88-90. Revision: El Ghoul, 1991, p. 1637-1655. Bouri: Bernasconi, et al., 1991, p. 1907-1928. Sedimentology: Racey, 2001, p. 79-100. Anketell and Mriheel, 2000, p. 425-447. Reali, et al. 2000, p. 76. Distribution: Sbeta, 1990. p. 51. Present-day *Nummulites:* Moody and Sandman, 2000, p. 61. Moody, et al. 2000. p. 70. Hasdrubal: Beavington-Penney, et al. 2000, p. 23.

50 Jirani Dolomite: Hammuda, et al. 1985, p. 91-92. El Ghoul. 1991, p. 1647. Giaj-Via, et al. 2000, p. 46. Tunisian equivalent of Jdeir Formation: Fournie, 1978, p. 112. Loucks, et al. 1998, p. 355-374. Tunisian outcrops: Jorry, et al. 2000, p. 54.

51 Hammuda, et al. 1985, p. 138-141. Sbeta, 1990, fig. 3.

52 Hammuda, et al. 1985, p. 134-137. Sbeta, 1990, p. 46.

53 Sbeta, 1991, p. 1929-1966.

54 Hammuda, et al. 1985, p. 82-83. Sbeta, 1991, p. 1937-1941.

55 Dahman Limestone: Hammuda et al. 1985, p. 58-59. Sbeta, 1991, p. 1937-1948. Reineche Member: Fournie, 1978, p. 112.

56 Samdun Formation: Hammuda, et al. 1985, p. 121-122. Sbeta, 1991, p. 1940-1965.

57 Hammuda, 1985, p. 75-78. Sbeta, 1991, p.1937-1966. Fournie, 1978, p. 112.

58 Barr and Weegar, 1972, p. 75-88.

59 Abugares, 1996, p. 45-63.

60 Elag, 1996, p. 99-114.

61 Dolomitization: Lashhab and West, 1996, p. 31-44. Sedimentology, Zallah Trough: Lashhab and West, 1991, p. 1855-1870.

62 Barr and Weegar, 1972, p. 70-74. Banerjee, 1980, p. 126-127.

63 Barr and Weegar, 1972, p. 64-69. Banerjee, 1980, p. 123.

64 Barr and Weegar, 1972, p. 28-33. Banerjee, 1980, p.48-49.

65 Gregory, 1911, p. 593-598. Pietersz, 1968, p. 125-130, Central belt: Rohlich, 1974, p. 32-34. Western outcrops: Klen, 1974, p. 26. Eastern outcrops: Zert, 1974, p.13-14. Age: Khoudary, 1980, p. 193-204. Wadi Bakur: Helmdach and Khoudary, 1980, p. 255-269. Tocra Limestone: Banerjee, 1980, p. 261-262. El Mehaghag and Daw, 2000, p. 253. Wadi Athrun: Haq and Aubrey, 1980, p. 279-282. Summary: Barr and Berggren, 1980, p. 167-169. Antlat Formation: Muftah

and Mohammed, 2000, p. 147. A1-NC 120: Duronio, et al. 1991, p. 1598-1599. El Hawat and Shelmani, 1993, p. 16-17. Diagenesis: Bausch and Schellhorn, 1991, p. 2671-2679.

66 Gregory, 1911, p. 579, 582. Pietersz, 1968, p. 127-128. Rohlich, 1974, p. 35-38. Zert, 1974, p. 14-17. Barr and Berggren, 1980, p. 169-171. Duronio, et al. 1991, p. 1606-1609. El Hawat and Shelmani, 1993, p. 33, 53, 54

67 Zallah: Vesely, 1985, p. 44-45. Vertebrates: Arnould-Saget and Magnier, 1961, p. 283-287. Arambourg and Magnier. 1961. p. 1181-1183. Maradah: Mastera, 1985, p. 36-38. Jabal Hasawnah: Jurak, 1978, p. 37-39.

68 Abu Na'im: Zikmund, 1985, p. 49-51. Dur at Talah: Bellair, et al. 1954, p. 1822-1824. Arambourg and Magnier, 1961, p. 1181-1183. Wight, 1980, p. 309-324.

69 Jordi and Lonfat, 1963, fig. 3. Megerisi and Mamgain, 1980a, p. 44-45. Type section: Shakoor and Shagroni. 1984, p. 66-70. Al Qaryat al Sharqiyah: Cepek, 1979, p. 67-68. Abu Njaym: Said, 1981, p. 51-54. Al Washkah: Woller, 1978, p. 52-55. Zallah: Vesely, 1985, p. 43-54. Hun Graben: Abadi and van Dijk, 1993 p.14-16. van Dijk and Eabadi, 1996, p. 155-166.

70 Burollet, 1960, p. 18. Megerisi and Mamgain, 1980a, p. 39-41. An Nuwfaliah: Srivastava, et al. 1980, p. 37-42. Wadi bu ash Shaykh: Jurak, 1985, p. 41-44. Maradah: Mastera, 1985, p. 30-33. Abu Na'im: Zikmund, 1985, p. 39-42. Patch reefs: Hladil, et al. 1991, p. 1401-1420.

71 Burollet, 1960, p. 29. Megerisi and Mamgain, 1980a, p. 41-42. An Nuwfaliah: Srivastava, et al. 1980, p. 42-46. Maradah: Mastera, 1985, p. 33-36. Abu Na'im: Zikmund, 1985, p. 43-46.

72 Hammuda, et al. 1985, p. 60-67. Karst: Belhaj, 2000, p. 16. Fournie, 1978, p. 106-109.

73 Hammuda, et al. 1985, p. 113-120. Fournie, 1978, p. 109-111.

74 Barr and Weegar, 1972, p. 23-25. Mansour and Magairhy, 1996, p. 519.

75 Barr and Weegar, 1972, p. 55-57.

76 Benfield and Wright, 1980, p. 463-478.

77 Bezan and Malak, 1996, p. 119-127. Gruenwald, 2001, p. 213-231. Meteorite craters: Underwood and Fisk, 1980, p. 894-900.

78 Gregory, 1911, p. 583. Kleinsmiede and van den Berg, 1968, p. 115-123. Pietersz, 1968, p. 125-130. Rohlich, 1974, p. 38-42. Megerisi and Mamgain, 1980a, p. 59-60. El Hawat and Shelmani, 1993, p. 17-18.

79 Rohlich, 1974, p.38-42. El Hawat and Shelmani, 1993, p. 18. Darnah: Zert, 1974, p. 15. Offshore: Duronio, et al. 1991, p. 1598-1607.

80 Kleinsmiede and van den Berg, 1968, p. 115-123. Al Bayda: Rohlich, 1974, p. 42-48. Soluq: Francis and Issawi, 1977, p. 21. Binghazi: Klen, 1977, p. 30-32. Darnah: Zert, 1974, p. 22-26. Megerisi and Mamgain 1980a, p. 62-65. Geological Map of Libya, 1985, scale

1:1,000,000. El Hawat and Shelmani, 1993, p. 17-18. Duronio, et al. 1991, p. 1598-1607.

81 Magnier, 1969, p. 119-130. Selley, 1969, p. 419-460. El Hawat, 1980, p. 427-448. Benfield and Wright, 1980, p. 463-500. Jabal as Sawda volcanics: Woller and Fediuk, 1980, p. 1081-1095. Busrewil and Esson, 1991, p. 2599-2603. Gharyan: Busrewil and Wadsworth, 1980b, p. 1095-1105. Al Haruj al Aswad: Busrewil and Wadsworth, 1980a, p. 1077-1180. Busrewil and Suwesi, 1993, p. 20-35. Busrewil, 1996, p. 331-345. Himmali and Oun, 1996, p. 347-356.

82 Desio, 1935, p. 110. Selley, 1966, p. 1-30. Selley, 1967, p. 215-234. Selley, 1968, p. 363-372. Selley, 1969, p. 419-460. Selley, 1971, p. 99-106. El Hawat, 1980, p. 427-448. Maradah: Mastera, 1985, p. 39-48.

83 Maradah: Mastera, 1985, p. 44-48. Al Aqaylah: Innocenti and Pertusati, 1984, p. 41-45.

84 Wadi bu ash Shaykh: Jurak, 1985, p. 59-63. Zallah: Vesely, 1985, p. 55-61. Abu Na'im: Zikmund, 1985, p. 55-65. Dur at Talah: Geological Map of Libya, 1985, scale 1:000,000.

85 Sabkhat Ghuzayil: Kodym, 1985, p. 21-32. Bi'r Zaltan: Domaci, 1985, p. 29-48. Mammals: Savage, 1971, p. 215-223. Pickford, 1991, p. 1483-1490.

86 Floridia, 1939, p. 245-260. Mann, 1975a, p. 35-38. Megerisi and Mamgain, 1980a, p. 42-44. Salem and Spreng, 1980, p. 97-116. Foraminifera: Sherif, 1991, p. 1421-1456. Ostracods: El Waer, 1991, p. 1457-1481.

87 Misratah: Mann, 1975b, p. 11-13. Bani Walid: Zivanovic, 1977, p. 40-42. Al Qaddahiyah: Mijalkovic, 1977, p. 21-24. Abu Njaym: Said, 1981, p. 55-57. An Nuwfaliyah: Srivastava, et al. 198, p. 48-52. Wadi bu ash Shaykh, Zallah, eastern margin Al Haruj al Aswad: Geological Map of Libya, 1985, scale 1:1,000,000.

88 Al Aqaylah: Innocenti and Pertusati, 1984, p. 53-66.

89 Maradah: Mastera, 1985, p. 49-52. Sabkhat Ghuzayil: Kodym, 1985, p. 34-38. Bi'r Zaltan: Domaci, 1985, p. 18.

90 Desio, 1935, p. 95. Benfield and Wright, 1980, p. 485-493. Giglia, 1984, p. 34-48.

91 Ajdabiya: Giglia, 1984, p. 34-48. Sabkhat Ghuzayil: Kodym, 1985, p. 39-47.

92 Sahabi fauna: de Heinzelin, 1980, p. 127-136. Boaz, et al. 1982, 141pp. Boaz, et al. 1987, 383pp. Boaz, 1996, p. 531- 539.

93 Sahabi channel: Barr and Walker, 1973, p. 1244-1251. Benfield and Wright, 1980, p. 491. Boaz, 1987, p. xii.

94 Geological Map of Libya, 1985, scale 1:1,000,000.

95 Hammuda, et al. 1985, p. 25-32. Fournie, 1978, p. 106.

96 Sidi Bannour Formation: Hammuda, et al. 1985, p.129-133.

97 Bi'r Sharuf Formation: Hammuda, et al. 1985, p. 49-53.

98 Tubtah Formation: Hammuda, et al. 1985, p. 149-153.

99 Hammuda, et al. 1985, p. 106-108. Mriheel and Alhnaish, 1995, p. 213-221. Fournie, 1978, p. 103.

100 B1-NC 35a: van Hinte, et al. 1980, p. 205-244.

101 Deep channels: Barr and Walker, 1973, p. 1244-1251. Tunisia: Wiman, 1980, p. 117-126. General reviews: Hsu, 1973, p. 1203-1231. Sonnenfeld, 1985, p. 323-346.

102 Benfield and Wright, 1980, p. 477-485.

103 Benfield and Wright, 1980, p. 485-493.

104 Pietersz, 1968, p. 129. Rohlich, 1974, p. 48-50.

105 Darnah, Bamba, Tubruq: Zert, 1974, p. 26-31. Al Burdi: El Deftar and Issawi, 1977, p. 32-33. Duronio, et al. 1991, p. 1597-1598.

106 Benghazi: Klen, 1974, p. 31-33. Zawayat Masus: Mazhar and Issawi, 1977, p.35. Bi'r Hakim: Swedan and Issawi, 1977, p.24. Wadi al Khali: Giammarino, 1984, p. 12-13. Geological Map of Libya, 1985, scale 1:1,000,000.

107 Eliagoubi, 1972. El Hawat and Shelmani, 1993, p. 19.

108 Desio, 1935, p. 77, 325. Klen, 1974, p. 33-39. Megerisi and Mamgain, 1980a, p. 65-69.

109 Klen, 1974, p. 35-36. Megerisi and Mamgain, 1980a, p. 68-69.

110 Francis and Issawi, 1977, p. 30-35. Megerisi and Mamgain, 1980a, p. 67-68.

111 Francis and Issawi, 1977, p. 36-39. Megerisi and Mamgain, 1980a, p. 68-69.

112 Klen, 1974, p. 38. Megerisi and Mamgain, 1980a, p. 69. El Hawat 1980, p. 449-461.

113 Desio, 1928, p. 94. Banerjee, 1980, p. 26-28. Al Burdi: El Deftar and Issawi, 1977, p. 39-43.

114 Qaret Meriem 'Formation': Swedan and Issawi, 1977, p. 36-38. Zawiyat Masus: Mazhar and Issawi, 1977, p. 40-46.

115 Wadi al Khali: Giammarino, 1984, p. 39-48. Wadi al Hamim: Carmignani, 1984, p. 35-48. Al Muffawaz: Manetti, 1984, p. 26-34.

116 Vita-Finzi, 1971, p. 409-430. Al Assah Formation: Smetana, 1975, p. 29-31. Assabria Formation: Hammuda et al. 1985, p. 33-35. Sbabil Formation: Hammuda et al. 1985, p. 123-128. Raf-Raf Formation: Fournie, 1978, p. 101.

117 Vita-Finzi, 1971, p. 409-430. Al Hishah Formation: Mijalkovic, 1977, p. 24-28. Al Aqaylah: Innocenti and Pertusati, 1984, p. 67-70. Maradah: Mastera, 1985, p. 55-58. Sabkhat Ghuzayil: Kodym, 1985, p. 49-52.

118 Vita-Finzi, 1971, p. 409-430. Qarat al Weddah Formation: Di Cesare, et al. 1963, p. 1344-1362. A1-NC 120: Duronio, et al. 1991, p. 1597-1598. Al Aqaylah: Innocenti and Pertusati, 1984, p. 71-74. Ajdabiyah: Giglia, 1984, p. 53-56. Wadi al Hamim: Carmignani, 1984, p. 62-66. Maradah: Mastera, 1985, p. 58-59. Sabkhat Ghuzayil: Kodym, 1985, p. 60-61. Bi'r Zaltan: Domaci, 1985, p. 48-50. Sahabi fauna: Boaz, et al. 1987, 383pp. Murzuq: Domaci, et al. 1991, p. 1785-1801.

119 Geological Map of Libya, 1985, scale 1:1,000,000. Anketell and Ghellali, 1991, p. 1987-2013. Kufrah lake: El Ramly, 1980, p. 659-670. Murzuq: Thiedig and El Chair, 1998, p. 31-32. Bagnold, 1935.

Chapter 6

1   The 9000 wells can be broken down roughly as follows. Wildcat wells 1700, appraisal and development wells 4500, water wells 2600, core holes and strat wells 200. Regional structure maps: Goudarzi and Smith, 1978. Goudarzi, 1980, p. 879-892. Mikbel, 1977, p. 19-34. Mikbel, 1979, p. 209-220. Mikbel, 1981, p. 547-554. Wennekers, et al. 1996, figure 1. Anketell, 1996, figure 4.

2   Burke and Dewey, 1974, p. 313-316. Anketell, 1996, p. 75-77.

3   Klitzsch, 1971, p. 255-262. Boote, et al. 1998, p. 15-17.

4   Klitzsch, 1971, plate 1. Selley, 1997b, p. 17-26. Sutcliffe, et al. 2000, p. 26-27. Burke and Kraus, 1998, p. 42.

5   Klitzsch, 1971, p. 255-262. Klerkx, 1980, p. 901-906. Boote, et al. 1998, p. 15-17. Murzuq: Adamson, et al. 2000, p. 443. Echikh and Sola, 2000, p. 191. Sutcliffe, et al. 2000, p. 60-63.

6   Klitzsch, 1971, p. 255-262, Burke and Dewey, 1974, p. 313-316. Geological Map of Libya, 1985, scale 1:1,000,000. Anketell, 1996, p. 57-89.

7   Geological Map of Libya, 1985, scale 1:1,000,000. Volcanics: Gharyan: Busrewil and Wadsworth, 1980b, p. 1095-1106. Jabal as Sawda: Busrewil and Esson, 1991, p. 2599-2604. Al Haruj al Aswad: Busrewil and Wadworth, 1980a, p. 1077-1080. Busrewil, 1996, p. 330-346. Jabal Awaynat: Bell and Sandford, 1971, p. 333-340. Flinn, et al. 1991, p. 2539-2557. Tibisti: Vincent, 1970, p. 301-319. Klitzsch, 1971, plate 1.

8   Lüning, et al. 1999, p. 695, 715. Grignani, et al. 1991, p. 1171. Bellini and Massa, 1980, p. 32-56. Bellini, et al. 1991, p. 2155-2184. Lüning, et al. 1999, p. 693-718.

9   Sub-basins: Lüning, et al. 1999, p. 715.

10 Bellini, et al. 1991, p. 2181, Lüning, et al. 1999, p. 715.

11 Lüning, et al. 1999, p. 713-714. Selley, 1997b, p. 17-26. Grignani, et al. 1991, p. 1163-1164. Bellini, et al. 1991, p. 2161.

12 Bellini and Massa, 1980, p. 46-54. Klitzsch, 1971, p. 255-262. Lüning, et al. 1999, p. 714. Selley, 1997b, p. 22. Grignani, et al. 1991, p. 1164-1166.

13 Klitzsch, 1971, fig. 6. Bellini and Massa, 1980, p. 53, Lüning, et al. 1999, p. 714. Carboniferous: see chapter 3, this publication. Cretaceous: Lüning, et al. 1999, fig. 3, fig. 5.

14 Selley, 1997b, p. 21-24. Boote, et al. 1998, p. 52-53.

15 Mikbel, 1977, p. 19-34. Meister, et al. 1991, p. 2725-2742. Pierobon, 1991, p. 1766-1784. Sutcliffe, et al. 2000, figures 29A and 31. Echikh and Sola, 2000, p. 193-203. Dur al Qussah: Klitzsch, 1966b, p. 29.

16 Echikh and Sola, 2000, p. 193-197. N1 to N4-1 wells: Montgomery, 1994a, p. 26.

17 Echikh and Sola, 2000, p. 193-197.

18 Echikh and Sola, 2000, p. 194-197.

19 Echikh and Sola, 2000, p. 197.

20 Echikh and Sola, 2000, p. 197-199.

21 Echikh and Sola, 2000, p. 197, 217. Klitzsch, 1971, p. 256-257.

22 Klitzsch, 1971, p. 256-257. Echikh and Sola, 2000, p. 197-199. Belhaj, 1996, p. 70.

23 Selley, 1997b, p. 22. Davidson, et al. 2000, p. 315-316. Aziz, 2000, p. 361.

24 Vail, 1991, p. 2262. Selley, 1997b, p. 21-22. Lüning, et al. 1999, p. 713-714. Glover, et al. 1998, p. 18.

25 O1-1: Abugares and Ramaekers, 1993, p. 59. Palaeozoic sequences: Sutcliffe, et al. 2000, p. 14-16. Klitzsch, 1966, p. 29. Echikh and Sola, 2000, p. 176-181, 200.

26 Silurian hot shales: Lüning, et al. 1999, p. 707. Al Fuqaha, C1-44: Belhaj, 1996, figure 8. Akakus Formation: Sutcliffe, et al. 2000, p. 54. Tihemboka Ridge: Abugares and Ramaekers, 1993, p. 59. Mid-Devonian inversion: Pierobon, 1991, p. 1772. Meister, et al. 1991, p. 2727. Sutcliffe, et al. 2000, figures 29B and 31. Echikh and Sola, 2000, p. 191.

27 Selley, 1997a, p. 21. See also chapters 4 and 5, this publication.

28 Oun, et al. 2000, p. 71.

29 See chapter 3, this volume. Blanpied, 1998, p. 43-44. Rubino, et al. 2000, p. 80. Deynoux, et al. 2000, p. 27.

30 Selley, 1997b, p. 23. Sutcliffe, et al. 2000, p. 22-24, 54-60. Carboniferous inliers: Wennekers, et al. 1996, p. 15.

31 Echikh, 1998, p. 109. Mikbel. 1977, plate 1.

32 Klitzsch, 1971, figure 1. Vail, 1991, p. 2262.

33 Vail, 1991, p. 2261-2263. Echikh, 1998, p. 111.

34 Echikh, 1998, p. 111-112. Boote, et al. 1998, p. 17.

35 Echikh, 1998, p. 112, Boote, et al. 1998, p. 17-19.

36 Sinha, 1980, figures 2-6. Pallas, 1980, figure 3. Boote, et al. 1998, p. 21. Echikh, 1998, p. 114.

37 Boote, et al. 1998, p. 21. Dercourt, et al. 1986, plate 9.

38 Wennekers, et al. 1996, figure 3. Boote, et al. 1998, figure 17b. Echikh, 1998, p. 110-112.

39 Wennekers, et al. 1996, figure 3. Belhaj, 1996, figure 6. Boote, et al. 1998, figure 9. Echikh, 1998, figure 4. Water well: Kruseman and Floegel, 1980, section E.

40 Singh, 1980, figure 7. Sinha, 1980, figure 5.

41 Anketell and Ghallali, 1991, p.2381-2405. Anketell, 1996, p. 57-87. Busrewil and Wadsworth, 1980b, p. 1095-1106.

42 Echikh, 1998, p. 109.

43 Kruseman and Floegel, 1980, sections G and N. Sbeta, 1991, p. 1932. Hammuda, et al. 1985, p. 43-48. Anketell and Ghellali, 1991, p. 2381-2406.

44 Kruseman and Floegel, 1980, sections A-N. Anketall and Ghellali, 1991, figure 3.

45 Gumati and Kanes, 1985, p. 39-52. Gumati and Nairn, 1991, p. 93-102. van der Meer and Cloetingh, 1996, p. 211-230. Baird, et al. 1996, p. 3-56. Suleiman, et al. 1991, p. 2461-2468.

46 Harding, 1984, p. 333-362. Baird, et al. 1996, p. 3-56. Gras and Thusu, 1998, p. 322. Wennekers, et al. 1996,

figure 22. Listric faults: Baird, et al. 1996, p. 36-38.

[47] Vail, 1991, p. 2262. Klitzsch, 1971, p. 256-257. Haruj al Aswad: Klitzsch, 1971, figure 2.

[48] See chapter 2, this volume. Hot spot: van Houten, 1991, p. 115-118. Fouad, 1991, p. 2451-2460.

[49] Dercourt, et al. 1986, plate 4. Anketell, 1996, p. 72-77. Granite: Wilson and Guiraud, 1998, p. 242.

[50] See chapter 2, this volume. Two plates: Burke and Dewey, 1974, p. 313-316. Anketell and Kumati, 1991b, p. 2353-2370. Gras and Thusu, 1998, p. 322.

[51] Anketell, 1996, p. 72-83. Baird, et al. 1996, p. 15.

[52] Gumati and Kanes, 1985, p. 39-52. Gumati and Nairn, 1991, p. 93-102, Anketell, 1996, p. 72-75. van der Meer and Cloetingh, 1996, p. 211-230.

[53] Anketell, 1996, p. 61-62. Mikbel, 1981, figure 1. Hallett and El Ghoul, 1996, figures 1-2. Wennekers, et al. 1996, figures 19, 23. Relay growth faulting: Abadi and van Dijk, 1993, figure 18. van Dijk and Eabadi, 1996, p. 155-166.

[54] Anketell, 1996, p. 62-63. Mikbel, 1981, figure 1. Hallett and El Ghoul, 1996, figures 1-2. Wennekers, et al. 1996, figure 19. Unnamed wrench fault: Anketell, 1996, figure 12.

[55] Anketell, 1996, p. 60. Mikbel, 1981, figure 1. Klitzsch, 1971, figure 2. Bender, et al. 1996, figure 1. Baird, et al. 1996, figure 9. Hallett and El Ghoul, 1996, figures 1-2.

[56] Schröter, 1996, p.123-135. Post rift phase: Gras and Thusu, 1998, p. 322.

[57] Knytl, et al. 1996, p. 167-199.

[58] Antetell, 1996, p.63, Hallett and El Ghoul, 1996, figures 1-2. Wennekers, et al. 1996, figure 19. Goudarzi, 1980, figure 6. Roohi, 1996b, figures 3 and 10.

[59] Johnson and Nicoud, 1996, figure 2. Sinha and Mriheel, 1996, p.153-195.

[60] Hallett and El Ghoul, 1996, figure 2. Wennekers, et al. 1996, figure 19. Goudarzi, 1980, figure 6. Anketell, 1996, p. 62-64. Johnson and Nicoud, 1996, p. 212-215. Schneiter, 2000, p. 23.

[61] Hallett and El Ghoul, 1996, figure 2. Wennekers, et al. 1996, figure 19. Goudarzi, 1980, figure 6. Baird, et al. 1996, figure 9. Anketell, 1996, p. 62-64. Anketell and Kumati, 1991, p. 2253-2370.

[62] Wennekers, et al. 1996, table 2. P1-16: Rasul, 2000, p. 64. Schröter, 1996, figure 4. Roohi, 1996b, p. 436-441.

[63] Roohi, 1996b, p. 435-454. Anketell, 1996, p.63, 66.

[64] Hallett and El Ghoul, 1996, figure 2. Wennekers, et al. 1996, figure 19. Goudarzi, 1980, figure 6. Anketell, 1996, p. 62-64. Sinha and Mriheel, 1996, p. 153-195.

[65] Roohi, 1996b, figure 3. Schröter, 1996, figure 4. Sinha and Mriheel, 1996, figures 4, 5.

[66] Sinha and Mriheel, 1996, p. 153-195. Hammuda, 1980, p. 513-517.

[67] Antetell, 1996, p. 63.

[68] Hallett and El Ghoul, 1996, figure 2. Wennekers, et al.

1996, figure 19. Goudarzi, 1980, figure 6. Baird, et al. 1996, figure 9. Bonnefous, 1972, p. 225-261. El Hawat, et al. 1996, p. 27. Hammuda, 1980, p. 510-520. Jones, 1996, p. 169-184. El Ghoul, 1996, figure 4. Bezan, 1996, p. 97-118. Abugares, 1996, p. 45-64. Roohi, 1996a, p. 329.

[69] El Batroukh and Zentani, 1980, p. 965-978.

[70] Anketell, 1996, p. 63. Baird, et al. 1996, figures 9, 18.

[71] Hallett and El Ghoul, 1996, figure 2. Rasul, 2000, p. 64. Wennekers, et al. 1996, table 2 and figure 6. Baird, et al. 1996, figure 18.

[72] Anketell, 1996, figures 4 and 12. Bezan, 1996, p. 101-102. Bezan and Malak, 1996, p. 121-122.

[73] Anketell, 1996, p. 63-64. Wennekers, et al. 1996, p. 35 and figure 19.

[74] Hallett and El Ghoul, 1996, figure 3. Thusu, et al. 1988, p. 197-198. Gras and Thusu, 1998, p. 322-323. Bonnefous, 1972, p.259. Wennekers, et al. p. 24, 43.

[75] Parsons, et al. 1980, p. 723-732. Wennekers, et al. 1996, p. 48 and figure 19. Anketell, 1996, figure 4. Gras and Thusu, 1998, figure 3. Ibrahim, 1991, figure 7. Ambrose, 2000, figure 1. Roohi, 1996a, figure 1.

[76] El Arnauti and Shelmani, 1985, p. 9 and figures 1, 4. Anketell, 1996, p. 64, figures 4, 12.

[77] Hallett and El Ghoul. 1996, figure 3. Brugman and Visscher, 1988, p. 159. Gardiner, 1988, p. 259-265. Thusu, et al. 1988, p. 197-198. Thusu, 1996, p.455-474. Gras and Thusu, 1998, p. 322. Benfield and Wright, 1980, p. 463-500. Baird, et al. 1996, p. 15-19. Ambrose, 2000, figure 4.

[78] Ibrahim, 1991, p. 2757-2779. Gras and Thusu, 1998, p. 317-334. Ambrose, 2000, p. 165-191.

[79] Anketell, 1996, p. 72-77.

[80] Anketell, 1996, p. 69-72.

[81] Vail, 1991, p. 2262. Baudet, 1988, p. 17-26. El Arnauti and Shelmani, 1988, figures 4-7. Klitzsch, 1971, p. 256

[82] Klitzsch, 1971, p. 258-261. El Arnauti and Shelmani, 1988, figures 2,8.

[83] Thusu, et al. 1988, figures 5 and 6. El Hawat and Shelmani, 1993, p. 9-12.

[84] Baird, et al. 1996, figure 4. Barr and Berggren, 1980, p. 163-192. Rohlich, 1978, p. 401-412. El Hawat and Shelmani, 1993, p. 13-20. Barr and Hammuda, 1971, p. 27-40. Pallas, 1980, figure 4. Duronio, et al. 1991, p. 1607.

[85] El Hawat and Shelmani, 1993, p. 17-20. Geological Map of Libya, 1985, scale 1:1,000,000

[86] Anketell,1996, p. 69-73 and figure 4. Pallas, 1980, figure 4. El Arnauti and Shelmani, 1985, figures 1-7. El Arnauti and Shelmani, 1988, figure 2. Bezan, 1996, p. 93-117.

[87] Anketell, 1996, p. 67-73. El Hawat and Shelmani, 1993, p. 8-20 and figure 5.

[88] El Arnauti and Shelmani, 1985, p. 10. El Arnauti and Shelmani, 1988, figure 1. El Hawat and Shelmani, p. 9. Ghori, 1991, p. 2750-2752. Duronio, et al. 1991, p. 1607. Anketell, 1996, p. 69-73 and figure 3.

89 Ghori, 1991, p. 2750-2752. Duronio, et al. 1991, p. 1598. El Hawat and Shelmani, 1993, p. 9. Del Ben and Finetti, 1991, p. 2417-2431.
90 El Arnauti and Shelmani, 1985, p. 9 and figure 6.
91 Baird, et al, 1996, p. 14-36. Sghair and El Alami, 1996, p. 263-274. Thusu, 1996, p. 455-474. Belazi, 1989, p. 356.
92 Roberts, 1970, p. 438-448. Belazi, 1989, p. 353-366. Thusu, 1996, p. 455-474.
93 Jongsma, 1991, figure 1. Ghori, 1991, p. 2752. Pallas, 1980, figure 4. El Arnauti and Shelmani, 1985, p. 10. Anketell, 1996, p. 73.
94 Dercourt, et al, 1986, plates 9, 10. Jongsma, et al. 1987, p. 87-106. Jongsma, 1991, p. 2331-2352. Burollet, 1978, p. 331-359. Finetti, 1982, plate 3.
95 Anketell, 1996, p. 67-73. Finetti, 1982, plate 3. Finetti, 1985, p. 216. Depth to base Cainozoic: El Ghoul, 1991, p. 1640. Jifarah Basin: Kruseman and Floegel, 1980, section G.
96 Sbeta, 1991, p. 1929-1937. El Ghoul, 1991, p. 1640.
97 Sbeta, 1991, p. 1929-1937. Hammuda, et al. 1985. Jongsma, 1991, p. 2331-2352. van Hinte, et al. 1980, p. 205-244.
98 El Ghoul, 1980, figure 4.
99 Del Ben and Finetti, 1991, figure 2. Bishop and Debono, 1993, p. 60-63. Bishop and Debono, 1996, p. 129-160.
100 Del Ben and Finetti, 1991, figure 2.
101 Smith and Karki, 1996, p. 129-137. Barr and Walker, 1973, p. 1244-1251.
102 Del Ben and Finetti, 1991, p. 2417-2431. Finetti and Del Ben, 2000, p. 42.
103 Del Ben and Finetti, 1991, p. 2424-2431. Smith and Karki, 1996, p. 131. Jongsma, 1991, figure 1.
104 Duronio, et al. 1991, p. 1598. Smith and Karki, 1996, p. 129-137.
105 Boote, et al. 1998, p. 9-21. Craig, et al. 2000, p. 29-31.
106 Baird, et al. 1996, p. 9-36.
107 Anketell, 1996, p. 64-73.

Chapter 7

1 Macgregor, 1996, p. 197-217
2 Klemme and Ulmishek, 1991, p. 1809-1851.
3 Algeria: particularly the works of Tissot, Espitalie and other workers from IFP.
4 Agip wells: Grignani, et al. 1991, p. 1159-1227. Lüning, et al. 2000c, p. 151-174.
5 Silurian hot shale: Lüning, et al. 2000a, p. 121-200. Kufrah Basin: Lüning, et al. 1999, p. 693-718. Lüning, et al. 2000c, p. 151-174. Depth to basement: Bellini, et al. 1991, figure 15. Well KW2: Grignani, et al. 1991, p. 1162-1163.
6 Thomas, 1995a, p. 46, 1995b, p. 100. Lüning, et al. 2000a, figure 32. Davidson, et al. 2000, p. 311-317.
7 Echikh and Sola, 2000, figure 22. Lüning, et al. in press.
7 C1-NC 174: Lüning, et al. (in press). Lüning, et al. 2000a, p. 166. Meister, et al. 1991, p. 2729-2733. Echikh and Sola, 2000, p. 210-213. Sikander, et al. 2000, p. 88. Davidson, et al. 2000, p. 314-316. Aziz, 2000, p. 360-362.
8 Echikh and Sola, 2000, p. 212. Meister, et al. 1991, tables 1and 8, p. 2735-2736. Davidson, et al. 2000, p. 314. Subfamilies: Himmali, et al. 1998.
9 Meister, et al. 1991, table 4 and p. 2728-2729.
10 Echikh and Sola, 2000, figure 22. Boote, et al. 1998, p. 25. Sikander, et al. 2000, p. 88. Daniels and Emme, 1995, p. 101-123. Lüning, et al. 2000a, p. 166. Lüning, 2000, p. 59. Lüning, et al. in press. Bracaccia, 1991, p. 1738. Aziez, et al. 1998. Ghori and Mohammed, 2000, p. 44. A1-NC 118: Pajola and Ricchiuto, 2000, p. 72.
11 Boote, et al. 1998, figure 17a. Daniels and Emme, 1995, p. 109-112. Hangari, et al. 2000, p. 50. Lüning, 2000, p. 59. Aziez, et al. 1998. Byramjee and Vasse, 1969, p. 319-330.
12 Dieb and Hrouda, 2000, p. 28. Daniels and Emme, 1995, p. 109. Hangari, et al. 2000, p. 50. Ghenima, 1995, p. 6. Lüning, 2000, p. 59. Sikander, et al. 2000, p. 88. Byramjee and Vasse, 1969, p. 327-329. Ghori and Mohammed, 2000, p. 44.
13 Hangari, et al. 2000, p. 50. Sikander, et al. 2000, p. 88. Davidson, et al. 2000, p. 314. Atshan: Thomas, 1995a, p. 44.
15 Parsons, et al. 1980, p. 729-730. Baird, et al. 1996, p. 44-46. El Alami, et al. 1989, p. 22. Geothermal gradient, Maradah Trough: Ibrahim, 1996, p. 419-433.
14 These figures must be qualified to the extent that the west Libya figures relate to TOC levels generally over 3%, whereas the Sirt Basin figures relate to TOC levels over 1%. Nevertheless there is still a very large difference between the effective thickness of the two source rocks.
16 El Alami, et al. 1989, p. 22-24.
17 El Alami, et al. 1989, p. 23-25. A1-NC 157: Baric, et al. 1996, p. 541-553.
18 El Alami, et al. 1989, p. 25-27.
19 Burwood, 1996, p. 6-12. Burwood, 1998, p. 1-4.
20 Ghori and Mohammed, 1996, p. 529-539.
21 El Alami, et al. 1989, p. 28-29. Ghori and Mohammed, 1996, p. 529-539. Bender, et al. 1996, p. 139-155.
22 El Alami, 1996, p. 337-348. Bu Argoub and Thusu, 1996. Burwood, 1998, p. 1-4. Turonian anoxic event: Kuhnt, et al. 2000, p. 51. Lüning, 2001.
23 Gumati and Schamel, 1988, p. 205-218.
24 Other potential source rocks: Baric, et al. 1996, p. 543. TOC values: Parsons, et al. 1980, p. 723-732. Intisar: Brady, et al. 1980, p. 564. Oil families: Burwood, et al. 2000, p. 51. Burwood, 1998, p. 1-4.
25 Ghori, 1991, p. 2743-2749. Thusu, personal communication.

26 Ghori, 1991, figure 2A and p. 2749.

27 Ghori, 1991, figure 2B and p. 2749-2750.

28 Ghori, 1991, figure 2C and p. 2750.

29 El Arnauti and Shelmani, 1988, figure 5 and p. 9-10. El Arnauti and Shelmani, 1985, figure 2. Silurian oil shows: Thusu, personal communication. Overall potential: Sola and Ozcicek, 1990, figure 19.

30 Pratsch, 1994, p. 5. Racey, et al. 2001, p. 30. Zaier, et al. 1998, p. 380-390. Loucks, et al. 1998, p. 355-374.

31 Racey, et al. 2001, p. 33. Hammuda, et al. 1985, p. 79-81. El Ghoul, 1991, p. 1647. Sbeta, 1990, p. 42-56.

32 Racey, et al. 2001, p. 36-42. Belina 1: Baric and Cota, 2000, p. 14.

33 Bailey, et al. 1989, p. 301-302.

34 Racey, et al. 2001, p. 36-42. Bernasconi, et al. 1991, p. 1927. Burwood, et al. 1995, p. 27. Anketell and Mriheel, 2000, p. 425-447.

35 El Ghoul, 1991, p. 1640. Bernasconi, et al. 1991, p. 1909.

36 Hammuda, et al. 1985, p. 24, 81. Fournie, 1978, p. 115. Sbeta, 1990, p. 55. Sbeta, 1991, figures 10, 21, 31. Hamyouni and Sadeegh, 1985, p. 27. Pratsch, 1994, figure 5. Isis: Ben Brahim, 1989, p. 309. Burwood, et al. 1995, p. 27.

37 Baair, et al. 2000, p. 12. Baair, et al. 2001.

38 A1-NC 120: BAAPG Petroleum Developments in north Africa for 1984, p. 1656. Duronio, et al. 1991, p. 1598, 1605.

Chapter 8

1 Petroleum Systems: Magoon, L.B. and Dow, W.G. 1994. *The Petroleum System*. In: The petroleum system – from source to trap. (Ed. L.B Magoon and W.G. Dow), Mem. Amer. Assoc. Pet. Geol., 40, p. 3-24. Demaison, G. and Huizinga, B.J. 1994. *Genetic classification of petroleum systems using three factors: charge, migration and entrapment.* In: The petroleum system – from source to trap. (Ed. L.B Magoon and W.G. Dow), Mem. Amer. Assoc. Pet. Geol., 40, p. 73-89.
Distribution of reserves: Parsons, et al. 1980, p. 723-732. Macgregor, 1996, p. 329-340. Macgregor, 1998, p. 201-216.
Hassi Messaoud: Djarnia and Fekirine, 1998, p. 167-174. Magregor, 1998, p. 83-84. Boote, et al. 1998, p. 29-34. Boutouaou and Larter, 2001.

2 Aziz, 2000, p. 349-368. Echikh and Sola, 2000, p. 175-222.

3 Echikh and Sola, 2000, p. 180-181, 204-205, 213. Burdon, 1980, p. 600-601. Meister, et al. 1991, p. 2728. Davidson, et al. 2000, p. 309. Elephant: Petzet, 1999, p. 65-66. Hogan and Davidson, 1998, p.1-12. Petroleum Exploration Society, GB., Newsletter, May 1999, p. 51. B1-NC 186: Petroleum Exploration Society, GB., Newsletter, June 2001.

4 Echikh and Sola, 2000, p. 181-182, 184-188, 208-213. B1-NC 174: Davidson, et al. 2000, p. 317. Aziz, 2000, p. 361. Petzet, 1999, p. 66.

5 Echikh and Sola, 2000, p. 183, 213, 215.

6 Davidson, et al. 2000, p. 304. Echikh and Sola, 2000, p. 206-208, 213-214.

7 Echikh and Sola, 2000, p. 175, 214. Meister, et al. 1991, p. 2729, 2739-2740. Aziz, 2000, p. 362. Water flushing: Burdon, 1980, p. 595-608.

8 Repsol discovery: Petrol. Explor. Soc. GB. Newsletter, February, 2001, p. 61. Aziz, 2000, p. 354-355, 364. Echikh and Sola, 2000, p. 180, 189. Atshan: Meister, et al. 1991, p. 2728-2829.

9 Daniels and Emme, 1995, p. 116-123. Boote, et al. 1998, p. 40-52. Echikh, 1998, p. 119-120. Macgregor, 1996, p. 329-340.

10 Echikh, 1998, p. 109-120. Boote, et al. 1998, p. 40-52. Jifarah Basin: Al Matmaty, et al. 2000, p. 9. Swire and Gashgesh, 2000, p. 96.

11 Echikh, 1998, p. 115. Boote, et al. 1998, p. 47-52

12 Echikh, 1998, p. 112, 115-116. F6 reservoir: Alem, et al. 1998, p. 175-186.

13 Echikh, 1998, p. 116-117, 124. Aliev, et al. 1971, p. 262. Shah, et al. 1993, p. 345-352. Poyntz, 1995, p. 45.

14 See chapter 7, this volume. Daniels and Emme, 1995, p. 101-123. Echkh and Sola, 2000, p. 209. Byramjee and Vasse, 1969, p. 319-330.

15 Echikh, 1998, p. 112-118, 124-128. Boote, et al. 1998, p. 48-49. Hammuda, 1980, p. 501-507.

16 Echikh, 1998, p. 113, 122-123. Hammuda, 1980, p. 501-507. Tiji: Byramjee and Vasse, 1969, p. 319-330.

17 Echikh, 1998, p. 119-121. Boote, et al. 1998, p. 48-49.

18 Echikh, 1998, p. 123-128. Boote, et al. 1998, p. 47-52.

19 Boote, et al. 1998, p. 47-52.

20 Daniels and Emme, 1995, p. 101-123.

21 Echikh, 1998, p. 117-118. Algerian fields: Aliev, et al. 1971, p. 193-194. Alrar: Chaouchi, et al. 1998, p. 187-200.

22 Aliev, et al. 1971, p. 194-196. Atshan: Meister, et al, 1991, p. 2728-2729. Hangari, et al. 2000, p. 50. Tahara: El Mehdawi, 2000, p. 273-294.

23 Southern Tunisia: Acheche, et al. 2001. Libya: Echikh, 1998, p. 118-119, 124-129. Ghenima, 1995, p. 3-15.

24 Parsons, et al. 1980, p. 723-732. Baird, et al. 1996, p. 43. Macgregor, 1996, p. 329-340. Macgregor and Moody, 1998, p. 212-213.

25 The figure of 0.25 barrels per m^3 of source rock is derived by calculating the volume of effective source rock from the map of El Alami, et al. 1989, p. 23. Oil in place: Thomas, 1995b, p. 100, 103.

26 Gumati and Nairn, 1991, p. 93-102, Baird, et al. 1996, p. 23-42. Roohi, 1996a, p. 323-336.

27 El Alami, et al. 1989, p. 23. Roohi, 1996a, p. 330-335. A1-10, B1-10: Baair, et al. 2000, p. 12. Baair, et al. 2001.

28 Abushagur and LeMone, 1991, p. 1871-1881. Abushagur, 1991, p. 1827-1840. Bezan, 1996, p. 97-118. Bezan, et al. 1996, p. 135-152.

29 Abugares, 1996, p. 45-63.

30 See chapter 7, this volume. Al Alami, et al. 1989, p. 29. Roohi, 1996b, p. 446.

31 Abugares, 1996, p. 56-57. Abushagur and LeMone, 1991, p. 1877. Knytl, et al. 1996, p. 184-187.

32 Schröter, 1996, p. 123-135, Knytl, et al. 1996, p. 167-199. Wennekers, et al. 1996, p. 44.

33 El Alami, et al. 1989, p. 19-29. Field data: AAPG Annual Petroleum Developments reports. Abu Alwan: Ojaley, 2000, p. 132.

34 B1-NC 177: Oil & Gas Journal, March 16th 1998, p. 40. Schneiter, 2000, p. 23.

35 Wennekers, et al. 1996, p. 44. Abugares, 1996, p. 56. Bezan et al. 1996, p. 146.

36 El Alami, et al. 1989, p. 29. Roohi, et al 1996a, p. 323-336. Roohi, 1996b, p. 435-454. Ibrahim and Al Mahruqi, 2000.

37 Roohi, 1996b, p. 442-444.

38 Roohi, 1996b, p. 442-445. Gardner, et al. 2000, p. 29.

39 Mriheel, 1994, p. 65-66. Sinha and Mriheel, 1996, p. 179-180. Bezan, et al. 1996, p. 135-152. Bezan, 1996, p. 97-117.

40 El Alami, et al. 1989, p. 23-24. Roohi, 1996b, p. 440. Gardner, et al. 2000, p. 29. Ibrahim, 1996, p. 430.

41 Roohi, 1996b, p. 441. Sinha and Mriheel, 1996, p. 179-180.

42 Roohi, 1996b, p. 448-452.

43 Brennan, 1992, p. 267-289. Broughton, 1996, p. 391-418. Gravity data Raqubah: El Batroukh and Zentani, 1980b, p. 1154-1163. Bazzuzi: Roohi, 1996b, p. 439, 448. El Hawat, et al. 1996, p. 8-12.

44 Bonnefous, 1972, p. 236-261. El Hawat, et al. 1996, p. 6-12. Gravity: El Batroukh and Zentani, 1980a, p. 972.

45 Bull. Am. Assoc. Pet. Geol. annual petroleum development reports.

46 Hammuda, 1980, p. 509-520. Jones, 1996, p. 169-184. El Ghoul. 1996, p. 137-154.

47 Parsons, et al. 1980, p.723-732. Ibrahim and Al Mahruqi, 2000, p. 26.

48 El Ghoul, 1996, p.137-154. Reservoir properties by analogy with Ar Raqubah.

49 El Alami, et al. 1989, p. 20-29.

50 Roohi, 1996a, p. 332.

51 El Ghoul, 1996, p. 137-154.

52 Baird, et al. 1996, p. 42-56. Fraser, 1967, p. 259-264. Bebout and Pendexter, 1975, p. 665-693. Thomas, 1995b, p. 100.

53 Bezan, et al. 1996, p. 143. Barr and Weegar, 1972, p. 51-53. Roohi, 1996a, p. 332. Halbouty, M.T. 1980 *Giant oilfields of the decade 1968-1978*. Mem. Amer. Ass. Pet. Geol. No. 30. Thomas, 1995b, p. 100.

54 El Alami, et al. 1989, p. 29.

55 Bebout and Pendexter, 1975, p. 665-693. Bezan, et al. 1996, p. 143.

56 Bebout and Pendexter, 1975, p. 665-693.

57 BAAPG Annual Petroleum Developments, 1968, 1969.

58 Jahamah Platform: BAAPG Annual Petroleum Developments, 1959, 1966. Zaltan Platform: Wennekers, et al. 1996, p. 39-40. Migration routes: El Alami, et al. 1989, p. 29. As Sahel, Assumud: Gurney, 1996, p. 178-180.

59 Attahadi gas in place, planned production rates: Gurney, 1996, p. 179. Cambrian age: Rasul, 2000, p. 64. Source: El Alami, et al. 1989, p. 29. Arshad: BAAPG Petroleum Developments, 1966, 1967.

60 Ibrahim, 1991, p. 2768. Ambrose, 2000, p. 169.

61 Rossi, et al. 1991, p. 2211-2226. El Hawat, et al. 1996, p. 3-30. Ibrahim, 1991, p. 2757-2779. Gras and Thusu, 1998, p. 317-334. Ambrose, 2000, p. 165-192.

62 Ambrose, 2000, p. 188.

63 Ambrose, 2000, p. 181-183, 188.

64 Ambrose, 2000, p. 170.

65 Ambrose, 2000, p. 170, 181-190.

66 Terry and Williams, 1969, p. 31-48. Brady, et al. 1980, p. 543-564. Hammer, 1987, p. 339-344.

67 Barr and Weegar, 1972, p. 154-155. Brady, et al. 1980, p. 564. Burwood, et al. 2000, p. 52.

68 Brady, et al. 1980, p. 547.

69 Brady, et al. 1980, p. 543, 564.

70 BAAPG Annual Petroleum Developments, 1960, 1961. Wennekers, et al. 1996, p. 39-40. Ghori, 1991, p. 2747-2748. Burwood, et al. 2000, p. 52.

71 Nubian reserves: Baird, et al. 1996, p.43. Thomas, 1995b, p. 100. Petroleum systems: Ambrose, 2000, p. 181-185. Burwood, et al. 2000, p. 52. PUC subdivision: Barasi, et al., 2000, p. 41.

72 Ambrose, 2000, p. 165-192. Jalu area: Mansour and Magairhy, 1996, p. 508-509. Sarir: Gillespie and Sanford, 1967, p. 181-193. Sanford, 1970, p. 449-476. Lewis, 1996, p. 253-267. Messlah: Clifford, et al. 1980, p. 507-524. Exploration staff of Agoco, 1980, p. 507-524. Abu Attiffel: Erba, et al. 1984, p. 89-99. Concesssion 82: Gras and Thusu, 1998, p. 327.

73 Lewis, 1990, p. 253-267. Sanford, 1970, p. 449-476. Gillespie and Sanford, 1967, p. 181-193.

74 Clifford, et al. 1980, p. 507-524. Exploration staff of Agoco, 1980, p. 521-536. Koscec and Gherryo, 1996, p. 365-389.

75 Erba, et al. 1984, p. 89-99. Reserves: Thomas, 1995b, p. 100. Campbell, 1991, p. 340.

76 Ambrose, 2000, p. 184. Burwood, et al. 2000, p. 52.

77 El Alami, et al. 1989, p. 19-30. Turonian: Bu Argoub, 1996, p. 450. Bender, et al. 1996, p. 139-155. Lewis, 1990, p. 264-267. Clifford, et al. 1980, p. 523. Koscec and Gherryo, 1996, p. 365-389. Carbon isotopes: Burwood, 1997, p. 1365. Gras and Thusu, 1998, p. 323-326. Abu Attiffel: Erba, et al. 1984, p. 89-99. El Alami, 1996b, p. 337-348. Ambrose, 2000, p. 183. Burwood, et al. 2000, p. 52.

78 Lewis, 1990, p. 260. Clifford, et al. 1980, p. 518-519.

Mansour and Magairhy, 1996, p. 503, 513-514. Gras and Thusu, 1998, p. 330. Bu Argoub, 1996, p. 419-453.

79 Gillespie and Sanford, 1967, p. 189. Gras and Thusu, 1998, p. 326-330. Ambrose, 2000. p. 169. Ibrahim, 1991, p. 2777. Faulted ramps, Messlah High: Gras, 1996, p. 201-210. Gras, 1998, p.329-342. Majid: Winnock and Mermod, 1979, p. 11-18.

80 Ibrahim, 1991, fig. 3. Bezan and Malak, 1996, p. 121, 125. Thomas, 1995b, p. 100. Campbell, 1991, p. 260. Rachmat Formation: Burwood, et al. 2000, p. 52. Latif area: Mansour and Maghairhy, 1996, p. 513-519. Possible source rocks: Ghori and Mohammed, 1996, p. 531. Maragh Graben: Gruenwald, 2001, p. 213-231.

81 Thusu, 1996, p. 455-461. Ghori and Mohammed, 1996, p. 537. Burwood, et al. 2000, p. 51-52. Kuehn, 1996, p. 251-261. Ball, et al. 1996, p. 349-363. Formation Microscanner: Haase, 1996, p. 275-285.

82 Gras and Thusu, 1998, p. 323-326. Ambrose, 2000, p. 181-185. Burwood, et al. 2000, p. 51-52.

83 Roberts, 1970, p. 438-448. Burwood, 1997, p. 1365. Burwood, et al. 2000, p. 51-52. Campbell, 1991, p. 260. Thomas, 1995b, p. 100.

84 Belazi, 1989, p. 353-366. Williams, 1972, p. 623-632. Burwood, et al. 2000, p. 51-52. Thomas, 1995b, p. 100.

85 Pratsch, 1994, p. 1-25. Misallati, 2000, p. 67-68.

86 Tmalla and Abutarruma, 2000, p. 97. Sbeta, 1990, p. 53. Bernasconi, et al. 1991, p. 1910-1915. Racey, 2001, p. 94-96. Mriheel, 1991, p. 44-52. Anketell and Mriheel, 2000, p. 429-446. Jirani Dolomite: Mriheel and Anketell, 2000, p. 449-474. Bouri reserves: Campbell, 1991, p. 340 (5 billion barrels recoverable). Gurney, 1996, p. 87 (2 billion barrels oil in place). Thomas, 1995b, p. 100 (5 billion barrels oil in place, 670 million barrels recoverable). Pratsch, 1994, p. 8 (700 million barrels recoverable). Gas while drilling: Barghut, 2001. Bishop and Debono, 1996, p. 152 quote reserves of 3.7 billion barrels and 1.9 TCF gas, including secondary recovery.

87 Sbeta, 1990, p. 53. Racey, et al, 2001, p. 48-51.

88 Ashtart: Bishop, 1988, p. 1042. Loucks, et al. 1998, p. 368-371. Bailey, et al. 1989, p. 281-307. Racey, et al 2001, p. 29-53.

89 Ricchiuto and Pajola, 2000, p. 78. Bailey, et al. 1989. p. 302. Bishop, 1988, p. 1052-1056. B1-NC 167: Baric and Cota, 2000, p. 14. Hasdrubal: Racey, et al. 2001, p. 33, 41. Anketell and Mriheel, 2000, p. 428. El Ghoul, 1991, p. 1637-1656. Sbeta, 1990, p. 42-56.

90 Bernasconi, et al. 1991, p. 1907-1927. Sbeta, 1991, p. 1929-1941. Hasdrubal: Racey, et al. 2001, p. 33.

91 Bernasconi, et al. 1991, p. 1909. El Ghoul, 1991, p. 1640. Hasdrubal: Racey, et al. 2001, p. 36.

92 Bailey, et al. 1989. p. 301-302.

93 Bishop, 1988, p. 1044-1052. Pratsch, 1994, p. 1-25. Burwood, et al. 1995, p. 27. Isis: Ben Brahim, 1989, p. 309-311

94 Pratsch, 1994, p. 1-25. Bishop, 1988, p. 1044-1052. Malta: Bishop and Debono, 1993, p. 60-63. Bishop and Debono, 1996, p. 129-160.

95 Pratsch, 1994, p. 1-25. Sicily: Bishop and Debono, 1993, p. 60-63.

96 Bishop. 1988, p. 1052.

97 Ricchiuto and Pajola, 2000, p. 78. Yukler, et al. 1994, p. 169-213. Pratsch, 1994, p. 1-25. Bishop, 1988, p. 1052-1056.

98 Burwood, et al. 1995, p. 27.

99 Del Ben and Finetti, 1991, p. 2421. Finetti, 1982, plate 3. Anketell, 1996, p. 73. A1-NC 42: Belhaj, 1996, p. 62. C1-87, A1-87: Baair, et al. 2000, p. 12. Baair, et al. 2001. Smith and Karki, 1996, p. 129-137.

100 R.C Michel, 1985, *Oil and Gas Developments in North Africa in 1984*, Bull. Amer. Assn. Pet. Geol., vol. 69, p. 1656. Duronio, et al. 1991, p. 1598. Ghori, 1991, p. 2754.

Chapter 9

1 Campbell, 1997, table 1.

2 Lüning, et al. 1999, p. 693-718. Lüning, et al. 2000c, p. 151-174.

3 Echikh and Sola, 2000, p. 175-222. Klitzsch, 1963, p. 97-113. Klitzsch, 1966b, p. 19-32. Belhaj, 1996, p. 68-72. Thusu, 1996, p. 463. Boote, et al. 1998, p. 50-53.

4 Echikh, 1998, p. 109-129. Boote, et al. 1998, p. 47-52.

5 Belhaj, 1996, p. 57-96. Belhaj, 2000, p. 117-142.

6 Al Matmaty, et al. 2000, p. 9.

7 El Alami, et al. 1989, p. 17-28. Roohi, 1996a, p. 323-336. Roohi, 1996b, p. 435-454. Knytl, et al. 1996, p. 167-199. Schröter, 1996, p. 123-135. Klitzsch, 1971, p. 257.

8 Johnson and Nicoud, 1996, p. 211-221. Abugares, 1996, p. 45-63. Schneiter, 2000, p. 23. Anonymous, Oil and Gas Journal, March 16th 1998, p. 40.

9 Roohi, 1996a, p. 323-336. Roohi, 1996b, p. 435-454. Bezan, 1996, p. 97-118. Bezan, et al. 1996, p. 135-152. Abugares, 1996, p. 45-63.

10 Hallett and El Ghoul, 1996, p. 455-484. Gravity: El Batroukh and Zentani, 1980, p. 965-978. Ibrahim, 1996, p. 419-434. Ibrahim and Al Mahruqi, 2000, p. 26. El Ghoul, 1996, p. 137-154. Jones, 1996, p. 169-184. El Hawat, et al. 1996, p. 3-20.

11 El Alami, et al. 1989, p. 19-29. Wennekers, et al. 1996, p. 2-56. Abugares, 1996, p. 45-63. Bezan and Malak, 1996, p. 119-128. Benfield and Wright, 1980, p. 463-500.

12 Ambrose, 2000, p. 165-192.

13 Terry and Williams, 1968, p. 31-48. Brady, et al. 1980, p. 843-861. Spring and Hansen, 1998, p. 335-354.

14 Ghori, 1991, p. 2743-2756. Ghori and Mohammed, 1996, p. 529-540. Burwood, 1997, p. 1365.

15 Ibrahim, 1991, p. 2757-2779. Thusu, et al. 1988, p. 171-213. Thusu, 1996, p. 455-474. Gras and Thusu,

448

1998, p. 317-334. El Hawat, et al. 1996, p. 3-30. Ambrose, 2000, p. 165-192. Mansour and Magairhy, 1996, p. 485-528.

16 Ghori, 1991, p. 2743-2753. Sola and Ozcicek, 1990, p. 25-41.

17 Bishop, 1988, p. 1040. Sbeta, 1990, p. 53.

18 Bishop, 1988, p. 1046. Hammuda, et al. 1985, p. 7-105.

19 Hammuda, et al. 1989, p. 154-162. Bishop, 1988, p. 1053-1055.

20 Hammuda, et al. 1989, p. 93-97. Del Ben and Finetti, 1991, p. 2420. Finetti, 1982, plate 3. Kruseman and Floegel, 1980, p. 539-594.

21 Del Ben and Finetti, 1991, p. 2420. Finetti, 1982, plate 3.

22 Del Ben and Finetti, 1991, p. 2421. Baair, et al, 2001.

23 R.C Michel, 1985, Oil and Gas Developments in North Africa in 1984, Bull. Amer. Assn. Pet. Geol., vol. 69, p. 1656. Duronio, et al. 1991, p. 1598. Ghori, 1991, p. 2754. Sola and Ozcicek, 1990, p. 40. B1-33: Thusu, et al. 1988, p. 179.

REFERENCES

ABADI, A.M. and VAN DIJK, P.M. 1993. Short notes and guidebook on the stratigraphy and tectonics of West Zallah Trough, Sirt Basin, Libya. *First Symposium on the Sedimentary Basins of Libya, Geology of the Sirt Basin*, Field Guide, Earth Sciences Society of Libya, 52p.

ABDELHAMEED, M.S. 2000. Study of late Cretaceous Rakb Group, southeastern Sirt Basin, Libya. (abstract only). *5th Intern. conf. on the geology of the Arab world*, Cairo University. Book of abstracts, p. 283.

ABDULGHADER, G.S. 1996. Sedimentology and reservoir heterogeneities of the Nubian Formation, southeastern Sirt Basin, Libya. *First Symposium on the Sedimentary Basins of Libya, Geology of the Sirt Basin*. vol. 2 *(eds. M.J. Salem, A.S. El Hawat, A.M. Sbeta)*, Elsevier, Amsterdam, p. 223-250.

ABUGARES, Y.I. and RAMAEKERS, P. 1993. Short notes and guidebook on the Palaeozoic geology of the Ghat area, SW Libya. *First Symposium on the Sedimentary Basins of Libya, Geology of the Sirt Basin*, Field Guide, Earth Sciences Society of Libya, 84p.

ABUGARES, Y.I. 1996. Sedimentology and hydrocarbon potential of the Gir Formation, Sirt Basin, Libya. *First Symposium on the Sedimentary Basins of Libya, Geology of the Sirt Basin*. vol. 2. *(eds. M.J. Salem, A.S. El Hawat, A.M. Sbeta)*, Elsevier, Amsterdam, p. 45-64.

ABUSHAGUR, S.A. 1991. Cyclic transgressive and regressive sequences and their association with hydrocarbons - Ghani field, Sirt Basin, Libya. *Third Symposium on the Geology of Libya*. vol. 5 *(eds. M.J. Salem and M.N. Belaid)*, Elsevier, Amsterdam, p. 1827-1840.

ABUSHAGUR, S.A. and LeMONE, D.V. 1991. Depositional facies and diagenesis as a guide to correlating porosity zones – Al Furud Formation, Ghani field, Sirt Basin, Libya. *Third Symposium on the Geology of Libya*. vol. 5 *(eds. M.J. Salem and M.N. Belaid)*, Elsevier, Amsterdam, p. 1871-1882.

ABUTARRUMA, Y.G. 2000. Palaeocene planktonic foraminifera from well D2-NC 41 in the Tarabulus-Gabes Basin, central Mediterranean Sea (abstract only). *Int. workshop on north African micropalaeo. for pet. expl., Univ. Coll. London, August 2000*. Book of abstracts, p. 19.

ACHECHE, M. H., M'RABET, A., GHARIANI, H., OUAHCHI, A. and MONTGOMERY, S.L. 2001. Ghadames Basin, southern Tunisia: a reappraisal of Triassic reservoirs and future prospectivity. *Bull. Amer. Assoc. Pet. Geol.*, vol. 85, p. 765-780.

ADAMSON, K., GLOVER, T., WHITTINGTON. R. and CRAIG, J. 2000. The Lower Devonian succession of the Murzuq Basin- possible indicators of eustatic and tectonic controls on sedimentation. *Symposium on Geological Exploration in Murzuq Basin (Eds. M.A. Sola and D. Worsley)*, Elsevier, Amsterdam, p. 431-448.

ADLORFF, M.C., DOUBINGER, J., MASSA, D. and VACHARD, D, 1986. Trias de Tripolitain (Libya), nouvelles donnees biostratigraphiques et paly-nologiques. *Rev. Inst. Fr. Petrole*, vol. 40, p. 27-63.

AL JUMAILY, W.A. 2000. Palynological study of Khashm az Zarzur Formation in a quarry in western Yifran, Libya (abstract only). Second *Symposium on the Sedimentary Basins of Libya, Geology of Northwest Libya*. Book of abstracts, p. 8.

AL MAGHRABI, I. and CHESHITEV, G. 1977. Geological Map of Libya, scale 1:2,000,000, Second edition. Industrial Research Centre, Tripoli.

AL MATMATY, I., GASHGESH, T.M., ERKMEN, U. and SWIRE, P. 2000. Hydrocarbon prospectivity of the northwest Ghadamis Basin and Al Jifarah Trough (abstract only). *Second Symposium on the Sedimentary Basins of Libya, Geology of Northwest Libya*. Book of abstracts, p. 9.

ALEM, N., ASSASSI, S., BENHEBOUCHE, S., and KADI, B. 1998. Controls on Hydrocarbon occurrence and productivity in the F6 reservoir, Tin Fouyé-Tabankort area, NW Illizi Basin. In: *Petroleum Geology of North Africa. (ed. D.S. Macgregor, R.T.J. Moody)*, Geol. Soc. Special Publication No. 132, p. 175-186.

ALEXEYEV, E.S. 1985a. Sheet Bi'r Amāsīn (NH 32-11),

450

Geological Map of Libya, scale 1:250,000, Explanatory Booklet, Industrial Research Centre, Tripoli.

ALEXEYEV, E.S. 1985b. Sheet Habayt al Ghardaqah (NH 33-9), *Geological Map of Libya, scale 1:250,000, Explanatory Booklet*, Industrial Research Centre, Tripoli.

ALEXEYEV, E.S. 1985c. Sheet Wādī Tanārūt (NH 32-12), *Geological Map of Libya, scale 1:250,000, Explanatory Booklet*, Industrial Research Centre, Tripoli.

ALIEV, M., AIT LAOUSSINE, N., AVROV, V., ALEKSINE, G., BAROULINE, G., IAKOVLEV, B., KORJ, M., KOUVYKINE, J., MAKAROV, V., MAZANOV, V., MEDVEDEV, E., MKRTCHIANE, O., MOUSTAFINOV, R., ORIEV, L. OROUDJEVA, D., OULMI, M., and SAID, A. 1971. Geological structures and estimation of oil and gas in the Sahara in Algeria. Sonatrach, Algiers, 265p.

ALMEHDI, B., GOJKOVIC, S., MEGERISI, M., OBRENOVIC, M., PURIC, D., ZELENKA, J. 1991. Radioactive elements in sedimentary rocks in the western part of the Murzuq Basin, Libya. *Third Symposium on the Geology of Libya*. vol. 7 (eds. M.J. Salem and M.T. Busrewil and A.M. Ben Ashour), Elsevier, Amsterdam, p. 2645-2658.

ALMOND, D.C., BUSREWIL, M.T. and WADSWORTH, 1974. The Ghirian Tertiary volcanic province of Tripolitania, Libya. *Geol. Journ.* vol. 9, p. 17-28.

ALMOND, D.C. 1991. Anorogenic magmatism in northeast Africa. *Third Symposium on the Geology of Libya*. vol. 7 (eds. M.J. Salem and M.T. Busrewil and A.M. Ben Ashour), Elsevier, Amsterdam, p. 2495-2512.

AMBROSE, G. 2000. The geology and hydrocarbon habitat of the Sarir Sandstone, SE Sirt Basin, Libya. *Journ. Pet. Geol.*, vol. 23, p. 165-192.

ANDRE, L., KLERKX, J., BUSREWIL, M.T. 1991. Geochemical and Rb-Sr isotopic data on felsic rocks from the Jabal Awaynat alkaline intrusive complex (SE Libya). 1991. *Third Symposium on the Geology of Libya*. vol. 7 (eds. M.J. Salem and M.T. Busrewil and A.M. Ben Ashour), Elsevier, Amsterdam, p. 2511-2528.

ANKETELL, J.M. and GHELLALI, S.M. 1991a. The Jifarah Formation – aeolian and fluvial deposits of Quaternary age, Jifarah Plain, GSPLAJ. A redefinition in terms of composite stratotype. *Third Symposium on the Geology of Libya*. vol. 5 (eds. M.J. Salem and M.N. Belaid), Elsevier, Amsterdam, p. 1967-1986.

ANKETELL, J.M. and GHELLALI, S.M. 1991b.

Quaternary sediments of the Jifarah Plain. *Third Symposium on the Geology of Libya*. vol. 5 (eds. M.J. Salem and M.N. Belaid), Elsevier, Amsterdam, p.1987-2016.

ANKETELL, J.M. and GHELLALI, S.M. 1991c. A palaeogeologic map of the pre-Tertiary surface in the region of the Jifarah Plain and its implications to the structural history of northern Libya. *Third Symposium on the Geology of Libya*. vol. 6 (eds. M.J. Salem, A.M. Sbeta and M.R. Bakbak), Elsevier, Amsterdam, p. 2381-2406.

ANKETELL, J.M. and KUMATI, S.M. 1991a. Sedimentary history of the Asiyah Formation-Upper Eocene of Al Hufrah area, western Sirt Basin, GSPLAJ. *Third Symposium on the Geology of Libya*. vol. 5 (eds. M.J. Salem and M.N. Belaid), Elsevier, Amsterdam, p. 1883-1906.

ANKETELL, J.M. and KUMATI, S.M. 1991b. Structure of Al Hufrah region – western Sirt Basin, GSPLAJ. *Third Symposium on the Geology of Libya*. vol. 6 (eds. M.J. Salem, A.M. Sbeta and M.R. Bakbak), Elsevier, Amsterdam, p. 2353-2370.

ANKETELL, J.M. 1996. Structural history of the Sirt Basin and its relationship to the Sabratah Basin and Cyrenaican Platform, northern Libya. *First Symposium on the Sedimentary Basins of Libya, Geology of the Sirt Basin*. vol. 3. (eds. M.J. Salem, M.T. Busrewil, A.A. Misallati, and M.A. Sola), Elsevier, Amsterdam, p. 57-89.

ANKETELL, J.M. and MRIHEEL, I.Y. 2000. Depositional environment and diagenesis of the Eocene Jdeir Formation, Gabes-Tripoli Basin, western offshore, Libya. *Journ. Petrol. Geology*, vol. 23, p. 425-447.

ANONYMOUS, 1970. Deep Sea Drilling Project: Leg 13. *Geotimes*, vol. 15, no. 10. p. 12-15.

ANONYMOUS, 1975. Glomar Challenger returns to the Mediterranean Sea. *Geotimes*, vol. 20. no. 8., p. 16-19.

ANONYMOUS, 1983. North American Stratigraphic Code. *North American Commission on Stratigraphic Nomenclature. Bull. Amer. Assoc. Pet. Geol.* vol. 67, p. 841-875.

ANONYMOUS, 1985. Geological Map of Libya, scale 1:1,000,000, (4 sheets). Industrial Research Centre, Tripoli.

ANONYMOUS, 1998. Red Sea Oil's Libyan venture marks success. *Oil & Gas Journ.* Mar. 16th, 1998, p. 40.

ANTONOVIC, A. 1977. Sheet Mizdah (NH 33-1), *Geological Map of Libya, scale 1:250,000, Explanatory Booklet*, Industrial Research Centre, Tripoli.

ARAMBOURG, C. and MAGNIER, P. 1961. Gisements de vertebres dans le bassin Tertiaire de Syrte (Libye). *C.R. Acad. Sci.* Paris, vol 252, p. 1181-1183.

ARNOULD-SAGET, S. and MAGNIER, P. 1961. Decouverte de dents de *Paleomastodontes* dans la region de Zella (Tripolitanie). *Bull. Geol. Soc.* Fr. vol 3, p. 283-287.

ASSAF, H.S. 1991. Folding structure in the Basement rocks of the Jabal Nuqay area, southern Libya. *Third Symposium on the Geology of Libya.* vol. 6 *(eds. M.J. Salem, A.M. Sbeta and M.R. Bakbak)*, Elsevier, Amsterdam, p. 2287-2294.

ASSERETO, R. and BENELLI, F. 1971. Sedimentology of the pre-Cenomanian formations of the Jebel Garian, Libya. *First Symposium on the Geology of Libya (ed. C. Gray).* Faculty of Science, University of Libya, Tripoli, p. 37-86.

ATHERTON, M.P., LAGHA, S., and FLINN, D. 1991. The geochemistry of the Jabal Arknu and Jabal al Awaynat alkaline ring complexes of SE Libya. *Third Symposium on the Geology of Libya.* vol. 7 *(eds. M.J. Salem and M.T. Busrewil and A.M. Ben Ashour)*, Elsevier, Amsterdam, p. 2559-2576.

AZIEZ, M., HROUDA, M. and M'MALLA, M.S. 1998. Geochemical study of potential Silurian source rocks and crude oils in the Ghadames Basin (Algeria-Libya) (abstract only). *Symposium on Geological Exploration in Murzuq Basin, Sabha,* 20-22nd September 1998. Book of abstracts.

AZIZ, A. 2000. Stratigraphy and hydrocarbon potential of the lower Palaeozoic succession of licence NC 115, Murzuq Basin, SW Libya. *Symposium on Geological Exploration in Murzuq Basin (Eds. M.A. Sola and D. Worsley)*, Elsevier, Amsterdam, p. 349-368.

BAAIR, M.Y., RABTI, I., JOHNSON, B., MILADI, N. and SWIRE, P.H. 2000. The regional geology of the northwestern edge of the Sirt Basin (abstract only). *Second Symposium on the Sedimentary Basins of Libya, Geology of Northwest Libya.* Book of abstracts, p. 12.

BAAIR, M.Y., RABTI, I, and SWIRE, P.H. 2001. Hydrocarbon source rock quality, distribution and migration in northwest Sirt basin, Libya (abstract only). *Abstracts, North Africa Research Workshop, Oxford Brookes University, 4th-5th September 2001.*

BAEGI M.B., ASSAF, H.S. and HANGARI, K.M. 1991. Al Awaynat surface uranium mineralisation – a new approach to its origin. *Third Symposium on the Geology of Libya.* vol. 7 *(eds. M.J. Salem and M.T. Busrewil and A.M. Ben Ashour)*, Elsevier, Amsterdam, p. 2619-2626.

BAGNOLD, R.A. 1935. Libyan sands. Hodder and Stoughton, London.

BAILEY, H.W., DUNGWORTH, G., HARDY, M., SCULL, D., and VAUGHAN, R.D. 1989. A fresh approach to the Metlaoui. *Actes des IIeme journees de geologie tunisienne appliquee a la recherche des hydrocarbures.* Mem. ETAP, No. 3, Tunis, p. 281-307.

BAIRD, D.W. 1967. The Permo-Carboniferous of southern Tunisia. In: *Guidebook to the geology and history of Tunisia (ed. Martin, L.).* Petroleum Exploration Society of Libya, Ninth Annual Field Conference, p. 85-108.

BAIRD, D.W., ABURAWI, R.M. and BAILEY, N.J.L. 1996. Geohistory and petroleum in the central Sirt Basin. *First Symposium on the Sedimentary Basins of Libya, Geology of the Sirt Basin.* vol. 3. *(eds. M.J. Salem, M.T. Busrewil, A.A. Misallati and M.J. Sola)*, Elsevier, Amsterdam, p. 3-56.

BALL, L., BOEBÉ, M., SADLER, P., CORBETT, P. and LEWIS, J. 1996. Permeability prediction in a braided fluvial reservoir: a probe permeameter study on the pre-Upper Cretaceous B sandstone, As Sarah field, Sirt Basin, Libya. *First Symposium on the Sedimentary Basins of Libya, Geology of the Sirt Basin.* vol. 2. *(eds. M.J. Salem, A.S. El-Hawat and A.M. Sbeta)*, Elsevier, Amsterdam, p. 349-364.

BANERJEE, S. 1980. Stratigraphic Lexicon of Libya. *Bulletin No. 13.* Industrial Research Centre, Tripoli, 300p.

BÄR, C.B. and KLITZSCH, E. 1964. Introduction to the geology of Egypt. In: *Guidebook to the geology and archaeology of Egypt (ed. Reilly F.A.).* Petroleum Exploration Society of Libya, Sixth Annual Field Conference, p. 71-98.

BARASI, S., MARIMI, A. and YANILMAZ, E. 2000. Tectono-facies of the hydrocarbon bearing pre-Upper Cretaceous clastics on the northeast margin of the Sirte Basin, Libya (abstract only). *Symposium on petroleum systems and evolving technologies in African E & P. Burlington House, London,* 16-18 May, 2000. Book of abstracts, p. 41.

BARBIERI, R. 1996. Micropalaeontology of the Rakb Group (Cenomanian to early Maastrichtian) in the Hameimat Basin, northern Libya. *First Symposium on the Sedimentary Basins of Libya, Geology of the Sirt Basin.* vol. 1. *(eds. M.J. Salem, A.J. Mouzughi and O.S. Hammuda)*, Elsevier, Amsterdam, p. 185-194.

BARGHUT, M and JALALH, A. 2001. Bouri field, NW offshore Libya. Gas while drilling data analysis (abstract only). *Abstracts, North Africa Research Workshop, Oxford Brookes University, 4th-5th*

452

September 2001.

BARIĆ, G., ŠPANIĆ, D. and MARIČIĆ, M. 1996. Geochemical characterization of source rocks in the NC-157 block (Zaltan Platform), Sirt Basin. *First Symposium on the Sedimentary Basins of Libya, Geology of the Sirt Basin.* vol. 2. *(eds. M.J. Salem, A.S. El-Hawat and A.M. Sbeta),* Elsevier, Amsterdam, p. 541-553.

BARIĆ, G. and COTA, L. 2000. A geochemical investigation of hydrocarbons in Belina 1 offshore well (abstract only). *Second Symposium on the Sedimentary Basins of Libya, Geology of Northwest Libya.* Book of abstracts, p. 14.

BARR, F.T. 1968. Upper Cretaceous stratigraphy of Jabal al Akhdar, northern Cyrenaica. In: *Geology and archaeology of northern Cyrenaica, Libya (ed. Barr, F.T.).* Petroleum Exploration Society of Libya, Tenth Annual Field Conference, p. 131-148.

BARR, F.T. and HAMMUDA, O.S. 1971. Biostratigraphy and planktonic zonation of the Upper Cretaceous Atrun Limestone and Hilal Shale, northeastern Libya. *In: Proc. 2nd Int. Conf. plankt. microfossils (1970)(Ed. A. Farinacci),* Rome, p. 27-40.

BARR, F.T. 1972. Cretaceous biostratigraphy and planktonic foraminifera of Libya. *Micropalaeontology,* vol. 18, p. 1-46.

BARR, F.T. and WEEGAR, A.A. 1972. Stratigraphic nomenclature of the Sirte Basin, Libya. Petroleum Exploration Society of Libya, Tripoli, 179p.

BARR, F.T. and WALKER, B.R. 1973. Late Tertiary channel system in northern Libya and its implication on Mediterranean sea level changes. In: *Initial Report Deep Sea Drilling Project, Leg 13. (Eds. W.B.F. Ryan and K.J. Hsu).* p. 1244-1251.

BARR, F.T. and BERGGREN, W.A. 1980. Lower Tertiary biostratigraphy and tectonics of northern Libya. *Second Symposium on the Geology of Libya.* vol. 1 *(eds. M.J. Salem and M.T. Busrewil),* Academic Press, London, p. 163-192.

BARY, E. von, 1880. Tagebuch gefuhrt auf seiner reise von Tripolis nach Ghat ind Aïr. *Zeit. Gesell. fur Erdkunde,* Berlin. vol. 15, no. III, XIV, XX.

BATTEN, D.J. and UWINS, P.J.R. 1985. Early-late Cretaceous (Aptian-Cenomanian) palynomorphs. In: *The palynostratigraphy of north-east Libya (eds. B. Thusu and B. Owens). Journ. Micropalaeontology,* vol. 4, p. 151-168.

BATTON, G. 1965. Contribution a l'etude anatomique et biostratigraphique de la flore du Continental intercalaire saharien. In: *Batton, G. et al. Paleobotanique saharien,* p. 7-97.

BAUDET, D. 1988. Precambrian palynomorphs from northeast Libya. In: *Subsurface palynostratigraphy of northeast Libya. (eds. A. El-Arnauti, B. Owens and B. Thusu).* Research Centre, Garyounis University, Benghazi, p. 17-26.

BAUSCH, W.M. 1980. Mineralogical composition of Jabal Nafusah phonolites. *Second Symposium on the Geology of Libya.* vol. 3 *(eds. M.J. Salem and M.T. Busrewil),* Academic Press, London, p. 1107-1115.

BAUSCH, W.M. and SCHELLHORN, H. 1991. Rhythmically alternating limestones of differing stages of diagenetic development (Eocene, Al Jabal al Akhdar). *Third Symposium on the Geology of Libya.* vol. 7 *(eds. M.J. Salem and M.T. Busrewil and A.M. Ben Ashour),* Elsevier, Amsterdam, p. 2671-2680.

BEAVINGTON-PENNEY, S.J., RACEY, A., WRIGHT, V.P. and BEN BRAHIM, A. 2000. Sedimentological and palaeoenvironmental analysis of nummulitic limestones from the Lower Eocene El Garia Formation, Tunisia (abstract only). *Int. workshop on north African micropalaeo. for pet. expl., Univ. Coll. London, August 2000.* Book of abstracts, p. 23.

BEBOUT D.G. and PENDEXTER, C. 1975. Secondary carbonate porosity as related to early Tertiary depositional facies, Zelten field, Libya. *Bull. Amer. Assoc. Pet. Geol.* vol. 59, p. 665-693.

BECKER, R.E. and FÜRST, M. 1991. Sedimentological time markers and ground-water dating: a study of the Quaternary evolution of the Al Kufrah area. *Third Symposium on the Geology of Libya.* vol. 5 *(eds. M.J. Salem and M.N. Belaid),* Elsevier, Amsterdam, p. 2017-2026.

BELAZI, H.S. 1989. The geology of the Nafoora oilfield, Sirte Basin. Libya. *Journ. Pet. Geol.* vol. 12, p. 353-366.

BELHAJ, F. 1996. Palaeozoic and Mesozoic stratigraphy of eastern Ghadamis and western Sirt Basins. *First Symposium on the Sedimentary Basins of Libya, Geology of the Sirt Basin.* vol. 1. *(eds. M.J. Salem, A.J. Mouzughi and O.S. Hammuda),* Elsevier, Amsterdam, p. 57-96.

BELHAJ, F. 2000. Evaluation of the Dirbal Formation (Ketatna) potential reservoir and a loss of circulation zone 'A' structure, concession NC 41, Sabratah Basin (abstract only). *Second Symposium on the Sedimentary Basins of Libya, Geology of Northwest Libya.* Book of abstracts, p. 16.

BELHAJ, F. 2000. Carboniferous and Devonian stratigraphy – the M'rar and Tadrart reservoirs, Ghadames Basin, Libya. *Symposium on Geological Exploration in Murzuq Basin (Eds. M.A. Sola and D. Worsley),* Elsevier, Amsterdam, p. 117-142.

BELL, J.D. and SANDFORD, K.S. 1971. Nodular masses

of manganese in volcanic rocks around Jebel 'Uweinat. *First Symposium on the Geology of Libya (ed. C. Gray).* Faculty of Science, University of Libya, Tripoli, p. 333-340.

BELLAIR, P., FREULON, J. and LEFRANC, J. 1954. Decouverte d'une formation a vertebres et vegetaux d'age tertiaire au bord occidental du desert Libyque (Sahara oriental). *C.R. Acad. Sci.* Paris, vol. 239, p. 1822-1824.

BELLINI, E. and MASSA, D. 1980. A stratigraphic contribution to the Palaeozoic of the southern basins of Libya. *Second Symposium on the Geology of Libya.* vol. 1 *(eds. M.J. Salem and M.T. Busrewil),* Academic Press, London,p. 3-56.

BELLINI, E., GIORI, I., ASHURI, O., and BENELLI, F. 1991. Geology of the Al Kufrah basin, Libya. *Third Symposium on the Geology of Libya.* vol. 6 *(eds. M.J. Salem, A.M. Sbeta and M.R. Bakbak),* Elsevier, Amsterdam, p. 2287-2294.

BEN BRAHIM, A. 1989. Etude du reservoir carbonate d'Isis. *Actes des IIeme journees de geologie tunisienne appliquee a la recherche des hydrocarbures.* Mem. ETAP, No. 3, Tunis, p. 309- 319.

BEN FERJANI, A., BUROLLET, P.F. and MEJRI, F. 1990. Petroleum Geology of Tunisia. Enterprise Tunisienne d'Activités Pétrolières, Tunis. 194p.

BENDER, A.A., COELHO, F.M., and BEDREGAL, R.P. 1996. A basin modelling study of the southeastern part of the Sirt basin. Libya. *First Symposium on the Sedimentary Basins of Libya, Geology of the Sirt Basin.* vol. 1. *(eds. M.J. Salem, A.J. Mouzughi and O,S. Hammuda),* Elsevier, Amsterdam, p. 139-156.

BENFIELD, A.C. and WRIGHT, E.P. 1980. Post-Eocene sedimentation in the eastern Sirt Basin, Libya. *Second Symposium on the Geology of Libya.* vol. 2 *(eds. M.J. Salem and M.T. Busrewil),* Academic Press, London, p. 463-500.

BENZARTI, R. 1993. Synthese biostratigraphique de la serie paleozoique de quelque sondages de l'ancien permis de Kirchaou. *Actes des IIIeme journees de geologie tunisienne appliquee a la recherche des hydrocarbures.* Mem. ETAP, No. 4, Tunis, p. 297-313.

BERENDEYEV, N.S. 1985a. Sheet Hamadat Tanghirt (NH 32-16), *Geological Map of Libya, scale 1:250,000, Explanatory Booklet,* Industrial Research Centre, Tripoli.

BERENDEYEV, N.S. 1985b. Sheet Wadi ain Armas (NH 32-15), *Geological Map of Libya, scale 1:250,000, Explanatory Booklet,* Industrial Research Centre, Tripoli.

BERGGREN, W.A. 1969. Biostratigraphy and foraminifera zonation of the Tertiary system of the Sirte Basin of Libya. In: *Proc, 1st Int. Conf. Plankt. Microfossils. (Eds. P. Bronnimann and H.H. Renz).* E.J. Brill, Leyden, p. 104-120.

BERGSTROM, S.M. and MASSA, D. 1991. Stratigraphic biogeographic significance of Upper Ordovician conodonts from northwestern Libya. *Third Symposium on the Geology of Libya.* vol. 4 *(eds. M.J. Salem, O.S. Hammuda and B.A. Eliagoubi),* Elsevier, Amsterdam, p. 1323-1342.

BERNASCONI, A., POLIANI, G. and DAKSHE, A. 1991. Sedimentology, petrography and diagenesis of Metlaoui Group in the offshore northwest of Tripoli. *Third Symposium on the Geology of Libya.* vol. 5 *(eds. M.J. Salem and M.N. Belaid),* Elsevier, Amsterdam, p. 1907-1928.

BEUF, S., BIJU-DUVAL, B., STEVAUX, J. and KULBICKI, G. 1969. Extent of 'Silurian' glaciation in the Sahara: its influences and consequences upon sedimentation. In: *Geology, archaeology and prehistory of the southwestern Fezzan, Libya (ed. Kanes, W.H.).* Petroleum Exploration Society of Libya, Eleventh Annual Field Conference, p. 103-116.

BEUF, S., BIJU-DUVAL, B., DE CHARPAL, O., de ROGNON, P. GABRIEL, O. and BENNACEF, A. 1971. Les gres du Paleozoique inferieur au Sahara. Editions Technip, Paris, 464p.

BEYRICH, E. 1852. Bericht uber die von Overweg auf der reise von Tripolis nach Mursuk und von Mursuk nach Ghat gefundenen versteinerungen. *Zeit. D. g. Gesell.,* vol. 4, p. 143-161.

BEZAN, A.M. 1996. The Palaeocene sequence in Sirt Basin. *First Symposium on the Sedimentary Basins of Libya, Geology of the Sirt Basin.* vol. 1. *(eds. M.J. Salem, A.J. Mouzughi and O,S. Hammuda),* Elsevier, Amsterdam, p. 97-118.

BEZAN, A.M., BELHAJ, F., and HAMMUDA, K. 1996. The Beda Formation in Sirt Basin. *First Symposium on the Sedimentary Basins of Libya, Geology of the Sirt Basin.* vol. 2. *(eds. M.J. Salem, A.S. El-Hawat and A.M. Sbeta),* Elsevier, Amsterdam, p. 135-152.

BEZAN, A.M. and MALAK, E.K. 1996. Oligocene sediments of Sirt Basin and their hydrocarbon potential. *First Symposium on the Sedimentary Basins of Libya, Geology of the Sirt Basin.* vol. 1. *(eds. M.J. Salem, A.J. Mouzughi and O.S. Hammuda),* Elsevier, Amsterdam, p. 119-128.

BHATTACHARYYA, D.P. and LORENZ, J.C. 1983. Different depositional settings of the Nubian lithofacies in Libya and southern Egypt. In: *Modern and ancient fluvial systems. Internat. Assoc.*

Sedimentologists, Spec.Publ., no. 6, p. 435-448.

BIJU-DUVAL, B. 1974. Exemples de depots fluvio-glaciaires dans l'Ordovicien superieur et le Precambrian superieur du Sahara central. *Bull. Cent. Rech, Pau.* vol. 8, p. 209-236.

BIJU-DUVAL, B., LETOUZEY, J., MONTADERT, L., COURRIER, P. MAGNIOT, J.F. and SANCHO, J. 1974. Geology of the Mediterranean Sea basins. In: *Geology of the continental margins (Ed. C.A. Burke and C.L. Drake)* Springer-Verlag, Berlin, p. 695-721.

BIJU-DUVAL, B. and MONTADERT, L. (Eds.) 1977. Structural history of the Mediterranean basins. Editions Technip, Paris, 448 p.

BIJU-DUVAL, B. DEYNOUX, M and ROGNON, P. 1981. Late Ordovician tillites of the central Sahara. In: *Earth's pre-Pleistocene glacial record (Eds. M.J. Hambrey and W.B. Harland)*. Cambridge University Press. p. 99-107.

BISHOP, W.F. 1988. Petroleum Geology of east-central Tunisia. *Bull. Amer. Assoc. Pet. Geol.* vol. 72, p. 1033-1058.

BISHOP, W.F. and DEBONO, G. 1993. Mediterranean Sea potential seen in area south of Malta. *Oil and Gas Journ.* July 5th, 1993, p. 60-63.

BISHOP, W.F. and DEBONO, G. 1996. The hydrocarbon geology of southern offshore Malta. *Journ. Petrol. Geol.* vol. 19, p. 129-160.

BISMUTH, H., BONNEFOUS, J. and DUFAURE, P. 1967. Mesozoic microfacies of Tunisia. In: *Guidebook to the geology and history of Tunisia (ed. Martin, L.)*. Petroleum Exploration Society of Libya, Ninth Annual Field Conference, p. 159-214.

BLANPIED, C., DEYNOUX, M., GHIENNE, J.F. and RUBINO, J.L. 2000. Late Ordovician glacially related depositional systems of the Gargaf Uplift (Libya) and comparisons with correlative deposits in the Taoudini Basin (Mauritania). *Symposium on Geological Exploration in Murzuq Basin (Eds. M.A. Sola and D. Worsley)*, Elsevier, Amsterdam, p. 485-508.

BOAZ, N.T., GAZIRY, A.W., DE HEINZELIN, J., and EL-ARNAUTI, A. (Eds.) 1982. Results from the International Sahabi research project. *Special Issue, Garyounis Scientific Bull. No. 4*. University of Garyounis, Benghazi, 141p.

BOAZ, N.T., EL-ARNAUTI, A., GAZIRY, A.W., DE HEINZELIN, J. and BOAZ, D.D. (Eds.). 1987. Neogene paleontology and geology of Sahabi. Alan Liss, New York, 396p.

BOAZ, N.T. 1996. Vertebrate palaeontology and terrestrial palaeoecology of As Sahabi and the Sirt basin. *First Symposium on the Sedimentary Basins of Libya,*

Geology of the Sirt Basin. vol. 1. *(eds. M.J. Salem, A.J. Mouzughi and O.S. Hammuda)*, Elsevier, Amsterdam, p. 531-539.

BOCCALETTI, M. and GUAZZONE, G. 1973. Plate tectonics in the Mediterranean region. In: *Geology of Italy*, vol. 1 *(ed. Squyres, C.H.)*. Petroleum Exploration Society of Libya, Fifteenth Annual Field Conference, p. 143-164.

BONNEFOUS, J. 1967. Jurassic stratigraphy of Tunisia: a tentative synthesis. In: *Guidebook to the geology and history of Tunisia (ed. Martin, L.)*. Petroleum Exploration Society of Libya, Ninth Annual Field Conference, p. 109-130.

BONNEFOUS, J. 1972. Geology of the quartzitic 'Gargaf Formation' in the Sirte Basin, Libya. *Bull. Centre. Rech. Pau-SNPA*, vol. 6, p. 225-261.

BOOTE, D.R.D., CLARK-LOWES, D.D., TRAUT, M.W. 1998. Palaeozoic petroleum systems of North Africa. In: *Petroleum Geology of North Africa. (ed. D.S. Macgregor, R.T.J. Moody, D.D. Clark-Lowes)*, Geol. Soc. Special Publication No. 132, p. 7-68.

BORGHI, P. 1939. Fossili Devonici del Fezzan. *Ann. Mus. Libico Storia Naturale*, vol. 1, p. 115-184.

BORGHI, P. and CHIESA, C. 1940. Cenni geologici e paleontologici sul Paleozoico dell'Egghidi Uan Caza nel deserto di Taita (Fezzan occidentale). *Ann. Mus. Libico Storia Nat.* vol. 2, p. 122-137.

BOUTOUAOU, D. and LARTER, S. 2001. Regional migration routes and field charging of the Hassi Messaoud field, Algeria (abstract only). *Abstracts, North Africa Research Workshop, Oxford Brookes University, 4th-5th September 2001.*

BRACACCIA, V., CARCANO, C., and DRERA, K. 1991. Sedimentology of the Silurian-Devonian Series in the southeastern part of the Ghadamis Basin. *Third Symposium on the Geology of Libya.* vol. 5 *(eds. M.J. Salem and M.N. Belaid)*, Elsevier, Amsterdam, p. 1727-1744.

BRADY, T.J., CAMPBELL, N.D.J., and MAHER, C.E. 1980. Intisar 'D' oil field, Libya. In: *Giant oil and gas fields of the decade 1968-1978. (ed. M.T. Halbouty)*. Mem. Amer. Assoc. Pet. Geol. No. 30, p. 843-861.

BRADY, T.J., CAMPBELL, N.J. and MAHER, C.E. 1981. Intisar 'D' oilfield, Libya. In: *Petroleum Geology of the continental shelf of north-west Europe (Ed. L.V. Illing and G.D. Hobson)*, Hayden, London, p. 543-564.

BRENCHLEY, P.J., MARSHALL, J.D., CARDEN, G.A.F. ROBERTSON, D.B.R., LONG, D.G.F., MEIDLA, T., HINGS, L. and ANDERSON, T.F. 1994. Bathymetric and isotopic evidence for a short-lived late Ordovician glaciation in a greenhouse period. *Geology*, vol 22, p. 295-298.

BRENNAN, P. 1992. Raguba Field – Libya. . In: *Treatise of Petroleum Geology. Atlas of oil and gas fields, Structural Traps VII (compiled by E.A. Beaumont and N.H. Foster)*. Amer. Assoc. Pet. Geol. p. 267-289.

BRICHANT, A.L. 1952. Sur la decouverte du Trias au pied du Djebel Gharian (Tripolitaine-Libye). *C.R. Hebd. Seanc. Acad. Sci.* Paris, vol. 234, p. 1456-1458.

BROUGHTON, P.L. 1996. Effect of permeability variations on water encroachment patterns at the Ar Raqubah field, Sirt Basin. *First Symposium on the Sedimentary Basins of Libya, Geology of the Sirt Basin*. vol. 2. *(eds. M.J. Salem, A.S. El-Hawat and A.M. Sbeta)*, Elsevier, Amsterdam, p. 391-418.

BRUGMAN, W.A., EGGINK, J.W. LOBOZIAK, S. and VISSCHER, H. 1985. Late Carboniferous-early Permian (Gzelian-Artinskian) palynomorphs. In: *The palynostratigraphy of north-east Libya (eds. B. Thusu and B. Owens). Journ. Micropalae-ontology*, vol. 4, p. 93-106.

BRUGMAN, W.A., EGGINK, J.W., and VISSCHER, H. 1985. Middle Triassic (Anisian-Ladinian) palynomorphs. In: *The palynostratigraphy of north-east Libya (eds. B. Thusu and B. Owens). Journ. Micropalae-ontology*, vol. 4, p. 107-112.

BRUGMAN, W.A., LOBOZIAK, S. and VISSCHER, H. 1988. The problem of the Carboniferous-Permian boundary in northeast Libya from a palynological point of view. In: *Subsurface palynostratigraphy of northeast Libya. (eds. A. El-Arnauti, B. Owens and B. Thusu)*. Research Centre, Garyounis University, Benghazi, p. 151-156.

BRUGMAN, W.A. and VISSCHER, H. 1988. Permian and Triassic palynostratigraphy of northeast Libya. In: *Subsurface palynostrati-graphy of northeast Libya. (eds. A. El-Arnauti, B. Owens and B. Thusu)*. Research Centre, Garyounis University, Benghazi, p. 157-170.

BU-ARGOUB, F.M. 1996. Palynological and palynofacies studies of the Upper Cretaceous sequence in well C'275-65, Sirt Basin, NE Libya. *First Symposium on the Sedimentary Basins of Libya, Geology of the Sirt Basin*. vol. 1. *(eds. M.J. Salem, A.J. Mouzughi and O.S. Hammuda)*, Elsevier, Amsterdam, p. 419-454.

BU-ARGOUB, F.M. and THUSU, B. 1996. Stratigraphy and palynofacies character of organic rich sediments in pre-Campanian sandstone in southeast Sirt Basin (abstract only). *The 2nd Middle East Geosciences Conference, Bahrain*, 1996. Book of abstracts.

BURDON, D.J. 1980. Infiltration conditions of a major sandstone aquifer around Ghat, Libya. *Second Symposium on the Geology of Libya*. vol. 2 *(eds. M.J. Salem and M.T. Busrewil)*, Academic Press, London, p. 595-609.

BURDON, D.J. and GONFIANTINI, R. 1991. Lakes in the Awbari Sand Sea of Fazzan, Libya. *Third Symposium on the Geology of Libya*. vol. 5 *(eds. M.J. Salem and M.N. Belaid)*, Elsevier, Amsterdam, p. 2027-2042.

BURKE, K., and DEWEY, J. 1974. Two plates in Africa during the Cretaceous? *Nature*, vol. 249, p. 313-316.

BURKE, K. 1977. Aulacogens and continental break-up. *Ann. Rev. Earth Planet. Sci.*, vol. 50, p. 371-396.

BURKE, K. and KRAUS, J.U. 1998. Are quartz-rich, Cambro-Ordovician sandstone sequences in northern Africa and Arabia products of the collapse and erosion of huge, Pan-African, Tibetan-style plateaux? *Journ. African Earth Sci.* vol. 27, p. 42.

BUROLLET, P.F. 1956. Contribution a l'etude strati-graphique de la Tunisie centrale. *Ann. Mines et Geol., Min. des Trav. Publ., Tunis*, No. 18.

BUROLLET, P.F. 1960. Libye. *Lexique Stratigraphique International, Afrique (dir. R. Furon) Fascicule IVa*. Congres Geologique International, Cent. Nat. Rech. Sci. Paris, 62p.

BUROLLET, P.F. 1963a. Reconnaissance geologique dans le sud-est du bassin de Kufra. *First Saharan Symposium*. Rev. Inst. Fran. Pet. special volume, p. 219-227.

BUROLLET, P.F. 1963b. Field trip guidebook of the excursions to Jebal Nefusa. Pet. Expl. Soc. Libya, 19p.

BUROLLET, P.F. 1967a. General geology of Tunisia. In: *Guidebook to the geology and history of Tunisia (ed. Martin, L.)*. Petroleum Exploration Society of Libya, Ninth Annual Field Conference, p. 51-58.

BUROLLET, P.F. 1967b. Tertiary geology of Tunisia. In: *Guidebook to the geology and history of Tunisia (ed. Martin, L.)*. Petroleum Exploration Society of Libya, Ninth Annual Field Conference, p. 215-226.

BUROLLET, P.F. and MANDERSCHEID, G. 1967. Le Devonien en Libye et en Tunisie. *Int. Symp. Devonian System*, Calgary, vol. 1. p. 285-302.

BUROLLET, P.F., and BYRAMJEE, R. 1969. Sedimentological remarks on Lower Palaeozoic sandstones of south Libya. In: *Geology, archaeology and prehistory of the southwestern Fezzan, Libya (ed. Kanes, W.H.)*. Petroleum Exploration Society of Libya, Eleventh Annual Field Conference, p. 91-102.

BUROLLET, P.F., MUGNIOT, J.M. and SWEENY, P. 1978. The geology of the Pelagian Block: the

456

margins and basins off southern Tunisia and Tripolitania. In: *The Ocean Basins and Margins,* vol. 4B: *The Western Mediterranean (eds. A.E.M. Nairn, W.H. Kanes, and F.G. Stehli).* Plenum Press, New York, p. 331-359.

BUROLLET, P.F. 1981. Structure and petroleum potential of the Ionian Sea. *Proc. Deep Offshore Tech. Conf.* (Oct. 1981), vol. 19/22, p. 1-11.

BURWOOD, R., HOCKEY, A., MYCKE, B., SAIDI, M., and GHENIMA, R. 1995. Geochemical evaluation of the Gabes-Tripoli Basin hydrocarbon habitat offshore Tunisia, (abstract only). *Proceedings of the Seminar on Source Rocks and hydrocarbon habitat in Tunisia.* ETAP, Tunis, Memoir No. 9, p. 27.

BURWOOD, R. 1996. Geochemical evaluation of east Sirte Basin. Petroleum systems and oil provenance. *First Magrebhian Conference on Petroleum Exploration, Benghazi, Libya, November 18th-20th, 1996,* 62p.

BURWOOD, R. 1997. Libya: Petroleum systems of the east Sirte Basin, (abstract only). *Bull. Amer. Assoc. Pet. Geol.* vol. 81, p. 1365.

BURWOOD, R. 1998. Petroleum System Assessment for the east Sirte Basin (Cyrenaica) oil province. *Extended abstracts, volume 1, AAPG Annual Convention, Salt Lake City, Utah.* Abstract A99, p. 1-4.

BURWOOD, R., COPE, M. and REDFERN, J. 2000. Petroleum systems of the eastern Sirte Basin, Libya, (abstract only). *Symp. on petrol. systems and evolving technol. in Afr. E&P, Geol. Soc. London, May 2000.* Book of abstracts p. 50-51.

BUSREWIL, M.T. and WADSWORTH, W.J. 1980a. Preliminary chemical data on the volcanic rocks of Al Haruj area, central Libya. *Second Symposium on the Geology of Libya.* vol. 3 *(eds. M.J. Salem and M.T. Busrewil),* Academic Press, London, p. 1077-1080.

BUSREWIL, M.T. and WADSWORTH, W.J. 1980b. The basanitic volcanoes of the Gharyan area, NW Libya. *Second Symposium on the Geology of Libya.* vol. 3 *(eds. M.J. Salem and M.T. Busrewil),* Academic Press, London, p. 1095-1106.

BUSREWIL, M.T. and OUN, K.M. 1991. Geochemistry of the Tertiary alkaline rocks of Jabal al Hasawinah, west central Libya. *Third Symposium on the Geology of Libya.* vol. 7 *(eds. M.J. Salem and M.T. Busrewil and A.M. Ben Ashour),* Elsevier, Amsterdam, p. 2587-2598.

BUSREWIL, M.T. and ESSON, J. 1991. Chronology and composition of the igneous rocks of Jabal as Sawda. *Third Symposium on the Geology of Libya.* vol. 7 *(eds. M.J. Salem and M.T. Busrewil and A.M. Ben*

Ashour), Elsevier, Amsterdam, p. 2599-2604.

BUSREWIL, M.T. and SUWESI, K.S. 1993. Sheet Al Haruj al Aswad (NG 33-4), *Geological Map of Libya, scale 1:250,000, Explanatory Booklet,* Industrial Research Centre, Tripoli.

BUSREWIL, M.T., MRIHEEL, I.Y. and AL FASATWI, Y.A. 1996. Volcanism, tectonism, and hydrocarbon potential of parts of Al Haruj area, SW Sirt Basin, Libya. *First Symposium on the Sedimentary Basins of Libya, Geology of the Sirt Basin.* vol. 3. *(eds. M.J. Salem, M.T. Busrewil, A.A. Misallati and M.J. Sola),* Elsevier, Amsterdam, p. 317-329.

BUSREWIL, M.T. 1996. The volcanology of central Jabal al Haruj al Aswad volcanic province, central Libya. *First Symposium on the Sedimentary Basins of Libya, Geology of the Sirt Basin.* vol. 3. *(eds. M.J. Salem, M.T. Busrewil, A.A. Misallati and M.J. Sola),* Elsevier, Amsterdam, p. 330-346.

BUSSON, G. 1967a. Mesozoic of southern Tunisia. In: *Guidebook to the geology and history of Tunisia (ed. Martin, L.).* Petroleum Exploration Society of Libya, Ninth Annual Field Conference, p. 131-152.

BUSSON, G. 1967b. Le Mesozoique en Djeffara libyenne et dans la falaise du Djebel Nefoussa. In: *Le Mesozoique Saharien. Prem. Partie: L'Extreme-Sud tunisien. Cent. Rech. Zones Arides. Ser. Geol.,* No. 8. Edits. Cent. Nat. Rech. Sci. Paris. Chap. XII, p. 137-159.

BUSSON, G. and CORNEE, A. 1991. The Sahara from the middle Jurassic to the middle Cretaceous: data on environments and climates based on outcrops in the Algerian Sahara. *Jour. African Earth Sci.,* vol 12, p. 85-195.

BUTT, A.A. 1986. Upper Cretaceous biostratigraphy of the Sirte Basin, northern Libya. *Rev. Paleobiol.* vol. 5, p. 175-191.

BYRAMJEE, R. and VASSA, L. 1969. Geochemical interpretation of Libyan and north Saharan crude oil analyses. In: *1968 Conference on Advances in Organic Geochemistry.* Pergamon Press, Oxford, p. 319-330.

CAHEN, L., SNELLING, N.J., DELHAL, J. and VAIL, J.R. 1984. North-east Africa and Arabia. In: *The geochronology and evolution of Africa.* Clarendon Press, Oxford. chap. 14, p. 254-269.

CAMPBELL, C.J. 1991. The golden century of oil 1950-2050. Kluwer Academic Publishers, Dordrecht, 345p.

CAMPBELL, C.J. 1997. The coming oil crisis. Multi-Science Publishing, Brentwood, 210p.

CAPUTO, M.V. and CROWELL, J.C. 1985. Migration of glacial centres across Gondwana during the Paleozoic era. *Bull. Geol. Soc. Am.* vol. 96, p. 1020-1036.

CARMIGNANI, L. 1984. Sheet Wadi al Hamim (NH 34-7), *Geological Map of Libya, scale 1:250,000, Explanatory Booklet,* Industrial Research Centre, Tripoli.

ČEPEK, P. 1979. Sheet Al Qaryat ash Sharqiyah (NH 33-6), *Geological Map of Libya, scale 1:250,000, Explanatory Booklet,* Industrial Research Centre, Tripoli.

ČEPEK, P. 1980. The sedimentology and facies development of the Hasawnah Formation in Libya. *Second Symposium on the Geology of Libya.* vol. 2 *(eds. M.J. Salem and M.T. Busrewil),* Academic Press, London, p. 375-382.

CHAABANI, F., EL MANAAI, M., SOUISSI, F., SASSI-SOUISSI, R. BEN MAMMOU, A. and SASSI, S. 2000. Dolomitization and calcitization stages in rocks of the Nalut Formation, Jabal Nafusah, NW Libya (abstract only). *Second Symposium on the Sedimentary Basins of Libya, Geology of Northwest Libya.* Book of abstracts, p. 25.

CHALOUPSKY, J. 1979. Sheet Shawa (NH 32-8), *Geological Map of Libya, scale 1:250,000, Explanatory Booklet,* Industrial Research Centre, Tripoli.

CHAOUCHI, R., MALLA, M.S. and KECHOU, F. 1998. Sedimentological evolution of the Givetian-Eifelian (F3) sand bar of the West Alrar field, Illizi Basin, Algeria. In: *Petroleum Geology of North Africa. (ed. D.S. Macgregor, R.T.J. Moody, D.D. Clark-Lowes),* Geol. Soc. Special Publication No. 132, p. 187-200.

CHATELIER, J. and SLEVIN, A. 1988. Review of African petroleum and gas deposits. *Journ. African Earth Sci.* vol 7, p. 561-578.

CHRISTIE, A.M. 1955. Geology of the Gharian map area, Tripolitania, with logs of the wells drilled by LATAS in Tripolitania between March 1953 and March 1954. *United Nations, Bull. No. 5, Ministry of Industry,* L.A.R., Tripoli, 64p.

CLARK-LOWES, D.D. 1978. Depositional environment of the lower and middle Devonian Tadrart and Uan Casa Formations of the southwest Fezzan and their relationship to the underlying Silurian deposits (abstract only). *Second Symposium on the Geology of Libya,* abstracts, p. 21-23.

CLARK-LOWES, D.D. and WARD, J. 1991. Palaeoenvironmental evidence from the Palaeozoic 'Nubian sandstones' of the Sahara. *Third Symposium on the Geology of Libya.* vol. 6 *(eds. M.J. Salem, A.M. Sbeta and M.R. Bakbak),* Elsevier, Amsterdam, p. 2099-2154.

CLAYTON, G. and LOBOZIAK, S. 1985. Early Carboniferous (Early Visean-Serpukhovian) palynomorphs. In: *The palynostratigraphy of north-east Libya (eds. B. Thusu and B. Owens).* Journ. Micropalaeontology, vol. 4, p. 83-92.

CLIFFORD, A.C. 1986. African oil – past, present and future. In: *Future petroleum provinces of the world (Ed. M.T. Halbouty).* Amer. Assoc. Pet. Geol. Mem. 40, p. 339-372.

CLIFFORD, H.J., GRUND, R. and MUSRATI, H. 1980. Geology of a stratigraphic giant: Messla oil field, Libya. In: *Giant oil and gas fields of the decade 1968-1978. (ed. M.T. Halbouty).* Mem. Amer. Assoc. Pet. Geol. No. 30, p. 507-524.

COLLOMB, G.R. 1962. Etude geologique du Jebel Fezzan et de sa bordure Palaeozoique. *Com. Fran. du Pet, Notes et Mem.* no. 1, p. 7-35.

CONANT, L.C. and GOUDARZI, G.H. 1964. Geological map of the Kingdom of Libya. *U.S. Geol. Surv. Misc. Geol. Inv.* Map 1-350A, scale 1:2,000,000. Washington.

CONANT, L.C. and GOUDARZI, G.H. 1967. Stratigraphic and tectonic framework of Libya. *Bull. Amer. Assoc. Pet. Geol.* vol. 51, p. 719-730.

CONLEY, C.D. 1971. Stratigraphy and lithofacies of Lower Paleocene rocks, Sirte Basin, L.A.R. *First Symposium on the Geology of Libya (ed. C. Gray).* Faculty of Science, University of Libya, Tripoli, p. 127-140.

CRAIG, J., LÜNING, S. and ZANELLA, E. 2000. Regional overview of the geology and petroleum prospectivity of north Africa (abstract only). *Int. workshop on north African micropalaeo. for pet. expl., Univ. Coll. London, August 2000.* Book of abstracts, p. 29-31.

CREMA, C. 1926. Sulle manifestazioni di idrocarburi del pozzo artesiano di Sidi-Mesri presso Tripoli. *La Miniera Italiano,* vol. 10, p. 49-50.

CROSSLEY, R. and McDOUGALL, N. 1998. Lower Palaeozoic reservoirs of North Africa. In: *Petroleum Geology of North Africa. (ed. D.S. Macgregor, R.T.J. Moody, D.D. Clark-Lowes),* Geol. Soc. Special Publication No. 132, p. 157-166.

DABER, R. 1971. *Cooksonia,* one of the most ancient Psilophyta, widely distributed but rare. *Botanique,* vol. 2, p. 35-40

DALLONI, M. 1934. Mission au Tibesti (1930-31). *Mem. Acad. Sci. Inst. Fr.,* Ser. 2, vol. 61, 372p.

DANIELS, R.P. and EMME, J.J. 1995. Petroleum system model, eastern Algeria from source rock to accumulation: when, where and how? Proc. *Seminar on source rocks and hydrocarbon habitat in Tunisia, Tunis,* 1995. ETAP Memoir 9, p. 101-123.

DAVIDSON, L., BESWETHERICK, S., CRAIG, J., EALES, M., FISHER, A., HIMMALI, A., JHO, J.,

MEJRAB, B. and SMART, J. 2000. The structure, stratigraphy and petroleum geology of the Murzuq Basin, southwest Libya. *Symposium on Geological Exploration in Murzuq Basin (Eds. M.A. Sola and D. Worsley)*, Elsevier, Amsterdam, p. 295-320.

DE CASTRO, J.C., DELLA FAVERA, J.C. and EL-JADI, M. 1991. Tempestite facies, Murzuq Basin, Great Socialist People's Libyan Arab Jamahiriya: their recognition and stratigraphic implications. *Third Symposium on the Geology of Libya.* vol. 5 *(eds. M.J. Salem and M.N. Belaid)*, Elsevier, Amsterdam, p. 1757-1766.

DE HEINZELIN, J., EL-ARNAUTI, A. and GAZIRY, W. 1980. A preliminary revision of the Sahabi Formation. *Second Symposium on the Geology of Libya.* vol. 1 *(eds. M.J. Salem and M.T. Busrewil)*, Academic Press, London, p. 127-136.

DE LAPPARENT, A.F. and LELUBRE, M. 1948. Interpretation stratigraphique des series continentales entre Ohanet et Bourarhet (Sahara central). *C.R. Acad. Sci. Paris.* vol. 227, p.1106-1108.

DE LAPPARENT, A.F. 1952. Stratigraphie du Trias de la Jeffara (extreme sud Tunisien et Tripolitaine). *Proc. 19th Int. Geol. Cong.*, Alger. No. 21, p. 129-134.

DE LAPPARENT, A.F. 1958. Sur les dinosauriens du 'Continental Intercalaire' du Sahara central. *C.R. Acad. Sci. Paris.* vol. 246, p. 1237-1240.

DE LESTANG, J. 1965. Das Paleozoikum am Rande des Afro-Arabischen Gondwana-Kintinents. Mitteilung ubar das Erdi-Becken (Republic Tschad). *Zeit. Deutsch. Geol. Ges.* Vol. 117, p. 479-488.

DE WIT, M., JEFFERY, M., BERGH, H. and NICOLAYSEN, L. 1988. Geological map of sectors of Gondwana reconstructed to their dispositions 150 Ma. *University of Witwatersrand.* Published by Amer. Assoc. Pet. Geol.

DEL BEN, A. and FINETTI, I. 1991. Geophysical study of the Sirt Rise. *Third Symposium on the Geology of Libya.* vol. 6 *(eds. M.J. Salem, A.M. Sbeta and M.R. Bakbak)*, Elsevier, Amsterdam, p. 2417-2432.

DERCOURT, J., ZONENHAIN, L.P., RICOU, L.E., KAZMIN, V.G., PICHON, X., KNIPPER. A.L., GRANDJACQUET, C., SBORTCHIKOV, I.M., GEYSSANT, J., LEPVRIER, D.H. BOULIN, J., SIBUET, J.C., SAVOSTIN, O., WESTPHAL, M., BAZHENOV, M.L., LAUER, J.P., and BIJU-DUVAL, B. 1986. Geological evolution of the Tethys belt from the Atlantic to the Pamirs since the Lias. *Tectonophysics*, vol 123. p. 241-315.

DERCOURT, J., RICOU, L.E. and VRIELYNCK, B., (Eds.) 1993. Atlas Tethys Palaeo-environmental maps. Gauthier-Villars, Paris.

DESIO, A. 1928. Resultati scientific della missione alla Oasi di Giarabub (1926-27), Pt. 2: La Geologia. *Pub. R. Soc. Geog. Ital,* p. 81-164.

DESIO, A. 1935. Studi geologici sulla Cirenaica, sul deserto Libico, sulla Tripolitania e sul Fezzan orientale. *Missione scientifica della R. Accad. d'Italie a Cufra, (1931)*, Roma., vol. 1. 480p.

DESIO, A. 1936a. Prime notizie sulla presenza del Silurico fossilifero nel Fezzan. *Boll. Soc. Geol. Ital.* vol. 55, p.116-120.

DESIO, A. 1936b, Riassunto sulla presenza del Silurico fossilifero nel Fezzan. *Boll. Soc. Geol. Ital.* vol. 55, p. 319-356.

DESIO, A. 1943. L'esplorazione mineraria della Libia. *Coll. Sci. Doc., Min. Afr. Ital.*, vol. 10. Milan. 333p.

DESIO, A., RONCHETTI, R.C. and INVERNIZZI, G. 1960a. Il Giurassico dei dintorni di Jefren in Tripolitania. *Riv. Ital. Paleont. Strat.* vol. 66, p. 65-118.

DESIO, A. and RONCHETTI, C. 1960. Sul Giurassico medio di Garet al-Bellaa (Tripolitania) e sulla posizione stratigrafica della formazione di Tachbal. Riv. *Ital. Paleont. Strat,* vol. 66, p. 173-196.

DESIO, A., RONCHETTI, R.C. and VIGANO, P.L. 1960b. Sulla stratigrafia del Trias in Tripolitania e nel sud-Tunisino. *Riv. Ital. Paleont. Strat,* vol. 66, p. 273-322.

DESIO, A., RONCHETTI, R.C., POZZI, R., CLERICI, F., INVERNIZZI, G., PISONI, C. and VIGANO, P.L. 1963. Stratigraphic studies in the Tripolitanian Jebel (Libya). *Mem. Riv. Ital. Paleont.* vol. 9, 126p.

DESIO, A. 1968. History of geologic exploration in Cyrenaica. In: *Geology and archaeology of northern Cyrenaica, Libya (ed. Barr, F.T.).* Petroleum Exploration Society of Libya, Tenth Annual Field Conference, p. 79-114.

DESIO, A. 1971. Outlines and problems of the geomorphological evolution of Libya from the Tertiary to the present day. *First Symposium on the Geology of Libya (ed. C. Gray).* Faculty of Science, University of Libya, Tripoli, p. 11-36.

DEUNFF, J.M.M. and MASSA, D, 1975. Palynologie et stratigraphie du Cambro-Ordovicien (Libye nord-occidentale). *C.R. Acad. Sci., Paris.* Ser. D, vol. 281, p.21-24.

DEWEY, J.F., HELMAN, M.L., TURCO, E., HUTTON, D.H.W. and KNOTT, S.D. 1989. Kinematics of the western Mediterranean. In: *Alpine Tectonics (ed. M.P. Coward, D. Dietrich and R.G. Park)*, Geol. Soc. Special Publication No. 45, p. 265-283.

DEYNOUX, M., GHIENNE, J.F., MANATSCHAL, G., RUBINO, J.L., BLANPIED, C., TAWENGI, K. and

GUILLANDE, R. 2000. Glacial valleys within the lower Mamuniyat unit in western Al Qarqaf Arch (abstract only). *Second Symposium on the Sedimentary Basins of Libya, Geology of Northwest Libya.* Book of abstracts, p. 27.

DI CESARE, F., FRANCHINO, A. and SOMM-ARUGA, C. 1963. The Pliocene-Quaternary of Giarabub Erg region. *First Saharan Symposium.* Rev. Inst. Fran. Pet. special volume, p. 30-48.

DIEB, M. and HROUDA, M. 2000. The hydrocarbon potential of Aounet Ouenine Formation, Ghadamis Basin, western Libya (abstract only). *Second Symposium on the Sedimentary Basins of Libya, Geology of Northwest Libya.* Book of abstracts, p. 28.

DJARNIA, M.R. and FEKIRINE, B. 1998. Sedimentological and diagenetic controls on Cambro-Ordovician reservoir quality in the southern Hassi Messaoud area (Saharan Platform, Algeria). In: *Petroleum Geology of North Africa. (ed. D.S. Macgregor, R.T.J. Moody, D.D. Clark-Lowes),* Geol. Soc. Special Publication No. 132, p. 167-175.

DOMÁCÍ, L. 1985. Sheet Bi'r Zaltan (NH 34-14), *Geological Map of Libya, scale 1:250,000, Explanatory Booklet,* Industrial Research Centre, Tripoli.

DOMÁCÍ, L., RÖHLICH, P., and BOSÁK, P. 1996. Neogene to Pleistocene continental deposits in northern Fazzan and the central Sirt Basin. *Third Symposium on the Geology of Libya.* vol. 5 *(eds. M.J. Salem and M.N. Belaid),* Elsevier, Amsterdam, p. 1785-1802.

DUBOIS, P., BEUF, S., and BIJU-DUVAL, B. 1969. Lithostratigraphy of Lower Devonian sandstones of the Tassili N'Ajjer. In: *Geology, archaeology and prehistory of the southwestern Fezzan, Libya (ed. Kanes, W.H.).* Petroleum Exploration Society of Libya, Eleventh Annual Field Conference, p. 125-130.

DURONIO, P., DAKSHE, A. and BELLINI, E. 1991. Stratigraphy of the offshore Cyrenaica (Libya). 1991. *Third Symposium on the Geology of Libya.* vol. 4 *(eds. M.J. Salem, O.S. Hammuda and B.A. Eliagoubi),* Elsevier, Amsterdam, p. 1589-1620.

ECHIKH, K. 1975. Geologie des provinces petroliferes de l'Algerie. S.N.E.D. Alger.

ECHIKH, K. 1998. Geology and hydrocarbon occurrences in the Ghadames Basin, Algeria, Tunisia, Libya. In: *Petroleum Geology of North Africa. (ed. D.S. Macgregor, R.T.J. Moody, D.D. Clark-Lowes),* Geol. Soc. Special Publication No. 132, p. 109-130.

ECHIKH, K. and SOLA, M.A. 2000. Geology and hydrocarbon occurrences in the Murzuq Basin, SW Libya. *Symposium on Geological Exploration in Murzuq Basin (eds. M.A. Sola and D. Worsley),* Elsevier, Amsterdam, p. 175-222.

EDMUNDS, W.M. 1980. The hydrogeochemical characterization of ground waters in the Sirt Basin using strontium and other elements. *Second Symposium on the Geology of Libya.* vol. 2 *(eds. M.J. Salem and M.T. Busrewil),* Academic Press, London, p. 703-714.

EL-ALAMI, M., RAHOUMA, S. and BUTT, A.A. 1989. Hydrocarbon habitat in the Sirte Basin northern Libya. *Pet. Res. Journ.* Tripoli, vol. 1. p. 17-28.

EL-ALAMI, M.A. 1996a. Petrography and reservoir quality of the Lower Cretaceous sandstone in the deep Mar Trough, Sirt Basin. *First Symposium on the Sedimentary Basins of Libya, Geology of the Sirt Basin.* vol. 2. *(eds. M.J. Salem, A.S. El-Hawat and A.M. Sbeta),* Elsevier, Amsterdam, p. 309-322.

EL-ALAMI, M.A. 1996b. Habitat of oil in Abu Attiffel area, Sirt Basin, Libya. *First Symposium on the Sedimentary Basins of Libya, Geology of the Sirt Basin.* vol. 2. *(eds. M.J. Salem, A.S. El-Hawat and A.M. Sbeta),* Elsevier, Amsterdam, p. 337-348.

EL-ARNAUTI, A. and SHELMANI, M. 1985. Stratigraphic and structural setting. In: *The palynostratigraphy of north-east Libya (eds. B. Thusu and B. Owens).* Journ. Micropalaeontology, vol. 4, p. 1-10.

EL-ARNOUTI, A and SHELMANI, M. 1988. A contribution to the northeast Libya subsurface stratigraphy with emphasis on pre-Mesozoic. In: *Subsurface palynostratigraphy of northeast Libya. (eds. A. El-Arnauti, B. Owens and B. Thusu).* Research Centre, Garyounis University, Benghazi, p. 1-17.

EL-BAKAI, M.T. 1996. Diagenesis and diagenetic history of the Lidam Formation in NW Sirt Basin. *First Symposium on the Sedimentary Basins of Libya, Geology of the Sirt Basin.* vol. 2. *(eds. M.J. Salem, A.S. El-Hawat and A.M. Sbeta),* Elsevier, Amsterdam, p. 83-96.

EL-BAKAI, M.T. 2000. Petrography, geochemistry and stable isotope constraints on the origin of the Cretaceous dolomite (Ain Tobi Member), NW Libya (abstract only). *Second Symposium on the Sedimentary Basins of Libya, Geology of Northwest Libya.* Book of abstracts, p. 30.

EL-BATROUKH, S.I. and ZENTANI, A.S. 1980a. The geological interpretation of a gravity map of the northern part of Maradah Graben (Sirt Basin, Libya). *Second Symposium on the Geology of Libya.* vol. 3 *(eds. M.J. Salem and M.T. Busrewil),* Academic Press, London, p. 965-978.

EL-BATROUKH, S.I. and ZENTANI, A.S. 1980b.

Gravity interpretation of Raguba field, Sirte Basin, Libya. *Geophysics*, vol. 45, p. 1154-1163.

EL-DEFTAR, T. and ISSAWI, B. 1977. Sheet Al Bardia (NH 35-1), *Geological Map of Libya, scale 1:250,000, Explanatory Booklet*, Industrial Research Centre, Tripoli.

EL-GHOUL, A. 1991. A modified Farwah Group type section and its application to understanding stratigraphy and sedimentation along an E-W section through NC 35A, Sabratah Basin. *Third Symposium on the Geology of Libya*. vol. 4 *(eds. M.J. Salem, O.S. Hammuda and B.A. Eliagoubi)*, Elsevier, Amsterdam, p. 1637-1656.

EL-GHOUL, A. 1996. An approach to locate subtle Waha structural traps on the Zaltan Platform: geology and geophysics. *First Symposium on the Sedimentary Basins of Libya, Geology of the Sirt Basin*. vol. 3. *(eds. M.J. Salem, M.T. Busrewil, A.A. Misallati and M.J. Sola)*, Elsevier, Amsterdam, p. 137-154.

EL-HADDAD, A., EL-HODAIRI, A. and EL-CHAIR, M. 2000. Sedimentology and Cu-U mineralisation of the upper Cretaceous Bin Affin Member, Dur Waddan, southwestern El Haruj, Murzuq Basin, Libya. *Symposium on Geological Exploration in Murzuq Basin (eds. M.A. Sola and D. Worsley)*, Elsevier, Amsterdam, p. 369-378.

EL-HADDAD, A., EL-HODAIRI, A. and ALI, O.E. 2000. Depositional environments of the Jurassic Bir el-Ghanum evaporites, Jabal Nafusa, NW Libya (abstract only). *5th Intern. conf. on the geology of the Arab world*, Cairo University. Book of abstracts, p. 156.

EL-HARBI, A. 2000. Lower Carboniferous miospores from surface samples of the Marar Formation, Ghadamis Basin, NW Libya (abstract only). *Second Symposium on the Sedimentary Basins of Libya, Geology of Northwest Libya*. Book of abstracts, p. 35.

EL-HARBI, A. and ABUHAMIDA, F. 2000. Palynostratigraphy of the uppermost Devonian strata in well A1-NC 40B, Ghadamis Basin, NW Libya (abstract only). *Second Symposium on the Sedimentary Basins of Libya, Geology of Northwest Libya*. Book of abstracts, p. 35.

EL-HAWAT, A. 1980a. Carbonate-terrigenous cyclic sedimentation and palaeogeography of the Marada Formation (Middle Miocene), Sirt Basin. *Second Symposium on the Geology of Libya*. vol. 2 *(eds. M.J. Salem and M.T. Busrewil)*, Academic Press, London, p. 427-448.

EL-HAWAT, A. 1980b. Intertidal and storm sedimentation from Wadi al Qattarah Member, Ar Rajmah Formation (Middle Miocene), Al Jabal al Akhdar.

Second Symposium on the Geology of Libya. vol. 2 *(eds. M.J. Salem and M.T. Busrewil)*, Academic Press, London, p. 449-462.

EL-HAWAT, A.S and SALEM, M.J. 1987. A case study of the stratigraphic subdivision of Ar Rajmah Formation and its implications on the Miocene of northern Libya. In: P*roc. VIIIth Cong. Med. Neogene Strat., Budapest*. Ann. Inst. Geol. Hungary, vol. 70, p. 173-184.

EL-HAWAT, A. 1992. The Nubian Sandstone sequence in Sirte Basin, Libya: sedimentary facies and events. In: *Geology of the Arab World*, vol. 1, *(ed. A. Sadek)*. Cairo Univ. p. 317-327.

EL-HAWAT, A. and SHELMANI, M.A. 1993. Short notes and guidebook on the geology of Al Jabal al Akhdar Cyrenaica, NE Libya. *First Symposium on the Sedimentary Basins of Libya, Geology of the Sirt Basin*, Field Guide, Earth Sciences Society of Libya, 69p.

EL-HAWAT, A., MISSALLATI, A.A., BEZAN, A.M. and TALEB, T.M. 1996. The Nubian Sandstone in Sirt Basin and its correlatives. *First Symposium on the Sedimentary Basins of Libya, Geology of the Sirt Basin*. vol. 2. *(eds. M.J. Salem, A.S. El-Hawat and A.M. Sbeta)*, Elsevier, Amsterdam, p. 3-20.

EL-HAWAT, A. 1997. Sedimentary basins of Egypt: an overview of dynamic stratigraphy. In: *Sedimentary Basins of the World (ed. K.J. Hsu), Part 4 African Basins (ed. R.C. Selley)*. Elsevier, Amsterdam, chap. 4, p. 39-85.

EL-HAWAT, A.S. and BEZAN, A.M. 1998. Early Palaeozoic event stratigraphy of western Libya: glacio-eustatic and tectonic signatures and their impact on hydrocarbon exploration (abstract only). *Symposium on Geological Exploration in Murzuq Basin, Sabha, 20-22nd September 1998*. Book of abstracts.

EL-HAWAT, A.S. BEZAN, A.M., OBEIDI, A and BARGHATHI, H. 2000. The Upper Ordovician-Lower Silurian succession in western Libya: sequence stratigraphy and glacioeustatic-tectonic scenario (abstract only). *Second Symposium on the Sedimentary Basins of Libya, Geology of Northwest Libya*. Book of abstracts, p. 36.

EL-HINNAWY, M. and CHESHITEV, G. 1975. Sheet Tarabulus (NI 33-13), Geological Map of Libya, scale 1:250,000, Explanatory Booklet, Industrial Research Centre, Tripoli.

EL-KHOUDARY, R.H. 1980. Planktonic foraminifera from the Middle Eocene of the northern escarpment of Al Jabal al Akhdar, NE Libya. *Second Symposium on the Geology of Libya*. vol. 1 *(eds. M.J. Salem and*

M.T. Busrewil), Academic Press, London, p. 193-204.

EL-MAKHROUF, A.A. 1996. The Tibisti-Sirt orogenic belt, Libya, G.S.P.L.A.J. *First Symposium on the Sedimentary Basins of Libya, Geology of the Sirt Basin.* vol. 3. *(eds. M.J. Salem, M.T. Busrewil, A.A. Misallati and M.J. Sola)*, Elsevier, Amsterdam, p. 137-154.

EL-MAKHROUF, A. A. and FULLAGAR, P.D. 2000. The rubidium-strontium geochronology of the Pan-African post-orogenic granites of the eastern Tibisti orogenic belt, Tibisti massif, south-central Libya: application to origin and tectonic evolution. *Symposium on Geological Exploration in Murzuq Basin (Eds. M.A. Sola and D. Worsley)*, Elsevier, Amsterdam, p. 379-396.

EL-MEHAGHAG, A.A. and MUFTAH, A.M. 1996. Calcareous nannofossils and foraminiferal biostratigraphy of Al Hilal Formation north-east Libya. *3rd Internat. Conf. on the geology of the Arab world.* Cairo University. Proceedings, p. 501-520.

EL-MEHAGHAG, A.A. 2000. Re-evaluation of Al Athrun and Apollonia formations based on calcareous nanno-fossils, Cyrenaica, north-east Libya (abstract only). *5th Intern. conf. on the geology of the Arab world*, Cairo University. Book of abstracts, p. 254.

EL-MEHAGHAG, A.A. and BADI, S. 2000. Nannofossil biostratigraphy of Al Uwayliah Formation at Jabal al Akhdar area, north east Libya (abstract only). *5th Intern. conf. on the geology of the Arab world*, Cairo University. Book of abstracts, p. 253.

EL-MEHAGHAG, A.A. and DAW, S. 2000. Nannifossil biostratigraphy of Tocra Formation, Cyrenaica, north-east Libya (abstract only). *5th Intern. conf. on the geology of the Arab world*, Cairo University. Book of abstracts, p. 253.

EL-MEHDAWI, A.D. 1991. Preliminary palynological study of the Upper Cretaceous Al Hilal Formation, Ra's al Hilal area, NE Libya. *Third Symposium on the Geology of Libya.* vol. 4 *(eds. M.J. Salem, O.S. Hammuda and B.A. Eliagoubi)*, Elsevier, Amsterdam, p. 1351-1356.

EL-MEHDAWI, A.D. 2000. Palynology of the upper Tahara Formation in concession NC 7A, Ghadames Basin. *Symposium on Geological Exploration in Murzuq Basin (eds. M.A. Sola and D. Worsley)*, Elsevier, Amsterdam, p. 273-294.

EL-MEHDAWI, A.D. 2000. Palynology of the Lower Acacus Member, Al Hamadah al Hamra area (Ghadamis Basin) (abstract only). *Second Symposium on the Sedimentary Basins of Libya, Geology of Northwest Libya.* Book of abstracts, p. 40.

EL-RAMLY, I.M. 1980. Al Kufrah Pleistocene lake – its evolution and role in present-day land reclamation. *Second Symposium on the Geology of Libya.* vol. 2 *(eds. M.J. Salem and M.T. Busrewil)*, Academic Press, London, p. 659-670.

EL-RWEIMI, W.S. 1991. Geology of the Aouinet Ouenine and Tahara Formations, Al Hamada al Hamra area, Ghadamis Basin. *Third Symposium on the Geology of Libya.* vol. 6 *(eds. M.J. Salem, A.M. Sbeta and M.R. Bakbak)*, Elsevier, Amsterdam, p. 2185-2194.

EL-SOGHER, A. 1996. Late Cretaceous and Palaeocene ostracods from the Waha Limestone and Hagfa Shale Formations of the Sirt Basin, Libya. *First Symposium on the Sedimentary Basins of Libya, Geology of the Sirt Basin.* vol. 1. *(eds. M.J. Salem, A.J. Mouzughi and O.S. Hammuda)*, Elsevier, Amsterdam, p. 287-382.

EL-WAER, A.A. 1991. Miocene ostracoda from Al Khums Formation, northwestern Libya. *Third Symposium on the Geology of Libya.* vol. 4 *(eds. M.J. Salem, O.S. Hammuda and B.A. Eliagoubi)*, Elsevier, Amsterdam, p. 1457-1482.

EL-WERFALLI, H.O. and STOW, D.A.V. 2000. Diagenesis of the lower Tertiary carbonates of NE Libya: origin of dolomites (abstract only). *5th Intern. conf. on the geology of the Arab world*, Cairo University. Book of abstracts, p. 158.

EL-ZOUKI, A.Y. 1980a. Stratigraphy and lithofacies of the continental clastics (Upper Jurassic and Lower Cretaceous) of Jabal Nafusah, NW Libya. *Second Symposium on the Geology of Libya.* vol. 2 *(eds. M.J. Salem and M.T. Busrewil)*, Academic Press, London, p. 393-418.

EL-ZOUKI, A.Y. 1980b. Depositional environment of the Chicla, Cabao, and Chameau Mort Formations as revealed by scanning electron microscopy. *Second Symposium on the Geology of Libya.* vol. 2 *(eds. M.J. Salem and M.T. Busrewil)*, Academic Press, London, p. 419-426.

EL-ZOUKI, A. and ALIAN, A. 1991. Application of neutron activation methods for trace element analysis in the Nubian Sandstone of southern Libya. *Third Symposium on the Geology of Libya.* vol. 5 *(eds. M.J. Salem and M.N. Belaid)*, Elsevier, Amsterdam, p. 1745-1756.

ELAG, M.O. 1996. Sedimentological study of the Facha Member in the southwest Sirt Basin, Libya. *First Symposium on the Sedimentary Basins of Libya, Geology of the Sirt Basin.* vol. 2. *(eds. M.J. Salem, A.S. El-Hawat and A.M. Sbeta)*, Elsevier, Amsterdam, p. 99-114.

ELAZEZI, M.M. and ASHCROFT, W.A. 1996. Modelling

of the seismic expression of the Upper Cretaceous and Precambrian reservoirs in the An Nafurah-Awjilah oil field, Sirt Basin, Libya. *First Symposium on the Sedimentary Basins of Libya, Geology of the Sirt Basin.* vol. 3. *(eds. M.J. Salem, M.T. Busrewil, A.A. Misallati and M.J. Sola)*, Elsevier, Amsterdam, p. 267-282.

ELFIGIH, O.B. 2000. Regional diagenesis and its relation to facies change in the Lower Acacus Formation, Al Hamadah Basin, NW Libya (abstract only). *Second Symposium on the Sedimentary Basins of Libya, Geology of Northwest Libya.* Book of abstracts, p. 33.

ELIAGOUBI, B.A. 1972. Vindobonian (Miocene) foraminifera from the Faidia Formation, Um Errzen region, N.E Libya. *M.S. thesis*, Idaho University.

ELIAGOUBI, B.A. 1980. Planktonic foraminifera of the Palaocene Al Uwaliah Formation at its type locality – northeastern Libya. *Second Symposium on the Geology of Libya.* vol. 1 *(eds. M.J. Salem and M.T. Busrewil)*, Academic Press, London, p. 155-162.

ELIAGOUBI, B.A. and POWELL, J.D. 1980. Biostratigraphy and palaeoenvironment of Upper Cretaceous (Maastrichtian) foraminifera of north-central and northwestern Libya. *Second Symposium on the Geology of Libya.* vol. 1 *(eds. M.J. Salem and M.T. Busrewil)*, Academic Press, London, p. 137-154.

ELZAROUG, R. and LASHHAB, M.I. 1998. Palynostratigraphy and palynofacies of subsurface Devonian (Middle-Upper) strata of Al Wafa field (abstract only). *Symposium on Geological Exploration in Murzuq Basin, Sabah, 20th-22nd September, 1998.* Book of abstracts, p. 38.

ERBA, M., ROVERE, A., TURRIANI, C. and SAFSAF, S. 1984. Bu Attifel field – a synergetic geological and engineering approach to reservoir management. *Proc. 11th World Petrol. Cong.* vol. 3, p. 89-99.

ETTALHI, J.A., KROKOVIC, D. and BANERJEE, S. 1978. Bibliography of the Geology of Libya. *Bulletin No. 11.* Industrial Research Centre, Tripoli, 135p.

EXPLORATION STAFF OF THE ARABIAN GULF OIL COMPANY, 1980. Geology of a stratigraphic giant – the Messlah oil field. *First Symposium on the Sedimentary Basins of Libya, Geology of the Sirt Basin.* vol. 2. *(eds. M.J. Salem, A.S. El-Hawat and A.M. Sbeta)*, Elsevier, Amsterdam, p. 521-538.

FATMI, A.N., SBETA, A.M., and ELIAGOUBI, B.A. 1978. Guide to the Mesozoic stratigraphy of Jabal Nefusa, Libyan Jamahiriya. *Publication No. 7*, Arab Development Inst., Tripoli. 35p.

FATMI, A.N., ELIAGOUBI, B.A. and HAMMUDA, O.S. 1980. Stratigraphic nomenclature of the pre-Upper Cretaceous Mesozoic rocks of Jabal Nafusah, NW. Libya. *Second Symposium on the Geology of Libya.* vol. 1 *(eds. M.J. Salem and M.T. Busrewil)*, Academic Press, London, p. 57-66.

FATMI, A.N. and SBETA, A.M. 1991. The significance of the occurrence of Abu Ghaylan and Kiklah Formations east of Wadi Ghan, eastern Jabal Nafusah. *Third Symposium on the Geology of Libya.* vol. 6 *(eds. M.J. Salem, A.M. Sbeta and M.R. Bakbak)*, Elsevier, Amsterdam, p. 2227-2234.

FEKIRINE, B. and ABDALLAH, H. 1998. Palaeozoic lithofacies correlatives and sequence stratigraphy of the Saharan Platform, Algeria. In: *Petroleum Geology of North Africa. (ed. D.S. Macgregor, R.T.J. Moody, D.D. Clark-Lowes)*, Geol. Soc. Special Publication No. 132, p. 97-106.

FELLO, N.M. and TURNER, B.R. 2001. Provenance analysis, tectonism and shifting depositional systems in the NW part of the Murzuq Basin, Libya: implications for hydrocarbon prospectivity (abstract only). *Abstracts, North Africa Research Workshop, Oxford Brookes University, 4th-5th September 2001.*

FINETTI, I. 1982. Structure, stratigraphy and evolution of central Mediterranean. *Boll. Geofis. Teor. Appl.* vol. 24, p. 247-312.

FINETTI, I. 1985. Structure and evolution of the central Mediterranean (Pelagian and Ionian Seas). In: *Geological Evolution of the Mediterranean Basin. (Raimondo Selli Commemorative Volume) (eds. D.J. Stanley and F-C. Wezel)*, Springer-Verlag, Berlin.

FINETTI, I.R. and DEL BEN, A. 2000. Crustal stratigraphy and tectono-dynamics of the Pelagian Sea region from new 'CROP' seismic data (abstract only). *Second Symposium on the Sedimentary Basins of Libya, Geology of Northwest Libya.* Book of abstracts, p. 42.

FISCHER, A.G. 1973. Tethys. In: *Geology of Italy, vol. 1 (ed. Squyres, C.H.)*. Petroleum Exploration Society of Libya, Fifteenth Annual Field Conference, p. 143-164.

FITCHES, R. 1998. The Saharan glaciation – an overview (abstract only). *Symposium on Geological Exploration in Murzuq Basin, Sabha, 20-22nd September 1998.* Book of abstracts, p. 45-46.

FLINN, D., LAGHA, S., ATHERTON, M.P. and CLIFF, R.A. 1991. The rock-forming minerals of the Jabal Arknu and Jabal al Awaynat alkaline ring complexes of SE Libya. *Third Symposium on the Geology of Libya.* vol. 7 *(eds. M.J. Salem and M.T. Busrewil and A.M. Ben Ashour)*, Elsevier, Amsterdam, p. 2539-2558.

FLORIDIA, G.B. 1939. Osservazioni sul Miocene del

dintorni di Homs. *Boll. Soc. Geol. Ital.* vol 58, p. 245-260.

FOUAD, K.M. 1991. Correlation of satellite geoid features and hot-spot activity with the tectonic framework of Libya. *Third Symposium on the Geology of Libya.* vol. 6 *(eds. M.J. Salem, A.M. Sbeta and M.R. Bakbak)*, Elsevier, Amsterdam, p. 2451-2460.

FOURNIÉ, D. 1978. Nomenclature lithostratigraphique des séries du Crétacé supérieur au Tertiaire de Tunisie. *Bull. Cent. Rech. Explor.-Prod. Elf-Aquitaine.* vol. 2, p. 97-148.

FRANCIS, M. and ISSAWI, B. 1977. Sheet Soluq (NH 34-2), *Geological Map of Libya, scale 1:250,000, Explanatory Booklet*, Industrial Research Centre, Tripoli.

FRASER, W.W. 1967. Geology of the Zelten field, Libya, north Africa. *Proc. 7th World Pet. Cong. Mexico*, Elsevier, Amsterdam, vol. 2, p. 259-264.

FREULON, J.M. 1953. Existence d'un niveau a stromatolithes *(Collenia)* dans le Carbonifere marin du Sahara oriental. *C.R. Somm. Seanc. Soc. Geol. Fr.* Nos. 11/12, p. 233-234.

FREULON, J.M. 1964. Etude geologique des series primaires du Sahara central (Tassili n'Ajjer et Fezzan). *Cent. Nat. Rech. Sci.* Paris. Ser. Geol. No. 3, 198p.

FÜRST, M. and KLITZSCH, E. 1963. Late Caledonian paleogeography of the Murzuk Basin. *First Saharan Symp.* Rev. Inst. Fran. Pet., special volume, p. 158-170.

FÜRST, M. 1965. Die Oberkreide-Paleozän transgression im ostlichen Fezzan. *Geol. Rundsch.* vol. 54, p. 1060-1088.

FÜRST, M. 1968. Die Paleozän-Eozän transgression in Sudlibyan. *Geol. Rundsch.* vol. 58, p. 296-313.

FUTYAN, A. and JAWZI, A.H. 1996. The hydrocarbon habitat of the oil and gas fields of north Africa with emphasis on the Sirt Basin. *First Symposium on the Sedimentary Basins of Libya, Geology of the Sirt Basin.* vol. 2. *(eds. M.J. Salem, A.S. El-Hawat and A.M. Sbeta)*, Elsevier, Amsterdam, p. 287-308.

GALEČIČ, M. 1984. Sheet Anay (NG 32-16), *Geological Map of Libya, scale 1:250,000, Explanatory Booklet*, Industrial Research Centre, Tripoli.

GARDINER, B. 1988. A new *Cleithrolepis* from the Triassic of central Cyrenaica, northeast Libya. In: *Subsurface palynostratigraphy of northeast Libya. (eds. A. El-Arnauti, B. Owens and B. Thusu)*. Research Centre, Garyounis University, Benghazi, p. 259-266.

GARDNER, R., HILL, A., KARKI, M., ERKMEN, U., SALEM, F.A. and ELAG, M. 2000. The regional geology and petroleum systems of the north Dahra and Jahama Platforms, north Sirte Basin, Libya (abstract only). *Symposium on petroleum systems and evolving technologies in African E & P.* Burlington House, London, 16-18 May, 2000. Book of abstracts, p. 29.

GAREA, B.B. 1996. Environment of deposition and diagenesis of the Beda reservoir in block NC 74F, SW Sirt Basin, Libya. *First Symposium on the Sedimentary Basins of Libya, Geology of the Sirt Basin.* vol. 2. *(eds. M.J. Salem, A.S. El-Hawat and A.M. Sbeta)*, Elsevier, Amsterdam, p. 115-134.

GAREA, B.B. and BRAITHWAITE, C.J.R. 1996. Geochemistry, isotopic composition and origin of the Beda dolomite, Block NC 74F, Sirte Basin, Libya. *Journ. Pet. Geol.* vol. 19, p. 289-304.

GEISS, E. 1991. Is there plate tectonics in the Mediterranean area? *Third Symposium on the Geology of Libya.* vol. 6 *(eds. M.J. Salem, A.M. Sbeta and M.R. Bakbak)*, Elsevier, Amsterdam, p. 2407-2416.

GENT, M.R. 1991. A preliminary photogeologic investigation of the basement rocks and structures of the Jabal Eghi area, Libya. *Third Symposium on the Geology of Libya.* vol. 6 *(eds. M.J. Salem, A.M. Sbeta and M.R. Bakbak)*, Elsevier, Amsterdam, p. 2295-2318.

GHENIMA, R. 1995. Hydrocarbon generation and migration in the Ghedames Basin application to the filling of the El Borma oil field. *Proceedings of the Seminar on Source Rocks and hydrocarbon habitat in Tunisia.* ETAP, Tunis, Memoir No. 9, p. 3-15.

GHORI, K.A.R. 1991. Petroleum geochemical aspects of Cyrenaica, NE Libya. *Third Symposium on the Geology of Libya.* vol. 7 *(eds. M.J. Salem and M.T. Busrewil and A.M. Ben Ashour)*, Elsevier, Amsterdam, p. 2743-2756.

GHORI, K.A.R. and MOHAMMED, R.A. 1996. The application of petroleum generation modelling to the eastern Sirt Basin, Libya. *First Symposium on the Sedimentary Basins of Libya, Geology of the Sirt Basin.* vol. 2. *(eds. M.J. Salem, A.S. El-Hawat and A.M. Sbeta)*, Elsevier, Amsterdam, p. 529-540.

GHORI, K.A.R. and MOHAMMED, R.A. 2000. Petroleum system modelling in Ghadamis Basin, NW Libya (abstract only). *Second Symposium on the Sedimentary Basins of Libya, Geology of Northwest Libya.* Book of abstracts, p. 44.

GHUMA, M.A. and ROGERS, J.W. 1978. Geology, geochemistry and tectonic setting of the Ben Ghnema batholith, Tibesti massif, southern Libya. *Bull. Amer. Soc. Pet. Geol.* vol. 89, p. 1351-1358.

GHUMA, M.A. and ROGERS, J.W. 1980. Pan-African

evolution in Jamahiriya and north Africa. *Second Symposium on the Geology of Libya*. vol. 3 *(eds. M.J. Salem and M.T. Busrewil)*, Academic Press, London, p. 1059-1064.

GIAJ-VIA, P., RONCHI, P. and RICCHIUTO, T. 2000. Sedimentary cyclicity and porous system variations in peritidal dolomite (Jianai Dolomite, Eocene, offshore Libya) (abstract only). *Second Symposium on the Sedimentary Basins of Libya, Geology of Northwest Libya*. Book of abstracts, p. 46.

GIAMMARINO, S. 1984. Sheet Wadi al Khali (NH 34-8), *Geological Map of Libya, scale 1:250,000, Explanatory Booklet*, Industrial Research Centre, Tripoli.

GIGLIA, G. 1984. Sheet Ajdabiya (NH 34-6), *Geological Map of Libya, scale 1:250,000, Explanatory Booklet*, Industrial Research Centre, Tripoli.

GILLESPIE, J. and SANFORD, R.M. 1967. The geology of the Sarir oilfield, Sirte Basin, Libya. *Proc. 7th World Pet. Cong. Mexico*, Elsevier, Amsterdam, vol. 2, p. 181-193.

GLOVER, T., ADAMSON, K., FITCHES, B., WHITTINGTON, R. and CRAIG, J. 1998. Intraplate deformation in the Murzuq intracratonic basin, S.W. Libya. (abstract only). *Geol. Conf. on Expl. in Murzuq Basin*, Sabha, book of abstracts, p.18.

GLOVER, T., ADAMSON, K., WHITTINGTON, R. FITCHES, B., and CRAIG, J. 2000. Evidence for soft-sediment deformation – the Duwaysah slide of the Gargaf Arch, central Libya. *Symposium on Geological Exploration in Murzuq Basin (eds. M.A. Sola and D. Worsley)*, Elsevier, Amsterdam, p. 417-430.

GOHRBANDT, K.H.A. 1966. Upper Cretaceous and Lower Tertiary stratigraphy along the western and southwestern edge of the Sirte basin, Libya. In: *South-central Libya and northern Chad (Conf. Chairm. J.J. Williams)*. Petroleum Exploration Society of Libya, Eighth Annual Field Conference, p. 33-42.

GOODELL, P.C. 1991. Uranium potential in and around the Tibisti and Hoggar Massifs. *Third Symposium on the Geology of Libya*. vol. 7 *(eds. M.J. Salem and M.T. Busrewil and A.M. Ben Ashour)*, Elsevier, Amsterdam, p. 2627-2636.

GOUDARZI, G.H. 1970. Geology and mineral resources of Libya: a reconnaissance. *U.S. Geol. Surv. Prof. Paper no. 660*, Washington, 104p.

GOUDARZI, G.H. 1971. Geology of the Shati valley area iron deposit, Fezzan, Libyan Arab Republic. *First Symposium on the Geology of Libya (ed. C. Gray)*. Faculty of Science, University of Libya, Tripoli, p.

489-500.

GOUDARZI, G.H. and SMITH, J.P. 1978. Preliminary structure-contour map of the Libyan Arab Republic and adjacent areas; 1:2,000,000. *U.S. Geol. Surv. Misc. Geol. Invest.*, Map I-350C.

GOUDARZI, G.H. 1980. Structure – Libya. *Second Symposium on the Geology of Libya*. vol. 3 *(eds. M.J. Salem and M.T. Busrewil)*, Academic Press, London, p. 879-892.

GRAS, R. 1996. Structural style on the southern margin of the Messlah High. *First Symposium on the Sedimentary Basins of Libya, Geology of the Sirt Basin*. vol. 3. *(eds. M.J. Salem, M.T. Busrewil, A.A. Misallati and M.J. Sola)*, Elsevier, Amsterdam, p. 201-210.

GRAS, R. 1998. Statistical analyses of syn-rift sediments: an example from the Sarir Sandstone, Messla field, Sirt Basin, Libya. *Journ. Pet. Geol.*, vol. 21, p. 329-342.

GRAS, R. and THUSU, B. 1998. Trap architecture of the early Cretaceous Sarir Sandstone in the eastern Sirt Basin, Libya. In: *Petroleum Geology of North Africa. (ed. D.S. Macgregor, R.T.J. Moody, D.D. Clark-Lowes)*, Geol. Soc. Special Publication No. 132, p. 317-334.

GRAY, C. 1971. Structure and origin of the Gharian domes. *First Symposium on the Geology of Libya (ed. C. Gray)*. Faculty of Science, University of Libya, Tripoli, p. 310-319.

GRAY, J., MASSA, D. and BOUCOT, A.J. 1982. Caradocian plant microfossils from Libya. *Geology*, vol. 10, p. 197-201.

GREGORY, J.W. 1911. The Geology of Cyrenaica. *Quart. Journ. Geol.* Soc. vol. 67, p. 572-615.

GRIGNANI, D., LANZONI, E. and ELATRASH, H. 1991. Palaeozoic and Mesozoic subsurface palynos-tratigraphy in the Al Kufrah Basin, Libya. *Third Symposium on the Geology of Libya*. vol. 4 *(eds. M.J. Salem, O.S. Hammuda and B.A. Eliagoubi)*, Elsevier, Amsterdam, p. 1159-1228.

GRUBIĆ, A. 1984a. Sheet Adrar in Yahia (NF 32-3), *Geological Map of Libya, scale 1:250,000, Explanatory Booklet*, Industrial Research Centre, Tripoli.

GRUBIĆ, A. 1984b. Sheet South Anay (NF 32-4), *Geological Map of Libya, scale 1:250,000, Explanatory Booklet*, Industrial Research Centre, Tripoli.

GRUBIĆ, A., DIMITRIJEVIĆ, M., GALEĆIĆ, M., JAKOVLJECĆ, Z., KOMARNICKI, S., PROTĆ, D., RADULOVIĆ, P. and RONČEVIĆ, G. 1991. Stratigraphy of western Fezzan (SW Libya). *Third*

Symposium on the Geology of Libya. vol. 4 *(eds. M.J. Salem, O.S. Hammuda and B.A. Eliagoubi)*, Elsevier, Amsterdam, p. 1529-1564.

GRUENWALD, R. 2001. The hydrocarbon prospectivity of Lower Oligocene deposits in the Maragh Trough, SE Sirt Basin, Libya. *Journ. Pet. Geol.* vol. 24, p.213-231.

GUALTIERI, J.F. 1959. Exploration of the Jefren gypsum-anhydrite deposit, Tripolitania. *US Mission to Libya. Min. of Ind. Geol. Sec.* Tripoli. Bull. No. 6, 64p.

GUERRAK, S. 1991. The Palaeozoic oolitic ironstone belt of north Africa; from the Zemmour to Libya. *Third Symposium on the Geology of Libya*. vol. 7 *(eds. M.J. Salem and M.T. Busrewil and A.M. Ben Ashour)*, Elsevier, Amsterdam, p. 2703-2725.

GUIRAUD, R. 1998. Mesozoic rifting and basin inversion along the northern African Tethyan margin: an overview. In: *Petroleum Geology of North Africa. (ed. D.S. Macgregor, R.T.J. Moody, D.D. Clark-Lowes)*, Geol. Soc. Special Publication No. 132, p. 217-230.

GUMATI, Y.D. 1985. Crustal extension, subsidence and thermal history of the Sirte Basin, Libya. *ESRI Occasional Pubs.* No. 3. Columbia, S. Carolina, 207p.

GUMATI, Y.D. and KANES, W.H. 1985. Early Tertiary subsidence and sedimentary facies, northern Sirte Basin, Libya. *Bull. Amer. Assoc. Pet. Geol.* vol. 69, p. 39-52.

GUMATI, Y.D. and SCHAMEL, S. 1988. Thermal maturation history of the Sirte Basin, Libya. *Journ. Pet. Geol.* vol. 11, p. 205-218.

GUMATI, Y.D. and NAIRN, A.E.M. 1991. Tectonic subsidence of the Sirte Basin, Libya. *Journ. Pet. Geol.* vol. 14, p. 93-102.

GUMATI, Y.D. 1992. Lithostratigraphy of oil bearing Tertiary bioherms in the Sirte Basin, Libya. *Jour. Petrol. Geol.* vol. 15, p. 305-318.

GUMATI, Y.D., KANES, W.H. and SCHAMEL, S. 1996. An evaluation of the hydrocarbon potential of the sedimentary basins of Libya. *Journ. Pet. Geol.* vol. 19, p. 39-52.

GUNDOBIN, V.M. 1985. Sheet Qararat al Marar (NH 33-13), *Geological Map of Libya, scale 1:250,000, Explanatory Booklet*, Industrial Research Centre, Tripoli.

GUNDOBIN V.M. and YEVDOKIMOV, S.V. 1985. Sheet Qasr ash Shwayrif (NH 33-10), *Geological Map of Libya, scale 1:250,000, Explanatory Booklet*, Industrial Research Centre, Tripoli.

GURNEY, J. 1996. Libya; the political economy of oil. *Oxford Institute of Energy Studies*. Oxford University Press. 241p.

HAASE, G.M. 1996. Formation MicroScanner interpretation and caliper measurements in Nubian Sandstones, SE Sirt Basin. *First Symposium on the Sedimentary Basins of Libya, Geology of the Sirt Basin*. vol. 2. *(eds. M.J. Salem, A.S. El-Hawat and A.M. Sbeta)*, Elsevier, Amsterdam, p. 275-286.

HALLETT, D. and EL GHOUL, A. 1996. Oil and gas potential of the deep trough areas in the Sirt Basin, Libya. *First Symposium on the Sedimentary Basins of Libya, Geology of the Sirt Basin*. vol. 2. *(eds. M.J. Salem, A.S. El-Hawat and A.M. Sbeta)*, Elsevier, Amsterdam, p. 455-484.

HAMBREY, M.J. and HARLAND, W.B. 1981. (Eds.) Earth's pre-Pleistocene glacial record. Cambridge University Press, 1019p.

HAMBREY, M. J. 1985. The late Ordovician-early Silurian glacial period. *Palaeogeog. Palaeoclimatol. Palaeoecol.* vol. 51, p. 273-289.

HAMMER, A. and LYNDON, N. 1987. Hammer. Putnam, New York. 544p.

HAMMUDA, O.S. 1969. Jurassic and Lower Cretaceous rocks of central Jabal Nefusa, NW Libya. Petroleum Exploration Society of Libya. 74p.

HAMMUDA, O.S. 1971. Nature and significance of the Lower Cretaceous unconformity in Jebel nefusa, northwest Libya. *First Symposium on the Geology of Libya (ed. C. Gray)*. Faculty of Science, University of Libya, Tripoli, p. 87-96.

HAMMUDA, O.S. 1980a. Geologic factors controlling fluid trapping and anomalous freshwater occurrences in the Tadrart Sandstone, Al Hamadah al Hamra area, Ghadamis Basin. *Second Symposium on the Geology of Libya*. vol. 2 *(eds. M.J. Salem and M.T. Busrewil)*, Academic Press, London, p. 501-508.

HAMMUDA, O.S. 1980b. Sediments and palaeogeography of the Lower Campanian sand bodies along the southern tip of Ad Daffah-Al Wahah Ridge. Sirt Basin. *Second Symposium on the Geology of Libya*. vol. 2 *(eds. M.J. Salem and M.T. Busrewil)*, Academic Press, London, p. 509-520.

HAMMUDA, O.S., SBETA, A.M., MOUZUGHI, A.J. and ELIAGOUBI, B.A. 1985. Stratigraphic nomenclature of the northwestern offshore of Libya. Earth Sciences Society of Libya. 166p.

HAMMUDA, O.S. VAN HINTE, J.E. and NEDERBRAGT, S. 1991. Geohistory analysis mapping in central and southern Tarabulus Basin, northwestern offshore of Libya. *Third Symposium on the Geology of Libya*. vol. 4 *(eds. M.J. Salem, O.S. Hammuda and B.A. Eliagoubi)*, Elsevier, Amsterdam, p. 1657-1680.

HAMYOUNI, E.A. and SADEEGH, H.E. 1985. Petroleum geochemistry of the offshore Sebrata Basin, north-western Libya. (abstract only). In: *Internat Conf. Petroleum geochemistry and exploration in the Afro-Asian region.* Oil and Natural Gas Comm., Dehra Dun, India, abstracts p. 27.

HANGARI, K., ELOUJALI, R., HROUDA, M., ELBADRI, A., ELKELANI, M., GHENIMA, R., SAIDI, M. and INOUBLI, H. 2000. Geochemical characterisation of the source rock potential of Ordovician to early Carboniferous sediments of eastern Ghadamis Basin (E Tunisia and NW Libya) (abstract only). *Second Symposium on the Sedimentary Basins of Libya. The Geology of northwest Libya.* Book of abstracts, p. 50.

HAQ, B.U. and AUBREY, M.P. 1980. Early Cenozoic calcareous nannoplankton bio-stratigraphy and palaeobiogeography of North Africa and the Middle East and Trans-Tethyan correlations. *Second Symposium on the Geology of Libya.* vol. 1 *(eds. M.J. Salem and M.T. Busrewil),* Academic Press, London, p. 271-305.

HAQ, B.U., HARDENBOL, J. and VAIL, P.R. 1987. Chronology of fluctuating sea levels since the Triassic. *Science,* vol. 235, p. 1156-1166.

HARDING, T. 1984. Graben hydrocarbon occurrence and structural style. *Bull. Am. Assoc. Pet. Geol.* vol. 68, p. 333-362.

HARGREAVES, R.B. and VAN HOUTEN, F.B. 1985. Palaeogeography of Africa in early-middle Palaeozoic; palaeomagnetic and stratigraphic constraints and tectonic implications. *Occas. Pubs. Int. Cent. Train. Exchs. Geosci.* vol. 3, p. 160-161.

HARLAND, W.B., ARMSTRONG, R.L., COX, A.V., CRAIG, L.E., SMITH, A.G. and SMITH, D.G. 1990. A geologic time scale, 1989. Cambridge Univ. Press, 263p.

HASSI, I.A. 1998. Sequence stratigraphic analysis of the Tahara in Hamada Basin, western Libya (abstract only). *Symposium on Geological Exploration in Murzuq Basin, Sabha, 20-22nd September 1998.* Book of abstracts.

HAVLICEC, V. and MASSA, D. 1973. Brachiopodes de l'Ordovicien superieur de Libye occidentale, implications stratigraphiques regionales. *Geobios,* vol. 6, p. 267-290.

HAY, W.W. 1968. Coccoliths and other calcareous nanno-fossils in marine sediments in Cyrenaica. In: *Geology and archaeology of northern Cyrenaica, Libya (ed. Barr, F.T.).* Petroleum Exploration Society of Libya, Tenth Annual Field Conference, p. 149-158.

HAYNES, J. 1962. *Operculina* and associated foraminifers from the Paleocene of NE Fezzan, Libya. *Contr. Cushman Found. Foram. Res.* vol. 13, p. 90-97.

HEA, J.P. 1971. Petrography of the Paleozoic-Mesozoic sandstones of the southern Sirte Basin, Libya. *First Symposium on the Geology of Libya (ed. C. Gray).* Faculty of Science, University of Libya, Tripoli, p. 107-126.

HELMDACH, F. and EL KHOUDARY, R.H. 1980. Ostracoda and planktonic foraminifera from the late Eocene of Al Jabal al Akhdar, northeastern Libya. *Second Symposium on the Geology of Libya.* vol. 1 *(eds. M.J. Salem and M.T. Busrewil),* Academic Press, London, p. 255-270.

HESELDEN, R.G.W., CUBITT, J.M., BORMAN, P. and MADI, F.M. 1996. Lithofacies study of the Lidam and Maragh Formations (late Cretaceous) of the Masrab field and adjacent areas, Sirt basin, Libya. *First Symposium on the Sedimentary Basins of Libya, Geology of the Sirt Basin.* vol. 2. *(eds. M.J. Salem, A.S. El-Hawat and A.M. Sbeta),* Elsevier, Amsterdam, p. 197-210.

HILL, P.J., PARIS, F. and RICHARDSON, J.B. 1985. Silurian Palynomorphs. In: *The palyno-stratigraphy of north-east Libya (eds. B. Thusu and B. Owens).* Journ. Micropalaeontology, vol. 4, p. 27-48.

HILL, P.J. and MOLYNEUX, S.G. 1988. Biostratigraphy, palynofacies and provincialism of late Ordovician-early Silurian acritarchs from northeast Libya. In: *Subsurface palynostratigraphy of northeast Libya. (eds. A. El-Arnauti, B. Owens and B. Thusu).* Research Centre, Garyounis University, Benghazi, p. 27-44.

HIMMALI, A.A. and OUN, K.M. 1996. The Tertiary volcanics of sheet Zallah, NH 33-16: a reappraisal. *First Symposium on the Sedimentary Basins of Libya, Geology of the Sirt Basin.* vol. 3. *(eds. M.J. Salem, M.T. Busrewil, A.A. Misallati and M.J. Sola),* Elsevier, Amsterdam, p. 347-356.

HIMMALI, A., MEJRAB, B. and SMART, J. 1998. Review of hydrocarbon migration models for the Murzuq Basin (abstract only). *Symposium on Geological Exploration in Murzuq Basin, Sabha, 20-22nd September 1998.* Book of abstracts.

HIRST, J., PHILIP, P., BENBAKIR, A., PAYNE, D. and WESTLAKE, I. 2001. Density flow processes and their influence on reservoir quality in the late Ordovician proglacial sequence, Algeria (abstract only). *Abstracts, North Africa Research Workshop, Oxford Brookes University, 4th-5th September 2001.*

HLADIL, J., OTAVA, J and GALLE, A. 1991. Oligocene carbonate buildups of the Sirt Basin, Libya. *Third Symposium on the Geology of Libya.* vol. 4 *(eds. M.J.*

Salem, O.S. Hammuda and B.A. Eliagoubi), Elsevier, Amsterdam, p. 1401-1420.

HLUŠTÍK, A. 1991. Late Palaeozoic floras of the Wadi ash Shati area, Libya. *Third Symposium on the Geology of Libya*. vol. 4 *(eds. M.J. Salem, O.S. Hammuda and B.A. Eliagoubi)*, Elsevier, Amsterdam, p. 1275-1284.

HOFFMEISTER, W.S. 1959. Lower Silurian plant spores from Libya. *Micropaleontology*, vol. 5, p. 331-334.

HOGAN, J.A. and DAVIDSON, L.M. 1998. Exploration in the Murzuk Basin Libya: a case history. *Algeria, Libya and Egypt Oil and Gas Summit, Café Royal, London*, 15p.

HSÜ, K.J., CITA, M.B. and RYAN, W.B.F. 1973. The origin of the Mediterranean evaporites. *Initial Reports of the Deep Sea Drilling Project*, vol. 13 *(Leg 13)*, pt. 2, chap. 43, p. 1203-1230.

HUGHES, W.D. and REED, J.D. 1995. Oil and source rock geochemistry and exploration implications in northern Tunisia. *Proc. Seminar on source rocks and hydrocarbon habitat in Tunisia, Tunis, 1995*. ETAP Memoir 9, p. 51-63.

IBRAHIM, M.W. 1991. Petroleum geology of the Sirt Group sandstones, eastern Sirt Basin. *Third Symposium on the Geology of Libya*. vol. 7 *(eds. M.J. Salem and M.T. Busrewil and A.M. Ben Ashour)*, Elsevier, Amsterdam, p. 2757-2779.

IBRAHIM, M.W. 1996. Geothermal gradient anomalies of hydrocarbon entrapment; Al Hagfah Trough, Sirt Basin, *First Symposium on the Sedimentary Basins of Libya, Geology of the Sirt Basin*. vol. 2. *(eds. M.J. Salem, A.S. El-Hawat and A.M. Sbeta)*, Elsevier, Amsterdam, p. 419-434.

IBRAHIM, M.W. and AL MAHRUQI, S. 2000. Petroleum systems of the Hagfa Trough, Libya (abstract only). *Symposium on petroleum systems and evolving technologies in African E & P. Burlington House, London, 16-18 May, 2000*. Book of abstracts, p. 26.

INNOCENTI, F. and PERTUSATI, P. 1984. Sheet Al Aqaylah (NH 34-5), *Geological Map of Libya, scale 1:250,000, Explanatory Booklet*, Industrial Research Centre, Tripoli.

JACQUÉ, M. 1962. Reconnaissance geologique du Fezzan oriental. *Notes et Mem, Comp. Fran. Pet.* No. 5, 44p.

JAEGER, H., BONNEFOUS, J. and MASSA, D. 1975. Le Silurien en Tunisie; ses relations avec le Silurien de Libye nord-occidentale. *Bull. Soc Geol. Fr.* ser. 7, vol. 17, p. 68-77

JAKOVLJEVIĆ, Ž. 1984. Sheet Al Awaynat (NG 32-12), *Geological Map of Libya, scale 1:250,000, Explanatory Booklet*, Industrial Research Centre, Tripoli.

JAKOVLJEVIĆ, Ž., GRUBIĆ, A. and PANTIĆ, N. 1991. The age of the Mesozoic continental formations on the western margin of the Murzuq Basin. *Third Symposium on the Geology of Libya*. vol. 4 *(eds. M.J. Salem, O.S. Hammuda and B.A. Eliagoubi)*, Elsevier, Amsterdam, p. 1583-1588.

JERZKIEWICZ, T., GHUMMED, M.A., TSHAKREEN, S.O. and ABUGARES, A.A. 2000. Sequence stratigraphic definition of the Ghadamis/Sirt boundary in late Cretaceous time (abstract only). *Second Symposium on the Sedimentary Basins of Libya. The Geology of northwest Libya*. Book of abstracts, p. 53.

JOHNSON, B.A. and NICOUD, D.A. 1996. Integrated exploration for Beda Formation reservoirs in the southern Zallah Trough (west Sirt basin, Libya). *First Symposium on the Sedimentary Basins of Libya, Geology of the Sirt Basin*. vol. 2. *(eds. M.J. Salem, A.S. El-Hawat and A.M. Sbeta)*, Elsevier, Amsterdam, p. 211-222.

JONES, J.R. 1971. Ground-water provinces of Libyan Arab Republic. *First Symposium on the Geology of Libya (ed. C. Gray)*. Faculty of Science, University of Libya, Tripoli, p. 449-458.

JONES, H.L. 1996. Late Cretaceous-early Tertiary stratigraphy of southern concession 6, Sirt Basin, Libya. *First Symposium on the Sedimentary Basins of Libya, Geology of the Sirt Basin*. vol. 1. *(eds. M.J. Salem, A.J. Mouzughi and O.S. Hammuda)*, Elsevier, Amsterdam, p. 169-184.

JONGSMA, D., WOODSIDE, J.M., KING, G.C.P. and VAN HINTE, J. 1987. The Medina wrench: a key to the kinematics of the central and eastern Mediterranean over the past 5 Ma. *Earth and Plan. Sci. Letters*, vol. 82, p. 87-106.

JONGSMA, D. 1991. The Medina and Sirt wrenches; a quantitative estimate of strike-slip deformation within the Sirt Rise over the past 5 Ma. *Third Symposium on the Geology of Libya*. vol. 6 *(eds. M.J. Salem, A.M. Sbeta and M.R. Bakbak)*, Elsevier, Amsterdam, p. 2331-2352.

JORDI, H.A. and LONFAT, F. 1963. Stratigraphic subdivision and problems in Upper Cretaceous-Lower Tertiary deposits in northwestern Libya. *First Saharan Symposium*. Rev. Inst. Fran. Pet. special volume, p. 114-122.

JORRY, S., BEN-AHMAD, M. and CANTALOUBE, S. 2000. Facies distribution and reservoir heterogeneity of El Gueria nummulitic formation in concession 137, NW offshore Libya: a field analog study from central Tunisia (Kesra Plateau) (abstract only). *Second Symposium on the Sedimentary Basins of Libya. The Geology of northwest Libya*. Book of

468

abstracts, p. 54.

JURÁK, L. 1978. Sheet Jabal al Hasawnah (NH 33-14), *Geological Map of Libya, scale 1:250,000, Explanatory Booklet*, Industrial Research Centre, Tripoli.

JURÁK, L. 1985. Sheet Wadi bu ash Shaykh (NH 33-12), *Geological Map of Libya, scale 1:250,000, Explanatory Booklet*, Industrial Research Centre, Tripoli.

KALLENBACH, H. 1972. Beitrage zur sedimentologie des kontinentalen Mesozoikums am westrand des Murzukbeckens (Libyen) *Geol. Rundschau. Deuts.* vol. 61, p. 302-322.

KAR, R.K., KAISER, G., and JAIN K.P. 1972. Permo-Triassic subsurface palynology from Libya. *Pollen et Spores*, vol 4, p. 389-453.

KARASEK, R.M. 1981. Structural and stratigraphic analysis of the Palaeozoic Murzuk and Ghadames Basins, western Libya. *ESRI Occasional Pubs.* No. 1. Columbia, S. Carolina, 146p.

KEELEY, M.L. and MASSOUD, M.S. 1998. Tectonic controls on the petroleum geology of NE Africa. In: *Petroleum Geology of North Africa. (ed. D.S. Macgregor, R.T.J. Moody, D.D. Clark-Lowes)*, Geol. Soc. Special Publication No. 132, p. 265-282.

KILANI-MAZRAOUDI, F., RAZGALLAH-GARGOURI, S. and MANNAI-TAYECH, B. 1990. The Permo-Triassic of southern Tunisia-biostratigraphy and palaeoenvironment. *Rev. Palaeobot. Palyn.*, vol. 66, p. 273-291.

KILIAN, C. 1931. Des principaux complexes continentaux du Sahara. *C.R. Soc. Geol. Fr.* No. 9, p. 109-111.

KLEINSMIEDE, W.F.J. and VAN DEN BERG, N.J. 1968. Surface geology of the Jabal al Akhdar, northern Cyrenaica, Libya. In: *Geology and archaeology of northern Cyrenaica, Libya (ed. Barr, F.T.)*. Petroleum Exploration Society of Libya, Tenth Annual Field Conference, p. 115-124.

KLEMME, H.D. and ULMISHEK, G.F. 1991. Effective petroleum source rocks of the world: stratigraphic distribution and controlling depositional factors. *Bull. Amer. Assoc. Pet. Geol.* vol. 75, p. 1809-1851.

KLEN. L. 1974. Sheet Benghazi (NI 34-14), *Geological Map of Libya, scale 1:250,000, Explanatory Booklet*, Industrial Research Centre, Tripoli.

KLERKX, J. 1980. Age and metamorphic evolution of the Basement complex around Jabal al Awaynat. *Second Symposium on the Geology of Libya*. vol. 3 *(eds. M.J. Salem and M.T. Busrewil)*, Academic Press, London, p. 901-906.

KLITZSCH, E. 1963. Geology of the north-east flank of the Murzuk Basin (Djebel ben Ghnema-Dor el Gussa area). *First Saharan Symposium. Rev.* Inst. Fran. Pet. special volume, p. 97-113.

KLITZSCH, E. 1965 Ein profil aus dem typusgebiet gotlandischer und devonische schichten der zentralsahara (westrand Murzukbecken, Libyen). *Erdol u. Kohle. Deut.* vol. 18, p. 605-607.

KLITZSCH, E. 1966a. Comments on the geology of the central parts of southern Libya and northern Chad. In: *South-central Libya and northern Chad (Conf. Chairm. J.J. Williams)*. Petroleum Exploration Society of Libya, Eighth Annual Field Conference, p. 1-18.

KLITZSCH, E. 1966b. Geology of the northeast flank of the Murzuk Basin. In: *South-central Libya and northern Chad (Conf. Chairm. J.J. Williams)*. Petroleum Exploration Society of Libya, Eighth Annual Field Conference, p. 19-32.

KLITZSCH, E. 1968a. Outline of the geology of Libya. In: *Geology and archaeology of northern Cyrenaica, Libya (ed. Barr, F.T.)*. Petroleum Exploration Society of Libya, Tenth Annual Field Conference, p. 71-78.

KLITZSCH, E. 1968b. Die Gothlandium Transgression in der Zentral-Sahara. *Zeit. Deut. Geol. Ges.* vol. 117, p. 492-501.

KLITZSCH, E. 1969. Stratigraphic section from the type areas of Silurian and Devonian strata at western Murzuk Basin (Libya). In: *Geology, archaeology and prehistory of the southwestern Fezzan, Libya (ed. Kanes, W.H.)*. Petroleum Exploration Society of Libya, Eleventh Annual Field Conference, p. 83-90.

KLITZSCH, E. and BAIRD, D.W. 1969. Stratigraphy and palaeohydrology of the Germa (Jarma) area southwest Libya. In: *Geology, archaeology and prehistory of the southwestern Fezzan, Libya (ed. Kanes, W.H.)*. Petroleum Exploration Society of Libya, Eleventh Annual Field Conference, p. 67-80.

KLITZSCH, E. 1970b. Die Strukturgeschichte der Zentralsahara: neue Erkenntnisse zum Bau und zur Paläogeographie eines Tafellandes. *Geol. Rundsch.* vol. 59, p. 459-527.

KLITZSCH, E. 1971. The structural development of parts of north Africa since Cambrian time. *First Symposium on the Geology of Libya (ed. C. Gray)*. Faculty of Science, University of Libya, Tripoli, p. 253-262.

KLITZSCH, E. 1972. Problems of continental Mesozoic strata of south-western Libya. In: *Proc. Conf. African Geology (Dec. 1970): Regional Geology, (Eds T.F.J. Dessauvagie and A.J. Whiteman)*, Dept. of Geology, Univ. Ibadan, Nigeria, publication 1, p. 483-494.

KLITZSCH, E., LEJAL-NICOL, A. and MASSA, D.

1973. Le Siluro-Devonien a Psilophytes et Lycophytes du bassin de Mourzouk, Libye. *C.R. Acad. Sci.* Paris. vol. 277, ser. D, p. 2465-2467.

KLITZSCH, E., HARMS, J.C., LEJAL-NICOL, A. and LIST, F.K. 1979. Major subdivisions and depositional environment of Nubian strata, southwestern Egypt. *Bull. Amer. Ass. Pet. Geol.* vol. 63, p. 967-974.

KLITZSCH, E. 1981. Lower Palaeozoic rocks of Libya, Egypt and Sudan. In: *Lower Palaeozoic of the Middle East, eastern and southern Africa and Antarctica. (ed. C.H. Holland).* John Wiley, New York, p. 131-163.

KLITZSCH, E. and WYCISK, P. 1987. Geology of sedimentary basins of northern Sudan and bordering area. *Berl. Geowiss. Abh,* vol. 75, p. 97-136.

KLITZSCH, E. and SQUYRES, C.H. 1990. Paleozoic and Mesozoic geological history of northeastern Africa based upon new interpretation of Nubian strata. *Bull. Amer. Assoc. Pet. Geol.* vol. 74, p. 1203-1211.

KLITZSCH, E. and SEMTNER, E. 1993. Silurian palaeogeography of NE Africa and Arabia – an updated interpretation. In: *Geoscientific research in northeast Africa (eds. U. Thornweihe and H. Schandelmeier).* Springer-Verlag, Berlin, p. 341-344.

KLITZSCH, E. 1996. Geological observations from Nubia and their structural interpretation. *First Symposium on the Sedimentary Basins of Libya, Geology of the Sirt Basin.* vol. 3. *(eds. M.J. Salem, M.T. Busrewil, A.A. Misallati and M.J. Sola),* Elsevier, Amsterdam, p. 101-106.

KLITZSCH, E. 2000. The structural development of the Murzuq and Kufra basins – significance for oil and mineral exploration. *Symposium on Geological Exploration in Murzuq Basin (Eds. M.A. Sola and D. Worsley),* Elsevier, Amsterdam, p. 143-150.

KNYTL, J., BRYDLE, G. and GREENWOOD, D. 1996. Tectonic history and structural development of the Kaf-Themar trend of the western Sirt Basin. *First Symposium on the Sedimentary Basins of Libya, Geology of the Sirt Basin.* vol. 3. *(eds. M.J. Salem, M.T. Busrewil, A.A. Misallati and M.J. Sola),* Elsevier, Amsterdam, p. 167-200.

KODYM, O. 1985. Sheet Sabkhat Ghuzayil (NH 34-10), *Geological Map of Libya, scale 1:250,000, Explanatory Booklet,* Industrial Research Centre, Tripoli.

KOGBE, C. 1980. The trans-Saharan seaway during the Cretaceous. *Second Symposium on the Geology of Libya.* vol. 1 *(eds. M.J. Salem and M.T. Busrewil),* Academic Press, London, p. 91-96.

KOMARNICKI, S. 1984. Sheet Wadi Irawan (NG 32-8), *Geological Map of Libya, scale 1:250,000, Explanatory Booklet,* Industrial Research Centre, Tripoli.

KONZALOVÁ, M. 1991. Palynological studies in the Mesak Formation (central Libya). *Third Symposium on the Geology of Libya.* vol. 4 *(eds. M.J. Salem, O.S. Hammuda and B.A. Eliagoubi),* Elsevier, Amsterdam, p. 1229-1242.

KOOPMANS, B.N. and AL FASATWI, Y.A. 1996. Synergistic use of SPOT and SIR data for geological mapping, Sirt Basin, Libya. *First Symposium on the Sedimentary Basins of Libya, Geology of the Sirt Basin.* vol. 3. *(eds. M.J. Salem, M.T. Busrewil, A.A. Misallati and M.J. Sola),* Elsevier, Amsterdam, p. 231-242.

KORÁB, T. 1984. Sheet Tmassah (NG 33-7), *Geological Map of Libya, scale 1:250,000, Explanatory Booklet,* Industrial Research Centre, Tripoli.

KOSCEC, B.G. and GHERRYO, Y.S. 1996. Geology and reservoir performance of Messlah oil field, Libya. *First Symposium on the Sedimentary Basins of Libya, Geology of the Sirt Basin.* vol. 2. *(eds. M.J. Salem, A.S. El-Hawat and A.M. Sbeta),* Elsevier, Amsterdam, p. 365-390.

KOSTANDI, A.B. 1963. Eocene facies maps and tectonic interpretation in the Western Desert, U.A.R. *First Saharan Symposium. Rev.* Inst. Fran. Pet. special volume, p. 17-28.

KRUMBECK, L. 1906. Beitrage zur geologie und palaeontologie von Tripolis. *Palaeontographica,* vol. 53, p. 51-136.

KRUSEMAN, G.P. and FLOEGEL, H. 1980. Hydrogeology of the Jifarah, NW Libya. *Second Symposium on the Geology of Libya.* vol. 2 *(eds. M.J. Salem and M.T. Busrewil),* Academic Press, London, p. 763-777.

KUEHN, M. 1996. Palaeosols as an additional tool in defining reservoir characteristics of pre-Upper Cretaceous sections in As Sarah and Tuama oil fields. *First Symposium on the Sedimentary Basins of Libya, Geology of the Sirt Basin.* vol. 2. *(eds. M.J. Salem, A.S. El-Hawat and A.M. Sbeta),* Elsevier, Amsterdam, p. 251-262.

KUHNT, W., WIEDMANN, J., HERBIN, J.P., MOULLADE, M. and THUROW, J. 2000. Paleogeographic distribution patterns of Cenomanian/Turonian source rocks: a north African perspective (abstract only). *Int. workshop on north African micropalaeo. for pet. expl., Univ. Coll. London, August 2000.* Book of abstracts, p. 51.

LABABIDI, M.M. and HAMDAN, A.N. 1985. Preliminary lithostratigraphic correlation study in OAPEC member countries. Energy Resources Dept,

Org. of Arab Pet. Exp. Countries, Kuwait, 171p.

LAGHA, S. 1991. The geology and geochemistry of the south Al Awaynat ultrabasic body, southeast Libya. *Third Symposium on the Geology of Libya.* vol. 7 *(eds. M.J. Salem and M.T. Busrewil and A.M. Ben Ashour),* Elsevier, Amsterdam, p. 2529-2538.

LASHHAB, M.I. and WEST, I.M. 1991. Sedimentology and geochemistry of the Jir Formation in Jabal al Jir and the western Sirt Basin. *Third Symposium on the Geology of Libya.* vol. 5 *(eds. M.J. Salem and M.N. Belaid),* Elsevier, Amsterdam, p. 1855-1870.

LASHHAB, M.I. and WEST, I.M. 1996. Dolomitization of the Jir and Rawaghah Formations in Jabal al Jir and the western Sirt Basin. *First Symposium on the Sedimentary Basins of Libya, Geology of the Sirt Basin.* vol. 2. *(eds. M.J. Salem, A.S. El-Hawat and A.M. Sbeta),* Elsevier, Amsterdam, p. 31-44.

LEFRANC, J.P. 1958. Stratigraphie des series continentales intercalaires au Fezzan nord-occidental (Libye). *C.R. Hebd. Seanc. Acad. Sci.* Paris. vol 247, p. 1360-1363.

LEFRANC, J.P. 1991. Alum deposits of Fazzan and the Sahara; their origin, distribution and importance *Third Symposium on the Geology of Libya.* vol. 7 *(eds. M.J. Salem and M.T. Busrewil and A.M. Ben Ashour),* Elsevier, Amsterdam, p. 2605-2618.

LEGRAND, P. 1985. Lower Palaeozoic rocks of Algeria. In: *Lower Palaeozoic of north-western and west-central Africa (Ed. C.H. Holland). J. Wiley,* Chichester, p. 5-90.

LEGRAND, P. 1988. The Ordovician-Silurian boundary in the Algerian Sahara. *Bull. Brit. Mus. Nat. Hist. (Geol).* vol. 43, p. 171-176.

LEHMANN, E.P. 1964. Tertiary-Cretaceous boundary facies in the Sirte Basin. *Proc. 22nd Int. Geol. Cong. New Delhi,* vol. 3, p. 56-73.

LEHMANN, E.P., ROSENBOOM, J.J., WALLER, H.O. and CONLEY, C.D. 1968. *Microfacies of Libya.* Pet. Expl. Soc. Libya. 80pp.

LEJAL-NICOL, A. 1991. The importance of Libyan Devono-Carboniferous palaeofloras: a comparison with Egyptian and new Nigerian ones. *Third Symposium on the Geology of Libya.* vol. 4 *(eds. M.J. Salem, O.S. Hammuda and B.A. Eliagoubi),* Elsevier, Amsterdam, p. 1343-1350.

LELUBRE, M. 1946a. Le Tibesti septentrionale. Esquisse morphologique et structurale. *C.R. Acad. Sci. Colon.* vol. 6. p. 337-357.

LELUBRE, M. 1946b. Sur la Paleozoique du Fezzan. *C.R. Hebd. Seanc. Acad. Sci.* vol. 222, p. 1403-1404

LELUBRE, M. 1946c. Sur les series antecambriennes du Tibesti septentrional. *C.R. Hebd. Seanc. Acad. Sci.*

vol 223, p. 429-431.

LELUBRE, M. 1948. Le Paleozoique du Fezzan sud-oriental. *C.R. Soc. Geol. Fr.* vol. 18, p. 79-81.

LELUBRE, M. 1952. Apercu sur la geologie du Fezzan. *Bull. Serv. Carte Geol. d'Algerie, Trav. Rec. Coll.* no. 3, p. 109-148.

LEWIS, C.J. 1990. Sarir Field. In: *Treatise of Petroleum Geology. Atlas of oil and gas fields, Structural Traps II (compiled by E.A. Beaumont and N.H. Foster).* Amer. Assoc. Pet. Geol. p. 253-267.

LINSENBARTH, A. 1991. Sand desert forms in Libya – remote sensing analysis. T*hird Symposium on the Geology of Libya.* vol. 6 *(eds. M.J. Salem, A.M. Sbeta and M.R. Bakbak),* Elsevier, Amsterdam, p. 2235-2258.

LIPPARINI, T. 1940. Tettonica e geomorfologia della Tripolitania. *Boll. Soc. Geol. Ital.* vol 59, p. 221-301.

LIVERMORE, R.A. and SMITH, A.G. 1985. Some boundary conditions for the evolution of the Mediterranean region. In: *Geological Evolution of the Mediterranean Basin. (Raimondo Selli Commemorative Volume) (eds. D.J. Stanley and F-C. Wezel),* Springer-Verlag, Berlin., chap. 5, p. 83-97.

LIVERMORE, R.A., SMITH, A.G. and VINE, F.J. 1986. Late Palaeozoic to early Mesozoic evolution of Pangea. *Nature,* vol. 322, p. 162-165.

LOBOZIAK, S. and CLAYTON, G. 1988. The Carboniferous palynostratigraphy of northeast Libya. In: *Subsurface palynostratigraphy of northeast Libya. (eds. A. El-Arnauti, B. Owens and B. Thusu).* Research Centre, Garyounis University, Benghazi, p. 129-150.

LORENZ, J. 1980. Late Jurassic-early Cretaceous sedimentation and tectonics of the Murzuq Basin, southwestern Libya. *Second Symposium on the Geology of Libya.* vol. 2 *(eds. M.J. Salem and M.T. Busrewil),* Academic Press, London, p. 383-392.

LORENZ, J.C. 1987. Mixed fluvial systems of the Messak Sandstone, a deposit of the Nubian lithofacies, south-western Libya. *Sedimentary Geology,* vol. 54, p. 245-264.

LOUCKS, R.G., MOODY, R.T.J., BELLIS, J.K. and BROWN, A.A. 1998. Regional depositional setting and pore network systems of the El Garia Formation (Metlaoui Group, Lower Eocene), offshore Tunisia. In: *Petroleum Geology of North Africa. (ed. D.S. Macgregor, R.T.J. Moody, D.D. Clark-Lowes),* Geol. Soc. Special Publication No. 132, p. 355-374.

LÜNING, S., CRAIG, J., FITCHES, B., MAYOUF, J., BUSREWIL, A., EL DIEB, M., GAMMUDI, A., LOYDELL, D. and McILROY, D. 1999. Re-evaluation of the petroleum potential of the Kufra

Basin, (SE Libya, NE Chad): does the source rock barrier fall? *Marine and Petrol. Geol.*, vol. 16, p. 693-718.

LÜNING, S., CRAIG, J., LOYDELL, D.K., STORCH, P. and FITCHES, B. 2000a. Lowermost Silurian 'hot shales' in north Africa and Arabia: regional distribution and depositional model. *Earth Science Reviews*, vol. 49, p. 121-200.

LÜNING, S., GRÄFE, K.U., BOSENCE, D., LUCIANI, V. and CRAIG, J. 2000b. Discovery of marine late Cretaceous carbonates and evaporites in the Kufra Basin (Libya) redefines the southern limit of the late Cretaceous transgression. *Cretaceous Research*, vol. 21, p. 721-731.

LÜNING, S., CRAIG, J., FITCHES, B., MAYOUF, J., BUSREWIL, A., EL DIEB, M., GAMMUDI, A, and LOYDELL, D.K. 2000c. Petroleum source and reservoir rock re-evaluation in the Kufra Basin (SE Libya, NE Chad, NW Sudan). *Symposium on Geological Exploration in Murzuq Basin (Eds. M.A. Sola and D. Worsley)*, Elsevier, Amsterdam, p. 151-174.

LÜNING, S., 2000. Source rocks and petroleum generation in the Palaeozoic of north Africa: their geological setting and fossil associations (abstract only). *Int. workshop on north African micropalaeo. for pet. expl., Univ. Coll. London, August 2000*. Book of abstracts, p. 54-55.

LÜNING, S., 2000. The role of microfossils in deciphering anoxic events in north Africa with emphasis on the Mesozoic (abstract only). *Int. workshop on north African micropalaeo. for pet. expl., Univ. Coll. London, August 2000*. Book of abstracts, p. 56-57.

LÜNING, S., 2000. Early Silurian and late Devonian organic-rich shales in western Libya and neighbouring areas (abstract only). *Second symposium on the sedimentary basins of Libya, Tripoli, Geology of northwest Libya, November 2000*. Book of abstracts, p. 59.

LÜNING, S., 2001. An integrated depositional model for the Cenomanian-Turonian organic-rich strata in north Africa (abstract only). *Abstracts, North Africa Research Workshop, Oxford Brookes University, 4th-5th September 2001*.

LÜNING, S., ARCHER, R., CRAIG, J. and LOYDELL, D.K. In the press. The Lower Silurian 'Hot Shales' and 'Double Hot Shales' in north Africa and Arabia. *Second symposium on the sedimentary basins of Libya, Tripoli, Geology of northwest Libya, November 2000*.

MACGREGOR, D.S. 1996a. Factors controlling the destruction or preservation of giant light oilfields.

Petroleum Geoscience, vol. 2, p. 197-217.

MACGREGOR, D.S. 1996b. The hydrocarbon systems of North Africa. *Marine and Petroleum Geology*, vol. 13, p. 329-340.

MACGREGOR, D.S. 1998. Giant fields, petroleum systems and exploration maturity in Algeria. In: *Petroleum Geology of North Africa. (ed. D.S. Macgregor, R.T.J. Moody, D.D. Clark-Lowes)*, Geol. Soc. Special Publication No. 132, p. 79-96.

MACGREGOR, D.S. and MOODY, R.T.J. 1998. Mesozoic and Cenozoic petroleum systems of North Africa. In: *Petroleum Geology of North Africa. (ed. D.S. Macgregor, R.T.J. Moody, D.D. Clark-Lowes)*, Geol. Soc. Special Publication No. 132, p. 201-216.

MAGNIER, P. 1962. Etude geologique du gisement de vertebres du Gebel Zelten (Libye). *C. R. Somm. Seanc. Soc. Geol. Fr.* no. 2, p. 55-57.

MAGNIER, P. 1963. Etude stratigraphique dans le Gebel Nefousa et le Gebel Garian (Tripolitaine, Libye). *Bull. Soc. Geol. Fr.* ser. 7, vol. 5, p. 89-94.

MAGNIER, P. 1964. Le Neogene du bassin de Syrte et du sud de la Cyrenaique (Libye). *Curs. Conf. Inst. Lucas Mallada*, Spain. no. 9, p. 193-198.

MAGNIER, P. 1969. Etude lithostratigraphique du miocene inferieur du bassin de Syrte (Libye). *Com. Mediterr. Neogene strat. (4th session). Proc. Bologna Mus. Geol.*, vol. 35, p. 119-130.

MAHRHOLZ, W.W. 1968. Geological exploration of the Kufrah region. *Ministry of Industry, Geol. Section.* Bull. 8, Tripoli, 76p.

MAISONNEUVE, J. 1991. CO2 emanations, alkaline magmatism and the relationship between rift structures and mineralizations in the African plate. *Third Symposium on the Geology of Libya.* vol. 7 *(eds. M.J. Salem and M.T. Busrewil and A.M. Ben Ashour)*, Elsevier, Amsterdam, p. 2577-2586.

MAMGAIN, V.D. 1980. The pre-Mesozoic (Precambrian to Palaeozoic) stratigraphy of Libya – a reappraisal. *Bulletin No. 14.* Industrial Research Centre, Tripoli, 104p.

MANETTI, P. 1984. Sheet Al Mufawwaz (NH 35-5), *Geological Map of Libya, scale 1:250,000, Explanatory Booklet*, Industrial Research Centre, Tripoli.

MANN, K. 1975a. Sheet Al Khums (NI 33-14), *Geological Map of Libya, scale 1:250,000, Explanatory Booklet*, Industrial Research Centre, Tripoli.

MANN, K. 1975b. Sheet Misratah (NI 33-15), *Geological Map of Libya, scale 1:250,000, Explanatory Booklet*, Industrial Research Centre, Tripoli.

MANSOUR, A.T. and MAGAIRHY, I.A. 1996. Petroleum

472

geology and stratigraphy of the southeastern part of the Sirt Basin, Libya. *First Symposium on the Sedimentary Basins of Libya, Geology of the Sirt Basin.* vol. 2. *(eds. M.J. Salem, A.S. El-Hawat and A.M. Sbeta)*, Elsevier, Amsterdam, p. 485-528.

MARCHETTI, M. 1934. Note illustrative per un abbozzo di carta geplogica della Cirenaica. *Boll. Soc. Geol. Ital.* vol. 53, p. 309-325.

MARTIN, D.L., MUNROE, H.D. and NAIRN, A.E.M. 1991. First results of the palaeomagnetic study of Libyan Mesozoic limestones; the Carnian Azizia Limestone Formation – preliminary data. *Third Symposium on the Geology of Libya.* vol. 6 *(eds. M.J. Salem, A.M. Sbeta and M.R. Bakbak)*, Elsevier, Amsterdam, p. 2433-2440.

MASSA, D. and COLLOMB, G.R. 1960. Observations nouvelles sur la region d'Aouinet Ouenine et du Djebel Fezzan (Libye). *Proc. 21st Int. Geol. Cong.* Copenhagen, pt. 12, p. 65-73.

MASSA, D. and JAEGER, H. 1971. Donnees stratigraphiques sur le Silurien de l'Ouest de la Libye. *Mem. Bur. Rech. Geol. Min.* vol. 73, p. 313-321.

MASSA, D., TERMIER, G. and TERMIER, H. 1974. Le Carbonifere de Libye occidentale; stratigraphie, paleontologie. *Com. Fran. de Petrole, Notes et Mem.* no. 11, p. 139-206.

MASSA, D. and BELTRANDI, M. 1975. Sedimentologie du Silurien de Libye occidentale. *9th Cong. Int. Sediment., Nice, 1975*, vol. 1, p. 113-118.

MASSA, D. and MOREAU-BENOIT, A. 1976. Essai de synthese stratigraphique et palynologique du System devonien en Libye occidentale. *Rev. Inst. Fr. Petr.* vol. 31, p. 287-332.

MASSA, D., HAVLICEK, V. and BONNEFOUS, J. 1977. Stratigraphic and faunal data on the Ordovician of the Rhadames (Ghadames) Basin (Libya and Tunisia). *Bull. Cent. Rech. Exp. & Prod.* Elf-Aquit. vol. 2, p. 3-27.

MASSA, D. and VACHARD D. 1979. Le Carbonifere de Libye occidentale: biostratigraphie et micropaleontologie. Position dans le domaine tethysien d'Afrique du nord. *Rev. Inst Fran. du Petrole*, vol. 34, p. 1-65.

MASSA, D., COQUEL, R., LOBOZIAK, S. and TAUGOURDEAU-LANTZ, J. 1980. Essai de synthese stratigraphique et palynologique du Carbonifere en Libye occidentale. *Ann. Soc. Geol. Nord.* vol. 99, p. 429-442.

MASSA D. and DELORT, T. 1984. Evolution du bassin de Syrte (Libye) du Cambrien au Crétacé basal. *Bull. Soc. Geol. Fr.* ser. 7, vol. 26, p. 1087-1096.

MASSA, D. and BOURROUILH, R. 2000. Upper Ordovician mud-mounds of northern Ghadamis Basin, Libya: a review (abstract only). *Second symposium on the sedimentary basins of Libya, Tripoli, Geology of northwest Libya, November 2000.* Book of abstracts, p. 63.

MAŠTERA, L. 1985. Sheet Maradah (NH 34-9), *Geological Map of Libya, scale 1:250,000, Explanatory Booklet*, Industrial Research Centre, Tripoli.

MAYOUF, G. and TURNER, B. 2000. Palaeohydraulic analysis of the lower part of the Hassaounah Formation, Libya (abstract only). *Second symposium on the sedimentary basins of Libya, Tripoli, Geology of northwest Libya, November 2000.* Book of abstracts, p. 65.

MAZHAR, A. and ISSAWI, B. 1977. Sheet Zawiyat Msus (NH 34-3), *Geological Map of Libya, scale 1:250,000, Explanatory Booklet*, Industrial Research Centre, Tripoli.

McDOUGALL, N. and MARTIN, M. 2000. Facies models and sequence stratigraphy of upper Ordovician outcrops in the Murzuk Basin, SW Libya. *Symposium on Geological Exploration in Murzuq Basin (Eds. M.A. Sola and D. Worsley)*, Elsevier, Amsterdam, p. 223-236.

MEGERISI, M. and MAMGAIN, V.D. 1980a. The Upper Cretaceous-Tertiary formations of northern Libya: a synthesis. *Bulletin No. 12.* Industrial Research Centre, Tripoli, 85p.

MEGERISI, M. and MAMGAIN, V.D. 1980b. The Upper Cretaceous-Tertiary formations of northern Libya. *Second Symposium on the Geology of Libya.* vol. 1 *(eds. M.J. Salem and M.T. Busrewil)*, Academic Press, London, p. 67-72.

MEGERISI, M. and MAMGAIN, V.D. 1980c. Al Khowaymat Formation – an enigma in the stratigraphy of northeastern Libya. *Second Symposium on the Geology of Libya.* vol. 1 *(eds. M.J. Salem and M.T. Busrewil)*, Academic Press, London, p. 73-90.

MEISTER, E.M., ORTIZ, E.F., PIEROBON, E.S.T., ARRUDA, A.A. and OLIVEIRA, M.A.M. 1991. The origin and migration fairways of petroleum in the Murzuq Basin, Libya: an alternative exploration model. *Third Symposium on the Geology of Libya.* vol. 7 *(eds. M.J. Salem and M.T. Busrewil and A.M. Ben Ashour)*, Elsevier, Amsterdam, p. 2725-2742.

MENCHIKOFF, N. 1927. Etude petrographique des roches cristalines et volcaniques de la region d'Ouenat. *Bull. Soc. Geol. de Fr.* Ser. 4, No. 27, p 337-354.

MENGOLI, S. and SPINICCI, G. 1984. Tectonic evolution of North Africa (from Algeria to Sinai). *Seminar on Source and Habitat of Petroleum in the Arab Countries*, Kuwait. Proc. p. 119-174.

MENNIG, J.J., VITTIMBERGA, P. and LEHMANN, P. 1963. Étude sédimentologique et pétrographique de la formation Ras Hamia (Trias moyen) du nord-ouest de la Libye. *First Saharan Symposium. Rev.* Inst. Fran. Pet. special volume, p. 186-201.

MERGL, M. and MASSA, D. 1992. Devonian and Lower Carboniferous brachiopods and bivalves from western Libya. *Collection 'Biostratigraphie du Paleozoique', No. 12.* Universite Claude Bernard, Lyon. 216pp.

MERGL, M. and MASSA, D. 2000. A palaeontological review of the Devonian and Carboniferous succession of the Murzuq Basin and the Djado sub-basin. *Symposium on Geological Exploration in Murzuq Basin (Eds. M.A. Sola and D. Worsley),* Elsevier, Amsterdam, p. 41-88.

METTER, R.E. and FRASER, W.W. 1972. Geochemical study of crude oils from Ghadames Basin, western Libya. (abstract only). *Bull. Am. Assoc. Pet. Geol.* vol. 56, p. 639.

MIJALKOVIĆ, N. 1977a. Sheet Al Qaddahiyah (NH 33-3), *Geological Map of Libya, scale 1:250,000, Explanatory Booklet,* Industrial Research Centre, Tripoli.

MIJALKOVIĆ, N. 1977b. Sheet Qasr Sirt (NH 33-4), *Geological Map of Libya, scale 1:250,000, Explanatory Booklet,* Industrial Research Centre, Tripoli.

MIKBEL, S.R. 1977. Basement configuration and structure of west Libya. *Libyan Jour. Sci.* vol. 7A, p. 19-34.

MIKBEL, S.R. 1979. Structural and configuration map of the basement of east and central Libya. *Abh. N. Jb. Geol. Paläont.,* vol. 158, p. 209-220.

MIKBEL, S.R. 1981. Major basement structures and their relation to some oil fields in Libya (north of Lat. 26°N). *Zeit. Deut. Geol. Ges.* vol 132, p. 547-554.

MILLER, V.C. 1971. A preliminary investigation of the geomorphology of the Jebel Nefusa. *First Symposium on the Geology of Libya (ed. C. Gray).* Faculty of Science, University of Libya, Tripoli, p. 365-386.

MISALLATI, A.A. 2000. Tectonic and geological setting of the Pelagian Basin, NW Libya and eastern Tynisia (abstract only). *Second symposium on the sedimentary basins of Libya, Tripoli, Geology of northwest Libya, November 2000.* Book of abstracts, p. 67-68.

MOLYNEUX, S.G. and PARIS, F. 1985. Late Ordovician Palynomorphs. In: *The palynostratigraphy of northeast Libya (eds. B. Thusu and B. Owens).* Journ. Micropalaeontology, vol. 4, p. 11-26.

MOLYNEUX, S.G. 1988. Late Ordovician acritarchs from northeast Libya. In: *Subsurface palynostratigraphy of northeast Libya. (eds. A. El-Arnauti, B. Owens and B. Thusu).* Research Centre, Garyounis University, Benghazi, p. 45-60.

MONTGOMERY, S.L. 1993. Ghadames Basin of north central Africa: stratigraphy, geologic history and drilling summary. *Petroleum Frontiers, Pet. Inf. Corp., Littleton, Colorado.* vol. 10, no. 3, 51p.

MONTGOMERY, S.L. 1994a. Ghadames Basin and surrounding areas, structure, tectonics, geochemistry and field summaries. *Petroleum Frontiers, Pet. Inf. Corp., Littleton, Colorado.* vol. 10, no. 4, 79p.

MONTGOMERY, S.L. 1994b. Sirte Basin, north-cetral Libya: prospects for the future. *Petroleum Frontiers, Pet. Inf. Corp., Littleton, Colorado.* vol. 11, no. 1, 94p.

MOODY, R.T.J. and SANDMAN, R.I. 2000. Nummulites as indicators of environmental and depositional conditions (abstract only). *Int. workshop on north African micropalaeo. for pet. expl., Univ. Coll. London, August 2000.* Book of abstracts, p. 61.

MOODY, R.T.J., LOUCKS, R.G., BROWN, A.A. and SANDMAN. R.I. 2000 Nummulite deposits of the Pelagian area – depositional models and diagenesis (abstract only). *Second symposium on the sedimentary basins of Libya, Tripoli, Geology of northwest Libya, November 2000.* Book of abstracts, p. 70.

MOREAU-BENOIT, A. 1988. Considerations nouvelles sur la palynozonation du Devonien moyen et superieur du bassin de Rhadames (Ghadames), Lybie occidental. *C.R. Acad. Sci. Paris.* vol. 307, p. 863-869.

MORGAN, M.A., GROCOTT, J. and MOODY, R.T.J. 1998. The structural evolution of the Zaghouan-Ressas structural belt, northern Tunisia. In: *Petroleum Geology of North Africa. (ed. D.S. Macgregor, R.T.J. Moody, D.D. Clark-Lowes),* Geol. Soc. Special Publication No. 132, p. 405-422.

MOUZUGHI, A.J. 1991. Petrography and lithofacies of the Satal Formation in the northwestern Sirt Basin, Libya. *Third Symposium on the Geology of Libya.* vol. 5 *(eds. M.J. Salem and M.N. Belaid),* Elsevier, Amsterdam, p. 1841-1854.

MRÁZEK, P. 1984. Sheet Tanahmu (NG 33-6), *Geological Map of Libya, scale 1:250,000, Explanatory Booklet,* Industrial Research Centre, Tripoli.

MRIHEEL, I. Y. 1991. Diagenetic history of early-middle Eocene Jdeir Formation, Farwah Group, northwest Libyan offshore. *Petroleum Res. Journ. (Tripoli).* vol. 3, p. 44-52.

MRIHEEL, I.Y. 1994. Hydrocarbon potential of the Beda

depositional facies over the new tectonic element, the inferred horst, south-central Sirt Basin, Libya. In: *Proc. 4th Tunisian Pet. Explor. Conf. (Ed. A. Chine and K. Ben Hassine)*. ETAP, Tunis, p. 65-66.

MRIHEEL, I.Y. and ALHNAISH, A.S. 1995. Study of the Messinian carbonate-evaporite lithofacies offshore western Libya. *Terra Nova*, vol. 7, p. 213-220.

MRIHEEL, I.Y. and ANKETELL, J.M. 2000. Dolomitization of the early Eocene Jirani Dolomite Formation, Gabes-Tripoli Basin, western offshore, Libya. *Journ. Petrol. Geol.* vol. 24, p. 449-474.

MUFTAH, A.M. and MOHAMMED, A.H. 2000. Palaeoenvironment and facies analysis of the Antelat Formation in western Cyrenaica region, NE Libya (abstract only). *5th Intern. conf. on the geology of the Arab world, Cairo University*. Book of abstracts, p. 147.

NAIRN, A.E.M. and SALAJ, J. 1991. Al Gharbiyah Formation, Upper Campanian-Upper Maastrichtian (northwest Libya). *Third Symposium on the Geology of Libya*. vol. 4 *(eds. M.J. Salem, O.S. Hammuda and B.A. Eliagoubi)*, Elsevier, Amsterdam, p. 1621-1636.

NOVOVIĆ, T. 1977a. Sheet Djeneien (NH 32-3), *Geological Map of Libya, scale 1:250,000, Explanatory Booklet*, Industrial Research Centre, Tripoli.

NOVOVIĆ, T. 1977b. Sheet Nalut (NH 32-4), *Geological Map of Libya, scale 1:250,000, Explanatory Booklet*, Industrial Research Centre, Tripoli.

OJALEY, R. 2000 Geochemical investigation of crude oils from concession 47 in the Sirte Basin (abstract only). *5th Intern. conf. on the geology of the Arab world, Cairo University*. Book of abstracts, p. 132.

OUN, K.M., LIEGEOIS, J.P. and DALY, S. 2000. Evolution of the Pan-African Jabal al Hasawinah granites (abstract only). *Second Symposium on the Sedimentary Basins of Libya, Geology of Northwest Libya*. Book of abstracts, p. 71.

PACHUR, H.J. 1996. Reconstruction of palaeodrainage systems in Sirt Basin and the area surrounding the Tibisti Mountains: implications for the hydrological history of the region. *First Symposium on the Sedimentary Basins of Libya, Geology of the Sirt Basin*. vol. 1. *(eds. M.J. Salem, A.J. Mouzughi and O.S. Hammuda)*, Elsevier, Amsterdam, p. 157-168.

PAJOLA, M. and RICCHIUTO, T. 2000. Reservoir geochemistry study: an integrated approach on A-NC 118 structure (abstract only). *Second symposium on the sedimentary basins of Libya, Tripoli, Geology of northwest Libya, November 2000*. Book of abstracts, p. 72.

PALLAS, P. 1980. Water resources of the Socialist People's Libyan Arab Jamahiriya. *Second Symposium on the Geology of Libya*. vol. 2 *(eds. M.J. Salem and M.T. Busrewil)*, Academic Press, London, p. 539-594.

PARIS, F., RICHARDSON, J.B., RIEGEL, W., STREEL, M. and VANGUESTAINE, M. 1985. Devonian (Emsian-Famennian Palynomorphs. In: *The palynostratigraphy of north-east Libya (eds. B. Thusu and B. Owens)*. Journ. Micropalaeontology, vol. 4, p. 49-82.

PARIS, F. 1988a. Late Ordovician and early Silurian chitinozoans from central and southern Cyrenaica. In: *Subsurface palynostratigraphy of northeast Libya. (eds. A. El-Arnauti, B. Owens and B. Thusu)*. Research Centre, Garyounis University, Benghazi, p. 61-72.

PARIS, F. 1988b. New chitinozoans from the late Ordovician-late Devonian of northeast Libya. In: *Subsurface palynostratigraphy of northeast Libya. (eds. A. El-Arnauti, B. Owens and B. Thusu)*. Research Centre, Garyounis University, Benghazi, p. 73-88.

PARIS, F., ELAOUAD-DEBBAJ, Z., JAGLIN, J.C., MASSA, D. and OULEBSIR, L. 1995. Chitinozoans and late Ordovician glacial events in Gondwana. In: *Ordovician Odyssey: Short Papers for the 7th Int. Symp. on the Ordovician System (Eds. J.D. Cooper, M.L. Drosier, and S.C. Finney)*. SEPM, Fullerton, California, p. 171-177.

PAŘIZEK, A., KLEN, L. and RÖHLICH, P. 1984. Sheet Idri (NG 33-1), *Geological Map of Libya, scale 1:250,000, Explanatory Booklet*, Industrial Research Centre, Tripoli.

PARONA, C.F. 1914. Per la geologia della Tripolitania. *Atti. R. Acad Sci, Torino*, vol. 50, p. 1-26.

PARSONS, M.G., ZAGAAR, A.M. and CURRY, J.J. 1980. Hydrocarbon occurrences in the Sirte Basin, Libya. In: *Facts and principles of World petroleum occurrence (Ed. A.D. Miall)*. Canadian Soc. Pet. Geol., Mem. 6, p. 723-732.

PEDLEY, H.M., HOUSE, M.R. and WAUGH, B. 1978. The geology of the Pelagian Block: the Maltese Islands. In: *The Ocean Basins and Margins*, vol. 4B: *The Western Mediterranean (eds. A.E.M. Nairn, W.H. Kanes, and F.G. Stehli)*. Plenum Press, New York, p. 417-435.

PESCE, A. 1966. Uau en Namus. In: *South-central Libya and northern Chad (Conf. Chairm. J.J. Williams)*. Petroleum Exploration Society of Libya, Eighth Annual Field Conference, p. 47-52.

PESCE, A. 1968. Gemini space photographs of Libya and Tibesti. A geological and geographical analysis. Petroleum Exploration Society of Libya, special publication. 81pp.

PESCE, A. 1971. Erg Idrisi and Hammada ibn Battutah: two 'new' geographical features in southeastern Libya. *First Symposium on the Geology of Libya (ed. C. Gray).* Faculty of Science, University of Libya, Tripoli, p. 351-364.

PETERSON, J.A. 1985. Geology and petroleum resources of north-central and northeastern Africa. *U.S. Dept of Int. Geol. Surv.* Open file rept. 85-709, 54p.

PETERSON, J.A. 1994. Regional geology and hydrocarbon resource potential, the Mediterranean Sea region. *U.S. Dept. of Int. Geol. Surv.* Open file rept. OF 94-0166, 118p.

PETZET, G.A. 1999. Patience, persistence preceded Elephant discovery in Libya. *Oil and Gas Journal,* vol. 97, pt, 20, p. 65-66.

PEYBERNÈS, B. 1991. The Jurassic of Tunisia: an attempt at reconstruction of the south Neotethyan margin during and after the rifting phase. *Third Symposium on the Geology of Libya.* vol. 4 *(eds. M.J. Salem, O.S. Hammuda and B.A. Eliagoubi),* Elsevier, Amsterdam, p. 1681-1709.

PFLUGER, F. 1998. Oxygenation facies as indicated by characteristic matground structures: the Akakus-Tanezzouft transition as an extraordinary window to inaccessible times and places (abstract only). *Symposium on Geological Exploration in Murzuq Basin, Sabha, 20-22nd September 1998.* Book of abstracts.

PHILLIPS, R.W. 1971. Biostratigraphic study in the Sirte Basin, Libya. *First Symposium on the Geology of Libya (ed. C. Gray).* Faculty of Science, University of Libya, Tripoli, p. 157-166.

PICCOLI, G. 1971. Outlines of volcanism in northern Tripolitania. *First Symposium on the Geology of Libya (ed. C. Gray).* Faculty of Science, University of Libya, Tripoli, p. 323-332.

PICKFORD, M. 1991. Biostratigraphic correlation of the Middle Miocene mammal locality of Jabal Zaltan, Libya. *Third Symposium on the Geology of Libya.* vol. 4 *(eds. M.J. Salem, O.S. Hammuda and B.A. Eliagoubi),* Elsevier, Amsterdam, p. 1483-1490.

PIEROBON, E.S.T. 1991. Contribution to the stratigraphy of the Murzuq Basin, SW Libya. *Third Symposium on the Geology of Libya.* vol. 5 *(eds. M.J. Salem and M.N. Belaid),* Elsevier, Amsterdam, p. 1766-1784.

PIETERSZ, C.R. 1968. Proposed nomenclature for rock units in northern Cyrenaica. In: *Geology and archaeology of northern Cyrenaica, Libya (ed. Barr, F.T.).* Petroleum Exploration Society of Libya, Tenth Annual Field Conference, p. 149-158.

PLAUCHUT, B. and FAURE, H. 1960. Notice explicative sur la carte geologique du bassin du Djado (feuilles Djado et Touro). *Bur. Rech. Geol. Min.* Dakar. 38p.

POMEYROL, R. 1969. A catharsis on the term Nubian Sandstone. In: *Geology, archaeology and prehistory of the southwestern Fezzan, Libya (ed. Kanes, W.H.).* Petroleum Exploration Society of Libya, Eleventh Annual Field Conference, p. 131-138.

POYNTZ, I. 1995. Hydrocarbon potential of the Tadrart and Ouan Kasa Formations (Lower Devonian), Ghadames Basin, NW Libya. (abstract). In: *Symposium on the Hydrocarbon geology of north Africa (Geol. Soc. London).* Abstracts volume, p. 45.

PRATSCH, J. 1994. Tunisia's oil and gas potential and future plays. In: *Proc. 4th Tunisian Pet. Explor. Conf. (eds. A. Chine and K. Ben Hassine).* ETAP, Tunis, p. 1-25.

PROTIĆ, D. 1984a. Sheet Bi'r Anzawa (NG 32-3), *Geological Map of Libya, scale 1:250,000, Explanatory Booklet,* Industrial Research Centre, Tripoli.

PROTIĆ, D. 1984b. Sheet Tikiumit (NG 32-7), *Geological Map of Libya, scale 1:250,000, Explanatory Booklet,* Industrial Research Centre, Tripoli.

RACEY, A., BAILEY, H.W., BECKETT, D., GALLAGHER, L.T., HAMPTON, M.J., and McQUILKEN, J. 2001. The petroleum geology of the early Eocene El Garia Formation, Hasdrubal field, offshore Tunisia. *Jour. Petrol. Geol.* vol. 24, p. 29-53.

RACEY, A. 2001. A review of Eocene nummulite accumulations: structure, formation and reservoir potential. *Journ. Petrol. Geol.* vol. 24, p. 79-100.

RADULOVIĆ, P. 1984a. Sheet Wadi Tanezzuft (NG 32-11), *Geological Map of Libya, scale 1:250,000, Explanatory Booklet,* Industrial Research Centre, Tripoli.

RADULOVIĆ, P. 1984b. Sheet Ghat (NG 32-15), *Geological Map of Libya, scale 1:250,000, Explanatory Booklet,* Industrial Research Centre, Tripoli.

RASUL, S. 2000. Cambrian microplankton from central Sirt Basin (abstract only). *Intern. Workshop on North African Micropalaeontology for Petroleum exploration. Univ. Coll. London, 21-25 August 2000.* Book of abstracts, p. 64.

RATSCHILLER, L.K. 1967. Sahara: correlazioni geologico-litostratigrafiche fra Sahara centrale e occidentale. *Mem. Mus. Tridentino di Sci. Nat.,* Trieste. vol. 15, p. 53-293.

REALI, S., RONCHI, P. and BORROMEO, O. 2000. Sedimentological model of El Garia Formation (NC 41 offshore Libya) (abstract only). *Second*

476

symposium on the sedimentary basins of Libya, Tripoli, Geology of northwest Libya, November 2000. Book of abstracts, p. 76.

REYMENT, R.A. and REYMENT, E.R. 1980. The Palaeocene Trans-Saharan transgression and its ostracod fauna. *First Symposium on the Geology of Libya (ed. C. Gray).* Faculty of Science, University of Libya, Tripoli, p. 245-254.

RIAD, S., EL-ETR, H.A. and MOHAMMED, M.A. 1980. Gravity-tectonic trend analysis in Siwa-Al Jaghbub region, NE Africa. *Second Symposium on the Geology of Libya.* vol. 3 *(eds. M.J. Salem and M.T. Busrewil),* Academic Press, London, p. 979-992.

RICCHIUTO, T. and PAJOLA, M. 2000. Libya NC 41: Integration of chemical and isotopic analyses (abstract only). *Second Symposium on the Sedimentary Basins of Libya. The Geology of northwest Libya.* Book of abstracts, p. 78.

RICHARDSON, J.B. and IOANNIDES, N. 1973. Silurian palynomorphs from the Tanezzuft and Acacus formations, Tripolitania. *Micropaleontology,* vol. 19, p. 257-307.

RICHARDSON, J.B. 1988. Late Ordovician and early Silurian cryptospores and miospores from northeast Libya. In: *Subsurface palynostratigraphy of northeast Libya. (eds. A. El-Arnauti, B. Owens and B. Thusu).* Research Centre, Garyounis University, Benghazi, p. 89-110.

RICOU, L.E. 1994. Tethys reconstructed: plates, continental fragments and their boundaries since 260 Ma from central America to south-eastern Asia. *Geodinamica Acta,* vol. 7, p. 169-218.

ROBERTS, J.M. 1970. Amal field, Libya. In: *Geology of giant petroleum fields. (ed. M.T. Halbouty).* Mem. Amer. Assoc. Pet. Geol. No. 14, p. 438-448.

RÖHLICH, P. 1974. Sheet Al Bayda (NI 34-15), *Geological Map of Libya, scale 1:250,000, Explanatory Booklet,* Industrial Research Centre, Tripoli.

RÖHLICH, P. 1978. Geological development of Jabal al Akhdar, Libya. *Geol Rundsch.* vol. 67, p. 401-412.

RÖHLICH, P. 1979. Sheet Ghadames (NH 32-7), *Geological Map of Libya, scale 1:250,000, Explanatory Booklet,* Industrial Research Centre, Tripoli.

RÖHLICH, P. 1980. Tectonic development of Al Jabal al Akhdar. *Second Symposium on the Geology of Libya.* vol. 3 *(eds. M.J. Salem and M.T. Busrewil),* Academic Press, London, p. 923-932.

RÖHLICH, P. and YOUSHAH, B.M. 1991. The Ghadamis fault – a disputed structure in NW Libya. *Third Symposium on the Geology of Libya.* vol. 6 *(eds. M.J.*

Salem, A.M. Sbeta and M.R. Bakbak), Elsevier, Amsterdam, p. 2371-2380.

RÖHLICH, P., SALAJ, J. and TRÖGER, K.A. 1996. Palaeontological dating of the pre-Campanian unconformity in the Gawt Sas area. *First Symposium on the Sedimentary Basins of Libya, Geology of the Sirt Basin.* vol. 1. *(eds. M.J. Salem, A.J. Mouzughi and O.S. Hammuda),* Elsevier, Amsterdam, p. 265-286.

ROLLAND, M. G. 1880. Sur le terrain cretacé du Sahara septentionale. *Bull. Soc. Geol. Fr.* vol. 9, p. 508-551.

RONČEVIĆ, G. 1984. Sheet Hasi Anjiwal (NG 32-4), *Geological Map of Libya, scale 1:250,000, Explanatory Booklet,* Industrial Research Centre, Tripoli.

ROOHI, M. 1996a. A geological view of source-reservoir relationships in the western Sirt Basin. *First Symposium on the Sedimentary Basins of Libya, Geology of the Sirt Basin.* vol. 2. *(eds. M.J. Salem, A.S. El-Hawat and A.M. Sbeta),* Elsevier, Amsterdam, p. 323-336.

ROOHI, M. 1996b. Geological history and hydrocarbon migration pattern of the central Az Zahrah-Al Hufrah Platform. *First Symposium on the Sedimentary Basins of Libya, Geology of the Sirt Basin.* vol. 2. *(eds. M.J. Salem, A.S. El-Hawat and A.M. Sbeta),* Elsevier, Amsterdam, p. 435-454.

ROSSI, M.E., TONNA, M. and LARBASH, M. 1991. Latest Jurassic-early Cretaceous deposits in the subsurface of the eastern Sirt Basin (Libya): facies and relationships with tectonics and sea-level changes. *Third Symposium on the Geology of Libya.* vol. 6 *(eds. M.J. Salem, A.M. Sbeta and M.R. Bakbak),* Elsevier, Amsterdam, p. 2211-2226.

RUBINO, J.L., GALEAZZI, A., BOUAZIZ, S., SBETA, A.M. and ASKER, A. 2000. Depositional facies and sequence stratigraphy of the Triassic series of the Libyan-Tunisian Jabal Nafusah: a preliminary report (abstract only). *Second Symposium on the Sedimentarty basins of Libya, Geology of Northwest Libya.* Book of abstracts, p. 79.

RUBINO, J.L., ANFRAY, R., BLANPIED, C., GHIENNE, J.F. and MANATSCHAL, G. 2000. Meander belt complex within the lower Mamuniyat Formation in western Al Qarqaf Arch (abstract only). *Second Symposium on the Sedimentarty basins of Libya, Geology of Northwest Libya.* Book of abstracts, p. 80.

RUBINO, J.L. and BLANPIED, C. 2000. Sedimentology and sequence stratigraphy of the Devonian to lowermost Carboniferous succession on the Gargaf Uplift (Murzuq Basin, Libya). *Symposium on Geological Exploration in Murzuq Basin (Eds. M.A.*

Sola and D. Worsley), Elsevier, Amsterdam, p. 321-348.

RUSSEGER, J. 1837. Kreide und sandstein. Einflus von granit auf leztern. *Neues Jb. Mineral.* Berlin. p. 665-669.

SABAOU, N. and TURNER, P. 2001. Triassic palaeosols in the Berkine Basin (abstract only). *Abstracts, North Africa Research Workshop, Oxford Brookes University, 4th-5th September 2001.*

SAID, F.M. 1974. Sedimentary history of the Palaeozoic rocks of the Ghadames Basin. *M.Sc thesis*, Univ. of South Carolina

SAID, M. 1981. Sheet Bunjim (NH 33-7), *Geological Map of Libya, scale 1:250,000, Explanatory Booklet*, Industrial Research Centre, Tripoli.

SAID, R. 1962. The Geology of Egypt. Elsevier, Amsterdam. 337p.

SALAJ, J. 1979. Sheet Al Qaryat al Gharbiyah (NH 33-5), *Geological Map of Libya, scale 1:250,000, Explanatory Booklet*, Industrial Research Centre, Tripoli.

SALAJ, J. 1984. Cretaceous/Tertiary boundary events in Tunisia and Libya. *Journ. Geol. Soc. Iraq.* vol. 16, p. 197-225.

SALEM, M.J. and SPRENG, A.C. 1980. Middle Miocene stratigraphy, Al Khums area, northwestern Libya. *Second Symposium on the Geology of Libya.* vol. 1 *(eds. M.J. Salem and M.T. Busrewil)*, Academic Press, London, p. 97-116.

SALVADOR, A. (Ed.) 1994. International stratigraphic guide. *International Union of Geological Sciences*, Geol. Soc. Amer.

SANDER, N.J. 1968. The pre-Mesozoic structural evolution of the Mediterranean region. In: *Geology and archaeology of northern Cyrenaica, Libya (ed. Barr, F.T.)*. Petroleum Exploration Society of Libya, Tenth Annual Field Conference, p. 47-70.

SANDER, N.J. 1970. Structural evolution of the Mediterranean region during the Mesozoic era. In: *Geology and history of Sicily. (eds W. Alvarez and K.H.A. Gohrbandt)*. Petroleum Exploration Society of Libya, Twelfth Annual Field Conference, p. 43-132.

SANDFORD, K.S. 1935. Geological observations on the north-west frontiers of the Anglo-Egytian Sudan and the adjoining part of the southern Libyan desert. *Quart. Journ. Geol. Soc.* London, vol. 91, p. 323-381.

SANFORD, R.M. 1970. Sarir oil field, Libya – desert surprise. *Geology of giant petroleum fields. (ed. M.T. Halbouty)*. Mem. Amer. Assoc. Pet. Geol. No. 14, p. 449-476.

SAVAGE, R.J.G. 1971. Review of the fossil mammals of Libya. *First Symposium on the Geology of Libya (ed. C. Gray)*. Faculty of Science, University of Libya, Tripoli, p. 215-226.

SBETA, A.M. 1990. Stratigraphy and lithofacies of Farwah Group and its equivalent: offshore NW Libya. *Petrol. Res. Jour. (Tripoli)*, vol. 2, p. 42-56.

SBETA, A.M. 1991. Petrography and facies of the Middle and Upper Eocene rocks (Tellil Group), offshore western Libya. *Third Symposium on the Geology of Libya.* vol. 5 *(eds. M.J. Salem and M.N. Belaid)*, Elsevier, Amsterdam, p. 1929-1966.

SCHÄFER, K., KRAFT, K.H., HÄUSLER, H. and ERDMANN, J. 1980. In situ stresses and palaeostresses in Libya. *Second Symposium on the Geology of Libya.* vol. 3 *(eds. M.J. Salem and M.T. Busrewil)*, Academic Press, London, p. 907-922.

SCHIETTECATTE, J.P. 1972. A new Cretaceous outcrop in northeastern Cyrenaica. *Libyan Jour. Sci*, vol. 2, p. 59-64.

SCHNEITER, A. 2000. Libya area NC 177. Hydrocarbon potential of the En Naga sub-basin, southwest Sirte Basin, Libya: the story of a successful partnership (abstract only). *Symposium on petroleum systems and evolving technologies in African E & P. Burlington House, London, 16-18 May, 2000*. Book of abstracts, p. 23.

SCHRÖTER, T. 1996. Tectonic and sedimentary development of the central Zallah Trough (west Sirt Basin, Libya). *First Symposium on the Sedimentary Basins of Libya, Geology of the Sirt Basin.* vol. 3. *(eds. M.J. Salem, M.T. Busrewil, A.A. Misallati and M.J. Sola)*, Elsevier, Amsterdam, p. 123-136.

SCHURMANN, H.M.E. 1974, The Pre-Cambrian in North Africa. Brill, Leyden. 352 pp.

SCOTESE, C.R., BOUCOT, A.J. and McKERROW, W.S. 1999. Gondwanan palaeogeography and palaeoclimatology. *Journ. Afr. Earth Sci.*, vol. 28, p. 99-114.

SEIDL, K. and RÖHLICH, P. 1984. Sheet Sabha (NG 33-2), *Geological Map of Libya, scale 1:250,000, Explanatory Booklet*, Industrial Research Centre, Tripoli.

SEILACHER, A. 1969. Sedimentary rhythms and trace fossils in Paleozoic sandstones in Libya. In: *Geology, archaeology and prehistory of the southwestern Fezzan, Libya (ed. Kanes, W.H.)*. Petroleum Exploration Society of Libya, Eleventh Annual Field Conference, p. 117-124.

SEILACHER, A. 1991. An updated *Cruziana* stratigraphy of Gonwanan Palaeozoic sand-stones. *Third Symposium on the Geology of Libya.* vol. 4 *(eds. M.J. Salem, O.S. Hammuda and B.A. Eliagoubi)*, Elsevier, Amsterdam, p. 1565-1582.

SEILACHER, A. 1993. Problems of correlation in the Nubian sandstone facies. In: *Geoscientific research in northeast Africa. (Eds. U. Thorweihe and H. Schandelmeier)*, Balkema, Rotterdam, p. 329-333.

SEILACHER, A. 1996. Evolution of burrowing behaviour in Silurian trilobites: ichnosubspecies of *Cruziana acacensis. First Symposium on the Sedimentary Basins of Libya, Geology of the Sirt Basin.* vol. 1. *(eds. M.J. Salem, A.J. Mouzughi and O.S. Hammuda)*, Elsevier, Amsterdam, p. 523-530.

SEILACHER, A. 2000. Ordovician and Silurian arthrophycid ichnostratigraphy. *Symposium on Geological Exploration in Murzuq Basin (Eds. M.A. Sola and D. Worsley)*, Elsevier, Amsterdam, p. 237-258.

SELLEY, R.C. 1966. The Miocene rocks of the Marada and Jebel Zelten area, central Libya: a study of shoreline sedimentation. *Pet. Explor. Soc. Libya.* Guidebook, 30p.

SELLEY, R.C. 1967. Palaocurrents and sediment transport in nearshore sediments of the Sirte Basin, Libya. *Jour. Geology*, vol. 75, p. 215-223.

SELLEY, R.C. 1968. Facies profiles and other new methods of graphic data presentation: application in a quantitative study of Libyan Tertiary shoreline deposits. *Journ. Sediment. Petrol.* vol. 35, p. 363-372.

SELLEY, R.C. 1969. Nearshore marine and continental sediments of the Sirte Basin, Libya. *Quart. Journ. Geol. Soc.* vol. 124, p. 419-460.

SELLEY, R.C. 1971. Structural control of Miocene sedimentation in the Sirte Basin. *First Symposium on the Geology of Libya (ed. C. Gray).* Faculty of Science, University of Libya, Tripoli, p. 99-106.

SELLEY, R.C. 1997a. The sedimentary basins of northwest Africa: stratigraphy and sedimentation. In: *Sedimentary basins of the world (series ed. K.J. Hsu)*, part 3: *African Basins (ed. R.C. Selley)*, Elsevier, Amsterdam, chap. 1, p. 3-16.

SELLEY, R.C. 1997b. The basins of northwest Africa: structural evolution. In: *Sedimentary basins of the world (series ed. K.J. Hsu)*, part 3: *African Basins (ed. R.C. Selley)*, Elsevier, Amsterdam, chap. 2, p. 17-26.

SELLEY, R.C. 1997c. The Sirte Basin of Libya. In: *Sedimentary basins of the world (series ed. K.J. Hsu)*, part 3: *African Basins (ed. R.C. Selley)*, Elsevier, Amsterdam, chap. 3, p. 27-37.

SEMTNER, A.K. and KLITZSCH, E. 1994. Early Paleozoic paleogeography of the northern Gondwanan margin: new evidence for Ordovician-Silurian glaciation. *Geol. Rundsch.* vol. 83, p. 743-751.

SGHAIR, A.M.A. 1996. Petrography, diagensesis and provenance of the Bahi Formation in the western part of the Sirt Basin, Libya. *First Symposium on the Sedimentary Basins of Libya, Geology of the Sirt Basin.* vol. 2. *(eds. M.J. Salem, A.S. El-Hawat and A.M. Sbeta)*, Elsevier, Amsterdam, p. 65-82.

SGHAIR, A.M.A. and El ALAMI, M.A. 1996. Depositional environment and diagenetic history of the Maragh Formation, NE Sirt Basin, Libya. *First Symposium on the Sedimentary Basins of Libya, Geology of the Sirt Basin.* vol. 2. *(eds. M.J. Salem, A.S. El-Hawat and A.M. Sbeta)*, Elsevier, Amsterdam, p. 263-274.

SGHAIR, A.M. and ANKETELL, J.M. 2001. Lithofacies, facies association and depositional environments of the Acacus Sandstone Formation (late Silurian), Ghadamis Basin, Libya (abstract only). *Abstracts, North Africa Research Workshop, Oxford Brookes University, 4th-5th September 2001.*

SHACKLETON, R.M. and GRANT, N.K. 1974. Nature and age of basement rocks from the Sarir oil field, Libya. *Ann. Rept. Inst. Afr. Geol. (ed. M.P. Coward).* Univ. Leeds. vol. 18, p. 3-6.

SHAH, S.H.A., MANSOURI, A. and EL GHOUL, M. 1993. Palaeozoic sandstone reservoirs of the Hamada Basin, NW Libya: effects of synsedimentary processes on porosity. *Journ. Pet. Geol.* vol. 16. p. 345-352.

SHAKOOR, A. and SHAGRONI, Y. 1984. Sheet Hun (NH 33-11), *Geological Map of Libya, scale 1:250,000, Explanatory Booklet*, Industrial Research Centre, Tripoli.

SHELMANI, M., THUSU, B. and EL-ARNAUTI, A. 1992. Subsurface occurrences of middle and upper Triassic sediments in eastern Libya. In: *Geology of the Arab World., vol 2., (ed. A. Sadek).* Cairo University, p. 233-240.

SHERIF, K.A.T. 1991. Biostratigraphy of the Miocene in Al Khums area, northwestern Libya. *Third Symposium on the Geology of Libya.* vol. 4 *(eds. M.J. Salem, O.S. Hammuda and B.A. Eliagoubi)*, Elsevier, Amsterdam, p. 1421-1456.

SIKANDER, A.H. 2000. The geology, structure and hydrocarbon potential of the Ghadamis and Murzuq Basins – an overview (abstract only). *Second Symposium on the Sedimentary Basins of Libya. The Geology of northwest Libya.* Book of abstracts, p. 87.

SIKANDER, A.H., BASU, S. and RASUL, S.M. 2000. Geochemical source-maturation and volumetric evaluation of lower Palaeozoic source rocks in the west Libyan basins (abstract only). *Second Symposium on the Sedimentary Basins of Libya. The*

Geology of northwest Libya. Book of abstracts, p. 88.

SINGH, G.D.S. 1980, Structural control of groundwater flow in the Mesozoic sandstone aquifers of the eastern part of the Jabal Nafusah, Libya. *Second Symposium on the Geology of Libya.* vol. 2 *(eds. M.J. Salem and M.T. Busrewil)*, Academic Press, London, p. 753-762.

SINHA, R.H. and BEN RAHUMA, M.M. 2000. Preliminary geological appraisal of Ghadamis Basin, Libya (abstract only). *Second Symposium on the Sedimentary Basins of Libya. The Geology of northwest Libya.* Book of abstracts, p. 89.

SINHA, R.N. and ELAND, H.B. 1996. The pre-Cretaceous (Triassic) sequence in the subsurface of the Maradah Trough, eastern Sirt Basin, Libya. *Pet. Res. Journ. Tripoli*, vol. 3, p. 49-60.

SINHA, R.N. and MRIHEEL, I.Y. 1996. Evolution of subsurface Palaeocene sequence and shoal carbonates, south-central Sirt Basin. *First Symposium on the Sedimentary Basins of Libya, Geology of the Sirt Basin.* vol. 2. *(eds. M.J. Salem, A.S. El-Hawat and A.M. Sbeta)*, Elsevier, Amsterdam, p. 153-196.

SINHA, S.C. and PANDEY, S.M. 1980. Hydrogeological studies in a part of the Murzuq Basin using geophysical logs. *Second Symposium on the Geology of Libya.* vol. 2 *(eds. M.J. Salem and M.T. Busrewil)*, Academic Press, London, p. 629-634.

SINHA, S.C. 1980. On the application of geophysical logging in the assessment of ground-water potential in Al Hamada al Hamra Basin. *Second Symposium on the Geology of Libya.* vol. 2 *(eds. M.J. Salem and M.T. Busrewil)*, Academic Press, London, p. 643-658.

SMART, J. 2000. Seismic expression of depositional processes in the upper Ordovician succession of the Murzuq Basin, SW Libya. *Symposium on Geological Exploration in Murzuq Basin (eds. M.A. Sola and D. Worsley)*, Elsevier, Amsterdam, p. 397-416.

SMETANA, R. 1975. Sheet Ra's Jdeir (NI 32-16), *Geological Map of Libya, scale 1:250,000, Explanatory Booklet*, Industrial Research Centre, Tripoli.

SMITH, D.N. and KARKI, M. 1996. Basin development of the offshore Sirt basin to the west of Binghazi, Libya. *First Symposium on the Sedimentary Basins of Libya, Geology of the Sirt Basin.* vol. 1. *(eds. M.J. Salem, A.J. Mouzughi and O.S. Hammuda)*, Elsevier, Amsterdam, p. 129-138.

SMITH, A.G. and LIVERMORE, R.A. 1991. Pangea in Permian to Jurassic time. *Tectonophysics*, vol. 187, p. 135-179.

SOLA, M. and OZCICEK, B. 1990. On the hydrocarbon prospectivity of northern Cyrenaica region, Libya. *Petroleum Research Journal (Tripoli)*, vol. 2, p. 25-41.

SONNENFELD, P. 1985. Models of Upper Miocene evaporite genesis in the Mediterranean region. In: *Geological Evolution of the Mediterranean Basin. (Raimondo Selli Commemorative Volume) (eds. D.J. Stanley and F-C. Wezel)*, Springer-Verlag, Berlin., chap. 16, p. 323-346.

SPRING, D. and HANSEN, O.P. 1998. The influence of platform morphology and sea level on the development of a carbonate sequence: the Harash Formation, eastern Sirt Basin, Libya. In: *Petroleum Geology of North Africa. (ed. D.S. Macgregor, R.T.J. Moody, D.D. Clark-Lowes)*, Geol. Soc. Special Publication No. 132, p. 335-354.

SQUYRES, C.H. and BRADLEY, W. 1964. Notes on the Western Desert of Egypt. In: *Guidebook to the geology archaeology of Egypt (ed. Reilly F.A..)*. Petroleum Exploration Society of Libya, Sixth Annual Field Conference, p. 99-106.

SRIVASTAVA, R.P., SHAGRONI, Y. and NJOMA, B. 1980. Sheet An Nuwfaliyah (NH 33-8), *Geological Map of Libya, scale 1:250,000, Explanatory Booklet*, Industrial Research Centre, Tripoli.

ŠTEFEK, V. and RÖHLICH, P. 1984. Sheet Awbari (NG 33-5), *Geological Map of Libya, scale 1:250,000, Explanatory Booklet*, Industrial Research Centre, Tripoli.

STREEL, M., PARIS, F., RIEGEL, W. and VANGUES-TAINE, M. 1988. Acritarch, chitinozoan and spore stratigraphy from the middle and late Devonian of northeast Libya. In: *Subsurface palynostratigraphy of northeast Libya. (eds. A. El-Arnauti, B. Owens and B. Thusu)*. Research Centre, Garyounis University, Benghazi, p. 111-128.

SULEIMAN, I.S., KELLER, G.R. and SULEIMAN, A.S. 1991. Gravity study of the Sirt basin, Libya. *Third Symposium on the Geology of Libya.* vol. 6 *(eds. M.J. Salem, A.M. Sbeta and M.R. Bakbak)*, Elsevier, Amsterdam, p. 2461-2468.

SUTCLIFFE, O.E., ADAMSON, K. and BEN RAHUMA, M.M. 2000. The geological evolution of the Palaeozoic rocks of western Libya: a review and fieldguide. *Second Symposium on the Sedimentary Basins of Libya, Geology of northwestern Libya.* Field Guide. Earth Sciences Society of Libya. 93pp.

SUTCLIFFE, O.E., CRAIG, J. and WHITTINGTON. R.J. 2001. Stratigraphic architecture of Upper Ordovician glacigenic rocks: understanding the glaciation of West Gondwana (abstract only). *Abstracts, North Africa Research Workshop, Oxford Brookes*

University, 4th-5th September 2001.

SWEDAN, A. and ISSAWI, B. 1977. Sheet Bir Hacheim (NH 34-4), *Geological Map of Libya, scale 1:250,000, Explanatory Booklet*, Industrial Research Centre, Tripoli.

SWIRE, P.H. and GASHGESH, T.M. 2000. The bio-chrono- and lithostratigraphy of Carboniferous, Permian and Triassic sediments of the Al Jifarah Trough, NW Libya (abstract only). *Second Symposium on the Sedimentary Basins of Libya. The Geology of northwest Libya.* Book of abstracts, p. 96.

TAWADROS, E.E. 2001. Geology of Egypt and Libya. Balkema, Rotterdam, 468p.

TEKBALI, A.O. 1994. Palynological observations on the 'Nubian Sandstone' of south-western Libya. *Rev. Palaeobot. Palyn.* vol. 81, p. 297-311.

TEKBALI, A.O. and WOOD, G.D. 1991. Silurian spores, acritarchs and chitinozoans from the Bani Walid borehole of the Ghadamis Basin, northwest Libya. *Third Symposium on the Geology of Libya.* vol. 4 *(eds. M.J. Salem, O.S. Hammuda and B.A. Eliagoubi)*, Elsevier, Amsterdam, p. 1243-1274.

TEKBALI, A.O. 2000. Pollen and spores from subsurface Albian of the Kiklah Formation, NW Libya (abstract only). *Intern. Workshop on North African Micropalaeontology for Petroleum exploration. Univ. Coll. London, 21-25 August 2000.* Book of abstracts, p. 70.

TERRY, C.E. and WILLIAMS, J. J. 1969. The Idris 'A' bioherm and oilfield, Sirte Basin, Libya – its commercial development, regional Palaeocene geologic setting and stratigraphy. In: *The Exploration for petroleum in Europe and North Africa (ed. P. Hepple)*. Institute of Petroleum, London, and Elsevier, Amsterdam, p. 31-48.

THIEDIG, F. and EL CHAIR, M.M. 1998. Quaternary limnic sediments in the Murzuq Basin (abstract only). *Symposium on Geological Exploration in Murzuq Basin, Sabah, 20th-22nd September, 1998.* Book of abstracts, p. 31-32.

THOMAS, D. 1995a. Geology, Murzuk oil development could boost SW Libya prospects. *Oil and Gas Journ.* vol. 93, no. 10, p. 41-46.

THOMAS, D. 1995b. Exploration limited since '70s in Libya's Sirte Basin. *Oil and Gas Journ.* vol. 93, no. 11, p. 99-104.

THUSU, B. and VIGRAN, J.O. 1985. Middle-late Jurassic (late Bathonian-Tithonian) palynomorphs. In: *The palynostratigraphy of north-east Libya (eds. B. Thusu and B. Owens)*. Journ. Micropalaeontology, vol. 4, p.113-130.

THUSU, B. and VAN DER EEM, J.G.L.A. 1985. Early

Cretaceous (Neocomian-Cenomanian) palynomorphs. In: *The palynostratigraphy of north-east Libya (eds. B. Thusu and B. Owens)*. Journ. Micropalaeontology, vol. 4, p. 131-150.

THUSU, B., VAN DER EEM, J.G.L.A., EL-MEHDAWI, A. and BU-ARGOUB, F. 1988. Jurassic-early Cretaceous palynostratigraphy in northeast Libya. In: *Subsurface palynostratigraphy of northeast Libya. (eds. A. El-Arnauti, B. Owens and B. Thusu)*. Research Centre, Garyounis University, Benghazi, p. 171-214.

THUSU, B. 1996. Implication of the discovery of reworked and in situ late Palaeozoic and Triassic palynomorphs on the evolution of Sirt Basin, Libya. *First Symposium on the Sedimentary Basins of Libya, Geology of the Sirt Basin.* vol. 1. *(eds. M.J. Salem, A.J. Mouzughi and O.S. Hammuda)*, Elsevier, Amsterdam, p. 455-474.

TISSOT, B., ESPITALIE, J., DEROO, G., TEMPERE, C. and JONATHAN, D. 1973. Origin and migration of hydrocarbons in the eastern Sahara (Algeria). *Sixth Internat. Mtg. Org. Geochem.* p, 315-334.

TMALLA, A.F.A. 1992. Stratigraphic position of the Cretaceous-Tertiary boundary in the northern Sirt Basin, Libya. *Marine and Petrol. Geol.* vol 9, p. 542-552.

TMALLA, A.F.A. and ABUTARRUMA, Y.G. 2000. The nummulitic reservoir of the Bouri field, offshore NW Libya (abstract only). *Second Symposium on the Sedimentary Basins of Libya. The Geology of northwest Libya.* Book of abstracts, p. 97.

TRAUT, M.W., BOOTE, D.R.D. and CLARK-LOWES, D.D. 1998. Exploration history of the Palaeozoic petroleum systems of North Africa. In: *Petroleum Geology of North Africa. (ed. D.S. Macgregor, R.T.J. Moody, D.D. Clark-Lowes)*, Geol. Soc. Special Publication No. 132, p. 69-78.

TRÖGER, K.A. and RÖHLICH, P. 1991. Campanian-Maastrichtian Inoceramid (Bivalvia) assemblages from NW Libya. *Third Symposium on the Geology of Libya.* vol. 4 *(eds. M.J. Salem, O.S. Hammuda and B.A. Eliagoubi)*, Elsevier, Amsterdam, p. 1357-1382.

TRÖGER, K.A. 2000. Upper Campanian and lower Maastrichtian Inoceramids of NW Libya and Europe – a comparison (abstract only). *Second Symposium on the Sedimentary Basins of Libya. The Geology of northwest Libya.* Book of abstracts, p. 98.

TURK, T.M., DOUGHRI, A.K. and BANERJEE, S. 1980. A review of the recent investigation on the Wadi ash Shati iron ore deposits, northern Fezzan, Libya. *Second Symposium on the Geology of Libya.* vol. 3 *(eds. M.J. Salem and M.T. Busrewil)*, Academic

Press, London, p. 1019-1044.

TURNER, B.R. 1980. Palaeozoic sedimentology of the southeastern part of the Al Kufrah Basin: a model for oil exploration. *Second Symposium on the Geology of Libya*. vol. 2 *(eds. M.J. Salem and M.T. Busrewil)*, Academic Press, London, p. 351-374.

TURNER, B.R. 1991. Palaeozoic deltaic sedimentation in the southeastern part of the Al Kufrah basin, Libya: a model for oil exploration. *Third Symposium on the Geology of Libya*. vol. 5 *(eds. M.J. Salem and M.N. Belaid)*, Elsevier, Amsterdam, p. 1713-1726.

TURNER, P., PILLING, D., WALKER, D., EXTON, J., BINNIE, J. and SABAOU, N. 2001. Sequence stratigraphy and sedimentology of the late Triassic TAG-I (Blocks 401/402), Berkine Basin, Algeria (abstract only). *Abstracts, North Africa Research Workshop, Oxford Brookes University, 4th-5th September 2001.*

UNDERWOOD, J.R. and FISK, E.P. 1980. Meterorite impact structures, southeast Libya. *Second Symposium on the Geology of Libya*. vol. 3 *(eds. M.J. Salem and M.T. Busrewil)*, Academic Press, London, p. 893-900.

UNIS, M.M. (Director) 1978. National Atlas of the Socialist People's Libyan Arab Jamahiriya. Sec. Planning, Survey Dept, Tripoli. 119p.

UNRUG, R. (Ed.). 1996. Geodynamic map of Gondwana supercontinent assembly, scale 1: 10,000,000. Bur. Rech. Geol. Min. Orleans.

UNRUG, R. 1997. Rodinia to Gondwana: the geodynamic map of Gondwana supercontinent assembly. *GSA Today*, vol. 7, p. 2-6

UNRUG, R., HARANCZYK, C. and CHOCYK-JAMINSKA, M. 1999. Easternmost Avalonian and Armorican-Cadomian terranes of central Europe and Caledonian-Variscan evolution of the polydeformed Krakow mobile belt: geological constraints. *Tectonophysics*, vol. 302, p. 133-157.

UWINS, P.J.R. and BATTEN, D.J. 1988. Early to mid-Cretaceous palynology of northeast Libya. In: *Subsurface palynostratigraphy of northeast Libya. (eds. A. El-Arnauti, B. Owens and B. Thusu).* Research Centre, Garyounis University, Benghazi, p. 215-258.

VACHARD D. and MASSA, D. 1984. The Carboniferous of western Libya. *9th Internat. Cong. on Carboniferous Strat. and Geol.*, vol. 2, p. 165-174.

VAIL, J.R. 1971. Dike swarms and volcanic activity in northeastern Africa. *First Symposium on the Geology of Libya (ed. C. Gray)*. Faculty of Science, University of Libya, Tripoli, p. 341-346.

VAIL, J. R. 1983. Pan-African crustal accretion in north-east Africa. *Journ. African Earth Sci.* vol 1, p. 285-294.

VAIL, J.R. 1991. The Precambrian tectonic structure of North Africa. *Third Symposium on the Geology of Libya*. vol. 6 *(eds. M.J. Salem, A.M. Sbeta and M.R. Bakbak)*, Elsevier, Amsterdam, p. 2259-2268.

VAN DE WEERD, A.A. and WARE, P.L.G. 1994. A review of the east Algerian Sahara oil and gas province (Triassic, Ghadames and Illizi Basins). *First Break*, vol. 12, p. 363-373.

VAN DER MEER, F. and CLOETINGH, S. 1996. Intraplate stresses and the subsidence history of the Sirt Basin. *First Symposium on the Sedimentary Basins of Libya, Geology of the Sirt Basin*. vol. 3. *(eds. M.J. Salem, M.T. Busrewil, A.A. Misallati and M.J. Sola)*, Elsevier, Amsterdam, p. 211-230.

VAN DIJK, P.M. and EABADI, A.M. 1996. Relay growth faulting and contemporaneous drape folding of the Ma'zul Ninah Formation in the southern extension of the Hun Graben, western Sirt basin, Libya. *First Symposium on the Sedimentary Basins of Libya, Geology of the Sirt Basin*. vol. 3. *(eds. M.J. Salem, M.T. Busrewil, A.A. Misallati and M.J. Sola)*, Elsevier, Amsterdam, p. 155-166.

VAN HINTE, J.E., COLIN, J.P. and LEHMANN, R. 1980. Micropalaeontologic record of the Messinian event at Esso Libya Inc. well B1-NC 35A on the Pelagian Platform. *Second Symposium on the Geology of Libya*. vol. 1 *(eds. M.J. Salem and M.T. Busrewil)*, Academic Press, London, p. 205-244.

VAN HOUTEN, F.B. 1980. Latest Jurassic-early Cretaceous regressive facies, northeast Africa craton. *Bull. Amer. Assoc. Pet. Geol.* vol 64, p. 857-867.

VAN HOUTEN, F.B. and KARASEK, R.M. 1981. Sedimentologic framework of late Devonian oolitic iron formation, Shatti valley, west-central Libya. *Jour. Sediment. Petrol.* vol. 51. p. 415-428.

VAN HOUTEN, F.B. 1983. Sirte Basin, north central Libya; Cretaceous rifting above a fixed mantle hotspot? *Geology*, vol. 11, p. 115-118.

VAN HOUTEN, F.B. and HARGREAVES, R.B. 1987. Palaeozoic drift of Gondwana: palaeomagnetic and stratigraphic constraints. *Geol. Journ.* vol. 22, p. 341-359.

VAVRDOVÁ, M. Latest Devonian miospores and acritarchs from the surface samples of the Ashkidah Formation. *Third Symposium on the Geology of Libya*. vol. 4 *(eds. M.J. Salem, O.S. Hammuda and B.A. Eliagoubi)*, Elsevier, Amsterdam, p. 1285-1296.

VESELY, J. 1985. Sheet Zallah (NH 33-16), *Geological Map of Libya, scale 1:250,000, Explanatory Booklet*, Industrial Research Centre, Tripoli.

482

VINASSA DE REGNY, P. 1912. Cenni geologici sulla Libia italiana, carta geol. scala 1:5,000,000. *Boll. Soc. Afr. Ital.* vol. 31. p. 3-35.

VINCENT, P.M. 1970. The evolution of the Tibesti volcanic province, eastern Sahara. In: *African Magmatism and Tectonics. (eds. T.N. Clifford and I.G. Gass)*, Oliver and Boyd, Edinburgh. chap. 14, p. 301-319.

VITA-FINZI, C. 1969. The Mediterranean valleys: geological changes in historical times. Cambridge University Press, 140p.

VITA-FINZI, C. 1971. Alluvial history of northern Libya since the last interglacial. *First Symposium on the Geology of Libya (ed. C. Gray).* Faculty of Science, University of Libya, Tripoli, p. 409-430.

VITERBO, I. 1968. Lower Cretaceous charophyta from the subsurface 'Nubian complex' of the Sirte Basin, Libya. *Proc. 3rd Afr. Micropal. Colloq*, Cairo. p. 393-402.

VITTIMBERGA, P. and CARDELLO, R. 1963. Sédimentologie et petrographie du Paleozoïque du bassin de Kufra. *First Saharan Symposium. Rev.* Inst. Fran. Pet. special volume, p. 228-240.

VOS, R.G. 1981a. Deltaic sedimentation in Devonian of western Libya. *Sedimentary Geology*, vol. 29, p. 67-88.

VOS, R.G. 1981b. Sedimentology of an Ordovician fan delta complex, western Libya. *Sedimentary Geology*, vol. 29, p. 153-170.

WACRENIER, P. 1959. Apercu sur l'antecambrien du Tibesti. *Proc. 20th Int. Geol. Cong., Mexico.* Ass. Serv. Geol. Afr. p. 281-288

WADDAMS, F.C. 1980. The Libyan oil industry. Croom Helm, London. 338p.

WENDT, J. 1991. Depositional and structural evolution of the middle and late Devonian on the northwestern margin of the Saharan craton (Morocco, Algeria, Libya). *Third Symposium on the Geology of Libya.* vol. 6 *(eds. M.J. Salem, A.M. Sbeta and M.R. Bakbak)*, Elsevier, Amsterdam, p. 2195-2210.

WENNEKERS, J.H.N., WALLACE, F.K. and ABUGARES, Y.I. 1996. The geology and hydrocarbons of the Sirt Basin: a synopsis. *First Symposium on the Sedimentary Basins of Libya, Geology of the Sirt Basin.* vol. 1. *(eds. M.J. Salem, A.J. Mouzughi and O.S. Hammuda)*, Elsevier, Amsterdam, p. 3-56.

WESTAWAY, R. 1996. Active tectonic deformation in the Sirt basin and its surroundings. *First Symposium on the Sedimentary Basins of Libya, Geology of the Sirt Basin.* vol. 3. *(eds. M.J. Salem, M.T. Busrewil, A.A. Misallati and M.J. Sola)*, Elsevier, Amsterdam, p. 89-100.

WEYANT, M. and MASSA, D. 1991. Contribution of conodonts to the Devonian biostratigraphy of western Libya. *Third Symposium on the Geology of Libya.* vol. 4 *(eds. M.J. Salem, O.S. Hammuda and B.A. Eliagoubi)*, Elsevier, Amsterdam, p. 1297-1322.

WHITBREAD, T. and KELLING, G. 1982. Mrar Formation of western Libya – evolution of an early Carboniferous delta system. *Bull. Amer. Assoc. Pet. Geol.* vol. 66, p. 1091-1107.

WHITTINGTON, R. and WALKER, R. 1998. Comparison of interpreted ancient and modern fluvio-glacial ice tunnel channels (abstract only). *Symposium on Geological Exploration in Murzuq Basin, Sabah, 20th-22nd September, 1998.* Book of abstracts, p. 42.

WIGHT, A.W.R. 1980. Paleogene vertebrate fauna and regressive sediments of Dur at Talhah, southern Sirt Basin, Libya. *Second Symposium on the Geology of Libya.* vol. 1 *(eds. M.J. Salem and M.T. Busrewil)*, Academic Press, London, p. 309-326.

WILLIAMS, J.J. 1968. The sedimentary and igneous reservoirs of the Augila oil field, Libya. In: *Geology and archaeology of northern Cyrenaica, Libya (ed. Barr, F.T.)*. Petroleum Exploration Society of Libya, Tenth Annual Field Conference, p. 197-206.

WILLIAMS, J.J. 1971. Igneous and sedimentary reservoir rocks, Augila field, Libyan Arab Republic. *First Symposium on the Geology of Libya (ed. C. Gray).* Faculty of Science, University of Libya, Tripoli, p. 501-512.

WILLIAMS, J.J. 1972. Augila field, Libya: depositional environment and diagenesis of sedimentary reservoir and description of igneous reservoir. In: *Stratigraphic oil and gas fields; classification exploration methods and case histories (ed. R.E. King)*. Amer. Assoc. Pet. Geol. Mem. 16, p. 623-632.

WILSON, M. and GUIRAUD, R. 1998. Late Permian to Recent magmatic activity on the African-Arabian margin of Tethys. In: *Petroleum Geology of North Africa. (ed. D.S. Macgregor, R.T.J. Moody, D.D. Clark-Lowes)*, Geol. Soc. Special Publication No. 132, p. 231-264.

WIMAN, S.K. 1980. Stratigraphic and micropalaeontologic expression of the Mediterranean late Miocene (Messinian) regression and early Pliocene (Zanclean) transgression in northern Tunisia. *Second Symposium on the Geology of Libya.* vol. 1 *(eds. M.J. Salem and M.T. Busrewil)*, Academic Press, London, p. 117-126.

WINNOCK, E. and MERMOD, P. 1979. Libye: le gisement de Magid A. *Bull Assn. Fran. des Tech. du Pet.* No. 234, p. 11-18.

WOLF, M., MOSER, H. and ZEINO, H. 1991. Isotope

hydrological investigations in the Tawargha area, Libya. *Third Symposium on the Geology of Libya.* vol. 5 *(eds. M.J. Salem and M.N. Belaid),* Elsevier, Amsterdam, p. 2043-2050.

WOLLER, F. 1978. Sheet Al Washkah (NH 33-15), *Geological Map of Libya, scale 1:250,000, Explanatory Booklet,* Industrial Research Centre, Tripoli.

WOLLER, F. and FEDIUK, F. 1980. Volcanic rocks of Jabal as Sawda. *Second Symposium on the Geology of Libya.* vol. 3 *(eds. M.J. Salem and M.T. Busrewil),* Academic Press, London, p. 1081-1095.

WOLLER, F. 1984. Sheet Al Fuqaha (NG 33-3), *Geological Map of Libya, scale 1:250,000, Explanatory Booklet,* Industrial Research Centre, Tripoli.

WOODSIDE, J.M. 1991. Disruption of the African plate margin in the eastern Mediterranean. *Third Symposium on the Geology of Libya.* vol. 6 *(eds. M.J. Salem, A.M. Sbeta and M.R. Bakbak),* Elsevier, Amsterdam, p. 2319-2330.

WORSLEY, D. 2000. A bibliography of the geology of the Murzuq Basin. *Symposium on Geological Exploration in Murzuq Basin (Eds. M.A. Sola and D. Worsley),* Elsevier, Amsterdam, p. 509-517.

WRIGHT, E.P. and EDMUNDS, W.M. 1971. Hydrogeological studies in central Cyrenaica. *First Symposium on the Geology of Libya (ed. C. Gray).* Faculty of Science, University of Libya, Tripoli, p. 459-482.

WRIGHT, J. 1981. Libya: a modern history. Croom Helm, London. 306p.

YERGIN, D. 1991. The Prize: the epic quest for oil, money and power. Simon and Schuster, London. 884p.

YUKLER, A., MESKINI, A., SAIDI, M., MOUMEN, I., DAADOUCH, I., BOUHLEL, H. and JARRAYA, H. 1994. Quantitive evaluation of the geologic evolution and hydrocarbon potential of the Gulf of Gabes. In: *Proc. 4th Tunisian Pet. Explor. Conf. (Ed. A. Chine and K. Ben Hassine).* ETAP, Tunis, p. 169-213.

ZACCAGNA, D. 1919. Itinerari geologici nella Tripolitania occidentale. *Mem. Descr. Cart. Geol. d'Ital.* vol. 18, p. 1-70.

ZAÏER, A., BEJI-SASSI, A., SASSI, S. and MOODY, R.T.J. 1998. Basin evolution and deposition during the early Paleogene in Tunisia. In: *Petroleum Geology of North Africa. (ed. D.S. Macgregor, R.T.J. Moody, D.D. Clark-Lowes),* Geol. Soc. Special Publication No. 132, p. 375-394.

ŽERT, B. 1974. Sheet Darnah (NI 34-16), *Geological Map of Libya, scale 1:250,000, Explanatory Booklet,* Industrial Research Centre, Tripoli.

ZIKMUND, J. 1985. Sheet Abu Na'im (NH 34-13), *Geological Map of Libya, scale 1:250,000, Explanatory Booklet,* Industrial Research Centre, Tripoli.

ŽIVANOVIĆ, M. 1977. Sheet Bani Walid (NH 33-2), *Geological Map of Libya, scale 1:250,000, Explanatory Booklet,* Industrial Research Centre, Tripoli.

Appendix

CURRENT TRANSLITERATION of LIBYAN GEOGRAPHIC NAMES

The present transliterations are taken from the National Atlas of SPLAJ, 1978, and from information kindly provided by Dr Mustafa Salem of Al Fatah University, Tripoli.

Only names which have changed from earlier styles are shown.

A few of the current transliterations (e.g. Tarabulus) have not been adopted in this book.

The list shows name changes in addition to transliteration changes.

FORMER STYLES	CURRENT STYLE	FORMER STYLES	CURRENT STYLE
Abrag field (NC 74F)	Al Abraq	Bardia	Al Burdi
Acacus	Akakus	Bazuzi field (Conc 16)	Bazzuzi
Achebyat	Ash Shabiyat	Beda field (Conc 47)	Al Bayda
Ad Daffah field (Conc 59)	Dayfah	Beida	Al Bayda
Agedabia	Ajdabiya	Bel Hedan field (Conc 59)	Bilhizan
Ain Jarbi field (conc 6)	Ayn Jarbi	Ben Afen	Bin Affin
Ain Tobi	Ayn Tobi	Ben Isa	Bin Isa
Al Awra field (Conc 13)	Tibisti	Ben Khashir	Bin Ghashir
Al Baida (Cyrenaica)	Al Bayda	Bengasi	Binghazi
Al Faidia	Al Fayidiyah	Benghazi	Binghazi
Al Fatah field (Conc 51)	Al Fatih	Beni Ulid	Bani Walid
Al Haleigh	Al Hulayq	Beni Walid	Bani Walid
Al Merg	Al Marj	Benia	Al Baniyah
Al Sceleidima	Ash Shulaydimah	Berguen	Birgin
Al Urah field (Conc 13)	Tibisti	Beshima	Bishimah
Al Wafa field (NC 169)	Al Wafaa	Bir el Ghnem	Bi'r al Ghanam
Al Watiyah	Al Watyah	Bir Hacheim	Bi'r Hakim
Amur	Ammur	Bir Zelten	Bi'r Zaltan
Antelat field (Conc 41)	Antlat	Bir Ziden	Bi'r Zaydan
Aouniat Ouenine	Awaynat Wanin	Bisher	Bishr
Apollonia	Susah	Bomba	Bamba
Areeda field (Conc 6)	Attahadi	Brega	Al Brayqah
Arknou, Jebel	Arknu	Bu Alawan field (Conc 47)	Abu Alwan
Assumood field (Conc 6)	Assumud	Bu Amud field (Conc 11)	Abu Amud
Attahaddy field (Conc 6)	Attahadi	Bu Attifel field (Conc 100)	Abu Attiffel
Auenat Uennin	Awaynat Wanin	Bu en Niran	Abu an Niran
Augila field (Conc 102)	Awjilah	Bu Gheilan	Abu Ghaylan
Ayadhar	Iyadar	Bu Grea field (Conc 5)	Abu Quray
Azba, Jebel	Azbah	Bu Gren	Abu Grayn
Azizia	Al Aziziyah	Bu Hashish	Abu Hashish
Barce	Al Marj	Bu Isa	Abu Isa

FORMER STYLES	CURRENT STYLE	FORMER STYLES	CURRENT STYLE
Bu Mras field (Conc 11)	Abu Maras	Farigh field (Conc 12)	Al Farigh
Bu Ngem	Abu Njaym	Farrud field (Conc 11)	Al Furud
Bu Njim	Abu Njaym	Garet Uedda	Qarat Waddah
Bu Ras	Abu Ra's	Gargaf, Jebel	Al Qarqaf
Bu Sceba	Abu Shaybah	Gargaresh	Qarqarish
Buerat el Hsun	Buwayrat al Hasun	Garian	Gharyan
Cabao	Kabaw	Gasr al Abid	Qasr al Ahrar
Calanscio field (EE-59)	Kalanshiyu	Gasr Tigrinna	Qasr Tagrinnah
Chicla	Kiklah	Gatrun	Al Qatrun
Cirenaica	Cyrenaica	Gattar field (Conc 11)	Qattar
Cufra	Al Kufrah	Gazeil field (Conc 26)	Ghziel
Currusc	Kurrush (Jabal)	Gazzun field	Qazzun
Cyrene	Shahhat	Gelta	Qaltah
Dahra field (Conc 32)	Az Zahrah	Gerdes el Abid	Qasr al Ahrar
Dalma	Az Zalmah	Germa	Jarmah
Defa field (Conc 59)	Dayfah	Ghadames	Ghadamis
Dembaba	Dimbabah	Ghazzun field (Conc 13)	Qazzun
Derna	Darnah	Ghemines	Qaminis
Djebel	Jabal	Gheriat	Al Qaryah
Djeffara	Jifarah	Giado	Jadu
Dor el Abd	Umm ad Dahiy	Gialo	Jalu
Dor el Gussa	Dur al Qussah	Gialo field (Conc 59)	Jalu
Dor field (Conc 47)	Dur	Gianzur	Janzur
Dor Marada field (Conc 13)	Dur Maradah	Giarabub	Al Jaghbub
Ed Dib field (Conc 11)	Adh Dhi'b	Giosc field (Conc 23)	Al Jawsh
Ed Duera field (Conc 23)	Ed Duerah	Gir	Al Jir
Edri	Idri	Greir bu Hascisc	Bu Hashish
Eghei, Jebel	Eghi	Gsur field (Conc 11)	Qusur
Eghei, Jebel	Nuqay	Gubba	Qubba
El Agheila	Al Aqaylah	Hagfa	Hagfah
El Beida	Al Bayda	Hagfa Trough	Maradah
El Fogaha	Al Fuqaha	Hamada al Homra	Al Hamadah al Hamra
El Gheriat ech Cherguia	Al Qaryat ash Sharqiyah	Hamama	Hamamah
El Giof	Al Jawf	Haouaz	Hawaz
El Hauria	Al Hawwari	Harsha	Harshah
El Magroun	Al Maqrun	Hasaouna	Hasawnah
El Meheiriga field (Conc LP3C)	Al Mheirigah	Hateiba field (Conc 6)	Hutaybah
El Merj	Al Marj	Hauaisc, Jebel	Al Hawa'ish
El Mlaghi field (Conc 61)	Al Malaqi	Hofra field (Conc 11)	Al Hufrah
El Uweinat	Al Awaynat	Homs	Al Khums
Emgayet field (Conc 66)	Emghayet	Hon	Hun
En Naga field (Conc 72)	'Ayn An Naqah	Idris fields (Conc 103)	Intisar
Es Sania field (Conc 23)	As Sania	In Ezzane	Ayn az Zan
Es Sider	As Sidrah	Jahama	Al Jahamah
Et Tallab	At Tullab	Jakhira field (Conc 96)	Jakharrah
Faidia	Al Faydiyah	Jardas al Abid	Qasr al Ahrar
Faregh field (Conc 59)	Kalanshiyu	Jebel Acacus	Jabal Akakus

FORMER STYLES	CURRENT STYLE	FORMER STYLES	CURRENT STYLE
Jebel Akhdar	Al Jabal al Akhdar	Mn Berber field (Conc 6)	Assumud
Jebel Archenu	Jabal Arknu	Mrar	Marar
Jebel Azba	Jabal Azbah	Msus	Masus
Jebel ben Ghnema	Jabal bin Ghanimah	Murizidie	Mourizidie
Jebel Dalma	Jabal az Zalmah	Murzuch	Murzuq
Jebel Eghei	Jabal Nuqay	Murzuk	Murzuq
Jebel el Akhdar	Jabal al Akhdar	Nafoora field (Conc 51)	An Nafurah
Jebel el Hauaisc	Jabal al Hawa'ish	Nafoora field (Conc 51)	Nafurah
Jebel field (Conc 6)	Al Jabal	Nefusa, Jebel	Nafusah
Jebel Gargaf	Jabal Qarqaf	Nofilia	An Nuwfaliyah
Jebel Hassaouna	Jabal al Hasawnah	Nufliah	An Nuwfaliyah
Jebel Nefusa	Jabal Nafusah	Ogle field (Conc 11)	Awjilah
Jebel Uweinat	Jabal Al Awaynat	Ora field (Conc 13)	Al Awra (now Tibisti)
Jebel Zalma	Jabal az Zalmah	Ouan Kasa	Wan Kasa
Jefren	Yifran	Oued Chebbi field (Conc 23)	Ouled Chebbi
Jerma	Jarmah	Oued Scarciuf field (Conc 23)	Sharcuf
Jiffara	Jifarah	Oued Zuzam field (Conc 23)	Wadi Zuzam
Kebir field (NC 7A)	Kabir	Ouled Chebbi Fm	Al Guidr
Khalifa	Khalifah	Pisida	Bu Kammash
Khalifa field (Conc 59)	Khalifah	Qararat al Jifah	Qrarat al Jifah
Khatt Graben	Maragh	Qarat Meriam	Qarat Mariem
Kheir	Khayir	Qarat Weddah	Qarat Waddah
Khuff field (Conc 93)	Al Kuf	Qayrat el M'rar	Qararat al Marar
Kotla field (Conc 47)	Al Kotlah	Rachmat	Rakhmat
Kufra	Al Kufrah	Rachmat field (Conc 13)	Rakhmat
Labrak	Al Abraq	Raguba field (Conc 20)	Ar Raqubah
Lehib field (Conc 6)	Lahib	Ras Hamia Fm	Kurrush
Les Abreghs	Abreghs	Ras Jdeir	Ra's Ajdir
Maazul Neina	Ma'zul Ninah	Rebiana	Rabyanah
Magid field (Conc 105)	Majid	Regima	Ar Rajmah
Mansour field (Conc 104)	Mansur	Rhnem	Bi'r al Ghanam
Mar Trough	Hameimat	Rimel field (Conc 82)	Ar Raml
Marada	Maradah	R'Mel	Ar Raml
Maregh	Maragh	Ruaga	Rawaghah
Marsa el Brega	Al Brayqah	Sabratha	Sabratah
Marsa el Hariga	Marsa al Hariqa	Sahl field (Conc 6)	As Sahel
Mazuza	Mazuzah	Sardalas	Al Awaynat
Melez Chograne	Melaz Shuqran	Scecsciuc	Shakshuk
Melita	Mellitah	Sceleidima	Ash Shulaydimah
Mellugh field (Conc 11)	Meulagh	Sciahhat	Shahhat
Memouniat	Mamuniyat (Jabal al)	Sciuduma	Ash Shulaydimah
Menzella	Manzilah	Sebha	Sabha
Messak	Msak	Sebkha	Sabkha
Messla field (Conc 65)	Messlah	Sebkhet al Kebir	As Sabkhah al Kabirah
Metem Depression	Hameimat	Serdeles	Al Awaynat
Misurata	Misratah	Sharara, El, field (NC 115)	Ash Shararah
Mizda	Mizdah	Sharma	Shammar

FORMER STYLES	CURRENT STYLE	FORMER STYLES	CURRENT STYLE
Shehabat field (Conc 13)	Shuhaybat	Toummo	Tumu
Sheterat	Shterat	Tripoli	Tarabulus
Sidi as Sid	Sidi Assayd	Tuareg	Touareg
Sidi Mesri	Sidi al Masri	Tumayyim	Abu Tumayam
Sinauen	Sinawan	Uaddan	Waddan
Sirte	Sirt	Uau al Kebir	Waw al Kabir
Slonta	Salantah	Uau an Namus	Waw an Namus
Socna	Suknah	Uazzen	Wazin
Soda	Sawda	Ubari	Awbari
Soffagin	Suf Ajjin	Umm Farrud field (Conc 92)	Umm al Furud
Soluch	Suluq	Uotia	Al Watyah
Sorra field (Conc 6)	As Surah	Wadi Ducchan	Wadi Dukhan
Surfa	Shurfah	Wadi esc Sciati	Wadi ash Shati
Tabet field (Conc 6)	Tabit	Wadi Etba	Wadi Atabah
Tagiura	Tajura	Wadi Tamet	Wadi Thamat
Tagrifet field (Conc 11)	Taqrifat	Wadi Tanezzuft	Tanzuft
Tanezzuft	Tanzuft	Wadi Thamet	Wadi Thamit
Tarhuna	Tarhunah	Wadi Younis	Wadi Yunis
Tazerbo	Tazirbu	Waha field (Conc 59)	Al Wahah
Tedjere	Tajihri	Wasat Ridge	Messlah
Tesaoua	Tsawah	Yosr field (Conc 47)	Al Yusr
Thala	Talah	Zaggut field (Conc 59)	Zaqqut
Tibesti	Tibisti	Zalma	Zalmah
Tigi field (Conc 23)	Tiji	Zauam field (Conc 11)	Az Zawwam
Tigrinna	Taghrinnah	Zawia	Az Zawiyah
Tihemboka	Tihimbukah	Zawiyat Msus	Zawiyat Masus
Tlacsin field (Conc 70)	Bi'r Tlakshin	Zellah field (NC 74B)	Zallah
Tmed al Ksour	Thmed al Qusur (well)	Zelten field (Conc 6)	Zaltan (now Nasser)
Tmed field (Conc 13)	Tmad	Zemzem	Zamzam
Tmessa	Tmassah	Zenad field (Conc 11)	Znad
Tobruch	Tubruq	Zmam	Zimam
Tobruk	Tubruq	Zuara	Zuwarah
Tocbal	Takbal	Zuetina	Azzuwaytinah
Tocra	Tukrah	Zuila	Zuwaylah
Tolmeita	Tulmaythah		

INDEX

map, 339, 382, burial history plot, 341, high geothermal gradient, 382, petroleum systems, 379-387, map, 384, amount of oil generated, 379, hydrocarbon potential, 420, 423, map, 418, deep trough potential, 423.

Maragh Formation, 174, description, 190, 191, map 176, columnar section, 177, oil reservoir, 406, 407, oil reserves, 374.

Maragh Graben, 267, 300, description, 306, map, 284, 301, cross-section, 302, petroleum systems, 404-6, 425, map, 405.

Marar Formation, 115, description, 130, 132, 133, map and columnar section, 131.

Marfeg Formation (Tunisia), 174, 189.

Marmarica Basin, 300, description, 305, map, 301, potential source rocks, 348, 416, map, 347, hydrocarbon potential, 420, 425, 426, map, 418.

Marsa Zouaghah Formation, 241, description, 259, map, 252, columnar section 253.

Masid Formation, 146, description, 165, 168, map, 166.

Masrab field, 22, 27, 29, 44, 403, map, 400, migration routes, 402, trap, 403, on production, 38.

Masus Member, 241, description, 262, map, 252, columnar section, 253, relationships, 255.

Maturity indicators, 323.

Ma'zul Ninah Formation, 241, description, 244, map, 242, columnar section, 243.

Mazuzah Member, 174, description, 184, map 182, columnar section, 183.

Medina Escarpment, 267, 310, map. 308, 409.

Medina Graben, map, 308, cross-section, 313.

Medina Plateau, map, 308, 409, 413, cross-section, 313.

Medina Wrench, 73, 307, 310, 319.

Mediterranean Sea, plate tectonic reconstruction, 72, 73, present-day subduction, 72, 73.

Meem Member, 215, oil reservoir, 381.

Meghil field, 389, map, 388, cross-section, 390, reserves, 389.

Meghil Formation, 220.

Melaz Shuqran Formation, 89, description, 97, 99, map and columnar section, 98, glacial features, 97, 99, land plants, 99, oil source potential, 325, 364.

Melita-Medina Plateau, 267, 310, description, 312, map, 308.

Melita Plateau, map, 308, 409, 413, 414, cross-section, 313.

Melitah gas processing plant, 46.

Melqart field (Tunisia), map, 349, 409, source rock, 350, 351.

Melqart Formation (Tunisia), 241, 259.

Mereksen field (Algeria), 333.

Mesdar Member, 204, description, 236, 237, map 222, columnar section, 223.

Mesogea (spreading axis), 64, 288, 317.

Messak Formation, 146, description, 168, 169, map, 166, columnar section, 167.

Messinian salinity event, 73, 257, 312, 314, 319, discussion, 259, 260.

Messlah field, 27, 44, discovery, 33, 34, giant field, 37, description, 401, map, 400, reservoir characteristics, 401, source rocks, 398, diagram, 345, stratigraphic trap, 403, oil characteristics, 402, reserves, 401, on production, 31, 38.

Messlah High, 298, 299, map, 284, 400, hydrocarbon potential, 420, 424, map, 418.

Messlah-Kalanshiyu High, 298.

Mestaoua Formation (Tunisia), 146.

Metem Trough, 298.

Meteorite craters, 247.

Methane, first discovery, 17.

Metlaoui Group (Tunisia), 204, 230, facies model, 410, oil reservoir, 350, 408, 410, heat flow, 350.

Meulagh field, 27.

Meulagh Graben, 291, 377, 422, map, 284, petroleum systems, 374, map,

375, hydrocarbon potential, 420, map, 418.

Mid-Devonian tectonism, 118, 316, 319, map 122.

Middle/Upper Devonian-Awaynat Wanin petroleum system (Ghadamis Basin), 371-372, map, 367, cross-section, 369.

Military bases (foreign) closed, 30.

Minerals Law, 17, 18.

Miocene, stratigraphy, 248, et seq.

Miocene tectonism, 73, 282, 283, 285, 290, 308-312, 319.

Miskar field (Tunisia), 350, 408, 412, 414, 415, map, 349, 409, 413.

Misratah Basin, 267, description, 312, map, 308, 409, 413, cross-section, 313, hydrocarbon source potential, 415, hydrocarbon potential, 420, 426, map, 418.

Misratah High, map, 409.

Mizdah Formation, 174, description, 179, 184, map 182, columnar section, 183.

Mn Berber field, 22.

Mobil (oil Co), 19, 20, 24, 26, 29, production cuts, 30, state participation, 32, production sharing awards, 34, 35, Ghani discovery, 37, withdrwal from Libya, 38.

Mourdi Basin (Sudan), 268.

Mourizidie Formation, 84, 85, 89.

Munchar Formation, 89, 102.

Murzuq Basin, 58, 267, description, 271, cross-section, 87, 273, 362, evolution, 275-277, source rock potential, 326-331, maturity, 328, 329, map, 329, burial history plot, 331, petroleum systems, 356-365, table, 358, map, 360, cross-section, 362, oil reserves, 326, 356, remaining hydrocarbon potential, 420, 421, map, 419.

Murzuq fields (NC 115), 44, on production, 31, 43, reservoir characteristics, 359, map, 360.

Murzuq-Jadu Trough, 56, map 272.

Nafusah Uplift, 58, 59, 267, description, 281, 282, map, 278, cross-section, 279, 280, 282.

Nakhla field, 27, map, 400, source